Units and conversion factors

Length:	$1 \text{ m} = 3.281$ feet
Mass:	$1 \text{ kg} = 2.2046$ pounds
Energy:	$1 \text{ J} = 1$ watt-second $= 10^7$ ergs $= 10^7$ dyne-cm $= 0.2390$ cal $= 0.948 \times 10^{-3}$ BTU $= 0.7376$ ft-lb_f $= 0.9868 \times 10^{-2}$ liter-atm
Pressure:	$1 \text{ Pa} = 1$ newton $\text{m}^{-2} = 1$ joule m^{-3} $= 0.9869 \times 10^{-5}$ atmospheres $= 0.750 \times 10^{-2}$ millimeters of mercury $= 1 \times 10^{-5}$ bars $= 1.450 \times 10^{-4}$ psi
Molar Heat Capacity:	$1 \text{ J mol}^{-1} \text{ K}^{-1} = 0.2390 \text{ cal(mol K)}^{-1}$ $= 0.2390 \text{ BTU(lb-mol °F)}^{-1}$
Specific Heat Capacity:	$1 \text{ J g}^{-1} \text{ K}^{-1} = 0.2390 \text{ cal(g K)}^{-1}$ $= 0.2390 \text{ BTU(lb °F)}^{-1}$ $= 1 \text{ kJ (kg K)}^{-1}$
Power:	$1 \text{ J sec}^{-1} = 1$ watt $= 0.1341 \times 10^{-2}$ horsepower (hp) $= 0.7356$ ft-lb_f/sec $= 0.948 \times 10^{-3}$ BTU/sec

Chemical and Engineering Thermodynamics

Wiley Series in Chemical Engineering

Bird, Armstrong and Hassager: DYNAMICS OF POLYMERIC LIQUIDS, Vol. I FLUID MECHANICS

Bird, Hassager, Armstrong and Curtiss: DYNAMICS OF POLYMERIC LIQUIDS, Vol. II KINETIC THEORY

Bird, Stewart and Lightfoot: TRANSPORT PHENOMENA

Brownell and Young: PROCESS EQUIPMENT DESIGN: VESSEL DESIGN

Davis: NUMERICAL METHODS AND MODELING FOR CHEMICAL ENGINEERS

Doraiswamy and Sharma: HETEROGENEOUS REACTIONS ANALYSIS, EXAMPLES AND REACTOR DESIGN, Vol. 1, Vol. 2

Felder and Rousseau: ELEMENTARY PRINCIPLES OF CHEMICAL PROCESSES, 2nd Edition

Foust, Wenzel, Clump, Maus and Andersen: PRINCIPLES OF UNIT OPERATIONS, 2nd Edition

Froment and Bischoff: CHEMICAL REACTOR ANALYSIS AND DESIGN

Franks: MODELING AND SIMULATION IN CHEMICAL ENGINEERING

Henley and Seader: EQUILIBRIUM–STAGE SEPARATION OPERATIONS IN CHEMICAL ENGINEERING

Hill: AN INTRODUCTION TO CHEMICAL ENGINEERING KINETICS AND REACTOR DESIGN

Jawad and Farr: STRUCTURAL ANALYSIS AND DESIGN OF PROCESS EQUIPMENT

Kellogg Company: DESIGN OF PIPING SYSTEMS, Revised 2nd Edition

Klein and Bischoff: CHEMICAL KINETICS AND REACTOR DESIGN

Levenspiel: CHEMICAL REACTION ENGINEERING, 2nd Edition

Nauman and Buffham: MIXING IN CONTINUOUS FLOW SYSTEMS

Nauman: CHEMICAL REACTOR DESIGN

Rase: CHEMICAL REACTOR DESIGN FOR PROCESS PLANTS PRINCIPLES AND TECHNIQUES, Vol. 1

Rase and Barrow: PIPING DESIGN FOR PROCESS PLANTS

Reklaitis: INTRODUCTION TO MATERIAL AND ENERGY BALANCES

Rudd, Fathi-Afshar, Trevino and Stadtherr: PETROCHEMICAL TECHNOLOGY ASSESSMENT

Rudd and Watson: STRATEGY OF PROCESS ENGINEERING

Sandler: CHEMICAL AND ENGINEERING THERMODYNAMICS, 2nd Edition

Seborg, Edgar and Mellichamp: PROCESS DYNAMICS AND CONTROL

Smith and Corripo: PRINCIPLES AND PRACTICE OF AUTOMATIC PROCESS CONTROL

Smith and Missen: CHEMICAL REACTION EQUILIBRIUM ANALYSIS

Ulrich: A GUIDE TO CHEMICAL ENGINEERING PROCESS DESIGN AND ECONOMICS

Welty, Wicks and Wilson: FUNDAMENTALS OF MOMENTUM, HEAT AND MASS TRANSFER, 3rd Edition

Chemical and Engineering Thermodynamics

Stanley I. Sandler

University of Delaware

JOHN WILEY & SONS

New York • Chichester • Brisbane • Toronto • Singapore

Library of Congress Cataloging in Publication Data:

Sandler, Stanley I., 1940–
Chemical and engineering thermodynamics/Stanley I. Sandler.—2nd ed.
p. cm.—(Wiley series in chemical engineering)
Includes index.
ISBN 0-471-83050-X
1. Thermodynamics. 2. Chemical engineering. I. Title.
II. Series.
QD504.S25 1989
541.3′69—dc19 88-10141
CIP

Printed in the United States of America

10 9 8 7 6 5 4 3 2 1

About the Author

STANLEY I. SANDLER earned the B.Ch.E. degree in 1962 from the City College of New York, and the Ph.D. in chemical engineering from the University of Minnesota in 1966. His research at Minnesota was on the kinetic theory of gases. He was then a National Science Foundation Postdoctoral Fellow at the Institute for Molecular Physics at the University of Maryland for the 1966–67 academic year where he worked on the theory of transport in ionized gases. He joined the faculty of the University of Delaware in 1967 as an assistant professor, and was promoted to associate professor in 1970, professor in 1973 and Henry Belin du Pont Professor of Chemical Engineering in 1982. He was department chairman from 1982 to 1986.

Professor Sandler is the author of 115 papers and the editor of five conference proceedings books. Among his many awards are a Faculty-Scholar Award (1971) from the Camille and Henry Dreyfus Foundation, a Research Fellowship (1980) and U.S. Senior Distinguished Scientist Award (1988) from the Alexander von Humboldt Foundation (West Germany), the 3M Chemical Engineering Lectureship Award (1988) from the American Society for Engineering Education and the Professional Progress Award (1984) from the American Institute of Chemical Engineers.

to Judith,
Catherine,
Joel,
and Michael

Preface

This book is intended to be the text for a course in thermodynamics for undergraduate students in chemical engineering. I had two objectives in writing the first edition of this book, which I retained in this edition. The first was to develop a modern thermodynamics course especially for chemical engineering students which was relevant to other parts of the curriculum, specifically courses in separations processes, chemical reactor analysis, and process design. The other objective was to organize and present material in sufficient detail, and in such a way that the student obtained a good understanding of the principles of thermodynamics, and proficiency in applying these principles to the solution of a large variety of energy flow and equilibrium problems.

Since the first edition largely met these goals, and the thermodynamic principles have not changed in the last decade, this edition is very similar in structure to the first. However, what has changed in engineering education is the availability of microcomputers, such as the IBM-PC. This has made it possible to bring engineering science, industrial practice, and undergraduate education much closer together. In particular, it is now possible for students to perform sophisticated thermodynamic calculations with microcomputers, including calculations of the type they will encounter in industry.

For this reason that I have included four BASIC language computer programs in this second edition on an accompanying diskette. These programs are for: (1) the calculation of the thermodynamic properties and vapor–liquid equilibrium of a pure fluid described by a cubic equation of state; (2) the calculation of the thermodynamic properties and phase equilibria for a multicomponent mixture described by a cubic equation of state; (3) the calculation of activity coefficients as a function of composition, and low pressure vapor–liquid equilibrium using a group contribution activity coefficient model; and (4) the calculation of the chemical equilibrium constant and the standard state heat of reaction as a function of temperature using a database of 100 compounds.

These programs permit students to obtain solutions to realistic problems by concentrating their efforts on the thermodynamics of the situation, rather than on numerical analysis and computer programming. Students can also use these programs elsewhere in the chemical engineering program, such as in the separations and the design courses. Further, since the BASIC language code is given, instructors and students can modify these programs for specific applications, if needed. (A version of these programs in BASIC, suitable for use on the Apple MacIntosh, is available from the author. Also, though I have not explicitly indicated so in the text, personal computer equation solving programs such as Eureka: The Solver or tk Solver are helpful in solving many of the illustrations and homework problems.

Another major change in the second edition is the almost complete conversion to SI units. The only exceptions occur where data are needed from other sources which are not in those units. For example, tables of heats and free energies of formation in chemical engineering handbooks are in units of (thermochemical) calories with a standard state of 1 atm. resulting in some use of these units, together with SI units, in Chapter 9.

As for changes within each chapter, a reader familiar with the first edition of the text will notice relatively little change in Chapters 1, 2, 3, and 6 of this edition, and significant changes in Chapters 4, 5, 7, 8, and 9. For example, Section 8.1 of the first edition, which dealt with vapor-liquid equilibrium, has now been split into two sections, the first using activity coefficients, and the second using equations of state. Further, in this latter section, there is additional emphasis on phase equilibrium at high pressure. Similarly, the section on the solubility of a solid in a liquid now also includes the solubility of a solid in a gas and a supercritical fluid, and the section on liquid–liquid equilibrium also considers vapor–liquid–liquid equilibrium. The concept of fugacity, which had previously been introduced in Chapter 7, now appears first in Chapter 5 and is the basis for pure fluid equation of state vapor–liquid equilibrium (vapor pressure) calculations in that chapter. Also, throughout this edition, there is more emphasis on the use of equations of state than was previously the case.

An important feature of this book is that the emphasis differs somewhat from other books on the same subject, based on my perceptions of the special needs of the chemical engineer. The student using this book will learn how to analyze and predict liquid–liquid and vapor–liquid–liquid equilibria, the solubility of gases and solids in liquids, the solubility of liquids and solids in gases and supercritical fluids, freezing point depressions and osmotic equilibria, as well as traditional vapor–liquid and chemical reaction equilibria. These topics are presented in Chapter 6 by first considering the criteria for multicomponent phase and chemical equilibria, and are then followed, in Chapter 7, by a discussion of equations of state, solution theories, and fugacity estimation techniques. This material is then combined, in Chapter 8, to treat the aforementioned types of phase equilibria in a unified manner and, in Chapter 9, to study chemical equilibria in one or more phases. Thus, the presentation separates the thermodynamic principles from questions of the estimation of thermodynamic properties, and illustrates how both are used to predict or correlate a variety of equilibrium phenomena.

Similarly, in Chapters 4 and 5, students will develop the skills to compute the thermodynamic properties of a fluid given a volumetric equation of state, and to prepare their own thermodynamic tables and charts (as is done in two illustrations). In contrast, while power cycles and energy transformation methods are considered in Chapter 3, they are not studied in the same detail as is found in mechanical engineering textbooks.

The illustrations and homework problems included in this book, many of which are new to this edition, are meant to expose the student to the broad range of applications of thermodynamics in the process industries. Most of the problems involve, in their solution, the use of physical properties data for real substances and mixtures; others require the estimation of such information when experimental data are not available. Thus, students using this text should develop the knowledge needed to apply thermodynamics to the scientific and technological problems they will encounter in their professional careers.

The first part of this book, Chapters 1 to 5, dealing with the thermodynamics of pure fluids, mass and energy flow problems, power cycles, and the properties and phase equilibria of one-component systems, can be covered in the first semester of a two-semester thermodynamics course. Chapters 6 to 9, dealing with mixtures, form the second part of the book, and the second semester of the course. Though designed for a two-semester undergraduate course, the first edition has also been used in junior colleges, and, with supplementary material, as the basis for graduate courses. This book is a demanding one, but I hope that students will feel repaid with knowledge for the demands made of them.

I want to thank my colleagues in the Department of Chemical Engineering at the University of Delaware, and elsewhere around the country, for their encouragement, comments, and suggestions over the years. Many of these suggestions have been incorporated into this edition. The most significant impact has been that of my students, whose proclivity to ask the question "Why?" or to point out that what seemed logical or obvious was neither, has made me rethink, reorganize, and rewrite many parts of this book over the seventeen years since I began. In fact, I am no longer sure whether the form of this book is more a result of my vision or their guidance.

STANLEY I. SANDLER

Note: Borland's equation solving software, Eureka: The Solver is available for use with this text. For approximately 15% of the current price, the complete software package with a specially prepared student manual can be obtained from Wiley. For more information write to Eureka, Software Publications Group, John Wiley and Sons, 605 Third Avenue, New York, NY 10158.

Contents

9

CHEMICAL EQUILIBRIUM AND THE BALANCE EQUATIONS FOR CHEMICALLY REACTING SYSTEMS 494

APPENDIXES 579

INDEX 611

Notation

Standard, generally accepted notation has been used throughout this text. This list contains the important symbols, their definition, and, when appropriate, the page of first occurrence (where a more detailed definition is given). Symbols used only once, or within only a single section are not listed. In a few cases it has been necessary to use the same symbol twice. These occurrences are rare and widely separated so, it is hoped, no confusion will result.

SPECIAL NOTATION

Symbol	*Designates*
^ (caret as in $\hat{H}$)	property per unit mass (enthalpy per unit mass)
_ (underscore as in $\underline{H}$)	property per mole (enthalpy per mole)
$\bar{}_i$ (overbar as in $\bar{H}_i$)	partial molar property (partial molar enthalpy)
± (as in $M_{\pm}$)	mean ionic property (mean ionic molality)
* (as in G_i^*)	property (Gibbs free energy) in hypothetical pure component state extrapolated from infinite dilution behavior
□ (as in $G_i^{\square}$)	ideal unit molal property (Gibbs free energy) extrapolated from infinite dilution behavior
°	standard state

GENERAL NOTATION

A	Helmholtz free energy (92)
a_i	activity of species i (501)
$a, b, c, \ldots$	constants in heat capacity equation, equation of state, etc.
$B(T), C(T), \ldots$	virial coefficients (149)
$\mathscr{C}$	number of components (251)
°C	degrees Celsius (12)
C_i	concentration of species i (472)
C_V, C_P	constant-volume and constant-pressure heat capacities (44)

C_V^*, C_P^*	ideal gas heat capacity (44)
∂, d, D	partial, total, and substantial derivative symbols
D	diffusion coefficient (21)
$\mathcal{F}$	degrees of freedom (234)
°F	degrees Fahrenheit (12)
f	pure component fugacity (220)
$\bar{f}_i$	fugacity of a species in a mixture (308)
F_{fr}	frictional forces (65)
g	acceleration of gravity (9)
G	Gibbs free energy (93)
ΔG^{fus}, ΔG_{rxn}, ΔG_{mix}	Gibbs free energy changes on fusion (461), reaction (275), and mixing (307)
$\Delta \underline{G}^\circ_{f,i}$	molar Gibbs free energy of formation of species i (276)
H	enthalpy (33)
H_i, $\mathcal{H}_i$	Henry's law constants (356, 357)
ΔH^{vap}, ΔH^{fus}	enthalpy changes on mixing (269), reaction (275)
ΔH_{mix}, ΔH_{rxn}	vaporization (461) and fusion (238)
$\Delta H^\circ_{c,i}$	standard heat of combustion of species i (277)
$\Delta \underline{H}^\circ_{f,i}$	molar heat of formation of species i (275)
$\boldsymbol{I}$	unit tensor (75)
K	number of flow streams (26)
K	degrees Kelvin (11)
K, K_c, K_x	distribution coefficients (395, 470, 472) in Chapter 8
K_a	chemical equilibrium constant (501)
K_c, K_p, K_x, K_y	chemical equilibrium ratios (513, 514) in Chapter 9
K_c°, K_s°	ideal solution ionization (520) and solubility (529) products
k_{fr}	coefficient of sliding friction (65)
K_i	K-factor, y_i/x_i (395)
K_s	solubility product (528)
K_γ	product of activity coefficients (514)
K_ν	product of fugacity coefficients (514)
L	moles of liquid phase (396)
M	mass (9)
$\dot{M}$	mass flow rate (26)
m	molecular weight (29)
$\mathcal{M}$	number of independent chemical reactions (266)
ΔM_k	amount of mass that entered from k^{th} flow stream (29)
M_1	mass of system in state 1 (30) or mass of species 1 (6) (267)
N	number of moles (29)
$N_{i,0}$	initial number of moles of species i (263)
P	pressure (9)
$\boldsymbol{P}$	pressure tensor (75)

$\mathcal{P}$	number of phases (293)
$P^{vap}, P^{sub}, P^{sat}$	vapor (238), sublimation (238), and saturation (238) pressures
P_{atm}	atmospheric pressure (10)
$P_c, P_{c,m}$	critical pressure (168) and mixture pseudocritical pressure (346)
P_g	gauge pressure (9)
P_i	partial pressure of species i (383)
P_r, $P_{r,m}$	reduced pressure for a pure component (172) and a mixture (346)
P_{rxn}	reaction pressure (510)
Q, $\dot{Q}$	heat flow (21) and heat flow rate
$\mathbf{q}$	heat flux vector (75)
q	volumetric flow rate (546)
R	gas constant (11)
r	specific reaction rate (547)
S	entropy (86)
S_{gen}, $\dot{S}_{gen}$	entropy generated (88) and entropy generation rate (86)
S_{UV}, S_{UN}, etc.	second partial derivatives of entropy (207)
T	temperature (10)
T_c, $T_{c,m}$	critical temperature (168) and mixture pseudocritical temperature (346)
T_f	mixture freezing temperature (479)
T_{lc}, T_{uc}	lower and upper consolute temperatures (446)
T_m	melting temperature (460)
T_r, $T_{r,m}$	reduced temperature of a pure fluid (172) and a mixture (346)
T_T	triple point temperature (536)
U	internal energy (18), and mixture internal energy (251)
V	volume (11), mixture volume (251) (also moles of vapor in Secs. 8.1 and 9.4 only)
ΔV^{fus}	volume change on melting or fusion (462)
v	velocity (21)
$\mathbf{v}$	velocity vector (73)
$\underline{V}_C$	molar critical volume (168)
ΔV_{mix}	volume change on mixing (279)
V_r	reduced volume (172)
w_i	mass fraction (21)
W, $\dot{W}$	work (35) and rate at which work is supplied to system (31)
W^{NET}	net work supplied from surroundings (66)
W_{fr}	work against function (67)
W_s, $\dot{W}_s$	shaft work (35), and rate of shaft work (31)
X	any thermodynamic variable in Chap. 4, molar extent of chemical reaction (263) elsewhere
x	coordinate direction
x^I, x^{II}	fraction of mass in a phase (46)
x^L, x^V	fraction of mass in vapor and liquid phases (214)

x_i	mole fraction of species i in vapor or liquid phase (251)
Y	any thermodynamic variable (132)
y	coordinate direction
y_i	vapor-phase mole fraction (316)
Z, Z_m	compressibility of a pure fluid (148) and a mixture (349)
z	coordinate direction
Z_C	critical compressibility (172)
z_+, z_-	ionic valence (359)
α	coefficient of thermal expansion (135) (in Chap. 4)
α, β	coefficients in van Laar and Debye–Hückel equations (Chaps. 7–9)
γ	specific heat ratio (60)
γ_i, γ_i^*, $\gamma_i^\square$	activity coefficients of species i (314, 356, 357)
$\gamma_\pm$	mean ionic activity coefficient (360)
δ	solubility parameter (339)
δ_{ij}	Kronecker delta function (118)
θ	general thermodynamic variable (46)
κ_s	adiabatic compressibility (195)
κ_T	isothermal compressibility (135)
μ	viscosity (21), Joule–Thomson coefficient (141), and ionic strength (361) in Chaps. 7–9
ν	stoichiometric coefficient (263)
ν_+, ν_-	ionic stoichiometric coefficient (359)
ρ	mass density (21)
$\dot{\sigma}_s$	rate of entropy generation per unit volume (117)
$\boldsymbol{\tau}$	stress tensor (75)
τ_{xy}	component of stress tensor (21)
ω	acentric factor (174)
ϕ	viscous dissipation function (118)
Φ	volume fraction (339)
ψ	potential energy (30)

SUBSCRIPTS

Symbol	*Designates*
A, B, C, . . . ,	species
AB, D	dissociated electrolyte AB
ad	adiabatic process
c	critical property
eq	equilibrium state
i	ith species $i = 1, \ldots, \mathscr{C}$
in	inlet conditions
j	generally denotes jth reaction $j = 1, \ldots, \mathscr{M}$
k	kth flow stream; $k = 1, \ldots, K$
m	mixture property

mix	mixing or mixture
R	reference property
rxn	reaction
sat	property along two-phase coexistence line
x, y, z	coordinate direction

SUPERSCRIPTS

Symbol	*Designates*
I, II	phase
ex	excess property on mixing
fus	property change on melting or fusion
i, f	initial and final states, respectively
IG	ideal gas property
IGM	ideal gas mixture property
IM	ideal mixture property
max	maximum
rev	reversible process
sat	property along vapor–liquid coexistence line
sub	property change on sublimation
V, L, S	vapor, liquid, and solid phase, respectively
vap	property change on vaporization
z_+, z_-	charge on an ionic species

Chemical and Engineering Thermodynamics

1

Introduction

A major objective of any field of pure or applied science is to summarize a large amount of experimental information with a few basic principles. The hope, then, is that any new experimental measurement or phenomenon can be easily understood in terms of the established principles, and that predictions based on these principles will be accurate. This book demonstrates how a collection of general experimental observations can be used to establish the principles of an area of science called thermodynamics and then shows how these principles can be used to study a wide variety of physical and chemical phenomena.

1.1 THE CENTRAL PROBLEMS OF THERMODYNAMICS

Thermodynamics is the study of the changes in the state or condition of a substance when changes in its **internal energy** are important. By internal energy we mean that energy of a substance associated with the motions, interactions, and bonding of its constituent molecules, rather than the **external energy** associated with the velocity and location of its center of mass, which is of primary interest in mechanics. Thermodynamics is, however, a macroscopic science; it deals with the average changes that occur among large numbers of molecules rather than the detailed changes that occur in a single molecule. Consequently, there will be no attempt in this book to quantitatively relate the internal energy of a substance to its molecular motions and interactions, but only to other macroscopic variables such as temperature, which is primarily related to the extent of molecular motions, and density, which is a measure of how closely the molecules are packed and thus largely determines the extent of molecular interactions. The total energy of any substance is the sum of its internal energy and its bulk potential and kinetic energy; that is, it is the sum of the internal and external energies.

Our interest in thermodynamics is mainly in changes that occur in some small part of the universe, for example, within a steam engine, a laboratory beaker, or a chemical reactor. The region under study, which may be a specified volume in space or a quantity of matter, will be called the **system**; the rest of the universe is its **surroundings**. Throughout this book the term **state** will refer to the thermodynamic state of a system as characterized by its density, refractive index, composition, pressure, temperature, or other variables to be introduced later. The state of agglomeration of the system, that is, whether it is a gas, liquid, or solid, will be called its **phase**.

A system is said to be in contact with its surroundings if a change in the surroundings can produce a change in the system. Thus a thermodynamic system is in **mechanical contact** with its surroundings if a change in pressure in the surroundings results in a pressure change in the system. Similarly, a system is in **thermal contact** with its surroundings if a temperature change in the surroundings of a system can produce a change in the system. If a system does not change as a result of changes in its surroundings, the system is said to be **isolated**. Systems may be partially isolated from their surroundings. An **adiabatic** system is one that is thermally isolated from its surroundings; that is, it is a system that is not in thermal contact, but may be in mechanical contact, with its surroundings. If mass can flow into or out of the thermodynamic system, the system is said to be **open**; if not, the system is **closed**. Similarly, if heat can be added to the system, or work done on it, we will say the system is open with respect to heat and work flows.[1]

An important concept in thermodynamics is the equilibrium state, which will be discussed in detail in the following sections. Here, we merely note that if a system is not subjected to a continual forced flow of mass, heat, or work, the system will eventually evolve to a time-invariant state in which there are no internal or external flows of heat or mass and no change in composition as a result of chemical reactions. This state of the system is the **equilibrium** state. The precise nature of the equilibrium state depends on both the character of the system and the **constraints** imposed on the system by its immediate surroundings and its container (e.g., a constant-volume container fixes the system volume, a thermostatic bath fixes the system temperature; see Problem 1.1).

Using these definitions we can identify the two general classes of problems that are of interest in thermodynamics. In the first class are problems concerned with computing the amount of work or the flow of heat either required or released to accomplish a specified change of state of a system or, alternatively, the prediction of the change in the thermodynamic state that occurs for given heat or work flows. We refer to these problems as **energy flow** problems.

The second class of thermodynamic problems are those involving equilibrium. Of particular interest here is the identification or prediction of the equilibrium state of a system that initially is not in equilibrium. The most common problem of this type is the prediction of the new equilibrium state of a system that has undergone a change in the constraints that had been maintaining it in a previous state. For example, we will want to predict whether a single liquid mixture or two partially miscible liquid phases will be the equilibrium state when two pure liquids (the initial equilibrium state) are mixed (the change of constraint; see Chapter 8). Similarly, we will be interested in predicting the final temperatures and pressures of gas in two gas cylinders after opening the connecting valve (change of constraint) between a cylinder that was initially filled and another that was empty (see Chapters 2 and 3).

It is useful to mention another class of problems, related to those referred to in the previous paragraphs, that will not be considered. We will not try to

[1]Both heat and work will be defined shortly.

answer the question of how fast a system will respond to a change in constraints; that is, we will not try to study system dynamics. The answers to such problems, depending on the system and its constraints, may involve chemical kinetics, heat or mass transfer, and fluid mechanics, all of which are studied elsewhere. Thus, in the foregoing example, we will be interested in the final state of the gas in each cylinder, but not in computing how long a valve of given size must be held open to allow the necessary amount of gas to pass from one cylinder to the other. Similarly, when, in Chapter 8, we study phase equilibrium and, in Chapter 9, chemical equilibrium, our interest will be in the prediction of the equilibrium state, not how long it will take to achieve this equilibrium state.

Shortly we will start the formal development of the principles of thermodynamics, first qualitatively and then, in the following chapters, in a quantitative manner. First, however, we make a short digression to discuss the system of units to be used in this text.

1.2 A SYSTEM OF UNITS

The study of thermodynamics involves mechanical variables such as force, pressure, and work, and thermal variables such as temperature and energy. Over the years many definitions and units for each of these variables have been proposed; thus, for example, there are several values of the calorie, British thermal unit, and horsepower. Also, whole systems of units, such as the English and cgs systems, have been used. Recently, the problem of units was reconsidered and the Système International d'Unités (abbreviated SI units) was agreed to at the Eleventh General Conference on Weights and Measures in 1960. This conference was one of a series convened periodically to obtain international agreement on questions of metrology. The SI unit system will be used whenever possible in this book (unfortunately, not all thermodynamic data we need are available in SI units).

In the SI system the seven basic units listed in Table 1.2-1 are identified and their values are assigned. From these seven basic, well-defined units, the units of other quantities can be derived. Also, certain quantities appear so frequently that they have been given special names and symbols in the SI system. Those of interest here are listed in Table 1.2-2. Some other derived units acceptable in the SI system are given in Table 1.2-3, and Table 1.2-4 gives the acceptable scaling prefixes. [It should be pointed out that, except at the end of a sentence, a period is never used after a symbol of an SI unit, and the degree symbol is not used. Also, capital letters are not used in units that are written out (e.g., pascals, joules, or meters) except at the beginning of a sentence. When the units are expressed in symbols, the first letter is capitalized only when the unit name is that of a person (e.g., Pa and J, but m not M).]

Appendix I contains approximate factors to convert from various common units to acceptable SI units. In the SI unit system, energy is expressed in joules, J, with 1 joule being the energy required to move an object 1 meter when it is opposed by a force of 1 newton. Thus, $1 \text{ J} = 1 \text{ N m} = 1 \text{ kg m}^2 \text{ s}^{-2}$. A pulse of the human heart, or lifting this book 0.1 meters, requires approximately 1 joule. Since this is such a small unit of energy, kilojoules (kJ = 1000 J) are frequently used. Similarly, we will frequently use bar = 10^5 Pa = 0.987 atm as the unit of pressure.

Table 1.2-1
THE SI UNIT SYSTEM

Unit	Name	Abbreviation	Basis of Definition
Length	meter	m	Proportional to the wavelength of one of krypton-86 radiative transitions
Mass	kilogram	kg	Platinum-iridium prototype at the International Bureau of Weights and Measures, Sèvres, France
Time	second	s	Proportional to the period of one of cesium-133 radiative transitions
Electric current	ampere	A	Current that would produce a specified force between two parallel conductors in a specified geometry
Temperature	kelvin	K	1/273.16 of the thermodynamic temperature (to be defined shortly) of water at its triple point (see Chapter 5)
Amount of substance	mole	mol	Amount of a substance that contains as many elementary entities as there are atoms in 0.012 kilograms of carbon-12 (6.022×10^{23}, which is Avogadro's number)
Luminous intensity	candela	cd	Related to the black-body radiation from freezing platinum (2045 K)

Table 1.2-2
DERIVED UNITS WITH SPECIAL NAMES AND SYMBOLS ACCEPTABLE IN SI UNITS

			Expression in	
Quantity	Name	Symbol	SI Units	Derived Units
Force	newton	N	$m\ kg\ s^{-2}$	$J\ m^{-1}$
Energy, work, or quantity of heat	joule	J	$m^2\ kg\ s^{-2}$	N m
Pressure or stress	pascal	Pa	$m^{-1}\ kg\ s^{-2}$	N/m^2
Power	watt	W	$m^2\ kg\ s^{-3}$	J/s
Frequency	hertz	Hz	s^{-1}	

Table 1.2-3
OTHER DERIVED UNITS IN TERMS OF ACCEPTABLE SI UNITS

Quantity	Expression in SI Units	Symbol
Concentration of substance	mol m^{-3}	mol/m^3
Mass density (ρ)	kg m^{-3}	kg/m^3
Heat capacity or entropy	m^2 kg s^{-1} K^{-1}	J/K
Heat flow rate ($\dot{Q}$)	m^2 kg s^{-3}	W or J/s
Molar energy	m^2 kg s^{-2} mol^{-1}	J/mol
Specific energy	m^2 s^{-2}	J/kg
Specific heat capacity or specific entropy	m^2 s^{-2} K^{-1}	J/(kg K)
Specific volume	m^3 kg^{-1}	m^3/kg
Viscosity (absolute or dynamic)	m^{-1} kg s^{-1}	Pa s
Volume	m^3	m^3
Work, energy (W)	m^2 kg s^{-2}	J or N m

Table 1.2-4
PREFIX FOR SI UNITS

Multiplication Factor	Prefix	Symbol	
10^{12}	tera	T	
10^9	giga	G	
10^6	mega	M	
10^3	kilo	k	(e.g., kilogram)
10^2	hecto	h	
10	deka	da	
10^{-1}	deci	d	
10^{-2}	centi	c	(e.g., centimeter)
10^{-3}	milli	m	
10^{-6}	micro	μ	
10^{-9}	nano	n	
10^{-12}	pico	p	
10^{-15}	femto	f	

1.3 THE EQUILIBRIUM STATE

As was indicated in Sec. 1.1, the equilibrium state plays a central role in thermodynamics. The general characteristics of the equilibrium state are that (1) it does not vary with time, (2) the system is uniform, or composed of several subsystems each of which is uniform, (3) all flows (of mass, heat, or work) both within the system and between the system and its surroundings are zero, and (4) the net rate of all chemical reactions is zero.

At first it might appear that the characteristics of the equilibrium state are so restrictive that such states rarely occur. In fact, just the opposite is true. The equilibrium state will always occur, given sufficient time, as the terminal state of a system closed to the flow of mass, heat, or work across its boundaries. In addition, systems open to such flows, depending on the nature of the interaction between the system and its surroundings, may also evolve to an equilibrium state. If the surroundings merely impose a value of temperature, pressure, or volume on the system, the system will evolve to an equilibrium state. If, on the other hand, the surroundings impose a mass flow into and out of the system (as a result of a pumping mechanism), or a heat flow (as would occur if one part of the system were exposed to one temperature and another part of the system to a different temperature), the system may evolve to a time-invariant state only if the flows are steady. The time-invariant states of these **driven** systems are not equilibrium states in that the systems may or may not be uniform (this will become clear when the continuous flow stirred tank and plug flow chemical reactors are considered in Chapter 9) and certainly do not satisfy part or all of criterion (3). Such time-invariant states are called **steady states** and occur frequently in continuous chemical and physical processing. Steady-state processes are of only minor interest in this book.

Nondriven systems reach equilibrium because all spontaneous flows that occur in nature tend to dissipate the driving forces that cause them. Thus, the flow of heat that arises in response to a temperature difference occurs in the direction that dissipates the temperature difference; the mass diffusion flux that arises in response to a concentration gradient occurs in such a way that a state of uniform concentration obtains; and the flux of momentum that occurs when a velocity gradient is present in a fluid tends to dissipate that gradient. Similarly, chemical reactions occur in a direction that drives the system toward equilibrium (Chapter 9). At various points throughout this book it will be useful to distinguish between the flows that arise naturally and drive the system to equilibrium, which we will call **natural flows**, and flows imposed on the system by its surroundings, which we term **forced flows**.

An important experimental observation in thermodynamics is that any system free from forced flows will, given sufficient time, evolve to an equilibrium state. This empirical fact will be used repeatedly in our discussion.

It is useful to distinguish between two types of equilibrium states according to their response to small disturbances. To be specific, suppose a system in equilibrium is subjected to a small disturbance that is then removed (e.g., temperature fluctuation or pressure pulse). If the system returns to its initial equilibrium state, this state of the system is said to be **stable** with respect to small disturbances. If, however, the system does not return to the initial state, that state is said to have been **unstable**.

There is a simple mechanical analogy, shown in Fig. 1.3-1, that can be used to illustrate the concept of stability. Figures 1.3-1*a*, *b*, and *c* each represent equilibrium positions of a block on a horizontal surface. The configuration in Fig. 1.3-1*c* is, however, precarious; an infinitesimal movement of the block in any direction (so that its center of gravity is not directly over the pivotal point) would cause the block to revert to the configuration of either Fig. 1.3-1*a* or *b*. Thus, Fig.

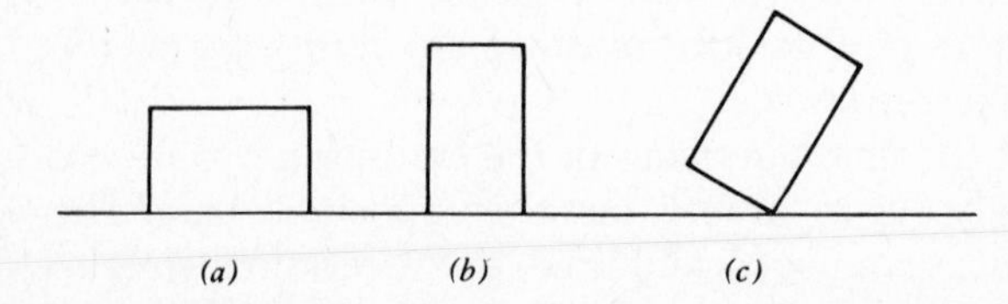

Figure 1.3-1
States (*a*) and (*b*) are both stable with respect to small mechanical disturbances, the delicately balanced block in state (*c*) is not.

1.3-1*c* represents an unstable equilibrium position. The configurations of both Figs. 1.3-1*a* and *b* are not affected by small disturbances, and these states are stable. Intuition suggests that the configuration of Fig. 1.3-1*a* is the most stable; clearly it has the lowest center of gravity and hence the lowest potential energy. To go from the configuration of Fig. 1.3-1*b* to that of Fig. 1.3-1*a*, the block must pass through the still higher potential energy state indicated in Fig. 1.3-1*c*. If we let $\Delta\varepsilon$ represent the potential energy difference between the configurations of Fig. 1.3-1*b* and *c*, we can clearly see the equilibrium state of Fig. 1.3-1*b* is stable to energy disturbances less than $\Delta\varepsilon$ in magnitude and unstable to larger disturbances.

Certain equilibrium states of thermodynamic systems are stable to small fluctuations; others are not. For example, the equilibrium state of a simple gas is stable to all fluctuations, as are most of the equilibrium states we will be concerned with. It is possible, however, to carefully prepare a subcooled liquid, that is, a liquid below its normal solidification temperature, which satisfies the equilibrium criteria. This is an unstable equilibrium state because the slightest disturbance, such as tapping on the side of the containing vessel, will cause the liquid to freeze. One sometimes encounters mixtures that, by the chemical reaction equilibrium criterion (see Chapter 9), should react; however, the chemical reaction rate is so small as to be immeasurable at the temperature of interest. Such a mixture can achieve a state of thermal equilibrium that is stable with respect to small fluctuations of temperature and pressure. If, however, there is a sufficiently large, but temporary, temperature fluctuation (so that the rate of the chemical reaction is appreciable for some period of time), a new thermal equilibrium state with a chemical composition that differs from the initial state will be obtained. The initial equilibrium state, like the mechanical state in Fig. 1.3-1*b*, is then said to be stable with respect to small disturbances, but not to large disturbances.

Unstable equilibrium states are rarely encountered in nature unless they have been specially prepared (e.g., the subcooled liquid mentioned earlier). The reason for this is that during the approach to equilibrium, temperature gradients, density gradients or other nonuniformities that exist within a system are of a sufficient magnitude to act as disturbances to unstable states and prevent their natural occurrence. In fact, the natural occurrence of an unstable thermodynamic equilibrium state is about as likely as the natural occurrence of the unstable mechanical equilibrium state of Fig. 1.3-1*c*. Consequently, our concern in this book will be mainly with stable equilibrium states.

If an equilibrium state is stable with respect to all disturbances, the properties of this state cannot depend on the past history of the system or, to be more specific, on the path followed during the approach to equilibrium. Similarly, if an equilibrium state is stable with respect to small disturbances, its properties do not depend on the path followed in the immediate vicinity of the equilibrium state. We can establish the validity of the latter statement by the following thought experiment (the validity of the first statement follows from a simple generalization of the argument). Suppose a system in a stable equilibrium state is subjected to a small temporary disturbance of a completely arbitrary nature. Since the initial state was one of stable equilibrium, the system will return to precisely that state after the removal of the disturbance. However, since any type of small disturbance is permitted, the return to the equilibrium state may be along a path that is different from the path followed in initially achieving the stable equilibrium state. The fact that the system is in exactly the same state as before means that all the properties of the system that characterize the equilibrium state must have their previous values; the fact that different paths were followed in obtaining this equilibrium state implies that none of these properties can depend on the path followed.

Another important experimental observation for the development of thermodynamics is that a system in a stable equilibrium state will never *spontaneously* evolve to a state of nonequilibrium. For example, any temperature gradients in a thermally conducting material free from a forced flow of heat will eventually dissipate so that a state of uniform temperature is achieved. Once this equilibrium state has been achieved, a macroscopic temperature gradient will never spontaneously occur in the material.

The two observations that (1) a system free from forced flows will evolve to an equilibrium state, and (2) once in equilibrium a system will never spontaneously evolve to a nonequilibrium state, are evidence for a unidirectional character of natural processes. Thus we can take as a general principle that the direction of natural processes is such that systems evolve toward an equilibrium state, not away from it.

1.4 PRESSURE, TEMPERATURE, AND EQUILIBRIUM

Most people have at least a primitive understanding of the notions of temperature, pressure, heat, and work, and we have, perhaps unfairly, relied on this understanding in previous sections. Since these concepts are important for the development of thermodynamics, each will be discussed in detail here and in the following sections.

The concept of pressure as being the total force exerted on an element of surface divided by the surface area should be familiar from courses in physics and chemistry. Pressure, or equivalently, force, is important in both mechanics and thermodynamics because it is closely related to the concept of mechanical equilibrium. This is simply illustrated by considering the two piston-and-cylinder devices shown in Figs. 1.4-1*a* and *b*. In each case we will assume that the piston and cylinder have been carefully machined so that there is no friction

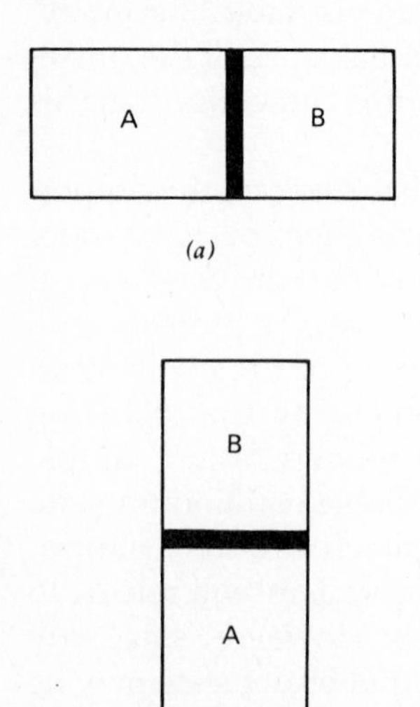

Figure 1.4-1
The piston separating gases A and B and the cylinder containing them have been carefully machined so that the piston moves freely in the cylinder.

between them. From elementary physics we know that for the systems in these figures to be in mechanical equilibrium (as recognized by an absence of movement of the piston) there must be no unbalanced forces; the pressure of gas A must equal that of gas B in the system of Fig. 1.4-1*a*, and, in the system of Fig. 1.4-1*b*, it must just be equal to the sum of the pressure of gas B and the force of gravity on the piston divided by its surface area. Thus, the requirement that a state of mechanical equilibrium exists is really a restriction on the pressure of the system.

Since pressure is a force per unit area, the direction of the pressure scale is evident; the greater the force per unit area, the greater the pressure. To measure pressure, one uses a pressure gauge, a device that produces a change in some indicator, such as the position of a pointer, the height of a column of liquid, or the electrical properties of a specially designed circuit, in response to a change in pressure. Pressure gauges are calibrated using devices such as that shown in Fig. 1.4-2. There, known pressures are created by placing weights on a frictionless piston of known weight. The pressure at the gauge P_g due to the metal weight and the piston is

$$P_g = \frac{M_w + M_P}{A} g \tag{1.4-1}$$

Here g is the local acceleration of gravity on an element of mass; the standard value is 9.80665 m/s^2. The position of the indicator at several known pressures is recorded, and the scale of the pressure gauge is completed by interpolation.

There is, however, a complication with this calibration procedure. It arises because the weight of the air of the earth's atmosphere produces an average pressure of 14.696 pounds force per square inch or 101.3 kilopascals at sea level. Since atmospheric pressure acts equally in all directions, we usually are not aware of its presence, so that in most nonscientific uses of pressure the zero of the pressure scale is the sea level atmospheric pressure (i.e., the pressure of the atmosphere is neglected in the pressure gauge calibration). Thus, when the recommended inflation pressure of an automobile tire is 200 kPa, what is really meant is 200 kPa above atmospheric pressure. We refer to pressures on such a

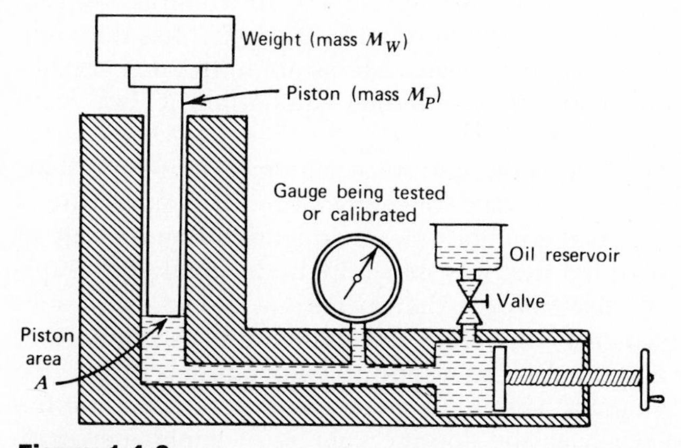

Figure 1.4-2
A simple deadweight pressure tester. (The purpose of the oil reservoir and the system volume adjustment are to maintain equal heights of the oil column in the cylinder and gauge sections, so that no corrections for the height of the liquid column need be made in the pressure calibration.)

scale as gauge pressures. Note that gauge pressures may be negative (in partially or completely evacuated systems), zero, or positive, and errors in pressure measurement result from changes in atmospheric pressure from the gauge calibration conditions (e.g., using gauge calibrated in New York for pressure measurements in Denver).

We define the total pressure P to be equal to the sum of the gauge pressure P_g and the ambient atmospheric pressure P_{atm}. By accounting for the atmospheric pressure in this way we have developed an **absolute pressure** scale; that is, a pressure scale with zero as the lowest pressure attainable (the pressure in a completely evacuated region of space). One advantage of such a scale is its simplicity; the pressure is always a positive quantity and measurements do not have to be corrected for either fluctuations in atmospheric pressure or its change with height above sea level. We will frequently be concerned with interrelationships between the temperature, pressure, and specific volume of fluids. These interrelationships are simplest if the absolute pressure is used. Consequently, unless otherwise indicated, the term pressure in this book will refer to absolute pressure.

Although pressure arises naturally from mechanics, the concept of temperature is more abstract. To the nonscientist, temperature is a measure of hotness or coldness and as such is not carefully defined, but rather is a quantity related to such things as physical comfort, cooking conditions, or the level of mercury or colored alcohol in a thermometer. To the scientist, temperature is a precisely defined quantity, deeply rooted in the concept of equilibrium, and related to the energy content of a substance.

The origin of the formal definition of temperature is in the concept of thermal equilibrium. Consider a thermodynamic system composed of two subsystems that are in thermal contact but that do not interchange mass (e.g., the two subsystems may be two solids in contact, or liquids or gases separated by a thin, impenetrable barrier or membrane) and are isolated from their surroundings. When this composite system achieves a state of equilibrium (detected by observing that the properties of each system are time invariant), it is found that that property measured by the height of fluid in a given thermometer is the same in each system, although the other properties of the subsystems, such as their density and chemical composition, may be different. In accord with this observation, **temperature** is defined to be that system property which, if it has the same value for any two systems, indicates that these systems are in thermal equilibrium if they are in contact, or would be in thermal equilibrium if they were placed in thermal contact.

Although this definition provides the link between temperature and thermal equilibrium, it does not suggest a scale for temperature. If temperature is used only as an indicator of thermal equilibrium, any quantification or scale of temperature is satisfactory provided that it is generally understood and reproducible, though the accepted convention is that increasing hotness of a substance should correspond to increasing values of temperature. An important consideration in developing a thermodynamic scale of temperature is that it, like all other aspects of thermodynamics, should be general and not depend on the properties of any one fluid (such as the specific volume of liquid mercury). Experimental evidence indicates that it should be possible to formulate a completely universal temperature scale. The first indication came from the study of gases at densities so low that intermolecular interactions are unimportant (such gases are called **ideal gases**), where it was found that the product of the absolute pressure P and the molar volume $\underline{V}$ of any low-density gas away from its condensation line (see Chapter 5) increases with increasing hotness. This observation has been used as the basis for a temperature scale by defining the temperature $\mathcal{T}$ to be linearly proportional to the product of $P\underline{V}$ for a particular low-density gas, that is, by choosing $\mathcal{T}$ so that

$$P\underline{V} = A + R\mathcal{T} \tag{1.4-2}$$

where A and R are constants. In fact, without any loss of generality one can define a new temperature $T = \mathcal{T} + (A/R)$, which differs from the choice of Eq. 1.4-2 only by an additive constant, to obtain

$$P\underline{V} = RT \tag{1.4-3}$$

Since neither the absolute pressure nor the molar volume of a gas can ever be negative, the temperature defined in this way must always be positive, the ideal gas temperature scale of Eq. 1.4-3 is an absolute scale (i.e., $T \geqslant 0$).

To complete this low-density gas temperature scale it still remains to specify the constant R, or equivalently the size of a unit of temperature. This can be done in two equivalent ways. The first is to specify the value of T for a given value of $P\underline{V}$ and thus determine the constant R; the second is to choose two reproducible points on a hotness scale and to decide arbitrarily how many units of T correspond to the difference in the product of $P\underline{V}$ at these two fixed points. In fact, it is the latter procedure that is used; the ice point temperature of water[2] and the boiling point temperature of water at atmospheric pressure (101.3 kPa) providing the two reproducible fixed-point temperatures. What is done, then, is to allow a low-density gas to achieve thermal equilibrium with water at its ice point and measure the product $P\underline{V}$, and then repeat the process at the boiling temperature. One then decides how many units of temperature correspond to this measured difference in the product $P\underline{V}$; the choice of 100 units or degrees leads to the Kelvin temperature scale, whereas the use of 180 degrees leads to the Rankine scale. With either of these choices, the constant R can be evaluated for a given low-density gas. *The important fact for the formulation of a universal temperature scale is that the constant R and hence the temperature scales determined in this way are the same for all low-density gases*! Values of the gas constant R in SI units are given in Table 1.4-1.

For the present we will assume this low-density or ideal gas Kelvin (denoted by K) temperature scale is equivalent to an absolute universal thermodynamic temperature scale; this will be proved in Chapter 4.

Table 1.4-1
THE GAS CONSTANT

$R = 8.314$	J/mol K
$= 8.314$	N m/mol K
$= 8.314 \times 10^{-3}$	kPa m^3/mol K
$= 8.314 \times 10^{-5}$	bar m^3/mol K
$= 8.314 \times 10^{-2}$	bar m^3/kmol K
$= 8.314 \times 10^{-6}$	MPa m^3/mol K

[2]The freezing temperature of water saturated with air at 101.3 kPa. On this scale the triple point of water is 0.01°C.

More common than the Kelvin temperature scale for nonscientific uses of temperature, are the closely related Fahrenheit and Celsius scales. The size of the degree is the same in both the Celsius (°C) and Kelvin temperature scales. However, the zero point of the Celsius scale is arbitrarily chosen to be the ice point temperature of water. Consequently, it is found that

$$T(\mathrm{K}) = T(^\circ\mathrm{C}) + 273.15 \tag{1.4-4a}$$

In the Fahrenheit (°F) temperature scale the ice point and boiling point of water (at 101.3 kPa) are 32°F and 212°F, respectively. Thus

$$T(\mathrm{K}) = \frac{T(^\circ\mathrm{F}) + 459.67}{1.8} \tag{1.4-4b}$$

Since we are assuming, for the present, that only the ideal gas Kelvin temperature scale has a firm thermodynamic basis, we will use it, rather than the Fahrenheit and Celsius scales, in all thermodynamic calculations.[3] (Another justification for the use of an absolute temperature scale is that the interrelation between the pressure, volume, and temperature for fluids is simplest when absolute temperature is used). Consequently, if the data for a thermodynamic calculation are not given in terms of absolute temperature, it will generally be necessary to convert these data to absolute temperatures using Eqs. 1.4-4.

The product of $P\underline{V}$ for a low-density gas is said to be a **thermometric property** in that to each value of $P\underline{V}$ there corresponds only a single value of temperature. The ideal gas thermometer is not convenient to use, however, because of both its mechanical construction (see Fig. 1.4-3) and the manipulation required to make a measurement. Therefore, common thermometers make use of thermometric properties of other materials; for example, the single-valued relation between temperature and the specific volume of liquid mercury (Problem 1.2) or the electrical resistance of platinum wire. There are two steps in the construction of thermometers based on these other thermometric properties. First, fabrication of the device, such as sealing liquid mercury in an otherwise evacuated tube, and then the calibration of the thermometric indicator with a known temperature scale. To calibrate a thermometer its readings (e.g., the height of a mercury column) are determined at a collection of known temperatures, and its scale is completed by interpolating between these fixed points. The calibration procedure for a common mercury thermometer is usually far simpler. The height of the mercury column is determined at only two fixed points (e.g., the temperature of an ice-water bath and the temperature of boiling water at atmospheric pressure), and the distance between these two heights is divided into equal units; the number of units depends on whether the Rankine or Kelvin degree is used, and whether a unit is to represent a fraction of a degree, a degree, or several degrees. Since only two fixed points are used in the calibration, intermediate temperatures recorded on such a thermometer may be different from those that would be obtained using an ideal gas thermometer because (1) the specific volume of liquid mercury has a slightly nonlinear dependence on temperature, and (2) the diameter of the capillary tube may not be completely uniform (so that the volume of mercury will not be simply related to its height in the tube; see Problem 1.2).

[3]Of course for calculations involving only temperature differences, any convenient temperature scale may be used, since a temperature difference is independent of the zero of the scale.

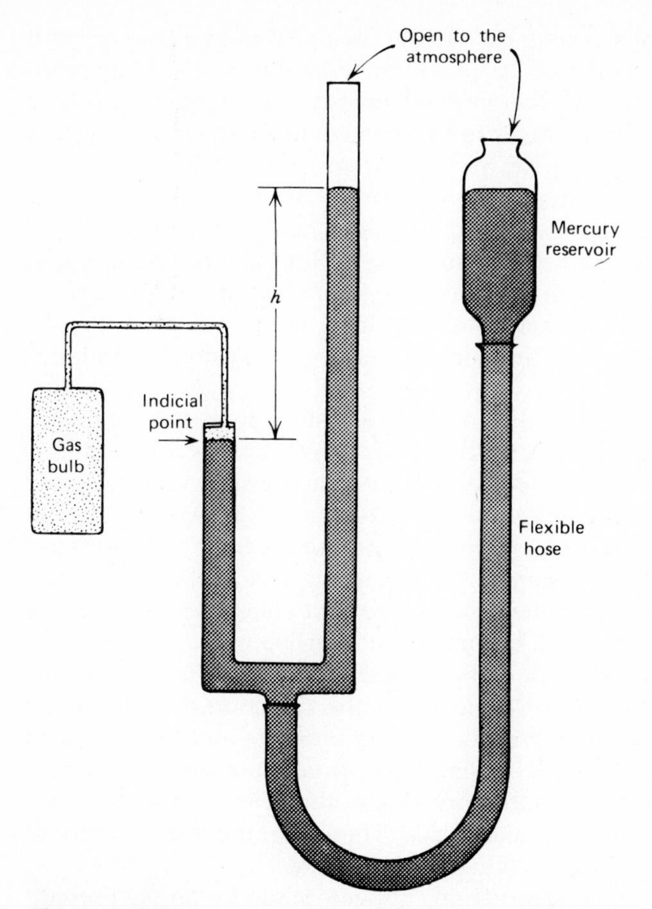

Figure 1.4-3
A simplified diagram of a constant-volume ideal gas thermometer. In this thermometer the product $P\underline{V}$ for a gas at various temperatures is found by measuring the pressure P at constant volume. For each measurement the mercury reservoir is raised or lowered until the mercury column at the left touches an index mark. The pressure of the gas in the bulb is then equal to the atmospheric pressure plus the pressure due to the height of the mercury column.

1.5 HEAT, WORK, AND THE CONSERVATION OF ENERGY

As we have already indicated, two systems in thermal contact but otherwise isolated from their surroundings will eventually reach an equilibrium state in which the systems have the same temperature. During the approach to this equilibrium state the temperature of the initially low-temperature system increases, while that of the initially high-temperature system decreases. We know that the temperature of a substance is directly related to its internal energy, especially the energy of molecular motion. Thus, in the approach to equilibrium, energy has been transferred from the high-temperature system to the one of lower temperature. This transfer of energy as a result of only a temperature difference is called a flow of **heat**.

It is also possible to increase the total energy (internal, potential, and kinetic) of a system by mechanical processes involving motion. In particular, the kinetic or potential energy of a system can change as a result of motion without deformation of the system boundaries, as in the movement of a solid body acted on by an external force, whereas the internal energy and temperature of a system may change when external forces result in the deformation of the system boundaries, as in the compression of a gas. Energy transfer by mechanical motion also occurs as a result of the motion of a drive shaft, push rod, or similar device across the system boundaries. For example, mechanical stirring of a fluid first results in fluid motion (evidence of an increase in fluid kinetic energy), and then, as this motion is damped by the action of the fluid viscosity, in an increase in the temperature (and internal energy) of the fluid. Energy transfer by any mechanism that involves mechanical motion of or across the system boundaries is called **work**.

Finally, it is possible to increase the energy of a system by supplying it with electrical energy in the form of an electrical current driven by a potential difference. This electrical energy can be converted into mechanical energy if the system contains an electric motor, it can increase the temperature of the system if it is dissipated through a resistor (resistive heating), or it can be used to cause an electrochemical change in the system (e.g., recharging a lead storage battery). Throughout this book we consider the flow of electrical energy to be a form of work. The reason for this choice will become clear shortly.

The amount of mechanical work is, from mechanics, equal to the product of the force exerted times the distance moved in the direction of the applied force, or, alternatively, to the product of the applied pressure and the displaced volume. Similarly, the electrical work is equal to the product of the current flow through the system, the potential difference across the system, and the time interval over which the current flow takes place. Therefore, the total amount of work supplied to a system is frequently easy to calculate.

An important experimental observation, initially made by James Prescott Joule between the years 1837 and 1847, is that a specified amount of energy can always be used in such a way as to produce the same temperature rise in a given mass of water, regardless of the precise mechanism or device used to supply the energy, and regardless of whether this energy is in the form of mechanical work, electrical work, or heat. Rather than describe Joule's experiments, consider how this hypothesis could be proved in the laboratory. Suppose a sample of water at temperature T_1 is placed in a well-insulated container (e.g., a Dewar flask) and, by the series of experiments in Table 1.5-1, the amount of energy expended in producing a final equilibrium temperature T_2 is measured. Based on the experiments of Joule and others, we would expect to find that this energy (determined by correcting the measurements of column 4 for the temperature rise of the container and the effects of column 5) to be precisely the same in all cases.

By comparing the first two experiments with the third and fourth, we conclude that there is an equivalence between mechanical energy (or work) and heat, in that precisely the same amount of energy was required to produce a given temperature rise, independent of whether this energy was delivered as heat or work. Furthermore, since both mechanical and electrical energy sources have been used (see column 3), there is similar equivalence between mechanical and electrical energy, and hence among all three energy forms. This conclusion is not specific to the experiments in Table 1.5-1, but is, in fact, a special case of a more general experimental observation; that is, any change of state in a system that occurs solely as a result of the addition of heat can also be produced by

Table 1.5-1
EXPERIMENTS DESIGNED TO PROVE THE ENERGY EQUIVALENCE OF HEAT AND WORK

Form in which Energy is Transferred to Water	Mechanism Used	Form of Energy Supplied to Mechanism	Method of Measuring Energy Input	Corrections that Must be Made to Energy Input Data
(1) Mechanical energy	Stirring: Paddlewheel driven by electric motor	Electrical energy	Product of voltage, current, and time	Electrical energy loss in motor and circuit, temperature rise of paddlewheel
(2) Mechanical energy	Stirring: Paddlewheel driven by pulley and falling weight	Mechanical energy	Change in potential energy of weight: product of mass of weight, change in height, and g	Temperature rise of paddlewheel
(3) Heat flow	Electrical energy converted to heat in a resistor	Electrical energy	Product of voltage, current, and time	Temperature rise of resistor and electrical losses in circuit
(4) Heat flow	Mechanical energy of falling weight is converted to heat through friction of rubbing two surfaces together as with a brake on the axle of a pulley	Mechanical energy	Change in potential energy of weight: product of mass of weight, change in height, and g	Temperature rise of mechanical brakes, etc.

adding the same amount of energy as work, electrical energy, or a combination of heat, work, and electrical energy.

Returning to the experiments of Table 1.5-1, we can now ask what has happened to the energy that was supplied to the water. The answer, of course, is that at the end of the experiment the temperature, and hence the molecular energy, of the water has increased. Consequently, the energy added to the water is now present as increased internal energy. It is possible to extract this increased internal energy by processes that return the water to its original temperature. One could, for example, use the warm water to heat a metal bar. The important experimental observation here is that if you measured the temperature rise in the metal, which occurred in returning the water to its initial state, and compared it with the electrical or mechanical energy required to cause the same temperature rise in the metal, you would find that all the energy added to the water in raising its temperature could be recovered as heat by returning the water to its initial state. Thus, total energy has been conserved in the process.

The observation that energy has been conserved in this experiment is only one example of a general energy conservation principle that is based on a much wider range of experiments. The more general principle is that in any change of state, the total energy, which is the sum of the internal, kinetic, and potential energy of the system, heat, and electrical and mechanical work, is conserved. A more succinct statement is that energy is neither created nor destroyed, but may change in form.

Although heat, mechanical work and electrical work are equivalent in that a given energy input, in any form, can be made to produce the same internal energy increase in a system, there is an equally important difference among the various energy forms. To see this, suppose that the internal energy of some system (perhaps the water sample in the experiments just considered) has been increased by increasing its temperature from T_1 to a higher temperature T_2, and we now wish to recover the added energy by returning the system to its initial state at temperature T_1. It is clear that we can recover the added energy completely as a heat flow merely by putting the system in contact with another system at a lower temperature. There is, however, no process or device by which it is possible to convert all the added internal energy of the system to mechanical energy and restore both the system and the surroundings to their initial states, even though the increased internal energy may have resulted from adding only mechanical energy to the system. In general, only a portion of the increased internal energy can be recovered as mechanical energy, the remainder appearing as heat. This situation is not specific to the experiments discussed here; it occurs in all similar efforts to convert both heat and internal energy to work or mechanical energy.

We use the term **thermal energy** to designate energy in the form of internal energy and heat, and **mechanical energy** to designate mechanical and electrical work and the external energy of a system. This distinction is based on the general experimental observation that while, in principle, any form of mechanical energy can be completely converted to other forms of mechanical energy or thermal energy, only a fraction of the thermal energy can be converted into mechanical energy in any **cyclic process** (that is, in a process which at the end of the cycle restores the system and surroundings to their states at the beginning of the cycle). In Chapter 3 we consider what this fraction is and on what it depends.

The units of mechanical work arise naturally from its definition as the product of a force and a distance. Typical units of mechanical work are foot-pounds force, dyne-centimeters, and ergs, though we will use newton-meters (N m), which are equal to joules; and from the formulation of work as pressure times displaced volume, pascal-meters^3, which also equal a joule. The unit of electrical work is the volt-ampere-second or, equivalently, the watt-second (again equal to one joule). Heat, however, not having a mechanical definition, has traditionally been defined experimentally. Thus, the heat unit calorie had been defined as the amount of heat required to raise the temperature of one gram of water from 14.5°C to 15.5°C, and the British thermal unit (Btu) was defined to be the amount of heat required to raise 1 pound of water from 59°F to 60°F. These experimental definitions of heat units have proved unsatisfactory because the amount of energy in both the calorie and Btu have been subject to continual change as caloric measurement techniques improved. Consequently, there are several different definitions of the Btu and calorie (e.g., the thermochemical calorie, the mean calorie, and the International Table calorie) that differ by less than one and one-half parts in a thousand. Current practice is to recognize the energy equivalence of heat and work and to use a common energy unit for both. We will use the joule, which is equal to 0.2390 calories (thermochemical) or 0.9485×10^{-3} Btu (thermochemical).

1.6 SPECIFICATION OF THE EQUILIBRIUM STATE; INTENSIVE AND EXTENSIVE VARIABLES; EQUATIONS OF STATE

Since our main interest throughout this book is with stable equilibrium states, it is important to consider how to characterize the equilibrium state and, especially, what minimum number of properties of a system in equilibrium must be specified to fix the values of all its remaining properties completely.[4] To be specific, suppose we had one kilogram of a pure gas in equilibrium, say oxygen, whose temperature is some value T, pressure some value P, volume V, refractive index R, electrical permitivity ε, and so on, and we wanted to adjust *some* of the equilibrium properties of a second sample of oxygen so that *all* the properties of the two samples would be identical. The questions we are asking then is what sort of properties, and how many properties, must correspond if all the properties of both systems are to be identical?

The fact that we are interested only in stable equilibrium states is sufficient to decide the type of properties needed to specify the equilibrium state. First, since gradients in velocity, pressure, and temperature cannot be present in the equilibrium state, they do not enter into its characterization. Next, since, as we saw in Sec. 1.3, the properties of a stable equilibrium state do not depend on the past history of the system or its approach to equilibrium, the *stable equilibrium state is characterized by only equilibrium properties of the system.*

The remaining question, that of how many equilibrium properties are necessary to specify the equilibrium state of the system, can only be answered by experiment. The important experimental observation here is that an equilibrium state of a single-phase, one-component system in the absence of external electric and magnetic fields is completely specified if its mass and two other thermodynamic properties are given. Thus, to go back to our example, if the second oxygen sample also weighs one kilogram, and if it were made to have the same

[4]Throughout this book, we will implicitly assume that a system contains large numbers of molecules (at least several tens of thousands), so surface effects of very small systems are unimportant.

temperature and pressure as the first sample, it would also be found to have the same volume, refractive index, and so forth. If, however, only the temperature of the second one kilogram sample was set equal to that of the first sample, neither its pressure nor any other physical property would necessarily be the same as those of the first sample. Consequently, the values of the density, refractive index, and, more generally, all thermodynamic properties of an equilibrium single-component, single-phase fluid are completely fixed once the mass of the system and the values of at least two other system parameters are given. (The specification of the equilibrium state of multiphase and multicomponent systems will be considered in Chapters 5 and 6.)

The specification of an equilibrium system can be made slightly simpler by recognizing that the variables used in thermodynamic descriptions are of two different types. To see this, consider a gas of mass M, which has a temperature T, pressure P, and is confined to a glass bulb of volume V. Suppose that an identical glass bulb is also filled with mass M of the same gas and heated to the same temperature T. Based on the previous discussion, the pressure in the second glass bulb is also P. If these two bulbs are now connected to form a new system, the temperature and pressure of this composite system are unchanged from those of the separated systems, although the volume and mass of this new system are clearly twice that of the original single glass bulb. The pressure and temperature, because of their size-independent property, are called **intensive variables**, whereas the mass, volume, and total energy are **extensive variables**, or variables dependent on the size of the system. Extensive variables can be transformed into intensive variables by dividing by the total mass or total number of moles so that a specific volume (volume per unit mass or volume per mole), a specific energy (energy per unit mass or per mole), and so forth are obtained. By definition, the term **state variable** will refer to any of the intensive variables of an equilibrium system: temperature, pressure, specific volume, specific internal energy, refractive index, and other variables introduced in the following chapters. Clearly, from the previous discussion, the value of any state variable depends only on the equilibrium state of the system, not on the path by which the equilibrium state was reached.

With the distinction now made between intensive and extensive variables, it is possible to rephrase the requirement for the complete specification of a thermodynamic state in a more coherent manner. The experimental observation is that the specification of two state variables uniquely determines the values of all other state variables of an equilibrium, single-component, single-phase system. [Remember, however, that to determine the size of the system, that is, its mass or total volume, one must also specify the mass of the system, or the value of one extensive parameter (total volume, total energy, etc.).] The implication of this statement is that for each substance there exists, in principle, equations relating each state variable to two others. For example

$$P = P(T, \hat{V})$$

and

$$\hat{U} = \hat{U}(T, \hat{V}) \tag{1.6-1}$$

Here P is the pressure, T the temperature, $\hat{U}$ the internal energy per unit mass, and $\hat{V}$ the volume per unit mass.[5] The first equation indicates that the pressure is

[5]In this book we will use a caret, as on $\hat{U}$ and $\hat{V}$, to indicate properties per unit mass, and an underline, as on $\underline{U}$ and $\underline{V}$, to indicate properties per mole, which will be referred to as molar properties.

a function of the temperature and volume, the second indicates that the internal energy is a function of temperature and volume. Also, there are relations of the form

$$\hat{U} = \hat{U}(T, P)$$
$$\hat{U} = \hat{U}(P, \hat{V})$$
$$P = P(\hat{U}, \hat{V}) \tag{1.6-2}$$

Similar equations will also be valid for the additional thermodynamic properties to be introduced later.

The interrelations of the form of Eqs. 1.6-1 and 2 are always obeyed in nature, though we may not have been sufficiently accurate in our experiments, or in other ways clever enough to have discovered them. In particular, Eq. 1.6-1 indicates that if we prepare a fluid such that it has specified values T and $\hat{V}$, it will always have the same pressure P. What is this value of the pressure P? To know this we have to have either done the experiment sometime in the past or know the exact functional relationship between T, $\hat{V}$, and P for the fluid being considered. What is frequently done for fluids of scientific or engineering interest is to make a large number of measurements of P, $\hat{V}$, and T, and then try to infer the **volumetric equation of state** for the fluid, that is, the mathematical relationship between the variables P, $\hat{V}$, and T. Similarly, measurements of $\hat{U}$, $\hat{V}$, and T are made to infer a **thermal equation of state** for the fluid. Alternatively, the data that have been obtained may be presented directly in graphical or tabular form.

There are some complications in the description of thermodynamic states of systems. For certain idealized fluids, such as the ideal gas and the incompressible liquid (both discussed in Sec. 2.4), the specification of any two state variables may not be sufficient to fix the thermodynamic state of the system. To be specific, the internal energy of the ideal gas is a function only of its temperature, and not of its pressure or density. Thus, the specification of the energy and temperature of an ideal gas contains no more information than specifying only its temperature and is insufficient to determine its pressure. Similarly, if a liquid is incompressible, its molar volume will depend on temperature but not on the pressure exerted on it. Consequently, specifying the temperature and the specific volume of an incompressible liquid contains no more information than specifying only its temperature. The ideal gas and the incompressible liquid are limiting cases of the behavior of real fluids, so that although the internal energy of a real gas depends on density and temperature, the density dependence may be weak and the densities of most liquids are only weakly dependent on their pressure. Therefore, although in principle any two state variables may be used to describe the thermodynamic state of a system, it is best to avoid the combinations of $\hat{U}$ and T in gases and $\hat{V}$ and T in liquids and solids.

As was pointed out in the previous paragraphs, two state variables are needed to fix the thermodynamic state of an equilibrium system. The obvious next question is how does one specify the thermodynamic state of a nonequilibrium system? This is clearly a much more complicated question to answer, and the detailed answer would involve a discussion of the relative time scales for changes imposed on the system and the changes that occur within the system (as a result of chemical reaction, internal energy flows, and fluid motion). Such a discussion is beyond the scope of this book. The important observation is that if we do not consider very fast system changes (as occur within a shock wave), nor systems that relax at a very slow, but perceptible rate (e.g., molten polymers), the equilibrium relationships between the fluid properties, such as the volumetric and thermal equations of state, are also satisfied in nonequilibrium flows on a point-by-point basis. That is, even though the temperature and pressure may vary in a flowing fluid, as long as the changes are not as sharp as in a shock wave and the fluid internal relaxation times are rapid,[6] the properties at each point in the fluid are interrelated by the same equations of state as for the equilibrium fluid. This situation is referred to a **local equilibrium**. This is an important concept, since it allows us to consider not only equilibrium phenomena in thermodynamics, but also energy flow problems involving distinctly nonequilibrium processes.

1.7 A SUMMARY OF IMPORTANT EXPERIMENTAL OBSERVATIONS

An objective of this book is to present the subject of thermodynamics in a logical, coherent manner. We will try to do this by demonstrating how the complete structure of thermodynamics can be built from a number of important experimental observations, some of which have been introduced in this chapter, some of which are familiar from mechanics, and others that will be introduced in the following chapters. For convenience, the most important of these observations are listed here.

From classical mechanics and chemistry we have the following observations.

Experimental Observation 1. In any change of state (except those involving nuclear reaction, which will not be considered in this book) total mass is conserved.

Experimental Observation 2. In any change of state total momentum is a conserved quantity.

In this chapter the following experimental facts have been mentioned.

Experimental Observation 3 (Section 1.5). In any change of state the total energy, which includes internal, potential, and kinetic energy, heat, and work, is a conserved quantity.

Experimental Observation 4 (Section 1.5). A flow of heat and a work flow are equivalent in that supplying a given amount of energy to a system in either of these forms can be made to result in the same increase in its internal energy. Heat and work, or more generally, thermal and mechanical energy, are not equivalent in the sense that mechanical energy can be completely converted into thermal energy, but thermal energy can be only partially converted into mechanical energy in a cyclic process.

Experimental Observation 5 (Section 1.3). A system that is not subject to forced flows of mass or energy from its surroundings will evolve to a time-invariant state that is uniform or composed of uniform subsystems. This is the equilibrium state.

Experimental Observation 6 (Section 1.3). A system in equilibrium with its surroundings will never spontaneously revert to a nonequilibrium state.

[6] The situation being considered here is not as restrictive as it appears. In fact, it is, by far, the most frequent case in engineering. It is the only case that will be considered in this book.

Experimental Observation 7 (Section 1.3). Equilibrium states that arise naturally are stable to small disturbances.

Experimental Observation 8 (Sections 1.3 and 1.6). The stable equilibrium state of a system is completely characterized by values of only equilibrium properties (and not properties that describe the approach to equilibrium). For a single-component, single-phase system the values of only two state variables are needed to fix the thermodynamic state of the equilibrium system completely; the further specification of one extensive variable of the system fixes its size.

Experimental Observation 9 (Section 1.6). The interrelationships that exist between the thermodynamic state variables for a fluid in equilibrium also prevail locally (i.e., at each point) for a fluid not in equilibrium, provided the internal relaxation processes are rapid with respect to the rate at which changes are imposed on the system. For fluids of interest in this book, this condition is satisfied.

Although we will not, in general, be interested in the detailed description of nonequilibrium systems, it is useful to note that the rates at which natural relaxation processes (i.e., heat fluxes, mass fluxes, etc.) occur are directly proportional to the magnitude of the driving forces (i.e., temperature gradients, concentration gradients, etc.) for their occurrence.

Experimental Observation 10. The flow of heat $\dot{Q}$ (J/s or W) that arises because of a temperature difference ΔT is linearly proportional to the magnitude of the temperature difference[7]

$$\dot{Q} = -h\Delta T \tag{1.7-1}$$

Here h is a positive constant, and the minus sign in the equation indicates that the heat flow is in the opposite direction to the temperature difference, that is, the flow of heat is from a region of high temperature to a region of low temperature. Similarly, on a microscopic scale, the heat flux in the x-coordinate direction, denoted by q_x (with units of J/m^2 s) is linearly related to the temperature gradient in that direction

$$q_x = -k\frac{\partial T}{\partial x} \tag{1.7-2}$$

The mass flux of species A in the x direction $j_A|_x$ (kg/m^2 s) relative to the fluid mass average velocity is linearly related to its concentration gradient

$$j_A|_x = -\rho D\frac{\partial w_A}{\partial x} \tag{1.7-3}$$

and, for many fluids, the flux of the x-component of momentum in the y-coordinate direction is

$$\tau_{yx} = -\mu\frac{\partial v_x}{\partial y} \tag{1.7-4}$$

In these equations T is the temperature, ρ the mass density, w_A is the mass fraction of species A, and v_x the x-component of fluid velocity vector. The

[7]Throughout this book we will use a dot, as on $\dot{Q}$, to indicate a flow term. Thus, $\dot{Q}$ is a flow of heat with units of J/s, and $\dot{M}$ is a flow of mass with units of kg/s. Also, radiative heat transfer will be neglected when considering heat flows.

parameter k is the thermal conductivity, D the diffusion coefficient for species A, and μ the fluid viscosity; from experiment these parameters are all greater than or equal to zero (this is, in fact, a requirement for the system to evolve toward equilibrium). Equation 1.7-2 is known as Fourier's law of heat conduction, Eq. 1.7-3 is called Fick's first law of diffusion; and Eq. 1.7-4 is Newton's law of viscosity.

1.8 A COMMENT ON THE DEVELOPMENT OF THERMODYNAMICS

The formulation of the principles of thermodynamics that will be used in this book is a reflection of the author's preference and experience, rather than an indication of the historical development of the subject. This is the case with most textbooks, as a good textbook should present its subject in an orderly, coherent fashion, even though most branches of science have developed in a disordered manner marked with both brilliant, and frequently unfounded, generalizations and, in retrospect, equally amazing blunders. It would serve little purpose to relate here the caloric or other theories of heat that have been proposed in the past, or to describe all the futile efforts that went into the construction of perpetual motion machines. Similarly, the energy equivalence of heat and work seems obvious now, though it was accepted by the scientific community only after 10 years of work by J. P. Joule. Historically, this equivalence was first pointed out by a medical doctor, J. R. Mayer. However, it would be foolish to reproduce in a textbook his development, which started from the observation that the venous blood of sailors being bled in Java was unusually red, made use of a theory of Lavoisier relating the rate of oxidation in animals to their heat losses, and ultimately led to the conclusion that heat and work were energetically equivalent.

The science of thermodynamics as we now know it is basically the work of the experimentalist, in that each of its principles represents the generalization of a large amount of varied experimental data and the life's work of many. We have tried to keep this flavor to the subject by basing our development of thermodynamics on a number of key experimental observations. However, the presentation of thermodynamics in this book, and especially in Chapter 3, certainly does not parallel its historical development.

PROBLEMS

1.1 For each of the cases that follow list as many properties of the equilibrium state as you can, especially the constraints placed on the equilibrium state of the system by its surroundings and/or its container.

a. The system is placed in thermal contact with a thermostatic bath maintained at the temperature T.

b. The system is contained in a constant-volume container and thermally and mechanically isolated from its surroundings.

c. The system is contained in a frictionless piston and cylinder exposed to an atmosphere at pressure P and thermally isolated from its surroundings.

d. The system is contained in a frictionless piston and cylinder exposed to an atmosphere at pressure P and is in thermal contact with a thermostatic bath maintained at the temperature T.

e. The system consists of two tanks of gas connected by tubing. A valve between the two tanks is fully opened for a short time and then closed.

1.2 The table lists the volumes of 1 gram of water and 1 gram of mercury as functions of temperature.

T(°C)	**Volume of 1 gram of H_2O (cm³)**	**Volume of 1 gram of Hg (cm³)**
0	1.0001329	0.0735560
1	1.0000733	0.0735694
2	1.0000321	0.0735828
3	1.0000078	0.0735961
4	1.0000000	0.0736095
5	1.0000081	0.0736228
6	1.0000318	0.0736362
7	1.0000704	0.0736496
8	1.0001236	0.0736629
9	1.0001909	0.0736763
10	1.0002719	0.0736893
20	1.0015678	0.0738233
30	1.0043408	0.0739572
40	1.0078108	0.0740910
50	1.012074	0.0742250
60	1.017046	0.0743592
70	1.022694	0.0744936
80	1.028987	0.0746282
90	1.035904	0.0747631
100	1.043427	0.0748981

*Based on data in R. H. Perry and D. Green, eds., *Chemical Engineers' Handbook*, 6th ed., McGraw–Hill, New York, 1984, pp. 3-75–3-77.

a. Discuss why water would not be an appropriate thermometer fluid between 0°C and 10°C.

b. Because of the slightly nonlinear temperature dependence of the specific volume of liquid mercury, there is an inherent error in using a mercury-filled thermometer that has been calibrated against an ideal gas thermometer at only 0°C and 100°C. Using the data in the table, prepare a graph of the error, ΔT, as a function of temperautre.

c. Why do you suppose a common mercury thermometer consists of a large volume mercury-filled bulb attached to a capillary tube?

2

Conservation of Mass and Energy

In this chapter we start the quantitative development of thermodynamics. Our objective is to use the qualitative observations of the previous chapter to develop the first two of three balance equations that will be used in the thermodynamic description of physical and chemical processes. These balance equations, together with experimental data and information about the process, will then be used to relate the change in system properties to a change in its thermodynamic state. In this context, physics, fluid mechanics, thermodynamics, and other physical sciences are all similar, in that the tools of each are the same: a set of conservation and balance equations, a collection of experimental observations (i.e., equation of state data in thermodynamics, viscosity data in fluid mechanics, etc.), and the initial and boundary conditions for each problem. The real distinction between these different subject areas is the class of problems, and in some cases the portion of a particular problem, that each deals with.

One important difference between thermodynamics, and, say, fluid mechanics and chemical reactor analysis, is the level of description used. In fluid mechanics one is usually interested in a very detailed microscopic description of flow phenomena and may try to determine, for example, the fluid velocity profile for flow in a pipe. Similarly, in chemical reactor analysis one is interested in determining the concentrations and rate of chemical reaction everywhere in the reactor. In thermodynamics the description is usually more primitive in that we choose either a region of space or an element of mass as the system and merely try to balance the change in the system with what is entering and leaving it. The advantage of such a description is that we can frequently make important predictions about certain types of processes for which a more detailed description might not be possible. Of course, a compromise is being made in that the thermodynamic description yields information only about certain overall properties of the system, though with relatively little labor and simple initial information. A detailed microscopic analysis can yield a much more comprehensive description of the system, but with considerably more work and more detailed initial data.

In this chapter we will be concerned with developing the equations of mass and energy conservation to be used in the thermodynamics of systems of pure substances. (The thermodynamics of mixtures will be considered later.) To emphasize both the generality of these equations and the lack of detail necessary for our description, we will write these equations for a general black-box system. For contrast, and also because a more detailed description will be useful in Chapter 3, the rudiments of the fluid mechanics description are provided in the final section of this chapter. Since the microscopic description is not central to our development of the thermodynamic principles, that section may be omitted.

2.1 A GENERAL BALANCE EQUATION AND CONSERVED QUANTITIES

The balance equations used in the study of thermodynamics are conceptually simple. Each is obtained by choosing a system, either a quantity of mass or a region of space (e.g., the contents of a tank), and equating the change of some property of this system to the amounts of the property that have entered and left the system and have been produced within it. We are interested both in the change of a system property over a time interval and with its instantaneous rate of change; therefore we will formulate equations of change for both time scales. Which of the two formulations of the equations of change is used for the description of a particular physical situation will largely depend on the type of information desired or available.

To illustrate the two types of descriptions and the relationship between them, as well as the idea of using balance equations, consider the problem of studying the total mass of water in Lake Mead (the lake behind the Hoover Dam on the Colorado River). If you were interested in determining, at any instant of time, whether the water level in this lake were rising or falling, you would have to ascertain whether the water flows into the lake were greater or less than the flows of water out of the lake. That is, at some instant of time you would determine the rates at which water was entering the lake (due to the flow of the Colorado River and rainfall) and leaving it (due to flow across the dam, evaporation from the lake surface, and seepage through the canyon walls), and then use the equation

$$\begin{pmatrix}\text{rate of change of}\\ \text{amount of water}\\ \text{in the lake}\end{pmatrix} = \begin{pmatrix}\text{rate at which}\\ \text{water flows}\\ \text{into the lake}\end{pmatrix} - \begin{pmatrix}\text{rate at which}\\ \text{water flows out}\\ \text{of the lake}\end{pmatrix} \tag{2.1-1}$$

to determine the precise rate of change of the amount of water in the lake.

If, on the other hand, you were interested in determining the change in amount of water for some period of time, say the month of January, you could use a balance equation in terms of the total amounts of water that entered and left the lake during this time. Thus

$$\begin{pmatrix}\text{change in amount}\\ \text{of water in the}\\ \text{lake during the}\\ \text{month of January}\end{pmatrix} = \begin{pmatrix}\text{amount of water that}\\ \text{flowed into the lake}\\ \text{during the month of}\\ \text{January}\end{pmatrix} - \begin{pmatrix}\text{amount of water}\\ \text{that flowed out of}\\ \text{the lake during the}\\ \text{month of January}\end{pmatrix} \tag{2.1-2}$$

Notice that Eq. 2.1-1 is concerned with an instantaneous rate of change, and it requires data on the rates at which flows occur. Equation 2.1-2, on the other hand, is for computing the total change that has occurred and requires data only on the total flows over the time interval. These two equations, one for the instantaneous rate of change of a system property (here the amount of water) and the other for the change over an interval of time, illustrate the two types of change of state problems that will be of interest in this book and the forms of the balance equations that will be used in their solution.

There is, of course, an interrelationship between the two balance equations. If you had information on each water flow rate at each instant of time for the whole month of January, you could integrate Eq. 2.1-1 over that period of time to obtain the same answer for the total change in the amount of water as would be obtained directly from Eq. 2.1-2 using only the much less detailed information on the total flows for the month.

The example used here to illustrate the balance equation concept is artificial in that although water flows into and out of a lake are difficult to measure, the amount of water in the lake can be determined directly from the water level. Thus, Eqs. 2.1-1 and 2 are not likely to be used. However, the system properties of interest in thermodynamics and, indeed, in most areas of engineering are frequently much more difficult to measure than flow rates of mass and energy, so that the balance equation approach may be the only practical way to proceed.

In the remainder of this section, the balance equations for an unspecified extensive property θ of a thermodynamic system are developed; specific choices for θ, such as mass and energy, will be considered in the following sections. These balance equations will be formulated in a general manner so that they are applicable (by appropriate simplification) to all systems studied in this book. In this way it will not be necessary to rederive the balance equations for each new problem. Thus, we consider a general system that may be moving or still, mass and energy may flow across its boundaries at one or more places, and the boundaries may distort. Since we are concerned with equating the total change within the system to flows across its boundaries, the details of the internal structure of the system will be left unspecified. This "black-box" system is illustrated in Fig. 2.1-1, and certain of its characteristics are listed here.

1. Mass may flow into one, several, all, or none of the K entry ports labeled 1, 2, . . . , K (i.e., the system may be either open or closed to the flow of mass). Since we are concerned with pure fluids here, only one molecular species will be involved, though its temperature and pressure may be different at each entry port. The mass flow rate *into* the system at the kth entry port will be $\dot{M}_k$, so that $\dot{M}_k > 0$ for flow into the system, and $\dot{M}_k < 0$ for flow out of the system.
2. The boundaries of the black-box system may be stationary, or they may be moving. If the system boundaries are moving, it can be either because the system is expanding or contracting, or because the system as a whole is moving, or both.

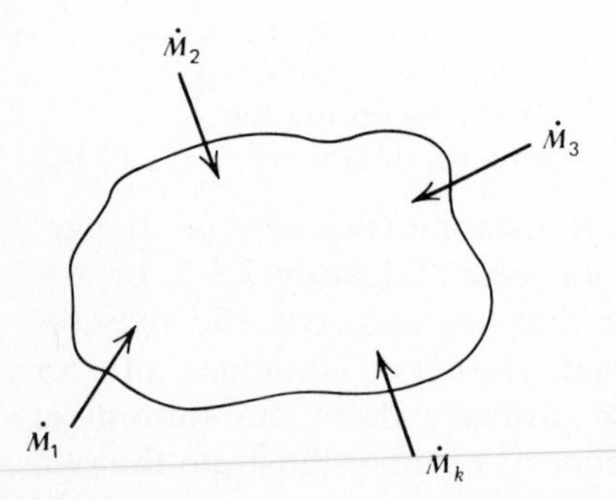

Figure 2.1-1
A system with several mass flows.

3. Energy in the form of heat may enter or leave the system across the system boundaries.
4. Energy in the form of work (mechanical shaft motion, electrical energy, etc.) may enter or leave the system across the system boundaries.

In 3 and 4, an energy flow into the system is positive and an energy flow out of the system is negative.

The balance equation for the total amount of any extensive quantity θ in this system is obtained by equating the change in the amount of θ in the system between the times t and $t + \Delta t$ to the flows of θ into and out of the system, and the generation of θ within the system, in the time interval Δt. Thus,

$$\begin{aligned}
&\begin{pmatrix}\text{amount of } \theta \text{ in the} \\ \text{system at time } t + \Delta t\end{pmatrix} - \begin{pmatrix}\text{amount of } \theta \text{ in the} \\ \text{system at time } t\end{pmatrix} \\
&\quad = \begin{pmatrix}\text{amount of } \theta \text{ that entered the system across} \\ \text{system boundaries between } t \text{ and } t + \Delta t\end{pmatrix} \\
&\quad - \begin{pmatrix}\text{amount of } \theta \text{ that left the system across} \\ \text{system boundaries between } t \text{ and } t + \Delta t\end{pmatrix} \\
&\quad + \begin{pmatrix}\text{amount of } \theta \text{ generated within the} \\ \text{system between } t \text{ and } t + \Delta t\end{pmatrix} \qquad (2.1\text{-}3)
\end{aligned}$$

The meaning of the first two terms on the right side of this equation is clear, but the last term deserves some discussion. If the extensive property θ is equal to the total mass, energy, or momentum, quantities that are conserved (Experimental Observations 1 to 3 of Sec. 1.7), then the internal generation of θ is equal to zero. This is easily seen for the special case of a system isolated from its environment (so that the flow terms across the system boundaries vanish); here Eq. 2.1-3 reduces to

$$\begin{pmatrix}\text{amount of } \theta \text{ in the} \\ \text{system at time} \\ t + \Delta t\end{pmatrix} - \begin{pmatrix}\text{amount of } \theta \text{ in the} \\ \text{system at time } t \\ \end{pmatrix} = \begin{pmatrix}\text{amount of } \theta \text{ generated} \\ \text{within the system} \\ \text{between } t \text{ and } t + \Delta t\end{pmatrix} \qquad (2.1\text{-}3a)$$

Since neither total mass, momentum, nor energy can be spontaneously produced, if θ is either of these quantities the internal generation term must be zero. If, however, θ is some other quantity, the internal generation term may be positive (if θ is produced within the system), negative (if θ is consumed within the system), or zero. For example, suppose the black-box system in Fig. 2.1-1 were a closed (batch) chemical reactor in which cyclohexane is partially dehydrogenated to benzene and hydrogen according to the reaction

$$C_6H_{12} \rightarrow C_6H_6 + 3H_2$$

If θ were set equal to the total mass, then, by the principle of conservation of mass, the internal generation term in Eq. 2.1-3a would be zero. If, however, θ were taken to be the mass of benzene in the system, the internal generation term for benzene would be positive, since benzene is produced by the chemical reaction. Conversely, if θ were taken to be the mass of cyclohexane in the system, the internal generation term would be negative. In either case the magnitude of the internal generation term would depend on the rate of reaction.

The balance equation (Eq. 2.1-3) is appropriate for computing the change in the extensive property θ over the time interval Δt. We can also obtain an equation for computing the instantaneous rate of change of θ by letting the time interval Δt go to zero. This is done as follows. First we use the symbol $\theta(t)$ to represent the amount of θ in the system at time t, and we recognize that for a very small time interval Δt (over which the flows into and out of the system are constant) we can write

$$\begin{pmatrix}\text{amount of } \theta \text{ that enters the}\\ \text{system across system boundaries}\\ \text{between } t \text{ and } t+\Delta t\end{pmatrix} \quad \text{as} \quad \begin{pmatrix}\text{rate at which } \theta \text{ enters}\\ \text{the system across system}\\ \text{boundaries}\end{pmatrix}\Delta t$$

with similar expressions for the outflow and generation terms. Next we rewrite Eq. 2.1-3 as

$$\frac{\theta(t+\Delta t)-\theta(t)}{\Delta t} = \begin{pmatrix}\text{rate at which } \theta \text{ enters the system}\\ \text{across system boundaries}\end{pmatrix} - \begin{pmatrix}\text{rate at which } \theta \text{ leaves the system}\\ \text{across system boundaries}\end{pmatrix} + \begin{pmatrix}\text{rate at which } \theta \text{ is generated}\\ \text{within the system}\end{pmatrix}$$

Finally, taking the limit as $\Delta t \to 0$ and using the definition of the derivative from calculus

$$\frac{d\theta}{dt} = \lim_{\Delta t \to 0} \frac{\theta(t+\Delta t)-\theta(t)}{\Delta t}$$

we obtain

$$\frac{d\theta}{dt} = \begin{pmatrix}\text{rate of change of}\\ \theta \text{ in the system}\end{pmatrix} = \begin{pmatrix}\text{rate at which } \theta \text{ enters the}\\ \text{system across system boundaries}\end{pmatrix} - \begin{pmatrix}\text{rate at which } \theta \text{ leaves the}\\ \text{system across system boundaries}\end{pmatrix} + \begin{pmatrix}\text{rate at which } \theta \text{ is generated}\\ \text{within the system}\end{pmatrix} \quad (2.1\text{-}4)$$

Balance equation 2.1-4 is general and applicable to conserved and nonconserved quantities. There is, however, the important advantage of not having to formulate an expression for the internal generation term when using this equation for conserved quantities. For example, to use the total mass balance to compute the rate of change of mass in the system, we need know only the mass flows into and out of the system. On the other hand, to compute the rate of change of the mass of cyclohexane undergoing a dehydrogenation reaction in a chemical reactor, we would need data on the rate of reaction in the system, which may be a function of concentration, temperature, catalyst activity, structure, and other internal characteristics of the system. Thus, a wealth of information may be needed to use the balance equation for the mass of cyclohexane, and, more generally, for any nonconserved quantity. The applications of thermodynamics sometimes require the use of balance equations for nonconserved quantities.

2.2 CONSERVATION OF MASS

The first balance equation of interest in thermodynamics is the conservation equation for total mass. If θ is taken to be the total mass in the system, designated by the symbol M, we have, from Eq. 2.1-3

$$M(t+\Delta t) - M(t) = \begin{pmatrix} \text{amount of mass that} \\ \text{entered the system} \\ \text{across the system} \\ \text{boundaries between} \\ t \text{ and } t+\Delta t \end{pmatrix} - \begin{pmatrix} \text{amount of mass} \\ \text{that left the system} \\ \text{across the system} \\ \text{boundaries between} \\ t \text{ and } t+\Delta t \end{pmatrix} \tag{2.2-1a}$$

where we have recognized that the total mass is a conserved quantity and that the only mechanism by which mass enters a system is by a mass flow. Using $\dot{M}_k$ to represent the mass flow rate into the system at the kth entry point, we have, from Eq. 2.1-4, the equation for the instantaneous rate of change of mass in the system

$$\frac{dM}{dt} = \sum_{k=1}^{K} \dot{M}_k \tag{2.2-1b}$$

Equations 2.2-1 are general and valid regardless of the details of the system and whether the system is stationary or moving.

Since we are interested only in pure fluids, we can divide Eqs. 2.2-1 by the molecular weight m of the fluid and use the fact that N, the number of moles in the system, is equal to M/m, and $\dot{N}_k$, the molar flow rate into the system at the kth entry port, is $\dot{M}_k/m$, to obtain instead of Eq. 2.2-1a a similar equation in which the term moles replaces the word mass and, instead of Eq. 2.2-1b

$$\frac{dN}{dt} = \sum_{k=1}^{K} \dot{N}_k \tag{2.2-2}$$

We introduce these equations here because it is frequently convenient to do calculations on a molar rather than mass basis.

In Sec. 2.1 it was indicated that the equation for the change of an extensive state variable of a system in the time interval Δt could be obtained by the integration over that time interval of the equation for the rate of change for that variable. Here we will demonstrate precisely how this integration is accomplished. For convenience, t_1 will represent the beginning of the time interval, and t_2 will represent the end of the time interval, so that $\Delta t = t_2 - t_1$. Integrating Eq. 2.2-1b between t_1 and t_2 yields

$$\int_{t_1}^{t_2} \frac{dM}{dt}\, dt = \sum_{k=1}^{K} \int_{t_1}^{t_2} \dot{M}_k\, dt \tag{2.2-3}$$

The left side of the equation is treated as follows:

$$\int_{t_1}^{t_2} \frac{dM}{dt}\, dt = \int_{M(t_1)}^{M(t_2)} dM = M(t_2) - M(t_1) = \begin{pmatrix} \text{change in total mass} \\ \text{of system between} \\ t_1 \text{ and } t_2 \end{pmatrix}$$

where $M(t)$ is the mass in the system at time t. The term on the right side of the equation may be simplified by observing that

$$\int_{t_1}^{t_2} \dot{M}_k\, dt = \begin{pmatrix} \text{mass that entered the} \\ \text{system at the } k\text{th entry} \\ \text{port between } t_1 \text{ and } t_2 \end{pmatrix} \equiv (\Delta M)_k$$

Thus

$$M(t_2) - M(t_1) = \sum_{k=1}^{K} (\Delta M)_k \tag{2.2-4}$$

This is the symbolic form of Eq. 2.2-1a.

Equation 2.2-4 may be written in a simpler form when the mass flow rates are steady, that is, independent of time. For this case

$$\int_{t_1}^{t_2} \dot{M}_k \, dt = \dot{M}_k \int_{t_1}^{t_2} dt = \dot{M}_k \Delta t$$

so that

$$M(t_2) - M(t_1) = \sum_{k=1}^{K} \dot{M}_k \Delta t \qquad \text{(steady flows)} \qquad (2.2\text{-}5)$$

The equations of this section that will be used throughout this book are listed in Table 2.2-1.

2.3 CONSERVATION OF ENERGY

The energy conservation equation for the black-box system can be obtained by starting from Eq. 2.1-3 and taking θ to be the sum of the internal, kinetic, and potential energy of the system

$$\theta = U + M(v^2/2 + \psi)$$

Here U is the total internal energy, $v^2/2$ is the kinetic energy per unit mass (where v is the center of mass velocity), and ψ is the potential energy per unit mass.[1] If gravity is the only force field present, then $\psi = gh$, where h is the height

Table 2.2-1
THE MASS CONSERVATION EQUATION

	Mass basis	Molar basis
Differential Form of the Mass Balance		
General equation:	$\dfrac{dM}{dt} = \sum_{k=1}^{K} \dot{M}_k$	$\dfrac{dN}{dt} = \sum_{k=1}^{K} \dot{N}_k$
Special case:		
Closed system:	$\dfrac{dM}{dt} = 0$	$\dfrac{dN}{dt} = 0$
	$M = \text{constant}$	$N = \text{constant}$
Difference Form of the Mass Balance*		
General equation:	$M_2 - M_1 = \sum_{k=1}^{K} (\Delta M)_k$	$N_2 - N_1 = \sum_{k=1}^{K} (\Delta N)_k$
Special cases:		
(i) Closed system:	$M_2 = M_1$	$N_2 = N_1$
(ii) Steady flow:	$M_2 - M_1 = \sum_{k=1}^{K} \dot{M}_k \Delta t$	$N_2 - N_1 = \sum_{k=1}^{K} \dot{N}_k \Delta t$

*Here we have used the abbreviated notation $M_i = M(t_i)$.

[1]In writing this form for the energy term, it has been assumed that the system consists of only one phase, that is, a gas, a liquid, or a solid. If the system consists of several distinct parts, for example, a gas and a liquid, or a gas and the piston and cylinder containing it, the total energy, which is an extensive property, is the sum of the energies of the constituent parts.

of the center of mass with respect to some reference, and g is the force of gravity. Since energy is a conserved quantity, we can write

$$\frac{d}{dt}\{U + M(v^2/2 + \psi)\} = \begin{pmatrix}\text{rate at which energy}\\ \text{enters the system}\end{pmatrix} - \begin{pmatrix}\text{rate at which energy}\\ \text{leaves the system}\end{pmatrix} \quad (2.3\text{-}1)$$

To complete the balance it only remains to identify the various mechanisms by which energy can enter and leave the system. These are

1. Energy flow accompanying mass flow. As a fluid element enters or leaves the system, it carries with it internal, potential, and kinetic energy. This energy flow accompanying the mass flow is simply the product of a mass flow and the energy per unit mass,

$$\sum_{k=1}^{K} \dot{M}_k(\hat{U} + v^2/2 + \psi)_k \quad (2.3\text{-}2)$$

where $\hat{U}_k$ is the internal energy per unit mass of the kth flow stream, and $\dot{M}_k$ its mass flow rate.

2. Heat. We will use $\dot{Q}$ to denote the total rate of flow of heat *into* the system, so that $\dot{Q}$ is positive if energy in the form of heat flows into the system and negative if heat flows from the system to its surroundings. If heat flows occur at several different places, the total rate of heat flow into the system is

$$\dot{Q} = \sum \dot{Q}_j$$

where $\dot{Q}_j$ is the heat flow at the jth heat flow port.

3. Work. The total energy flow into the system due to work will be divided into several parts. The first part, called shaft work and denoted by the symbol W_s, is the mechanical energy flow that occurs without a deformation of the system boundaries. For example, if the system under consideration were a steam turbine or an internal combustion engine, the rate of shaft work $\dot{W}_s$ would be equal to the rate at which energy was transferred across the stationary system boundaries by the drive shaft or push rod. Following the convention that energy flow into the system is positive, $\dot{W}_s$ is positive if the surroundings do work on the system and negative if the system does work on its surroundings.

   For convenience, the flow of electrical energy into or out of the system will be included in the shaft work term. In this case, $\dot{W}_s = \pm EI$, where E is the electrical potential difference across the system and I is the current flow through the system. The positive sign applies if electrical energy is being supplied to the system, and the negative sign applies if the system is the source of electrical energy.

   Work also results from the movement of the system boundaries. The rate at which work is done when a force F is moved through a distance in the direction of the applied force dL in the time interval dt is

$$\dot{W}_F = F\frac{dL}{dt}$$

   Here we recognize that pressure is a force per unit area and write

$$\dot{W}_F = -P\frac{dV}{dt} \quad (2.3\text{-}3)$$

where P is the pressure exerted by the system at its boundaries.[2] The negative sign in this equation arises from the convention that work done on a system in a compression (for which dV/dt is negative) is positive, and work done by the system on its surroundings in an expansion (for which dV/dt is positive) is negative. The pressure at the boundaries of a nonstationary system will be opposed by (1) the pressure of the environment surrounding the system, (2) inertial forces if the expansion or compression of the system results in an acceleration or deceleration of the surroundings, and (3) other external forces such as gravity. As we will see in Illustration 2.5-7, the contribution to the energy balance of the first of these forces is a term corresponding to the work done against the atmosphere, the second is a work term corresponding to the change in kinetic energy of the surroundings, and the last is the work done that changes the potential energy of the surroundings.

4. One additional flow of energy for systems open to the flow of mass must be included in the energy balance equation; it is more subtle than the energy flows just considered. This is the energy flow that arises from the fact that as an element of fluid moves it does work on the fluid ahead of it and the fluid behind it does work on it. Clearly, each of these work terms is of the $P\,\Delta V$ type. To evaluate this energy flow term, which occurs only in systems open to the flow of mass, we will compute the net work done as one fluid element of mass ΔM_1 enters a system, such as the valve in Fig. 2.3-1, and another fluid element of mass ΔM_2 leaves the system. The pressure of the fluid at the inlet side of the valve is P_1 and the fluid pressure at the outlet side is P_2, so that we have

$$\begin{pmatrix}\text{work done } by \text{ surrounding fluid in} \\ \text{pushing fluid element of mass } \Delta M_1 \\ \text{into the valve}\end{pmatrix} = P_1\hat{V}_1\Delta M_1$$

$$\begin{pmatrix}\text{work done } on \text{ surrounding fluid by} \\ \text{movement of fluid element of mass} \\ \Delta M_2 \text{ out of valve (since this fluid} \\ \text{element is pushing the surrounding} \\ \text{fluid)}\end{pmatrix} = -P_2\hat{V}_2\Delta M_2$$

$$\begin{pmatrix}\text{net work done on the system due to} \\ \text{movement of fluid}\end{pmatrix} = P_1\hat{V}_1\Delta M_1 - P_2\hat{V}_2\Delta M_2$$

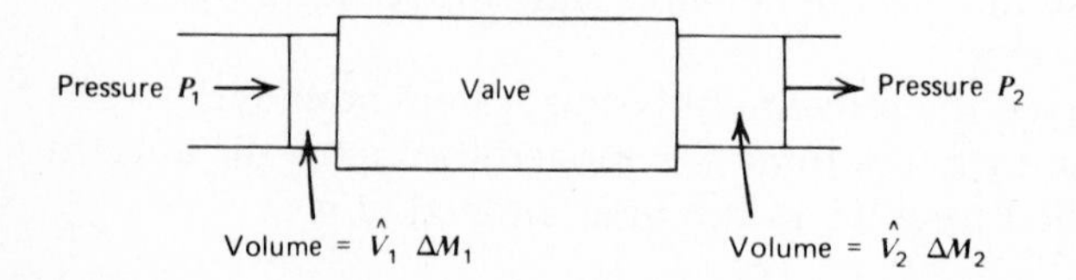

Figure 2.3-1 A schematic representation of flow through a valve.

[2] In writing this form for the work term, we have assumed the pressure to be uniform at the system boundary. If this is not the case, Eq. 2.3-3 is to be replaced with an integral over the surface of the system.

For a more general system, with numerous mass flow ports, we have

$$\begin{pmatrix}\text{net work done on the system due}\\ \text{to the pressure forces acting on}\\ \text{the fluids moving into and out of}\\ \text{the system}\end{pmatrix} = \sum_{k=1}^{K} \Delta M_k (P\hat{V})_k$$

Finally, to obtain the net *rate* at which work is done, we replace each mass flow ΔM_k with a mass flow rate $\dot{M}_k$, so that

$$\begin{pmatrix}\text{net } rate \text{ at which work is done on}\\ \text{the system due to pressure forces}\\ \text{acting on fluids moving into and out}\\ \text{of the system}\end{pmatrix} = \sum_{k=1}^{K} \dot{M}_k (P\hat{V})_k$$

where the sign of each term of this energy flow is the same as $\dot{M}_k$.

An important application of this pressure-induced energy flow accompanying a mass flow is hydroelectric power generation, schematically indicated in Fig. 2.3-2. Here a water turbine is being used to obtain mechanical energy from the flow of water through the base of a dam. Since the water velocity, height, and temperature are approximately the same at both sides of the turbine (even though there are large velocity changes within the turbine), the mechanical (or electrical) energy obtained is a result of only the mass flow across the pressure difference at the turbine.

Now, collecting all the energy terms discussed gives

$$\frac{d}{dt}\{U + M(v^2/2 + \psi)\} = \sum_{k=1}^{K} \dot{M}_k(\hat{U} + v^2/2 + \psi)_k + \dot{Q} + \dot{W}_s - P\frac{dV}{dt} + \sum_{k=1}^{K} \dot{M}_k(P\hat{V})_k \tag{2.3-4}$$

This equation can be written in a more compact form by combining the first and last terms on the right side and introducing the notation

$$H = U + PV$$

where the function H is called the **enthalpy**, and by using the symbol $\dot{W}$ to represent the combination of shaft work $\dot{W}_s$ and expansion work $-P(dV/dt)$.

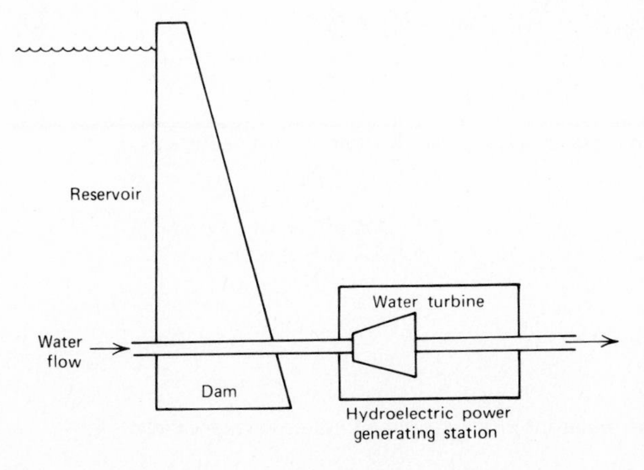

Figure 2.3-2
A hydroelectric power generating station: a device for obtaining work from a fluid flowing across a large pressure drop.

Thus we have

$$\frac{d}{dt}\{U + M(v^2/2 + \psi)\} = \sum_{k=1}^{K} \dot{M}_k(\hat{H} + v^2/2 + \psi)_k + \dot{Q} + \dot{W} \tag{2.3-4a}$$

It is also convenient to have the energy balance on a molar rather than a mass basis. This change is easily accomplished by recognizing that $\dot{M}_k\hat{H}_k$ can equally well be written as $\dot{N}_k\underline{H}_k$, where $\underline{H}$ is the enthalpy per mole or molar enthalpy,[3] and $M(v^2/2 + \psi) = Nm(v^2/2 + \psi)$, where m is the molecular weight. Therefore, we can write the energy balance:

$$\frac{d}{dt}\{U + Nm(v^2/2 + \psi)\} = \sum_{k=}^{K} \dot{N}_k\{\underline{H} + m(v^2/2 + \psi)\}_k + \dot{Q} + \dot{W} \tag{2.3-4b}$$

Several special cases of Eq. 2.3-4 are listed in Table 2.3-1.

Table 2.3-1
DIFFERENTIAL FORM OF THE ENERGY BALANCE

General equation:

$$\frac{d}{dt}\{U + M(v^2/2 + \psi)\} = \sum_{k=1}^{K} \dot{M}_k(\hat{H} + v^2/2 + \psi)_k + \dot{Q} + \dot{W} \tag{a}$$

Special cases:

(i) Closed system: $\dot{M}_k = 0, \quad \dfrac{dM}{dt} = 0$

so

$$\frac{dU}{dt} + M\frac{d}{dt}(v^2/2 + \psi) = \dot{Q} + \dot{W} \tag{b}$$

(ii) Adiabatic process: $\dot{Q} = 0$ in Eq. a (c)

(iii) Open steady-state system:

$$\frac{dM}{dt} = 0, \quad \frac{dV}{dt} = 0, \quad \frac{d}{dt}\{U + M(v^2/2 + \psi)\} = 0$$

so

$$0 = \sum_{k=1}^{K} \dot{M}_k(\hat{H} + v^2/2 + \psi)_k + \dot{Q} + \dot{W}_s \tag{d}$$

(iv) Uniform system: $U = M\hat{U}$ in Eq. a (e)

Note: To obtain the energy balance on a molar basis, make the following substitutions:

Replace	*with*
$M(v^2/2 + \psi)$	$Nm(v^2/2 + \psi)$
$\dot{M}_k(\hat{H} + v^2/2 + \psi)_k$	$\dot{N}_k\{\underline{H} + m(v^2/2 + \psi)\}_k$
$M\hat{U}$	$N\underline{U}$

[3] $\underline{H} = \underline{U} + P\underline{V}$, where $\underline{U}$ and $\underline{V}$ are the molar internal energy and volume, respectively.

The changes in energy associated with either the kinetic energy or potential energy terms, especially for gases, are usually very small compared to the thermal (internal) energy terms, unless the fluid velocity is near the velocity of sound, the change in height is very large, or the system temperature is nearly constant. This point will become evident in some of the examples and problems (see particularly Illustration 2.5-2). Therefore, it is frequently possible to approximate Eqs. 2.3-4 by

$$\frac{dU}{dt} = \sum_{k=1}^{K} (\dot{M}\hat{H})_k + \dot{Q} + \dot{W} \qquad \text{(mass basis)} \qquad (2.3\text{-}5a)$$

$$\frac{dU}{dt} = \sum_{k=1}^{K} (\dot{N}\underline{H})_k + \dot{Q} + \dot{W} \qquad \text{(molar basis)} \qquad (2.3\text{-}5b)$$

As with the mass balance, it is useful to have a form of the energy balance applicable to a change from state 1 to state 2. This is easily obtained by integrating Eq. 2.3-4 over the time interval t_1 to t_2, the time required for the system to go from state 1 to state 2. The result is

$$\{U + M(v^2/2 + \psi)\}_{t_2} - \{U + M(v^2/2 + \psi)\}_{t_1} = \sum_{k=1}^{K} \int_{t_1}^{t_2} \dot{M}_k(\hat{H} + v^2/2 + \psi)_k \, dt + Q + W \qquad (2.3\text{-}6)$$

where

$$Q = \int_{t_1}^{t_2} \dot{Q}\, dt, \qquad W_s = \int_{t_1}^{t_2} \dot{W}_s\, dt, \qquad \int_{V(t_1)}^{V(t_2)} P\, dV = \int_{t_1}^{t_2} P \frac{dV}{dt}\, dt,$$

and

$$W = W_s - \int_{V(t_1)}^{V(t_2)} P\, dV$$

The first term on the right side of Eq. 2.3-6 is usually the most troublesome to evaluate because the mass flow rate and the thermodynamic properties of the flowing fluid may change with time. If the thermodynamic properties of the fluid entering the system are independent of time (even though the mass flow rate may depend on time), we have

$$\sum_{k=1}^{K} \int_{t_1}^{t_2} \dot{M}_k(\hat{H} + v^2/2 + \psi)_k \, dt = \sum_{k=1}^{K} (\hat{H} + v^2/2 + \psi)_k \int_{t_1}^{t_2} \dot{M}_k \, dt$$

$$= \sum_{k=1}^{K} \Delta M_k(\hat{H} + v^2/2 + \psi)_k \qquad (2.3\text{-}7)$$

If, on the other hand, the thermodynamic properties of the flow streams change with time in some arbitrary way, the energy balance of Eq. 2.3-6 may not be useful. The usual procedure, then, is to try to choose a new system (or subsystem) for the description of the process in which these time-dependent flows do not occur or are more easily handled (see Illustration 2.5-5).

Table 2.3-2 contains various special cases of Eq. 2.3-6 that will be useful in solving thermodynamic problems.

Table 2.3-2
DIFFERENCE FORM OF THE ENERGY BALANCE

General equation:

$$\{U + M(v^2/2 + \psi)\}_{t_2} - \{U + M(v^2/2 + \psi)\}_{t_1} = \sum_{k=1}^{K} \int_{t_1}^{t_2} \dot{M}_k(\hat{H} + v^2/2 + \psi)_k \, dt + Q + W \quad \text{(a)}$$

Special cases:

(i) Closed system:

$$\{U + M(v^2/2 + \psi)\}_{t_2} - \{U + M(v^2/2 + \psi)\}_{t_1} = Q + W \quad \text{(b)}$$

and

$$M(t_1) = M(t_2)$$

(ii) Adiabatic process: $Q = 0$ in Eq. a (c)

(iii) Open system, flow of fluids of constant thermodynamic properties:

$$\sum_{k=1}^{K} \int_{t_1}^{t_2} \dot{M}_k(\hat{H} + v^2/2 + \psi)_k \, dt = \sum_{k=1}^{K} \Delta M_k(\hat{H} + v^2/2 + \psi)_k \quad \text{in Eq. a} \quad \text{(d)}$$

(iv) Uniform system:

$$\{U + M(v^2/2 + \psi)\} = M(\hat{U} + v^2/2 + \psi) \quad \text{in Eq. a} \quad \text{(e)}$$

Note: To obtain the energy balance on a molar basis, make the following substitutions:

Replace	*with*
$M(v^2/2 + \psi)$	$Nm(v^2/2 + \psi)$
$\int_{t_1}^{t_2} \dot{M}_k(\hat{H} + v^2/2 + \psi)_k \, dt$	$\int_{t_1}^{t_2} \dot{N}_k\{\underline{H} + m(v^2/2 + \psi)\}_k \, dt$
$M(\hat{U} + v^2/2 + \psi)$	$N\{\underline{U} + m(v^2/2 + \psi)\}$

For the study of thermodynamics it will be useful to have equations that relate the differential change in certain thermodynamic variables of the system to differential changes in other system properties. Such equations can be obtained from the differential form of the mass and energy balances. For process in which the kinetic and potential energy terms are unimportant, there is no shaft work, and there is only a single mass flow stream, these equations reduce to

$$\frac{dM}{dt} = \dot{M}$$

and

$$\frac{dU}{dt} = \dot{M}\hat{H} + \dot{Q} - P\frac{dV}{dt}$$

which can be combined to give

$$\frac{dU}{dt} = \hat{H}\frac{dM}{dt} + \dot{Q} - P\frac{dV}{dt} \quad \text{(2.3-8)}$$

where $\hat{H}$ is the enthalpy per unit mass entering or leaving the system. (Note that for a system closed to the flow of mass, $dM/dt = \dot{M} = 0$.) Defining $Q = \dot{Q}\,dt$ to be equal to the heat flow into the system in the differential time interval dt, and $dM = \dot{M}\,dt = (dM/dt)\,dt$ to be equal to the mass flow in that time interval, we obtain the following expression for the change of the internal energy in the time interval dt

$$dU = \hat{H}\,dM + Q - P\,dV \tag{2.3-9a}$$

For a closed system this equation reduces to

$$dU = Q - P\,dV \tag{2.3-9b}$$

Since the time derivative operator d/dt is mathematically well defined, and the operator d is not, it is important to remember when using these equations that they are abbreviations of Eq. 2.3-8. It is part of the traditional notation of thermodynamics to use $d\theta$ to indicate a differential change in the property θ.

The mass and energy balance equations developed so far in this chapter can be used for the description of any process. As the first step in using these equations, it is necessary to choose a black-box system. The important fact for the student of thermodynamics to recognize is that processes occurring in nature are in no way influenced by our mathematical description of them. Therefore, if our descriptions are correct, they must lead to the same final result for the system and its surroundings regardless of which system choice is made. This is demonstrated in the example below, where the same result is obtained by choosing for the system first a given mass of material and then a specified region in space. Since the first system choice is closed, and the second open, this illustration also establishes the way in which the open system energy flow $PV\dot{M}$ is related to the closed system work term $P(dV/dt)$.

ILLUSTRATION 2.3-1

A compressor is operating in a continuous, steady-state manner to produce a gas at temperature T_2 and pressure P_2 from one at T_1 and P_1. Show that for the time interval Δt

$$Q + W_s = (\hat{H}_2 - \hat{H}_1)\Delta M$$

where ΔM is the mass of gas that has flowed into or out of the system in Δt. Establish this result by (a) first writing the balance equations for a closed system consisting of some convenient element of mass, and then (b) by writing the balance equations for the compressor and its contents, which is an open system.

Solution

a. The Closed System Analysis

Here we take the gas in the compressor and the mass of gas ΔM, which will enter the compressor in the time interval Δt as the system. This system is enclosed by the dotted lines in the figure.

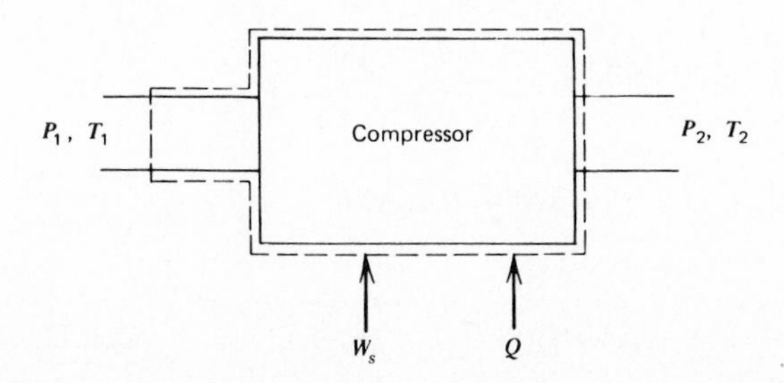

At the later time, $t + \Delta t$, the mass of gas we have chosen as the system is as shown here.

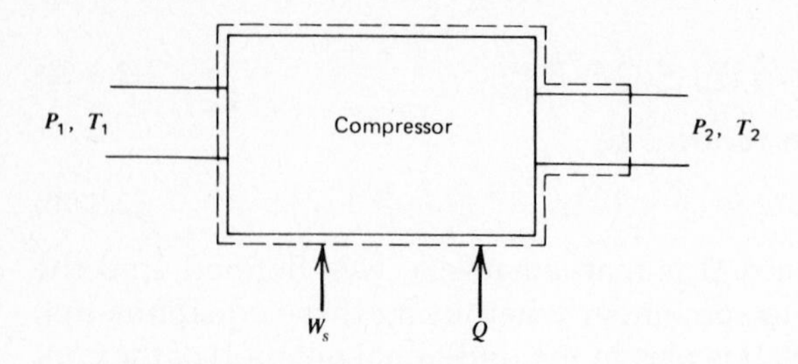

We use the subscript c to denote the characteristics of the fluid in the compressor, the subscript 1 for the gas contained in the system that is in the inlet pipe at time t, and the subscript 2 for the gas in the system that is in the exit pipe at time $t + \Delta t$. With this notation the mass balance for the closed system is

$$M_2(t + \Delta t) + M_c(t + \Delta t) = M_1(t) + M_c(t)$$

Since the compressor is in steady-state operation, the amount of gas contained within it and the properties of this gas are constant. Thus, $M_c(t + \Delta t) = M_c(t)$, and

$$M_2(t + \Delta t) = M_1(t) = \Delta M$$

The energy balance for this system, neglecting the potential and kinetic energy terms (which, if retained, would largely cancel) is

$$M_2\hat{U}_2|_{t+\Delta t} + M_c\hat{U}_c|_{t+\Delta t} - M_1\hat{U}_1|_t - M_c\hat{U}_c|_t = W_s + Q + P_1\hat{V}_1M_1 - P_2\hat{V}_2M_2$$

In writing this equation we have recognized that the flow terms vanish for the closed system and that there are two contributions of the $\int P\,dV$ type, one due to the deformation of the system boundary at the compressor inlet and another at the compressor outlet. Since the inlet and exit pressures are constant at P_1 and P_2, these terms arise as

$$-\int P\,dV = -P_1 \int dV\Big|_{\text{inlet}} \quad -P_2 \int dV\Big|_{\text{outlet}}$$

$$= -P_1\{V_1(t + \Delta t) - V_1(t)\} - P_2\{V_2(t + \Delta t) - V_2(t)\}$$

However, $V_1(t + \Delta t) = 0$ and $V_2(t) = 0$, so that

$$-\int P\,dV = +P_1V_1 - P_2V_2 = P_1\hat{V}_1M_1 - P_2\hat{V}_2M_2$$

Now using Eq. a and recognizing that since the compressor is in steady-state operation,

$$M_c\hat{U}_c|_{t+\Delta t} = M_c\hat{U}_c|_t$$

we obtain

$$\Delta M(\hat{U}_2 - \hat{U}_1) = W_s + Q + P_1\hat{V}_1\Delta M - P_2\hat{V}_2\Delta M$$

or

$$\Delta M(\hat{U}_2 + P_2\hat{V}_2 - \hat{U}_1 - P_1\hat{V}_1) = \Delta M(\hat{H}_2 - \hat{H}_1) = W_s + Q \qquad \text{Q.E.D.}$$

b. The Open System Analysis

Here we take the contents of the compressor at any time to be the system. The mass balance for this system over the time interval Δt is

$$M(t + \Delta t) - M(t) = \int_t^{t+\Delta t} \dot{M}_1 \, dt + \int_t^{t+\Delta t} \dot{M}_2 \, dt = \Delta M_1 + \Delta M_2$$

and the energy balance is

$$\{U + M(v^2/2 + \psi)\}_{t+\Delta t} - \{U + M(v^2/2 + \psi)\}_t$$

$$= \int_t^{t+\Delta t} \dot{M}_1(\hat{H}_1 + v_1^2/2 + \psi_1) \, dt + \int_t^{t+\Delta t} \dot{M}_2(\hat{H}_2 + v_2^2/2 + \psi_2) \, dt + Q + W$$

These equations may be simplified:

1. Since the compressor is operating continuously in a steady-state manner, its contents must, by definition, have the same mass and thermodynamic properties at all times. Therefore

   $$M(t + \Delta t) = M(t)$$

   and

   $$\{U + M(v^2/2 + \psi)\}_{t+\Delta t} = \{U + M(v^2/2 + \psi)\}_t$$

2. Since the thermodynamic properties of the fluids entering and leaving the turbine do not change in time, we can write

   $$\int_t^{t+\Delta t} \dot{M}_1(\hat{H}_1 + v_1^2/2 + \psi_1) \, dt = (\hat{H}_1 + v_1^2/2 + \psi_1) \int_t^{t+\Delta t} \dot{M}_1 \, dt$$

   $$= (\hat{H}_1 + v_1^2/2 + \psi_1)\Delta M_1$$

   with a similar expression for the compressor exit stream.

3. Since the volume of the system, the contents of the compressor, is constant

   $$\int_{V_1}^{V_2} P \, dV = 0$$

   so that

   $$W = W_s$$

4. Finally, we will neglect the potential and kinetic energy changes of the entering and exiting fluids.

With these simplifications we have

$$0 = \Delta M_1 + \Delta M_2 \qquad \text{or} \qquad \Delta M_1 = -\Delta M_2 = \Delta M$$

and

$$0 = \Delta M_1 \hat{H}_1 + \Delta M_2 \hat{H}_2 + Q + W_s$$

Combining these two equations, we obtain

$$Q + W_s = (\hat{H}_2 - \hat{H}_1)\Delta M$$

Comment

Notice that in the closed system analysis the surroundings are doing work on the system (the mass element) at the inlet to the compressor, while the system is doing work on its surroundings at the outlet pipe. Each of these terms is calculable as a $\int P\,dV$ type work term. For the open system this work term has been included in the energy balance as a $P\hat{V}\,\Delta M$ term so that it is the enthalpy, rather than the internal energy, of the flow streams that appear in the equation. The explicit $\int P\,dV$ term that does appear in the open system energy balance represents only the work done as the system boundaries deform; this term is zero here unless the compressor (the boundary of our system) explodes. ■[4]

This illustration demonstrates that the sum $Q + W_s$ is the same for a fluid undergoing some change in a continuous process regardless of whether we choose to compute this sum from the closed system analysis on a mass of gas or from an open system analysis on a given volume in space. In Illustration 2.3-2 we consider another problem, the compression of a gas by two *different* processes, the first being a closed system piston and cylinder process and the second being a flow compressor process. Here we will find that the sum $Q + W$ is different in both processes, but that the origin of this difference is easily understood.

ILLUSTRATION 2.3-2

A mass M of gas is to be compressed from a temperature T_1 and a pressure P_1 to T_2 and P_2 in (a) a one-step process in a frictionless piston and cylinder[5] and (b) as part of a continuous process in which the mass M of gas is part of the feed stream to the compressor of the previous illustration. Compute the sum $Q + W$ for each process.

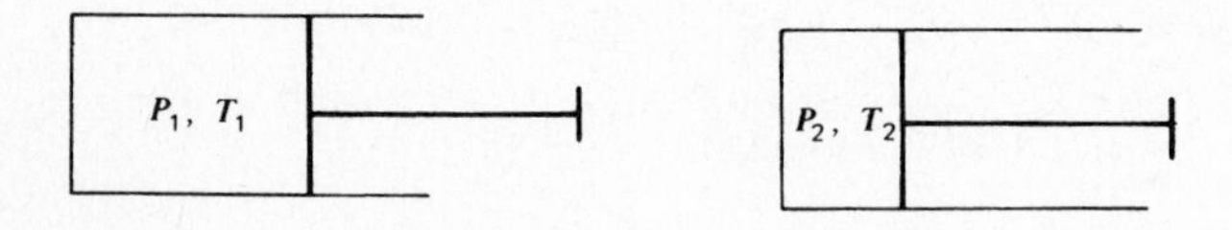

Solution

a. The Piston and Cylinder Process

Here we take the gas within the piston and cylinder as the system. The energy balance for this closed system is

$$M(\hat{U}_2 - \hat{U}_1) = Q + W \qquad \text{(piston-cylinder process)}$$

It is useful to note that $W_s = 0$ and $W = -\int P\,dV$.

b. The Flow Compressor Process

If we take the contents of the compressor as the system, and follow the analysis of the previous illustration, we obtain

$$M(\hat{H}_2 - \hat{H}_1) = Q + W \qquad \text{(flow compressor)}$$

where, since $\int P\,dV = 0$, $W = W_s$.

Comment

From these results it is evident that the sum $Q + W$ is different in the two cases. The origin of the difference in the flow and nonflow energy changes accompanying a change of state is easily identified by considering two different ways of

[4]Throughout this text the symbol ■ will be used to indicate the end of an illustration.

[5]Since the piston is frictionless, the pressure of the gas is equal to the pressure applied by the piston.

compressing a mass M of gas in a piston and cylinder from (T_1, P_1) to (T_2, P_2). The first way is merely to compress the gas in situ. The sum of heat and work flows needed to accomplish the change of state is, from the preceding computations,

$$Q + W = M(\hat{U}_2 - \hat{U}_1)$$

A second way to accomplish the compression is to open a valve at the side of the cylinder and use the piston movement (at constant pressure P_1) to inject the gas into the compressor inlet stream, use the compressor to compress the gas, and then withdraw the gas from the compressor exit stream by moving the piston against a constant external pressure P_2. The energy required in the compressor stage is just that found above

$$(Q + W)_c = M(\hat{H}_2 - \hat{H}_1)$$

To this we must add the work in using the piston movement to pump the fluid into the compressor inlet stream

$$W_1 = \int P\,dV = P_1 V_1 = P_1 \hat{V}_1 M$$

(this is the work done by the system on the gas in the inlet pipe to the compressor) and subtract the work obtained as a result of the piston movement as the cylinder is refilled

$$W_2 = -\int P\,dV = -P_2 V_2 = -P_2 \hat{V}_2 M$$

(this is the work done on the system by the gas in the compressor exit stream). Thus the total energy change in the process is

$$Q + W = (Q + W)_c + W_1 + W_2 = M(\hat{H}_2 - \hat{H}_1) + P_1 \hat{V}_1 M - P_2 \hat{V}_2 M = M(\hat{U}_2 - \hat{U}_1)$$

which is what we found in part a. Here, however, it results from an energy requirement of $M(\hat{H}_2 - \hat{H}_1)$ in the flow compressor and the two pumping terms. ■

As the final illustration of this section, the problem of relating the downstream temperature and pressure of a gas in steady flow across a flow constriction (e.g., a valve, orifice, or porous plug) to its upstream temperature and pressure is considered.

ILLUSTRATION 2.3-3

A gas at pressure and temperature of P_1 and T_1, respectively, is steadily exhausted to the atmosphere at pressure P_2 through a pressure-reducing valve. Find an expression relating the downstream gas temperature T_2 to P_1, P_2, and T_1. Since the gas flows through the valve rapidly, one can assume that there is no heat transfer to the gas. Also, the potential and kinetic energy terms can be neglected.

Solution

The flow process is schematically shown here. We will consider the region of space that includes the flow obstruction (indicated by the dashed line) to be the system, though, as in Illustration 2.3-1, a fixed mass of gas could have been chosen as well. The pressure of the gas exiting the reducing valve will be P_2, the

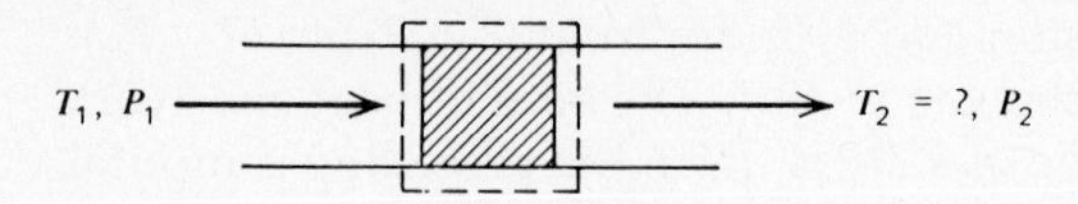

pressure of the surrounding atmosphere. (It is not completely obvious that these two pressures should be the same. However, in the laboratory we find that the velocity of the flowing fluid will always adjust in such a way that the fluid exit pressure and the pressure of the surroundings are equal.) Now recognizing that our system is of constant volume, that the flow is steady, that there are no heat or work flows and negligible kinetic and potential energy changes, the mass and energy balances yield

$$0 = \dot{N}_1 + \dot{N}_2, \qquad \text{or} \qquad \dot{N}_2 = -\dot{N}_1$$

and

$$0 = \dot{N}_1 \underline{H}_1 + \dot{N}_2 \underline{H}_2 = \dot{N}_1(\underline{H}_1 - \underline{H}_2)$$

Thus

$$\underline{H}_1 = \underline{H}_2$$

or, to be explicit,

$$\underline{H}(T_1, P_1) = \underline{H}(T_2, P_2)$$

so that the initial and final states of the gas have the same enthalpy. Consequently, if we knew how the enthalpy of the gas depended on its temperature and pressure, we could use the known values of T_1, P_1, and P_2 to determine the unknown downstream temperature T_2.

Comments

1. The equality of enthalpies in the upstream and downstream states is the only information we get from the thermodynamic balance equations. To proceed further we need constitutive information (i.e., an equation of state or experimental data) interrelating $\underline{H}$, T, and P. Equations of state are discussed in the following section and much of Chapter 4.
2. The experiment discussed in this illustration was devised by William Thomson (later Lord Kelvin) and performed by J. P. Joule to study departures from ideal gas behavior. The **Joule–Thomson expansion**, as it is called, is still used in the liquifaction of gases and in refrigeration processes. (See Problems 3.21 and 22.) ■

2.4 THE THERMODYNAMIC PROPERTIES OF MATTER

The balance equations of this chapter allow one to relate the mass, work, and heat flows of a system to the change in its thermodynamic state. From the discussion in Chapter 1 it is evident that a change of state for a single-component, single-phase system can be described by specifying the initial and final values of any two independent intensive variables. However, certain intensive variables, especially temperature and pressure, are far easier to measure than others. Consequently, for most problems we will want to specify the state of a system by its temperature and pressure, rather than its specific volume, internal energy, and enthalpy, which appear in the energy balance. What are needed, then, are interrelations between the fluid properties that allow one to eliminate some thermodynamic variables in terms of more easily measured ones. Of particular interest is the volumetric equation of state, which is a relation between temperature, pressure, and specific volume, and the thermal equation of state, which is usually in the form of relationship between internal energy, temperature, and specific volume, or enthalpy, temperature, and pressure. Such information may be available in either of two forms. First, there are analytic equations of state, which provide an algebraic relation between the thermodynamic state variables. Second, experimental data, usually in graphical or tabular form, may be available to provide the needed interrelationships between the fluid properties.

Equations of state of fluids are examined in detail in Chapter 4. To illustrate the use of the mass and energy balance equations, we briefly consider here the equation of state for the ideal gas and the graphical and tabular display of the thermodynamic properties of several real fluids.

An **ideal gas** is a gas that is so dilute that there are no interactions among its molecules. For such gases it is possible to show, either experimentally or by the methods of statistical mechanics, that at all absolute temperatures and pressures the volumetric equation of state is

$$P \underline{V} = RT \tag{2.4-1}$$

(as was indicated in Section 1.4), and that the enthalpy and internal energy are functions of temperature only (and not pressure or specific volume). We denote this latter fact by $\underline{H} = \underline{H}(T)$ and $\underline{U} = \underline{U}(T)$. This simple behavior is to be compared with the enthalpy for real fluids, which is a function of temperature and pressure [i.e., $\underline{H} = \underline{H}(T, P)$] and the internal energy, which is a function of temperature and specific volume [$\underline{U} = \underline{U}(T, \underline{V})$].

The temperature dependence of the internal energy and enthalpy of all substances (not merely ideal gases) can be found by measuring the temperature rise that accompanies a heat flow into a closed stationary system. If a sufficiently small quantity of heat is added to such a system, it is observed that the temperature rise produced, ΔT, is linearly related to the heat flow and inversely proportional to N, the number of moles in the system

$$\frac{Q}{N} = C\Delta T = C\{T(t_2) - T(t_1)\}$$

where C is a parameter and Q is the heat added to the system between the times t_1 and t_2. The object of the experiment is to accurately measure the parameter C for as small a temperature rise as possible since C generally is a function of temperature; if the measurement is made at constant volume and with $\dot{W}_s = 0$, we have, from the energy balance and the foregoing equation

$$U(t_2) - U(t_1) = Q = NC_v\{T(t_2) - T(t_1)\}$$

Thus

$$C_v = \frac{U(t_2) - U(t_1)}{N\{T(t_2) - T(t_1)\}} = \frac{\underline{U}(t_2) - \underline{U}(t_1)}{T(t_2) - T(t_1)}$$

where the subscript v has been introduced to remind us that the parameter C was determined in a constant-volume experiment. In the limit of a very small temperature difference we have

$$C_V = \lim_{T(t_2)-T(t_1)\to 0} \frac{\underline{U}(t_2) - \underline{U}(t_1)}{T(t_2) - T(t_1)} = \left(\frac{\partial \underline{U}}{\partial T}\right)_V = \left(\frac{\partial \underline{U}(T, V)}{\partial T}\right)_V \tag{2.4-2}$$

so that the measured parameter C_V is, in fact, equal to the temperature derivative of the internal energy at constant volume. Similarly, if the parameter C is determined in a constant-pressure experiment, we have

$$\begin{aligned} Q &= U(t_2) - U(t_1) + P\{V(t_2) - V(t_1)\} \\ &= U(t_2) + P(t_2)V(t_2) - U(t_1) - P(t_1)V(t_1) = H(t_2) - H(t_1) \\ &= NC_P\{T(t_2) - T(t_1)\} \end{aligned}$$

where we have used the fact that since pressure is constant, $P = P(t_1) = P(t_2)$. Then

$$C_P = \left(\frac{\partial \underline{H}}{\partial T}\right)_P = \left(\frac{\partial \underline{H}(T, P)}{\partial T}\right)_P \tag{2.4-3}$$

so that the measured parameter here is equal to the temperature derivative of the enthalpy at constant pressure.

The quantity C_V is called the **constant-volume heat capacity**, and C_P the **constant-pressure heat capacity**; both will appear frequently throughout this book. Partial derivatives have been used in Eqs. 2.4-2 and 3 to indicate that although the internal energy is a function of temperature and density or specific volume, C_V has been measured along a path of constant volume, and that although the enthalpy is a function of temperature and pressure, C_P has been evaluated in an experiment in which the pressure was held constant.

For the special case of the ideal gas where the enthalpy and internal energy of the fluid are only functions of temperature we have

$$C_P(T) = \frac{d\underline{H}(T)}{dT} \quad \text{and} \quad C_V(T) = \frac{d\underline{U}(T)}{dT} \tag{2.4-4}$$

so that C_P and C_V are only functions of T as well, and the derivatives are total derivatives. The temperature dependence of the ideal gas heat capacity can be measured or, in some cases, computed using the methods of statistical mechanics and detailed information about molecular structure, bond lengths, vibrational frequencies, and so forth. For our purposes C_P for the ideal gas will either be considered to be a constant or be written as a function of temperature in the form

$$C_P = a + bT + cT^2 + dT^3 + \cdots \tag{2.4-5}$$

Since $\underline{H} = \underline{U} + P\underline{V}$, and for the ideal gas $P\underline{V} = RT$, we have $\underline{H} = \underline{U} + RT$, and

$$C_P(T) = \frac{d\underline{H}}{dT} = \frac{d(\underline{U} + RT)}{dT} = C_V(T) + R$$

so that $C_V(T) = C_P(T) - R = (a - R) + bT + cT^2 + dT^3 + \cdots$. The constants in Eq. 2.4-5 for various gases are given in Appendix II. [In this Appendix and from Chapter 4 onwards, the ideal gas constant-pressure and constant-volume heat capacities will be denoted by C_P^* and C_V^*, respectively.]

The enthalpy and internal energy of an ideal gas at a temperature T_2 can be related to its values at T_1 by integration of Eqs. 2.4-4 to obtain

$$\underline{H}(T_2) = \underline{H}(T_1) + \int_{T_1}^{T_2} C_P(T)\, dt$$

and

$$\underline{U}(T_2) = \underline{U}(T_1) + \int_{T_1}^{T_2} C_V(T)\, dt \tag{2.4-6}$$

Our interest in the first part of this book is with energy flow problems in single-component systems. Since the only energy information needed in solving these problems is the change in internal energy and enthalpy of a substance between two states, and since the determination of the absolute energy of a substance is not possible, what is done is to choose an easily accessible state of a substance to be the **reference state**, for which $\underline{H}$ is arbitrarily set equal to zero, and then to report the enthalpy and internal energy of all other states relative to this reference state. The fact that there is a state for each substance for which the enthalpy has been arbitrarily set to zero does lead to difficulties when chemical reactions occur. Consequently, another energy convention will be introduced in Chapter 6.

If, for the ideal gas, the temperature T_R is chosen as the reference state (i.e., $\underline{H}$ at T_R is set equal to zero), the enthalpy at temperature T is

$$\underline{H}(T) = \int_{T_R}^{T} C_P(T)\, dT \tag{2.4-7}$$

Similarly, the internal energy at T is

$$\begin{aligned}\underline{U}(T) &= \underline{U}(T_R) + \int_{T_R}^{T} C_V(T)\, dT \\ &= \{\underline{H}(T_R) - RT_R\} + \int_{T_R}^{T} C_V(T)\, dT \\ &= \int_{T_R}^{T} C_V(T)\, dT - RT_R\end{aligned} \tag{2.4-8}$$

A convenient choice for the reference temperature T_R is absolute zero. In this case

$$\underline{H} = \int_0^T C_P\, dT \qquad \text{and} \qquad \underline{U} = \int_0^T C_V\, dT$$

For the special case in which the constant-pressure and constant-volume heat capacities are independent of temperature, we have, from Eqs. 2.4-7 and 8

$$\underline{H}(T) = C_P(T - T_R)$$

and

$$\underline{U}(T) = C_V(T - T_R) - RT_R = C_VT - C_PT_R$$

which, when T_R is taken to be absolute zero, simplify further to

$$\underline{H} = C_PT \qquad \text{and} \qquad \underline{U} = C_VT$$

As we will see in Chapters 4 and 5, very few fluids are ideal gases and the mathematics of relating the enthalpy and internal energy to the temperature and pressure of a real fluid are much more complicated than indicated here. Therefore, for fluids of industrial and scientific importance detailed experimental thermodynamic data have been collected. These data can be presented in tabular form (see Appendix III for a table of the thermodynamic properties of steam), or in graphical form, as in Figs. 2.4-1, 2, and 3 for steam, methane, and nitrogen. (Can you identify the reference state for the construction of the steam tables?) With these detailed data one can, given values of temperature and pressure, easily find the enthalpy, specific volume, and entropy (a thermodynamic quantity that will be introduced in the next chapter). More generally, given any two intensive variables of a single-component, single-phase system, the rest may be found.

You should notice that different choices have been made for the independent variables in these figures. Although the independent variables may be chosen arbitrarily,[6] some choices are especially convenient when solving certain problems. Thus, as we will see, an enthalpy-entropy (*H-S*) or Mollier diagram, such as Fig. 2.4-1, is useful in turbine and compressor problems; enthalpy-pressure (*H-P*) diagrams, for example Figs. 2.4-2*a* and 3, are useful in solving refrigeration problems; and temperature-entropy (*T-S*) diagrams, of which Figs. 2.4-1*b* and 2*b* are examples, are used in the analysis of engines and cycles (see Sec. 3.3).

An important characteristic of real fluids is that at sufficiently low temperatures they condense to form liquids and solids. Also, many applications of thermodynamics of interest to engineers involve either a range of thermodynamic states for which the fluid of interest undergoes a phase change or equilibrium multiphase mixtures (e.g., steam and water at 100°C and 101.3 kPa). Since the energy balance equation is in terms of the internal energy and enthalpy per unit mass of the system, this equation is valid regardless of which phase or mixture of phases is present. Consequently, there is no difficulty, in principle, in using the energy balance (or other equations to be introduced later) for multiphase or phase-change problems, provided thermodynamic information is available for each of the phases present. Figures 2.4-1, 2, and 3 and the steam tables in Appendix III, provide such information for the vapor and liquid phases and, within the dome-shaped region, for vapor-liquid mixtures. Similar information for many other fluids is also available. Thus, you should not hesitate to apply the equations of thermodynamics to the solution of problems involving gases, liquids, solids, and mixtures thereof.

There is a simple relationship between the thermodynamic properties of a two-phase mixture (e.g., a mixture of water and steam), the properties of the individual phases, and the mass distribution between the phases. If $\hat{\theta}$ is any intensive property, such as internal energy per unit mass or volume per unit mass, its value in a two-phase mixture is

$$\hat{\theta} = x^{\mathrm{I}}\hat{\theta}^{\mathrm{I}} + x^{\mathrm{II}}\hat{\theta}^{\mathrm{II}} = x^{\mathrm{I}}\hat{\theta}^{\mathrm{I}} + (1 - x^{\mathrm{I}})\hat{\theta}^{\mathrm{II}} \tag{2.4-9}$$

Here x^{I} is the mass fraction of phase I, and $\hat{\theta}^{\mathrm{I}}$ is the value of the variable in that phase. Also, by definition of the mass fraction, $x^{\mathrm{I}} + x^{\mathrm{II}} = 1$. (For mixtures of steam and water, the mass fraction of steam is termed the **quality** and is frequently expressed as a percent, rather than as a fraction; for example, a steam-water mixture containing 0.02 kg of water for each kilogram of mixture is referred to as steam of 98% quality.)

[6]See, however, the comments made in Section 1.6 concerning the use of the combinations *U* and *T* and *V* and *T* as the independent thermodynamic variables.

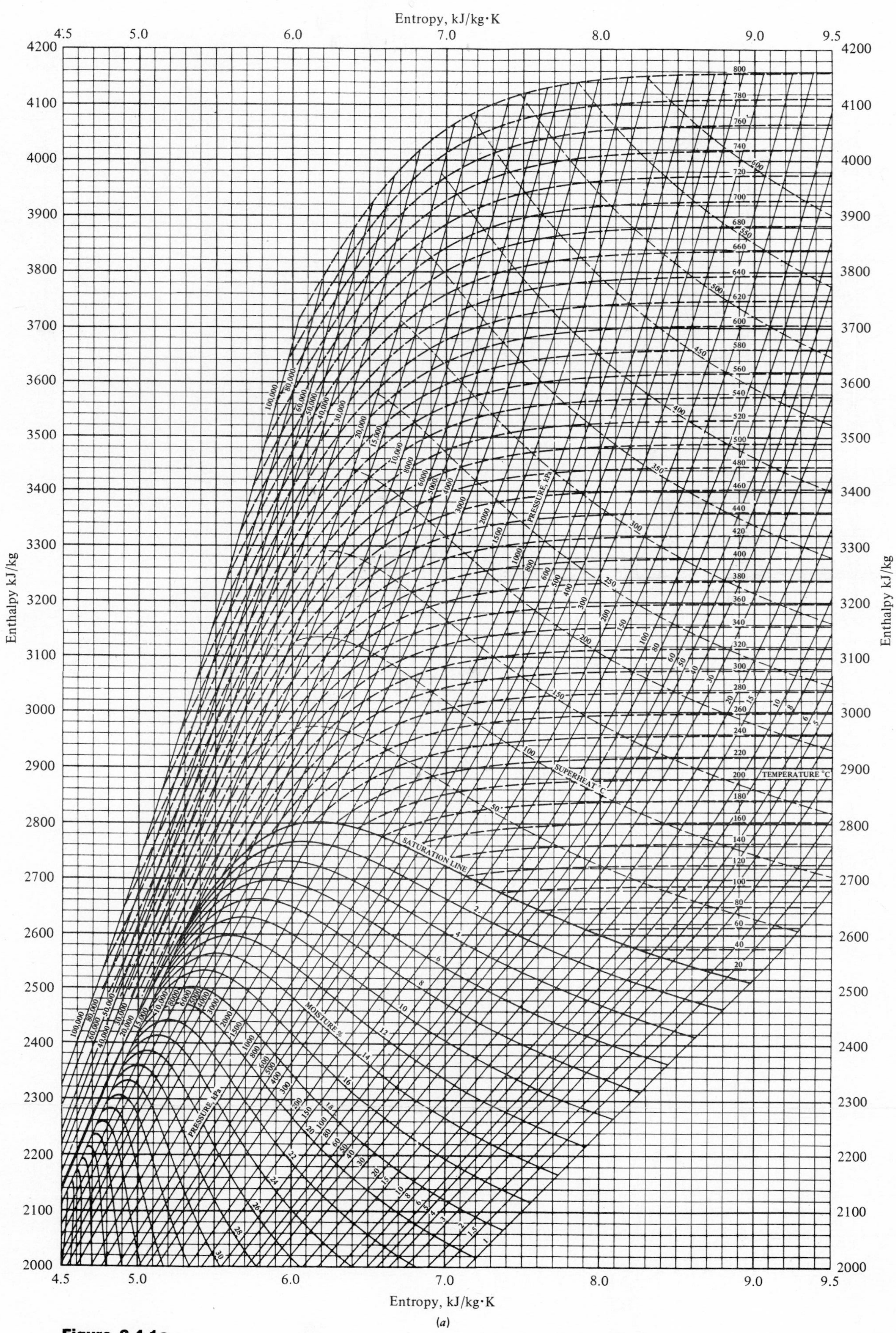

Figure 2.4-1a
Enthalpy–entropy or Mollier diagram for steam. (*Source:* ASME Steam Tables in SI (Metric) Units for Instructional Use, American Society of Mechanical Engineers, New York, 1967. Used with permission.)

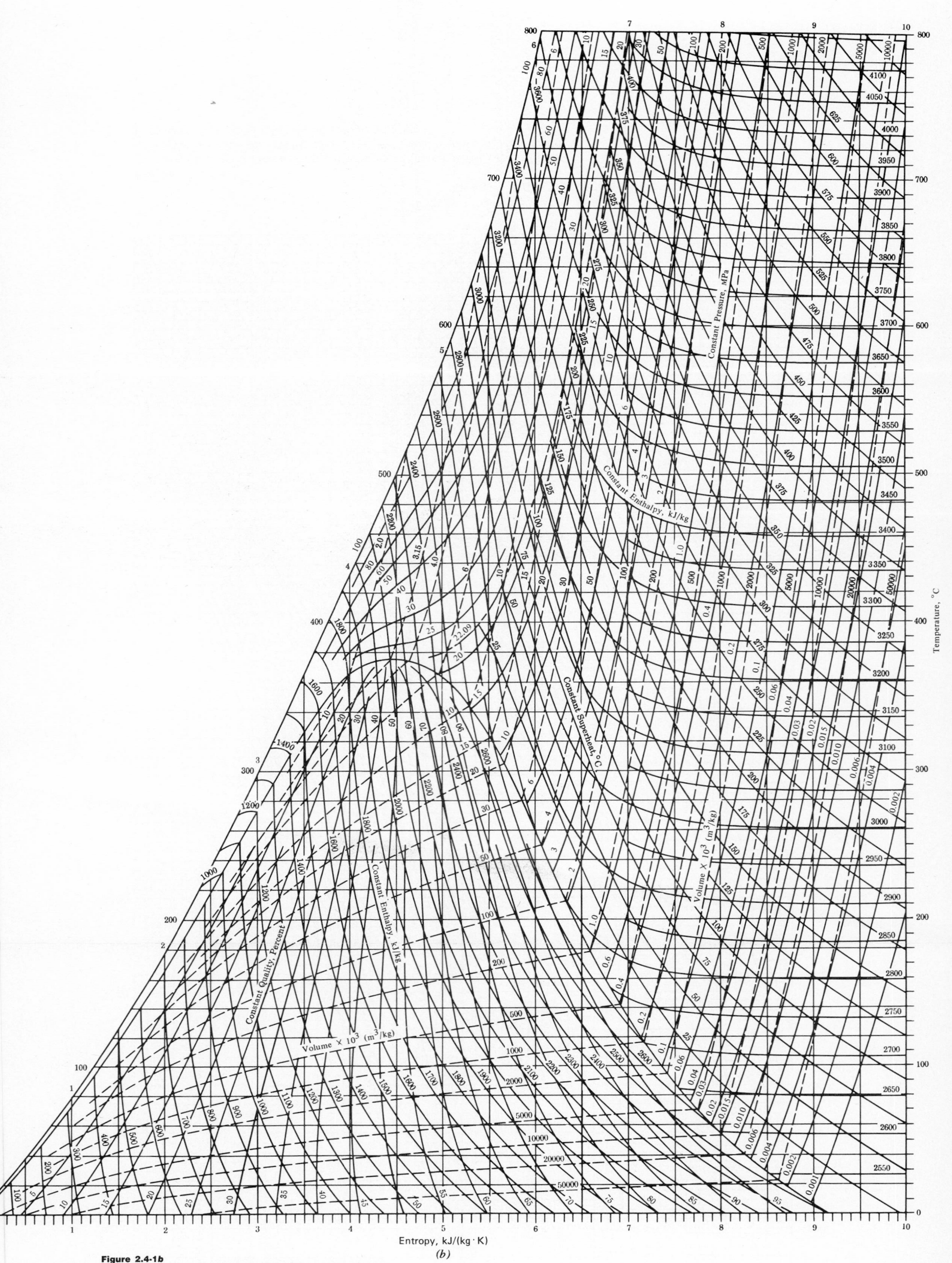

Figure 2.4-1b
Temperature–entropy diagram for steam. (*Source:* J. H. Keenan, F. G. Keyes, P. G. Hill and J. G. Moore, *Steam Tables* (International Edition–Metric Units). Copyright © 1969, John Wiley & Sons, Inc., New York. Used with permission.)

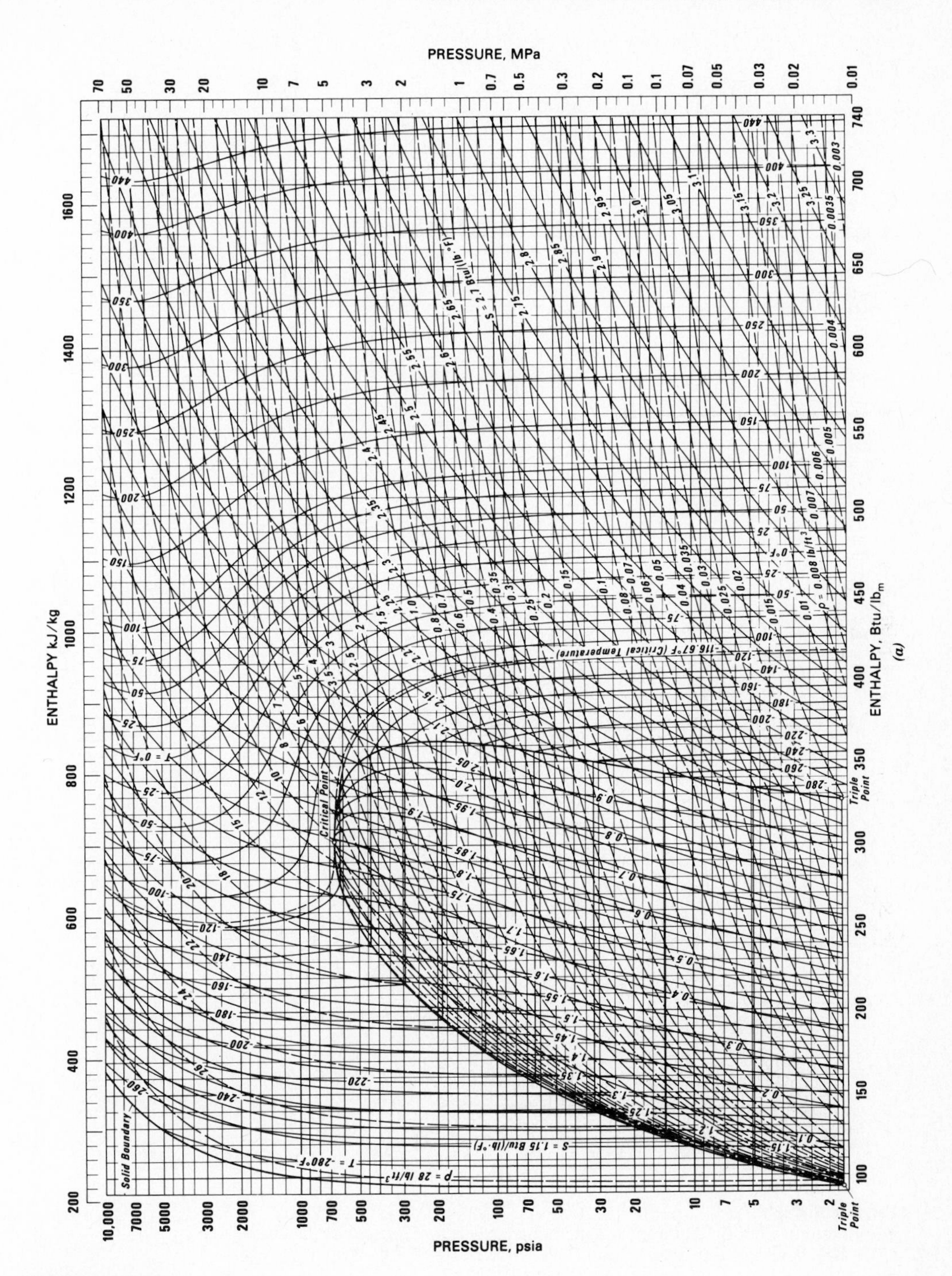

Figure 2.4-2a
Pressure–enthalpy diagram for methane. (*Source:* Cryogenics Division, NBS-IBS, Boulder, Colorado, Chart 2029, April 1, 1977. Used with permission.)

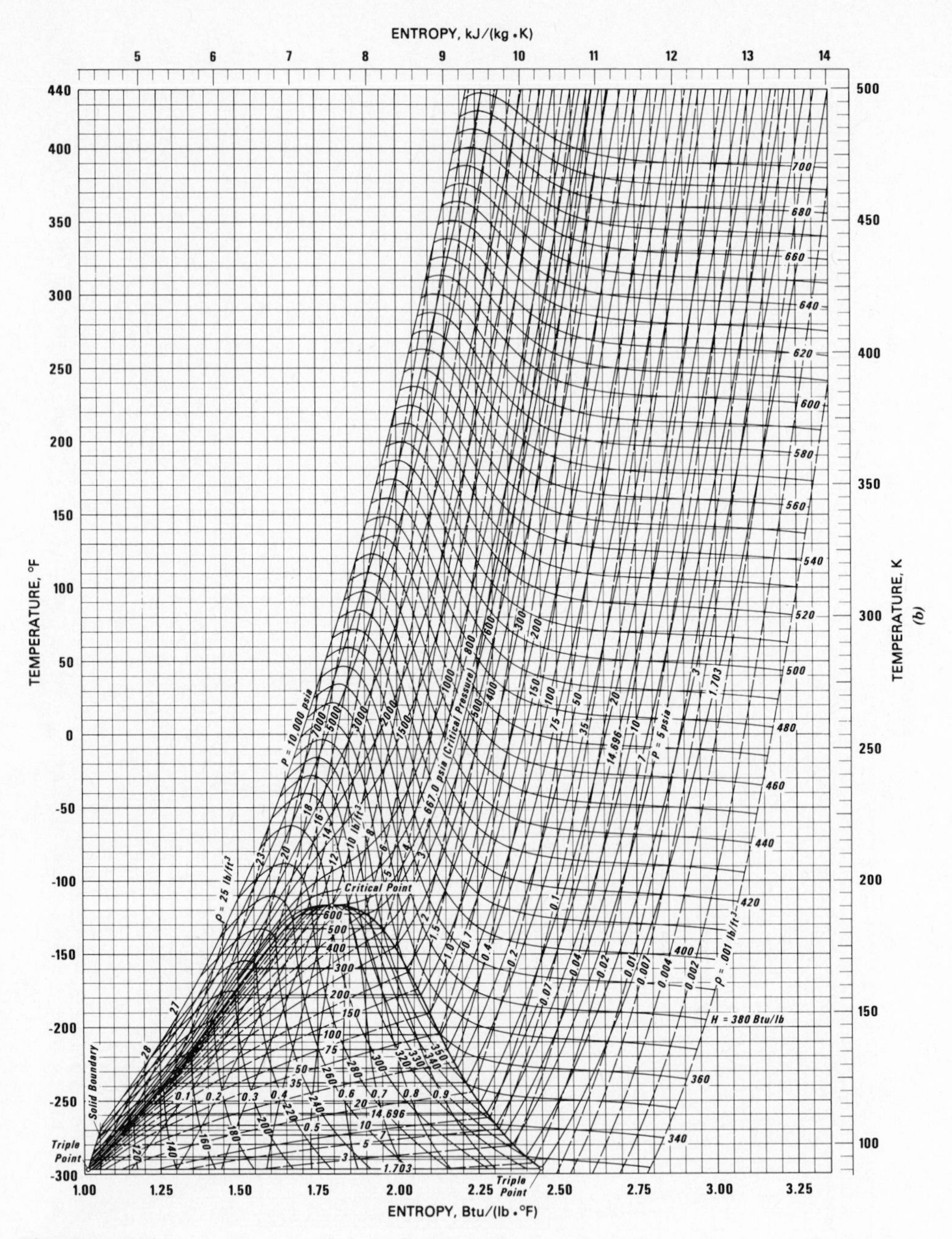

Figure 2.4-2b

Temperature–entropy diagram for methane. (*Source:* Cryogenics Division, NBS–IBS, Boulder, Colorado, Chart 2026, April 1, 1977. Used with permission.)

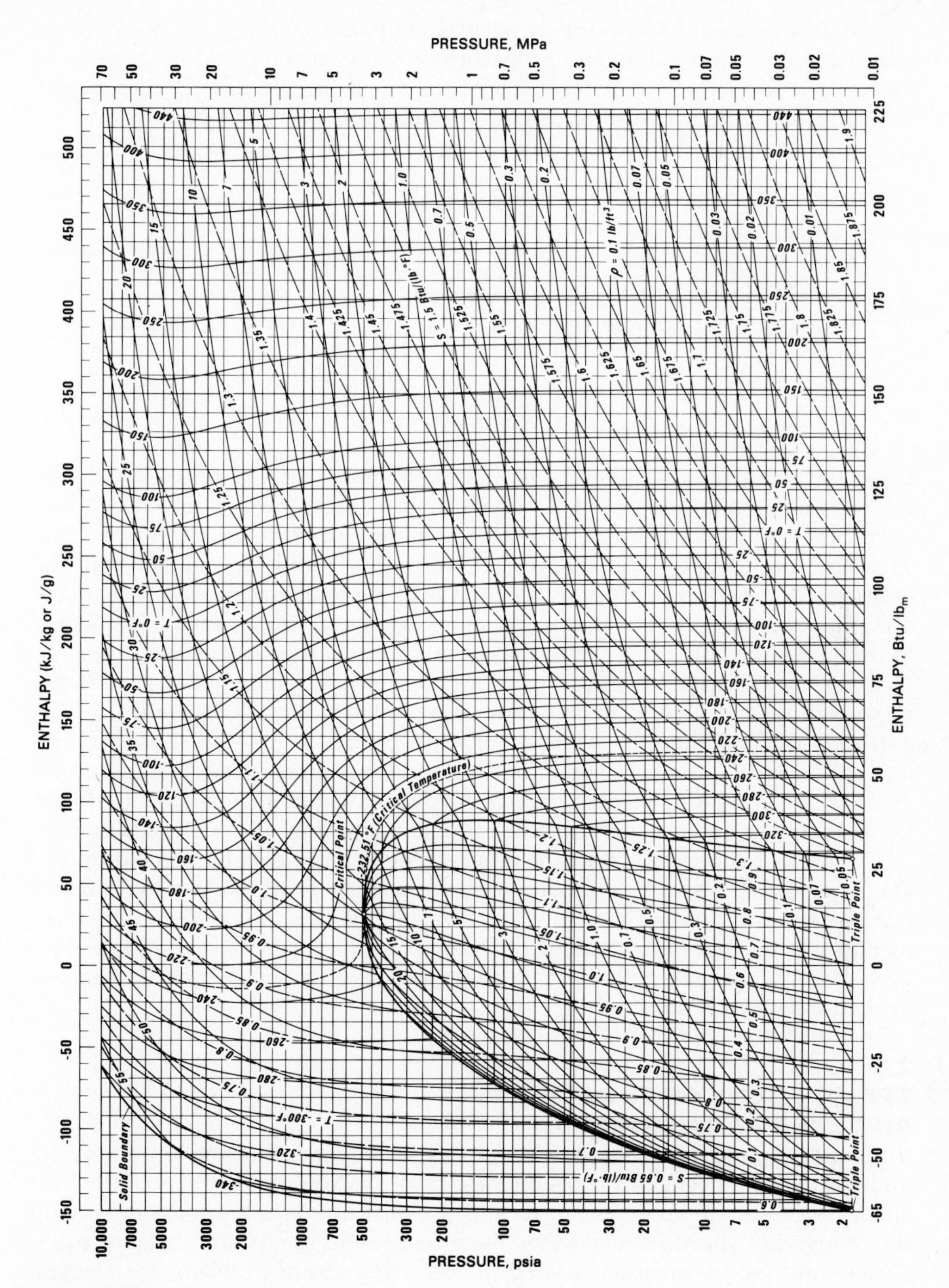

Figure 2.4-3
Pressure–enthalpy diagram for nitrogen (*Source:* Cryogenics Division, NBS-IBS, Boulder, Colorado, Chart 2028, April 1, 1977. Used with permission.)

It is also useful to note that several simplifications can be made when computing the thermodynamic properties of solids and liquids. First, because the molar volumes of condensed phases are small, the product $P\underline{V}$ can be neglected, unless the pressure is very high. Thus, for solids and liquids,

$$\underline{H} \approx \underline{U} \tag{2.4-10}$$

and

$$C_P \approx C_V \tag{2.4-11}$$

A further simplification commonly made for liquids and solids is to assume that they are also incompressible; that is, their volume is only a function of temperature so that

$$\left(\frac{\partial \underline{V}}{\partial P}\right)_T = 0 \tag{2.4-12}$$

Using the techniques of Chapter 4 it can be shown that for incompressible fluids, the thermodynamic properties $\underline{H}$, $\underline{U}$, C_P, and C_V are functions of temperature only. Since, in fact, solids, and most liquids away from their critical point (see Chapter 5), are relatively incompressible, Eqs. 2.4-10 and 11, together with the assumption that these properties depend only on temperature, are reasonably accurate and often used in thermodynamic studies of liquids and solids. Thus, for example, the enthalpy of liquid water at a temperature T_1 and pressure P_1 is, to a very good approximation, equal to the enthalpy of liquid water at the temperature T_1 and any convenient pressure. Consequently, the entries for the enthalpy of liquid water for a variety of temperatures (at pressures corresponding to the vapor-liquid coexistence or saturation pressures for each temperature) given in the saturation steam tables of Appendix III can also be used to obtain the enthalpy of liquid water at these same temperatures and higher pressures.

Finally, although we have formulated the balance equations for pure fluids, these equations can also be applied to mixtures, provided that the mixture composition is the same throughout the system. Thus, for example, using information on the thermodynamic properties of air, we could solve energy flow problems involving air in the same manner as pure fluid problems will be solved in the following section.

2.5 APPLICATIONS OF THE MASS AND ENERGY BALANCES

In many thermodynamics problems one is given some information about the initial equilibrium state of a substance and asked to find the final state if the heat and work flows are specified, or to find the heat or work flows accompanying the change to a specified final state. Since we use thermodynamic balance equations to get the information needed to solve this sort of problem, the starting point is always the same, the identification of a convenient thermodynamic system. The main restriction on the choice of the system is that the flow terms into and out of it must be of a simple form, for example, time-invariant or perhaps even zero. Next, the forms of the mass and energy balance equations appropriate to the system choice are written, and any information about the initial and final states of the system and the flow terms is used. Finally, the thermal equation of state is used to replace the internal energy and enthalpy in the balance equations with temperature, pressure, and volume; the volumetric equation of state may then be used to eliminate the volume in terms of temperature and pressure. In this way equations are obtained that contain temperature and pressure as the only state variables.

The volumetric equation of state may also provide another relationship between the temperature, pressure, mass, and volume when the information about the final state of the system is presented in terms of total volume, rather than volume per unit mass or molar volume (see Illustration 2.5-5).

By using the balance equations and the equation of state information, we will frequently be left with equations that contain only temperature, pressure, mass, shaft work W_s, and the heat flow Q. If the number of equations equals the number of unknowns, the problem can be solved. The mass and energy balance equations, together with equation of state information, are sufficient to solve many, but not all, energy flow problems. In some situations we are left with more unknowns than equations. In fact, we can readily identify a class of problems of this sort. The mass and energy balance equations together can, at most, yield new information about only one intensive variable of the system (the internal energy or enthalpy per unit mass), or about the sum of the heat and work flows if only the state variables are specified. Therefore, we will not, at present, be able to solve problems in which (1) there is no information about any intensive variable of the final state of the system, or (2) both the heat flow Q and the shaft work W_s are unspecified, or, as in Illustration 2.5-4, in which (3) one intensive variable of the final state and either Q or W_s are unknown. To solve these problems, an additional balance equation is needed; such an equation will be developed in Chapter 3.

The seemingly most arbitrary step in thermodynamic problem solving is the choice of the system. Since the mass and energy balances were formulated with great generality, they apply to any choice of system, and, as was demonstrated in Illustration 2.3-1, the solution of a problem is independent of the system chosen in obtaining the solution. However, some system choices may result in less effort being required to obtain a solution than others. This will be demonstrated here and again in Chapter 3.

ILLUSTRATION 2.5-1

Steam at 400 bar and 500°C undergoes a Joule-Thomson expansion to 1 bar. Determine the temperature of the steam after the expansion using

a. Figure 2.4-1a
b. Figure 2.4-1b
c. The steam tables in Appendix III

Solution

From Illustration 2.3-3 we have that

$$\hat{H}_2 = \hat{H}(T_1, P_1) = \hat{H}(T_2, P_2) = \hat{H}_2$$

for a Joule-Thomson expansion. Since T_1 and P_1 are known, $\hat{H}_1$ can be found from either Figs. 2.4-1 or the steam tables. Then, since $\hat{H}_2$ (= $\hat{H}_1$, from the foregoing) and P_2 are known, T_2 can be found.

a. Using Fig. 2.4-1a we first locate the point P = 400 bar = 40000 kPa and T = 500°C, which corresponds to $\hat{H}_1$ = 2900 kJ/kg. Now following a line of constant enthalpy (horizontal line on the Mollier diagram) to P = 1 bar = 100 kPa, we find that the final temperature is about 212°C.

b. Using Fig. 2.4-1b we locate the point $P = 400$ bar and $T = 500°C$ (which is somewhat easier to do than using Fig. 2.4-1a) and follow the curved line of constant enthalpy to a pressure of 1 bar and see that $T_2 = 212°C$.

c. Using the steam tables of Appendix III we have that at $P = 400$ bar $=$ 40 MPa and T $= 500°C$, $\hat{H} = 2903.3$ kJ/kg. At $P = 1$ bar $= 0.1$ MPa, $\hat{H} =$ 2879.5 kJ/kg at $T = 200°C$ and $\hat{H} = 2977.3$ kJ/kg at $T = 250°C$. Assuming that the enthalpy varies linearly with temperature between 200 and 250°C at $P = 1$ bar we have by interpolation

$$T = 200 + (250 - 200) \times \frac{(2903.3 - 2879.5)}{(2977.3 - 2879.5)} = 212.2 \text{ K}$$

Comment

For many problems a graphical representation of thermodynamic data, such as Figures 2.4-1, is easiest to use although the answers obtained are approximate and certain parts of the graphs may be difficult to read accurately. Use of tables of thermodynamic data, such as the steam tables, generally leads to the most accurate answers; however, one or more interpolations may be required. For example, if the initial condition of the steam had been 475 bar and 530°C instead of 400 bar and 500°C the method of solution using Figs. 2.4-1 would be unchanged; however, using the steam tables, we would have to interpolate with respect to both temperature and pressure to get the initial enthalpy of the steam.

[One way to do this is to first, by interpolation with temperature, to obtain the enthalpy of steam at 530°C at both 400 bar = 40 MPa and at 500 bar. Then, by interpolation with respect to pressure between these two values, to obtain the enthalpy at 475 bar. That is, from

$$\hat{H}(40 \text{ MPa}, 500°\text{C}) = 2903.3 \text{ kJ/kg} \qquad \hat{H}(50 \text{ MPa}, 500°\text{C}) = 2720.1 \text{ kJ/kg}$$

$$\hat{H}(40 \text{ MPa}, 550°\text{C}) = 3149.1 \text{ kJ/kg} \qquad \hat{H}(50 \text{ MPa}, 550°\text{C}) = 3019.5 \text{ kJ/kg}$$

and the interpolation formula

$$\Theta(x + \Delta) = \Theta(x) + \Delta \frac{\Theta(y) - \Theta(x)}{y - x}$$

where Θ is any tabulated function and x and y are two adjacent values at which Θ is available, we have that

$$\begin{aligned}
\hat{H}&(40 \text{ MPa}, 530°\text{C}) \\
&= \hat{H}(40 \text{ MPa}, 500°\text{C}) + 30 \times \frac{(\hat{H}(40 \text{ MPa}, 550°\text{C}) - \hat{H}(40 \text{ MPa}, 500°\text{C}))}{550 - 500} \\
&= 2903.3 + 30 \times \frac{(3149.1 - 2903.3)}{50} = 2903.3 + 147.5 \\
&= 3050.8 \text{ kJ/kg}
\end{aligned}$$

and

$$\hat{H}(50 \text{ MPa}, 530°\text{C}) = 2702.1 + 30 \times \frac{(3019.5 - 2720.1)}{50} = 2881.7 \text{ kJ/kg}$$

then

$\hat{H}(47.5 \text{ mPa}, 530°\text{C})$

$$= \hat{H}(40 \text{ mPa}, 530°\text{C}) + 7.5 \times \frac{(\hat{H}(50 \text{ MPa}, 530°\text{C}) - \hat{H}(40 \text{ mPa}, 530°\text{C}))}{50 - 40}$$

$$= 3050.8 + \frac{7.5}{10} \times (2881.7 - 3050.8) = 2924.0 \text{ kJ/kg} \quad \blacksquare$$

ILLUSTRATION 2.5-2

An adiabatic steady-state steam turbine is being designed to serve as an energy source for a small electrical generator. The inlet to the turbine will be steam at 600°C and 10 bar, with a velocity of 100 m/s and a flow rate of 2.5 kg/s. The conditions at the turbine exit are $T = 400°\text{C}$, $P = 1$ bar, and a gas velocity of 30 m/s. Estimate the rate at which work can be obtained from this turbine.

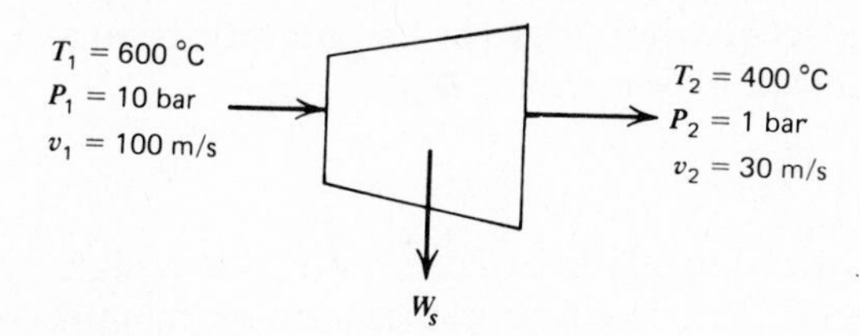

Solution

The first step in solving any energy flow problem is to choose the thermodynamic system; the second step is to write the balance equations for the system. Here we will take the turbine and its contents to be the system. The mass and energy balance equations for this adiabatic, steady-state system are

$$\frac{dM}{dt} = 0 = \dot{M}_1 + \dot{M}_2 \tag{a}$$

and

$$\frac{d}{dt}\{U + M(v^2/2 + gh)\} = 0 = \dot{M}_1(\hat{H}_1 + v_1^2/2) + \dot{M}_2(\hat{H}_2 + v_2^2/2) + \dot{W}_s \tag{b}$$

In writing these equations we have set the rate of change of mass and energy equal to zero because the turbine is in steady-state operation; Q is equal to zero because the process is adiabatic; and $P(dV/dt)$ is equal to zero because the volume of the system is constant (unless the turbine explodes). Finally, since the schematic diagram indicates that the turbine is positioned horizontally, we have assumed there is no potential energy change in the flowing steam.

There are four unknowns, $\dot{M}_2$, $\hat{H}_1$, $\hat{H}_2$ and $\dot{W}_s$ in Eqs. a and b. However, thermal equation of state information (here the steam tables in Appendix III) relates the enthalpies to temperature and pressure, both of which are known. Thus, $\dot{M}_2$ and $\dot{W}_s$ are the only true unknowns, and these may be found from the balance equations. From the mass balance equation we have

$$\dot{M}_2 = -\dot{M}_1 = -2.5 \text{ kg/s}$$

Also from the steam tables, or less accurately from Figs. 2.4-1, we have

$$\hat{H}_1 = 3697.9 \text{ kJ/kg} \quad \text{and} \quad \hat{H}_2 = 3278.2 \text{ kJ/kg}$$

so that the energy balance yields

$$\dot{W}_s = -\dot{M}_2(\hat{H}_2 + v_1^2/2) - \dot{M}_1(\hat{H}_1 + v_2^2/2)$$

$$= -2.5\,\frac{\text{kg}}{\text{s}}\,\{(\hat{H}_1 - \hat{H}_2) + \frac{1}{2}\,(v_1^2 - v_2^2)\}$$

$$= -2.5\,\frac{\text{kg}}{\text{s}}\left\{419.7\,\frac{\text{kJ}}{\text{kg}} + \frac{1}{2}\,(100^2 - 30^2)\,\frac{\text{m}^2}{\text{s}^2} \times \frac{1\text{J/kg}}{\text{m}^2/\text{s}^2} \times \frac{1\text{kJ}}{1000\text{J}}\right\}$$

$$= -2.5\,\frac{\text{kg}}{\text{s}}\,\{419.7 + 4.55\}\,\frac{\text{kJ}}{\text{kg}} = -1060.6\ \text{kJ/s} \quad (= -1422\ \text{hp})$$

Here the minus sign indicates that work is being done *by* the turbine; that is, shaft work is being supplied by the system to its surroundings.

Comment

If we had completely neglected the kinetic energy terms in this calculation the error in the work term would be about 1%. Generally, the contribution of kinetic and potential energy terms can be neglected when there is a significant change in the fluid temperature, as was suggested in Sec. 2.3. ■

ILLUSTRATION 2.5-3

A compressed air tank is to be repressurized to 40 bar by connecting it to a high-pressure line containing air at 50 bar and 20°C. The repressurization of the tank occurs so quickly that the process may be assumed to be adiabatic; also there is no heat transfer from the air to the tank. If the tank initially contains air at 1 bar and 20°C, what will be the temperature of the air in the tank at the end of the filling process? After a sufficiently long period of time, the gas in the tank is found to be at room temperature (20°C) because of heat exchange with the tank and the atmosphere. What is the new pressure of air in the tank? You may assume air to be an ideal gas with C_v = 21 J/mol K.

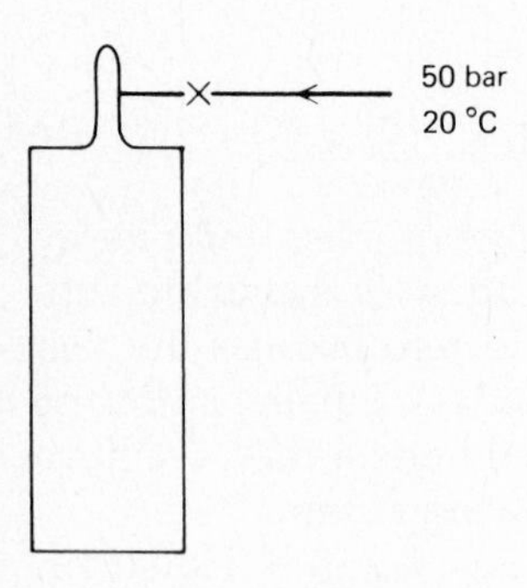

Solution

We will take the contents of the tank to be the system. The difference form of the mass (or rather mole) and energy balances for this open system are

$$N_2 - N_1 = \Delta N \tag{a}$$

$$N_2\underline{U}_2 - N_1\underline{U}_1 = (\Delta N)\underline{H}_{\text{in}} \tag{b}$$

In writing the energy balance we have made the following observations:

1. The kinetic and potential energy terms are small and may be neglected.
2. Since the tank is connected to a source of gas at constant temperature and pressure, $\underline{H}_{\text{in}}$ is constant.
3. The initial process is adiabatic, so $Q = 0$, and the system (the contents of the tank) is of constant volume, so that $\Delta V = 0$.

Substituting Eq. a in Eq. b, and recognizing that for the ideal gas $\underline{H}(T) = C_P(T - T_R)$ and $\underline{U}(T) = C_V(T - T_R) - RT_R$, yields

$$N_2\{C_V(T_2 - T_R) - RT_R\} - N_1\{C_V(T_1 - T_R) - RT_R\} = (N_2 - N_1)C_P(T_{\text{in}} - T_R)$$

or

$$N_2C_VT_2 - N_1C_VT_1 = (N_2 - N_1)C_PT_{\text{in}}$$

[Note that the reference temperature T_R cancels out of the equation.] Finally, using the ideal gas equation of state to eliminate N_1 and N_2, and recognizing that $V_1 = V_2$, yields

$$\frac{P_2}{T_2} = \frac{P_1}{T_1} + \frac{C_V}{C_P}\left(\frac{P_2 - P_1}{T_{\text{in}}}\right) \quad \text{or} \quad T_2 = \frac{P_2}{\dfrac{P_1}{T_1} + \dfrac{C_V}{C_P}\left(\dfrac{P_2 - P_1}{T_{\text{in}}}\right)}$$

The only unknown in this equation is T_2, so, formally, the problem is solved. The answer is $T = 405.2$ K $= 132.05$°C.

Before proceeding to the second part of the problem, it is interesting to consider the case in which the tank is initially evacuated. Here $P_1 = 0$, and so

$$T_2 = \frac{C_P}{C_V} T_{\text{in}}$$

independent of the final pressure. Since C_P is always greater than C_V for a gas, this implies that the temperature of the gas in the tank will be greater than the temperature of gas in the line. Why is this so?

To find the pressure in the tank after the heat transfer process, we use the mass balance and the equation of state. Again, choosing the contents of the tank as the system, the mass (mole) balance is $N_2 = N_1$, since there is no transfer of mass into or out of the system during the heat transfer process (unless, of course, the tank is leaking; we will not consider this complication here). Now using the ideal gas equation of state we have

$$\frac{P_2}{T_2} = \frac{P_1}{T_1} \quad \text{or} \quad P_2 = P_1\frac{T_2}{T_1}$$

Thus, $P_2 = 28.94$ bar. ■

ILLUSTRATION 2.5-4

A compressor is a gas pumping device that takes in gas at a low pressure and discharges it at a higher pressure. Since this process occurs quickly compared to heat transfer, it is usually assumed to be adiabatic; that is, there is no heat transfer to or from the gas during its compression. Assuming that the inlet to the compressor is air (which we will take to be an ideal gas with $C_P = 29.3$ J/mol K) at 1 bar and 290 K and that the discharge is at a pressure of 10 bar, estimate

a. The temperature of the exit gas

b. The rate at which work is done on the gas (i.e., the power requirement) for a gas flow of 2.5 mol/s

Solution

The system will be taken to be the gas contained in the compressor. The differential form of the molar mass and energy balances for this open system are

$$\frac{dN}{dt} = \dot{N}_1 + \dot{N}_2$$

$$\frac{dU}{dt} = \dot{N}_1\underline{H}_1 + \dot{N}\underline{H}_2 + \dot{Q} + \dot{W}$$

where we have used the subscript 1 to indicate the flow stream into the compressor and 2 to indicate the flow stream out of the compressor.

Since the compressor operates continuously, the process may be assumed to be in a steady state,

$$\frac{dN}{dt} = 0 \quad \text{or} \quad \dot{N}_1 = -\dot{N}_2$$

$$\frac{dU}{dt} = 0$$

that is, the time variations of the mass of the gas contained in the compressor and of the energy content of this gas are both zero. Also, $\dot{Q} = 0$ since there is no heat transfer to the gas, and $\dot{W} = \dot{W}_s$ since the system boundaries (the compressor) are not changing with time. Thus we have

$$\dot{W}_s = \dot{N}_1\underline{H}_2 - \dot{N}_1\underline{H}_1 = \dot{N}_1 C_P(T_2 - T_1)$$

or

$$\underline{W}_s = C_P(T_2 - T_1)$$

where $\underline{W}_s = \dot{W}_s/\dot{N}_1$ is the work done per mole of gas. Therefore, the power necessary to drive the compressor can be computed once the outlet temperature of the gas is known, or the outlet temperature can be determined if the power input is known.

We are at an impasse; we need more information before a solution can be obtained. It is clear, by comparison with the previous examples, why we cannot obtain a solution here. In the previous cases, the mass balance and the energy balance, together with the equation of state of the fluid and the problem statement provided the information necessary to determine the final state of the system. However, here we have a situation where the energy balance contains two unknowns, the final temperature and $\underline{W}_s$. Since neither is specified, we will need additional information about the system or process before we can solve the problem. This additional information will be obtained in the next chapter. ■

ILLUSTRATION 2.5-5

A gas cylinder of 1 m^3 volume containing nitrogen initially at a pressure of 40 bar and a temperature of 200 K is connected to another cylinder of 1 m^3 volume that is evacuated. A valve between the two cylinders is opened until the pressures in the cylinders equalize. Find the final temperature and pressure in each cylinder if there is no heat flow into or out of the cylinders or between the gas and the cylinder. You may assume that the gas is ideal with a constant pressure heat capacity of 29.3 J/mol K.

Solution

This problem is more complicated than the previous ones because we are interested in changes that occur in two separate cylinders. We can try to obtain a solution to this problem in two different ways. First, we could consider each tank to be a separate system, and so obtain two mass balance and two energy balance equations, which are coupled by the facts that the mass flow rate and enthalpy of the gas leaving the first cylinder are equal to the like quantities entering the second cylinder.[7] Alternatively, we could obtain an equivalent set of equations by choosing a composite system of the two interconnected gas cylinders to be the first system and the second system to be either one of the cylinders. In this way the first (composite) system is closed and the second system is open. We will use the second system choice here; you are encouraged to explore the first system choice independently and to verify that the same solution is obtained.

The difference form of the mass and energy balance equations (on a molar basis) for the two-cylinder composite system are

$$N_1^i = N_1^f + N_2^f \tag{a}$$

and

$$N_1^i \underline{U}_1^i = N_1^f \underline{U}_1^f + N_2^f \underline{U}_2^f \tag{b}$$

Here the subscripts 1 and 2 refer to the cylinders, and the superscripts i and f refer to the initial and final states. In writing the energy balance equation we have recognized that for the system consisting of both cylinders there is no mass flow, heat flow, or change in volume.

Now using, in Eq. a, the ideal gas equation of state written as $N = PV/RT$ and the fact that the volumes of both cylinders are equal yields

$$\frac{P_1^i}{T_1^i} = \frac{P_1^f}{T_1^f} + \frac{P_2^f}{T_2^f} \tag{a'}$$

Using the same observations in Eq. b, and further recognizing that for a constant heat capacity gas we have, from Eq. 2.4-8, that

$$\underline{U}(T) = C_V T - C_P T_R$$

yields

$$\frac{P_1^i}{T_1^i}\{C_V T_1^i - C_P T_R\} = \frac{P_1^f}{T_1^f}\{C_V T_1^f - C_P T_R\} + \frac{P_2^f}{T_2^f}\{C_V T_2^f - C_P T_R\}$$

which, on rearrangement, gives

$$-\left\{\frac{P_1^i}{T_2^i} - \frac{P_1^f}{T_1^f} - \frac{P_2^f}{T_2^f}\right\} C_P T_R + C_V\{P_1^i - P_1^f - P_2^f\} = 0$$

[7]That the enthalpy of the gas leaving the first cylinder is equal to that entering the second even though the two cylinders are at different pressures follows from the fact that the plumbing between the two can be thought of as a flow constriction, as in the Joule–Thomson expansion. Thus the analysis of Illustration 2.3-3 applies to this part of the total process.

Since the bracketed quantity in the first term is identically zero (see Eq. a′), we obtain

$$P_1^i = P_1^f + P_2^f \tag{c}$$

(Note that the properties of reference state have canceled. This is to be expected, since the solution to a change of state problem must be independent of the arbitrarily chosen reference state.)

Next, we observe that from the problem statement $P_1^f = P_2^f$; thus

$$P_1^f = P_2^f = \tfrac{1}{2}P_1^i = 20 \text{ bar}$$

and

$$\frac{1}{T_1^f} + \frac{1}{T_2^f} = \frac{2}{T_1^i} \tag{c'}$$

Thus we have one equation for the two unknowns, the two final temperatures. We cannot assume that the final gas temperatures in both cylinders are the same because nothing in the problem statement indicates that a transfer of heat between the two cylinders necessary to equalize the gas temperatures has occurred.

To get the additional information necessary to solve this problem, we next write the mass and energy balance equations for the initially filled cylinder. The rate of change form of these equations for this system are

$$\frac{dN_1}{dt} = \dot{N} \tag{d}$$

and

$$\frac{d(N_1\underline{U}_1)}{dt} = \dot{N}\underline{H}_1 \tag{e}$$

In writing the energy balance equation, we have made use of the fact that $\dot{Q}$, $\dot{W}_s$, and dV/dt are all zero. Also, we have assumed that while the gas temperature is changing with time, it is spatially uniform within the cylinder, so that at any instant in time the temperature and pressure of the gas leaving the cylinder are identical with those properties of the gas in the cylinder. Thus, the molar enthalpy of the gas leaving the cylinder is

$$\underline{H} = \underline{H}(T_1, P_1) = \underline{H}_1$$

Since our interest is in the change in temperature of the gas that occurs as its pressure drops from 40 bar to 20 bar due to the escaping gas, you may ask why the balance equations here have been written in the rate of change rather than the change over a time interval form. The answer is that since the properties of the gas within the cylinder (i.e., its temperature and pressure) are changing with time, so is $\underline{H}_1$, the enthalpy of the exiting gas. Thus, if we were to use the form of Eq. e integrated over a time interval (i.e., the difference form of the energy balance equation)

$$N_1^f\underline{U}_1^f - N_1^i\underline{U}_1^i = \int \dot{N}\underline{H}_1\,dt$$

we would have no way of evaluating the integral on the right side. Consequently, the difference equation provides no useful information for the solution of the problem. However, by starting with Eqs. d and e, it is possible to obtain a solution, as will be evident shortly.

To proceed with the solution, we first combine and rearrange the mass and energy balances to obtain

$$\frac{d(N_1\underline{U}_1)}{dt} \equiv N_1\frac{d\underline{U}_1}{dt} + \underline{U}_1\frac{dN_1}{dt} = \dot{N}\underline{H}_1 = \underline{H}_1\frac{dN_1}{dt}$$

so that we have

$$N\frac{d\underline{U}_1}{dt} = (\underline{H}_1 - \underline{U}_1)\frac{dN_1}{dt}$$

Now we use the following properties of the ideal gas (see Eqs. 2.4-7 and 8)

$$N = PV/RT, \qquad \underline{H} = C_P(T - T_R)$$

and

$$\underline{U} = C_V(T - T_R) - RT_R$$

to obtain

$$\frac{P_1V}{RT_1}C_V\frac{dT_1}{dt} = RT_1\frac{d}{dt}\left(\frac{P_1V}{RT_1}\right)$$

Simplifying this equation yields

$$\frac{C_V}{R}\frac{1}{T_1}\frac{dT_1}{dt} = \frac{T_1}{P_1}\frac{d}{dt}\left(\frac{P_1}{T_1}\right)$$

or

$$\frac{C_V}{R}\frac{d\ln T_1}{dt} = \frac{d}{dt}\ln\left(\frac{P_1}{T_1}\right)$$

Now integrating between the initial and final states we obtain

$$\left(\frac{T_1^f}{T_1^i}\right)^{C_V/R} = \left(\frac{P_1^f}{P_1^i}\right)\left(\frac{T_1^i}{T_1^f}\right)$$

or

$$\left(\frac{T_1^f}{T_1^i}\right)^{C_P/R} = \left(\frac{P_1^f}{P_1^i}\right) \tag{f}$$

where we have used the fact that for the ideal gas $C_P = C_V + R$. Equation f provides the means to compute T_1^f, and T_2^f can then be found from Eq. c′. Finally, using the ideal gas equation of state we can also compute the final number of moles of gas in each cylinder using the relation

$$N_I^f = \frac{V_{\text{cylI}}P_I^f}{RT_I^f} \tag{g}$$

where the subscript I refers to the cylinder number. The answers are

$$T_1^f = 164.3\text{K} \qquad N_1^f = 1.464 \text{ kmol}$$

$$T_2^f = 255.6\text{K} \qquad N_2^f = 0.941 \text{ kmol}$$

Comments

The solution of this problem for real fluids is considerably more complicated than for the ideal gas. The starting points are again

$$P_1^f = P_2^f \tag{h}$$

and

$$N_1^f + N_2^f = N_1^i \tag{i}$$

and Eqs. d and e. However, instead of Eq. g we now have

$$N_1^f = \frac{V_{\text{cyl1}}}{\underline{V}_1^f} \tag{j}$$

and

$$N_2^f = \frac{V_{\text{cyl2}}}{\underline{V}_2^f} \tag{k}$$

where $\underline{V}_1^f$ and $\underline{V}_2^f$ are related to (T_1^f, P_1^f) and (T_2^f, P_2^f), respectively, through the equation of state or tabular PVT data of the form

$$\underline{V}_1^f = \underline{V}_1^f(T_1^f, P_1^f) \tag{l}$$

$$\underline{V}_2^f = \underline{V}_2^f(T_2^f, P_2^f) \tag{m}$$

The energy balance for the two-cylinder composite system is

$$N_1^i \underline{U}_1^i = N_1^f \underline{U}_1^f + N_2^f \underline{U}_2^f \tag{n}$$

Since a thermal equation of state or tabular data of the form $\underline{U} = \underline{U}(T, \underline{V})$ are presumed available, Eq. n introduces no new variables.

Thus we have seven equations among eight unknowns (N_1^f, N_2^f, T_1^f, T_2^f, P_1^f, P_2^f, $\underline{V}_1^f$, and $\underline{V}_2^f$). The final equation needed to solve this problem can, in principle, be obtained by the manipulation and integration of Eq. e, as in the ideal gas case, but now using the real fluid equation of state or tabular data and numerical integration techniques. Since this analysis is difficult, and a simpler method of solution (discussed in Chapter 3) is available, the solution of this problem for the real fluid case will be postponed until Sec. 3.5. ■

ILLUSTRATION 2.5-6

It is possible to go from a given initial equilibrium state of a system to a given final equilibrium state by a number of different paths. Since the internal energy of a system is a state property, its change between any two states must be independent of the path chosen (see Sec. 1.3). The heat and work flows are, however, path-dependent quantities. This assertion is established here by example. One gram mole of a gas at a temperature of 25°C and a pressure of 1 bar (the initial state) is to be heated and compressed in a frictionless piston and cylinder to 300°C and 10 bar (the final state). Compute the heat and work required along each of the following paths.

Path A Isothermal (constant temperature) compression to 10 bar, and then isobaric (constant pressure) heating to 300°C.

Path B Isobaric heating to 300°C followed by isothermal compression to 10 bar.

Path C A compression in which PV^{γ} = constant, where $\gamma = C_P/C_V$, followed by an isobaric cooling or heating, if necessary, to 300°C.

For simplicity, the gas will be assumed ideal, with C_P = 38 J/mol K.

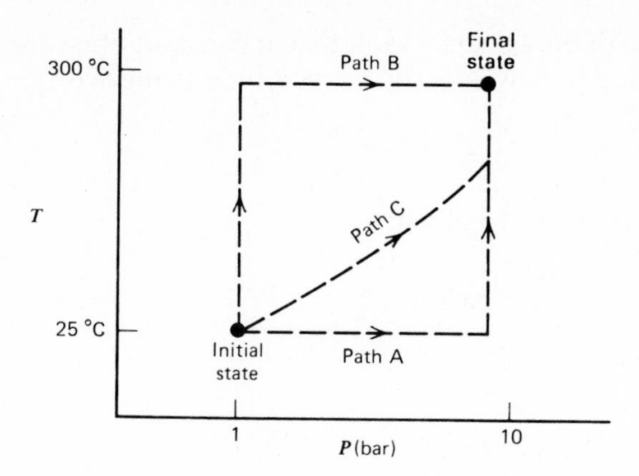

Solution

The 1 mol sample of gas will be taken as the thermodynamic system. The difference form of the mass balance for this closed, deforming volume of gas is

$$N = \text{constant} = 1 \text{ mol}$$

and the difference form of the energy balance is

$$\Delta U = Q - \int P\, dV = Q + W$$

Path A

i. ***For the isothermal compression***

$$W_i = -\int_{\underline{V}_1}^{\underline{V}_2} P\, d\underline{V} = -\int_{\underline{V}_1}^{\underline{V}_2} RT \frac{d\underline{V}}{\underline{V}} = -RT \int_{\underline{V}_1}^{\underline{V}_2} \frac{d\underline{V}}{\underline{V}} = -RT \ln \frac{\underline{V}_2}{\underline{V}_1} = RT \ln \frac{P_2}{P_1}$$

$$= 8.314 \text{ J/mol K} \times 298.15 \text{ K} \times \ln \frac{10}{1} = 5707.7 \text{ J/mol}$$

Since

$$\Delta \underline{U} = \int_{T_1}^{T_2} C_v\, dT = C_v(T_2 - T_1) \qquad \text{and} \qquad T_2 = T_1 = 25°\text{C}$$

we have

$$\Delta \underline{U} = 0 \qquad \text{and} \qquad Q_i = -W_i = -5707.7 \text{ J/mol}$$

ii. ***Isobaric heating***

$$W_{ii} = -\int_{\underline{V}_2}^{\underline{V}_3} P_2\, d\underline{V} = -P_2 \int_{\underline{V}_2}^{\underline{V}_3} d\underline{V} = -P_2(\underline{V}_3 - \underline{V}_2) = -R(T_3 - T_2)$$

$$\Delta \underline{U} = \int_{T_2}^{T_3} C_v\, dT = C_v(T_3 - T_2)$$

and

$$Q_{ii} = \Delta \underline{U} - W_{ii} = C_v(T_3 - T_2) + R(T_3 - T_2) = (C_v + R)(T_3 - T_2)$$
$$= C_P(T_3 - T_2)$$

[This is, in fact, a special case of the general result that at constant pressure for a closed system $Q = \int C_P \, dT$. This is easily proved by starting with

$$\dot{Q} = \frac{d\underline{U}}{dt} + P\frac{d\underline{V}}{dt}$$

and using the fact that P is constant to obtain

$$\dot{Q} = \frac{d\underline{U}}{dt} + \frac{d}{dt}(P\underline{V}) = \frac{d}{dt}(\underline{U} + P\underline{V}) = \frac{d\underline{H}}{dt} = C_P\frac{dT}{dt}$$

Now setting $Q = \int \dot{Q} \, dt$ yields $Q = \int C_P \, dT$.]

Therefore

$$W_{ii} = -8.314 \text{ J/mol K} \times 275 \text{ K} = -2286.3 \text{ J/mol}$$

$$Q_{ii} = 38 \text{ J/mol K} \times 275 \text{ K} = 10450 \text{ J/mol}$$

$$Q = Q_i + Q_{ii} = -5707.7 + 10450 = 4742.3 \text{ J/mol}$$

$$W = W_i + W_{ii} = 5707.7 - 2286.3 = 3421.4 \text{ J/mol}$$

Path B

i. *Isobaric heating*

$$Q_i = C_p(T_2 - T_1) = 10450 \text{ J/mol}$$

$$W_i = -R(T_2 - T_1) = -2286.3 \text{ J/mol}$$

ii. *Isothermal compression*

$$W_{ii} = RT \ln\frac{P_2}{P_1} = 8.314 \times 573.15 \ln\left(\frac{10}{1}\right) = 10972.2 \text{ J/mol}$$

$$Q_{ii} = -W_{ii} = -10972.2 \text{ J/mol}$$

$$Q = 10450 - 10972.2 = -522.2 \text{ J/mol}$$

$$W = -2286.3 + 10972.2 = 8685.9 \text{ J/mol}$$

Path C

i. *Compression with PV^γ = constant*

$$W_i = -\int_{\underline{V}_1}^{\underline{V}_2} P \, d\underline{V} = -\int_{\underline{V}_1}^{\underline{V}_2} \frac{\text{constant}}{V^\gamma} \, dV = -\frac{\text{constant}}{1-\gamma}(\underline{V}_2^{1-\gamma} - \underline{V}_1^{1-\gamma})$$

$$= -\frac{1}{1-\gamma}(P_2\underline{V}_2 - P_1\underline{V}_1) = \frac{-R(T_2 - T_1)}{1-\gamma} = \frac{-R(T_2 - T_1)}{1 - (C_P/C_V)} = C_V(T_2 - T_1)$$

where T_2 can be computed from

$$P_1\underline{V}_1^\gamma = P_1\left(\frac{RT_1}{P_1}\right)^\gamma = P_2\underline{V}_2^\gamma = P_2\left(\frac{RT_2}{P_2}\right)^\gamma$$

or

$$\frac{T_2}{T_1} = \left(\frac{P_2}{P_1}\right)^{(\gamma-1)/\gamma}$$

Now

$$\gamma = \frac{C_P}{C_V} = \frac{38}{38 - 8.314} = 1.280$$

so that

$$T_2 = 298.15 \text{ K } (10)^{(0.280/1.280)} = 493.38 \text{ K}$$

and

$$W_i = C_v(T_2 - T_1) = (38 - 8.314) \text{ J/mol K} \times (493.38 - 298.15)\text{K}$$

$$= 5795.6 \text{ J/mol}$$

$$\Delta U_i = C_v(T_2 - T_1) = 5795.6 \text{ J/mol}$$

$$Q_i = \Delta U_i - W_i = 0$$

ii. ***Isobaric heating***

$$Q_{ii} = C_P(T_3 - T_2) = 38 \text{ J/mol K} \times (573.15 - 493.38)\text{K} = 3031.3 \text{ J/mol}$$

$$W_{ii} = -R(T_3 - T_2) = -8.314 \text{ J/mol K} \times (573.15 - 493.38)\text{K} = -663.2 \text{ J/mol}$$

and

$$Q = 0 + 3031.3 = 3031.3 \text{ J/mol}$$

$$W = 5795.6 - 663.2 = 5132.4 \text{ J/mol}$$

Summary

Path	Q(J/mol)	W(J/mol)	$Q + W = \Delta U$(J/mol)
A	4742.3	3421.4	8163.7
B	−522.2	8685.9	8163.7
C	3031.3	5132.4	8163.7

Comment

It should be noticed that along each of the three paths considered (and, in fact, any other path between the initial and final states), the sum of Q and W is 8163.7 J/mol, even though Q and W separately are different along the different paths. This illustrates that whereas the internal energy is a state property and path independent (i.e., its change in going from state 1 to state 2 depends only on these states and not on the path between them), the heat and work flows depend on the path and are therefore path functions.

ILLUSTRATION 2.5-7

An initial pressure of 2.043 bar is maintained on 1 mol of air contained in a piston and cylinder system by a set of weights $\mathscr{W}$, the weight of the piston, and the surrounding atmosphere. Work is obtained from the air by removing some of the weights and allowing the air to isothermally expand at 25°C, thus lifting the piston and the remaining weights. The process will be repeated until all the weights have been removed. The piston has a mass of 5 kg and an area of 0.01 m^2. For simplicity, the air can be considered to be an ideal gas. Assume that as a result of sliding friction between the piston and the cylinder wall, all oscillatory motions of the piston after the removal of a weight will eventually be damped.

Compute the work obtained from the isothermal expansion and the heat required from external sources for each of the following:

a. The weight is taken off in one step.

b. The weight is taken off in two steps, $\mathcal{W}/2$ removed each time.
c. The weight is taken off in four steps, $\mathcal{W}/4$ removed each time.
d. The weight is replaced by a pile of sand (of total weight $\mathcal{W}$), and the grains of sand are removed one at a time.

Processes b and d are illustrated in the following figure.[8]

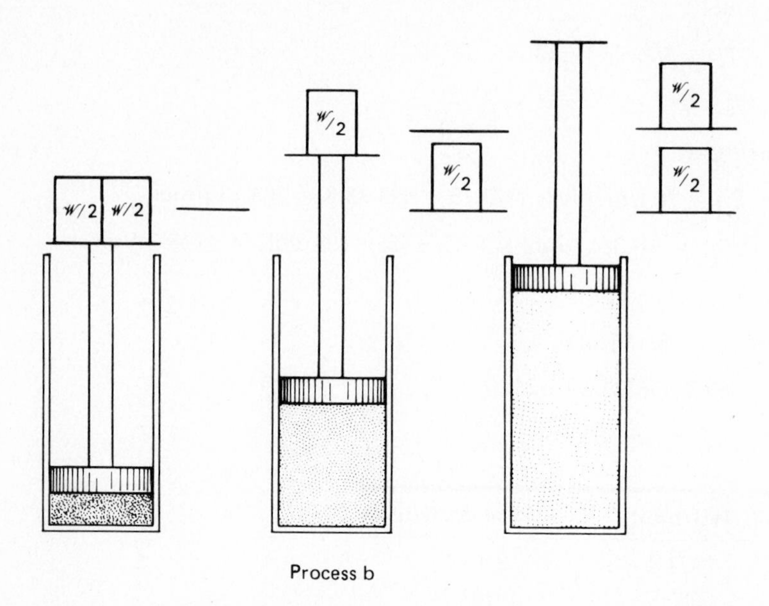

Process b

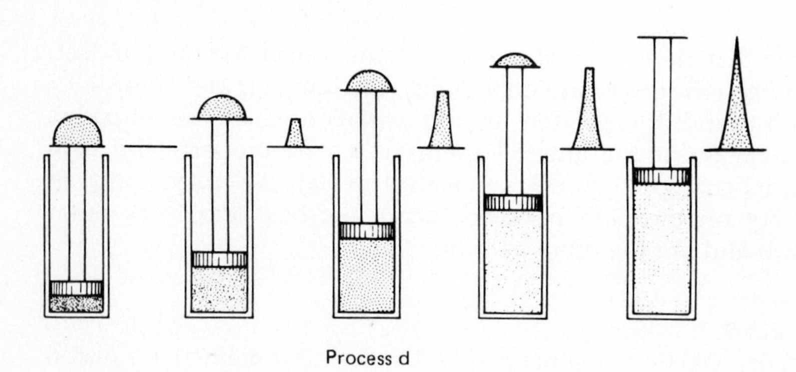

Process d

Solution

I. Analysis of the Problem

Choosing the air in the cylinder to be the system, recognizing that for an ideal gas at constant temperature U is constant so that $\Delta U = 0$, and neglecting the kinetic and potential energy terms for the gas (since the mass of 1 mol of air is only 29 g), we obtain the following energy balance equation:

$$0 = Q - \int P\,dV \qquad \text{(a)}$$

[8]From H. C. Van Ness, *Understanding Thermodynamics*. Copyright 1969 by McGraw–Hill, Inc. Used with permission of the McGraw–Hill Book Co.

The total work done by the gas in lifting and accelerating the piston and the weights, against the frictional forces, and in expanding the system volume against atmospheric pressure are contained in the $-\int P\,dV$ term. To see this we recognize that the laws of classical mechanics apply to the piston and weights, and equate, at each instant of time, all the forces on the piston and weights to their acceleration

$$\text{forces on piston and weights} = \begin{pmatrix}\text{mass of piston}\\ \text{and weights}\end{pmatrix} \times \text{acceleration}$$

and obtain

$$[P \times A - P_{\text{atm}} \times A - (\mathcal{W} + \omega)g + F_{\text{fr}}] = (\mathcal{W} + \omega)\frac{dv}{dt} \tag{b}$$

Here we have taken the vertical upward $(+z)$ direction as being positive, used P and P_{atm} to represent the pressure of the gas and atmosphere, respectively, A to represent the piston area, ω its mass, and $\mathcal{W}$ the mass of the weights at any time; v is the piston velocity and F_{fr} is the frictional force, which is proportional to the piston velocity. Recognizing that the piston velocity v is equal to the rate of change of the piston height h or the gas volume V, we have

$$v = \frac{dh}{dt} = \frac{1}{A}\frac{dV}{dt}$$

Also we can solve Eq. b for the gas pressure

$$P = P_{\text{atm}} + \frac{(\mathcal{W} + \omega)}{A}g - \frac{F_{\text{fr}}}{A} + \frac{(\mathcal{W} + \omega)}{A}\frac{dv}{dt} \tag{c}$$

At mechanical and thermodynamic equilibrium (i.e., when $dv/dt = 0$ and $v = 0$), we have

$$P = P_{\text{atm}} + \frac{(\mathcal{W} + \omega)}{A}g \tag{d}$$

With these results, the total work done by the gas can be computed. In particular

$$\int P\,dV = \int \left[P_{\text{atm}} + \left(\frac{\mathcal{W} + \omega}{A}\right)g - \frac{F_{\text{fr}}}{A} + \frac{(\mathcal{W} + \omega)}{A}\frac{dv}{dt}\right] dV$$

$$= \left[P_{\text{atm}} + \frac{(\mathcal{W} + \omega)}{A}g\right]\Delta V - \frac{1}{A}\int F_{\text{fr}}\,dV + \frac{(\mathcal{W} + \omega)}{A}\int \frac{dv}{dt}\,dV$$

This equation can be simplified by rewriting the last integral as follows

$$\frac{1}{A}\int \frac{dv}{dt}\,dV = \frac{1}{A}\int \frac{dv}{dt}\frac{dV}{dt}\,dt = \int \frac{dv}{dt}\,v\,dt = \frac{1}{2}\int \frac{dv^2}{dt}\,dt = \Delta\left(\frac{1}{2}v^2\right)$$

where the symbol Δ indicates the change between the initial and final states. Next, we recall from mechanics that the force due to sliding friction, here F_{fr}, is in the direction opposite to the relative velocity of the moving surfaces and can be written as

$$F_{\text{fr}} = -k_{\text{fr}}v$$

where k_{fr} is the coefficient of sliding friction. Thus, the remaining integral can be written as

$$\frac{1}{A}\int F_{\text{fr}}\,dV = -\frac{1}{A}\int k_{\text{fr}}v\frac{dV}{dt}\,dt = -k_{\text{fr}}\int v^2\,dt$$

The energy balance, Eq. a, then becomes

$$Q = \int P\,dV = P_{\text{atm}}\Delta V + (\mathcal{W} + \omega)g\,\Delta h + k_{\text{fr}}\int v^2\,dt + (\mathcal{W} + \omega)\,\Delta(\tfrac{1}{2}v^2) \tag{e}$$

This equation relates the heat flow into the gas (to maintain its temperature constant) to the work the gas does against the atmosphere, in lifting the piston and weights (and hence increasing their potential energy), against friction, and in accelerating the piston and weights (thus increasing their kinetic energy).

The work done against frictional forces is dissipated into thermal energy, resulting in a higher temperature at the piston and cylinder wall. This thermal energy is then absorbed by the gas (and appears as part of Q). Consequently, the net flow of heat from a temperature bath to the gas is

$$Q^{\text{NET}} = Q - k_{\text{fr}}\int v^2\,dt \tag{f}$$

Since heat will be transferred to the gas, and the integral is always positive, this equation establishes that less heat will be needed to keep the gas at constant temperature if the expansion occurs with friction than in a frictionless process.

Also, since the expansion occurs isothermally, the total heat flow to the gas is, from Eq. a,

$$Q = \int_{V_i}^{V_f} P\,dV = \int \frac{NRT}{V}\,dV = NRT \ln \frac{V_f}{V_i} \tag{g}$$

Combining Eqs. e and g, and recognizing that our interest here will be in computing the heat and work flows between states for which the piston has come to rest, yields

$$Q^{\text{NET}} = Q - k_{\text{fr}}\int v^2\,dt = P_{\text{atm}}\Delta V + (\mathcal{W} + \omega)g\,\Delta h = P\Delta V = -W^{\text{NET}} \tag{h}$$

where P is the equilibrium final pressure given by Eq. d. Also from Eqs. f, g, and h we have

$$Q^{\text{NET}} = -W^{\text{NET}} = NRT \ln\left(\frac{V_2}{V_1}\right) - k_{\text{fr}}\int v^2\,dt \tag{i}$$

Here W^{NET} represents the NET work obtained by the expansion of the gas (i.e., the work obtained in raising the piston and weights and in doing work against the atmosphere).

The foregoing equations can now be used in the solution of the problem. In particular, as a weight is removed, the new equilibrium gas pressure is computed from Eq. d, the resulting volume change from the ideal gas law, W^{NET} and Q^{NET} from Eq. h, and the work against friction from Eq. i. There is, however, one point that should be mentioned before proceeding with this calculation. If there were no mechanism for the dissipation of kinetic energy to thermal energy (here sliding friction between the piston and cylinder wall, but which could also include viscous dissipation on expansion and compression of the gas due to its bulk viscosity), when a weight was removed the piston would be put into a perpetual oscillatory motion. The presence of a dissipative mechanism will damp the oscillatory motion. (As will be seen, the value of the coefficient of sliding friction, k_{fr}, does not affect the amount of kinetic energy ultimately dissipated as heat. Its value does, however, affect the dynamics of the system and thus determines how quickly the oscillatory motion is damped.)

II. The Numerical Solution

First, the mass $\mathcal{W}$ of the weights will be computed using Eq. d and the fact that the initial pressure is 2.043 bar. Thus,

$$P = 2.043 \text{ bar} = 1.013 \text{ bar} + \frac{(5 + \mathcal{W})\text{kg}}{0.01 \text{ m}^2} \times 9.807 \frac{\text{m}}{\text{s}^2} \times \frac{1 \text{ Pa}}{\text{kg/m s}^2} \times \frac{1 \text{ bar}}{10^5 \text{ Pa}}$$

or

$$(5 + \mathcal{W})\text{kg} = 105.0 \text{ kg}$$

so that $\mathcal{W} = 100$ kg

The ideal gas equation of state for 1 mol of air at 25°C is

$$PV = NRT = 1 \text{ mol} \times 8.314 \times 10^{-5} \frac{\text{bar m}^3}{\text{mol K}} \times (25 + 273.15)\text{K}$$

$$= 2.479 \times 10^{-2} \text{ bar m}^3 = 2479 \text{ J} \qquad \text{(j)}$$

and the initial volume of the gas is

$$V = \frac{2.479 \times 10^{-2} \text{ bar m}^3}{2.043 \text{ bar}} = 1.213 \times 10^{-2} \text{ m}^3$$

Process a

The 100-kg weight is removed. The equilibrium pressure of the gas (after the piston has stopped oscillating) is

$$P_1 = 1.013 \text{ bar} + \frac{5 \text{ kg} \times 9.807 \text{ m/s}^2}{0.01 \text{ m}^2} \times \frac{10^{-5} \text{ bar}}{\text{kg/m s}^2} = 1.062 \text{ bar}$$

and the gas volume is

$$V_1 = \frac{2.479 \times 10^{-2} \text{ bar m}^3}{1.062 \text{ bar}} = 2.334 \times 10^{-2} \text{ m}^3$$

thus

$$\Delta V = (2.334 - 1.213) \times 10^{-2} \text{ m}^3 = 1.121 \times 10^{-2} \text{ m}^3$$

$$-W^{\text{NET}} = 1.062 \text{ bar} \times 1.121 \times 10^{-2} \text{ m}^3 \times 10^5 \frac{\text{J}}{\text{bar m}^3}$$

$$= 1190.5 \text{ J} = Q^{\text{NET}}$$

and

$$Q = NRT \ln \frac{V_1}{V_0} = 2479 \text{ J} \ln \frac{2.334 \times 10^{-2}}{1.213 \times 10^{-2}} = 1622.5 \text{ J}$$

(since $NRT = 2479$ J from Eq. j). Consequently, the work done against frictional forces (and converted to thermal energy), which we denote by W_{fr}, is

$$-W_{\text{fr}} = Q - Q^{\text{NET}} = (1622.5 - 1190.5) \text{ J} = 432 \text{ J}$$

The total useful work obtained, the net heat supplied, and the work against frictional forces are given in Table 1. Also, the net work, $P\,\Delta V$, is shown as the shaded area in the accompanying figure, together with the line representing the isothermal equation of state, Eq. j.

Process b

The situation here is similar to that of process a, except that the weight is removed in two 50-kg increments. The pressure, volume, work, and heat flows for both steps of the process are given in Table 2, and $-W_i^{\text{NET}} = P_i(\Delta V)_i$, the net work for each step, is given in the figure.

Table 1

	$-W^{NET} = Q^{NET}$ (J)	Q (J)	$-W_{fr}$ (J)
Process a	1190.5	1622.5	432.0
Process b	1378.7	1622.5	243.8
Process c	1493.0	1622.5	129.5
Process d	1622.5	1622.5	0

Process c

Here the weight is removed in four 25-kg increments. The pressure volume, work, and heat flows for each step are given in Table 2 and summarized in Table 1. Also, the net work for each step is given in the figure.

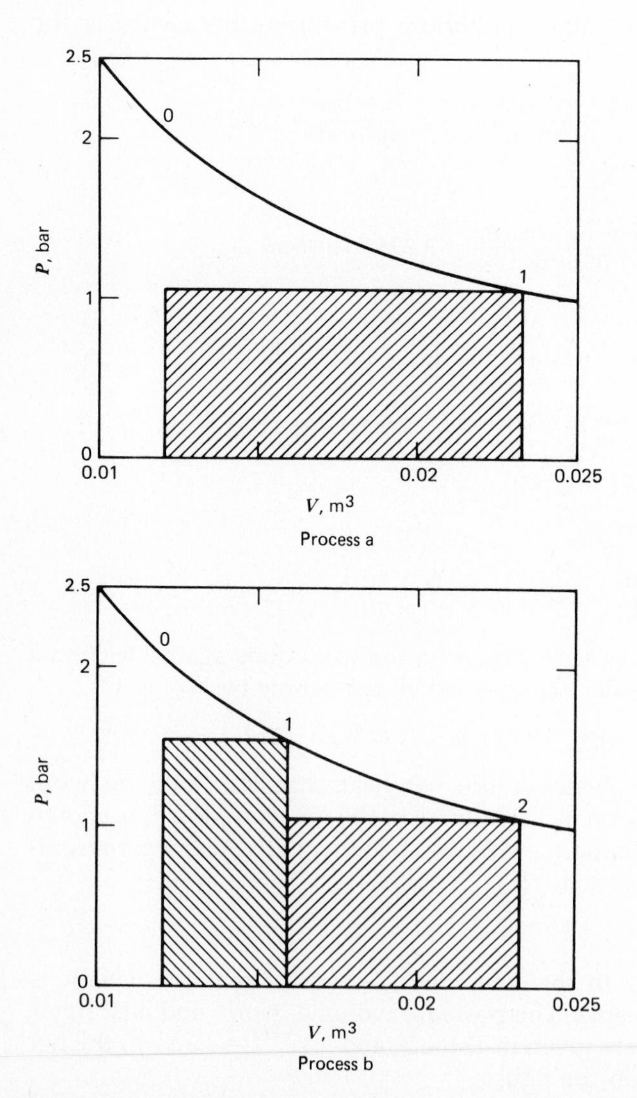

Table 2

Process b

Stage	P (bar)	V (m³)	$-W_i^{NET} = P_i(\Delta V)_i$ (J)	$Q = NRT \ln \frac{V_i}{V_{i-1}}$ (J)	$-W_{fr}$ (J)
0	2.043	1.213×10^{-2}			
1	1.552	1.597×10^{-2}	596.0	681.8	85.8
2	1.062	2.334×10^{-2}	782.7	940.7	158.0
Total			1378.7	1622.5	243.8

Process c

Stage	P (bar)	V (m³)	$-W_i^{NET} = P_i(\Delta V)_i$ (J)	$Q = NRT \ln \frac{V_i}{V_{i-1}}$ (J)	$-W_{fr}$ (J)
0	2.043	1.213×10^{-2}			
1	1.798	1.379×10^{-2}	298.5	318.0	19.5
2	1.552	1.597×10^{-2}	338.3	363.8	25.5
3	1.307	1.897×10^{-2}	392.1	426.8	34.7
4	1.062	2.334×10^{-2}	464.1	513.9	49.8
Total			1493.0	1622.5	129.5

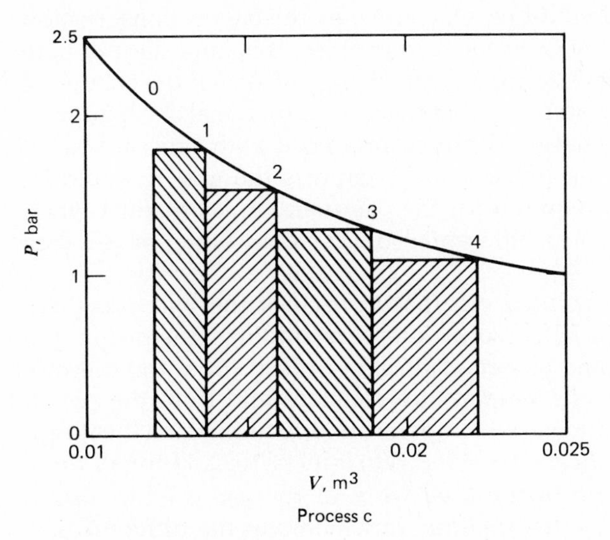

Process c

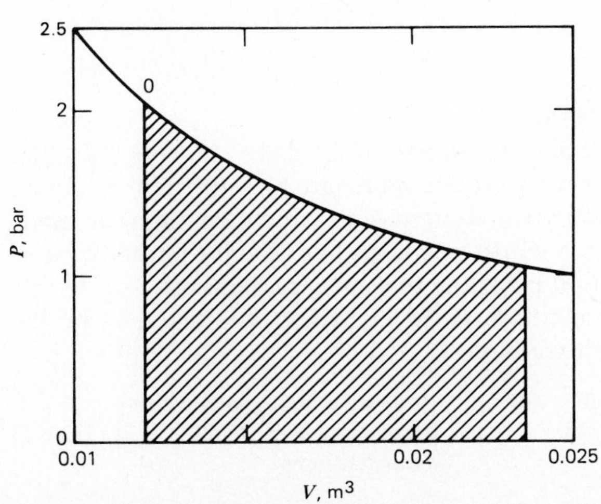

Process d

Process d

The computation here is somewhat more difficult since there are almost an infinite number of stages to the calculation. However, recognizing that in the limit of the mass of a grain of sand going to zero, there is only a differential change in the pressure of the gas and negligible velocity or acceleration of the piston, we have

$$-W^{\text{NET}} = \sum_i P_i(\Delta V_i) \to \int P\,dV = +NRT \ln \frac{V_f}{V_i} = Q^{\text{NET}}$$

and $W_{\text{fr}} = 0$, since the piston velocity is essentially zero at all times. Thus,

$$-W^{\text{NET}} = Q^{\text{NET}} = Q = 1 \text{ mol} \times 8.314 \frac{\text{J}}{\text{mol K}} \times 298.15 \text{ K} \times \ln \frac{2.334 \times 10^{-2}}{1.213 \times 10^{-2}}$$

$$= 1622.5 \text{ J}$$

This result is given in Table 1 and the figure.

Comments

Several points are worth noting in this illustration. First, although the initial and final states of the gas are the same in all three processes, the useful work obtained and the net heat required differ. Of course, by the energy conservation principle, it is true that $-W^{\text{NET}} = Q^{\text{NET}}$ for each process. It is important to note that the most useful work is obtained for a given change of state if the change of state is carried out in differential steps, so that there is no frictional dissipation of mechanical energy to thermal energy (compare process d with processes a, b, and c). If we were to reverse the process and compress the gas, it would be found that the minimum work required for the compression is obtained when weights are added to the piston in differential (rather than finite) steps. (See Problem 2.22.)

Finally, it should also be pointed out that in each of the four processes considered here the gas did 1622.5 J of work on its surroundings (the piston, the weights, and the atmosphere) and absorbed 1622.5 J of heat (from the thermostatic bath maintaining the system temperature constant and from the piston and cylinder as a result of their increased temperature due to frictional heating). We can see this from Table 1, since $-(W^{\text{NET}} + W_{\text{fr}}) = Q = 1622.5$ J for all three processes. However, the fraction of the total work of the gas used as useful work, and that used in work against friction, varies among the different processes. ■

2.6 CONSERVATION OF MOMENTUM

Based on the discussion of Sec. 2.5 and Illustration 2.5-4 we can conclude that the equations of mass and energy conservation alone are not sufficient to obtain the solution to all the problems of thermodynamics in which we might be interested. What is needed is a balance equation for an additional thermodynamic state variable. The one conservation principle that has not been used up to this point is the conservation of momentum. If, in Eq. 2.1-4, θ is taken to be the momentum of a black-box system we have

$$\frac{d}{dt}(M\mathbf{v}) = \begin{pmatrix}\text{rate at which momentum}\\ \text{enters system by all}\\ \text{mechanisms}\end{pmatrix} - \begin{pmatrix}\text{rate at which momentum}\\ \text{leaves system by all}\\ \text{mechanisms}\end{pmatrix} \quad (2.6\text{-}1)$$

where **v** is the center of mass velocity vector of the system, and M its total mass. We could now continue the derivation by evaluating all the momentum flows; however, it is clear by looking at the left side of this equation that we will get an equation for the rate of change of the center of mass velocity of the system, *not* an equation of change for a thermodynamic state variable. Consequently, the conservation of momentum equation will not lead to the additional balance equation we need, and this derivation will not be completed. The development of an additional, useful balance equation is not a straightforward task, and will be delayed until Chapter 3.

2.7 THE MICROSCOPIC EQUATIONS OF CHANGE FOR THERMODYNAMICS AND FLUID MECHANICS (OPTIONAL)

The balance equations we have developed so far are those commonly used in engineering thermodynamics. An important characteristic of these equations is that they are balances for large, black-box systems and therefore yield information about the total mass or total energy of the system. Frequently it is useful to have information not about the total mass and total energy, but about the mass density and energy density at each point in a fluid. To get balance equations for these density functions, a thermodynamic system of microscopic (or differential) size must be used. There is some advantage to developing the microscopic balance equations here. First, it exposes the essential similarity between thermodynamics and fluid mechanics, and, second, these equations will be of use in the following chapters. However, since the formulation of these equations requires a greater degree of mathematical complexity than used heretofore, and since these equations are of only peripheral interest in our development of thermodynamics, you may wish to proceed directly to Chapter 3.

To obtain the microscopic equations of fluid mechanics and thermodynamics, we apply the general balance equation, Eq. 2.1-4, to the small stationary volume element of Fig. 2.7-1. This volume element, of dimensions Δx, Δy, and Δz, is part of a much larger fluid system, so the boundaries are not physical boundaries and mass may flow across each of its faces. Finally, since this volume

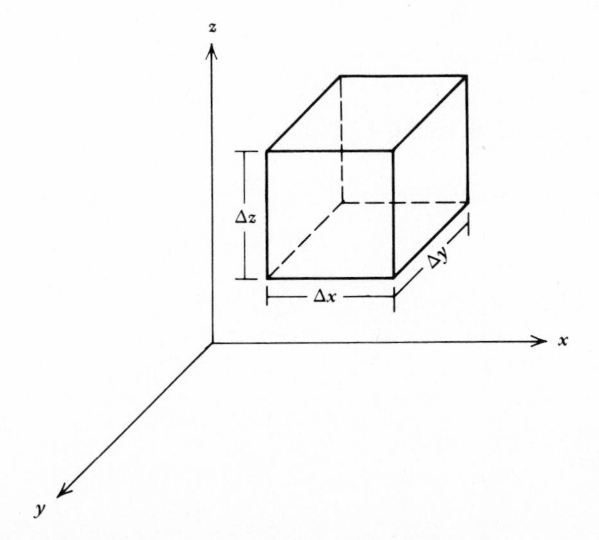

Figure 2.7-1
Volume element for a microscopic balance equation.

element is of infinitesimal size (our interest is in the case where Δx, Δy, and Δz simultaneously go to zero), the properties within it can be assumed to be uniform. Thus the mass contained within the volume element is $\rho\ \Delta x\ \Delta y\ \Delta z$, where ρ is the mass density within the infinitesimal volume element. With this introduction we can now make the following identifications in Eq. 2.1-4:

$$\begin{pmatrix}\text{rate of change of mass in}\\ \text{the volume element}\end{pmatrix} = \Delta x\ \Delta y\ \Delta z\ \frac{\partial \rho}{\partial t} \tag{2.7-1a}$$

$$\begin{aligned}\begin{pmatrix}\text{net rate at which}\\ \text{mass enters the}\\ \text{volume element}\end{pmatrix} &= \rho v_x\ \Delta y\ \Delta z|_x - \rho v_x\ \Delta y\ \Delta z|_{x+\Delta x}\\ &+ \rho v_y\ \Delta x\ \Delta z|_y - \rho v_y\ \Delta x\ \Delta z|_{y+\Delta y}\\ &+ \rho v_z\ \Delta x\ \Delta y|_z - \rho v_z\ \Delta x\ \Delta y|_{z+\Delta z}\end{aligned} \tag{2.7-1b}$$

where v_i is the fluid velocity in the ith coordinate direction. The interpretation of the terms in Eq. 2.7-1b is as follows. The mass flow into the volume element across the face of the volume element perpendicular to the x-axis at x is

$$\rho v_x\ \Delta y\ \Delta z|_x$$

where $\Delta y\ \Delta z$ is the area of the face and ρv_x is the flow rate per unit area. Similarly, the term

$$-\rho v_x\ \Delta y\ \Delta z|_{x+\Delta x}$$

is the mass flow out of the volume element at the face perpendicular to the x-axis at $x + \Delta x$ and therefore has a negative sign. The remaining terms in Eq. 2.7-1b represent the mass flows into and out of the other faces of the volume element.

Using Eqs. 2.7-1 in Eq. 2.1-4 yields

$$\begin{aligned}\Delta x\ \Delta y\ \Delta z\ \frac{\partial \rho}{\partial t} &= \rho v_x\ \Delta y\ \Delta z|_x - \rho v_x\ \Delta y\ \Delta z|_{x+\Delta x} + \rho v_y\ \Delta x\ \Delta z|_y - \rho v_y\ \Delta x\ \Delta z|_{y+\Delta y}\\ &+ \rho v_z\ \Delta x\ \Delta y|_z - \rho v_z\ \Delta x\ \Delta y|_{z+\Delta z}\end{aligned}$$

Now, dividing by $\Delta x\ \Delta y\ \Delta z$, and taking the limit as Δx, Δy, and Δz go to zero gives

$$\begin{aligned}\frac{\partial \rho}{\partial t} &= \lim_{\Delta x\to 0} \frac{\rho v_x|_x - \rho \mathrm{v}_x|_{x+\Delta x}}{\Delta x} + \lim_{\Delta y\to 0} \frac{\rho v_y|_y - \rho \mathrm{v}_y|_{y+\Delta y}}{\Delta y}\\ &+ \lim_{\Delta z\to 0} \frac{\rho v_z|_z - \rho v_z|_{z+\Delta z}}{\Delta z}\end{aligned} \tag{2.7-2}$$

Finally, using the definition of the partial derivative, that is

$$\frac{\partial F(x,\ t)}{\partial x} = \lim_{\Delta x\to 0} \frac{F(x + \Delta x,\ t) - F(x,\ t)}{\Delta x} \tag{2.7-3}$$

we obtain

$$\frac{\partial \rho}{\partial t} = -\frac{\partial(\rho v_x)}{\partial x} - \frac{\partial(\rho v_y)}{\partial y} - \frac{\partial(\rho v_z)}{\partial z}$$

or, in vector notation,

$$\frac{\partial \rho}{\partial t} = -\nabla \cdot (\rho \mathbf{v}) \tag{2.7-4}$$

Equation 2.7-4 is the mass conservation equation for a stationary differential volume element; in fluid mechanics it is called the continuity equation. This equation may be rearranged to yield

$$\frac{\partial \rho}{\partial t} = -\rho \nabla \cdot \mathbf{v} - \mathbf{v} \cdot \nabla \rho$$

or

$$\frac{\partial \rho}{\partial t} + \mathbf{v} \cdot \nabla \rho \equiv \frac{D\rho}{Dt} = -\rho \nabla \cdot \mathbf{v} \tag{2.7-5}$$

Here we have introduced the notation $D/Dt = \partial/\partial t + \mathbf{v} \cdot \nabla$, where D/Dt is the convected derivative; it is the derivative with respect to time in a volume element moving with the fluid velocity $\mathbf{v}$. To see this, consider the moving volume element shown in Fig. 2.7-2. The position vector of the center of this volume element is $\mathbf{r}$ at time t, and $\mathbf{r} + \Delta\mathbf{r}$ at time $t + \Delta t$, where $\Delta\mathbf{r} = \mathbf{v}\Delta t$. The time derivative of any function F taken in this moving differential volume, DF/Dt, is defined, in analogy with Eq. 2.7-3, as

$$\begin{aligned}
\frac{DF}{Dt} &= \lim_{\Delta t \to 0} \left\{ \frac{F(\mathbf{r} + \Delta\mathbf{r}, t + \Delta t) - F(\mathbf{r}, t)}{\Delta t} \right\} \\
&= \lim_{\Delta t \to 0} \left\{ \frac{F(\mathbf{r} + \Delta\mathbf{r}, t + \Delta t) - F(\mathbf{r} + \Delta\mathbf{r}, t)}{\Delta t} + \frac{F(\mathbf{r} + \Delta\mathbf{r}, t) - F(\mathbf{r}, t)}{\Delta t} \right\} \\
&= \lim_{\Delta t \to 0} \left\{ \frac{F(\mathbf{r} + \Delta\mathbf{r}, t + \Delta t) - F(\mathbf{r} + \Delta\mathbf{r}, t)}{\Delta t} + \frac{\Delta\mathbf{r}}{\Delta t} \cdot \frac{F(\mathbf{r} + \Delta\mathbf{r}, t) - F(\mathbf{r}, t)}{\Delta\mathbf{r}} \right\}
\end{aligned}$$

or

$$\frac{DF}{Dt} = \frac{\partial F}{\partial t} + \mathbf{v} \cdot \nabla F \tag{2.7-6}$$

where we have used the fact that $\lim_{\Delta t \to 0} \Delta\mathbf{r}/\Delta t = \mathbf{v}$, and $\Delta\mathbf{r} \to 0$ as $\Delta t \to 0$.

Equation 2.7-4 can be integrated over a finite volume element to obtain Eq. 2.2-1b. Since this requires the use of certain mathematical theorems that may be

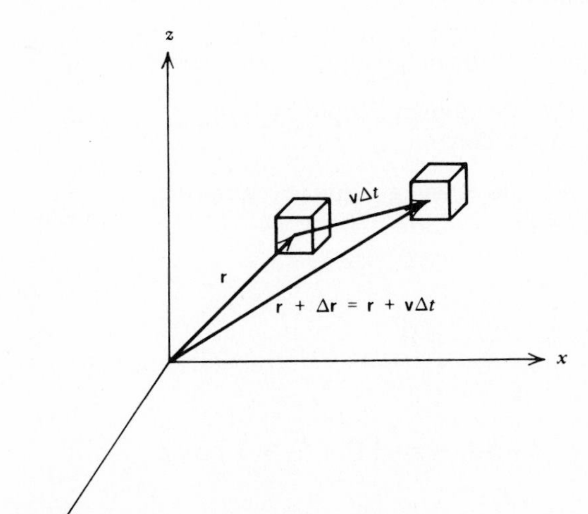

Figure 2.7-2
Positions of a moving volume element at t and at $t + \Delta t$.

unfamiliar to you, this integration will not be pursued here except to point out that for a stationary system ($v = 0$) we can integrate the left side of Eq. 2.7-4 over the volume V to obtain

$$\int \frac{\partial \rho}{\partial t} \, dV = \frac{d}{dt} \int \rho \, dV = \frac{dM}{dt}$$

which is equal to the left side of Eq. 2.2-1b. A similar correspondence exists between the right sides of Eqs. 2.2-1b and 2.7-4, but this is more difficult to prove. This analysis does establish a very important general relationship; the black-box thermodynamic equations developed in Secs. 2.2 and 3 can be gotten by an integration over volume of the more detailed microscopic equations.

At this point, you should reflect on the three levels of description used in this book. The microscopic equations, of which the mass conservation equation developed here is only the first, require detailed information about the internal structure and internal flows in any macroscopic system in order to be in use. Next, by integration over a finite volume element these microscopic equations yield equations for the time rate of change in a black-box system. For these equations to be of use, only information about flow rates into and out of the system as a function of time are needed. Finally, by integration over time of the time rate of change equations, we obtain equations for the change over an interval of time of various properties of the black-box system. For these last equations to be useful one merely needs information about the total flows into and out of the system over the time interval and not a history of how these flow rates varied with time.

The microscopic form of the energy conservation equation can be obtained in a manner similar to the mass conservation equation, so the development will only be sketched here. The starting point of the derivation is, of course, Eq. 2.1-4 with θ being the total energy of the fluid, that is, the sum of the internal, potential, and kinetic energies. Energy can enter the differential volume element by energy flows accompanying mass flows (convection), by heat flows across the faces of the volume element, and, since there may be both forces and fluid movement at the faces of the volume element, by the work being done by surface forces. Since the volume element is only of differential size, we will not include a shaft work term. Thus we have

$$\begin{pmatrix}\text{rate of change of energy}\\ \text{in the volume element}\end{pmatrix} = (\text{net rate of energy input by convection}) + \begin{pmatrix}\text{net rate of energy input by heat}\\ \text{conduction}\end{pmatrix} + \begin{pmatrix}\text{net rate of energy input by surface}\\ \text{forces}\end{pmatrix} \tag{2.7-7}$$

where

$$\begin{pmatrix}\text{rate of change of}\\ \text{energy in the}\\ \text{volume element}\end{pmatrix} = \Delta x \, \Delta y \, \Delta z \, \frac{\partial}{\partial t}\left\{\rho\left(\hat{U} + \frac{v^2}{2} + \psi\right)\right\}$$

$$\begin{aligned}\begin{pmatrix}\text{net rate of energy}\\ \text{input by}\\ \text{convection}\end{pmatrix} &= \rho v_x\left(\hat{U} + \frac{v^2}{2} + \psi\right) \Delta y \, \Delta z\Big|_x - \rho v_x\left(\hat{U} + \frac{v^2}{2} + \psi\right) \Delta y \, \Delta z\Big|_{x+\Delta x} \\ &+ \rho v_y\left(\hat{U} + \frac{v^2}{2} + \psi\right) \Delta x \, \Delta z\Big|_y - \rho v_y\left(\hat{U} + \frac{v^2}{2} + \psi\right) \Delta x \, \Delta z\Big|_{y+\Delta y} \\ &+ \rho v_z\left(\hat{U} + \frac{v^2}{2} + \psi\right) \Delta x \, \Delta y\Big|_z - \rho v_z\left(\hat{U} + \frac{v^2}{2} + \psi\right) \Delta x \, \Delta y\Big|_{z+\Delta z}\end{aligned}$$

$$\begin{pmatrix}\text{net rate of energy}\\ \text{input by heat}\\ \text{conduction}\end{pmatrix} = q_x\, \Delta y\, \Delta z|_x - q_x\, \Delta y\, \Delta z|_{x+\Delta x} + q_y\, \Delta x\, \Delta z|_y - q_y\, \Delta x\, \Delta z|_{y+\Delta y} + q_z\, \Delta x\, \Delta y|_z - q_z\, \Delta x\, \Delta y|_{z+\Delta z}$$

$$\begin{pmatrix}\text{net rate of energy}\\ \text{input by surface}\\ \text{forces}\end{pmatrix} = (\boldsymbol{P}\cdot\mathbf{v})_x\, \Delta y\, \Delta z|_x - (\boldsymbol{P}\cdot\mathbf{v})_x\, \Delta y\, \Delta z|_{x+\Delta x} + (\boldsymbol{P}\cdot\mathbf{v})_y\, \Delta x\, \Delta z|_y - (\boldsymbol{P}\cdot\mathbf{v})_y\, \Delta x\, \Delta z|_{y+\Delta y} + (\boldsymbol{P}\cdot\mathbf{v})_z\, \Delta x\, \Delta y|_z - (\boldsymbol{P}\cdot\mathbf{v})_z\, \Delta x\, \Delta y|_{z+\Delta z}$$

Here ρ is the mass density, $\mathbf{q}$ the heat flux per unit area, q_i is its component in the ith coordinate direction and $(\boldsymbol{P}\cdot\mathbf{v})_i$ is the component of the vector $\boldsymbol{P}\cdot\mathbf{v}$ in the ith coordinate direction. In writing the contribution to the energy balance from surface forces we have to recognize that the forces are of two different types; the first is merely a hydrostatic pressure term that acts perpendicular to the surface of the differential volume element and is uniform and isotropic; this term is written as $P\boldsymbol{I}$, where $\boldsymbol{I}$ is the unit tensor. The second surface force is more subtle; it is a drag force that occurs whenever a fluid element flows by either an adjacent surface or a fluid element of different velocity. The magnitude of the surface force or drag force depends on both the viscosity of the fluid μ and the velocity gradient $\nabla\mathbf{v}$ in the fluid at the surface of the fluid element. For Newtonian fluids the components of the stress tensor τ for a general three-dimensional flow are

$$\tau_{xx} = -2\mu\,\frac{\partial v_x}{\partial x} + \frac{2}{3}\,\mu\nabla\cdot\mathbf{v}, \qquad \tau_{xy} = \tau_{yx} = \mu\left(\frac{\partial v_x}{\partial y} + \frac{\partial v_y}{\partial x}\right)$$

with similar expressions for the other components. In the derivation of the energy balance we have combined the hydrostatic pressure (P) and the shear stress tensor (τ) contributions to the surface forces by defining a pressure tensor

$$\boldsymbol{P} = P\boldsymbol{I} + \boldsymbol{\tau} \tag{2.7-8}$$

Using these expressions in Eq. 2.7-7, dividing by $\Delta x\, \Delta y\, \Delta z$, and taking the limit as each of these goes to zero gives

$$\begin{aligned}\frac{\partial}{\partial t}\{\rho(\hat{U} + v^2/2 + \psi)\} &= -\nabla\cdot\{\rho\mathbf{v}(\hat{U} + v^2/2 + \psi)\} - \nabla\cdot\mathbf{q} - \nabla\cdot(\boldsymbol{P}\cdot\mathbf{v}) \\ &= -\nabla\cdot\{\rho\mathbf{v}(\hat{U} + v^2/2 + \psi)\} - \nabla\cdot\mathbf{q} - \nabla\cdot P\mathbf{v} - \nabla\cdot(\boldsymbol{\tau}\cdot\mathbf{v})\end{aligned} \tag{2.7-9}$$

The integration of this equation over a finite volume to obtain Eq. 2.3-4 can be accomplished, again establishing that the thermodynamic equations are volume integrals of the microscopic equations. This tedious process will not be done here.

The momentum balance equation for the microscopic volume element of Fig. 2.7-1 is nothing other than Newton's second law of motion for a fluid element. In deriving this equation one must remember that a change of momentum (an acceleration) results from an applied force. Consequently, in the momentum balance equation we have to include the forces acting at the surface of the volume element, since these forces result in a flow of momentum across the system boundaries. Considering first only the x-component of momentum, we make the following identifications:

$$\begin{pmatrix}\text{rate of change of} \\ x\text{-component of momentum} \\ \text{in volume element}\end{pmatrix} = \Delta x\, \Delta y\, \Delta z \frac{\partial}{\partial t}(\rho v_x)$$

$$\begin{pmatrix}\text{net rate of change of} \\ x\text{-component of} \\ \text{momentum into the volume} \\ \text{element by convection}\end{pmatrix} = (\rho v_x v_x|_x - \rho v_x v_x|_{x+\Delta x})\Delta y\, \Delta z$$
$$+ (\rho v_y v_x|_y - \rho v_y v_x|_{y+\Delta y})\Delta x\, \Delta z$$
$$+ (\rho v_z v_x|_z - \rho v_z v_x|_{z+\Delta z})\Delta x\, \Delta y$$

$$\begin{pmatrix}\text{net rate of change of} \\ x\text{-component of momentum} \\ \text{due to forces acting on} \\ \text{the volume element}\end{pmatrix} = \rho\, \Delta x\, \Delta y\, \Delta z g_x + (P|_x - P|_{x+\Delta x})\, \Delta y\, \Delta z$$
$$+ (\tau_{xx}|_x - \tau_{xx}|_{x+\Delta x})\, \Delta y\, \Delta z$$
$$+ (\tau_{yx}|_y - \tau_{yx}|_{y+\Delta y})\, \Delta x\, \Delta z$$
$$+ (\tau_{zx}|_z - \tau_{zx}|_{z+\Delta z})\, \Delta x\, \Delta y$$

Here the first force term is that due to gravity in the x direction, the second term results from pressure forces acting on the volume element, and the remaining terms result from viscous forces.

Now using these terms in Eq. 2.1-4, dividing by $\Delta x\, \Delta y\, \Delta z$, and taking the limit as Δx, Δy, and Δz go to zero yields

$$\frac{\partial}{\partial t}(\rho v_x) = -\nabla \cdot (\rho \mathbf{v} v_x) - \left(\frac{\partial \tau_{xx}}{\partial x} + \frac{\partial \tau_{yx}}{\partial y} + \frac{\partial \tau_{zx}}{\partial z}\right) - \frac{\partial P}{\partial x} + \rho g_x \tag{2.7-10}$$

In a similar fashion, one could obtain balance equations for the y and z components of momentum. Vectorially adding the equations for each of the coordinate directions yields

$$\frac{\partial(\rho \mathbf{v})}{\partial t} = -\nabla \cdot (\rho \mathbf{v}\mathbf{v}) - \nabla \cdot \boldsymbol{\tau} - \nabla P + \rho \mathbf{g} \tag{2.7-11}$$

This is the conservation equation for momentum, or the equation of motion. The only real use we have for this equation is in the development of the mechanical energy equation or Bernouilli equation of fluid mechanics, which is obtained by scalar multiplication of the momentum equation with the fluid velocity $\mathbf{v}$

$$\mathbf{v} \cdot \frac{\partial(\rho \mathbf{v})}{\partial t} = -\mathbf{v} \cdot \nabla \cdot (\rho \mathbf{v}\mathbf{v}) - \mathbf{v} \cdot (\nabla \cdot \boldsymbol{\tau}) - \mathbf{v} \cdot \nabla P + \rho \mathbf{v} \cdot \mathbf{g} \tag{2.7-12}$$

and then rearrangement to give

$$\frac{\partial}{\partial t}\left(\frac{\rho}{2} v^2\right) = -\nabla \cdot \left(\frac{\rho}{2} v^2 \mathbf{v}\right) - \mathbf{v} \cdot (\nabla \cdot \boldsymbol{\tau}) - \mathbf{v} \cdot \nabla P + \rho \mathbf{v} \cdot \mathbf{g} \tag{2.7-13}$$

This equation looks like a conservation equation for kinetic energy. However, kinetic energy is not a conserved quantity, and this equation was not obtained from an energy conservation principle, but merely from the velocity moment of the momentum conservation equation.

It is also possible to derive an equation that looks like a conservation equation for potential energy. For simplicity, we will assume that the only potential energy contribution is that due to the gravitational field, so that

$$\psi = -\mathbf{h} \cdot \mathbf{g} \tag{2.7-14}$$

where $\mathbf{h}$ is the vector to the center of mass of the volume element with respect to a convenient frame of reference, and $\mathbf{g}$ is the directed force of gravity. The minus sign appears because, by convention, the gravitational force is in the negative direction. Now, multiplying the continuity equation, Eq. 2.7-4, by ψ and rearranging gives

$$\frac{\partial(\rho\psi)}{\partial t} - \rho\frac{\partial\psi}{\partial t} = -\nabla\cdot(\rho\mathbf{v}\psi) + \rho\mathbf{v}\cdot\nabla\psi \tag{2.7-15}$$

For a stationary volume element we have $\partial\mathbf{h}/\partial t = 0$, and since the force of gravity is time invariant, $\partial\mathbf{g}/\partial t = 0$. Also, $\nabla\mathbf{h} = \boldsymbol{I}$, where $\boldsymbol{I}$ is the unit tensor, and, to an excellent approximation $\nabla\cdot\mathbf{g} = 0$. Using these observations in Eq. 2.7-15 yields

$$\frac{\partial}{\partial t}(\rho\psi) = -\nabla\cdot(\rho\mathbf{v}\psi) - \rho\mathbf{v}\cdot\mathbf{g} \tag{2.7-16}$$

Adding this to Eq. 2.7-13 gives a result that looks like a conservation equation for mechanical energy

$$\frac{\partial}{\partial t}[\rho(v^2/2 + \psi)] = -\nabla\cdot[\rho\mathbf{v}(v^2/2 + \psi)] - \mathbf{v}\cdot(\nabla\cdot\boldsymbol{\tau}) - \mathbf{v}\cdot\nabla P \tag{2.7-17}$$

Finally, we notice that this equation can be subtracted from the total energy balance, Eq. 2.7-9, to obtain a "thermal energy equation"

$$\frac{\partial}{\partial t}(\rho\hat{U}) = -\nabla\cdot(\rho\mathbf{v}\hat{U}) - \nabla\cdot\mathbf{q} - P\nabla\cdot\mathbf{v} - \boldsymbol{\tau}:\nabla\mathbf{v} \tag{2.7-18}$$

or

$$\rho\frac{D\hat{U}}{Dt} = -\nabla\cdot\mathbf{q} - P\nabla\cdot\mathbf{v} - \boldsymbol{\tau}:\nabla\mathbf{v} \tag{2.7-19}$$

where we have again used $D/Dt = \partial/\partial t + \mathbf{v}\cdot\boldsymbol{\nabla}$.

We will return to some of these equations in Section 3.6.

PROBLEMS

2.1 Steam at 500 bar and 600°C is to undergo a Joule–Thomson expansion to atmospheric pressure. What will the temperature of the steam be after the expansion? What would be the downstream temperature if the steam were replaced by an ideal gas?

2.2 Water in an open metal drum is to be heated from room temperature (25°C) to 80°C by adding steam slowly enough that all the steam condenses. The drum initially contains 100 kg of water, and steam is supplied at 3.0 bar and 300°C. How many kilograms of steam should be added so that the final temperature of the water in the tank is exactly 80°C? Neglect all heat losses from the water in this calculation.

2.3 Consider the following statement:
"The adiabatic work necessary to cause a given change of state in a closed system is independent of the path by which that change occurs."

a Is this statement true or false? Why? Does this statement contradict Illustration 2.5-6, which establishes the path dependence of work?

b Show that if the statement is false, it would be possible to construct a machine that would generate energy.

2.4 A nonconducting tank of negligible heat capacity and 1 m^3 volume is connected to a pipeline containing steam at 5 bar and 370°C, filled with steam to a pressure of 5 bar, and disconnected from the pipeline.

a If the tank is initially evacuated, how much steam is in the tank at the end of the filling process, and what is its temperature?

b If the tank initially contains steam at 1 bar and 150°C, how much steam is in the tank at the end of the filling process and what is its temperature?

2.5 The voltage drop across an electrical resistor is 10 volts and the current through it is 1 ampere. The total heat capacity of the resistor is 20 J/K, and heat is dissipated from the resistor to the surrounding air according to the relation

$$\dot{Q} = -h(T - T_{am})$$

where T_{am} is the ambient air temperature, 25°C, T is the temperature of the resistor, and h, the heat transfer coefficient, is equal to 0.2 J/K s. Compute the steady-state temperature of the resistor; that is, the temperature of the resistor when the energy loss from the resistor is just equal to the electrical energy input.

2.6 The frictionless piston and cylinder shown here is subjected to 1.013 bar external pressure. The piston mass is 200 kg, it has an area of 0.15 m^2, and the initial volume of the entrapped ideal gas is 0.12 m^3. The piston and cylinder do not conduct heat, but heat may be added to the gas by a heating coil. The gas has a constant volume heat capacity of 30.1 J/mol K, an initial temperature of 298 K, and 10.5 kJ of energy are to be supplied to the gas through the heating coil.

a If stops placed at the initial equilibrium position of the piston prevent it from rising, what will be the final temperature and pressure of the gas?

b If the piston is allowed to move freely, what is the final temperature and volume of the gas?

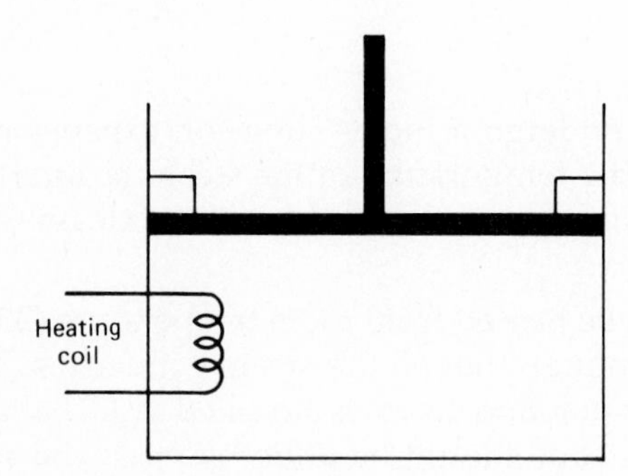

2.7 As an energy conservation measure in a chemical plant, a 40-m^3 tank will be used for temporary storage of exhaust process steam. This steam is then used in a later stage of the processing. The storage tank is well insulated, and initially contains 0.02 m^3 of liquid water at 50°C; the remainder of the tank contains water vapor in equilibrium with this liquid. Process steam at 1.013 bar and 90% quality enters the storage tank until the pressure in the tank is 1.013 bar. How many kilograms of wet steam enter the tank during the filling process, and how much liquid water is present at the end of the process? Assume that there is no heat transfer between the steam or water and the tank walls.

2.8 An isolated chamber with rigid walls is divided into two equal compartments, one containing gas and the other evacuated. The partition between the compartments ruptures. After the passage of a sufficiently long period of time the temperature and pressure are found to be uniform throughout the chamber.

a If the filled compartment initially contains an ideal gas at 1 MPa and 500 K, what is the final temperature and pressure in the chamber?

b If the filled chamber initially contains steam at 1 MPa at 500 K, what is the final temperature and pressure in the chamber?

c and **d** Repeat parts a and b if the second chamber initially contains the same fluid as the first chamber but at half the pressure and 100 K higher temperature.

2.9 **a** An adiabatic turbine expands steam from 500°C, 3.5 MPa to 200°C and 0.3 MPa. If the turbine generates 750 kW, what is the flow rate of steam through the turbine?

b If a breakdown of the thermal insulation around the turbine allows a heat loss of 60 kJ per kg of steam, and the exiting steam is at 150°C and 0.3 MPa, what will be the power developed by the turbine if the inlet steam conditions and flow rate are unchanged?

2.10 Intermolecular forces play an important role in determining the thermodynamic properties of fluids. To see this, consider the vaporization (boiling) of a liquid such as water in the frictionless piston and cylinder device shown here.

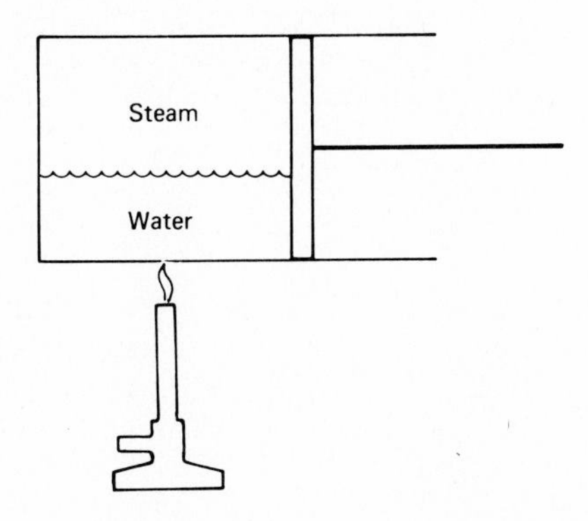

a Compute the work obtained from the piston when 1 kg of water is vaporized to steam at 100°C (the vapor and liquid volumes of steam at the boiling point can be found in the steam tables).

b Show that the heat required for the vaporization of the steam is considerably greater than the work done. (Note that the enthalpy change for the vaporization is given as 2257 kJ/kg in the steam tables in Appendix III.)

2.11 It is sometimes necessary to produce saturated steam from superheated steam (steam at a temperature higher than the vapor–liquid coexistence temperature at the given pressure). This change can be accomplished in a desuperheater, a device in which just the right amount of water is sprayed into superheated steam to produce dry saturated steam. If superheated steam at 3.0 MPa and 500°C enters the desuperheater at a rate of 500 kg/hr, at what rate should liquid water at 2.5 MPa and 25°C be added to the desuperheater to produce saturated steam at 2.25 MPa?

2.12 Nitrogen gas leaves a compressor at 2.0 MPa and 120°C and is collected in three different cylinders, each of 0.3 m^3. In each case the cylinder is to be filled to a pressure of 2.0 MPa. Cylinder 1 is initially evacuated, cylinder 2 contains nitrogen gas at 0.1 MPa and 20°C, and cylinder 3 contains nitrogen at 1 MPa and 20°C. Find the final temperature of nitrogen in each of the cylinders, assuming nitrogen to be an ideal gas with $C_P = 29.3$ J/mol K. In each case assume the gas does not exchange heat with the cylinder walls.

2.13 A clever (?) chemical engineer has devised the thermally operated elevator shown here. The elevator compartment is made to rise by electrically heating the air contained in the piston and cylinder drive mechanism, and the elevator is lowered by opening a valve at the side of the cylinder, allowing the air in the cylinder to slowly escape. Once the elevator compartment is back to the lower level, a small compressor forces out the air remaining in the cylinder and replaces it with air at 20°C and a pressure just sufficient to support the elevator compartment. The cycle can then be repeated. There is no heat transfer between the piston, cylinder, and the gas; the weight of the piston, elevator, and elevator contents is 4000 kg; the piston has a surface area of 2.5 m^2; and the volume contained in the cylinder when the elevator is at its lowest level is 25 m^3. There is no friction between the piston and the cylinder, and the air in the cylinder is assumed to be an ideal gas with $C_P = 30$ J/mol K.

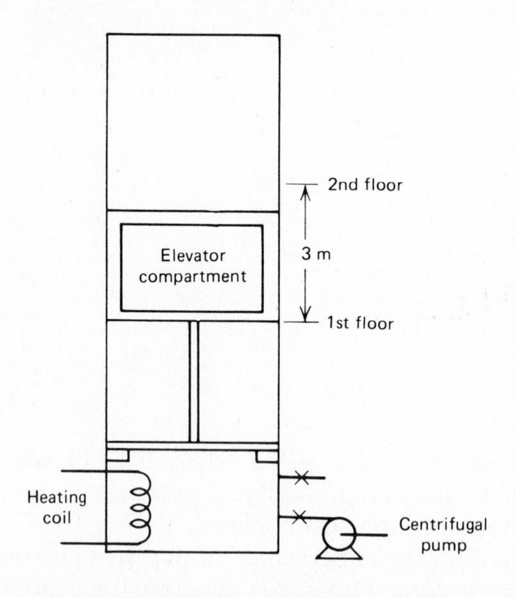

a What is the pressure in the cylinder throughout the process?

b How much heat must be added to the air during the process of raising the elevator 3 m, and what is the final temperature of the gas?

c What fraction of the heat added is used in doing work, and what fraction is used in raising the temperature of the gas?

d How many moles of air must be allowed to escape in order for the elevator to return to the lowest level?

2.14 The elevator in the previous problem is to be designed to ascend and descend at the rate of 0.2 m/s, and to rise a total of 3 meters.

a At what rate should heat be added to the cylinder during the ascent?

b How many kmol per second of air should be removed from the cylinder during the descent?

2.15 Nitrogen gas is being withdrawn at the rate of 4.5 g/s from a 0.15 m^3 cylinder, initially containing the gas at a pressure of 10 bar and 320 K. The cylinder does not conduct heat, nor does its temperature change during the emptying process. What will be the temperature and pressure of the gas in the cylinder after 5 minutes? What will be the rate of change of the gas temperature at this time? Nitrogen can be considered to be an ideal gas with $C_P = 30$ J/mol K.

2.16 In Illustration 2.5-6 we considered the compression of an ideal gas in which PV^γ = constant, where $\gamma = C_P/C_V$. Show that such a pressure-volume relationship is obtained in the adiabatic compression of an ideal gas of constant heat capacity.

2.17 Air in a 0.3 m^3 cylinder is initially at a pressure of 10 bar and a temperature of 330 K. The cylinder is to be emptied by opening a valve and letting the pressure drop to that of the atmosphere. What will be the temperature and mass of gas in the cylinder if this is accomplished

a In a manner that maintains the temperature of the gas at 330 K?

b In a well-insulated cylinder?

For simplicity assume, in part b, that the process occurs sufficiently rapidly that there is no heat transfer between the cylinder walls and the gas. The gas is ideal and $C_P = 29$ J/mol K.

2.18 A 0.01 m^3 cylinder containing nitrogen gas initially at a pressure of 200 bar and 250 K is connected to another cylinder 0.005 m^3 in volume, which is initially evacuated. A valve between the two cylinders is opened until the pressure in the cylinders equalize. Find the final temperature and pressure in each cylinder if there is no heat flow into or out of the cylinder. You may assume that there is no heat transfer between the gas and the cylinder walls and that the gas is ideal with a constant pressure heat capacity of 30 J/mol K.

2.19 Repeat the calculation of Problem 2.18, but now assume that sufficient heat transfer occurs between the gas in the two cylinders so that both final temperatures and both final pressures are the same.

2.20 Repeat the calculation in Problem 2.18, but now assume that the second cylinder, instead of being evacuated, is filled with nitrogen gas at 20 bar and 160 K.

2.21 A 1.5 kilowatt heater is to be used to heat a room with dimensions 3.5 m × 5.0 m × 3.0 m. There are no heat losses from the room, but the room is not airtight, so that the pressure in the room is always atmospheric. Consider the air in the room to be an ideal gas with $C_P = 29$ J/mol K, and its initial temperature is 10°C.

a Assuming that the rate of heat transfer from the air to the walls is small, what will be the rate of increase of temperature in the room when the heater is turned on?

b What would be the rate of increase in the room temperature if the room were hermetically sealed?

2.22 The piston and cylinder device of Illustration 2.5-7 is to be operated in reverse to isothermally compress the 1 mol of air. Assume that weights in the illustration have been left at the heights at which they were removed from the piston (i.e., in process *b* the first 50-kg weight is at the initial piston height and the second at $\Delta h = \Delta V/A = 0.384$ m above the initial piston height). Compute the minimum work that must be done by the surroundings, and the net heat that must be withdrawn, to return the gas, piston, and weights to their initial state. Also compute the total heat and the total work for each of the four expansion and compression cycles and comment on the results.

2.23 The piston and cylinder device shown here contains an ideal gas at 20 bar and 25°C. The piston has a mass of 300 kg, and a cross-sectional area of 0.05 m^2. The initial volume of the gas in the cylinder is 0.03 m^3, the piston is initially held in place by a pin, and the external pressure on the piston and cylinder is 1 bar. The pin suddenly breaks, and the piston moves 0.6 m further up the cylinder where it is stopped by another pin. Assuming that the gas is ideal with a constant pressure heat capacity of 30 kJ/kmol K, and that there is no heat transfer between the gas and the cylinder walls or piston, estimate the piston velocity, and the temperature and pressure of the gas just before the piston hits the second pin. Do this calculation assuming:

a No friction between the piston and the cylinder.
b Friction between the piston and the cylinder.

List and defend all assumptions you make in solving this problem.

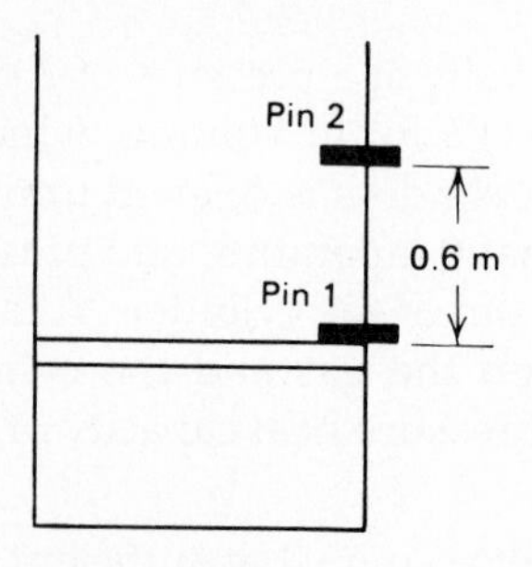

2.24 The mixing tank shown here initially contains 50 kg water at 25°C. Suddenly the two inlet valves and the single outlet valve are opened, so that

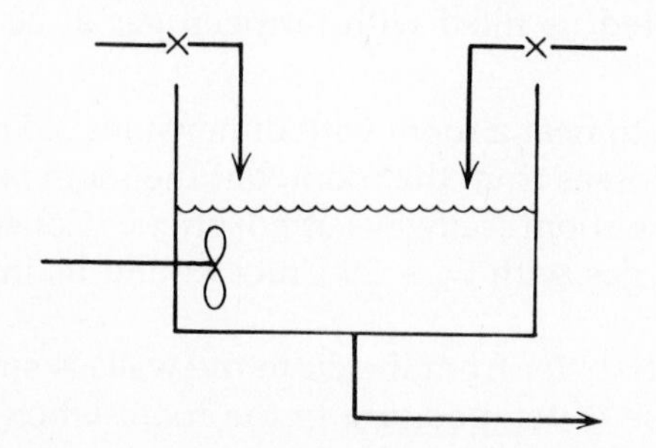

two water streams, each with a flow rate of 5 kg/min flow into the tank, and a single exit stream with a flow rate of 10 kg/min leaves the tank. The temperature of one inlet stream is 80°C, and that of the other is 50°C. The tank is well mixed, so that the temperature of the outlet stream is always the same as the temperature of the water in the tank.

a Compute the steady-state temperature that will finally be obtained in the tank.

b Develop an expression for the temperature of the fluid in the tank at any time.

2.25 A 0.6-m diameter gas pipeline is being used for the long-distance transport of natural gas. Just past a pumping station, the gas is found to be at a temperature of 25°C and a pressure of 3.0 MPa. The mass flow rate is 125 kg/s, and the gas flow is adiabatic. Forty miles down the pipeline is another pumping station. At this point the pressure is found to be 2.0 MPa. At the pumping station the gas is first adiabatically compressed to a pressure of 3.0 MPa, and then isobarically (i.e., at constant pressure) cooled to 25°C.

a Find the temperature and velocity of the gas just before entering the pumping station.

b Find the rate at which the gas compressor in the pumping station does work on the gas, the gas temperature leaving the compressor, and the heat load on the gas cooler. You may assume that the compressor exhaust is also a 0.6-m pipe. (Explain why you cannot solve this problem. You will have another chance in Chapter 3.)

Natural gas may be assumed to be pure methane (molecular weight = 16, C_P = 36.8 J/mol K), and an ideal gas at the conditions being considered here. Note that the mass flow rate M is $\rho v A$, where ρ is the mass density of the gas, v is the average gas velocity, and A is the area of the pipe.

3

Entropy: An Additional Balance Equation

The illustrations and problems in Chapter 2 make it clear that the equations of mass and energy conservation alone are not sufficient to solve all the thermodynamic energy flow problems in which we might be interested. To be more specific, these two equations are not always sufficient to determine the final values of two state variables, or the heat and work flows for a system undergoing a change of state. What is needed is a balance equation for an additional state variable. As we have seen, the principle of momentum conservation does not provide this additional equation. Although a large number of additional state variables could be defined, and could serve as the basis for a new balance equation, these variables would have the common feature that they are not conserved quantities. Thus the internal generation term for each of these variables would, in general, be nonzero and would have to be evaluated if the balance equation were to be of use. Clearly, the most useful variable to introduce as the basis for a new balance equation is one that has an internal generation rate that can be best specified and has some physical significance.

Another defect in our present development of thermodynamics has to do with the unidirectional character of natural processes considered in Section 1.3. There it was pointed out that all spontaneous or natural processes proceed only in the direction that tends to dissipate the gradients in the system and thus lead to a state of equilibrium, and never in the reverse direction. This unidirectional character of natural processes has not yet been included in our thermodynamic description.

To complete our thermodynamic description of pure component systems, it is therefore necessary that we (1) develop an additional balance equation for a state variable; and (2) incorporate into our description the unidirectional character of natural processes. In Section 3.1 we show that both these objectives can be accomplished by introducing a single new thermodynamic function. The remaining sections of this chapter are concerned with illustrating the properties and utility of this new variable and its balance equation.

3.1 ENTROPY—A NEW CONCEPT

We will take as the starting point for the identification of an additional thermodynamic variable the experimental observation that all spontaneous processes that occur in an isolated constant-volume system result in the evolution of the system to a state of equilibrium (this is a special case of Observation 5, Sect. 1.7). The problem is to quantify this qualitative observation. We can obtain some insight into how to do this by considering the general balance equation (Eq. 2.1-4) for any extensive variable θ of a closed, isolated, constant-volume system

$$\frac{d\theta}{dt} = \begin{pmatrix}\text{rate of change of}\\ \theta \text{ in the system}\end{pmatrix} = \begin{pmatrix}\text{rate at which } \theta \text{ is generated}\\ \text{within the system}\end{pmatrix} \tag{3.1-1}$$

Alternatively, we can write Eq. 3.1-1 as

$$\frac{d\theta}{dt} = \dot{\theta}_{\text{gen}} \tag{3.1-2}$$

where $\dot{\theta}_{\text{gen}}$ is the rate of internal generation of the yet unspecified state variable θ. Now, if the system under consideration were in a true time-invariant equilibrium state, $d\theta/dt = 0$ (since, by definition of a time-invariant state, no state variable can change with time). Thus

$$\dot{\theta}_{\text{gen}} = 0 \qquad \text{at equilibrium} \tag{3.1-3}$$

Equations 3.1-2 and 3 suggest a way of quantifying the qualitative observation of the undirectional evolution of an isolated system to an equilibrium state. In particular, suppose we could identify a thermodynamic variable θ whose rate of internal generation $\dot{\theta}_{\text{gen}}$ was positive,[1] except at equilibrium where $\dot{\theta}_{\text{gen}} = 0$. For this variable

$$\frac{d\theta}{dt} > 0 \qquad \text{away from equilibrium}$$

and

$$\left.\begin{array}{ll} & \dfrac{d\theta}{dt} = 0 \\ \text{or} & \theta = \text{constant}\end{array}\right\} \quad \text{at equilibrium} \tag{3.1-4}$$

Furthermore, since the function θ is increasing in the approach to equilibrium, θ must be a maximum at equilibrium, subject to the constraints of constant mass, energy, and volume for the isolated, constant-volume system.[2] Thus, if we could find a thermodynamic function with the properties given in Eq. 3.1-4, the experimental observation of the undirectional evolution to the equilibrium state would be built into the thermodynamic description through the properties of the function θ. That is, the unidirectional evolution to the equilibrium state would correspond to the monotonically increasing nature of the function θ, and the occurrence of equilibrium in an isolated, constant-volume system to the attainment of a maximum value of the function θ.

[1]If we chose the other possibility, $\dot{\theta}_{\text{gen}}$ being less than zero, the discussion here would still be valid except that θ would monotonically decrease to a minimum, rather than increase to a maximum, in the evolution to the equilibrium state. The positive choice is made here in agreement with standard thermodynamic convention.

[2]Since an isolated constant-volume system has fixed mass, internal energy, and volume, you might ask how θ can vary if M, U, and V, or alternatively, the two state variables $\underline{U}$ and $\underline{V}$ are fixed. The answer is that the discussion of Sections 1.3 and 1.6 established that two state variables completely fix the state of a *uniform* one-component, one-phase system. Consequently, θ (or any other state variable) is free to vary for fixed $\underline{U}$ and $\underline{V}$ in (1) a nonuniform system, (2) a multicomponent system, or (3) a multiphase system. The first case is of importance in the approach to equilibrium in the presence of internal relaxation processes, the second and third cases for chemical reaction equilibrium and phase equilibrium, which are discussed later in this book.

The problem then is to identify a thermodynamic state function θ with a rate of internal generation, $\dot{\theta}_{gen}$, that is always greater than or equal to zero. Before searching for the variable θ, it should be noted that the property we are looking for is that $\dot{\theta}_{gen} \geq 0$; this is clearly not as strong a statement as $\dot{\theta}_{gen} = 0$ always, as occurs if θ were a conserved variable such as total mass or total energy, *but it is as strong a general statement as we can expect for a nonconserved variable.*

We could now institute an extensive search of possible thermodynamic functions in the hope of finding a function that is a state variable and also has the property that its rate of internal generation is a positive quantity. Instead, we will introduce this new thermodynamic property by its definition and then show that the property so defined has the desired characteristics.

DEFINITION

The **entropy** (denoted by the symbol S) is a state function. In a system in which there are flows of both heat ($\dot{Q}$) and work [$\dot{W}_s$ and $P(dV/dt)$] across the system boundaries, the heat flow, but not the work flow, causes a change in the entropy of the system; this rate of entropy change is $\dot{Q}/T$, where T is the *absolute thermodynamic temperature* of the system at the point of the heat flow. If, in addition, there are mass flows across the system boundaries, the total entropy of the system will also change due to this convected flow.

Using this definition and Eq. 2.1-4, we have the following as the balance equation for entropy[3]

$$\frac{dS}{dt} = \sum_{k=1}^{K} \dot{M}_k \hat{S}_k + \frac{\dot{Q}}{T} + \dot{S}_{gen} \tag{3.1-5}$$

where

$\sum_{k=1}^{K} \dot{M}_k \hat{S}_k$ = the net rate of entropy flow due to the flows of mass into and out of the system ($\hat{S}$ = entropy per unit mass)

$\frac{\dot{Q}}{T}$ = the rate of entropy flow due to the flow of heat across the system boundary

$\dot{S}_{gen}$ = the rate of internal generation of entropy within the system.

Before we consider how the entropy balance, Eq. 3.1-5, will be used in problem solving, we should establish (1) that the entropy function is a state variable, and (2) that it has a positive rate of internal generation. Clearly, Eq. 3.1-5 cannot provide general information about the internal generation of entropy since it is an equation for the black-box description of a system, whereas $\dot{S}_{gen}$ depends on the detailed internal relaxation processes that occur within the system. In certain special cases, however, one can cleverly use Eq. 3.1-5 to get some insight into the form of $\dot{S}_{gen}$. To see this, consider the thermodynamic system of Fig. 3.1-1, which is a composite of two subsystems, A and B. These subsystems are well-insulated, except at their interface, so that the only heat transfer that occurs is a

[3]For simplicity, we have assumed that there is only a single heat flow into the system. If there are multiple heat flows, the term $\dot{Q}/T$ is to be replaced by a $\Sigma\, \dot{Q}_j/T_j$, where $\dot{Q}_j$ is the heat flow and T_j the temperature at the jth heat flow port into the system.

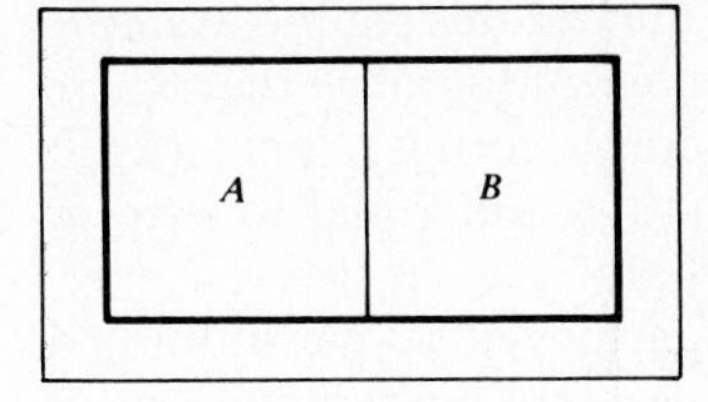

Figure 3.1-1
Systems A and B are free to interchange energy, but the composite system ($A + B$) is isolated from the environment.

flow of heat from the high temperature subsystem A to the low temperature subsystem B. We will assume that the resistance to heat transfer at this interface is large relative to the internal resistances of the subsystems (this would occur if, for example, the subsystems were well-mixed liquids or highly conducting solids), so that the temperature of each subsystem is uniform, but varying with time. In this situation the heat transfer process occurs in such a way that, at any instant of time, each subsystem is in a state of internal thermal equilibrium (i.e., if the two subsystems were suddenly separated, each would be of uniform, but different temperature and would no longer vary with time), though the composite system consisting of both subsystems (at different temperatures) is not in thermal equilibrium, as evidenced by the fact that though the composite system is isolated from the environment, its properties are changing with time (as heat is transferred from A to B and their temperatures are changing). The differential form of the entropy balances for subsystems A and B, which are passing through a succession of equilibrium states,[4] and therefore have no internal generation of entropy, are

$$\frac{dS_A}{dt} = \frac{\dot{Q}_A}{T_A} = -h\left(\frac{T_A - T_B}{T_A}\right) \tag{3.1-6a}$$

and

$$\frac{dS_B}{dt} = \frac{\dot{Q}_B}{T_B} = +h\left(\frac{T_A - T_B}{T_B}\right) \tag{3.1-6b}$$

In writing this equation we have recognized that the amount of heat that leaves subsystem A enters subsystem B, and that the heat flow from A to B is proportional to the temperature difference between the two systems, that is,

$$\dot{Q}_A = -\dot{Q}_B = -h(T_A - T_B)$$

where h is the heat transfer coefficient (Experimental Observation 10 of Sec. 1.7).

The entropy balance for the isolated ($\dot{Q} = 0$), nonequilibrium composite system is

$$\frac{dS}{dt} = \dot{S}_{\text{gen}} \tag{3.1-7}$$

Since the total entropy is an extensive property, $S = S_A + S_B$, and Eqs. 3.1-6 and 7 can be combined to yield

$$\dot{S}_{\text{gen}} = \frac{h(T_A - T_B)^2}{T_A T_B} = \frac{h(\Delta T)^2}{T_A T_B} \tag{3.1-8}$$

[4]A process in which a system goes through a succession of equilibrium states is termed a **quasistatic** process.

Since h, T_A, and T_B are positive, Eq. 3.1-8 establishes that for this simple example the entropy generation term is positive. It is also important to notice that $\dot{S}_{gen}$ is proportional to the second power of the system nonuniformity, here $(\Delta T)^2$. Thus, $\dot{S}_{gen}$ is positive away from equilibrium ($T_A \neq T_B$) and equal to zero at equilibrium.

This result for the entropy generation rate was achieved by partitioning a black-box system into two subsystems, thereby obtaining a limited amount of information about processes internal to the overall system. In Sec. 3.6 a more general derivation of the entropy generation term is given, based on the detailed microscopic description introduced in Sec. 2.7, and it is shown that the entropy is indeed a state function and that $\dot{S}_{gen}$ is positive, except in the equilibrium state where it is equal to zero. It is also established that $\dot{S}_{gen}$ is proportional to the second power of the gradients in each of temperature and velocity, thus the rate of generation of entropy is related to the square of the departure from the equilibrium state.

Table 3.1-1 gives several special cases of the entropy balance equation, on both a mass and molar basis, for situations similar to those considered for the mass and energy balance equations in Tables 2.2-1 and 2.3-1.

Frequently, one is interested in the change in entropy of a system in going from state 1 to state 2, rather than the rate of change of entropy with time. This entropy change can be gotten by integrating Eq. 3.1-5 over the time interval t_1 to t_2, where $(t_2 - t_1)$ is the (perhaps unknown) time required to go between the two states. The result is

$$S_2 - S_1 = \sum_k \int_{t_1}^{t_2} \dot{M}_k \hat{S}_k \, dt + \int_{t_1}^{t_2} \frac{\dot{Q}}{T} \, dt + S_{gen} \tag{3.1-9}$$

Table 3.1-1

DIFFERENTIAL FORM OF THE ENTROPY BALANCE

General equation: $$\frac{dS}{dt} = \sum_{k=1}^{K} \dot{M}_k \hat{S}_k + \frac{\dot{Q}}{T} + \dot{S}_{gen} \tag{a}$$

Special cases

(i) Closed system: set $\dot{M}_k = 0$

so $$\frac{dS}{dt} = \frac{\dot{Q}}{T} + \dot{S}_{gen} \tag{b}$$

(ii) Adiabatic process set $\dot{Q} = 0$ in Eq. a (c)

(iii) Reversible process: set $\dot{S}_{gen} = 0$ in Eq. a (d)

(iv) Open steady-state system:

$$\frac{dS}{dt} = 0$$

so $$0 = \sum_{k=1}^{K} \dot{M}_k \hat{S}_k + \frac{\dot{Q}}{T} + \dot{S}_{gen} \tag{e}$$

(v) Uniform system: $S = M\hat{S}$ in Eq. a (f)

Note: To obtain the entropy balance on a molar basis, replace $\dot{M}_k\hat{S}_k$ by $\dot{N}_k\underline{S}_k$, and $M\hat{S}$ by $N\underline{S}$, where $\underline{S}$ is the entropy per mole of fluid.

where

$$S_{gen} = \text{total entropy generated} = \int_{t_1}^{t_2} \dot{S}_{gen}\, dt$$

There are two important simplifications of Eq. 3.1-9. First, if the entropy per unit mass of each stream entering and leaving the system is constant in time (even though the flow rates may vary), we have

$$\sum_k \int_{t_1}^{t_2} \dot{M}_k \hat{S}_k\, dt = \sum_k \hat{S}_k \int_{t_1}^{t_2} \dot{M}_k\, dt = \sum_k (\Delta M)_k \hat{S}_k$$

where $(\Delta M)_k = \int_{t_1}^{t_2} \dot{M}_k\, dt$ is the total mass that has entered the system from the kth stream. Next, if the system is isothermal (i.e., at constant temperature), then

$$\int_{t_1}^{t_2} \frac{\dot{Q}}{T}\, dt = \frac{1}{T} \int_{t_1}^{t_2} \dot{Q}\, dt = \frac{Q}{T}$$

where $Q = \int_{t_1}^{t_2} \dot{Q}\, dt$ is the total heat flow into the system between t_1 and t_2. If either of these simplifications is not valid, the respective integrals must be evaluated if Eq. 3.1-9 is to be used. This may be a difficult or impossible task so that, as with the energy balance, the system must be chosen with care, as illustrated later in this chapter. Table 3.1-2 summarizes various forms of the integrated entropy balance.

It should be pointed out that we have introduced the entropy function in an axiomatic and mathematical fashion. Over the history of thermodynamics, entropy has been presented in many different ways, and it is interesting to read about these alternative approaches. One interesting source is *The Second Law* by P. W. Atkins (Copyright © 1984 by W. H. Freeman and Company, New York).

Table 3.1-2
DIFFERENCE FORM OF THE ENTROPY BALANCE

General equation: $$S_2 - S_1 = \sum_{k=1}^{K} \int_{t_1}^{t_2} \dot{M}_k \hat{S}_k\, dt + \int_{t_1}^{t_2} \frac{\dot{Q}}{T}\, dt + S_{gen} \qquad \text{(a)}$$

Special cases:

(i) Closed system: set $\dot{M}_k = 0$ in Eq. a

so $$S_2 - S_1 = \int_{t_1}^{t_2} \frac{\dot{Q}}{T}\, dt + S_{gen} \qquad \text{(b)}$$

(ii) Adiabatic process:

set $\int_{t_1}^{t_2} \frac{\dot{Q}}{T}\, dt = 0$ in Eq. a (c)

(iii) Reversible process: set $S_{gen} = 0$ in Eq. a (d)

(iv) Open system: Flow of fluids of constant thermodynamic properties

set $\sum_{k=1}^{K} \int_{t_1}^{t_2} \dot{M}_k \hat{S}_k\, dt = \sum_{k=1}^{K} (\Delta M)_k \hat{S}_k$ in Eq. a (e)

(v) Uniform system: $S = M\hat{S}$ in Eq. a (f)

Note: To obtain the entropy balance on a molar basis replace $M\hat{S}$ by $N\underline{S}$, $\int_{t_1}^{t_2} \dot{M}_k \hat{S}_k\, dt$ by $\int_{t_1}^{t_2} \dot{N}_k \underline{S}_k\, dt$ and $(\Delta M)_k \hat{S}_k$ with $(\Delta N)_k \underline{S}_k$.

3.2 THE ENTROPY BALANCE AND REVERSIBILITY

An important class of processes are those for which the rate of generation of entropy is always zero. Such processes are called **reversible** processes and are of special interest in thermodynamics. Since $\dot{S}_{gen}$ is proportional to the square of the system temperature and velocity gradients, such gradients must vanish in a process in which $\dot{S}_{gen}$ is zero. Notice, however, that although the rate of entropy generation is second order in the system gradients, the internal relaxation processes that occur in the approach to equilibrium are linearly proportional to these gradients (i.e., the heat flux $\mathbf{q}$ is proportional to the temperature gradient ∇T, the stress tensor is proportional to the velocity gradient, etc.). Therefore, if there is a very small temperature gradient in the system the heat flux $\mathbf{q}$, which depends on ∇T, will be small, and $\dot{S}_{gen}$, which depends on $(\nabla T)^2$, may be so small as to be negligible. Similarly, $\dot{S}_{gen}$ may be negligible for very small velocity gradients. Processes that occur with such small gradients in temperature and velocity that $\dot{S}_{gen}$ is essentially zero can also be considered to be reversible.

The designation reversible arises from the following observation. Consider the change in state of a general system open to the flow of mass, heat, and work, between two equal time intervals, 0 to t_1, and t_1 to t_2, where $t_2 = 2t_1$. The mass, energy, and entropy balances for this system are, from Eqs. 2.2-4, 2.3-6, and 3.1-9

$$M_2 - M_0 = \sum_k \int_0^{t_1} \dot{M}_k \, dt + \sum_k \int_{t_1}^{t_2} \dot{M}_k \, dt$$

$$U_2 - U_0 = \int_0^{t_1} \left[\sum_k \dot{M}_k \hat{H}_k - P \frac{dV}{dt} + \dot{W}_s + \dot{Q} \right] dt + \int_{t_1}^{t_2} \left[\sum_k \dot{M}_k \hat{H}_k - P \frac{dV}{dt} + \dot{W}_s + \dot{Q} \right] dt$$

and

$$S_2 - S_0 = \int_0^{t_1} \left[\sum \dot{M}_k \hat{S}_k + \frac{\dot{Q}}{T} \right] dt + \int_{t_1}^{t_2} \left[\sum \dot{M}_k \hat{S}_k + \frac{\dot{Q}}{T} \right] dt + \int_0^{t_1} \dot{S}_{gen} \, dt + \int_{t_1}^{t_2} \dot{S}_{gen} \, dt$$

Now, suppose that all the mass, heat, and work flows are just reversed between t_1 and t_2 from what they were between 0 and t_1, so that

$$\int_0^{t_1} \dot{M}_k \, dt = -\int_{t_1}^{t_2} \dot{M}_k \, dt \qquad \int_0^{t_1} \dot{Q} \, dt = -\int_{t_1}^{t_2} \dot{Q} \, dt$$

$$\int_0^{t_1} \dot{M}_k \hat{H}_k \, dt = -\int_{t_1}^{t_2} \dot{M}_k \hat{H}_k \, dt \qquad \int_0^{t_1} \frac{\dot{Q}}{T} \, dt = -\int_{t_1}^{t_2} \frac{\dot{Q}}{T} \, dt$$

$$\int_0^{t_1} \dot{M}_k \hat{S}_k \, dt = -\int_{t_1}^{t_2} \dot{M}_k \hat{S}_k \, dt \qquad \int_0^{t_1} \dot{W}_s \, dt = -\int_{t_1}^{t_2} \dot{W}_s \, dt$$

$$\int_0^{t_1} P \frac{dV}{dt} \, dt = -\int_{t_1}^{t_2} P \frac{dV}{dt} \, dt \tag{3.2-1}$$

In this case the equations reduce to

$$M_2 = M_0 \tag{3.2-2a}$$

$$U_2 = U_0 \tag{3.2-2b}$$

$$S_2 = S_0 + \int_0^{t_1} \dot{S}_{gen}\, dt + \int_{t_1}^{t_2} \dot{S}_{gen}\, dt \tag{3.2-2c}$$

In general, $\dot{S}_{gen} \geqslant 0$, so that the two integrals in Eq. 3.2-2c may be positive, and the entropy of the initial and final states will differ. Thus the initial and final states of the system must be different. If, however, the changes were accomplished in such a manner that the gradients in the system over the whole time interval are infinitesimal, then $\dot{S}_{gen} = 0$, and $S_2 = S_0$. In this case the system has been returned to its initial state from its state at t_1 by a process in which the work and each of the flows were reversed. Such a process is said to be reversible. If $\dot{S}_{gen}$ had not been equal to zero, then $S_2 > S_0$, and the system could not be returned to its initial state by simply reversing the work and other flows; the process is then said to be **irreversible**.

The main characteristic of a reversible process is that it proceeds with infinitesimal gradients within the system. Since transport processes are linearly related to the gradients in the system, this requirement implies that a reversible change occurs slowly on the time scale of macroscopic relaxation times. Changes of state in real systems can be approximated as being reversible if there is no appreciable internal heat flows or viscous dissipation; they are irreversible if these relaxation processes occur. Consequently, expansions and compressions that occur uniformly throughout a fluid, or in well-designed turbines, compressors, and nozzles for which viscous dissipation and internal heat flows are unimportant, can generally be considered to occur reversibly (i.e., $\dot{S}_{gen} = 0$). Flows through pipes, through flow constrictions (e.g., a valve or a porous plug), and through shock waves all involve velocity gradients and viscous dissipation and hence are irreversible. Table 3.2-1 contains some examples of reversible and irreversible processes.

Table 3.2-1
EXAMPLES OF REVERSIBLE AND IRREVERSIBLE PROCESSES

Reversible process: one with no (appreciable) internal heat flows or viscous dissipation.

Examples:

- Fluid flow in a well-designed turbine, compressor, or nozzle.
- Uniform and slow expansion or compression of a fluid.
- Many processes in which changes occur sufficiently slowly that gradients do not appear in the system.

Irreversible process: one with internal heat flow and/or viscous dissipation.

Examples:

- Flow of fluid in a pipe or duct where viscous forces are present.
- Flow of fluid through a constriction such as a partially open valve or porous plug (i.e., the Joule-Thomson expansion)
- Flow of fluid through a sharp gradient such as a shock wave.
- Heat conduction process in which temperature gradients exist.
- Any process in which friction is important.
- Mixing of fluids of different temperature or pressure.

Another characteristic of a reversible process is that if the surroundings are extracting work from the system, the maximum amount of work is obtained for a given change of state if the process is carried out reversibly (i.e., so that $\dot{S}_{gen} = 0$). A corollary to this statement is that if the surroundings are doing work on the system, a minimum amount of work is needed for a given change of state if the change occurs reversibly. The first of these statements is evident from Illustration 2.5-7 where it was found that the maximum work (W^{NET}) was extracted from an expansion of a gas between given initial and final states if the expansion was carried out reversibly, that is, so that only differential changes were occurring, and there was no frictional dissipation of kinetic energy to thermal energy in the work-producing device. It will be shown, in Illustration 3.5-8, that such a process is also one for which $S_{gen} = 0$.

Although few processes are truly reversible, it is sometimes useful to model them to be so. When this is done it is clear that any computations made based on Eqs. 3.1-5 or 9, with $\dot{S}_{gen} = S_{gen} = 0$, will only be approximate. However, these approximate results may be very useful since the term neglected (the entropy generation) is of known sign, so that we will know whether our estimate for the heat, work, or any state variable is an upper or lower bound to the true value. To see this, consider the energy and entropy balances for a closed, isothermal, constant-volume system

$$U_2 = U_1 + Q + W_s \tag{3.2-3}$$

and

$$S_2 = S_1 + \frac{Q}{T} + S_{gen} \tag{3.2-4}$$

Eliminating Q between these two equations and using $T_1 = T_2 = T$ gives

$$\begin{aligned} W_s &= (U_2 - T_2S_2) - (U_1 - T_1S_1) + TS_{gen} \\ &= A_2 - A_1 + TS_{gen} \end{aligned} \tag{3.2-5}$$

Here we have defined a new thermodynamic state variable, the **Helmholtz free energy**, by

$$A = U - TS \tag{3.2-6}$$

(Note that A must be a state variable since it is a combination of state variables.) The work required to bring the system from state 1 to state 2 by a reversible (i.e., $S_{gen} = 0$), isothermal, constant-volume process is

$$W_s^{rev} = A_2 - A_1 \tag{3.2-7}$$

while the work in an irreversible (i.e., $S_{gen} > 0$), isothermal, constant-volume process between the same initial and final states is

$$W_s = A_2 - A_1 + TS_{gen} = W_s^{rev} + TS_{gen}$$

Since $TS_{gen} > 0$, this equation establishes that more work is required to drive the system from state 1 to state 2 if the process is carried out irreversibly, than if it were carried out reversibly. Conversely, if we are interested in the amount of work the system can do on its surroundings at constant temperature and volume in going from state 1 to state 2 (so that W_s is negative), we find, by the same argument, that more work is obtained if the process is carried out reversibly than if it is carried out irreversibly.

One should not conclude from Eq. 3.2-7 that the reversible work for any process is equal to the change in Helmholtz free energy, since this result was derived only for an isothermal, constant-volume process. The value of W_s^{rev}, and the thermodynamic functions to which it is related, depends on the constraints placed on the system during the change of state (see Problem 3.2). For example, consider a process occurring in a closed system at fixed temperature and pressure. Here we have

$$U_2 = U_1 + Q + W_s - (P_2V_2 - P_1V_1)$$

where $P_2 = P_1$, $T_2 = T_1$, and

$$S_2 = S_1 + \frac{Q}{T} + S_{gen}$$

Thus

$$W_s^{rev} = G_2 - G_1$$

where

$$G \equiv U + PV - TS = H - TS \tag{3.2-8}$$

G is called the **Gibbs free energy**. Therefore, for the case of a closed system change at constant temperature and pressure, we have

$$W_s = G_2 - G_1 + TS_{gen} = W_s^{rev} + TS_{gen}$$

The quantity TS_{gen} in either process can be interpreted as the amount of mechanical energy that has been converted into thermal energy by viscous dissipation and other system irreversibilities. To see this, consider a reversible process ($S_{gen} = 0$) in a closed system. The energy balance is

$$Q^{rev} = U_2 - U_1 - W^{rev}$$

(here W is the sum of the shaft work and $P \Delta V$ work). If, on the other hand, the process were carried out so that macroscopic gradients arose in the system (that is, irreversibly, so $S_{gen} > 0$), then

$$Q = U_2 - U_1 - W = U_2 - U_1 - W^{rev} - TS_{gen}$$

or

$$Q = Q^{rev} - TS_{gen} \tag{3.2-9a}$$

and

$$W = W^{rev} + TS_{gen} \tag{3.2-9b}$$

Since TS_{gen} is greater than zero, less heat and more work are required to accomplish a given change of state in the second (irreversible) process than in the first.[5] This is because the additional mechanical energy supplied in the second case has been converted to thermal energy. It is generally true that system gradients that lead to heat flows, mass flows, and viscous dissipation also result in the conversion of mechanical energy into thermal energy and in a decrease in the amount of work that can be obtained from the system.

[5]The argument used here assumes that work is required to drive the system from state 1 to state 2. You should verify that the same conclusions would be reached if work were obtained in going from state 1 to state 2.

Finally, since the evaluation of the changes in the thermodynamic properties of a system accompanying its change of state are important in thermodynamics, it is useful to have an expression relating the entropy change to changes in other state variables. To obtain such an equation we start with the differential form of the mass, energy, and entropy balances for a system in which the kinetic and potential energy terms are unimportant, there is only one mass flow stream, and the mass and heat flows occur at the common temperature T:

$$\frac{dM}{dt} = \dot{M}$$

$$\frac{dU}{dt} = \dot{M}\hat{H} + \dot{Q} - P\frac{dV}{dt} + \dot{W}_s$$

and

$$\frac{dS}{dt} = \dot{M}\hat{S} + \frac{\dot{Q}}{T} + \dot{S}_{\text{gen}}$$

where $\hat{H}$ and $\hat{S}$ are the enthalpy per unit mass and entropy per unit mass of the fluid entering or leaving the system. Eliminating $\dot{M}$ between these equations and multiplying the third equation by T yields

$$\frac{dU}{dt} = \hat{H}\frac{dM}{dt} + \dot{Q} - P\frac{dV}{dt} + \dot{W}_s$$

and

$$T\frac{dS}{dt} = T\hat{S}\frac{dM}{dt} + \dot{Q} + T\dot{S}_{\text{gen}}$$

Of particular interest is the form of these equations applicable to the change of state occurring in the differential time interval dt. As in Sec. 2.3 we write

$Q = \dot{Q}\,dt$ = heat flow into the system in the time interval dt

$W_s = \dot{W}_s\,dt$ = shaft work into the system in the time interval dt

$S_{\text{gen}} = \dot{S}_{\text{gen}}\,dt$ = entropy generated in the system in the time interval dt

and obtain the following balance equations for the time interval dt

$$dU = \hat{H}\,dM + Q - P\,dV + W_s \tag{3.2-10a}$$

and

$$T\,dS = T\hat{S}\,dM + Q + TS_{\text{gen}} \tag{3.2-10b}$$

Equations 3.2-10 can be used to interrelate the differential changes in internal energy, entropy, and volume that occur between fixed initial and final states that are only slightly different. This is accomplished by first solving Eq. 3.2-10b for Q

$$Q = T\,dS - TS_{gen} - T\hat{S}\,dM \tag{3.2-11a}$$

and then using this result to eliminate the heat flow from Eq. 3.2-10a to obtain

$$\begin{aligned} dU &= T\,dS - TS_{\text{gen}} - P\,dV + W_s + (\hat{H} - T\hat{S})\,dM \\ &= T\,dS - TS_{\text{gen}} - P\,dV + W_s + \hat{G}\,dM \end{aligned} \tag{3.2-11b}$$ [6]

[6]Alternatively, this equation and those that follow can be written in terms of the molar Gibbs free energy and the mole number change by replacing $\hat{G}\,dM$ by $\underline{G}\,dN$.

For a reversible process ($S_{gen} = 0$) these equations reduce to

$$Q^{rev} = T\,dS - T\hat{S}\,dM \tag{3.2-12a}$$

and

$$dU = T\,dS + (-P\,dV + W_s)^{rev} + \hat{G}\,dM \tag{3.2-12b}$$

In writing these equations we have recognized that since the initial and final states of the system are fixed, the changes in the path-independent functions are the same for reversible and irreversible processes: The heat and work terms that depend on the path followed are denoted by the superscript rev.

Equating the changes in the (state variables) internal energy and entropy for the reversible process (Eqs. 3.2-12) and the irreversible process (Eqs. 3.2-11) yields

$$Q^{rev} = Q^{irrev} + TS_{gen}$$

and

$$(-P\,dV + W_s)^{rev} = (-P\,dV + W_s)^{irrev} - TS_{gen}$$

Substituting these results into Eqs. 3.2-11 again gives Eqs. 3.2-12, which are now seen to apply for any process (reversible or irreversible) for which kinetic and potential changes are negligible. There is a subtle point here. That is, Eq. 3.2-12b can be used to interrelate the internal energy and entropy changes for any process, reversible or irreversible. However, to relate the heat flow Q and the entropy, Eq. 3.2-11a is always valid whereas Eq. 3.2-12a can only be used for reversible processes.

Although both Eqs. 3.2-11b and 12b provide relationships between the changes in internal energy, entropy, mass, and the work flow for any real process, we will, in fact, use only Eq. 3.2-12b to interrelate these changes since, in many cases, this simplifies the computation. Finally, we note that for a system with no shaft work

$$dU = T\,dS - P\,dV + \hat{G}\,dM \tag{3.2-13a}$$[7]

and, further, if the system is closed to the flow of mass

$$dU = T\,dS - P\,dV$$

or

$$d\underline{U} = T\,d\underline{S} - P\,d\underline{V} \tag{3.2-13b}$$

Since we are always free to choose the system for a given change of state such that there is no shaft work (only $P\,\Delta\,V$ work), Eq. 3.2-13a can be generally used to compute entropy changes in open systems. Similarly, Eq. 3.2-13b can generally be used to compute entropy changes in closed systems. This last point should be emphasized. In many cases it is necessary to compute the change in thermodynamic properties between two states of a substance. Since, for this computation, we can choose the system in such a way that it is closed and shaft work is excluded, Eq. 3.2-13b can be used in the calculation of the change in thermodynamic properties of the substance between the given initial and final states, regardless of the device or path used to accomplish this change.

[7]Equivalently, on a molar basis, we have $dU = T\,dS - P\,dV + \underline{G}\,dN$.

3.3 HEAT, WORK, ENGINES, AND ENTROPY

The discussion of Sec. 3.2, and especially Eqs. 3.2-9, reveals an interesting distinction between mechanical energy (and work) and heat or thermal energy. In particular, while both heat and work are forms of energy, the relaxation processes (i.e., heat flow and viscous dissipation) that act naturally to reduce any temperature and velocity gradients in the system result in the conversion of mechanical energy in the form of work or the potential to do work, into heat or thermal energy. Friction in any moving object, which reduces its speed and increases its temperature, is the most common example of this phenomenon.

A problem of great concern to scientists and engineers since the late eighteenth century has been the development of devices (engines) to accomplish the reverse transformation, the conversion of heat (from the combustion of wood, coal, oil and other fuels) into mechanical energy or work. Much effort has been spent on developing engines of high efficiency, that is, engines that convert a large fraction of the heat supplied to useful work. Such engines are schematically represented in Fig. 3.3-1*a*, where $\dot{Q}_1$ is the heat flow rate into the engine from the surroundings at temperature T_1, $\dot{Q}_2$ is the heat flow rate from the engine to its surroundings at temperature T_2, and $\dot{W}$ is the rate at which work is done by the engine. The engine depicted in this figure may operate either in a steady-state fashion, in which case $\dot{Q}_1$, $\dot{Q}_2$, and $\dot{W}$ are independent of time, or cyclically. If the energy and entropy balances for the engine are integrated over a time interval Δt, which is the period of one cycle of the cyclic engine, or any convenient time interval for the steady-state engine, one obtains

$$0 = Q_1 - Q_2 + W \tag{3.3-1}$$

$$0 = \frac{Q_1}{T_1} - \frac{Q_2}{T_2} + S_{\text{gen}} \tag{3.3-2}$$

where $Q = \int_{t_1}^{t_2} \dot{Q}\, dt$ and $W = \int_{t_1}^{t_2} (\dot{W}_s - P(dV/dt))\, dt$. The left side of Eqs. 3.3-1 and 2 are zero since, for both the cyclic and steady-state engine, the engine is in the

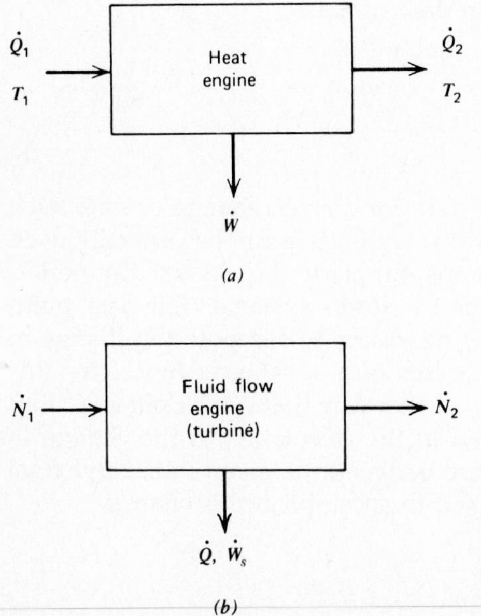

Figure 3.3-1
(*a*) Schematic drawing of a simple heat engine.
(*b*) Schematic drawing of a fluid flow engine.

same state at time t_2 as it was at t_1. Eliminating Q_2 between Eqs. 3.3-1 and 2 yields

$$\text{work done by the engine} = -W = Q_1\left(\frac{T_1 - T_2}{T_1}\right) - T_2 S_{\text{gen}} \tag{3.3-3}$$

Clearly, to obtain the maximum work from an engine operating between fixed temperatures T_1 and T_2, it is necessary that all processes be carried out reversibly, so that $S_{\text{gen}} = 0$. In this case

$$\text{maximum work done by the engine} = -W = Q_1\left(\frac{T_1 - T_2}{T_1}\right) \tag{3.3-4}$$

and

$$\text{engine efficiency} = \begin{pmatrix}\text{fraction of heat}\\ \text{supplied that is}\\ \text{converted to work}\end{pmatrix} = \frac{-W}{Q_1} = \frac{T_1 - T_2}{T_1} \tag{3.3-5}$$

This is a surprising result since it establishes that, independent of the engine design, there is a maximum engine efficiency that depends only on the temperature levels between which the engine operates. Less than ideal design or operation of the engine will, of course, result in the efficiency of an engine being less than the maximum efficiency given in Eq. 3.3-5. Industrial heat engines, due to design and operating limitations, heat losses, and friction, typically operate at about half this efficiency.

Another aspect of the conversion of heat to work can be illustrated by solving for Q_2 from Eqs. 3.3-1 and 3 to get

$$Q_2 = Q_1 \frac{T_2}{T_1} + T_2 S_{\text{gen}} \tag{3.3-6}$$

This equation establishes that it is impossible to convert all the heat supplied to an engine into work (that is, for Q_2 to equal zero) in a continuous or cyclic process, unless the engine has a lower operating temperature (T_2) of absolute zero. In contrast, the inverse process, that of converting work or mechanical energy completely to heat, can be accomplished completely at any temperature, and unfortunately occurs frequently in natural processes (e.g., frictional heating).

Therefore, it is clear that although heat and work are equivalent in the sense that both are forms of energy (Experimental Observations 3 and 4 of Sec. 1.7), there is a real distinction between them in that work or mechanical energy can spontaneously (naturally) be converted completely to heat or thermal energy, but thermal energy can, with some effort, be only partially converted to mechanical energy. In this sense, mechanical energy can be regarded as a higher form of energy than thermal energy.

At this point you might ask if it is possible to construct an engine having the efficiency given by Eq. 3.3-5. Nicolas Léonard Sadi Carnot described such a cyclic engine in 1824. The **Carnot engine** consists of a fluid enclosed in a frictionless piston and cylinder device, schematically shown in Fig. 3.3-2. Work is extracted from this engine by the movement of the piston. In the first step of the four-part cycle, the fluid is isothermally and reversibly expanded from volume V_a to volume V_b at a constant temperature T_1 by adding an amount of heat Q_1 from the first heat source. The mechanical work obtained in this expansion is $\int_{V_a}^{V_b} P\,dV$. The next part of the cycle is a reversible adiabatic expansion of the

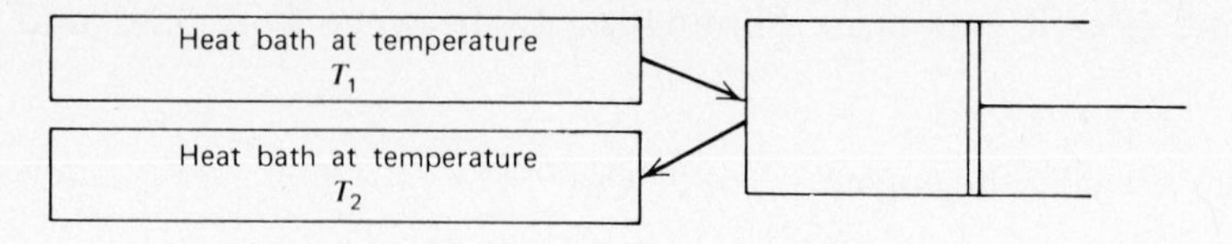

(a) Schematic diagram of a Carnot engine

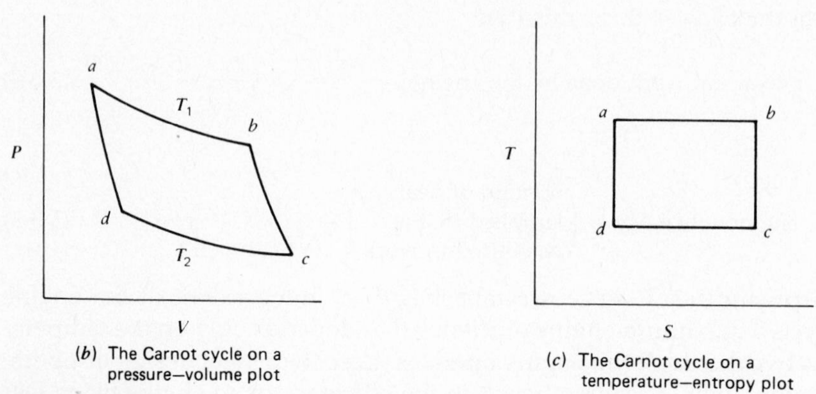

(b) The Carnot cycle on a pressure–volume plot

(c) The Carnot cycle on a temperature–entropy plot

Figure 3.3-2
The Carnot cycle.

fluid from the state (P_b, V_b, T_1) to the state (P_c, V_c, T_2). The work obtained in this expansion is $\int_{V_b}^{V_c} P\,dV$, and is gotten by directly converting the internal energy of the fluid into work. The next step in the cycle is to reversibly and isothermally compress the fluid to the state (P_d, V_d, T_2). The work done on the fluid in this process is $\int_{V_c}^{V_d} P\,dV$, and the heat *removed* is Q_2. The final step in the cycle is a reversible adiabatic compression to the initial state (P_a, V_a, T_1). The work done on the fluid in this part of the process is $\int_{V_d}^{V_a} P\,dV$. The complete work cycle is summarized in the following table.

Path	Work Done on the Fluid	Heat Added to the Fluid
$(P_a, V_a, T_1) \xrightarrow[\text{expansion}]{\text{reversible isothermal}} (P_b, V_b, T_1)$	$-\int_{V_a}^{V_b} P\,dV$	Q_1
$(P_b, V_b, T_1) \xrightarrow[\text{adiabatic expansion}]{\text{reversible}} (P_c, V_c, T_2)$	$-\int_{V_b}^{V_c} P\,dV$	0
$(P_c, V_c, T_2) \xrightarrow[\text{compression}]{\text{reversible isothermal}} (P_d, V_d, T_2)$	$\int_{V_c}^{V_d} P\,dV$	$-Q_2$
$(P_d, V_d, T_2) \xrightarrow[\text{adiabatic compression}]{\text{reversible}} (P_a, V_a, T_1)$	$\int_{V_d}^{V_a} P\,dV$	0

The energy and entropy balances for one complete cycle are

$$0 = W + Q_1 - Q_2 \tag{3.3-7}$$

$$0 = \frac{Q_1}{T_1} - \frac{Q_2}{T_2} \tag{3.3-8}$$

where

$$W = -\int_{V_a}^{V_b} P\,dV - \int_{V_b}^{V_c} P\,dV + \int_{V_c}^{V_d} P\,dV + \int_{V_d}^{V_a} P\,dV$$

Now, using Eq. 3.3-8 to eliminate Q_2 from Eq. 3.3-7 yields

$$-W = Q_1 - \left(\frac{T_2}{T_1} Q_1\right) = \frac{T_1 - T_2}{T_1} Q_1 \tag{3.3-9}$$

which is exactly the result of Eq. 3.3-4. Equations 3.3-5 and 6 then follow directly. Thus we conclude that the Carnot engine is the most efficient possible in the sense of extracting the most work from a given flow of heat between temperature baths at T_1 and T_2 in a cyclic or continuous manner.

Perhaps the most surprising aspect of the Carnot cycle engine (or Eq. 3.3-4 for that matter) is that the work obtained depends only on T_1, T_2, and the heat flow Q_1, and is completely independent of the working fluid.

Since the work supplied or obtained in each step of the Carnot cycle is expressible in the form $-\int P\,dV$, the enclosed area on the P–V diagram of Fig. 3.3-2*b* is equal to the total work *supplied* by the Carnot engine to its surroundings in one complete cycle. (You should verify that if the Carnot engine is driven in reverse, so that the cycle in Fig. 3.3-2*b* is traversed counterclockwise, the enclosed area is equal to the work *absorbed* by the engine from its surroundings in one cycle.)[8] Similarly, since the differential entropy change dS and the differential heat flow Q for a reversible process in a closed system are related as follows

$$dS = \frac{Q}{T} \qquad \text{or} \qquad Q = T\,dS$$

the enclosed area on the T–S diagram of Fig. 3.3-2*c* is equal to the net heat flow to the Carnot engine in one cycle. This heat flow is positive (i.e., into the engine) if the engine is operated as described and the cycle is traversed clockwise, and negative if the engine is operated in the reverse manner and the cycle is traversed counterclockwise. Thus, for reversible cycles, the P–V diagram supplies information about the net work flow, and the T–S diagram information about the net heat flow. For irreversible processes (i.e., processes for which $S_{\text{gen}} \neq 0$), the heat flow and entropy change are not simply related as above, and the area on a T–S diagram is not directly related to the heat flow.

In addition to the Carnot heat engine, other cycles and devices may be used for the conversion of thermal energy to mechanical energy or work; though the conversion efficiencies of these other cycles, owing to the paths followed, are less than that of the Carnot engine. Despite their decreased efficiency, these other engines offer certain design and operating advantages over the Carnot cycle, and hence are more widely used. The efficiency of some other cycles is considered in Problem 3.17.

Another class of work-producing devices are engines that convert the thermal energy of a *flowing fluid* into mechanical energy. Examples of this type of engine are the nozzle-turbine systems of Fig. 3.3-3. Here a high-pressure, high-

[8]Note that if the Carnot heat engine is operated as shown in Fig. 3.3-2 it absorbs heat from the high temperature bath, exhausts heat to the low temperature bath, and produces work. However, if the engine is operated in reverse, it accepts work, absorbs heat from the low temperature bath, and exhausts heat to the high temperature bath. In this mode it is operating as a refrigerator or heat pump.

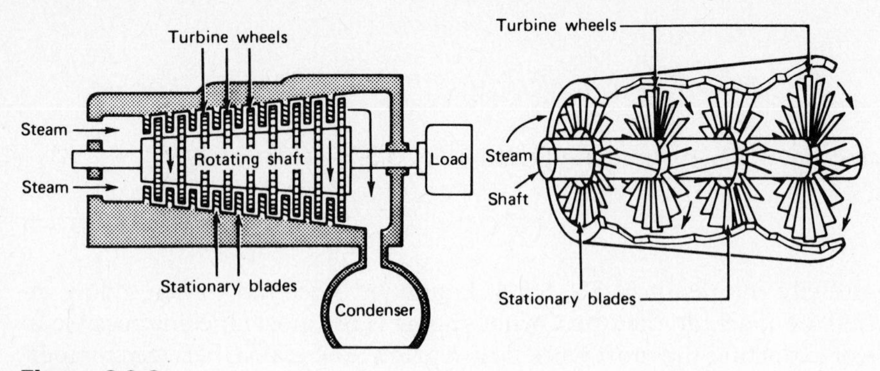

Figure 3.3-3
Sketch of a steam turbine. (Adapted from *The World Book Encyclopedia*, © 1976 Field Educational Enterprises.)

temperature fluid, frequently steam, is expanded through a nozzle to obtain a low-pressure, high-velocity gas. This gas then impinges on turbine blades where the kinetic energy of the gas is transferred to the turbine rotor, and thus is available as shaft work. The resulting low pressure, low velocity gas leaves the turbine. Of course, many other devices can also be used to accomplish the same energy transformation. All these devices may be schematically represented as shown in Fig. 3.3-1*b*. The steady-state mass, energy, and entropy balances for such heat engines are

$$0 = \dot{N}_1 + \dot{N}_2 \qquad \text{or} \qquad \dot{N}_2 = -\dot{N}_1$$

$$0 = (\underline{H}_1 - \underline{H}_2)\dot{N}_1 + \dot{Q} + \dot{W}_s \tag{3.3-10}$$

and

$$0 = (\underline{S}_1 - \underline{S}_2)\dot{N}_1 + \frac{\dot{Q}}{T} + \dot{S}_{\text{gen}} \tag{3.3-11}$$

Here we have assumed that the kinetic and potential energy changes of the fluid entering and leaving the device cancel or are negligible (regardless of what happens internally) and that the heat flow $\dot{Q}$ can be identified as occurring at a single temperature T.[9] Solving these equations for the heat flow $\dot{Q}$ and the work flow $\dot{W}_s$ yields

$$\dot{Q} = -T(\underline{S}_1 - \underline{S}_2)\dot{N}_1 - T\dot{S}_{\text{gen}} \tag{3.3-12}$$

and

$$\dot{W}_s = -\dot{N}_1[(\underline{H}_1 - T\underline{S}_1) - (\underline{H}_2 - T\underline{S}_2)] + T\dot{S}_{\text{gen}} \tag{3.3-13}$$

Note that the quantity $(\underline{H}_1 - T\underline{S}_1)$ is *not* equal to the Gibbs free energy unless the temperature T at which heat transfer occurs is equal to the inlet fluid temperature [i.e., $\underline{G}_1 = \underline{H}_1 - T_1\underline{S}_1$]; a similar comment applies to the term $(\underline{H}_2 - T\underline{S}_2)$.

Several special cases of these equations are important. First, for the isothermal flow engine (i.e., for an engine in which the inlet temperature T_1, the outlet

[9]If this is not the case, a sum of $\dot{Q}/T$ terms or an integral of the heat flux divided by the temperature over the surface of the system is needed.

temperature T_2, and the operating temperature T are all equal), Eq. 3.3-13 reduces to

$$\dot{W}_s = -\dot{N}_1(\underline{G}_1 - \underline{G}_2) + T\dot{S}_{gen}$$

The maximum rate at which work can be obtained from such an engine for fixed inlet and exit pressures and fixed temperature occurs when the engine is reversible; in this case

$$\dot{W}_s^{rev} = -\dot{N}_1(\underline{G}_1 - \underline{G}_2)$$

and the heat load for reversible, isothermal operation is

$$\dot{Q}^{rev} = -T\dot{N}_1(\underline{S}_1 - \underline{S}_2)$$

For adiabatic operation ($\dot{Q} = 0$) of a flow engine we have from Eq. 3.3-10 that

$$\dot{W}_s = -\dot{N}_1(\underline{H}_1 - \underline{H}_2) \tag{3.3-14}$$

and

$$0 = \dot{N}_1(\underline{S}_1 - \underline{S}_2) + \dot{S}_{gen} \tag{3.3-15}$$

so that the work flow is proportional to the difference in enthalpies of the inlet and exiting fluids. Usually, the inlet temperature and pressure and exit pressure of the adiabatic flow engine can be specified by the design engineer; the exit temperature cannot be specified, but instead adjusts so that Eq. 3.3-15 is satisfied. Thus, although the entropy generation term does not explicitly appear in the work term of Eq. 3.3-14, it is contained implicitly through the exit temperature and enthalpy $\underline{H}_2$. By example (Problems 3.6 and 3.15), one can establish that a reversible adiabatic engine has the lowest exit temperature and enthalpy for fixed inlet and exit pressures, and thereby achieves the best conversion of fluid thermal energy to work.

Finally, we want to develop an expression in terms of the pressure and volume, for the maximum rate at which work is obtained, or the minimum rate at which work must be added, to accomplish a given change of state in continuous flow systems such as turbines and compressors. Figure 3.3-4 is a generic diagram of a device through which fluid is flowing continuously. The volume element in the figure contained within the dashed lines is a very small region of length ΔL in which the temperature and pressure of the fluid can be taken to be approximately constant (in fact, shortly we will consider the limit in which $\Delta L \to 0$). The mass, energy, and entropy balances for this steady state system are

$$\frac{dN}{dt} = 0 = \dot{N}|_L - \dot{N}|_{L+\Delta L} \tag{3.3-16a}$$

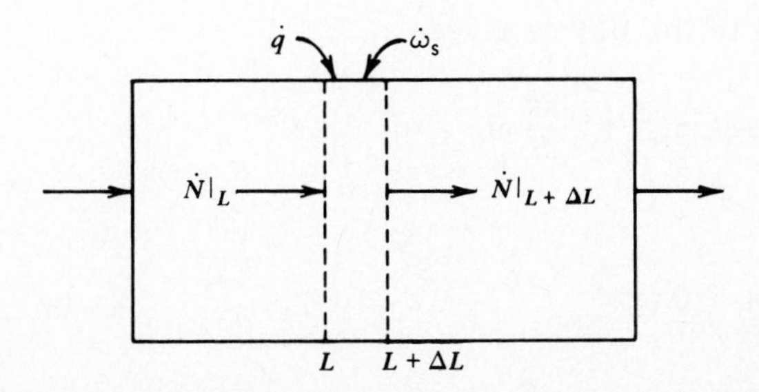

Figure 3.3-4
Device with fluid, heat, and work flows.

$$\frac{dU}{dt} = 0 = \dot{N}\underline{H}|_L - \dot{N}\underline{H}|_{L+\Delta L} + \dot{q}\,\Delta L + \dot{\omega}_S\,\Delta L \tag{3.3-16b}$$

and

$$\frac{dS}{dt} = 0 = \dot{N}\underline{S}|_L - \dot{N}\underline{S}|_{L+\Delta L} + \frac{\dot{q}}{T}\,\Delta L + \dot{\rho}_{gen}\,\Delta L \tag{3.3-16c}$$

In these equations, $\dot{q}$, $\dot{\omega}_S$ and $\dot{\rho}_{gen}$ are, respectively, the heat and work flows, and the rate of entropy generation per unit length of the device.

Dividing by ΔL, taking the limit as $\Delta L \rightarrow 0$, and using the definition of the total derivative from calculus, gives

$$\lim_{\Delta L \to 0} \frac{\dot{N}|_{L+\Delta L} - \dot{N}|_L}{\Delta L} = \frac{d\dot{N}}{dL} = 0 \quad \text{or} \quad \dot{N} = \text{constant} \tag{3.3-17a}$$

$$\dot{N} \lim_{\Delta L \to 0} \left(\frac{\underline{H}|_{L+\Delta L} - \underline{H}|_L}{\Delta L} \right) = \dot{N} \frac{d\underline{H}}{dL} = \dot{q} + \dot{\omega}_S \tag{3.3-17b}$$

$$\dot{N} \lim_{\Delta L \to 0} \left(\frac{\underline{S}|_{L+\Delta L} - \underline{S}|_L}{\Delta L} \right) = \dot{N} \frac{d\underline{S}}{dL} = \frac{\dot{q}}{T} + \dot{\rho}_{gen} \tag{3.3-17c}$$

From the discussion of the previous section, the maximum work that can be obtained, or the minimum work required, in a given change of state occurs in a reversible process. Setting $\dot{\rho}_{gen} = 0$ yields

$$\dot{N} \frac{d\underline{S}}{dL} = \frac{\dot{q}}{T}$$

$$\dot{\omega}_S = \dot{N} \left(\frac{d\underline{H}}{dL} - T \frac{d\underline{S}}{dL} \right)$$

Now, from $\underline{H} = \underline{U} + P\underline{V}$

$$d\underline{H} = d\underline{U} + d(P\underline{V}) = d\underline{U} + P\,d\underline{V} + \underline{V}\,dP$$

and Eq. 3.2-13b

$$d\underline{U} = T\,d\underline{S} - P\,d\underline{V}$$

we have that

$$d\underline{H} - T\,d\underline{S} = T\,d\underline{S} - P\,d\underline{V} + P\,d\underline{V} + \underline{V}\,dP - T\,d\underline{S} = \underline{V}\,dP$$

$$\left(\text{or} \quad \frac{d\underline{H}}{dL} - T \frac{d\underline{S}}{dL} = \underline{V} \frac{dP}{dL} \right)$$

and

$$\dot{\omega}_S = \dot{N} \underline{V} \frac{dP}{dL}$$

Further, integrating over the length of the device gives

$$\dot{W}_S = \int \dot{\omega}_S\, dL = \dot{N} \int \underline{V} \frac{dP}{dL}\, dL$$

or

$$\underline{W}_S = \frac{\dot{W}_S}{\dot{N}} = \int \underline{V}\, dP \tag{3.3-18}$$

Equation 3.3-18 is the desired result.

Finally, it is useful to consider several applications of Eq. 3.3-18. For the ideal gas undergoing an isothermal change, so that $P\underline{V} = RT =$ constant

$$\underline{W}_S = \int \underline{V}\, dP = \int \frac{RT}{P}\, dP = RT \int \frac{dP}{P} = RT \ln \frac{P_2}{P_1} \tag{3.3-19}$$

For an expansion or compression for which

$$P_1\underline{V}_1^{\gamma} = P_2\underline{V}_2^{\gamma} = \text{constant} \tag{3.3-20}$$

$$\underline{W}_S = \int \underline{V}\, dP = (\text{constant})^{(1/\gamma)} \int \frac{dP}{P^{(1/\gamma)}}$$

$$= \frac{(\text{constant})^{(1/\gamma)}(P_2^{(1-1/\gamma)} - P_1^{(1-1/\gamma)})}{\left(1 - \dfrac{1}{\gamma}\right)}$$

$$= \frac{[(P_2\underline{V}_2^{\gamma})^{(1/\gamma)}P_2^{(1-1/\gamma)} - (P_1\underline{V}_1^{\gamma})^{(1/\gamma)}P_1^{(1-1/\gamma)}]}{\left(1 - \dfrac{1}{\gamma}\right)}$$

$$= \frac{\gamma(P_2\underline{V}_2 - P_1\underline{V}_1)}{\gamma - 1} \tag{3.3-21}$$

A process that obeys Eq. 3.3-20 is referred to as a **polytropic process**. For an ideal gas it is easy to show that

$\gamma = 0$ for an isobaric process

$\gamma = 1$ for an isothermal process

$\gamma = \infty$ for a constant volume (isochoric) process

Also, we will show later that

$\gamma = C_p/C_v$ for a constant entropy (isentropic) process in an ideal gas of constant heat capacity

3.4 ENTROPY CHANGES OF MATTER

Equation 3.2-13b provides the basis for computing entropy changes for real fluids, and it will be used in that manner in Chapter 4. However, to illustrate the use of the entropy balance, we consider here the calculation of the entropy change accompanying a change of state for 1 mole of an ideal gas, and for incompressible liquids and solids.

From the discussion of Sec. 2.4 the internal energy change and pressure of an ideal gas are

$$d\underline{U} = C_v\, dT \qquad \text{and} \qquad P = RT/\underline{V}$$

respectively, so that for 1 mole of a fluid we have

$$d\underline{S} = \frac{1}{T}\, d\underline{U} + \frac{P}{T}\, d\underline{V} = \frac{C_v}{T}\, dT + \frac{R}{\underline{V}}\, d\underline{V} \tag{3.4-1}$$

If C_v is independent of temperature, we can immediately integrate this equation to obtain

$$\underline{S}(T_2, \underline{V}_2) - \underline{S}(T_1, \underline{V}_1) = C_v \int_{T_1}^{T_2} \frac{dT}{T} + R \int_{\underline{V}_1}^{\underline{V}_2} \frac{d\underline{V}}{\underline{V}}$$

$$= C_v \ln\left(\frac{T_2}{T_1}\right) + R \ln\left(\frac{\underline{V}_2}{\underline{V}_1}\right) \tag{3.4-2}$$

Using the ideal gas law, we can eliminate either the temperature or the volume in this equation and obtain expressions for the change in entropy with changes in temperature and pressure

$$\begin{aligned}\underline{S}(T_2, P_2) - \underline{S}(T_1, P_1) &= C_V \ln\left(\frac{T_2}{T_1}\right) + R \ln \frac{RT_2/P_2}{RT_1/P_1} \\ &= (C_V + R)\ln\left(\frac{T_2}{T_1}\right) - R\ln\left(\frac{P_2}{P_1}\right) \\ &= C_P \ln\left(\frac{T_2}{T_1}\right) - R\ln\left(\frac{P_2}{P_1}\right)\end{aligned} \tag{3.4-3}$$

and pressure and volume

$$\begin{aligned}\underline{S}(P_2, \underline{V}_2) - \underline{S}(P_1, \underline{V}_1) &= C_V \ln\frac{P_2\underline{V}_2/R}{P_1\underline{V}_1/R} + R\ln\left(\frac{\underline{V}_2}{\underline{V}_1}\right) \\ &= (C_V + R)\ln\left(\frac{\underline{V}_2}{\underline{V}_1}\right) + C_V\ln\left(\frac{P_2}{P_1}\right) \\ &= C_P\ln\left(\frac{\underline{V}_2}{\underline{V}_1}\right) + C_V\ln\left(\frac{P_2}{P_1}\right)\end{aligned} \tag{3.4-4}$$

The evaluation of the entropy change for an ideal gas in which the heat capacity is a function of temperature (see Eq. 2.4-5) leads to more complicated equations than those given here. It is left to you to work out the appropriate expressions (Problem 3.10).

For liquids or solids we can generally write

$$d\underline{S} = \frac{1}{T}\,d\underline{U} + \frac{P}{T}\,d\underline{V} \approx \frac{1}{T}\,d\underline{U} \tag{3.4-5}$$

since the molar volume is very weakly dependent on either temperature or pressure (i.e., $d\underline{V}$ is generally small). Furthermore, since for a liquid or a solid $C_V \approx C_P$, we have

$$d\underline{U} = C_V\,dT \approx C_P\,dT$$

for these substances, so that

$$d\underline{S} = C_P\frac{dT}{T}$$

and

$$\underline{S}(T_2) - \underline{S}(T_1) = \int_{T_1}^{T_2} C_P\frac{dT}{T} \tag{3.4-6}$$

Finally, the entropy changes accompanying a change in state can easily be computed for real fluids for which thermodynamic tables and charts have been prepared. In this way the entropy changes for methane, nitrogen, and steam can be found using the figures of Chapter 2 and the appendixes.

3.5 APPLICATIONS OF THE ENTROPY BALANCE

In this section we show, by example, that the entropy balance provides a useful additional equation for the analysis of thermodynamic problems. In fact, many of the examples considered here are continuations of the illustrations of the previous chapter, to emphasize that the entropy balance can provide the information needed to solve problems that were unsolvable using only the mass and energy balance equations, or, in some cases, to simplify the solutions of problems that were solvable.

ILLUSTRATION 3.5-1 (Illustration 2.5-4 continued)

In Illustration 2.5-4 we attempted to estimate the exit temperature and power requirements for a gas compressor. From the steady-state mass balance we found that

$$\dot{N}_1 = -\dot{N}_2 = \dot{N} \tag{a}$$

and from the steady-state energy balance we had

$$\dot{W}_s = \dot{N}C_p(T_2 - T_1) \tag{b}$$

Now, writing a molar entropy balance for the same system yields

$$0 = (\underline{S}_1 - \underline{S}_2)\dot{N} + \dot{S}_{\text{gen}} \tag{c}$$

To obtain an estimate of the exit temperature and the power requirements, we will assume that the compressor is well-designed and operates reversibly, that is

$$\dot{S}_{\text{gen}} = 0 \tag{d}$$

Thus, we have

$$\underline{S}_1 = \underline{S}_2 \tag{e}$$

which is the additional relation for a state variable needed to solve the problem. Now using Eq. 3.4-3

$$\underline{S}(P_2, T_2) - \underline{S}(P_1, T_1) = C_P \ln\left(\frac{T_2}{T_1}\right) - R \ln\left(\frac{P_2}{P_1}\right)$$

and recognizing that $\underline{S}(P_2, T_2) = \underline{S}(P_1, T_1)$ yields

$$\left(\frac{T_2}{T_1}\right) = \left(\frac{P_2}{P_1}\right)^{R/C_P} \tag{f}$$

or

$$T_2 = T_1\left(\frac{P_2}{P_1}\right)^{R/C_P} = 290\text{ K}\left(\frac{10}{1}\right)^{8.314/29.3} = 557.4\text{ K or }284.2^\circ\text{C} \tag{g}$$

Thus T_2 is known, and hence $\underline{W}_s$ can be computed

$$\underline{W}_s = C_P(T_2 - T_1) = 7834.8\text{ J/mol}$$

and

$$\dot{W}_s = \dot{N}\,\underline{W}_s = 2.5\,\frac{\text{mol}}{\text{s}} \times 7834.8\,\frac{\text{J}}{\text{mol}} = 19.59\,\frac{\text{kJ}}{\text{s}}$$

Before considering the problem to be solved, we should try to assess the validity of the assumption $\dot{S}_{\text{gen}} = 0$. Unfortunately, this can only be done by experiment. One method is to measure the inlet and exit temperatures and pressures for an adiabatic turbine and see if Eq. e is satisfied. Experiments of this type indicate that Eq. e is reasonably accurate, so that reversible operation is a reasonable approximation for a gas compressor. ■

ILLUSTRATION 3.5-2

Frequently it is possible to solve a thermodynamic problem several ways, based on different choices of the system. To see this, consider Illustration 2.5-5, which was concerned with the partial evacuation of a compressed gas cylinder into an evacuated cylinder of equal volume. Suppose we now choose for the system of interest only that portion of the contents of the first cylinder that remains in the cylinder when the pressures have equalized (see Fig. 3.5-1, where the thermodynamic system of interest is indicated by the dashed lines). Note that with this choice the system is closed, but of changing volume. Furthermore, since the gas on one side of the imaginary boundary has precisely the same temperature as the gas at the other side, we can assume there is no heat transfer across the boundary, so that the system is adiabatic. Also, with the exception of the region near the valve (which is *outside* the system), the gas in the cylinder is undergoing a uniform expansion so there will be no pressure, velocity, or temperature gradients in the cylinder. Therefore, we will assume that the changes taking place in the system occur reversibly.

The mass, energy, and entropy balances (on a molar basis) for this system are

$$N_1^f = N_1^i \tag{a}$$

$$N_1^f \underline{U}_1^f = N_1^i \underline{U}_1^i - \int_{V_1^i}^{V_1^f} P\, dV \tag{b}$$

and

$$N_1^f \underline{S}_1^f = N_1^i \underline{S}_1^i \tag{c}$$

Now the important observation is that combining Eqs. a and c we obtain

$$\underline{S}_1^f = \underline{S}_1^i \tag{d}$$

so that the process is **isentropic** (i.e., occurs at constant entropy) for the system we have chosen. Using Eq. 3.4-3

$$\underline{S}_1(P^f, T^f) - \underline{S}_1(P^i, T^i) = C_P \ln\left(\frac{T_1^f}{T_1^i}\right) - R \ln\left(\frac{P_1^f}{P_1^i}\right)$$

and Eq. d yields

$$\left(\frac{T_1^f}{T_1^i}\right)^{C_P/R} = \left(\frac{P_1^f}{P_1^i}\right)$$

This is precisely the result obtained in Eq. f of Illustration 2.5-5 using the energy balance on the open system consisting of the total contents of cylinder 1. The remainder of the problem can now be solved in exactly the same manner used in Illustration 2.5-5.

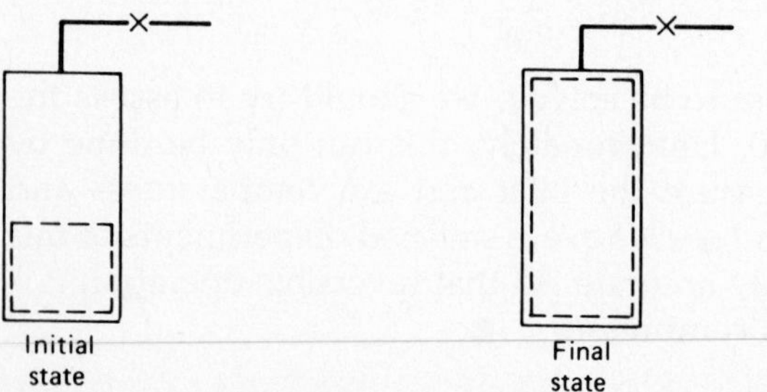

Figure 3.5-1

Although the system choice used in this illustration is an unusual one, it is one that leads quickly to a useful result. This demonstrates that sometimes a clever choice for the thermodynamic system can be the key to solving a thermodynamic problem with the minimum effort. However, one also has to be careful about the assumptions in unusual system choices. For example, consider the two cylinders connected as in Fig. 3.5-2, where the second cylinder is not initially evaluated. Here we have chosen to treat that part of the initial contents of cylinder 1 that will be in that cylinder finally as one system and the total initial contents of cylinder 2 as a second system. The change that occurs in the first system, as we already discussed, is adiabatic and reversible, so that

$$\left(\frac{T_1^f}{T_1^i}\right)^{C_P/R} = \left(\frac{P_1^f}{P_1^i}\right) \tag{e}$$

One might expect that a similar relation would hold for the system in cylinder 2.

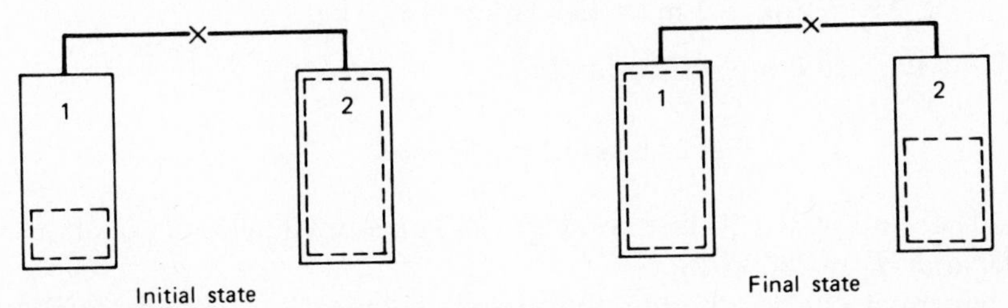

Figure 3.5-2

This is not the case, however, since the gas entering cylinder 2 is not necessarily at the same temperature as the gas already there (hydrodynamics will ensure that the pressures are the same). Therefore, temperature gradients will exist within the cylinder, and our system, the partial contents of cylinder 2, will not be adiabatic. Consequently, Eq. e will not apply. ■

ILLUSTRATION 3.5-3 (Illustration 2.5-5 continued)

A gas cylinder of 1-m³ volume containing nitrogen initially at a pressure of 40 bar and a temperature of 200 K is connected to another cylinder of 1-m³ volume, which is evacuated. A valve between the two cylinders is opened only until the pressure in both cylinders equalizes and then is closed. Find the final temperature and pressure in each cylinder if there is no heat flow into or out of the cylinders, or between the gas and the cylinder walls. The properties of nitrogen gas are given in Fig. 2.4-3.

Solution

Except for the fact that nitrogen is now being considered to be a real rather than ideal gas, the problem here is the same as in Illustration 2.5-5. In fact, Eqs. h–n in the comments to that example apply here. The additional equation needed to solve this problem is obtained in the same manner as in Illustration 3.5-2. Thus, for the cylinder initially filled we have

$$\underline{S}_1^i = \underline{S}_1^f \tag{o}$$

For an ideal gas of constant heat capacity (which is not the case here) this reduces to

$$\left(\frac{P_1^f}{P_1^i}\right) = \left(\frac{T_1^f}{T_1^i}\right)^{C_P/R}$$

by Eq. 3.4-3. For real nitrogen gas, Eq. o requires that the initial and final states in cylinder 1 be connected by a line of constant entropy in Fig. 2.4-3.

With eight equations, Eqs. h–o, and eight unknowns, this problem can be solved, though the solution is a tedious trial-and-error process. In general, a reasonable first guess for the pressure in the nonideal gas problem is the ideal gas solution, which was

$$P_1^f = P_2^f = 20 \text{ bar}$$

Locating the initial conditions ($P = 40$ bar $= 4$ MPa and $T = 200$ K $= -99.6$°F) on Fig. 2.4-3 yields[10] $\hat{H}_1^i \approx 191$ kJ/kg and $\rho = 4.59$ lb/ft$^3 = 73.5$ kg/m^3 so that

$$M_1^i = V_1\rho_1^i = 1 \text{ m}^3 \times 73.5 \text{ kg/m}^3 = 73.5 \text{ kg}$$

Now using Eq. (o) (in the form $\hat{S}_1^i = \hat{S}_1^f$) and Fig. 2.4-3, we find $T_1^f \sim 163$ K, $H_1^f \approx$ 15.6 kJ/kg and $\rho_1^f = 2.84$ lb/ft$^3 = 45.5$ kg/m^3 so that

$$M_1^f = V_1\rho_1^f = 1 \text{ m}^3 \times 45.5 \text{ kg/m}^3 = 45.5 \text{ kg}$$

and $M_2^f = M_1^i - M_1^f = 28.0$ kg, which implies

$$\rho_2^f = \frac{M_2^f}{V_2} = 28.0 \text{ kg/m}^3 = 1.75 \text{ lb/ft}^3$$

Locating $P = 20$ bar and $\rho_2^f = 1.75$ lb/ft^3 on Fig. 2.4-3 gives a value for T_2^f of about -18°F $= 245$ K and $\hat{H}_2^f \approx 250$ kJ/kg.

Finally, we must check whether the energy balance is satisfied for the conditions computed here based on the final pressure *we have assumed*. To do this we first must compute the internal energies of the initial and final states of the system as follows

$$\hat{U}_1^i = \hat{H}_1^i - P_1^i\hat{V}_1^i = \hat{H}_1^i - P_1^i/\rho_1^i$$

$$= 191\,\frac{\text{kJ}}{\text{kg}} - \frac{40 \text{ bar}}{73.5 \text{ kg m}^{-3}} \times 10^5\,\frac{\text{Pa}}{\text{bar}} \times \frac{1 \text{ J}}{\text{m}^3 \text{ Pa}} \times \frac{1 \text{ kJ}}{1000 \text{ J}} = 136.6\,\frac{\text{kJ}}{\text{kg}}$$

Similarly

$$\hat{U}_1^f = 112.0 \text{ kJ/kg}$$

and

$$\hat{U}_2^f = 178.6 \text{ kJ/kg}$$

The energy balance (Eq. n of Illustration 2.5-5) on a mass basis is

$$M_1^i\hat{U}_1^i = M_1^f\hat{U}_1^f + M_2^f\hat{U}_2^f$$

or

$$73.5 \times 136.6 \text{ kJ} = 45.5 \times 112.0 + 28.0 \times 178.6 \text{ kJ}$$

$$10040 \text{ kJ} \approx 10097 \text{ kJ}$$

So that, to the accuracy of our calculations, the energy balance will be considered to be satisfied and the problem solved. Had the energy balance not been satisfied it would have been necessary to make another guess for the final pressures and repeat the calculation.

[10]Since the thermodynamic properties in Fig. 2.4-3 are on a mass basis, all calculations here will be on a mass, rather than molar, basis.

It is interesting to note that the solution obtained here is essentially the same as that for the ideal gas case. This is not generally true, but occurs here because the initial and final pressures are sufficiently low, and the temperature sufficiently high, that nitrogen behaves like an ideal gas. Had we chosen the initial pressure to be higher, say several hundred bars, the ideal gas and real gas solutions would have been significantly different (see Problem 3.13). ■

ILLUSTRATION 3.5-4 (Illustration 2.5-6 continued)
Show that the entropy S is a state function by computing ΔS for each of the three paths of Illustration 2.5-6.

Solution
Since the piston and cylinder device is frictionless (see Illustration 2.5-6), each of the expansion processes will be reversible (see also Illustration 3.5-8). Thus, the entropy balance for the gas within the piston and cylinder reduces to

$$\frac{dS}{dt} = \frac{\dot{Q}}{T}$$

Path A

i. ***Isothermal compression***
Since T is constant

$$\Delta S_A = \frac{Q_A}{T} = -\frac{5707.7 \text{ J/mol}}{298.15 \text{ K}} = -19.14 \text{ J/mol K}$$

ii. ***Isobaric heating***

$$\dot{Q} = C_P \frac{dT}{dt}$$

so

$$\frac{d\underline{S}}{dt} = \frac{\dot{Q}}{T} = \frac{C_P}{T}\frac{dT}{dt}$$

and

$$\Delta \underline{S}_B = C_P \ln \frac{T_2}{T_1} = 38 \frac{\text{J}}{\text{mol K}} \times \ln \frac{573.15}{298.15} = 24.83 \text{ J/mol K}$$

$$\Delta \underline{S} = \Delta \underline{S}_A + \Delta \underline{S}_B = -19.14 + 24.83 = 5.69 \text{ J/mol K}$$

Path B

i. ***Isobaric heating***

$$\Delta \underline{S}_A = C_P \ln \frac{T_2}{T_1} = 24.83 \text{ J/mol K}$$

ii. ***Isothermal compression***

$$\Delta \underline{S}_B = \frac{\text{Q}}{\text{T}} = -\frac{10972.2}{573.15} = -19.14 \text{ J/mol K}$$

$$\Delta \underline{S} = 24.83 - 19.14 = 5.69 \text{ J/mol K}$$

Path C

i. ***Compression with PV^γ = constant***

$$\Delta \underline{S}_A = \frac{Q}{T} = 0$$

ii. ***Isobaric heating***

$$\Delta \underline{S}_B = C_P \ln \frac{T_3}{T_2} = 38 \frac{\text{J}}{\text{mol K}} \times \ln \frac{573.15}{493.38} = 5.69 \text{ J/mol K}$$

$$\Delta \underline{S} = 5.69 \text{ J/mol K}$$

Comment

This example verifies, at least for the paths considered here, that the entropy is a state function. For reversible processes in closed systems, the rate of change of entropy and the ratio $\dot{Q}/T$ are equal. Thus, for reversible changes, $\dot{Q}/T$ is also a state function, even though the total heat flow $\dot{Q}$ is a path function. ■

ILLUSTRATION 3.5-5

In Section 3.1 we established that the entropy function will be a maximum at equilibrium in an isolated system. This point will be illustrated by example for the system shown here.

Figure 3.5-3 shows a well-insulated box of volume 3 m³ divided into two equal volumes. The left-hand cell is initially filled with air at 100°C and 2 bar, and the right-hand cell is initially evacuated. The valve connecting the two cells will be opened so that gas will slowly pass from cell 1 to cell 2. The wall connecting the two cells conducts heat sufficiently well that the temperature of the gas in both cells will always be the same. Plot on the same graph (1) the pressure in the second tank versus the pressure in the first tank, and (2) the change in the total entropy of the system versus the pressure in tank 1. At these temperatures and pressures, air can be considered to be an ideal gas.

Solution

For this system

$$\text{total mass} = N = N_1 + N_2$$

$$\text{total energy} = U = U_1 + U_2$$

$$\text{total entropy} = S = S_1 + S_2 = N_1\underline{S}_1 + N_2\underline{S}_2$$

From the ideal gas equation of state and the fact that $V = N\underline{V}$ we have

$$N_1^i = \frac{PV}{RT} = \frac{2 \text{ bar} \times 3 \text{ m}^3}{8.314 \times 10^{-5} \frac{\text{bar m}^3}{\text{mol K}} \times 373.15 \text{ K}} = 193.4 \text{ mol} = 0.1934 \text{ kmol}$$

Now since U = constant, $T_1 = T_2$, and for the ideal gas $\underline{U}$ is a function of temperature only, we conclude that $T_1 = T_2 = 100°\text{C}$ at all times. This result greatly simplifies the computation. Suppose that the pressure in cell 1 is decreased from 2 bar to 1.9 bar by transferring some gas from cell 1 to cell 2. Since the temperature in cell 1 is constant, we have, from the ideal gas law, that

$$N_1 = 0.95 N_1^i$$

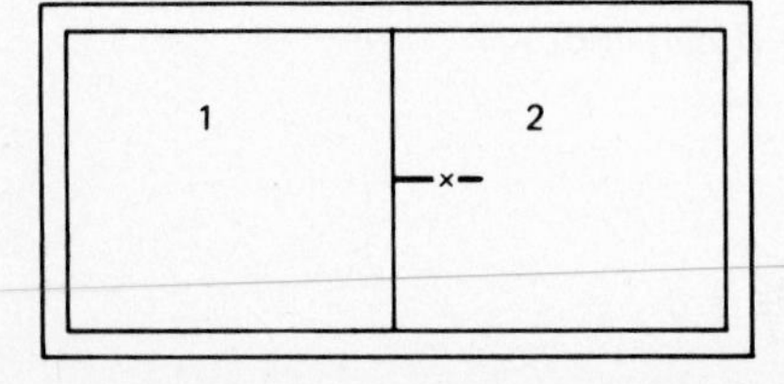

Figure 3.5-3
A well-insulated box divided into two equal compartments.

and by mass conservation that $N_2 = 0.05N_1^i$. Applying the ideal gas relation, we obtain $P_2 = 0.1$ bar.

For any element of gas, we have, from Eq. 3.4-3, that

$$\underline{S}^f - \underline{S}^i = C_P \ln \frac{T^f}{T^i} - R \ln \frac{P^f}{P^i} = -R \ln \frac{P^f}{P^i}$$

since temperature is constant. Therefore, to compute the change in entropy of the system, we visualize the process of transferring 0.05 N_1^i moles of gas from cell 1 to cell 2 as having two effects:

1. To decrease the pressure of the 0.95 N_1^i moles of gas remaining in cell 1 from 2 bar to 1.9 bar.
2. To decrease the pressure of the 0.05 N_1^i moles of gas that have been transferred from cell 1 to cell 2 from 2 bar to 0.1 bar.

Thus

$$S^f - S^i = -0.95\, N_1^i R \ln (1.9/2) - 0.05\, N_1^i R \ln (0.1/2)$$

or

$$\frac{\Delta S}{N_1^i R} = -0.95 \ln 0.95 - 0.05 \ln 0.05 = 0.199 \qquad \begin{cases} P_1 = 1.9 \text{ bar} \\ P_2 = 0.1 \text{ bar} \end{cases}$$

Similarly, if $P_1 = 1.8$ bar

$$\frac{\Delta S}{N_1^i R} = -0.9 \ln 0.9 - 0.1 \ln 0.1 = 0.325 \qquad \begin{cases} P_1 = 1.8 \text{ bar} \\ P_2 = 0.2 \text{ bar} \end{cases}$$

and so forth. The results are plotted in Fig. 3.5-4.

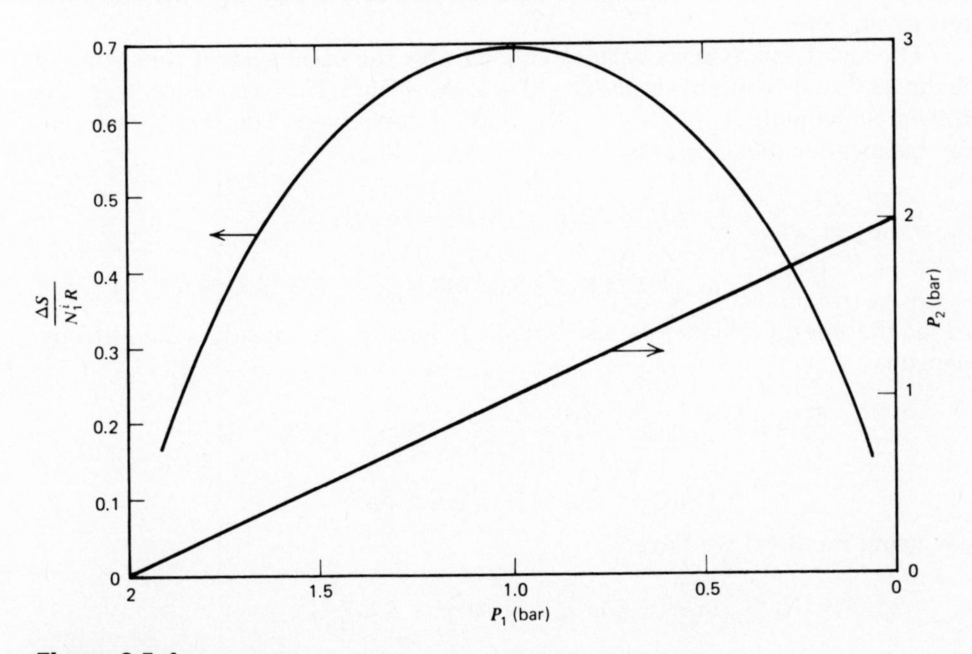

Figure 3.5-4
The system entropy change and the pressure in cell 2 as a function of the pressure in cell 1 (Illustration 3.5-5).

From this figure it is clear that ΔS, the change in entropy from the initial state, and therefore the total entropy of the system, reaches a maximum value when $P_1 = P_2 = 1$ bar. Consequently, the equilibrium state of the system under consideration is the state in which the pressure in both cells is the same, as one would expect. (Since the use of the entropy function leads to a solution that agrees with one's intuition, this example should reinforce confidence in the use of the entropy function as a criterion for equilibrium in an isolated constant-volume system.) ■

ILLUSTRATION 3.5-6

An engineer claims to have invented a steady flow device that will take air at 4 bar and 20°C and separate it into two streams of equal mass, one at 1 bar and −20°C and the second at 1 bar and 60°C. Furthermore, the inventor states that his device operates adiabatically and does not require (or produce) work. Is such a device possible? (Air can be assumed to be an ideal gas with a constant heat capacity of $C_P = 29.3$ J/mol K).

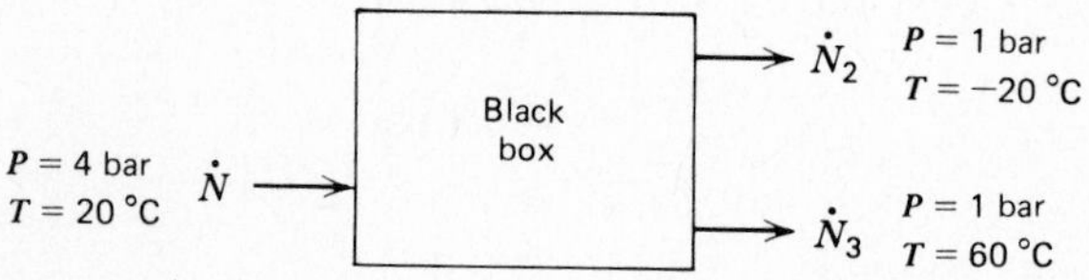

$P = 4$ bar, $T = 20$ °C, $\dot{N}$ → Black box → $\dot{N}_2$: $P = 1$ bar, $T = -20$ °C; $\dot{N}_3$: $P = 1$ bar, $T = 60$ °C

Solution

The three principles of thermodynamics, (1) conservation of mass, (2) conservation of energy, and (3) $\dot{S}_{\text{gen}} \geqslant 0$, must be satisfied for this or any other device. These principles can be used to test whether any device can meet the specifications given here.

The steady-state mass balance equation for the open system consisting of the device and its contents is $dN/dt = 0 = \Sigma_k \dot{N}_k = \dot{N}_1 + \dot{N}_2 + \dot{N}_3$. Since, from the problem statement, $\dot{N}_2 = \dot{N}_3 = -\frac{1}{2}\dot{N}_1$, mass is conserved. The steady-state energy balance for this device is

$$\frac{dU}{dt} = 0 = \sum_i \dot{N}_k \underline{H}_k = \dot{N}_1 \underline{H}_1 - \tfrac{1}{2}\dot{N}_1 \underline{H}_2 - \tfrac{1}{2}\dot{N}_1 \underline{H}_3$$

$$= \dot{N}_1 C_P (293.15 \text{ K} - \tfrac{1}{2} \times 253.15 \text{ K} - \tfrac{1}{2} \times 333.15 \text{ K}) = 0$$

so that the energy balance is also satisfied. Finally, the steady-state entropy balance is

$$\frac{dS}{dt} = 0 = \sum_k \dot{N}_k \underline{S}_k + \dot{S}_{\text{gen}} = \dot{N}_1 \underline{S}_1 - \tfrac{1}{2}\dot{N}_1 \underline{S}_2 - \tfrac{1}{2}\dot{N}_1 \underline{S}_3 + \dot{S}_{\text{gen}}$$

$$= \tfrac{1}{2}\dot{N}_1[(\underline{S}_1 - \underline{S}_2) + (\underline{S}_1 - \underline{S}_3)] + \dot{S}_{\text{gen}}$$

Now using Eq. 3.4-3 we have

$$\dot{S}_{\text{gen}} = -\tfrac{1}{2}\dot{N}_1 \left(C_P \ln \frac{T_1}{T_2} - R \ln \frac{P_1}{P_2} + C_P \ln \frac{T_1}{T_3} - R \ln \frac{P_1}{P_3} \right)$$

$$= -\tfrac{1}{2}\dot{N}_1 \left(29.3 \ln \frac{293.15 \times 293.15}{253.15 \times 333.15} - 8.314 \ln \frac{4 \times 4}{1 \times 1} \right) = 11.25\, \dot{N}_1 \frac{\text{J}}{\text{K s}}$$

Therefore we conclude, on the basis of thermodynamics, that it *is* possible to construct a device with the specifications claimed by the inventor. Thermodynamics, however, gives us no insight into how to design such a device.

Two possible devices are indicated in Fig. 3.5-5. The first device consists of an air-driven turbine that extracts work from the flowing gas. This work is then used to drive a heat pump (an air conditioner or refrigerator) to cool part of the

(*a*) Turbine-heat pump system

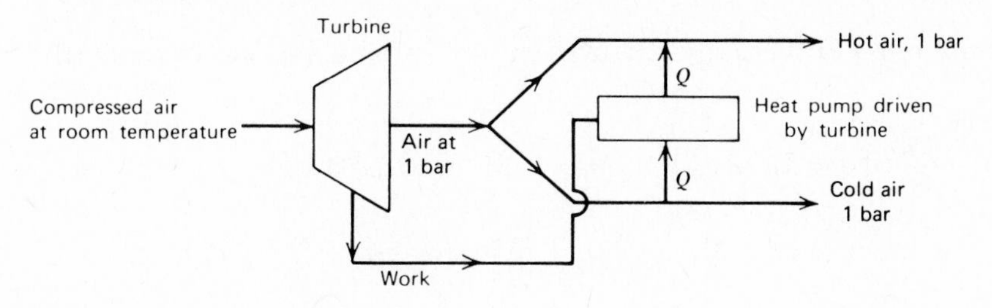

(*b*) Hilsch-Ranque vortex tube

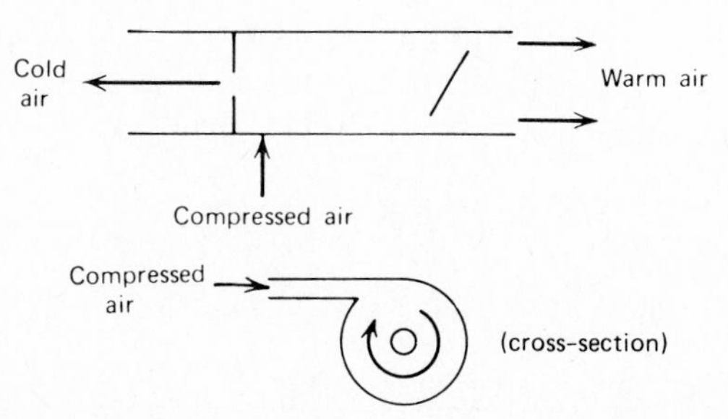

Figure 3.5-5
Two devices to separate compressed air into two low pressure air streams of different temperature.

gas and heat the rest. The second device, the Hilsch–Ranque vortex tube, is somewhat more interesting in that it accomplishes the desired change of state with only a butterfly valve. In this device the air expands as it enters the tube, thus gaining kinetic energy at the expense of internal energy (i.e., at the end of the expansion process we have high velocity air of both lower pressure and temperature than the incoming air). Some of this cooled air is withdrawn from the center of the vortex tube. The rest of the air swirls down the tube, where, as a result of viscous dissipation, the kinetic energy is dissipated into heat, which increases the internal energy (temperature) of the air. Thus, the air near the butterfly valve is warmer than the incoming air. ■

ILLUSTRATION 3.5-7

A steam turbine operates at the following conditions:

	Inlet		Outlet
Velocity (m/min)	2000		7500
T (K)	800		440
P (MPa)	3.5		0.15
Flow rate (kg/h)		10,000	
Heat loss (kJ/h)		125,000	

a. Compute the horsepower developed by the turbine, and the entropy change of the steam.

b. Suppose the turbine is now replaced with one that is well insulated, so that the heat loss is eliminated, and well designed, so that the expansion is reversible. If the exit pressure and velocity are maintained at the previous values, what is the outlet steam temperature and the horsepower developed by the turbine?

Solution

The steady-state mass and energy balances on the turbine and its contents (the system) yield

$$\frac{dM}{dt} = 0 = \dot{M}_1 + \dot{M}_2; \qquad \dot{M}_2 = -\dot{M}_1 = -10{,}000 \text{ kg/h}$$

$$\frac{d}{dt}\left[U + M\left(\frac{v^2}{2} + gh\right)\right] = 0 = \dot{M}_1\left(\hat{H}_1 + \frac{v_1^2}{2}\right) + \dot{M}_2\left(\hat{H}_2 + \frac{v_2^2}{2}\right) + \dot{W}_s + \dot{Q}$$

a. From the Mollier diagram of Fig. 2.4-1 (or Appendix III)

$$\hat{H}_1 \simeq 3510 \text{ J/g}$$

and

$$\hat{S}_1 \simeq 7.23 \text{ J/g K}$$

Also

$$\frac{v_1^2}{2} = \frac{1}{2}\left(\frac{2000 \text{ m/min}}{60 \text{ s/min}}\right)^2 \times \frac{1 \text{ J/kg}}{\text{m}^2/\text{s}^2} \times \frac{1 \text{ kg}}{1000 \text{ g}} = 0.56 \text{ J/g}$$

so that

$$\hat{H}_1 + \frac{v_1^2}{2} = 3510.6 \text{ J/g}$$

Similarly

$$\hat{H}_2 \simeq 2805 \text{ J/g}; \qquad \hat{H}_2 + \frac{v_2^2}{2} = 2805 + 7.8 = 2812.8 \text{ J/g}$$

and

$$\hat{S}_2 \simeq 7.50 \text{ J/g K}$$

Therefore

$$-\dot{W}_S = (3510.6 - 2812.8)\frac{\text{J}}{\text{g}} \times 10{,}000\frac{\text{kg}}{\text{h}} \times 1000\frac{\text{g}}{\text{kg}} - 12.5 \times 10^4 \frac{\text{kJ}}{\text{h}} \times 1000\frac{\text{J}}{\text{kJ}}$$

$$= 6.853 \times 10^9 \frac{\text{J}}{\text{h}} \times \frac{1 \text{ kJ}}{1000 \text{ J}} \times \frac{1 \text{ hr}}{3600 \text{ s}} = 1903.6 \text{ kJ/s} = 1.9036 \times 10^6 \text{ W}$$

$$= 2553 \text{ hp}$$

also

$$\Delta\hat{S} = (\hat{S}_2 - \hat{S}_1) = 0.27 \text{ J/g K}$$

b. The steady-state entropy balance for the turbine and its contents is

$$\frac{dS}{dt} = 0 = \dot{M}_1 \hat{S}_1 - \dot{M}_1 \hat{S}_2 + \dot{S}_{gen}$$

since $\dot{Q} = 0$, and $\dot{M}_2 = -\dot{M}_1$. Also, the turbine operates reversibly so that $\dot{S}_{gen} = 0$, and $\hat{S}_1 = \hat{S}_2$, that is, the expansion is isentropic. We now use Fig. 2.4-1, the entropy–enthalpy plot (Mollier diagram) for steam to solve this problem. In particular, we locate the initial steam conditions ($T = 800$ K, $P = 3.5$ MPa) on the chart, and follow a line of constant entropy (a vertical line on the Mollier diagram) to the exit pressure (0.15 MPa), to obtain the enthalpy of the exiting steam ($\hat{H}_2 \approx 2690$ J/g) and its final temperature ($T \approx 373$ K). Since the exit velocity is known, we can immediately compute the horsepower generated by the turbine

$$-\dot{W}_S = [(3510 + 0.6) - (2690 + 7.8)] \frac{\text{J}}{\text{g}} \times 10000 \frac{\text{kg}}{\text{h}} \times \frac{1000 \text{ g/kg}}{1000 \text{ J/kJ}} \times \frac{1 \text{ h}}{3600 \text{ s}}$$

$$= 2257.8 \text{ kJ/s} = 2.2578 \times 10^6 \ W = 3028 \text{ hp}$$

Comments

1. Here, as before, the kinetic energy term is of negligible importance compared to the internal energy term.
2. Notice, from the Mollier diagram, that the turbine exit steam is right at the boundary of a two-phase mixture of vapor and liquid. For the solution of this problem, no difficulties arise if the exit steam is a vapor, a liquid, or a two-phase vapor–liquid mixture since our mass, energy and entropy balances are of general applicability. In particular, the information required to use these balance equations are the internal energy, enthalpy, and entropy per unit mass of each of the flow streams. Provided we have this information, the balance equations can be used independent of whether the flow streams consist of single or multiple phases, or, in fact, single or multiple components (see Chapter 6). Here the Mollier diagram provides the necessary thermodynamic information, and the solution of this problem is straightforward. ■

ILLUSTRATION 3.5-8

1. By considering the gas contained within the piston and cylinder device of Illustration 2.5-7 to be the system, show that the gas undergoes a reversible expansion in each of the four processes considered in that illustration. That is, show that $S_{gen} = 0$ for each process.
2. By considering the gas and the piston and cylinder to be the system, show that processes (a), (b) and (c) of Illustration 2.5-7 are not reversible (i.e., $S_{gen} > 0$), and that process (d) is reversible.

Solution

1. The entropy balance for the 1 mol of gas contained in the piston and cylinder is

$$\underline{S}_f - \underline{S}_i = \frac{Q}{T} + S_{gen}$$

where T is the constant temperature of this isothermal system and Q is the total heat flow (from both the thermostatic bath and the cylinder walls) to the gas.

From Eq. g of Illustration 2.5-7 we have, for the 1 mol of gas, that

$$Q = RT \ln \frac{\underline{V}_f}{\underline{V}_i}$$

and from Eq. 3.4-2 we have

$$\underline{S}_f - \underline{S}_i = R \ln \frac{\underline{V}_f}{\underline{V}_i}$$

since the temperature of the gas is constant. Thus

$$S_{\text{gen}} = \underline{S}_f - \underline{S}_i - \frac{Q}{T} = R \ln \frac{\underline{V}_f}{\underline{V}_i} - \frac{1}{T}\left\{RT \ln \frac{\underline{V}_f}{\underline{V}_i}\right\} = 0$$

so that the gas undergoes a reversible expansion in all four processes.

2. The entropy balance for the isothermal system consisting of 1 mol of gas and the piston and cylinder is

$$S_f - S_i = \frac{Q}{T} + S_{\text{gen}}$$

where Q is the heat flow to the piston, cylinder, and gas, Q^{NET} of Illustration 2.5-7, and $S_f - S_i$ is the entropy change for that composite system,

$$S_f - S_i = (\underline{S}_f - \underline{S}_i)_{\text{gas}} + (S_f - S_i)_{\text{piston-cylinder}}$$

Since the system is isothermal

$$(\underline{S}_f - \underline{S}_i)_{\text{gas}} = R \ln \frac{\underline{V}_f}{\underline{V}_i} \qquad \text{(see Eq. 3.4-2)}$$

and

$$(S_f - S_i)_{\text{piston-cylinder}} = 0 \qquad \text{(see Eq. 3.4-6)}$$

Consequently

$$S_{\text{gen}} = R \ln \frac{\underline{V}_f}{\underline{V}_i} - \frac{Q}{T} = \frac{1622.5 - Q^{\text{NET}}}{298.15} \text{ J/K}$$

so we find, using the entries in Table 1 of Illustration 2.5-7, that

$$S_{\text{gen}} = \begin{cases} 1.4473 \text{ J/K} & \text{for process (a)} \\ 0.8177 \text{ J/K} & \text{for process (b)} \\ 0.4343 \text{ J/K} & \text{for process (c)} \\ 0 & \text{for process (d)} \end{cases}$$

Thus, we conclude that for the piston, cylinder, and gas system, processes (a), (b), and (c) are not reversible, whereas process (d) is reversible.

Comment

From the results of part 1 we find that for the gas all expansion processes are reversible (i.e., there are no dissipative mechanisms in the gas). However, from part 2, we see that when the piston, cylinder, and gas are taken to be the system, the expansion process is irreversible unless the expansion occurs in differential steps. The conclusion, then, is that the irreversibility, or the dissipation of mechanical energy to thermal energy, occurs in the piston and cylinder. This is, of course, obvious from the fact that the only source of dissipation in this problem is the friction between the piston and the cylinder wall. ■

3.6 THE MICROSCOPIC ENTROPY BALANCE (OPTIONAL)

As indicated in Sec. 3.1, in order to get general information about the rate of internal generation of entropy, one has to use a detailed, rather than black-box, description of thermodynamic systems. Based on the simple analysis of Sec. 3.1, we expect $\dot{S}_{gen}$ to be proportional to the square of the gradients of temperature and velocity, a result that is established here using the differential element description introduced in Sec. 2.7.

We start by writing each of the contributions to the entropy balance for the differential volume element of Fig. 2.7-1

$$\begin{pmatrix}\text{rate of change of}\\ \text{entropy in the}\\ \text{volume element}\end{pmatrix} = \Delta x \, \Delta y \, \Delta z \frac{\partial}{\partial t}(\rho \hat{S})$$

$$\begin{pmatrix}\text{net rate at which}\\ \text{entropy of the volume}\\ \text{element increases due to}\\ \text{mass flow across the}\\ \text{system boundaries}\end{pmatrix} = \begin{aligned}&\rho v_x \hat{S} \, \Delta y \, \Delta z|_x - \rho v_x \hat{S} \, \Delta y \, \Delta z|_{x+\Delta x}\\ &+ \rho v_y \hat{S} \, \Delta x \, \Delta z|_y - \rho v_y \hat{S} \, \Delta x \, \Delta z|_{y+\Delta y}\\ &+ \rho v_z \hat{S} \, \Delta x \, \Delta y|_z - \rho v_z \hat{S} \, \Delta x \, \Delta y|_{z+\Delta z}\end{aligned}$$

$$\begin{pmatrix}\text{net rate at which entropy}\\ \text{of the volume element}\\ \text{increases by conduction}\\ \text{across the system}\\ \text{boundaries}\end{pmatrix} = \begin{aligned}&\frac{1}{T} q_x \, \Delta y \, \Delta z|_x - \frac{1}{T} q_x \, \Delta y \, \Delta z|_{x+\Delta x}\\ &+ \frac{1}{T} q_y \, \Delta x \, \Delta z|_y - \frac{1}{T} q_y \, \Delta x \, \Delta z|_{y+\Delta y}\\ &+ \frac{1}{T} q_z \, \Delta x \, \Delta y|_z - \frac{1}{T} q_z \, \Delta x \, \Delta y|_{z+\Delta z}\end{aligned}$$

$$\begin{pmatrix}\text{rate of production}\\ \text{of entropy in the}\\ \text{volume element}\end{pmatrix} = \dot{\sigma}_S \, \Delta x \, \Delta y \, \Delta z$$

Here $\dot{\sigma}_S$ is the rate of internal generation of entropy per unit volume ($\dot{S}_{gen} = \int \dot{\sigma}_S \, dV$). Now putting all these terms into Eq. 2.1-4, dividing by $\Delta x \, \Delta y \, \Delta z$, taking the limit as each of these goes to zero, and using the definition of derivative given in Eq. 2.7-3 yields

$$\frac{\partial}{\partial t}(\rho \hat{S}) = -\nabla \cdot (\rho \mathbf{v} \hat{S}) - \nabla \cdot \frac{\mathbf{q}}{T} + \dot{\sigma}_S \tag{3.6-1}$$

This equation can be rewritten, using Eqs. 2.7-5 and 6, as

$$\rho \left(T \frac{D\hat{S}}{Dt} \right) = -T\nabla \cdot \frac{\mathbf{q}}{T} + T\dot{\sigma}_S \tag{3.6-2}$$

To establish that the entropy function has the properties suggested in Sec. 3.1, Eq. 3.6-2 is subtracted from the mechanical energy equation developed in the last chapter (Eq. 2.7-19)

$$\rho \frac{D\hat{U}}{Dt} = -\nabla \cdot \mathbf{q} - P\nabla \cdot \mathbf{v} - \boldsymbol{\tau} : \nabla \mathbf{v}$$

to obtain

$$\rho \left(\frac{D\hat{U}}{Dt} - T \frac{D\hat{S}}{Dt} \right) = -P\nabla \cdot \mathbf{v} - \frac{\mathbf{q}}{T} \cdot \nabla T - \boldsymbol{\tau} : \nabla \mathbf{v} - T\dot{\sigma}_S \tag{3.6-3}$$

Finally, rewriting the continuity equation

$$\nabla \cdot \mathbf{v} = -\frac{1}{\rho}\frac{D\rho}{Dt} = +\rho\frac{D\hat{V}}{Dt} \tag{3.6-4}$$

and using this result in Eq. 3.6-3 yields

$$\rho\left(\frac{D\hat{U}}{Dt} + P\frac{D\hat{V}}{Dt} - T\frac{D\hat{S}}{Dt}\right) = -\frac{\mathbf{q}}{T}\cdot\nabla T - \boldsymbol{\tau} : \nabla\mathbf{v} - T\dot{\sigma}_S \tag{3.6-5}$$

Now, using Fourier's law of heat conduction

$$\mathbf{q} = -\lambda\nabla T \qquad (\text{where } \lambda > 0)$$

and assuming that the fluid is Newtonian so that the stress tensor is given by Eq. 2.7-8 with $\mu > 0$ we have

$$\rho\left(\frac{D\hat{U}}{Dt} + P\frac{D\hat{V}}{Dt} - T\frac{D\hat{S}}{Dt}\right) = +\left(\frac{\lambda}{T}\right)(\nabla T)^2 + \mu\phi^2 - T\dot{\sigma}_S \tag{3.6-6}$$

where

$$\phi^2 = \sum_i\sum_j\left[\left(\frac{\partial v_i}{\partial j}\right) + \left(\frac{\partial v_j}{\partial i}\right) - \frac{2}{3}\nabla\cdot\mathbf{v}\delta_{ij}\right]^2$$

Here i and j are the coordinate directions x, y, and z, and δ_{ij} equals 1 if $i = j$, and is zero otherwise. (ϕ^2 has been used to emphasize that this term is a sum of squared terms.) On the basis of the simple example in Sec. 3.1, and the observation that $\dot{S}_{gen}$ and thus $\dot{\sigma}_S$ are related only to the gradients in the system (since $\dot{\sigma}_S$ must vanish in the uniform equilibrium state), the state variable and gradient terms can be separately equated Eq. 3.6-6 to yield

$$\frac{D\hat{S}}{Dt} = \frac{1}{T}\left(\frac{D\hat{U}}{Dt} + P\frac{D\hat{V}}{Dt}\right) \tag{3.6-7}$$

and

$$\dot{\sigma}_S = (\lambda/T^2)(\nabla T)^2 + (\mu T)\phi^2 \tag{3.6-8}$$

The first of these equations indicates that the entropy is a state function since its rate of change is related only to other state functions and their rates of change. The second equation indicates that $\dot{\sigma}_S$ (and hence $\dot{S}_{gen}$) is proportional to the squares of the gradients in the system and therefore is a positive definite quantity.

One can establish that more complicated constitutive relations for the heat flux and stress tensor will also lead to $\dot{\sigma}_S \geqslant 0$. In multicomponent systems chemical reactions and species diffusion can occur, as well as heat flows and viscous dissipation. These additional processes also lead to positive contributions to the entropy generation term.

PROBLEMS

3.1 A 5-kg copper ball at 75°C is dropped into 12 kg of water, initially at 5°C, in a well-insulated container.

a Find the common temperature of the water and copper ball after the passage of a long period of time.

b What is the entropy change of the water in going from its initial to final state? Of the ball? Of the composite system of water and ball?

Data C_P(copper) = 0.5 J/g K

C_P(water) = 4.2 J/g K

3.2 **a** Show that the rate at which shaft work is obtained or required for a reversible change of state in a closed system at constant internal energy and volume is equal to the negative of the product of the temperature and the rate of change of the entropy for the system.

b Show that the rate at which shaft work is obtained or required for a reversible change of state in a closed system at constant entropy and pressure is equal to the rate of change of enthalpy of the system.

3.3 Steam at 700 bar and 600°C is withdrawn from a steam line and adiabatically expanded to 10 bar at the rate of 2 kg/min. What is the temperature of the steam that was expanded, and what is the rate of entropy generation in this process?

3.4 Two metal blocks of equal mass M of the same substance, one at an initial temperature T_1^i and the other at an initial temperature of T_2^i are placed in a well-insulated (adiabatic) box of constant volume. A device that can produce work from a flow of heat across a temperature difference (i.e., a heat engine) is connected between the two blocks. Develop expressions for the maximum amount of work that can be obtained from this process and the common final temperature of the blocks when this amount of work is obtained. You may assume that the heat capacity of the blocks do not vary with temperature.

3.5 The compressor discussed in Illustrations 2.5-4 and 3.5-1 is being used to compress air from 1 bar and 290 K to 10 bar. The compression can be assumed to be adiabatic, and the compressed air is found to have an outlet temperature of 575 K.

a What is the value of ΔS for this process?

b How much work, W_s, is needed per mole of air for the compression?

c The temperature of the air leaving the compressor here is higher than in Illustration 3.5-1. How do you account for this?

In your calculations you may assume air is an ideal gas with C_P = 29.3 J/mol K.

3.6 **a** A steam turbine in a small electric power plant is designed to accept 4500 kg/h of steam at 60 bar and 500°C and exhaust the steam at 10 bar. Assuming that the turbine is adiabatic and has been well designed (so that $\dot{S}_{gen} = 0$), compute the exit temperature of the steam and the power generated by the turbine.

b The efficiency of a turbine is defined to be the ratio of the work actually obtained from the turbine to the work that would be obtained if the turbine operated isentropically between the same inlet and exit pressures. If the turbine in part a is adiabatic but only 80% efficient, what would be the exit temperature of the steam? At what rate would entropy be generated within the turbine?

c In off-peak hours the power output of the turbine in part a (100% efficient) is decreased by adjusting a throttling valve that reduces the turbine inlet steam pressure to 30 bar (see the diagram). Compute T_1, the steam temperature to the turbine, T_2, the steam temperature at the turbine exit, and the power output of the turbine.

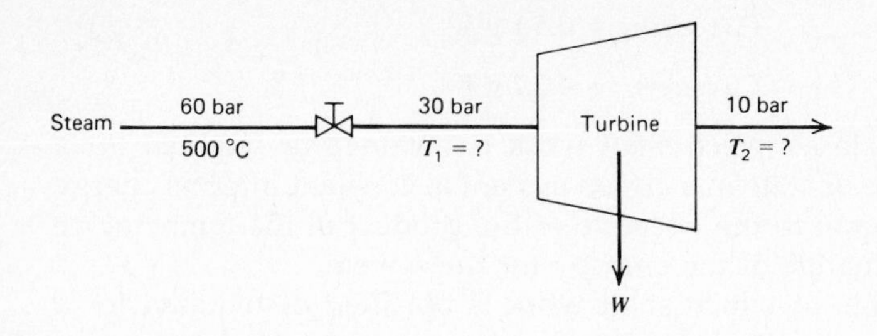

3.7 Complete part b of Problem 2.25, assuming the compressor operates reversibly and adiabatically.

3.8 Steam is produced at 70 bar and some unknown temperature. A small amount of steam is bled off just before entering a turbine and goes through an adiabatic throttling valve to atmospheric pressure. The temperature of the steam exiting the throttling valve is 400°C. The unthrottled steam is fed into the turbine, where it is adiabatically expanded to atmospheric pressure.

a What is the temperature of the steam entering the turbine?

b What is the maximum work per kilogram of steam that can be obtained using the turbine in its present mode of operation?

c Tests on the turbine exhaust indicate that the steam leaving is a saturated vapor. What is the efficiency of the turbine and the entropy generated per kilogram of steam?

d If the ambient temperature is 25°C and the ambient pressure is 1 bar, what is the maximum possible work that could be obtained per kilogram of steam in any continuous process?

3.9 A well-insulated, 0.7-m^3 gas cylinder containing natural gas (which you can consider to be pure methane) at 70 bar and 300 K is exhausted until the pressure drops to 3.5 bar. This process occurs fast enough that there is no heat transfer between the cylinder walls and the gas, but not so rapidly as to produce large velocity or temperature gradients in the gas within the cylinder. Compute the number of moles of gas withdrawn and the final temperature of the gas in the cylinder, if

a Methane gas is assumed to be ideal with $C_P = 36$ J/mol K.

b Methane is considered to be a real gas with properties given in Figs. 2.4-2.

3.10 If the heat capacity of an ideal gas is given by

$$C_V = (a - R) + bT + cT^2 + dT^3 + e/T^2$$

show that

$$\underline{S}(T_2, \underline{V}_2) - \underline{S}(T_1, \underline{V}_1) = (a - R)\ln\left(\frac{T_2}{T_1}\right) + b(T_2 - T_1) + \frac{c}{2}(T_2^2 - T_1^2) + \frac{d}{3}(T_2^3 - T_1^3) - \frac{e}{2}(T_2^{-2} - T_1^{-2}) + R\ln\left(\frac{\underline{V}_2}{\underline{V}_1}\right)$$

Also develop expressions for this fluid that replace Eqs. 3.4-3 and 3.4-4.

3.11 **a** Steam at 35 bar and 600 K enters a throttling valve that reduces the steam pressure to 7 bar. Assuming there is no heat loss from the valve, what is the exit temperature of the steam and its change in entropy?

b If air (assumed to be an ideal gas with $C_P = 29.3$ J/mol K) entered the valve at 35 bar and 600 K and left at 7 bar, what would be its exit temperature and entropy change?

3.12 In a large refrigeration plant it is necessary to compress a fluid, which we will assume to be an ideal gas with constant heat capacity, from a low pressure P_1 to a much higher pressure P_2.

a If the compression is done in a single compressor that operates reversibly and adiabatically, obtain an expression for the work needed for the compression in terms of the mass flow rate, P_1, P_2, and the initial temperature, T_1.

b If the compression is to be done in two stages, first compressing the gas from P_1 to P^*, then cooling the gas at constant pressure down to the compressor inlet temperature T_1, and then compressing the gas to P_2, develop an expression for the work needed for the compression. What should the value of the intermediate pressure be to accomplish the compression with minimum work?

3.13 Recompute Problem 2.18, now considering nitrogen to be a real gas with thermodynamic properties given in Fig. 2.4-3.

3.14 An isolated chamber with rigid walls is divided into two equal compartments, one containing steam at 10 bar and 370°C, and the other evacuated. A valve between the compartments is opened to permit steam to pass from one chamber to the other.

a After the pressures (but not the temperatures) in the two chambers have equalized, the valve is closed, isolating the two systems. What is the temperature and pressure in each cylinder?

b If the valve were left open, an equilibrium state is obtained in which each chamber has the same temperature and pressure. What is this temperature and pressure?

(*Note*: Steam is not an ideal gas under the conditions here.)

3.15 An adiabatic turbine is operating with an ideal gas working fluid of fixed inlet temperature and pressure, T_1 and P_1, respectively, and a fixed exit pressure P_2. Show that

a The minimum outlet temperature T_2 occurs when the turbine operates reversibly, that is, when $S_{gen} = 0$.

b The maximum work that can be extracted from the turbine is obtained when $S_{gen} = 0$.

3.16 **a** Consider the following statement: "Although the entropy of a given system may increase, decrease, or remain constant, the entropy of the universe cannot decrease." Is this statement true? Why?

b Consider any two states, labeled 1 and 2. Show that if state 1 is accessible from state 2 by a real (irreversible) adiabatic process, then state 2 is inaccessible from state 1 by a real adiabatic process.

3.17 Though the Carnot cycle is the most efficient for converting heat to work, it is rarely used because of the large amount of work that must be supplied to a Carnot engine during the isothermal compression step. There are, however, other cyclic processes used in commercial power generation; several are indicated here. Assuming that the working fluid is an ideal gas with constant heat capacity, draw pressure–volume and temperature–entropy diagrams for each of these cycles, and develop expressions for their maximum thermodynamic efficiency.

I. Stirling Cycle
(a) Isothermal compression
(b) Constant-volume heating
(c) Isothermal expansion
(d) Constant-volume cooling

II. Ericsson Cycle
(a) Isothermal compression
(b) Isobaric heating
(c) Isothermal expansion
(d) Isobaric cooling

III. Brayton Cycle
(a) Isobaric compression
(b) Reversible adiabatic compression
(c) Isobaric expansion
(d) Reversible adiabatic expansion

3.18 Electrical power is to be produced from a steam turbine connected to a nuclear reactor. Steam is obtained from the reactor at 540 K and 36 bar, the turbine exit pressure is 1.0 bar, and the turbine is adiabatic.

a Compute the maximum work per kilogram of steam that can be obtained from the turbine.

A clever (?) chemical engineer has suggested that the single-stage turbine considered here be replaced by a two-stage adiabatic turbine, and that the steam exiting from the first stage be returned to the reactor and reheated, at constant pressure, to 540 K, and then fed to the second stage of the turbine. (See the figure.)

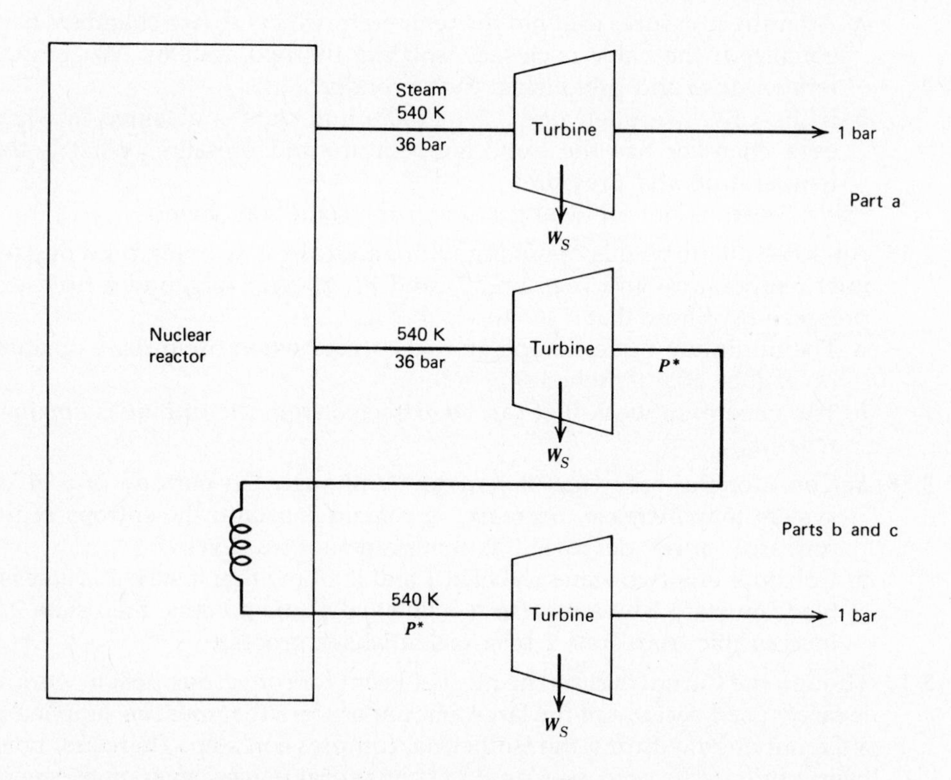

b Compute the maximum work obtained per kilogram of steam if the two-stage turbine is used and the exhaust pressure of the first stage is $P^* = \frac{1}{2}(36 + 1.0) = 18.5$ bar.

c Compute the maximum work obtained per kilogram of steam if the two-stage turbine is used and the exhaust pressure of the first stage is $P^* = \sqrt{36 \times 1} = 6.0$ bar.

3.19 It is necessary to estimate how rapidly a piece of equipment can be evacuated. The equipment, which is 0.7 m^3 in volume, initially contains carbon dioxide at 340 K and 1 bar pressure. The equipment will be evacuated by connecting it to a reciprocating constant displacement vacuum pump that will pump out 0.14 m^3/min of gas at any conditions. At the conditions here carbon dioxide can be considered to be an ideal gas with $C_P = 39$ J/mol K.

a What will be the temperature and pressure of the carbon dioxide inside the tank after 5 minutes of pumping if there is no exchange of heat between the gas and the process equipment?

b The gas exiting the pump is always at 1 bar pressure, and the pump operates in a reversible adiabatic manner. Compute the temperature of the gas exiting the pump after 5 minutes of operation.

3.20 A 0.2 m^3 tank containing helium at 15 bar and 22°C will be used to supply 4.5 moles per minute of helium at atmospheric pressure using a controlled adiabatic throttling valve.

a If the tank is well insulated, what will be the pressure in the tank and the temperature of the gas stream leaving the throttling valve at any later time t?

b If the tank is isothermal, what will be the pressure in the tank as a function of time?

You may assume helium to be an ideal gas with $C_P = 22$ J/mol K, and that there is no heat transfer between the tank and the gas.

3.21 Here is a schematic diagram of a vapor-compression refrigeration cycle or Rankine cycle used in household and commercial refrigerators.

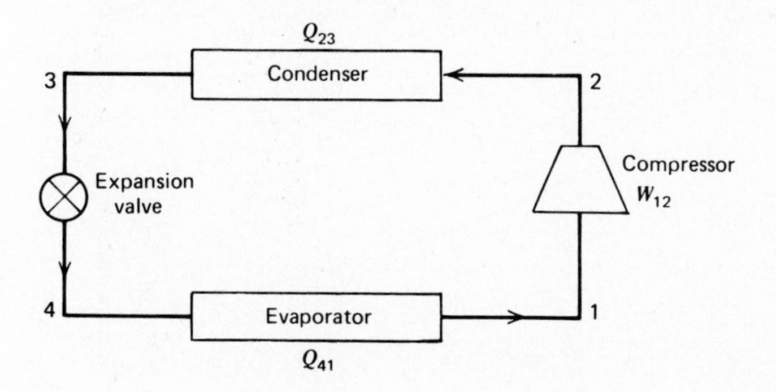

Between points 1 and 2 the vapor is compressed, the compressed vapor is then cooled and condensed to a liquid between points 2 and 3, and between points 3 and 4 the liquid undergoes an expansion and vaporization to form a cool gas–liquid mixture. Between points 4 and 1 the gas–liquid mixture absorbs heat and the remaining liquid vaporizes. In the home refrigerator the condenser is the air-cooled coil usually at the back of the refrigerator, and the evaporator is the coil in the freezer section.

The compressor will be assumed to operate adiabatically and reversibly, we will assume that there are no pressure changes in the system other than across the compressor and the expansion valve, and the refrigeration fluid is freon-12 (see the chart for its thermodynamic properties[11]). The following data are available for the process

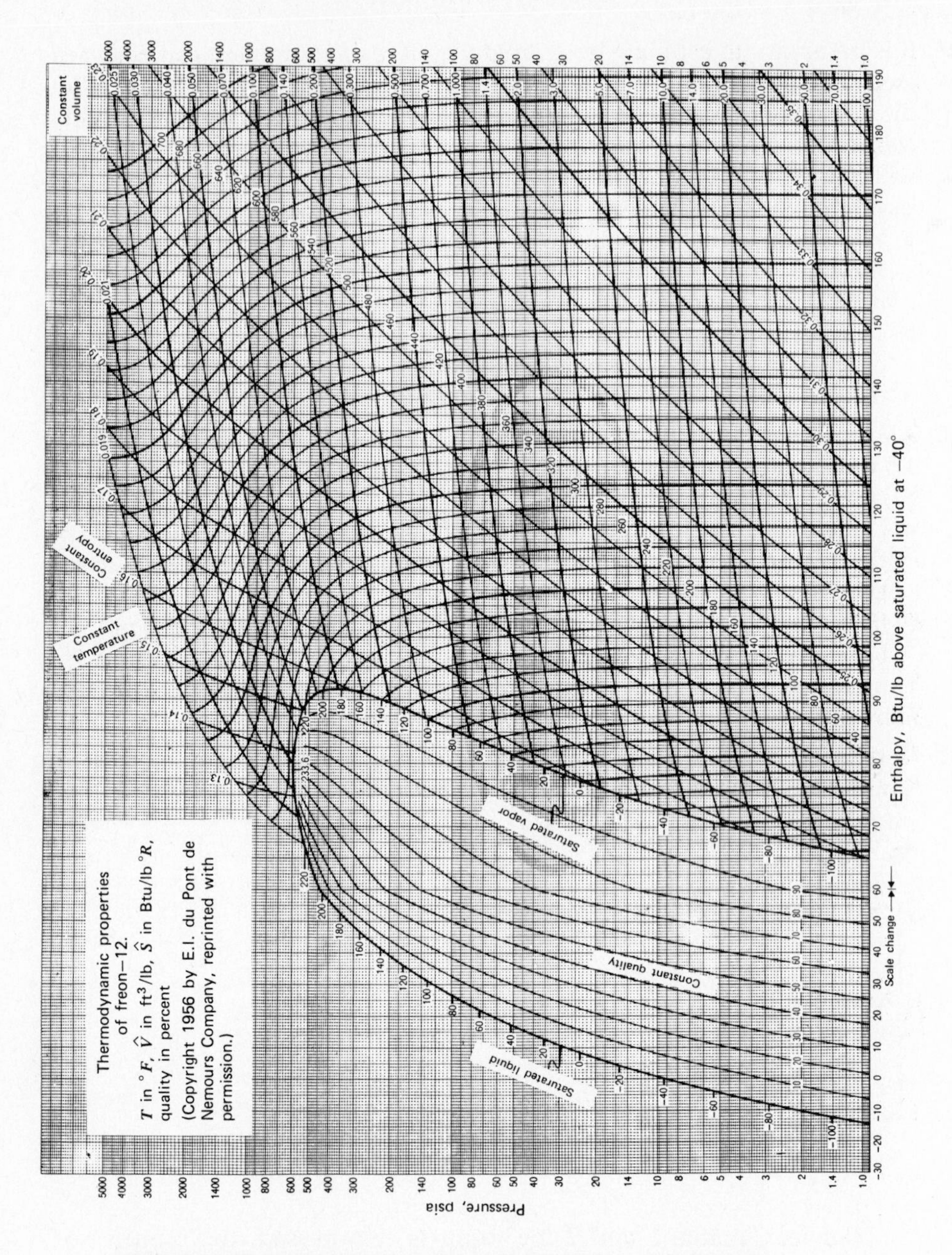

Location	Fluid State	Temperature	Pressure
1	Saturated vapor	0°F	
2	Vapor		
3	Saturated liquid	120°F	
4	Vapor–liquid mixture		

[11]Since the thermodynamic properties chart for this fluid is only available in British units, these units will be used for Problems 3.21 and 22.

a Supply the missing temperatures and pressures in the table.

b The coefficient of performance (C.O.P.) for a refrigeration system is defined to be the ratio of the heat absorbed by the evaporator to the work supplied to the compressor,

$$\text{C.O.P.} = \frac{Q_{41}}{W_{12}}$$

Evaluate the coefficient of performance for the refrigeration cycle described in this problem.

c One ton of refrigeration capacity corresponds to a heat removal rate of 12,000 BTU/h (roughly equivalent to the amount of energy that must be removed to freeze one ton of ice in a day). If the compressor in the cycle were driven by a 1 horsepower motor, what would be the tonnage rating of the refrigeration unit?

3.22 A Carnot or other engine can be operated in reverse so that mechanical energy (work) is used to pump heat from a low-temperature source to a high-temperature sink. Refrigerators and air conditioners are examples of heat pumps. Heat pump devices are now being installed in residential housing, with appropriate valving, and used for both winter heating (by pumping heat to the house from its surroundings) and summer cooling (by pumping heat from the house to the surroundings). The surroundings may be either the atmosphere, a lake, or, by using underground coils, the earth.

The coefficient of performance for a heat pump is defined to be the ratio of the heat flow from the low-temperature source to the work required to produce that flow. (The heat supplied to the high-temperature sink is the sum of the heat and work flows.)

Consider a heat pump that uses a lake as a heat source in the winter and as a heat sink in the summer. The house is to be maintained at a winter temperature of 65°F and a summer temperature of 78°F. To do this efficiently, it is found that the indoor coil temperature should be 100°F in the winter, 40°F in the summer. The outdoor coil temperature can be assumed to be 40°F during the winter months and 65°F during the summer.

a Compute the coefficients of performance for a Carnot cycle heat pump for winter and for summer operation.

b Instead of a Carnot cycle, the vapor compression (or Rankine) cycle shown in the figure will be used in the heat pump with freon-12 as the working fluid (see previous problem for thermodynamic data on freon-12).

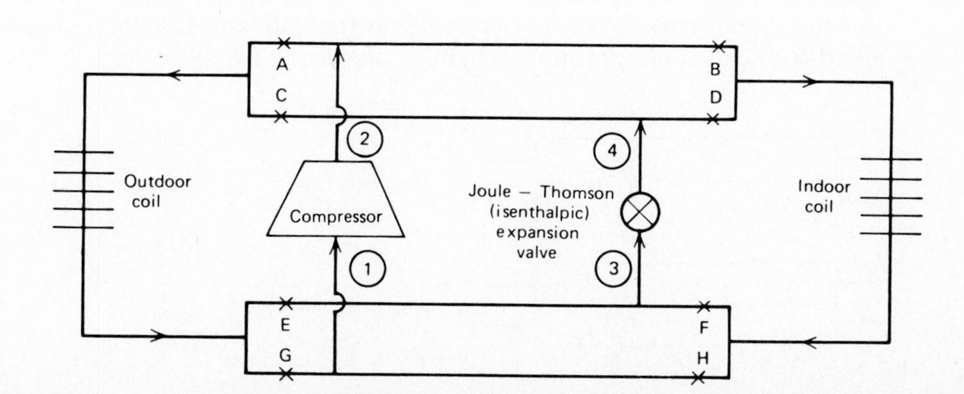

Compute the winter and summer coefficients of performance for this heat pump. You can assume that the only pressure changes in the cycle occur across the compressor and the expansion valve, and that the only heat transfer to or from the gas occurs in the indoor and outdoor coils.

Location	Heating Temperature	Cooling Temperature	Fluid State
1	40°F	40°F	Saturated vapor
2			Vapor
3	100°F	65°F	Saturated liquid
4			Vapor–liquid mixture

Valve	Heating Position	Cooling Position
A	Closed	Open
B	Open	Closed
C	Open	Closed
D	Closed	Open
E	Closed	Open
F	Open	Closed
G	Open	Closed
H	Closed	Open

3.23 A portable engine of nineteenth-century design used a tank of compressed air and an "evacuated" tank as its power source. The first tank had a capacity of 0.3 m^3 and was initially filled with air at 14 bar and a temperature of 700°C. The "evacuated" tank had a capacity of 0.75 m^3. Unfortunately, nineteenth-century vacuum techniques were not very efficient, so that the "evacuated" tank contained air at 0.35 bar and 25°C. What is the maximum total work that could be obtained from an air-driven engine connected between the two tanks if the process is adiabatic? What would be the temperature and pressure in each tank at the end of the process? You may assume that air is an ideal gas with $C_P = 7R/2$.

3.24 A heat exchanger is a device in which heat flows between two fluid streams brought into thermal contact through a barrier, such as a pipe wall. Heat exchangers may be operated in either the cocurrent (both fluid streams flowing in the same direction) or countercurrent (streams flowing in opposite directions) configurations; schematic diagrams are given here.

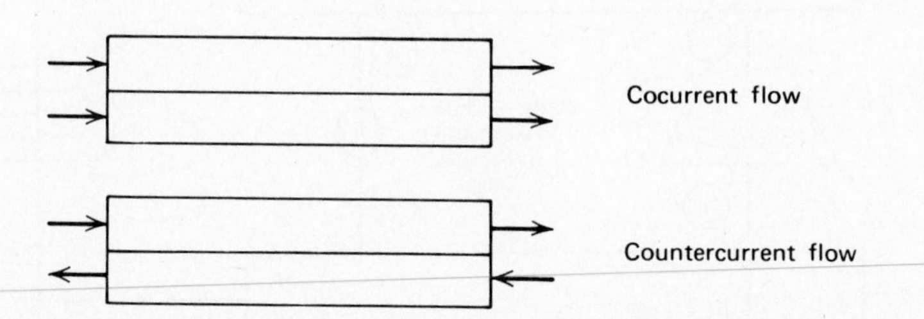

The heat flow rate from fluid 1 to fluid 2 per unit length of the heat exchanger, $\dot{\mathcal{Q}}$, is proportional to the temperature difference $(T_1 - T_2)$,

$$\dot{\mathcal{Q}} = \begin{pmatrix} \text{heat flow rate from fluid 1} \\ \text{to fluid 2 per unit length of} \\ \text{heat exchanger} \end{pmatrix} = \kappa(T_1 - T_2)$$

where κ is a constant of proportionality with units of J/m s K. The fluids in both streams are the same and their flow rates are equal. The initial and final temperatures of stream 1 will be 35°C and 15°C, respectively, and those for stream 2 will be −15°C and 5°C.

a Write the balance equations for each fluid stream in a portion of the heat exchanger of length dL and obtain differential equations by letting $dL \rightarrow 0$.

b Integrate the energy balance equations over the length of the exchanger to obtain expressions for the temperature of each stream at any point in the exchanger for each flow configuration. Also compute the length of the exchanger, in units of $L_0 = \dot{M}C_P/2\kappa$, (where $\dot{M}$ is the mass flow rate of either stream) needed to accomplish the desired heat transfer.

c Write an expression for the change of entropy of stream 1 with distance for any point in the exchanger.

3.25 One measure of the thermodynamic efficiency of a complex process (e.g., an electrical generating station or an automobile engine) is the ratio of the useful work obtained for a specified change of state to the maximum useful work obtainable with the ambient temperature T_0 and pressure P_0. Here by **useful work** we mean the total work done by the system less the work done in the expansion of the system boundaries against the ambient pressure.

Clearly, to obtain the maximum useful work all processes should occur reversibly, and all heat transferred to the surroundings should leave the system at T_0, since heat available at any other temperature could be used with a Carnot or similar engine to obtain additional work. For a similar reason, the feed and exit streams in an open process should also be at the ambient conditions.

a Show that the maximum useful work for a closed system change of state is

$$W_u^{\max} = \mathcal{A}_2 - \mathcal{A}_1$$

where $\mathcal{A} = U + P_0V - T_0S$ is the closed system **availability function**.

b For a steady-state or cyclic flow system show that the maximum useful work is

$$W_u^{\max} = M(\hat{\mathcal{B}}_2 - \hat{\mathcal{B}}_1)$$

where $\hat{\mathcal{B}} = \hat{H} - T_0\hat{S}$ is the flow availability function and M is the mass that has entered the system in any convenient time interval in a steady-flow process or in one complete cycle in a cyclic process. (Note that since the inlet and exit temperatures and pressures are equal, $\hat{\mathcal{B}}_1$ will equal $\hat{\mathcal{B}}_2$ and $W_u^{\max}$ will equal zero unless a chemical reaction, for example the burning of coal or gasoline, a phase change, or some other change in the composition of the inlet and exit streams occurs.)

c Compute the maximum useful work that can be obtained when 1 kg of steam undergoes a closed system change of state from 30 bar and 600°C to 5 bar and 300°C when the atmospheric temperature and pressure are as given in the data.

d A coal-fired power-generation station generates approximately 2.2 kilowatt-hours of electricity for each kilogram of coal (composition given below) burned. Using the availability concept, compute the thermodynamic efficiency of the station.

Data

Ambient conditions: 25°C and 1.013 bar

Chemical composition of coal (simplified)

Carbon	70 weight percent
Water (liquid)	15 weight percent
Inorganic matter	15 weight percent

THERMODYNAMIC DATA

	Enthalpy at 25°C, 1.013 bar (kcal/mol)*	Entropy at 25°C, 1.013 bar (kcal/mol K)
O_2	0.0	0.0
C	0.0	0.0
CO_2	−94.052	0.0
H_2O (liq)	−68.317	−0.039
H_2O (vap)	−57.800	−0.0106

* 1 kcal = 4.184 kJ.

Assume that the inorganic matter (ash) passes through the power station unchanged, that all the water leaves as vapor in the flue gas, and that all the carbon is converted into carbon dioxide.

3.26 The following closed-loop steam cycle has been proposed to generate work from burning fuel.

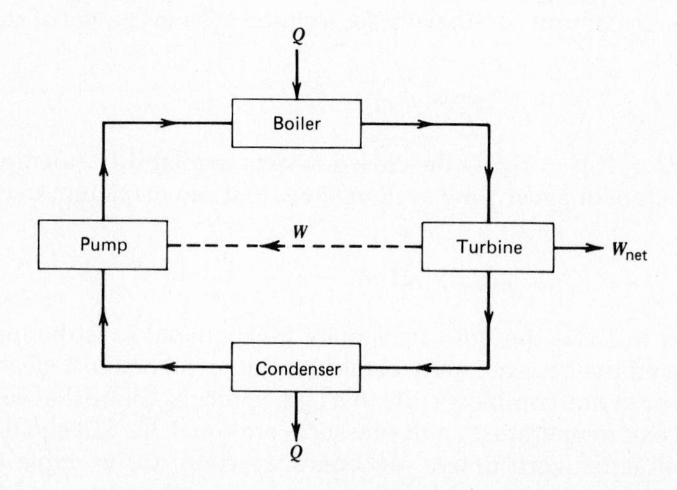

The temperature of the burning fuel is 1100°C, and cooling water is available at 15°C. The steam leaving the boiler is at 20 bar and 700°C, and the condenser produces a saturated liquid at 0.2 bar. The steam lines are well insulated, the turbine and pump operate reversibly and adiabatically, and some of the mechanical work generated by the turbine is used to drive the pump.

a What is the net work obtained in the cycle per kilogram of steam generated in the boiler?

b How much heat is discarded in the condenser per kilogram of steam generated in the boiler?

c What fraction of the work generated by the turbine is used to operate the pump?

d How much heat is absorbed in the boiler per kilogram of steam generated?

e Calculate the engine efficiency and compare it with the efficiency of a Carnot cycle receiving heat as 1100°C and discharging heat at 15°C.

3.27 Two tanks are connected as shown here. Tank 1 initially contains an ideal gas at 10 bar and 20°C, and both parts of Tank 2 contain the same gas at 1 bar and 20°C. The valve connecting the two tanks is opened long enough to allow the pressures in the tanks to equilibrate and then shut. There is no transfer of heat from the gas to the tanks or the frictionless piston, and the constant pressure heat capacity of the gas is $4R$.

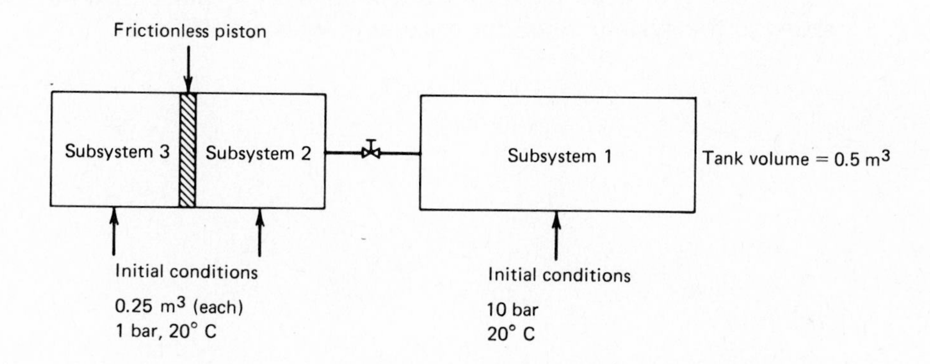

Compute the temperature and pressure of the gas in each part of the system at the end of the process and the work done on the gas behind the piston (i.e., the gas in subsystem 3).

3.28 A small can containing compressed air at a sufficiently high pressure can be used to reinflate a flat automobile tire. If the initial temperature of both the compressed air and the air in the tire is 295 K, estimate the initial pressure in the compressed air can necessary to reinflate one tire from 1.0 bar to 2.6 bar. Also, estimate the final air temperature in the tire and in the can. For the purposes of this calculation you may assume:

1 Air is an ideal gas with C_P = 30.0 J/mol K.

2 The tire does not change its size or shape during the inflation process.

3 The inner tube of the tire has a volume of 4×10^{-2} m^3.

4 The compressed air can has a volume of 6×10^{-4} m^3.

5 The final pressure in the can is the same as that of the tire.

3.29 A 0.5-m^3 tank containing air at 17.5 bar and 22°C is to be exhausted to the atmosphere through a cleverly designed black-box engine. The engine is directly connected to the tank, and the air exiting from the engine is always at 1 bar and 22°C. The process will be continued until the pressure of the air in the tank is 1 bar (see the figure). There is no transfer of heat between the air and the walls of the tank or into or out of the engine. Air can be considered to be an ideal gas with $C_P = 30.0$ J/mol K.

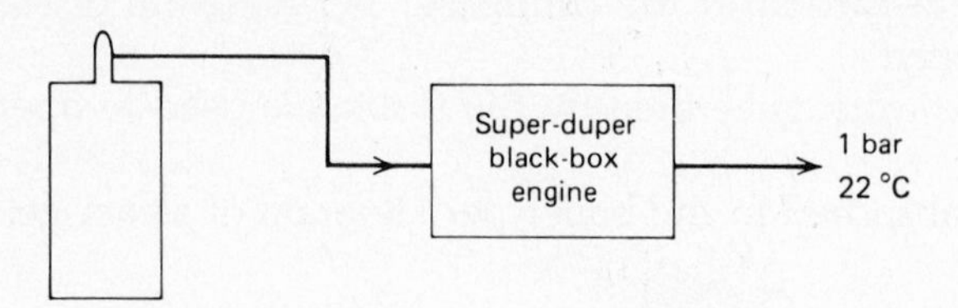

a Develop an expression relating the temperature and pressure of the air in the tank at any time during the process.
b Compute the final temperature of the air in the tank.
c Compute the maximum work obtained from the engine during the complete exhaustion process.
d Suppose the process were carried out isothermally and the temperature of the air throughout the system maintained at 22°C. Compute the maximum amount of work that could be obtained. How much heat must be added to the system when the maximum work is obtained?

4

The Thermodynamic Properties of Real Substances

In Chapters 2 and 3 we derived a general set of balance equations for mass, energy, and entropy that can be used to compute energy changes, heat or work requirements, and so forth for the changes of state of any substance. However, these balance equations are in terms of the internal energy, enthalpy, and entropy, rather than the pressure and temperature, the variables most easily measured and thus most often used to specify the thermodynamic state of the system. To illustrate the use of the balance equations in the simplest manner, examples were given using either ideal gases or fluids whose thermodynamic properties were available in graphical and tabular form. Unfortunately, no gas is ideal over the whole range of pressure and temperature, and thermodynamic properties tables are not always available, so that a necessary ingredient of many thermodynamic computations is the calculation of the thermodynamic properties of real substances in any state. The main concern of this chapter is to establish how to solve thermodynamic problems for real substances given heat capacity data, and data on the relationship between pressure, volume, and temperature. The problem of constructing a thermodynamic properties chart from such data is also considered. The discussion of the relationship between the ideal gas and absolute temperature scales, which began in Chapter 1, is completed here, and finally, the principle of corresponding states and generalized equations of state are considered, as is their application.

4.1 SOME MATHEMATICAL PRELIMINARIES

In the last two chapters eight thermodynamic state variables (P, T, $\underline{V}$, $\underline{S}$, $\underline{U}$, $\underline{H}$, $\underline{A}$, and $\underline{G}$), which frequently appear in thermodynamic calculations, were introduced. If values of any two of these variables are given, the thermodynamic state of a pure, single-phase system is fixed, as are the values of the remaining six variables (Experimental Observation 8 of Sec. 1.7). Mathematically we describe this situation by saying that any two variables may be chosen as the independent variables for the single-component, one-phase system, and the remaining six variables are dependent variables. If, for example, T and $\underline{V}$ are taken as the independent variables, then any other variable, such as the internal energy $\underline{U}$, is a dependent variable; this is denoted by $\underline{U} = \underline{U}(T, \underline{V})$ to indicate that the internal energy is a function of temperature and specific volume. The change in internal energy $d\underline{U}$, which results from differential changes in T and $\underline{V}$, can be computed using the chain rule of differentiation

$$d\underline{U} = \left(\frac{\partial \underline{U}}{\partial T}\right)_{\underline{V}} dT + \left(\frac{\partial \underline{U}}{\partial \underline{V}}\right)_T d\underline{V}$$

where the subscript on each derivative indicates the variable being held constant; that is, $(\partial \underline{U}/\partial T)_{\underline{V}}$ denotes the differential change in molar internal energy accompanying a differential change in temperature in a process in which the molar volume is constant. Note that both $(\partial \underline{U}/\partial T)_{\underline{V}}$ and $(\partial \underline{U}/\partial \underline{V})_T$ are partial derivatives of the type

$$\left(\frac{\partial \underline{X}}{\partial \underline{Y}}\right)_{\underline{Z}} \tag{4.1-1}$$

where $\underline{X}$, $\underline{Y}$, and $\underline{Z}$ are used here to denote the state variables P, T, $\underline{V}$, $\underline{S}$, $\underline{U}$, $\underline{H}$, $\underline{A}$, and $\underline{G}$. Since derivatives of this type occur whenever one tries to relate a change in one thermodynamic function (here $\underline{U}$) to changes in two others (here T and $\underline{V}$), one of the goals of this chapter is to develop methods for computing numerical values for these derivatives, as these will be needed in many thermodynamic calculations.

In open systems extensive properties, such as the total internal energy U, the total enthalpy H, and the total entropy S, are functions of three variables, usually taken to be two thermodynamic variables (either intensive or extensive) and the total mass (M) or mole number (N). Thus, for example, the total internal energy can be considered to be a function of temperature, total volume, and number of moles, so that

$$dU = \left(\frac{\partial U}{\partial T}\right)_{V,N} dT + \left(\frac{\partial U}{\partial V}\right)_{T,N} dV + \left(\frac{\partial U}{\partial N}\right)_{T,V} dN$$

where now two subscripts are needed to indicate the variables being held constant for each differential change.

The derivatives of extensive properties at constant mole number or mass are simply related to the analogous derivatives among the state variables, that is, to derivatives of the form of Eq. 4.1-1. To see this we note that since any extensive property X can be written as $N\underline{X}$ where $\underline{X}$ is an intensive (or molar) property, the derivative of an extensive property with respect to an intensive property (e.g., temperature, pressure, or specific volume) at constant mole number is

$$\left(\frac{\partial X}{\partial \underline{Y}}\right)_{Z,N} = \left(\frac{\partial (N\underline{X})}{\partial \underline{Y}}\right)_{Z,N} = N\left(\frac{\partial \underline{X}}{\partial \underline{Y}}\right)_{\underline{Z}}$$

Thus,

$$\left(\frac{\partial U}{\partial T}\right)_{V,N} = N\left(\frac{\partial \underline{U}}{\partial T}\right)_{\underline{V}} = NC_v$$

where C_v is the constant-volume heat capacity. Similarly, the derivative of an extensive property with respect to an extensive property at constant mole number is found from the observation that

$$\left(\frac{\partial X}{\partial Y}\right)_{Z,N} = \left(\frac{\partial (N\underline{X})}{\partial (N\underline{Y})}\right)_{Z,N} = \frac{N}{N}\left(\frac{\partial \underline{X}}{\partial \underline{Y}}\right)_{\underline{Z}} = \left(\frac{\partial \underline{X}}{\partial \underline{Y}}\right)_{\underline{Z}}$$

so that, for example,

$$\left(\frac{\partial U}{\partial V}\right)_{T,N} = \left(\frac{\partial \underline{U}}{\partial \underline{V}}\right)_T$$

Consequently, once a method is developed to obtain numerical values for derivatives in the form of Eq. 4.1-1,

$$\left(\frac{\partial \underline{X}}{\partial \underline{Y}}\right)_{\underline{Z}}$$

it can also be used to evaluate derivatives of the form

$$\left(\frac{\partial X}{\partial Y}\right)_{Z,N} \quad \text{and} \quad \left(\frac{\partial X}{\partial \underline{Y}}\right)_{Z,N}$$

The derivative of an extensive property with respect to mole number [e.g., the derivative $(\partial U/\partial N)_{T,V}$ in this discussion] is not of the form of Eq. 4.1-1. Such derivatives are considered in both the following sections and in Sec. 4.8.

We start the analysis of partial derivatives of the type indicated in Eq. 4.1-1 by listing several of their important properties. First, their numerical value depends on the path followed, that is, on which variable is being held constant. Thus,

$$\left(\frac{\partial \underline{U}}{\partial T}\right)_{\underline{V}} \neq \left(\frac{\partial \underline{U}}{\partial T}\right)_P$$

as will be verified shortly (Illustration 4.2-1). If two intensive variables are held constant in a one-component, single-phase system, all derivatives, such as

$$\left(\frac{\partial \underline{U}}{\partial T}\right)_{\underline{V},P}$$

must equal zero, since by fixing the values of two intensive variables, one has also fixed the values of all the remaining variables.

We will assume that the thermodynamic variables in which we are interested exist and are well-behaved in some mathematical sense that we will leave unspecified. In this case, the derivative of Eq. 4.1-1 has the properties that

$$\left(\frac{\partial \underline{X}}{\partial \underline{Y}}\right)_{\underline{Z}} = \frac{1}{(\partial \underline{Y}/\partial \underline{X})_{\underline{Z}}} \tag{4.1-2}$$

and

$$\frac{\partial}{\partial \underline{Z}}\bigg|_{\underline{Y}} \left(\frac{\partial \underline{X}}{\partial \underline{Y}}\right)_{\underline{Z}} = \frac{\partial}{\partial \underline{Y}}\bigg|_{\underline{Z}} \left(\frac{\partial \underline{X}}{\partial \underline{Z}}\right)_{\underline{Y}} \tag{4.1-3}$$

This last equation states that in a mixed second derivative the order of taking derivatives is unimportant. Also

$$\left(\frac{\partial \underline{X}}{\partial \underline{Y}}\right)_{\underline{X}} = 0 \tag{4.1-4a}$$

and

$$\left(\frac{\partial \underline{X}}{\partial \underline{X}}\right)_{\underline{Z}} = 1 \tag{4.1-4b}$$

since the first derivative is the change in $\underline{X}$ along a path of constant $\underline{X}$, and the second derivative is the change in $\underline{X}$ in response to an imposed change in $\underline{X}$.

If $\underline{X}$ is any dependent variable, and $\underline{Y}$ and $\underline{Z}$ the two independent variables, we can write $\underline{X} = \underline{X}(\underline{Y}, \underline{Z})$, and

$$d\underline{X} = \left(\frac{\partial \underline{X}}{\partial \underline{Y}}\right)_{\underline{Z}} d\underline{Y} + \left(\frac{\partial \underline{X}}{\partial \underline{Z}}\right)_{\underline{Y}} d\underline{Z} \tag{4.1-5}$$

Now from Eqs. 4.1-4a and 5 we have

$$\left(\frac{\partial \underline{X}}{\partial \underline{Y}}\right)_{\underline{X}} = 0 = \left(\frac{\partial \underline{X}}{\partial \underline{Y}}\right)_{\underline{Z}} \left(\frac{\partial \underline{Y}}{\partial \underline{Y}}\right)_{\underline{X}} + \left(\frac{\partial \underline{X}}{\partial \underline{Z}}\right)_{\underline{Y}} \left(\frac{\partial \underline{Z}}{\partial \underline{Y}}\right)_{\underline{X}}$$

and using Eq. 4.1-4b we obtain

$$\left(\frac{\partial \underline{X}}{\partial \underline{Y}}\right)_{\underline{Z}} \left(\frac{\partial \underline{Z}}{\partial \underline{X}}\right)_{\underline{Y}} \left(\frac{\partial \underline{Y}}{\partial \underline{Z}}\right)_{\underline{X}} = -1 \tag{4.1-6a}$$

or

$$\left(\frac{\partial \underline{X}}{\partial \underline{Y}}\right)_{\underline{Z}} \left(\frac{\partial \underline{Z}}{\partial \underline{X}}\right)_{\underline{Y}} = -\left(\frac{\partial \underline{Z}}{\partial \underline{Y}}\right)_{\underline{X}} \tag{4.1-6b}$$

Equations 4.1-6 are known as the triple product rule and will be used frequently in this book; it is easily remembered by noting the symmetric form of Eq. 4.1-6a in that each variable appears in each derivative position only once.

There are two other important results to be gotten from Eq. 4.1-5. The first is the expansion rule, obtained by introducing any two additional thermodynamic properties $\underline{K}$ and $\underline{L}$

$$\left(\frac{\partial \underline{X}}{\partial \underline{K}}\right)_{\underline{L}} = \left(\frac{\partial \underline{X}}{\partial \underline{Y}}\right)_{\underline{Z}} \left(\frac{\partial \underline{Y}}{\partial \underline{K}}\right)_{\underline{L}} + \left(\frac{\partial \underline{X}}{\partial \underline{Z}}\right)_{\underline{Y}} \left(\frac{\partial \underline{Z}}{\partial \underline{K}}\right)_{\underline{L}} \tag{4.1-7}$$

and the second is a special case of the first in which $\underline{L} = \underline{Z}$, so that, by Eq. 4.1-4a

$$\left(\frac{\partial \underline{X}}{\partial \underline{K}}\right)_{\underline{Z}} = \left(\frac{\partial \underline{X}}{\partial \underline{Y}}\right)_{\underline{Z}} \left(\frac{\partial \underline{Y}}{\partial \underline{K}}\right)_{\underline{Z}} \tag{4.1-8}$$

[You should compare Eq. 4.1-6b with Eq. 4.1-8, and note the difference between them.] Thus, if for some reason it is convenient, we can interpose a new variable in evaluating a partial derivative, as is the case with the variable $\underline{Y}$ in Eq. 4.1-8.

Finally, we note that for an open system it is usually convenient to use the mass M or mole number N, and two variables from among T, P, and the extensive variables U, V, S, G, H, and A, as the independent variables. Letting X, Y, and Z represent variables from among the set (U, V, S, G, H, A, T, and P), we have that $X = X(Y, Z, N)$ and

$$dX = \left(\frac{\partial X}{\partial Y}\right)_{Z,N} dY + \left(\frac{\partial X}{\partial Z}\right)_{Y,N} dZ + \left(\frac{\partial X}{\partial N}\right)_{Y,Z} dN \tag{4.1-9}$$

4.2 THE EVALUATION OF THERMODYNAMIC PARTIAL DERIVATIVES

Whenever the change in a thermodynamic property is related to changes in two others, derivatives of the type of Eq. 4.1-1 occur. Four of these partial derivatives occur so frequently in both experiment and calculation that they have been given special designations:

$$\left(\frac{\partial \underline{U}}{\partial T}\right)_{\underline{V}} = C_{\mathrm{v}} = \text{constant-volume heat capacity (see Sec. 2.4)} \tag{4.2-1}$$

$$\left(\frac{\partial \underline{H}}{\partial T}\right)_{P} = C_{\mathrm{P}} = \text{constant-pressure heat capacity (see Sec. 2.4)} \tag{4.2-2}$$

$$\frac{1}{\underline{V}}\left(\frac{\partial \underline{V}}{\partial T}\right)_P = \alpha = \text{coefficient of thermal expansion} \tag{4.2-3}$$

$$-\frac{1}{\underline{V}}\left(\frac{\partial \underline{V}}{\partial P}\right)_T = \kappa_T = \text{isothermal compressibility} \tag{4.2-4}$$

The starting point of the analysis of other thermodynamic derivatives is Eq. 3.2-13a for open systems

$$dU = T\,dS - P\,dV + \underline{G}\,dN \tag{4.2-5a}$$

and Eq. 3.2-13b for closed systems

$$d\underline{U} = T\,d\underline{S} - P\,d\underline{V} \tag{4.2-5b}$$

Alternatively, these equations can be rearranged to

$$dS = \frac{1}{T}\,dU + \frac{P}{T}\,dV - \frac{\underline{G}}{T}\,dN \tag{4.2-5c}$$

and

$$d\underline{S} = \frac{1}{T}\,d\underline{U} + \frac{P}{T}\,d\underline{V} \tag{4.2-5d}$$

By definition $H = U + PV$, so that

$$dH = dU + V\,dP + P\,dV = T\,dS - P\,dV + \underline{G}\,dN + V\,dP + P\,dV$$

$$= T\,dS + V\,dP + \underline{G}\,dN \tag{4.2-6a}$$

and, for the closed system

$$d\underline{H} = T\,d\underline{S} + \underline{V}\,dP \tag{4.2-6b}$$

Similarly, from Eq. 3.2-6 we have $A = U - TS$, so that $dA = dU - S\,dT - T\,dS$. Using Eq. 4.2-5 yields

$$dA = -P\,dV - S\,dT + \underline{G}\,dN \tag{4.2-7a}$$

which, for the closed system becomes

$$d\underline{A} = -P\,d\underline{V} - \underline{S}\,dT \tag{4.2-7b}$$

Finally, from Eq. 3.2-8, we have $G = H - TS$, so that

$$dG = V\,dP - S\,dT + \underline{G}\,dN \tag{4.2-8a}$$

and

$$d\underline{G} = \underline{V}\,dP - \underline{S}\,dT \tag{4.2-8b}$$

Next, we note the analogy between Eqs. 4.1-5 and 9 and Eqs. 4.2-5 through 8 and obtain the following relations

$$\left(\frac{\partial U}{\partial S}\right)_{V,N} = \left(\frac{\partial \underline{U}}{\partial \underline{S}}\right)_{\underline{V}} = T \tag{4.2-9a}$$

$$\left(\frac{\partial U}{\partial V}\right)_{S,N} = \left(\frac{\partial \underline{U}}{\partial \underline{V}}\right)_{\underline{S}} = -P \tag{4.2-9b}$$

$$\left(\frac{\partial U}{\partial N}\right)_{S,V} = \underline{G} \tag{4.2-9c}$$

$$\left(\frac{\partial S}{\partial N}\right)_{U,V} = -\frac{\underline{G}}{T} \tag{4.2-9d}$$

$$\left(\frac{\partial H}{\partial S}\right)_{P,N} = \left(\frac{\partial \underline{H}}{\partial \underline{S}}\right)_{P} = T \tag{4.2-10a}$$

$$\frac{1}{N}\left(\frac{\partial H}{\partial P}\right)_{S,N} = \left(\frac{\partial \underline{H}}{\partial P}\right)_{\underline{S}} = \underline{V} \tag{4.2-10b}$$

$$\left(\frac{\partial H}{\partial N}\right)_{P,S} = \underline{G} \tag{4.2-10c}$$

$$\left(\frac{\partial A}{\partial V}\right)_{T,N} = \left(\frac{\partial \underline{A}}{\partial \underline{V}}\right)_{T} = -P \tag{4.2-11a}$$

$$\frac{1}{N}\left(\frac{\partial A}{\partial T}\right)_{V,N} = \left(\frac{\partial \underline{A}}{\partial T}\right)_{\underline{V}} = -\underline{S} \tag{4.2-11b}$$

$$\left(\frac{\partial A}{\partial N}\right)_{T,V} = \underline{G} \tag{4.2-11c}$$

$$\frac{1}{N}\left(\frac{\partial G}{\partial P}\right)_{T,N} = \left(\frac{\partial \underline{G}}{\partial P}\right)_{T} = \underline{V} \tag{4.2-12a}$$

$$\frac{1}{N}\left(\frac{\partial G}{\partial T}\right)_{P,N} = \left(\frac{\partial \underline{G}}{\partial T}\right)_{P} = -\underline{S} \tag{4.2-12b}$$

and

$$\left(\frac{\partial G}{\partial N}\right)_{T,P} = \underline{G} \qquad \text{(4.2-12c)[1]}$$

To get expressions for several additional thermodynamic derivatives we next use Eq. 4.1-3, the commutative property of mixed second derivatives. In this way starting with Eqs. 4.2-9a and b we obtain

$$\frac{\partial}{\partial \underline{V}}\bigg|_{\underline{S}} \left(\frac{\partial \underline{U}}{\partial \underline{S}}\right)_{\underline{V}} = \left(\frac{\partial T}{\partial \underline{V}}\right)_{\underline{S}}$$

$$\frac{\partial}{\partial \underline{S}}\bigg|_{\underline{V}} \left(\frac{\partial \underline{U}}{\partial \underline{V}}\right)_{\underline{S}} = -\left(\frac{\partial P}{\partial \underline{S}}\right)_{\underline{V}}$$

so that from Eq. 4.1-3 we have

$$\left(\frac{\partial T}{\partial \underline{V}}\right)_{\underline{S}} = -\left(\frac{\partial P}{\partial \underline{S}}\right)_{\underline{V}} \tag{4.2-13}$$

Similarly, from Eqs. 4.2-10, 11, and 12 we obtain

$$\left(\frac{\partial T}{\partial P}\right)_{\underline{S}} = \left(\frac{\partial \underline{V}}{\partial \underline{S}}\right)_{P} \tag{4.2-14}$$

$$\left(\frac{\partial P}{\partial T}\right)_{\underline{V}} = \left(\frac{\partial \underline{S}}{\partial \underline{V}}\right)_{T} \tag{4.2-15}$$

[1] Note that comparing Eqs. 4.2-9c and d, 10c, 11c, and 12c we have

$$\left(\frac{\partial U}{\partial N}\right)_{S,V} = \left(\frac{\partial H}{\partial N}\right)_{P,S} = \left(\frac{\partial A}{\partial N}\right)_{T,V} = \left(\frac{\partial G}{\partial N}\right)_{T,P} = -T\left(\frac{\partial S}{\partial N}\right)_{U,V} = \underline{G}$$

The multicomponent analogs of these equations are given in Sec. 6.2.

and

$$\left(\frac{\partial \underline{V}}{\partial T}\right)_P = -\left(\frac{\partial \underline{S}}{\partial P}\right)_T \tag{4.2-16}$$

Equations 4.2-13 through 16 are known as the **Maxwell relations**. (It is left to you to derive the Maxwell relations for open systems; see Problem 4.27.)

Equation 4.2-5d relates the change in entropy to changes in internal energy and volume. Since temperature and pressure, or temperature and volume, are, because of the ease with which they can be measured, more common choices for the independent variables than $\underline{U}$ and $\underline{V}$, it would be useful to have expressions relating $d\underline{S}$ to dT and $d\underline{V}$, or dT and dP. We can derive such expressions by first writing $\underline{S}$ as a function of T and $\underline{V}$,

$$\underline{S} = \underline{S}(T, \underline{V})$$

and then using the chain rule of partial differentiation to obtain

$$d\underline{S} = \left(\frac{\partial \underline{S}}{\partial T}\right)_{\underline{V}} dT + \left(\frac{\partial \underline{S}}{\partial \underline{V}}\right)_T d\underline{V} \tag{4.2-17}$$

From the application of first Eq. 4.1-8, then Eqs. 4.1-2, and finally Eqs. 4.2-1 and 9a, we obtain

$$\left(\frac{\partial \underline{S}}{\partial T}\right)_{\underline{V}} = \left(\frac{\partial \underline{S}}{\partial \underline{U}}\right)_{\underline{V}} \left(\frac{\partial \underline{U}}{\partial T}\right)_{\underline{V}} = \left(\frac{\partial \underline{U}}{\partial T}\right)_{\underline{V}} \left[\left(\frac{\partial \underline{U}}{\partial \underline{S}}\right)_{\underline{V}}\right]^{-1}$$
$$= C_V/T \tag{4.2-18}$$

and, from Eq. 4.2-15, we have

$$\left(\frac{\partial \underline{S}}{\partial \underline{V}}\right)_T = \left(\frac{\partial P}{\partial T}\right)_{\underline{V}}$$

Thus

$$d\underline{S} = \frac{C_V}{T} dT + \left(\frac{\partial P}{\partial T}\right)_{\underline{V}} d\underline{V} \tag{4.2-19}$$

Consequently, given heat capacity data as a function of T and P or T and $\underline{V}$, volumetric equation of state information, and a value of the entropy at some value of T and $\underline{V}$, it is possible to compute the entropy at any other value of T and $\underline{V}$ by integration of Eq. 4.2-19. Similarly, starting from $\underline{S} = \underline{S}(T, P)$, one can easily show that

$$\left(\frac{\partial \underline{S}}{\partial T}\right)_P = \frac{C_P}{T}, \qquad \left(\frac{\partial \underline{S}}{\partial P}\right)_T = -\left(\frac{\partial \underline{V}}{\partial T}\right)_P$$

and

$$d\underline{S} = \frac{C_P}{T} dT - \left(\frac{\partial \underline{V}}{\partial T}\right)_P dP \tag{4.2-20}$$

Equations 4.2-19 and 20 can now be used in Eqs. 4.2-5 and 6 to get

$$d\underline{U} = T\,d\underline{S} - P\,d\underline{V}$$
$$= T\left[\frac{C_V}{T} dT + \left(\frac{\partial P}{\partial T}\right)_{\underline{V}} d\underline{V}\right] - P\,d\underline{V}$$
$$= C_V\,dT + \left[T\left(\frac{\partial P}{\partial T}\right)_{\underline{V}} - P\right] d\underline{V} \tag{4.2-21}$$

and

$$d\underline{H} = C_P\, dT + \left[\underline{V} - T\left(\frac{\partial \underline{V}}{\partial T}\right)_P\right] dP \tag{4.2-22}$$

From these last two equations we obtain

$$\left(\frac{\partial \underline{U}}{\partial \underline{V}}\right)_T = T\left(\frac{\partial P}{\partial T}\right)_{\underline{V}} - P \tag{4.2-23}$$

and

$$\left(\frac{\partial \underline{H}}{\partial P}\right)_T = \underline{V} - T\left(\frac{\partial \underline{V}}{\partial T}\right)_P \tag{4.2-24}$$

Table 4.2-1 summarizes the definitions used and some of the thermodynamic identities developed so far in this chapter. The equations in this table can be useful in obtaining information about some thermodynamic derivatives, as indicated in Illustration 4.2-1.

ILLUSTRATION 4.2-1

Obtain expressions for the two derivatives $(\partial \underline{U}/\partial T)_{\underline{V}}$ and $(\partial \underline{U}/\partial T)_P$, and show that they are not equal.

Solution

Starting from Eq. 4.2-21

$$d\underline{U} = C_V\, dT + \left[T\left(\frac{\partial P}{\partial T}\right)_{\underline{V}} - P\right] d\underline{V}$$

and using Eq. 4.1-7 yields

$$\left(\frac{\partial \underline{U}}{\partial T}\right)_{\underline{V}} = C_V\left(\frac{\partial T}{\partial T}\right)_{\underline{V}} + \left[T\left(\frac{\partial P}{\partial T}\right)_{\underline{V}} - P\right]\left(\frac{\partial \underline{V}}{\partial T}\right)_{\underline{V}}$$

which, with Eq. 4.1-4, reduces to

$$\left(\frac{\partial \underline{U}}{\partial T}\right)_{\underline{V}} = C_V$$

Similarly, again starting with Eq. 4.2-21, we obtain

$$\left(\frac{\partial \underline{U}}{\partial T}\right)_P = C_V + \left[T\left(\frac{\partial P}{\partial T}\right)_{\underline{V}} - P\right]\left(\frac{\partial \underline{V}}{\partial T}\right)_P \tag{4.2-25}$$

Clearly, then

$$\left(\frac{\partial \underline{U}}{\partial T}\right)_{\underline{V}} \neq \left(\frac{\partial \underline{U}}{\partial T}\right)_P$$

(see Problem 4.3). ■

The form of Eqs. 4.2-19 through 22 is nice for two reasons. First, the equations relate the change in entropy, internal energy, and enthalpy to changes in only P, $\underline{V}$, and T. Next, the right sides of these equations contain only C_P, C_V, P, $\underline{V}$, and T, and partial derivatives involving P, $\underline{V}$, and T. Thus, given heat capacity data and the volumetric equation of state for the fluid, the change in $\underline{S}$, $\underline{U}$, or $\underline{H}$ accompanying a change in system temperature, pressure, or volume can be computed.

Table 4.2-1
SOME USEFUL DEFINITIONS AND THERMODYNAMIC IDENTITIES

Definitions

Constant-volume heat capacity $= C_V = \left(\frac{\partial \underline{U}}{\partial T}\right)_{\underline{V}} = T\left(\frac{\partial \underline{S}}{\partial T}\right)_{\underline{V}}$

Constant-pressure heat capacity $= C_P = \left(\frac{\partial \underline{H}}{\partial T}\right)_P = T\left(\frac{\partial \underline{S}}{\partial T}\right)_P$

Isothermal compressibility $= \kappa_T = -\frac{1}{\underline{V}}\left(\frac{\partial \underline{V}}{\partial P}\right)_T$

Coefficient of thermal expansion $= \alpha = \frac{1}{\underline{V}}\left(\frac{\partial \underline{V}}{\partial T}\right)_P$

Maxwell relations

$$\left(\frac{\partial T}{\partial \underline{V}}\right)_{\underline{S}} = -\left(\frac{\partial P}{\partial \underline{S}}\right)_{\underline{V}} \qquad \left(\frac{\partial T}{\partial P}\right)_{\underline{S}} = \left(\frac{\partial \underline{V}}{\partial \underline{S}}\right)_P$$

$$\left(\frac{\partial P}{\partial T}\right)_{\underline{V}} = \left(\frac{\partial \underline{S}}{\partial \underline{V}}\right)_T \qquad \left(\frac{\partial \underline{V}}{\partial T}\right)_P = -\left(\frac{\partial \underline{S}}{\partial P}\right)_T$$

Thermodynamic identities

$$\left(\frac{\partial \underline{H}}{\partial \underline{S}}\right)_P = \left(\frac{\partial \underline{U}}{\partial \underline{S}}\right)_{\underline{V}} = T \qquad \left(\frac{\partial \underline{G}}{\partial P}\right)_T = \left(\frac{\partial \underline{H}}{\partial P}\right)_{\underline{S}} = \underline{V}$$

$$\left(\frac{\partial \underline{U}}{\partial \underline{V}}\right)_{\underline{S}} = \left(\frac{\partial \underline{A}}{\partial \underline{V}}\right)_T = -P \qquad \left(\frac{\partial \underline{A}}{\partial T}\right)_{\underline{V}} = \left(\frac{\partial \underline{G}}{\partial T}\right)_P = -\underline{S}$$

Thermodynamic functions

$$d\underline{U} = T\,d\underline{S} - P\,d\underline{V} = C_V\,dT + \left[T\left(\frac{\partial P}{\partial T}\right)_{\underline{V}} - P\right]d\underline{V}$$

$$d\underline{H} = T\,d\underline{S} + \underline{V}\,dP = C_P\,dT + \left[\underline{V} - T\left(\frac{\partial \underline{V}}{\partial T}\right)_P\right]dP$$

$$d\underline{A} = -P\,d\underline{V} - \underline{S}\,dT$$

$$d\underline{G} = \underline{V}\,dP - \underline{S}\,dT$$

Miscellaneous

$$\left(\frac{\partial \underline{U}}{\partial \underline{V}}\right)_T = T\left(\frac{\partial P}{\partial T}\right)_{\underline{V}} - P = \frac{T\alpha}{\kappa_T} - P$$

$$\left(\frac{\partial \underline{H}}{\partial P}\right)_T = \underline{V} - T\left(\frac{\partial \underline{V}}{\partial T}\right)_P = \underline{V}(1 - T\alpha)$$

Ideally, we would like to develop equations similar to Eqs. 4.2-19 through 22 for all the thermodynamic variables of interest, and more generally, to be able to relate numerically the change in any thermodynamic property to the changes in any two others. To do this we must be able to obtain a numerical value for any derivative of the form $(\partial \underline{X}/\partial \underline{Y})_{\underline{Z}}$. Since engineers generally use two variables from among pressure, temperature, and volume as the independent variables, and also have most information about the interrelationship between these variables, the discussion that follows centers on reducing all partial derivatives to functions of P, $\underline{V}$, and T, their mutual derivatives, and the heat capacity, as in the equations already derived. Unfortunately, it is not possible to reduce all partial derivatives to functions of only these variables because certain partial derivatives introduce the entropy (see Eqs. 4.2-11 and 12). Since, using Eqs. 4.2-19 and 20, entropy can be evaluated from heat capacity and volumetric equation of state data, its inclusion introduces no real difficulty. Thus, we will be satisfied if we can reduce any partial derivative to a form containing P, $\underline{V}$, T, $\underline{S}$, and C_P or C_V, and derivatives containing only P, $\underline{V}$, and T. In fact, as we shall see later in this section (Eq. 4.2-30), there are only three independent partial derivatives from among the four in Eqs. 4.2-1 through 4, for example C_P, α, and κ_T, so that it is possible to reduce the partial derivatives encountered in this chapter to functions of only P, $\underline{V}$, T, $\underline{S}$, α, κ_T, and C_P or C_V.

Since eight different variables (T, P, $\underline{V}$, $\underline{U}$, $\underline{H}$, $\underline{S}$, $\underline{A}$, and $\underline{G}$) may be used in the thermodynamic description of a one-component system, there are $8 \times 7 \times 6 = 336$ possible nontrivial derivatives of the form of Eq. 4.1-1 to be considered. (In a binary system there is a ninth variable, composition, and three independent variables—temperature, specific volume, and composition, so that the thermodynamic partial derivatives of possible interest number in the thousands.) Therefore, it is necessary that a systematic procedure be developed for reducing any such derivative. Although not needed for the discussion that follows, such a procedure is introduced in Sec. 4.8.

ILLUSTRATION 4.2-2

For the discussion of the difference between the constant-pressure heat capacity C_P and the constant-volume heat capacity C_V, it is useful to have an expression for the derivative $(\partial \underline{S}/\partial T)_P$ in which T and $\underline{V}$ are the independent variables. Derive such an expression.

Solution

Starting from Eq. 4.2-19

$$d\underline{S} = \frac{C_V}{T}\, dT + \left(\frac{\partial P}{\partial T}\right)_{\underline{V}} d\underline{V}$$

we have

$$\left(\frac{\partial \underline{S}}{\partial T}\right)_P = \frac{C_V}{T}\left(\frac{\partial T}{\partial T}\right)_P + \left(\frac{\partial P}{\partial T}\right)_{\underline{V}}\left(\frac{\partial \underline{V}}{\partial T}\right)_P$$

$$= \frac{C_V}{T} + \left(\frac{\partial P}{\partial T}\right)_{\underline{V}}\left(\frac{\partial \underline{V}}{\partial T}\right)_P \qquad (4.2\text{-}26)$$

Now using the triple product rule

$$\left(\frac{\partial P}{\partial T}\right)_{\underline{V}}\left(\frac{\partial \underline{V}}{\partial P}\right)_T\left(\frac{\partial T}{\partial \underline{V}}\right)_P = -1 \qquad (4.2\text{-}27)$$

we get

$$\left(\frac{\partial \underline{S}}{\partial T}\right)_P = \frac{C_V}{T} - \left(\frac{\partial P}{\partial \underline{V}}\right)_T \left(\frac{\partial \underline{V}}{\partial T}\right)_P^2 \qquad (4.2\text{-}28a)$$

$$= \frac{C_V}{T} - \left(\frac{\partial \underline{V}}{\partial P}\right)_T \left(\frac{\partial P}{\partial T}\right)_{\underline{V}}^2 \qquad (4.2\text{-}28b)$$

and finally

$$\left(\frac{\partial \underline{S}}{\partial T}\right)_P = \frac{C_V}{T} + \frac{\underline{V}\,\alpha^2}{\kappa_T} \qquad (4.2\text{-}29)$$

■

In Eqs. 4.2-1 through 4, four partial derivatives that frequently occur were introduced. As has already been indicated, only three of these derivatives are independent in that given values of three derivatives, the value of the fourth is easily computed. We establish this here by deriving an equation that relates C_P to C_V, α, and κ_T. The starting point is the relation $C_P = T(\partial \underline{S}/\partial T)_P$ and the results developed in Illustration 4.2-2, which yield

$$C_P = T\left(\frac{\partial \underline{S}}{\partial T}\right)_P = C_V + T\left(\frac{\partial P}{\partial T}\right)_{\underline{V}} \left(\frac{\partial \underline{V}}{\partial T}\right)_P$$

$$= C_V - T\left(\frac{\partial P}{\partial \underline{V}}\right)_T \left(\frac{\partial \underline{V}}{\partial T}\right)_P^2$$

$$= C_V - T\left(\frac{\partial \underline{V}}{\partial P}\right)_T \left(\frac{\partial P}{\partial T}\right)_{\underline{V}}^2$$

$$= C_V + T\underline{V}\alpha^2/\kappa_T \qquad (4.2\text{-}30)$$

establishing that C_P, C_V, α, and κ_T are all interrelated.

For the discussion of the following section, we need to know the dependence of the constant-volume heat capacity on specific volume (or density) at constant temperature. To obtain $(\partial C_V/\partial \underline{V})_T$ we start with

$$d\underline{U} = C_V\,dT + \left[T\left(\frac{\partial P}{\partial T}\right)_{\underline{V}} - P\right] d\underline{V}$$

and note that

$$\left(\frac{\partial \underline{U}}{\partial T}\right)_{\underline{V}} = C_V \quad \text{and} \quad \left(\frac{\partial \underline{U}}{\partial \underline{V}}\right)_T = T\left(\frac{\partial P}{\partial T}\right)_{\underline{V}} - P$$

The use of Eq. 4.1-3 yields the desired result

$$\left.\frac{\partial}{\partial \underline{V}}\right|_T \left(\frac{\partial \underline{U}}{\partial T}\right)_{\underline{V}} = \left(\frac{\partial C_V}{\partial \underline{V}}\right)_T = T\left(\frac{\partial^2 P}{\partial T^2}\right)_{\underline{V}} = \left.\frac{\partial}{\partial T}\right|_{\underline{V}} \left(\frac{\partial \underline{U}}{\partial \underline{V}}\right)_T \qquad (4.2\text{-}31)$$

In a similar fashion, starting with Eq. 4.2-22, one obtains

$$\left(\frac{\partial C_P}{\partial P}\right)_T = -T\left(\frac{\partial^2 \underline{V}}{\partial T^2}\right)_P \qquad (4.2\text{-}32)$$

ILLUSTRATION 4.2-3

Develop expressions for the coefficient of thermal expansion α, the isothermal compressibility κ_T, the Joule–Thomson coefficient μ, and the difference $C_P - C_V$, for (a) the ideal gas and (b) a gas that obeys the equation of state

$$\left(P + \frac{a}{\underline{V}^2}\right)(\underline{V} - b) = RT \qquad (4.2\text{-}33a)$$

where a and b are constants. (The equation of state was developed by J. D. van der Waals in 1873, and fluids that obey this equation of state are called van der Waals fluids.)

Solution

a. For the ideal gas $P\underline{V} = RT$, thus

$$\left(\frac{\partial \underline{V}}{\partial T}\right)_P = \frac{R}{P} = \frac{\underline{V}}{T} \qquad \text{so that} \qquad \alpha = \frac{1}{\underline{V}}\left(\frac{\partial \underline{V}}{\partial T}\right)_P = \frac{1}{T}$$

and

$$\left(\frac{\partial \underline{V}}{\partial P}\right)_T = -\frac{\underline{V}}{P} \qquad \text{so that} \qquad \kappa_T = -\frac{1}{\underline{V}}\left(\frac{\partial \underline{V}}{\partial P}\right)_T = \frac{1}{P}$$

From Eq. 4.2-22 we have

$$d\underline{H} = C_P\, dT + \left[\underline{V} - T\left(\frac{\partial \underline{V}}{\partial T}\right)_P\right] dP$$

so that for $\underline{H}$ = constant ($d\underline{H} = 0$)

$$0 = C_P\, dT|_{\underline{H}} + \left[\underline{V} - T\left(\frac{\partial \underline{V}}{\partial T}\right)_P\right] dP|_{\underline{H}}$$

or

$$\left(\frac{\partial P}{\partial T}\right)_{\underline{H}} = \mu = -\frac{\left[\underline{V} - T\left(\frac{\partial \underline{V}}{\partial T}\right)_P\right]}{C_P} = -\frac{\underline{V}}{C_P}[1 - T\alpha]$$

(Note: The procedure followed here can also be used with Eqs. 4.2-5 through 8, 4.2-19 through 22, and elsewhere in the evaluation of thermodynamic partial derivatives.)

For the ideal gas

$$\mu = \left(\frac{\partial \underline{V}}{\partial T}\right)_{\underline{H}} = -\frac{\underline{V}}{C_P}\left[1 - T \cdot \frac{1}{T}\right] = 0$$

and

$$C_P = C_V + \frac{T\underline{V}\alpha^2}{\kappa_T} = C_V + T\underline{V}\frac{1}{T^2}P = C_V + \frac{P\underline{V}}{T} = C_V + R$$

b. For the van der Waals gas we first rewrite the equation of state as

$$T = \frac{P\underline{V}}{R} - \frac{Pb}{R} + \frac{a}{\underline{V}R} - \frac{ab}{R\underline{V}^2}$$

so that

$$\left(\frac{\partial T}{\partial \underline{V}}\right)_P = \frac{P}{R} - \frac{a}{R\underline{V}^2} + \frac{2ab}{R\underline{V}^3}$$

and

$$(\alpha)^{-1} = \underline{V}\left(\frac{\partial T}{\partial \underline{V}}\right)_P = \frac{P\underline{V}}{R} - \frac{a}{R\underline{V}} + \frac{2ab}{R\underline{V}^2}$$

Now rewriting the van der Waals equation as

$$P = \frac{RT}{\underline{V} - b} - \frac{a}{\underline{V}^2} \tag{4.2-33b}$$

allows us to eliminate P from the expression for α to obtain

$$(\alpha)^{-1} = \frac{T\underline{V}}{(\underline{V} - b)} - \frac{2a}{R\underline{V}} + \frac{2ab}{R\underline{V}^2} = \frac{T\underline{V}}{(\underline{V} - b)} - \frac{2a}{R\underline{V}^2}(\underline{V} - b)$$

An expression for κ_T is obtained as follows

$$\left(\frac{\partial P}{\partial \underline{V}}\right)_T = -\frac{RT}{(\underline{V} - b)^2} + \frac{2a}{\underline{V}^3}$$

or

$$(\kappa_T)^{-1} = \frac{RT\underline{V}}{(\underline{V} - b)^2} - \frac{2a}{\underline{V}^2} = \frac{R}{(\underline{V} - b)}\left[\frac{T\underline{V}}{(\underline{V} - b)} - \frac{2a}{R\underline{V}^2}(\underline{V} - b)\right]$$

$$= \frac{R}{(\underline{V} - b)}\alpha^{-1}$$

Consequently,

$$\mu = \left(\frac{\partial T}{\partial P}\right)_{\underline{H}} = -\frac{\underline{V}}{\underline{C}_P}(1 - T\alpha) = -\frac{\underline{V}}{\underline{C}_P}\left[1 - \frac{1}{\dfrac{\underline{V}}{\underline{V} - b} - \dfrac{2a}{RT}\left(\dfrac{\underline{V} - b}{\underline{V}^2}\right)}\right]$$

and

$$\underline{C}_P = \underline{C}_V + \frac{T\underline{V}\alpha^2}{\kappa_T} = \underline{C}_V + T\underline{V}\alpha^2\left(\frac{R}{(\underline{V} - b)\alpha}\right)$$

$$= \underline{C}_V + \frac{R}{1 - \dfrac{2a}{RT}\dfrac{(\underline{V} - b)^2}{\underline{V}^3}}$$ ■

Finally, it is useful to note that although the molar internal energy can be considered to be a function of any two state variables, an equation of state that gives the internal energy as a function of entropy and volume is, in principle, more useful than an equation of state for the internal energy in terms of temperature and volume or any other pair of state variables. To see that this is so, suppose we had equations of state of the form $\underline{U} = \underline{U}(\underline{S}, \underline{V})$ and $\underline{U} = \underline{U}(T, \underline{V})$. From Eq. 4.2-5b it is evident that we could differentiate the first equation of state to get other thermodynamic functions directly. For example,

$$T = \left(\frac{\partial \underline{U}}{\partial \underline{S}}\right)_{\underline{V}} \qquad P = -\left(\frac{\partial \underline{U}}{\partial \underline{V}}\right)_{\underline{S}}$$

However, using the second equation of state and Eq. 4.2-21 we obtain

$$\left(\frac{\partial \underline{U}}{\partial T}\right)_{\underline{V}} = \underline{C}_V \quad \text{and} \quad \left(\frac{\partial \underline{U}}{\partial \underline{V}}\right)_V = T\left(\frac{\partial P}{\partial T}\right)_{\underline{V}} - P$$

In this case, on differentiation, we do not obtain thermodynamic state functions directly, but rather derivatives of state functions or combinations of state functions and their derivatives.

In a similar fashion it is possible to show that an equation of state that relates $\underline{H}$, $\underline{S}$, and P, or $\underline{A}$, $\underline{V}$ and T, or $\underline{G}$, P and T is more useful than other equations of state. Equations of state relating ($\underline{S}$, $\underline{U}$, and $\underline{V}$), ($\underline{H}$, $\underline{S}$, and P), ($\underline{A}$, $\underline{V}$, and T), or ($\underline{G}$, T, and P) are called **fundamental equations of state,** a term first used by the American physicist Josiah Willard Gibbs in 1873. Unfortunately, in general we do not have the information to obtain or construct a fundamental equation of state. More commonly, we have only a volumetric equation of state, that is, an equation relating P, $\underline{V}$, and T.

ILLUSTRATION 4.2-4

Show that from an equation of state relating the Gibbs free energy, temperature, and pressure, equations of state for all other state functions (and their derivatives, as well) can be obtained by appropriate differentiation.

Solution

Suppose we had an equation of state of the form $\underline{G} = \underline{G}(T, P)$. The entropy and volume, as a function of temperature and pressure, are then immediately obtained using Eqs. 4.2-12

$$\underline{S}(T, P) = -\left(\frac{\partial \underline{G}}{\partial T}\right)_P \quad \text{and} \quad \underline{V}(T, P) = \left(\frac{\partial \underline{G}}{\partial P}\right)_T$$

Next, the enthalpy and internal energy can be found as follows:

$$\underline{H}(T, P) = \underline{G}(T, P) + T\underline{S}(T, P) = \underline{G}(T, P) - T\left(\frac{\partial \underline{G}}{\partial T}\right)_P$$

and

$$\underline{U}(T, P) = \underline{G}(T, P) + T\underline{S}(T, P) - P\underline{V}(T, P) = \underline{G}(T, P) - T\left(\frac{\partial \underline{G}}{\partial T}\right)_P - P\left(\frac{\partial \underline{G}}{\partial P}\right)_T$$

The Helmholtz free energy is obtained from

$$\underline{A}(T, P) = \underline{G}(T, P) - P\underline{V} = \underline{G}(T, P) - P\left(\frac{\partial \underline{G}}{\partial P}\right)_T$$

and the constant-pressure and constant-volume heat capacities can then be found as follows:

$$C_P(T, P) = \left(\frac{\partial \underline{H}}{\partial T}\right)_P = \left.\frac{\partial}{\partial T}\right|_P \left[\underline{G}(T, P) - T\left(\frac{\partial \underline{G}}{\partial T}\right)_P\right] = -T\left(\frac{\partial^2 \underline{G}}{\partial T^2}\right)_P$$

and

$$C_V(T, P) = C_P + \frac{T(\partial \underline{V}/\partial T)_P^2}{(\partial \underline{V}/\partial P)_T} = -T\left(\frac{\partial^2 \underline{G}}{\partial T^2}\right)_P - T\left(\frac{\partial^2 \underline{G}}{\partial T \partial P}\right)^2 \left(\frac{\partial^2 \underline{G}}{\partial P^2}\right)_T^{-1}$$

Finally, the isothermal compressibility κ_T and coefficient of thermal expansion α are found as indicated here:

$$\kappa_T = -\frac{1}{\underline{V}}\left(\frac{\partial \underline{V}}{\partial P}\right)_T = -\frac{(\partial^2 \underline{G}/\partial P^2)_T}{(\partial \underline{G}/\partial P)_T}$$

and

$$\alpha = \frac{1}{\underline{V}}\left(\frac{\partial \underline{V}}{\partial T}\right)_P = \frac{(\partial^2 \underline{G}/\partial P \partial T)}{(\partial \underline{G}/\partial P)_T}$$

Thus, if the fundamental equation of state for a substance in the form $\underline{G} = \underline{G}(T, P)$ were available, we could, using these relations, obtain equations relating all other state variables for this substance to temperature and pressure by taking the appropriate derivatives of the fundamental equation.

Questions

1 Why do we need two equations, a volumetric equation of state $P = P(T, \underline{V})$ and a thermal equation of state $\underline{U} = \underline{U}(T, \underline{V})$, to define an ideal gas?

2 Can you develop a single equation of state that would completely specify all the properties of an ideal gas? (See Problem 5.7.) ■

Although fundamental equations of state are, in principle, the most useful thermodynamic descriptions of any substance, it is unlikely that such equations will be available for fluids of interest to engineers. In fact, frequently only heat capacity and $P\underline{V}T$ data are available; Section 4.4 examines how these data are used in thermodynamic problem solving.

4.3 THE IDEAL GAS AND ABSOLUTE TEMPERATURE SCALES

In Chapters 1 and 2 we introduced the concept of the ideal gas and suggested, without proof, that if the ideal gas were used to establish a scale of temperature, an absolute and universal or thermodynamic scale would be obtained. We now have developed sufficient thermodynamic theory to prove this to be the case.

We take as the starting point for this discussion the facts that the product $P\underline{V}$ and the internal energy $\underline{U}$ of an ideal gas are both unspecified, but increasing functions of the absolute temperature T and independent of density, pressure, or specific volume (see Sec. 2.4). To be perfectly general at this point we denote these characteristics by

$$P\underline{V} = RT^{IG} = \Theta_1(T) \tag{4.3-1}$$

and

$$\underline{U} = \Theta_2(T) \tag{4.3-2}$$

where T^{IG} is the temperature on the ideal gas temperature scale. To prove the equality of the ideal gas and thermodynamic temperature scales it is necessary to establish that $\Theta_1(T)$ is a linear function T as was suggested in Eqs. 1.4-2 and 2.4-1.

It is important to note that in the balance equations the thermodynamic temperature first appears in the introduction of the entropy function; in particular, in the $\dot{Q}/T$ term in Eq. 3.1-5. Thus, it is the thermodynamic temperature T that appears in all equations derived from Eq. 3.1-5, and therefore in the equations of Sec. 4.2. From Eq. 4.2-21 we have, in general, that

$$d\underline{U} = C_v\, dT + \left[T\left(\frac{\partial P}{\partial T}\right)_{\underline{V}} - P\right] d\underline{V}$$

whereas from Eq. 4.3-2 we have, for the ideal gas

$$d\underline{U} = \frac{d\Theta_2(T)}{dT}\, dT \tag{4.3-3}$$

The only way to reconcile these two equations is for the coefficient of the $d\underline{V}$ term in Eq. 4.2-21 to be zero for the ideal gas, that is, for

$$P = T\left(\frac{\partial P}{\partial T}\right)_{\underline{V}}$$

This implies that for changes at constant volume

$$\left.\frac{dP}{P}\right|_{\underline{V}} = \left.\frac{dT}{T}\right|_{\underline{V}}$$

or

$$\frac{P_2}{P_1} = \frac{T_2}{T_1} \tag{4.3-4}$$

for any two states 1 and 2 with the same specific volume. However, from Eq. 4.3-1 we have, under these circumstances, that

$$\frac{P_2}{P_1} = \frac{\Theta_1(T_2)}{\Theta_1(T_1)} \tag{4.3-5}$$

To satisfy Eqs. 4.3-4 and 5, $\Theta_1(T)$ must be a linear function of temperature, that is

$$\Theta_1(T) = RT \tag{4.3-6}$$

where R is a constant related to the unit of a degree (see Sec. 1.4). Using Eq. 4.3-6 in Eq. 4.3-1 yields

$$P\underline{V} = RT$$

which establishes that the ideal gas temperature scale is also a thermodynamic temperature scale.

4.4 THE EVALUATION OF CHANGES IN THE THERMODYNAMIC PROPERTIES OF REAL SUBSTANCES ACCOMPANYING A CHANGE OF STATE

The Necessary Data

In order to use the energy and entropy balances for any real substance for which thermodynamic tables are not available, we must be able to compute the changes in its internal energy, enthalpy, and entropy for any change of state. The equations of Sec. 4.2 provide the basis for such computations. However, before we discuss these calculations it is worthwhile to consider the minimum amount of information needed and the form in which this information is likely to be available.

VOLUMETRIC EQUATION OF STATE INFORMATION

Clearly, to use Eqs. 4.2-19 through 22 we need volumetric equation of state data, that is, information on the interrelationship between P, $\underline{V}$, and T. This information may be available as tables of experimental data, or, more frequently, as equations with parameters that have been fitted to experimental data. Hundreds of analytic equations of state have been suggested for the correlation of $P\underline{V}T$ data.

The equation

$$\left(P + \frac{a}{\underline{V}^2}\right)(\underline{V} - b) = RT \tag{4.2-33a}$$

or equivalently

$$P = \frac{RT}{\underline{V} - b} - \frac{a}{\underline{V}^2} \tag{4.2-33b}$$

with a and b being constants, was proposed by J. D. van der Waals in 1873 to describe the volumetric or $P\underline{V}T$ behavior of both vapors and liquids, work for which he was awarded the Nobel Prize in physics in 1910. The constants a and b can be determined either by fitting this equation to experimental data or, more commonly, from critical-point data as described in Sec. 4.6. Values for the parameters for several gases appear in Table 4.4-1.

The van der Waals equation of state is not very accurate and is mainly of historic interest in that it was the first equation capable of predicting the transition between vapor and liquid; this will be discussed in Sec. 5.3. It has also been the prototype for modern, more accurate equations of state such as, for example, those of Redlich–Kwong (1949),[2]

$$P = \frac{RT}{\underline{V} - b} - \frac{a}{T^{1/2}\underline{V}(\underline{V} + b)} \tag{4.4-1}$$

of Soave (1972),[3] in which the $a/T^{1/2}$ term in Eq. 4.4-1 is replaced with $a(T)$, a function of temperature, and of Peng and Robinson (1976),[4]

$$P = \frac{RT}{\underline{V} - b} - \frac{a(T)}{\underline{V}(\underline{V} + b) + b(\underline{V} - b)} \tag{4.4-2}$$

Table 4.4-1
PARAMETERS FOR THE VAN DER WAALS EQUATION OF STATE

Gas	a, Pa m^6/mol^2	b, (m^3/mol) × 10^5
O_2	0.1381	3.184
N_2	0.1368	3.864
H_2O	0.5542	3.051
CH_4	0.2303	4.306
CO	0.1473	3.951
CO_2	0.3658	4.286
NH_3	0.4253	3.737
H_2	0.0248	2.660
He	0.00346	2.376

These parameters were computed from critical-point data as described in Sec. 4.6.

[2] O. Redlich and J. N. S. Kwong, *Chem. Rev.* **44,** 233 (1949).
[3] G. Soave, *Chem. Eng. Sci.* **27,** 1197 (1972).
[4] D.-Y. Peng and D. B. Robinson, *I.E.C. Fund.* **15,** 59 (1976).

Equations 4.2-33b, 4.4-1, and 4.4-2 are special cases of the general class of equations of state

$$P = \frac{RT}{\underline{V} - b} - \frac{(\underline{V} - \eta)\theta}{(\underline{V} - b)(\underline{V}^2 + \delta\underline{V} + \varepsilon)} \tag{4.4-3}$$

where each of the five parameters b, θ, δ, ε, and η can depend on temperature. In practice, however, generally only θ is taken to be a function of temperature, and it is adjusted to give the correct boiling temperature as a function of pressure; this will be discussed in Sec. 5.5. Table 4.4-2 gives the parameters of Eq. 4.4-3 for some common equations of state from among the hundreds of this class that have been published. Clearly, many other choices are possible.

Numeric values for equation of state parameters are commonly obtained in one of two ways. First, parameters can be obtained by fitting the equation to $P\underline{V}T$ and other data for the fluid of interest; this leads to the most accurate values, but is very tedious. Second, as will be discussed in Sec. 4.7, general relations can be obtained between the equation of state parameters and critical-point properties. From these equations, somewhat less accurate parameter values are easily obtained from only critical-point properties.

Each of the equations of state discussed here can be written in the form

$$Z^3 + \alpha Z^2 + \beta Z + \gamma = 0 \tag{4.4-4}$$

where $Z = P\underline{V}/RT$ is **compressibility factor,** and the parameters α, β, and γ for some representative equations of state are given in Table 4.4-3. Consequently, these equations are said to be cubic equations of state. Many such equations

Table 4.4-2
PARAMETERS FOR CUBIC EQUATIONS OF STATE

Author	Year	θ	η	δ	ϵ	$P = \frac{RT}{\underline{V} - b} - \Delta$ Δ
van der Waals	1873	a	b	0	0	$\frac{a}{\underline{V}^2}$
Clausius	1880	a/T	b	$2c$	c^2	$\frac{a/T}{(\underline{V} + c)^2}$
Berthelot	1899	a/T	b	0	0	$\frac{a/T}{\underline{V}^2}$
Redlich–Kwong	1949	$a/\sqrt{T}$	b	b	0	$\frac{a/\sqrt{T}}{\underline{V}(\underline{V} + b)}$
Soave	1972	$\theta_S(T)$	b	b	0	$\frac{\theta_S(T)}{\underline{V}(\underline{V} + b)}$
Lee–Erbar–Edmister	1973	$\theta_L(T)$	$\eta(T)$	b	0	$\frac{\theta_L(T)[V - \eta(T)]}{(\underline{V} - b)(\underline{V} + b)}$
Peng–Robinson	1976	$\theta_{PR}(T)$	b	$2b$	$-b^2$	$\frac{\theta_{PR}(T)}{\underline{V}(\underline{V} + b) + b(\underline{V} - b)}$
Patel–Teja	1981	$\theta_{PT}(T)$	b	$b + c$	$-cb$	$\frac{\theta_{PT}(T)}{\underline{V}(\underline{V} + b) + c(\underline{V} - b)}$

Note: If $\eta = b$, Eq. 4.4-3 reduces to

$$P = \frac{RT}{\underline{V} - b} - \frac{\theta}{\underline{V}^2 + \delta\underline{V} + \varepsilon}$$

Table 4.4-3
PARAMETERS IN EQ. 4.4-4 FOR THREE EQUATIONS OF STATE

	van der Waals	Redlich–Kwong and Soave	Peng–Robinson
α	$-1 - B$	-1	$-1 + B$
β	A	$A - B - B^2$	$A - 3B^2 - 2B$
γ	$-AB$	$-AB$	$-AB + B^2 + B^3$

with

$$Z = P\underline{V}/RT$$

$$B = bP/RT$$

and

$$A = \begin{cases} aP/(RT)^2 & \text{in the van der Waals, Soave and Peng–Robinson equations of state} \\ aP/(RT)^2\sqrt{T} & \text{in the Redlich–Kwong equation of state} \end{cases}$$

have been suggested in the scientific literature. One should remember that all cubic equations of state are approximate; generally they provide a reasonable description of the $P\underline{V}T$ behavior in both the vapor and liquid regions for hydrocarbons, and of the vapor region only for many other pure fluids. Equations 4.4-1 and 2 are, at present, the most commonly used cubic equations of state.

A different type of equation of state is the virial equation

$$\frac{P\underline{V}}{RT} = 1 + \frac{B(T)}{\underline{V}} + \frac{C(T)}{\underline{V}^2} + \cdots \tag{4.4-5}$$

where $B(T)$ and $C(T)$ are the temperature-dependent second and third virial coefficients. Although higher order terms can be defined in a similar fashion, data generally are available only for the second virial coefficient.[5] The virial equation was first used by H. Kamerlingh Onnes in 1901 and is of theoretical interest since it can be derived from statistical mechanics with explicit expressions obtained for the virial coefficients in terms of the potential function between molecules. The virial equation of state is a power series expansion in density (or reciprocal volume) about the ideal gas result ($P\underline{V}/RT = 1$). With a sufficient number of coefficients, the virial equation can give excellent vapor-phase predictions, but it is not applicable to the liquid phase. When truncated at the $B(T)$ term, as is usually the case because of lack of higher virial coefficient data, the virial equation of state can be used only at low densities; as a general rule it should not be used at pressures above 10 bar for most fluids.

There are other more complicated equations of state that accurately predict the $P\underline{V}T$ behavior in most of the vapor and liquid regions; such equations con-

[5]A recent tabulation *The Virial Coefficients of Pure Gases and Mixtures. A Critical Evaluation* by J. H. Dymond and E. B. Smith, Clarendon Press, Oxford, 1980, provides second virial coefficient data for more than 250 compounds.

tain the reciprocal volume in both integral powers (like the virial equation) and exponential functions. One example is the equation of Benedict, Webb, and Rubin (1940)[6]

$$\frac{P\underline{V}}{RT} = 1 + \left(B - \frac{A}{RT} - \frac{C}{RT^3}\right)\frac{1}{\underline{V}} + \left(b - \frac{a}{RT}\right)\frac{1}{\underline{V}^2} + \frac{a\alpha}{RT\underline{V}^5} + \frac{\beta}{RT^3\underline{V}}\left(1 + \frac{\gamma}{\underline{V}^2}\right)\exp(-\gamma/\underline{V}^2) \tag{4.4-6a}$$

where the eight constants a, b, A, B, C, α, β, and γ are specific to each fluid and are obtained by fitting the equation of state to a variety of experimental data. The exponential term in this equation (and others of its class) is meant to compensate for the truncation of the virial series since, if expanded in a Taylor series around $\underline{V} = \infty$, the exponential function generates an infinite number of terms. The 20-constant Bender equation (1970)[7]

$$P = \frac{T}{\underline{V}}\left[R + \frac{B}{\underline{V}} + \frac{C}{\underline{V}^2} + \frac{D}{\underline{V}^3} + \frac{E}{\underline{V}^4} + \frac{F}{\underline{V}^5} + \left(G + \frac{H}{\underline{V}^2}\right)\frac{1}{\underline{V}^2}\exp(-a_{20}/\underline{V}^2)\right] \tag{4.4-6b}$$

with

$$B = a_1 - a_2/T - a_3/T^2 - a_4/T^3 - a_5/T^4$$

$$C = a_6 + a_7/T + a_8/T^2$$

$$D = a_9 + a_{10}/T$$

$$E = a_{11} + a_{12}/T$$

$$F = a_{13}/T$$

$$G = a_{14}/T^3 + a_{15}/T^4 + a_{16}/T^5$$

and

$$H = a_{17}/T^3 + a_{18}/T^4 + a_{19}/T^5$$

is another example of an equation of this type. Although such equations provide more accurate descriptions of fluid behavior, including the vapor–liquid phase transition, than simple cubic equations of state, they are only useful if digitial computation facilities are available. Furthermore, the coefficients that appear in these equations known only for light hydrocarbons and a few other substances.

For an evaluation of volumetric equations of state important in engineering, refer to R. C. Reid, J. M. Prausnitz, and B. Poling, *The Properties of Gases and Liquids*, 4th ed. (McGraw–Hill, New York, 1987). In the discussion that follows, we will generally assume that a volumetric equation of state in analytic form is available.

HEAT CAPACITY DATA

It is also evident from Eqs. 4.2-19 through 22 that data for C_P and C_V are needed to compute changes in thermodynamic properties. At first glance it might appear that we need data for C_V as a function of T and $\underline{V}$, and C_P as a function of T and P for each fluid over the complete range of conditions of interest. However,

[6]M. Benedict, G. B. Webb, and L. C. Rubin, *J. Chem. Phys.* **8**, 334 (1940), and later papers by the same authors.

[7]E. Bender, *5th Symp. Thermophys. Prop.* (ASME, New York, 1970), p. 227, and later papers by the same author.

from Eqs. 4.2-31 and 32 it is clear that our need for heat capacity data is much more modest once we have volumetric equation of state information. To see this, consider the situation in which we have data for C_P as a function of temperature at a pressure P_1, and want C_P as a function of temperature at another pressure P_2. At each temperature we can integrate Eq. 4.2-32 to obtain the desired result

$$\int_{P_1,T}^{P_2,T} dC_P = C_P(P_2, T) - C_P(P_1, T) = -T \int_{P_1,T}^{P_2,T} \left(\frac{\partial^2 \underline{V}}{\partial T^2}\right)_P dP \tag{4.4-7}$$

or

$$C_P(P_2, T) = C_P(P_1, T) - T \int_{P_1,T}^{P_2,T} \left(\frac{\partial^2 \underline{V}}{\partial T^2}\right)_P dP \tag{4.4-8}$$

Here we have included T in the limits of integration to stress that the integration is carried out over pressure at a fixed value of temperature. Similarly, for the constant-volume heat capacity one obtains (from Eq. 4.2-31)

$$C_V(\underline{V}_2, T) = C_V(\underline{V}_1, T) + T \int_{\underline{V}_1,T}^{\underline{V}_2,T} \left(\frac{\partial^2 P}{\partial T^2}\right)_{\underline{V}} d\underline{V} \tag{4.4-9}$$

Therefore, given the volumetric equation of state (or, equivalently, a numerical tabulation of the volumetric data for a fluid) and heat capacity data as a function of temperature at a single pressure or volume, the value of the heat capacity in any other state can be computed.

In practice, heat capacity data are tabulated only for states of very low pressure or, equivalently, large specific volume, where all fluids are ideal gases.[8] Therefore, if P_1 and $\underline{V}_1$ are taken as 0 and ∞, respectively, in Eqs. 4.4-8 and 9, we obtain

$$C_P(P, T) = C_P^*(T) - T \int_{P=0,T}^{P,T} \left(\frac{\partial^2 \underline{V}}{\partial T^2}\right)_P dP \tag{4.4-10}$$

and

$$C_V(\underline{V}, T) = C_V^*(T) + T \int_{\underline{V}=\infty,T}^{\underline{V},T} \left(\frac{\partial^2 P}{\partial T^2}\right)_{\underline{V}} d\underline{V} \tag{4.4-11}$$

where we have used the notation that

$$C_P^*(T) = C_P(P = 0, T)$$

and

$$C_V^*(T) = C_V(\underline{V} = \infty, T)$$

Data for C_P^* and C_V^* are available in most data reference books, such as the *Chemical Engineers' Handbook*[9] or *The Handbook of Chemistry and Physics.*[10] This information is frequently presented in the form

$$C_P^*(T) = a + bT + cT^2 + dT^3 + \cdots$$

See Appendix II for C_P^* data for some compounds.

[8]That all fluids become ideal gases at large specific volumes is easily verified by observing that all volumetric equations of state (e.g., Eqs. 4.2-33, 4.4-1, 2, and 3) reduce to $P\underline{V} = RT$ in the limit of $\underline{V} \to \infty$.

[9]R. H. Perry and D. Green, eds., *Chemical Engineers' Handbook*, 6th ed., McGraw–Hill, New York (1984).

[10]R.C. Weast, ed., *The Handbook of Chemistry and Physics*, Cleveland Chemical Rubber Co. This handbook is updated annually.

The Evaluation of $\Delta \underline{H}$, $\Delta \underline{U}$, and $\Delta \underline{S}$

To compute the change in enthalpy in going from the state (T_1, P_1) to the state (T_2, P_2), we start from

$$\Delta \underline{H} = \underline{H}(T_2, P_2) - \underline{H}(T_1, P_1) = \int_{T_1,P_1}^{T_2,P_2} d\underline{H} \tag{4.4-12}$$

and note that since enthalpy is a state function, we can compute its change between two states by evaluating the integral along any convenient path. In particular, if the path indicated by the solid line in Fig. 4.4-1 is used (isothermal expansion followed by isobaric heating and isothermal compression), we have, from Eq. 4.2-22,

$$\Delta \underline{H} = \int_{P_1,T_1}^{P=0,T_1} \left[\underline{V} - T\left(\frac{\partial \underline{V}}{\partial T}\right)_P\right] dP + \int_{T_1,P=0}^{T_2,P=0} C_P^* \, dT + \int_{P=0,T_2}^{P_2,T_2} \left[\underline{V} - T\left(\frac{\partial \underline{V}}{\partial T}\right)_P\right] dP \tag{4.4-13}$$

Alternatively, we could compute the enthalpy change using the path indicated by the dashed line in Fig. 4.4-1; isobaric heating followed by isothermal compression. For this path

$$\begin{aligned}\Delta \underline{H} &= \int_{P_1,T_1}^{P_1,T_2} C_P \, dT + \int_{P_1,T_2}^{P_2,T_2} \left[\underline{V} - T\left(\frac{\partial \underline{V}}{\partial T}\right)_P\right] dP \\ &= \int_{T_1}^{T_2} C_P^* \, dT - \int_{T_1}^{T_2} T\left\{\int_0^{P_1} \left(\frac{\partial^2 \underline{V}}{\partial T^2}\right)_P dP\right\} dT \\ &\quad + \int_{P_1,T_2}^{P_2,T_2} \left[\underline{V} - T\left(\frac{\partial \underline{V}}{\partial T}\right)_P\right] dP \end{aligned} \tag{4.4-14a}$$

where in going from the first to the second of these equations we have used Eq. 4.4-10.

The equality of Eqs. 4.4-13 and 14 is easily established as follows. First we note that

$$\begin{aligned}\int_{T_1}^{T_2} T\left\{\int_0^{P_1} \left(\frac{\partial^2 \underline{V}}{\partial T^2}\right)_P dP\right\} dT &= \int_0^{P_1} \left\{\int_{T_1}^{T_2} T\left(\frac{\partial^2 \underline{V}}{\partial T^2}\right)_P dT\right\} dP \\ &= \int_{P=0}^{P_1} \left\{\int_{T_1}^{T_2} \frac{\partial}{\partial T}\bigg|_P \left[T\left(\frac{\partial \underline{V}}{\partial T}\right)_P - \underline{V}\right] dT\right\} dP \\ &= \int_{P=0,T_2}^{P_1,T_2} \left[T\left(\frac{\partial \underline{V}}{\partial T}\right)_P - \underline{V}\right] dP - \int_{P=0,T_1}^{P_1,T_1} \left[T\left(\frac{\partial \underline{V}}{\partial T}\right)_P - \underline{V}\right] dP \end{aligned} \tag{4.4-15}$$

where we have used the fact that the order integration with respect to T and P can be interchanged, and then recognized that $T(\partial^2 \underline{V}/\partial T^2)$ has an exact differential. Next, substituting Eq. 4.4-15 in Eq. 4.4-14 yields Eq. 4.4-13, verifying that the enthalpy change between given initial and final states is independent of the path used in its computation.

Using the solid-line path in Fig. 4.4-1 and Eq. 4.2-20 we obtain

$$\Delta \underline{S} = -\int_{P_1,T_1}^{P=0,T_1} \left(\frac{\partial \underline{V}}{\partial T}\right)_P dP + \int_{T_1}^{T_2} \frac{C_P^*}{T} \, dT - \int_{P=0,T_2}^{P_2,T_2} \left(\frac{\partial \underline{V}}{\partial T}\right)_P dP \tag{4.4-16a}$$

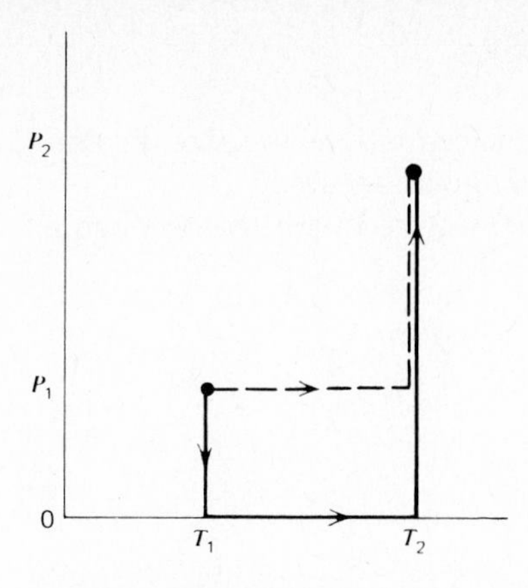

Figure 4.4-1
Two paths for the integration of Eq. 4.4-12.

By following the same argument, we can show that the entropy function is also path independent.

The path in the $\underline{V}$–T plane analogous to the solid-line path in the P–T plane of Fig. 4.4-1 is shown in Fig. 4.4-2. Here the gas is first isothermally expanded to zero pressure (and, hence, infinite volume), heated (at $\underline{V} = \infty$) from T_1 to T_2, and then compressed to a specific volume $\underline{V}_2$. The entropy and internal energy changes from Eqs. 4.2-19 and 21 are

$$\Delta \underline{S} = \int_{\underline{V}_1, T_1}^{\underline{V}=\infty, T_1} \left(\frac{\partial P}{\partial T}\right)_{\underline{V}} d\underline{V} + \int_{T_1}^{T_2} \frac{C_V^*}{T}\, dT + \int_{\underline{V}=\infty, T_2}^{\underline{V}_2, T_2} \left(\frac{\partial P}{\partial T}\right)_{\underline{V}} d\underline{V} \tag{4.4-17a}$$

and

$$\Delta \underline{U} = \int_{\underline{V}_1, T_1}^{\underline{V}=\infty, T_1} \left[T\left(\frac{\partial P}{\partial T}\right)_{\underline{V}} - P\right] d\underline{V} + \int_{T_1}^{T_2} C_V^*\, dT + \int_{\underline{V}=\infty, T_2}^{\underline{V}_2, T_2} \left[T\left(\frac{\partial P}{\partial T}\right)_{\underline{V}} - P\right] d\underline{V} \tag{4.4-18}$$

It can be easily shown that alternative paths lead to identical results for $\Delta \underline{S}$ and $\Delta \underline{U}$.

Note that only C_P^*, C_V^* and terms related to the volumetric equation of state appear in Eqs. 4.4-13, 14a, 16a, 17a, and 18. Thus, as has already been pointed out, we do not need heat capacity data at all densities, but merely in the low density limit and volumetric equation of state information.

Given C_P^* or C_V^* data, and volumetric equation of state data (in either analytic or tabular form), it is possible to compute $\Delta \underline{H}$, $\Delta \underline{U}$, and $\Delta \underline{S}$ for any two states of a fluid; given the value of $\underline{S}$ in any one state, it is also possible to compute $\Delta \underline{G}$ and $\Delta \underline{A}$. Thus, we now have the equations necessary to construct complete tables or charts interrelating $\underline{H}$, $\underline{U}$, $\underline{S}$, T, P, and $\underline{V}$, such as those given in Chapter 2. Although the process of constructing thermodynamic properties tables and charts is tedious (see Illustration 4.4-1), their availability, as we saw in

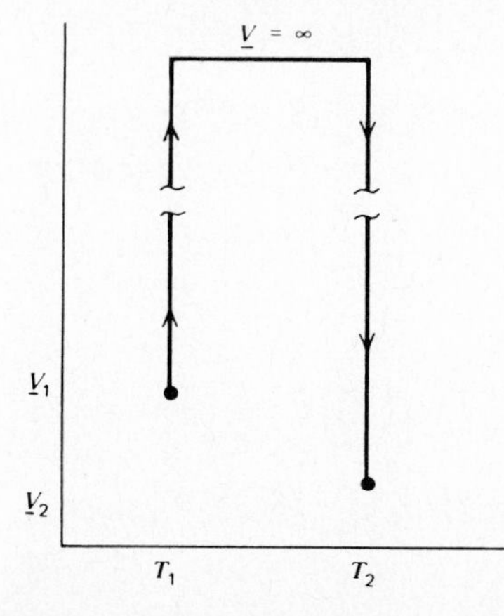

Figure 4.4-2
Integration path in the $\underline{V}$–T Plane.

Chapters 2 and 3, makes it possible to use the balance equations to solve thermodynamic problems for real fluids quickly and with good accuracy.

It is interesting to compare the equations just derived with the analogous results for an ideal gas. Since

$$\left(\frac{\partial \underline{V}}{\partial T}\right)_P^{IG} = \frac{\underline{V}}{T} = \frac{R}{P} \qquad \text{so} \qquad \underline{V}^{IG} - T\left(\frac{\partial \underline{V}}{\partial T}\right)_P^{IG} = 0$$

and

$$\left(\frac{\partial P}{\partial T}\right)_{\underline{V}}^{IG} = \frac{R}{\underline{V}} = \frac{P}{T}$$

we have that

$$\underline{H}^{IG}(T_2, P_2) - \underline{H}^{IG}(T_1, P_1) = \int_{T_1,P=0}^{T_2,P=0} C_P^* \, dT \tag{4.4-14b}$$

$$\begin{aligned}\underline{S}^{IG}(T_2, P_2) - \underline{S}^{IG}(T_1, P_1) &= -\int_{T_1,P_1}^{T_1,P=0} \frac{R}{P} \, dP + \int_{T_1}^{T_2} \frac{C_P^*}{T} \, dT - \int_{T_2,P=0}^{T_2,P_2} \frac{R}{P} \, dP \\ &= \int_{T_1}^{T_2} \frac{C_P^*}{T} \, dT - R \ln \frac{P_2}{P_1}\end{aligned} \tag{4.4-16b}$$

$$\begin{aligned}\underline{S}^{IG}(T_2, \underline{V}_2) - \underline{S}^{IG}(T_1, \underline{V}_1) &= +\int_{T_1,\underline{V}_1}^{T_1,\underline{V}=\infty} \frac{R}{\underline{V}} \, d\underline{V} + \int_{T_1}^{T_2} \frac{C_V^*}{T} \, dT + \int_{T_2,\underline{V}=\infty}^{T_2,\underline{V}_2} \frac{R}{\underline{V}} \, d\underline{V} \\ &= \int_{T_1}^{T_2} \frac{C_V^*}{T} \, dT + R \ln \frac{\underline{V}_2}{\underline{V}_1}\end{aligned} \tag{4.4-17b}$$

Thus, comparing Eqs. 4.4-14a and b, Eqs. 4.4-17a and b, and Eqs. 4.4-16a and b, we have for the real fluid that

$$\begin{aligned}\underline{H}(T_2, P_2) - \underline{H}(T_1, P_1) &= \underline{H}^{IG}(T_2, P_2) - \underline{H}^{IG}(T_1, P_1) \\ &\quad + \int_{T_1,P_1}^{T_1,P=0} \left[\underline{V} - T\left(\frac{\partial \underline{V}}{\partial T}\right)_P\right] dP + \int_{T_2,P=0}^{T_2,P_2} \left[\underline{V} - T\left(\frac{\partial \underline{V}}{\partial T}\right)_P\right] dP \\ &= \underline{H}^{IG}(T_2, P_2) - \underline{H}^{IG}(T_1, P_1) + (\underline{H} - \underline{H}^{IG})_{T_2,P_2} - (\underline{H} - \underline{H}^{IG})_{T_1,P_1}\end{aligned} \tag{4.4-19}$$

$$\begin{aligned}\underline{S}(T_2, P_2) - \underline{S}(T_1, P_1) &= \underline{S}^{IG}(T_2, P_2) - \underline{S}^{IG}(T_1, P_1) \\ &\quad - \int_{T_1,P_1}^{T_1,P=0} \left[\left(\frac{\partial \underline{V}}{\partial T}\right)_P - \frac{R}{P}\right] dP - \int_{T_2,P=0}^{T_2,P_2} \left[\left(\frac{\partial \underline{V}}{\partial T}\right)_P - \frac{R}{P}\right] dP \\ &= \underline{S}^{IG}(T_2, P_2) - \underline{S}^{IG}(T_1, P_1) + (\underline{S} - \underline{S}^{IG})_{T_2,P_2} - (\underline{S} - \underline{S}^{IG})_{T_1,P_1}\end{aligned} \tag{4.4-20}$$

and

$$\begin{aligned}\underline{S}(T_2, \underline{V}_2) - \underline{S}(T_1, \underline{V}_1) &= \underline{S}^{IG}(T_2, \underline{V}_2) - \underline{S}^{IG}(T_1, \underline{V}_1) \\ &\quad + \int_{T_1,\underline{V}_1}^{T_1,\underline{V}_1=\infty} \left[\left(\frac{\partial P}{\partial T}\right)_{\underline{V}} - \frac{R}{\underline{V}}\right] d\underline{V} + \int_{T_2,\underline{V}=\infty}^{T_2,\underline{V}_2} \left[\left(\frac{\partial P}{\partial T}\right)_{\underline{V}} - \frac{R}{\underline{V}}\right] d\underline{V} \\ &= \underline{S}^{IG}(T_2, \underline{V}_2) - \underline{S}^{IG}(T_1, \underline{V}_1) + (\underline{S} - \underline{S}^{IG})_{T_2,\underline{V}_2} - (\underline{S} - \underline{S}^{IG})_{T_1,\underline{V}_1}\end{aligned} \tag{4.4-21}$$

where

$$(\underline{H} - \underline{H}^{IG})_{T,P} = \int_{T,P=0}^{T,P} \left[\underline{V} - T\left(\frac{\partial \underline{V}}{\partial T}\right)_P \right] dP \tag{4.4-22}$$

$$(\underline{S} - \underline{S}^{IG})_{T,P} = -\int_{T,P=0}^{T,P} \left[\left(\frac{\partial \underline{V}}{\partial T}\right)_P - \frac{R}{P} \right] dP \tag{4.4-23}$$

and

$$(\underline{S} - \underline{S}^{IG})_{T,\underline{V}} = S(T,\underline{V}) - \underline{S}^{IG}(T,\underline{V}) = \int_{T,\underline{V}=\infty}^{T,V} \left[\left(\frac{\partial P}{\partial T}\right)_{\underline{V}} - \frac{R}{\underline{V}} \right] d\underline{V} \tag{4.4-24}$$

The interpretation of Eqs. 4.4-19, 20, and 21 is clear; the changes in enthalpy and entropy of a real fluid are equal to that for an ideal gas undergoing the same change of state plus the departure of the fluid from ideal gas behavior at the end state less the departure from ideal gas behavior of the initial state. These **departure functions,** given by Eqs. 4.4-22, 23, and 24, can be computed once the fluid equation of state is known.

Before leaving this subject, we note that although Eqs. 4.4-22, 23, and 24 are useful for calculating the enthalpy and entropy departures from ideal gas behavior for some equations of state, their form is less helpful for the Peng–Robinson and other equations of state considered in this section in which $\underline{V}$ and T are the convenient independent variables.[11] In such cases it is useful to have alternative expressions for the departure functions at fixed temperature and pressure. To obtain such expressions, we start with Eqs. 4.4-22 and 23 and use

$$dP = \frac{1}{\underline{V}} d(P\underline{V}) - \frac{P}{\underline{V}} d\underline{V} \tag{4.4-25}$$

and the triple product rule (Eq. 4.1-6a) in the form

$$\left(\frac{\partial \underline{V}}{\partial T}\right)_P \left(\frac{\partial P}{\partial \underline{V}}\right)_T \left(\frac{\partial T}{\partial P}\right)_{\underline{V}} = -1 \quad \text{or} \quad \left(\frac{\partial \underline{V}}{\partial T}\right)_P dP\bigg|_T = -\left(\frac{\partial P}{\partial T}\right)_{\underline{V}} d\underline{V}\bigg|_T \tag{4.4-26}$$

to obtain

$$\int_{P=0}^{P} \left[\underline{V} - T\left(\frac{\partial \underline{V}}{\partial T}\right)_P \right] dP = \int_{P\underline{V}=RT}^{P\underline{V}(T,P)} d(P\underline{V}) + \int_{\underline{V}=\infty}^{\underline{V}=\underline{V}(T,P)} \left[T\left(\frac{\partial P}{\partial T}\right)_{\underline{V}} - P \right] d\underline{V}$$

$$= (P\underline{V} - RT) + \int_{\underline{V}=\infty}^{\underline{V}=\underline{V}(T,P)} \left[T\left(\frac{\partial P}{\partial T}\right)_{\underline{V}} - P \right] d\underline{V}$$

and

$$\underline{H}(T, P) - \underline{H}^{IG}(T, P) = RT(Z - 1) + \int_{\underline{V}=\infty}^{\underline{V}=\underline{V}(T,P)} \left[T\left(\frac{\partial P}{\partial T}\right)_{\underline{V}} - P \right] d\underline{V} \tag{4.4-27}$$

Similarly, from

$$\frac{R}{P} dP = R\frac{d(P\underline{V})}{P\underline{V}} - R\frac{d\underline{V}}{\underline{V}} = Rd \ln (P\underline{V}) - \frac{R}{\underline{V}} d\underline{V}$$

[11]That is, with such equations of state, it is easier to solve for P given $\underline{V}$ and T, than for $\underline{V}$ given P and T. Consequently, derivatives of P with respect to $\underline{V}$ or T are more easily found in terms of $\underline{V}$ and T, than are derivatives of $\underline{V}$ with respect to P and T (try it!).

and Eq. 4.4-23 we obtain

$$\int_{P=0}^{P}\left[\frac{R}{P}-\left(\frac{\partial \underline{V}}{\partial T}\right)_P\right]dP = R\int_{P\underline{V}=RT}^{P\underline{V}(T,P)} d\ln(P\underline{V}) + \int_{\underline{V}=\infty}^{\underline{V}=\underline{V}(T,P)}\left[\left(\frac{\partial P}{\partial T}\right)_{\underline{V}}-\frac{R}{\underline{V}}\right]d\underline{V}$$

$$= R\ln\left(\frac{P\underline{V}}{RT}\right) + \int_{\underline{V}=\infty}^{\underline{V}=\underline{V}(T,P)}\left[\left(\frac{\partial P}{\partial T}\right)_{\underline{V}}-\frac{R}{\underline{V}}\right]d\underline{V}$$

and

$$\underline{S}(T,P)-\underline{S}^{IG}(T,P) = R\ln Z + \int_{\underline{V}=\infty}^{\underline{V}=\underline{V}(T,P)}\left[\left(\frac{\partial P}{\partial T}\right)_{\underline{V}}-\frac{R}{\underline{V}}\right]d\underline{V} \tag{4.4-28}$$

The desired equations, Eqs. 4.4-27 and 28, are general and can be used with any equation of state. Now using, for example, the Peng–Robinson equation of state, one obtains (Problem 4.2)

$$\underline{H}(T,P)-\underline{H}^{IG}(T,P) = RT(Z-1) + \frac{T\left(\frac{da}{dT}\right)-a}{2\sqrt{2}b}\ln\left[\frac{Z+(1+\sqrt{2})B}{Z+(1-\sqrt{2})B}\right] \tag{4.4-29}$$

and

$$\underline{S}(T,P)-\underline{S}^{IG}(T,P) = R\ln(Z-B) + \frac{\frac{da}{dT}}{2\sqrt{2}b}\ln\left[\frac{Z+(1+\sqrt{2})B}{Z+(1-\sqrt{2})B}\right] \tag{4.4-30}$$

where

$$Z = P\underline{V}/RT \text{ and } B = Pb/RT$$

ILLUSTRATION 4.4-1 CONSTRUCTION OF A THERMODYNAMIC PROPERTIES CHART

As an introduction to the problem of constructing a chart and/or table of the thermodynamic properties of a real fluid, develop a thermodynamic properties chart for oxygen over the temperature range of −100°C to +150°C and a pressure range of 1 to 100 bar. (A larger temperature range, including the vapor–liquid two-phase region, will be considered in Chapter 5.) In particular, calculate the compressibility factor, specific volume, molar enthalpy and entropy as a function of temperature and pressure. Also, prepare a pressure–volume plot, a pressure–enthalpy plot (see Fig. 2.4-2) and a temperature–entropy plot (see Fig. 2.4-3) for oxygen.

Data

For simplicity we will assume oxygen obeys the Peng–Robinson equation of state and has an ideal gas heat capacity given by

$$C_P^*\left(\frac{\text{J}}{\text{mol K}}\right) = 25.46 + 1.519\times10^{-2}T - 0.7151\times10^{-5}T^2 + 1.311\times10^{-9}T^3$$

we will choose the reference state of oxygen to be

$$\underline{H}^{IG}(T = 25°\text{C}, P = 1\text{ bar}) = 0$$

and

$$\underline{S}^{IG}(T = 25°\text{C}, P = 1\text{ bar}) = 0$$

As will be explained shortly, the Peng–Robinson parameters for oxygen are

$$a(T) = 0.45724 \frac{R^2T_c^2}{P_c} \alpha(T)$$

$$b(T) = 0.07780 \frac{RT_c}{P_c}$$

$$[\alpha(T)]^{1/2} = 1 + \kappa\left(1 - \sqrt{\frac{T}{T_c}}\right)$$

where $\kappa = 0.4069$, and $T_c = 154.6$ K is the critical temperature of oxygen, and $P_c = 5.046$ MPa is its critical pressure.

Solution

i. Volume

The volume is the easiest of the properties to calculate. For each value of temperature we compute values for $a(T)$ and b, and then at each pressure solve the equation

$$P = \frac{RT}{\underline{V} - b} - \frac{a(T)}{\underline{V}(\underline{V} + b) + b(\underline{V} - b)}$$

for the volume, or equivalently and preferably, solve the equation

$$Z^3 + (-1 + B)Z^2 + (A - 3B^2 - 2B)Z + (-AB + B^2 + B^3) = 0$$

with $B = bP/RT$ and $A = aP/(RT)^2$ (see Eq. 4.4-4 and Table 4.4-3) for the compressibility factor $Z = P\underline{V}/RT$, from which the volume is easily calculated $\underline{V} = ZRT/P$. Repeating the calculation a number of times leads to the entries in Table 4.4-4. We have done these calculations using the BASIC language program PR1.BAS described in Appendix A4.1.

Since the pressure and specific volume vary by several orders of magnitude over the range of interest, these results have been plotted in Fig. 4.4-3 as the ln P versus ln $\underline{V}$. Remember that for the ideal gas $P\underline{V} = RT$, so that an ideal gas isotherm on a log-log plot is a straight line. (Note that Fig. 4.4-3 also contains isotherms for temperatures below −100°C. The calculation of these is discussed in Sec. 5.5.)

ii. Enthalpy

To compute the difference in the enthalpy of oxygen between a state of temperature T and pressure P and the ideal gas reference state at 25°C and 1 bar we use

$$\begin{aligned}
&\underline{H}(T, P) - \underline{H}^{IG}(T = 25°\text{C}, P = 1 \text{ bar}) \\
&= \underline{H}(T, P) - \underline{H}^{IG}(T, P) + \underline{H}^{IG}(T, P) - \underline{H}^{IG}(T = 25°\text{C}, P = 1 \text{ bar}) \\
&= (\underline{H} - \underline{H}^{IG})_{T,P} + \int_{T=298.15\text{K}}^{T} C_P^* \, dT \\
&= (\underline{H} - \underline{H}^{IG})_{T,P} + 25.46(T - 298.15) + \frac{1.519 \times 10^{-2}}{2}(T^2 - 298.15^2) \\
&\quad - \frac{0.7151 \times 10^{-5}}{3}(T^3 - 298.15^3) + \frac{1.311 \times 10^{-9}}{4}(T^4 - 298.15^4)
\end{aligned}$$

Table 4.4-4
THE THERMODYNAMIC PROPERTIES OF OXYGEN CALCULATED USING THE PENG–ROBINSON EQUATION OF STATE

T(°C)	−100	−75	−50	−25	0	25	50	75	100	125	150
p = 1 bar											
Z	0.9945	0.9963	0.9974	0.9982	0.9987	0.9991	0.9994	0.9996	0.9997	0.9998	0.9999
$\underline{V}$	14.3200	16.4200	18.5100	20.5900	22.6800	24.7700	26.8500	28.9300	31.0200	33.1000	35.1800
$\underline{H}$	−3603.75	−2898.69	−2186.80	−1480.83	−742.14	−9.44	730.04	1476.18	2228.83	2987.86	3753.10
$\underline{S}$	−15.59	−11.81	−8.43	−5.38	−2.59	−0.02	2.36	4.58	6.67	8.64	10.50
P = 2 bar											
Z	0.9889	0.9925	0.9948	0.9963	0.9974	0.9982	0.9987	0.9991	0.9994	0.9996	0.9998
$\underline{V}$	7.1190	8.1760	9.2290	10.2800	11.3300	12.3700	13.4200	14.4600	15.5000	16.5500	17.5900
$\underline{H}$	−3626.18	−2917.04	−2202.08	−1480.83	−753.13	−18.88	721.89	1469.10	2222.68	2982.49	3748.42
$\underline{S}$	−21.38	−17.63	−14.23	−11.17	−8.38	−5.81	−3.43	−1.19	0.87	2.87	4.73
P = 5 bar											
Z	0.9722	0.9813	0.9871	0.9909	0.9936	0.9954	0.9968	0.9978	0.9986	0.9991	0.9996
$\underline{V}$	2.7990	3.2330	3.6630	4.0890	4.5130	4.9350	5.3560	5.7760	6.1960	6.6150	7.0330
$\underline{H}$	−3694.40	−2972.47	−2248.04	−1519.52	−786.05	−47.12	697.51	1447.97	2204.29	2966.37	3734.44
$\underline{S}$	−29.33	−25.44	−22.00	−18.90	−16.09	−13.50	−11.10	−8.86	−6.77	−4.79	−2.92
P = 10 bar											
Z	0.9439	0.9626	0.9743	0.9820	0.9872	0.9910	0.9937	0.9957	0.9972	0.9983	0.9992
$\underline{V}$	1.3590	1.5860	1.8080	2.0260	2.2420	2.4570	2.6700	2.8820	3.0940	3.3050	3.5150
$\underline{H}$	−3811.36	−3066.16	−2325.11	−1584.04	−840.75	−93.94	657.16	1413.02	2173.92	2940.01	3711.36
$\underline{S}$	−35.55	−31.53	−28.00	−24.85	−21.99	−19.38	−16.97	−14.71	−12.60	−10.61	−8.73
P = 20 bar											
Z	0.8856	0.9252	0.9491	0.9646	0.9751	0.9825	0.9878	0.9918	0.9947	0.9969	0.9987
$\underline{V}$	0.6375	0.7621	0.8804	0.9951	1.1070	1.2180	1.3270	1.4350	1.5430	1.6500	1.7570
$\underline{H}$	−4059.21	−3258.71	−2480.87	−1713.17	−949.56	−186.68	577.48	1344.15	2114.16	2888.07	3666.08
$\underline{S}$	−42.27	−37.95	−34.25	−30.99	−28.06	−25.39	−22.92	−20.64	−18.50	−16.50	−14.60
P = 30 bar											
Z	0.8245	0.8878	0.9246	0.9479	0.9636	0.9745	0.9824	0.9882	0.9925	0.9958	0.9983
$\underline{V}$	0.3957	0.4876	0.5718	0.6519	0.7295	0.8053	0.8798	0.9535	1.0260	1.0990	1.1710
$\underline{H}$	−4329.71	−3458.44	−2638.59	−1842.18	−1057.37	−278.07	499.24	1276.72	2055.76	2837.29	3621.95
$\underline{S}$	−46.72	−42.01	−38.11	−34.73	−31.72	−28.89	−26.48	−24.17	−22.00	−20.00	−18.07

P = 40 bar											
Z	0.7602	0.8508	0.9008	0.9320	0.9528	0.9671	0.9774	0.9850	0.9906	0.9949	0.9981
$\underline{V}$	0.2736	0.3504	0.4178	0.4807	0.5409	0.5994	0.6565	0.7128	0.7683	0.8233	0.8779
$\underline{H}$	−4629.51	−3665.63	−2797.93	−1970.75	−1163.97	−367.98	422.53	1210.75	1998.73	2787.83	3578.97
$\underline{S}$	−50.35	−45.14	−41.01	−37.50	−34.40	−31.61	29.06	−26.72	−24.54	−22.50	−20.56
P = 50 bar											
Z	0.6922	0.8143	0.8781	0.9170	0.9426	0.9603	0.9729	0.9821	0.9890	0.9942	0.9981
$\underline{V}$	0.1993	0.2683	0.3258	0.3784	0.4282	0.4761	0.5228	0.5686	0.6137	0.6582	0.7023
$\underline{H}$	−4968.26	−3880.25	−2958.41	−2098.53	−1269.13	−456.28	347.92	1146.30	1943.10	2739.64	3537.15
$\underline{S}$	−53.66	−47.77	−43.39	−39.72	−36.54	−33.70	−31.11	−28.73	−26.52	−24.45	−22.51
P = 60 bar											
Z	0.6202	0.7788	0.8564	0.9029	0.9332	0.9541	0.9698	0.9796	0.9877	0.9937	0.9983
$\underline{V}$	0.1448	0.2139	0.2648	0.3105	0.3532	0.3942	0.4338	0.4726	0.5107	0.5483	0.5854
$\underline{H}$	−5358.88	−4101.78	−3119.39	−2225.13	−1372.64	−542.81	273.98	1083.39	1888.88	2692.73	3496.47
$\underline{S}$	−56.90	−50.09	−45.42	−41.62	−38.34	−35.44	−32.80	−30.39	−28.16	−26.07	−24.12
P = 70 bar											
Z	0.5471	0.7450	0.8361	0.8898	0.9246	0.9484	0.9652	0.9775	0.9866	0.9935	0.9987
$\underline{V}$	0.1125	0.1753	0.2216	0.2623	0.3000	0.3358	0.3705	0.4042	0.4373	0.4698	0.5020
$\underline{H}$	−5878.46	−4328.89	−3280.08	−2350.13	−1472.25	−627.54	202.27	1022.07	1836.08	2647.08	3456.93
$\underline{S}$	−60.27	−52.21	−47.23	−43.27	−39.90	−36.94	−34.26	−31.82	−29.57	−27.46	−25.49
P = 80 bar											
Z	0.4833	0.7135	0.8173	0.8778	0.9167	0.9433	0.9621	0.9757	0.9859	0.9935	0.9993
$\underline{V}$	0.0870	0.1469	0.1896	0.2264	0.2602	0.2923	0.3231	0.3530	0.3823	0.4111	0.4395
$\underline{H}$	−6296.72	−4559.08	−3439.53	−2473.08	−1573.76	−710.26	132.35	962.34	1784.71	2602.71	3418.53
$\underline{S}$	−63.65	−54.19	−48.86	−44.75	−41.29	−38.27	−35.55	−33.08	−30.80	−28.68	−26.69
P = 90 bar											
Z	0.4444	0.6852	0.8003	0.8669	0.9096	0.9387	0.9593	0.9743	0.9854	0.9937	1.0001
$\underline{V}$	0.0711	0.1254	0.1650	0.1987	0.2295	0.2586	0.2864	0.3134	0.3397	0.3655	0.3909
$\underline{H}$	−6719.52	−4788.49	−3596.64	−2593.54	−1670.95	−790.91	64.27	904.24	1734.77	2559.61	3381.24
$\underline{S}$	−66.53	−56.03	−50.35	−46.09	−42.54	−39.46	−36.70	−34.20	−31.90	−29.76	−37.76
P = 100 bar											
Z	0.4297	0.6611	0.7851	0.8571	0.9034	0.9348	0.9570	0.9732	0.9851	0.9941	1.0010
$\underline{V}$	0.0619	0.1089	0.1457	0.1768	0.2052	0.2317	0.2571	0.2817	0.3056	0.3291	0.3522
$\underline{H}$	−7026.48	−5012.01	−3750.23	−2711.04	−1765.61	−869.40	−1.92	847.77	1686.27	2517.77	3345.07
$\underline{S}$	−68.69	−57.75	−51.74	−47.32	−43.69	−40.55	−37.75	−35.22	−32.89	−30.74	−28.72

$\underline{V}$ [=] m^3/kmol $\underline{H}$ [=] J/mol = kJ/kmol $\underline{S}$ [=] J/mol K = kJ/kmol K

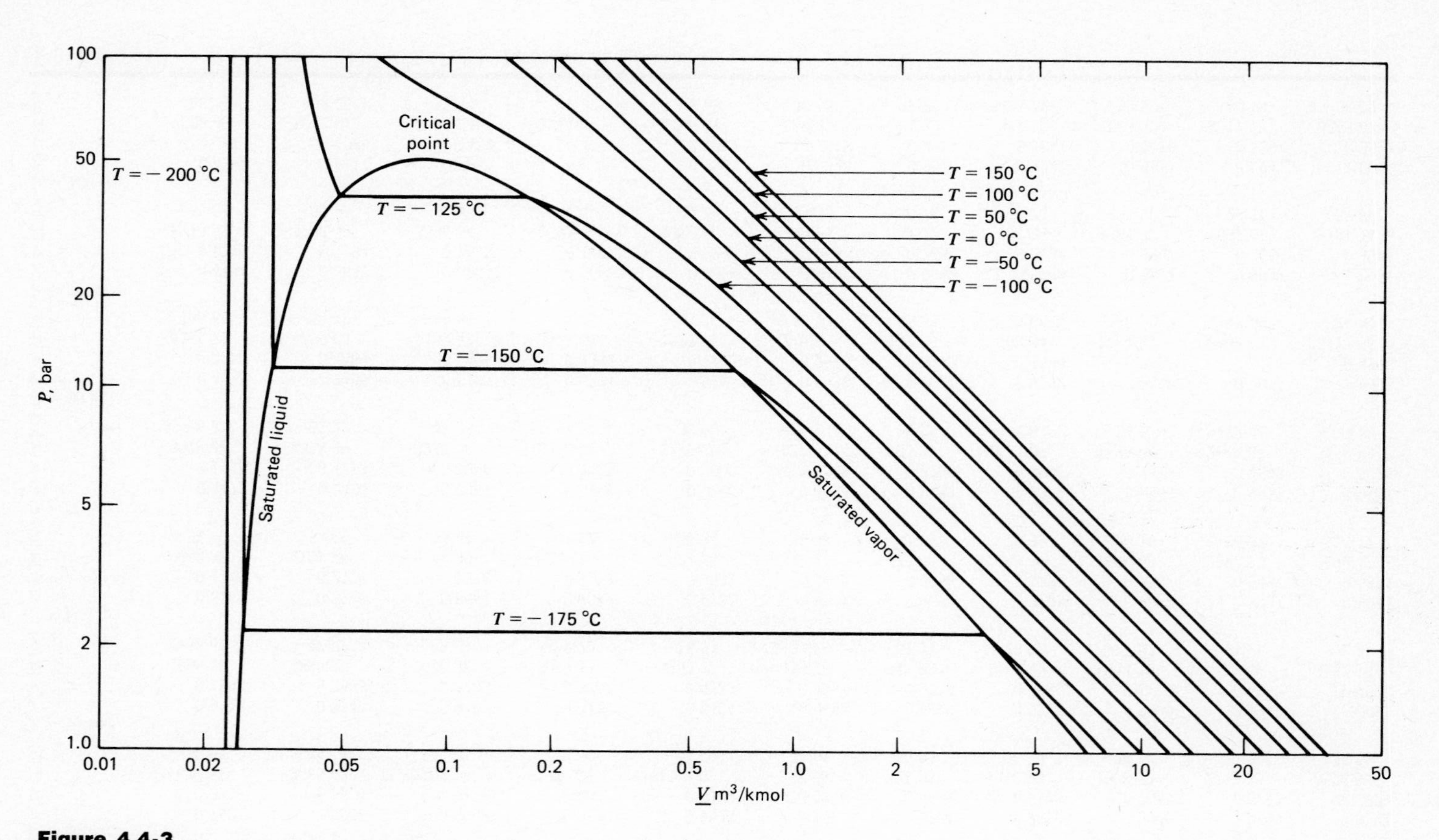

Figure 4.4-3
Pressure–volume diagram for oxygen calculated using the Peng–Robinson equation of state.

The quantity $(\underline{H} - \underline{H}^{IG})_{T,P}$ is computed using Eq. 4.4-29 and the value of the compressibility factor Z found earlier at each set of (T, P) values. The values of enthalpy computed in this manner appear in Table 4.4-4 and have been plotted in Fig. 4.4-4 as a pressure–enthalpy diagram. The BASIC program of Appendix A4.1 was used for this calculation as well. [Note that at $T = 25°C$ and $P = 1$ bar, oxygen is not quite an ideal gas; $Z = 0.9991$, not unity. Consequently $(\underline{H} - H^{IG})_{T=25°C, P=1\text{ bar}} = -9.44$ J/mol, so that the enthalpy of oxygen at $T = 25°C$, $P = 1$ bar is -9.44 J/mol, whereas if it were an ideal gas at these conditions its enthalpy would be zero.] Once again, lower temperature results, including the two-phase region, appear in Fig. 4.4-4. The basis for those calculations will be discussed in Sec. 5.5.

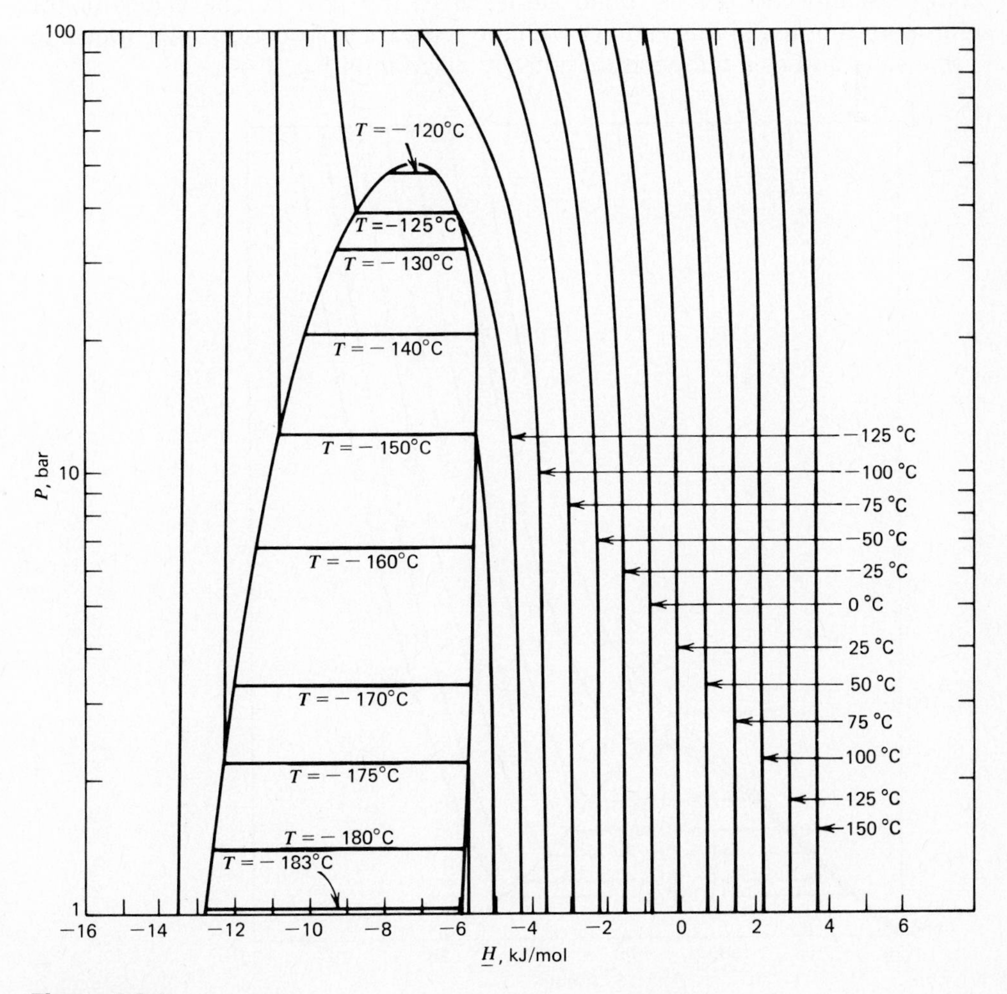

Figure 4.4-4
A pressure–enthalpy diagram for oxygen calculated using the Peng–Robinson equation of state.

iii. Entropy

To compute the difference in the entropy of oxygen between the state (T, P) and the ideal gas reference state at 25°C and 1 bar we use

$$\underline{S}(T, P) - \underline{S}^{IG}(T = 25°\text{C}, P = 1 \text{ bar})$$
$$= \underline{S}(T, P) - \underline{S}^{IG}(T, P) + \underline{S}^{IG}(T, P) - \underline{S}^{IG}(T = 25°\text{C}, P = 1 \text{ bar})$$
$$= (\underline{S} - \underline{S}^{IG})_{T,P} + \int_{T=298.15\text{K}}^{T} \frac{C_P^*}{T}\, dT - R \ln\left(\frac{P}{1 \text{ bar}}\right)$$
$$= (\underline{S} - \underline{S}^{IG})_{T,P} + 25.46 \ln \frac{T}{298.15} + 1.519 \times 10^{-2} (T - 298.15)$$
$$- \frac{0.7151 \times 10^{-5}}{2} (T^2 - 298.15^2) + \frac{1.311 \times 10^{-9}}{3} (T^3 - 298.15^3)$$
$$- R \ln\left(\frac{P}{1 \text{ bar}}\right)$$

The quantity $(\underline{S} - \underline{S}^{IG})_{T,P}$ is computed using Eq. 4.4-30, and the value of the compressibility factor Z is found earlier at each T and P. The values of the entropy computed in this manner with the program in Appendix A4.1 appear in Table 4.4-4 and as a temperature-entropy diagram in Fig. 4.4-5.

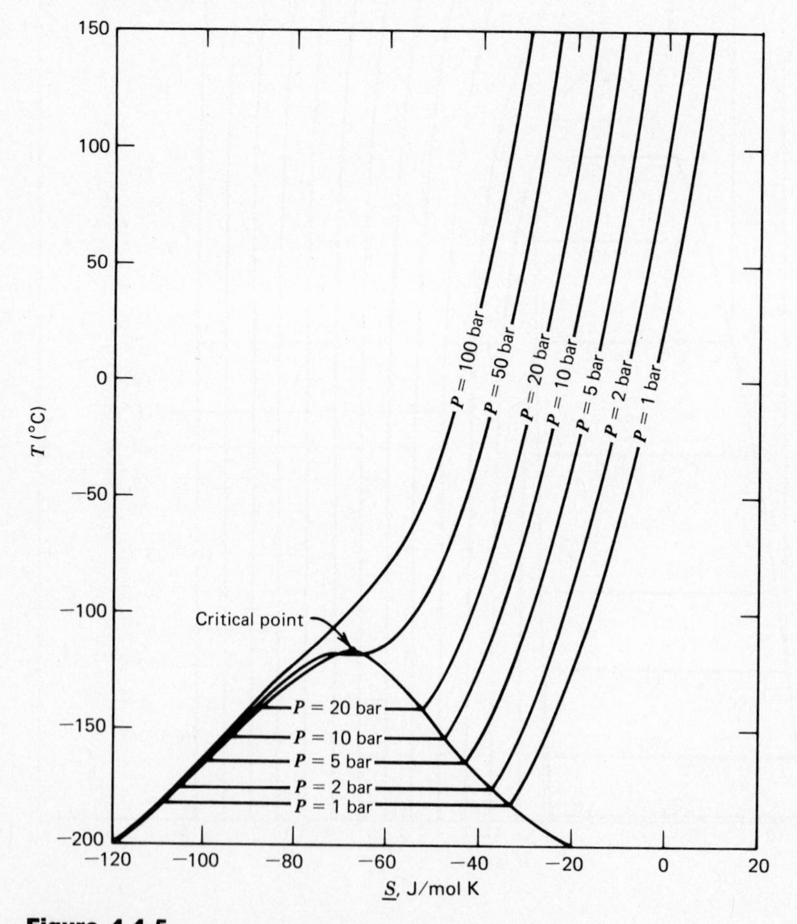

Figure 4.4-5
A temperature–entropy diagram for oxygen calculated using the Peng–Robinson equation of state.

Comments
For illustrative purposes, the Peng–Robinson equation of state was used here. Generally, if one were going to take the time and effort to construct tables or plots such as those developed here, much more complicated equations of state, for example the Bender equation (Eq. 4.4-6b), would be used. ■

4.5 AN EXAMPLE

Given low-density heat capacity data and volumetric equation of state information for a fluid, we can use the procedures developed in this chapter to calculate the change in the thermodynamic properties of the fluid accompanying any change in its state. Thus, at least in principle, we can use the mass, energy, and entropy balances to solve energy flow problems for all fluids. The starting point for solving such problems is exactly the same here as that used in Chapters 2 and 3, in that the balance equations are written for a convenient choice of the thermodynamic system. For real fluids, tedious calculations may be necessary to eliminate the internal energy, enthalpy, and entropy, which appear in the balance equations in terms of the temperature, pressure, and specific volume. If the fluid under study is likely to be of continual interest, it is probably worthwhile constructing a complete thermodynamic properties chart, as in Figs. 2.4-1 through 3 and in the previous illustration, so that all problems can be solved rapidly, as was demonstrated in Chapters 2 and 3. If, on the other hand, the fluid is of limited interest, it is more logical to use the heat capacity and volumetric equation of state data only to solve the problem at hand. This sort of calculation is illustrated in the following example. The example here uses the van der Waals equation of state merely as a prototype for the equations of state of real fluids. The use of the Benedict–Webb–Rubin equation would lead to predictions of greater accuracy, but with a great deal more computational effort. This example is also considered, and enlarged on, in the following two sections.

ILLUSTRATION 4.5-1

Nitrogen gas is being withdrawn from a 0.15-m^3 cylinder at the rate of 10 mol/min. The cylinder initially contains the gas at a pressure of 100 bar and 170 K. The cylinder is well insulated and there is a negligible heat transfer between the cylinder walls and the gas. How many moles of gas will be in the cylinder at any time? What will be the temperature and pressure of the gas in the cylinder after 50 minutes?

a. Assume that nitrogen is an ideal gas.
b. Use the nitrogen data in Fig. 2.4-3.
c. Assume that nitrogen is a van der Waals fluid.

Data
For parts (a) and (c) use

$$C_P^* \text{ (J/mol K)} = 27.2 + 4.2 \times 10^{-3} \, (T \text{ in K})$$

Solution
The mass and energy balances for the contents of the cylinder are

$$\frac{dN}{dt} = \dot{N} \qquad \text{or} \qquad N(t) = N(t = 0) + \dot{N}t \tag{a}$$

and

$$\frac{d(N\underline{U})}{dt} = \dot{N}\underline{H} \tag{b}$$

where $\dot{N} = -10$ mol/min.

Following Illustration 3.5-2, the result of writing an entropy balance for that portion of the gas that always remains in the cylinder is

$$\frac{d\underline{S}}{dt} = 0 \quad \text{or} \quad \underline{S}(t=0) = \underline{S}(t) \tag{c}$$

Also, from $V = N\underline{V}$, we have that

$$N(t=0) = \frac{V_{\text{cyl}}}{\underline{V}(t=0)} = \frac{0.15 \text{ m}^3}{\underline{V}(t=0)} \tag{d}$$

Equations a, b, c, and d apply to both ideal and nonideal gases.

Computation of $N(t)$

a. ***Using the ideal gas equation of state***

$$\underline{V}(t=0) = \frac{RT(t=0)}{P(t=0)} = \frac{8.314 \times 10^{-5} \text{ bar m}^3\text{/mol K} \times 170 \text{ K}}{100 \text{ bar}}$$

$$= 1.4134 \times 10^{-4} \text{ m}^3\text{/mol}$$

so that

$$N(t=0) = 1061.3 \text{ mol}$$

and

$$N(t) = 1061.3 - 10t \text{ mol}$$

b. ***Using Figure 2.4-3***

From Fig. 2.4-3,

$$\rho(T = 170 \text{ K}, P = 100 \text{ bar}) = \rho(T = -153.7°\text{F}, P = 10 \text{ MPa}) \approx 18.2 \text{ lb/ft}^3$$

$$\underline{V}(T = 170 \text{ K}, P = 10 \text{ MPa}) = \frac{28 \dfrac{\text{g}}{\text{mol}}}{18.2 \dfrac{\text{lb}}{\text{ft}^3} \times 16.02 \dfrac{\text{kg/m}^3}{\text{lb/ft}^3} \times 1000 \dfrac{\text{g}}{\text{kg}}}$$

$$= 9.60 \times 10^{-5} \text{ m}^3\text{/mol}$$

so that

$$N(t=0) = \frac{0.15 \text{ m}^3}{9.60 \times 10^{-5} \text{ m}^3\text{/mol}} = 1562.5 \text{ mol}$$

and

$$N(t) = 1562.5 - 10t \text{ mol}$$

c. ***Using the van der Waals equation of state***

The van der Waals equation is

$$P = \frac{RT}{\underline{V} - b} - \frac{a}{\underline{V}^2}$$

With the constants given in Table 4.4-1 we have

$$100 \text{ bar} = 1 \times 10^7 \text{ Pa} = \frac{8.314 \times 170}{\underline{V}(t=0) - 3.864 \times 10^{-5}} - \frac{0.1368}{[\underline{V}(t=0)]^2}$$

Solving this cubic equation, we obtain

$$\underline{V}(t = 0) = 9.437 \times 10^{-5}\ \mathrm{m^3/mol}$$

so that

$$N(t = 0) = 1589.5\ \mathrm{mol}$$

and

$$N(t) = 1589.5 - 10t\ \mathrm{mol}$$

Note also that since the volume of the cylinder is constant ($0.15\ \mathrm{m^3}$), and as $N(t)$ is known, we can compute the molar volume of the gas at any later time from

$$\underline{V}(t) = \frac{0.15\ \mathrm{m^3}}{N(t)}$$

In particular, we can use this equation to compute $\underline{V}(t = 50\ \mathrm{min})$. These results, together with other information gathered so far, and some results gotten from the following sections, are listed in Table 4.5-1.

Since we know the specific volume after 50 minutes, we need to determine only one further state property to have the final state of the system completely specified. In principle, either the energy or entropy balance could be used to find this property. The entropy balance is more convenient to use, especially for the nonideal gas calculations. Thus, all the calculations here are based on the fact (see Eq. c) that

$$\underline{S}(t = 0) = \underline{S}(t = 50\ \mathrm{min})$$

Computation of $T(t = 50\ \mathrm{min})$ and $P(t = 50\ \mathrm{min})$

a. ***Using the ideal gas equation of state***

From Eq. 3.4-1

$$d\underline{S} = C_V \frac{dT}{T} + R\frac{d\underline{V}}{\underline{V}}$$

Now for the ideal gas

$$C_V = C_V^* = C_P^* - R = (27.2 - 8.314) + 4.2 \times 10^{-3}\,T$$

$$\simeq 18.9 + 4.2 \times 10^{-3}\,T\ \mathrm{J/mol\ K}$$

Also, since $\underline{S}$ = constant or $\Delta\underline{S} = 0$, we have

$$\int_{T=170}^{T} \frac{(18.9 + 4.2 \times 10^{-3}\,T)}{T}\,dT + 8.314 \int_{\underline{V}=1.413\times10^{-4}}^{\underline{V}=2.672\times10^{-4}} \frac{d\underline{V}}{\underline{V}} = 0$$

Table 4.5-1

Equation of state	**$N(t = 0)$**	**$N(t)$**	**$\underline{V}(t = 50\ \mathrm{min})\ \mathrm{m^3/mol}$**
Ideal gas	1061.3	$1061.3 - 10t$	2.672×10^{-4}
Fig. 2.4-3	1562.5	$1562.5 - 10t$	1.412×10^{-4}
van der Waals	1589.5	$1589.5 - 10t$	1.377×10^{-4}
Corresponding states (Illustration 4.6-2)	1499.0	$1499.0 - 10t$	1.5015×10^{-4}
Peng–Robinson (Illustration 4.7-1)	1567.9	$1567.9 - 10t$	1.4046×10^{-4}

or

$$18.9 \ln \frac{T}{170} + 4.2 \times 10^{-3} (T - 170) + 8.314 \ln \frac{2.672 \times 10^{-4}}{1.413 \times 10^{-4}} = 0$$

The solution to this equation is 129.6 K, so that

$$P(t = 50 \text{ min}) = \frac{RT(t = 50 \text{ min})}{\underline{V}(t = 50 \text{ min})} = 40.3 \text{ bar}$$

b. ***Using Fig. 2.4-3***

$$\rho(t = 50 \text{ min}) = 28 \frac{\text{g}}{\text{mol}} \left(1.412 \times 10^{-4} \frac{\text{m}^3}{\text{mol}} \times 16.02 \frac{\text{kg/m}^3}{\text{lb/ft}^3} \times 1000 \frac{\text{g}}{\text{kg}}\right)^{-1}$$

$$= 12.38 \text{ lb/ft}^3$$

To find the final temperature and pressure using Fig. 2.4-3 we locate the initial point ($T = 170$ K $= -153.6$°F, $P = 10$ MPa) and follow a line of constant entropy (dashed line) through this point to the intersection with a line of constant density (solid line) through $\rho = 12.38$ lb/ft³. This intersection gives the pressure and temperature of the $t = 50$-minute state. We find

$$T(t = 50 \text{ min}) \approx -212°\text{F} = 138 \text{ K}$$

$$P(t = 50 \text{ min}) \approx 4.1 \text{ MPa} = 41 \text{ bar}$$

c. ***Using the van der Waals equation of state***
Here we start with Eq. 4.2-19

$$d\underline{S} = \frac{C_V}{T} dT + \left(\frac{\partial P}{\partial T}\right)_{\underline{V}} d\underline{V}$$

and note that for the van der Waals gas

$$\left(\frac{\partial P}{\partial T}\right)_{\underline{V}} = \frac{R}{\underline{V} - b}$$

The integration path to be followed is

i. $T(t = 0), \underline{V}(t = 0) \rightarrow T(t = 0), \underline{V} = \infty$
ii. $T(t = 0), \underline{V} = \infty \rightarrow T(t = 50 \text{ min}), \underline{V} = \infty$ ($C_V = C_V^*$ for this step)
iii. $T(t = 50 \text{ min}), \underline{V} = \infty \rightarrow T(t = 50 \text{ min}), \underline{V}(t = 50 \text{ min})$

so that

$$\underline{S}(t = 50 \text{ min}) - \underline{S}(t = 0) = 0$$
$$= R \int_{\underline{V}(t=0)}^{\underline{V}=\infty} \frac{d\underline{V}}{\underline{V} - b} + \int_{T=170\text{K}}^{T(t=50)} \frac{C_V^*}{T} dT + \int_{\underline{V}=\infty}^{\underline{V}(t=50)} \frac{d\underline{V}}{\underline{V} - b}$$

or

$$0 = R \int_{\underline{V}(t=0)}^{\underline{V}(t=50)} \frac{d\underline{V}}{\underline{V} - b} + \int_{T=170\text{K}}^{T(t=50)} \frac{18.9 + 4.2 \times 10^{-3} T}{T} dT$$

Thus, using $b = 3.864 \times 10^{-5}$ m³/mol we have

$$0 = 8.314 \ln \left[\frac{(13.77 - 3.864) \times 10^{-5}}{(9.437 - 3.864) \times 10^{-5}}\right] + 18.9 \ln \frac{T}{170} + 4.2 \times 10^{-3} (T - 170)$$

The solution to this equation is

$$T(t = 50 \text{ min}) = 133.1 \text{ K}$$

Now, using the van der Waals equation of state with $T = 133.1$ K and $\underline{V} = 13.77 \times 10^{-5}$ m^3/mol, we find

$$P(t = 50 \text{ min}) = 39.6 \text{ bar}$$

Summary

Equation of state	**$V(t = 50$ min) m^3/mol**	**$T(t = 50$ min) K**	**$P(t = 50$ min) bar**
Ideal gas	2.672×10^{-4}	129.6	40.3
Fig. 2.4-3	1.582×10^{-4}	138	41
van der Waals	1.377×10^{-4}	133.1	39.6
Corresponding states (Illustration 4.6-2)	1.502×10^{-4}	145	49.2
Peng–Robinson (Illustration 4.7-1)	1.4046×10^{-4}	143.6	48.6

Comments

It is clear from the results of this problem that one cannot assume that a high pressure gas is ideal and expect to make useful predictions. Of the techniques that have been considered for the thermodynamic calculations involving real fluids, the use of a previously prepared thermodynamic properties chart for the fluid is the most rapid way to proceed. The alternatives, the use of a simple volumetric equation of state for the fluid or of corresponding states correlations (discussed in the following section), are always tedious, and their accuracy is hard to assess. However, if you do not have a thermodynamic properties chart available for the working fluid, you have no choice but to use the more tedious methods. ■

4.6 THE PRINCIPLE OF CORRESPONDING STATES

The analysis presented in Sec. 4.4 makes it possible to solve thermodynamic problems for real substances or to construct tables and charts of their thermodynamic properties given only the low-density (ideal gas) heat capacity and analytic or tabular information on the volumetric equation of state. Unfortunately, these can be tedious tasks, and the necessary volumetric equation of state information is not available for all fluids. Thus, we consider here the principle of corresponding states, which allows one to predict some thermodynamic properties of fluids from generalized property correlations based on available experimental data for similar fluids.

Before we introduce the concept of generalized fluid properties correlations, it is useful to consider which properties we could hope to get from this sort of correlation scheme, and which we cannot. The two types of information needed in thermodynamic calculations are the volumetric equation of state and the ideal gas heat capacity. The ideal gas heat capacity is determined solely by the intramolecular structure (e.g., bond lengths, vibration frequencies, configuration of constituent atoms) of only a single molecule, as there is no intermolecular interaction energy in an ideal gas. As the structure of each molecular species is sufficiently different from others, and since experimental low-density heat capacity data are frequently available, we will not attempt to develop correlations for heat capacity data or for the ideal gas part of the enthalpy, internal energy, or entropy, which can be computed directly from C_P^* and C_V^*.

Conversely, the volumetric equation of state of a fluid is determined solely by the interactions of each molecule with its neighbors. An interesting fact that has emerged from the study of molecular behavior (i.e., statistical mechanics) is that as far as molecular interactions are concerned, molecules can be grouped into classes, such as spherical molecules, nonspherical molecules, molecules with permanent dipole moments, and so forth, and that within any one class molecular interactions are similar. It is also found that the volumetric equation of state obeyed by all members of a class are similar in the sense that if a given equation of state (e.g., Peng-Robinson, Benedict–Webb–Rubin) fits the volumetric data for one member of a class, the same equation of state, with different parameter values, is likely to fit the data for other molecular species in the same class.

The fact that a number of different molecular species may be represented by a volumetric equation of state of the same form suggests that it might be possible to construct generalized correlations, that is, correlations applicable to many different molecular species, for both the volumetric equation of state and the density (or pressure) dependent contribution to the enthalpy, entropy, or other thermodynamic properties. Historically, the first generalized correlation arose from the study of the van der Waals equation of state. Although *present correlations are largely based on experimental data,* it is useful to review the van der Waals generalized correlation scheme since, although not very accurate, it does indicate both the structure and correlative parameters used in modern fluid property correlations.

A plot of P versus $\underline{V}$ for various values of temperature for a van der Waals fluid is given in Fig. 4.6-1. An interesting characteristic of this figure is that isotherms below the one labeled T_C exhibit considerable structure over part of the specific volume–pressure range. As we will see in Chapter 5, this behavior is associated with a vapor–liquid phase transition. For $T > T_C$ this structure in the P–$\underline{V}$ plot has vanished, whereas at $T = T_C$ the only vestige of this structure that remains is an inflection point in the P–$\underline{V}$ plot. For the present discussion, this inflection point will be taken to be the **critical point** of the fluid, that is, the point of highest temperature at which a liquid can exist (justification for this identification will be given in Chapter 5). A brief list of the experimentally measured critical temperatures, critical pressures P_C, and critical volumes $\underline{V}_C$ (i.e., the pressure and volume of the highest temperature liquid) for various fluids is given in Table 4.6-1.

To identify the critical point for the van der Waals fluid analytically we make use of the following mathematical requirements for the occurrence of an inflection point on an isotherm in the P–$\underline{V}$ plane:

$$\left(\frac{\partial P}{\partial \underline{V}}\right)_{T_C} = 0 \quad \text{and} \quad \left(\frac{\partial^2 P}{\partial \underline{V}^2}\right)_{T_C} = 0 \quad \text{at } P_C \text{ and } \underline{V}_C \tag{4.6-1}$$

(also, the first nonzero derivative should be odd and negative in value, but this will not be used here). Using Eqs. 4.6-1 together with Eq. 4.2-33 we obtain that at

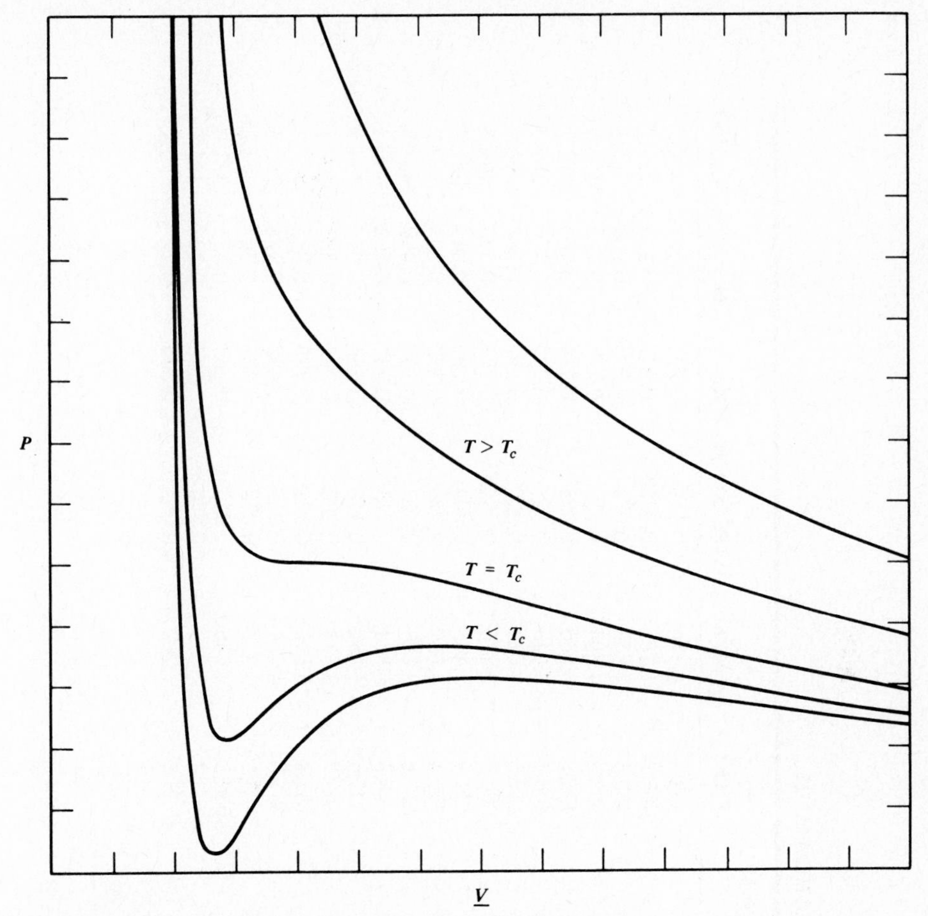

Figure 4.6-1
The van der Waals equation of state.

critical point

$$P_C = \frac{RT_C}{\underline{V}_C - b} - \frac{a}{\underline{V}_C^2} \tag{4.6-2a}$$

$$\left(\frac{\partial P}{\partial \underline{V}}\right)_T \Bigg|_{\substack{\text{evaluated at}\\ T_C, P_C, \underline{V}_C}} = 0 = -\frac{RT_C}{(\underline{V}_C - b)^2} + \frac{2a}{\underline{V}_C^3} \tag{4.6-2b}$$

and

$$\left(\frac{\partial^2 P}{\partial \underline{V}^2}\right)_T \Bigg|_{\substack{\text{evaluated at}\\ T_C, P_C, \underline{V}_C}} = 0 = \frac{2RT_C}{(\underline{V}_C - b)^3} - \frac{6a}{\underline{V}_C^4} \tag{4.6-2c}$$

Thus, three equations interrelate the two unknowns a and b. Using Eqs. 4.6-2b and c to ensure that the critical-point conditions of Eqs. 4.6-1 are satisfied, we obtain (Problem 4.5)

$$a = \frac{9\underline{V}_C RT_C}{8} \tag{4.6-3a}$$

Table 4.6-1
THE CRITICAL AND OTHER CONSTANTS FOR SELECTED FLUIDS

Substance	Symbol	Molecular Weight (g mol^{-1})	$T_C(K)$	P_C(MPa)	V_C(m^3/kmol)	Z_C	ω	T_{boil}(K)
Acetylene	C_2H_2	26.038	308.3	6.140	0.113	0.271	0.184	189.2
Ammonia	NH_3	17.031	405.6	11.28	0.0724	0.242	0.250	239.7
Argon	Ar	39.948	150.8	4.874	0.0749	0.291	−0.004	87.3
Benzene	C_6H_6	78.114	562.1	4.894	0.259	0.271	0.212	353.3
n-Butane	C_4H_{10}	58.124	425.2	3.800	0.255	0.274	0.193	272.7
Isobutane	C_4H_{10}	58.124	408.1	3.648	0.263	0.283	0.176	261.3
1-Butene	C_4H_8	56.108	419.6	4.023	0.240	0.277	0.187	266.9
Carbon dioxide	CO_2	44.010	304.2	7.376	0.0940	0.274	0.225	194.7
Carbon monoxide	CO	28.010	132.9	3.496	0.0931	0.295	0.049	81.7
Carbon tetrachloride	CCl_4	153.823	556.4	4.560	0.276	0.272	0.194	349.7
n-Decane	$C_{10}H_{22}$	142.286	617.6	2.108	0.603	0.247	0.490	447.3
n-Dodecane	$C_{12}H_{26}$	170.340	658.3	1.824	0.713	0.24	0.562	489.5
Ethane	C_2H_6	30.070	305.4	4.884	0.148	0.285	0.098	184.5
Ethyl ether	$C_4H_{10}O$	74.123	466.7	3.638	0.280	0.262	0.281	307.7
Ethylene	C_2H_4	28.054	282.4	5.036	0.129	0.276	0.085	169.4
Freon, F-12	CCl_2F_2	120.914	385.0	4.124	0.217	0.280	0.176	243.4
Helium	He	4.003	5.19	0.227	0.0573	0.301	−0.387	4.21
n-Heptane	C_7H_{16}	100.205	540.2	2.736	0.0432	0.263	0.351	371.6
n-Hexane	C_6H_{14}	86.178	507.4	2.969	0.370	0.260	0.296	341.9
Hydrogen	H_2	2.016	33.2	1.297	0.065	0.305	−0.22	20.4

Hydrogen fluoride	HF	20.006	461.0	6.488	0.069	0.12	0.372	292.7
Hydrogen sulfide	H_2S	34.080	373.2	8.942	0.0985	0.284	0.100	212.8
Methane	CH_4	16.043	190.6	4.600	0.099	0.288	0.008	111.7
Naphthalene	$C_{10}H_8$	128.174	748.4	4.05	0.410	0.267	0.302	491.1
Neon	Ne	20.183	44.4	2.756	0.0417	0.311	0.	27.0
Nitric oxide	NO	30.006	180.0	6.485	0.058	0.250	0.607	121.4
Nitrogen	N_2	28.013	126.2	3.394	0.0895	0.290	0.040	77.4
n-Octane	C_8H_{18}	114.232	568.8	2.482	0.492	0.259	0.394	398.8
Oxygen	O_2	31.999	154.6	5.046	0.0732	0.288	0.021	90.2
n-Pentane	C_5H_{12}	72.151	469.6	3.374	0.304	0.262	0.251	309.2
Isopentane	C_5H_{12}	72.151	460.4	3.384	0.306	0.271	0.227	301.0
Propane	C_3H_8	44.097	369.8	4.246	0.203	0.281	0.152	231.1
Propylene	C_3H_6	42.081	365.0	4.620	0.181	0.275	0.148	225.4
Sulfur dioxide	SO_2	64.063	430.8	7.883	0.122	0.268	0.251	263.
Toluene	C_7H_8	92.141	591.7	4.113	0.316	0.264	0.257	383.8
Water	H_2O	18.015	647.3	22.048	0.056	0.229	0.344	373.2
Xenon	Xe	131.300	289.7	5.836	0.118	0.286	0.002	165.0

Source: Adapted from R. C. Reid, J. M. Prausnitz, and B. E. Poling, *The Properties of Gases and Liquids*, 4th ed., McGraw–Hill, New York, 1986, Appendix A and other sources.

and

$$b = \frac{\underline{V}_C}{3}$$

Using these results in Eq. 4.6-2a yields

$$P_C = \frac{a}{27b^2} \qquad (4.6\text{-}3b)$$

By direct substitution we find that the compressibility at the critical point is,

$$Z_C = \frac{P_C \underline{V}_C}{RT_C} = \frac{3}{8} = 0.375 \qquad (4.6\text{-}3c)$$

This last equation can be used to obtain expressions for the van der Waals parameters in terms of (1) the critical temperature and pressure

$$a = \frac{27R^2T_C^2}{64P_C} \quad \text{and} \quad b = \frac{RT_C}{8P_C} \qquad (4.6\text{-}4a)$$

or (2) the critical pressure and volume

$$a = 3P_C\underline{V}_C^2 \quad \text{and} \quad b = \frac{\underline{V}_C}{3} \qquad (4.6\text{-}4b)$$

Substituting either of these sets of parameters into Eq. 4.2-32 yields

$$\left[\frac{P}{P_C} + 3\left(\frac{\underline{V}_C}{\underline{V}}\right)^2\right]\left[3\left(\frac{\underline{V}}{\underline{V}_C}\right) - 1\right] = 8\frac{T}{T_C}$$

Now defining a dimensionless or reduced temperature T_r, pressure P_r, and volume V_r by

$$T_r = \frac{T}{T_C} \quad P_r = \frac{P}{P_C} \quad \text{and} \quad V_r = \frac{\underline{V}}{\underline{V}_C} \qquad (4.6\text{-}5)$$

we obtain

$$\left[P_r + \frac{3}{V_r^2}\right][3V_r - 1] = 8T_r \qquad (4.6\text{-}6)$$

The form of Eq. 4.6-6 is very interesting since it suggests that for all fluids that obey the van der Waals equation, V_r is the same function of T_r and P_r. That is, at given values of the **reduced temperature** ($T_r = T/T_C$) and **reduced pressure** ($P_r = P/P_C$), all van der Waals fluids will have the same numerical value of **reduced volume** ($V_r = \underline{V}/\underline{V}_C$). This does not mean that at the same value of T and P all van der Waals fluids have the same value of $\underline{V}$, since this is *certainly not* the case, as can be seen from the illustration that follows. Two fluids, which have the same values of reduced temperature and pressure and therefore the same reduced volume, are said to be in **corresponding states**.

ILLUSTRATION 4.6-1

Assume that oxygen ($T_C = 154.6$ K, $P_C = 5.046 \times 10^6$ Pa, and $\underline{V}_C = 7.32 \times 10^{-5}$ m^3/mol) and water ($T_C = 647.3$ K, $P_C = 2.205 \times 10^7$ Pa, and $\underline{V}_C = 5.6 \times 10^{-5}$ m^3/mol) can be considered van der Waals fluids.

a. Find the value of reduced volume both fluids would have at $T_r = 3/2$ and $P_r = 3$.

b. Find the temperature, pressure, and volume of each gas at $T_r = 3/2$ and $P_r = 3$.

c. If O_2 and H_2O are both at a temperature of 200°C, and a pressure of 2.5×10^6 Pa, find their specific volumes.

Solution

a. Using $T_r = 3/2$ and $P_r = 3$ in Eq. 4.6-6, one finds that $V_r = 1$ for both fluids.

b. Since $V_r = 1$ at these conditions (see part a), the specific volume of each fluid is equal to the critical volume of that fluid. So

$$\underline{V}_{O_2} = 7.32 \times 10^{-5} \text{ m}^3/\text{mol} \qquad \underline{V}_{H_2O} = 5.6 \times 10^{-5} \text{ m}^3/\text{mol}$$

at at

$$P_{O_2} = 3 \times P_{C,O_2} = 1.514 \times 10^7 \text{ Pa} \qquad P_{H_2O} = 3 \times P_{C,H_2O} = 6.615 \times 10^7 \text{ Pa}$$

and and

$$T_{O_2} = 1.5 \times T_{C,O_2} = 232.2 \text{ K} \qquad T_{H_2O} = 1.5 \times T_{C,H_2O} = 971 \text{ K}$$

c. For oxygen we have $P_r = \dfrac{2.5 \times 10^6}{5.046 \times 10^6} = 0.495$, and $T_r = 473.2/154.6 = 3.061$. Therefore, $V_r = 16.5$ and $\underline{V} = 1.208 \times 10^{-3}$ m³/mol. Similarly for water, $P_r = \dfrac{2.5 \times 10^6}{2.205 \times 10^7} = 0.113$, $T_r = 473.2/647.3 = 0.731$, so that $V_r = 15.95$ and $\underline{V} = 8.932 \times 10^{-4}$ m³/mol. ■

As already indicated, the accuracy of the van der Waals equation is not very good. This may be verified by comparing the results of the previous example with experimental data, or by comparing the compressibility factor $Z = P\underline{V}/RT$ for the van der Waals equation of state with experimental data for a variety of gases. In particular, at the critical point (see Eq. 4.6-3c)

$$Z_C|_{\text{van der Waals}} = P_C\underline{V}_C/RT_C = 0.375$$

while for most fluids the critical compressibility Z_C is in the range 0.23 to 0.31 (Table 4.6-1), so that the van der Waals equation fails to predict accurate critical-point behavior. (It is, however, a great improvement over the ideal gas equation of state, which predicts that $Z = 1$ for all conditions.)

The fact that the critical compressibility of the van der Waals fluid is not equal to that for most real fluids also means that different values for the van der Waals parameters are obtained for any one fluid, depending on whether Eqs. 4.6-3a, Eqs. 4.6-4a, or Eqs. 4.6-4b are used to relate these parameters to the critical properties. In practice, the critical volume of a fluid is known with less experimental accuracy than either the critical temperature or critical pressure, so that Eqs. 4.6-4a and critical-point data are most frequently used to obtain the van der Waals parameters of a gas. Indeed, the entries in Tables 4.4-1 and 4.6-1 are related in this way. Thus, if the parameters in Table 4.4-1 are used in the van der Waals equation, the critical temperature and pressure will be correctly predicted, but the critical volume will be too high by the factor

$$\frac{Z_C|_{\text{van der Waals}}}{Z_C} = \frac{3}{8Z_C}$$

where Z_C is the real fluid critical compressibility.

Although the van der Waals equation is not accurate, the idea of a correspondence of states to which it historically led is both appealing and, as we will see, useful. Attempts at using the corresponding states concept over the last 40 years have been directed toward representing the compressibility factor Z as a function of the reduced pressure and temperature; that is

$$Z = P\underline{V}/RT = Z(P_r, T_r) \tag{4.6-7}$$

where the functional relationship between T_r, P_r, and Z is *determined by experiment*, rather than being based on an analytic equation of state. That such a procedure has some merit is evident from Fig. 4.6-2, where the compressibility data for different fluids have been made to almost superimpose by plotting each as a function of its reduced temperature and pressure.

A close study of Fig. 4.6-2 indicates that there are systematic deviations from the simple corresponding states relation of Eq. 4.6-7. In particular, the compressibility factors for the inorganic fluids are almost always below those for the hydrocarbons. Furthermore, if Eq. 4.6-7 were universally valid, all fluids would have the same value of the critical compressibility $Z_C = Z(P_r = 1, T_r = 1)$; however, from Table 4.6-1, it is clear that Z_C for most real fluids ranges from 0.23 to 0.31. These failings of Eq. 4.6-7 have led to the development of more complicated corresponding states principles. The simplest generalization is the suggestion that there should not be a single $Z = Z(P_r, T_r)$ relationship for all fluids, but rather a family of relationships for different values of Z_C. Thus, to find the value of the compressibility factor for given values T_r and P_r, it would be necessary to use a chart of $Z = Z(P_r, T_r)$ prepared from experimental data for fluids with the same value of Z_C. This is equivalent to saying that Eq. 4.6-7 is to be replaced by

$$Z = Z(T_r, P_r, Z_C)$$

Alternatively, fluid characteristics other than Z_C can be used as the additional parameter in the generalization of the simple corresponding states principle. In fact, since for many substances the critical density, and hence Z_C, is not known with great accuracy, if at all, there is some advantage in avoiding the use of Z_C. Pitzer[12] has suggested that for nonspherical molecules the **acentric factor** ω be used as the third correlative parameter, where ω is defined to be

$$\omega = -1.0 - \log_{10}[P^{\text{vap}}(T_r = 0.7)/P_C]$$

Here $P^{\text{vap}}(T_r = 0.7)$ is the vapor pressure of the fluid at $T_r = 0.7$, a temperature near the normal boiling point. In this case the corresponding states relation would be of the form

$$Z = Z(T_r, P_r, \omega)$$

Even these extensions of the corresponding states concept, which are meant to account for molecular structure, cannot be expected to be applicable to fluids with permanent dipoles and quadrupoles. Since molecules with strong permanent dipoles interact differently than molecules without dipoles, or than molecules with weak dipoles, one would expect the volumetric equation of state for polar fluids to be a function of the dipole moment. In principle, the corresponding states concept could be further generalized to include this new param-

[12]See Appendix 1 of *Thermodynamics*, 2 ed., by G. N. Lewis, M. Randall, K. S. Pitzer, and L. Brewer, McGraw–Hill, New York, 1961, and R. C. Reid, J. M. Prausnitz, and B. Poling, *Properties of Gases and Liquids*, 4th ed. McGraw–Hill, New York, 1987.

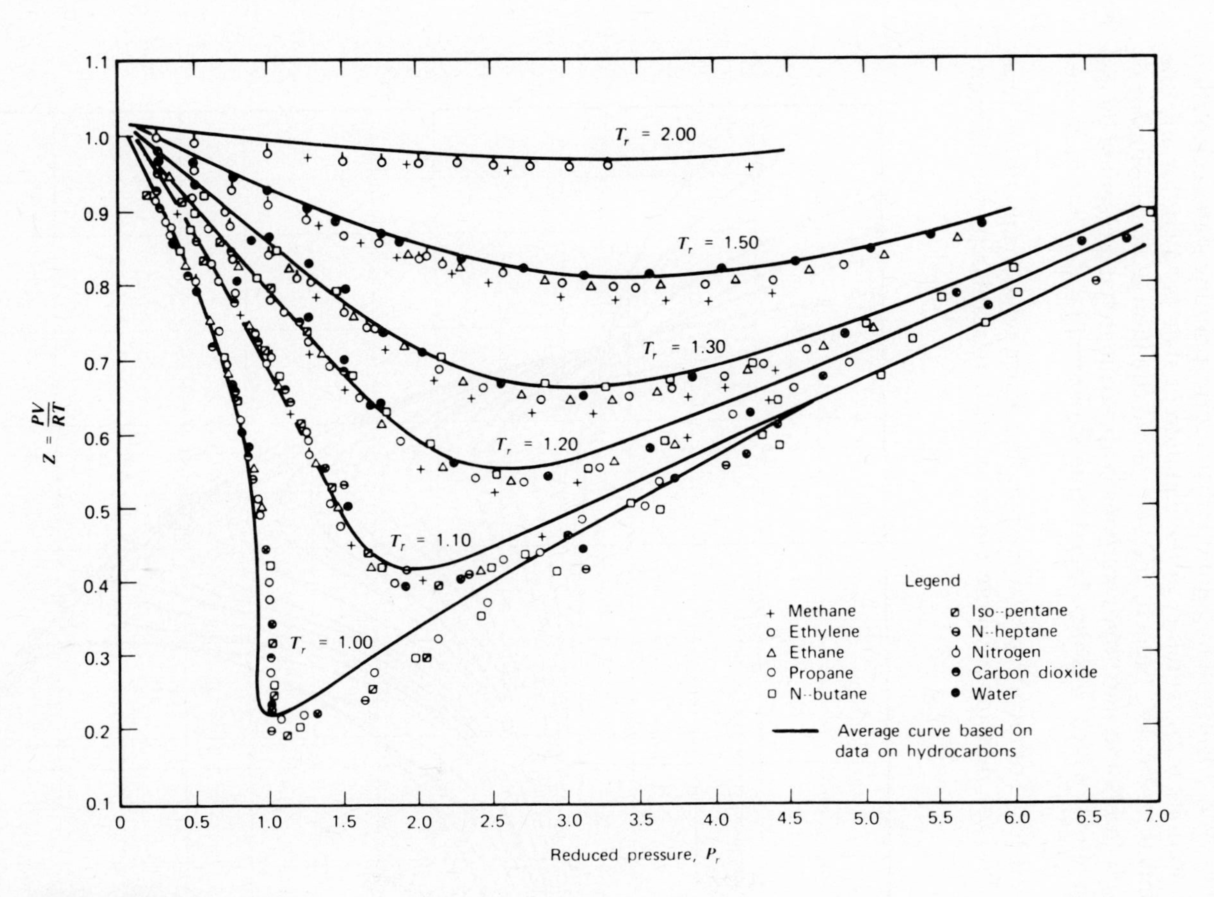

Figure 4.6-2
Compressibility factors for different fluids as a function of the reduced temperature and pressure. (Reprinted with permission from G.-J. Su, *Ind. Engr. Chem.* **38,** 803 (1946). Copyright American Chemical Society.)

eter, but we will not do so here. Instead, we refer you to the book by Reid, Prausnitz, and Poling for a detailed discussion of the corresponding states correlations commonly used by engineers.

The last several paragraphs have emphasized the shortcomings of a single corresponding states principle when dealing with fluids of different molecular classes. However, it is useful to point out that a corresponding states correlation can be an accurate representation of the equation of state behavior for any one class of similar molecules. Indeed, the volumetric equation of state behavior of many simple fluids and most hydrocarbons is approximately represented by the plot in Fig. 4.6-3, which was developed from experimental data for molecules with $Z_C = 0.27$.

The existence of an accurate corresponding states relationship of the type $Z = Z(T_r, P_r)$ (or perhaps a whole family of such relationships for different

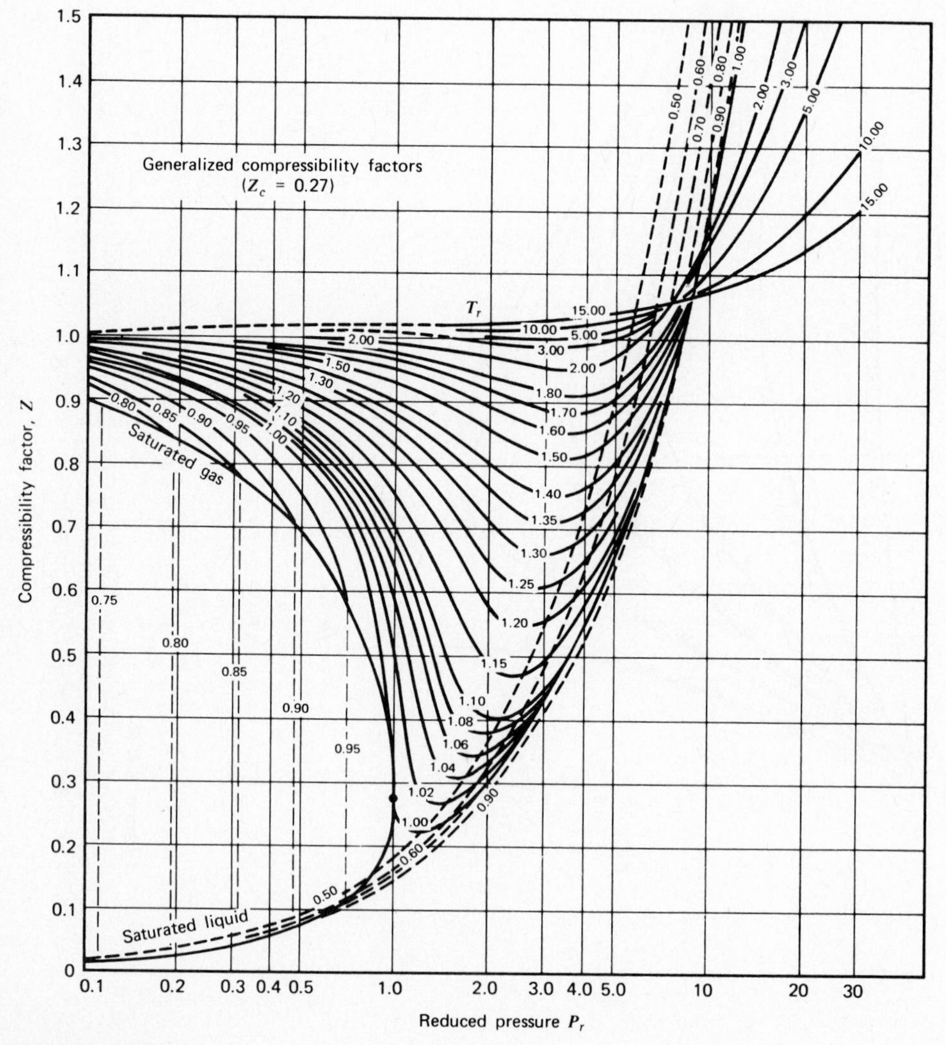

Figure 4.6-3
(Reprinted with permission from O. A. Hougen, K. M. Watson, and R. A. Ragatz, *Chemical Process Principles Charts*, 2nd ed., John Wiley & Sons, New York, 1960.)

values of Z_C or ω) allows one to also develop corresponding states correlations for that contribution to the thermodynamic properties of the fluid that results from molecular interactions, or nonideal behavior, that is, the departure functions of Sec. 4.4. For example, starting with Eq. 4.4-22 we have

$$\underline{H}(T, P) - \underline{H}^{IG}(T, P) = \int_{P=0,T}^{P,T} \left[\underline{V} - T\left(\frac{\partial \underline{V}}{\partial T}\right)_P\right] dP$$

and using the corresponding states relation, Eq. 4.6-7, yields

$$\underline{V} = \frac{RT}{P} Z(T_r, P_r)$$

and

$$\underline{V} - T\left(\frac{\partial \underline{V}}{\partial T}\right)_P = -\frac{RT^2}{P}\left(\frac{\partial Z}{\partial T}\right)_P$$

so that

$$\underline{H}(T, P) - \underline{H}^{IG}(T, P) = -\int_{P=0,T}^{P,T} \frac{RT^2}{P}\left(\frac{\partial Z}{\partial T}\right)_P dP = -RT_C \int_{P_r=0,T_r}^{P_r,T_r} \frac{T_r^2}{P_r}\left(\frac{\partial Z}{\partial T_r}\right)_{P_r} dP_r$$

or

$$\frac{\underline{H}(T, P) - \underline{H}^{IG}(T, P)}{T_C} = -RT_r^2 \int_{P_r=0,T_r}^{P_r,T_r} \frac{1}{P_r}\left(\frac{\partial Z}{\partial T_r}\right)_{P_r} dP_r \tag{4.6-8}$$

The important thing to notice about this equation is that nothing in the integral depends on the properties of a specific fluid, so that when the integral is evaluated using the corresponding states equation of state, the result will be applicable to *all* corresponding states fluids.

Figure 4.6-4 contains in detailed graphical form, the corresponding states prediction for the enthalpy departure from ideal gas behavior computed from Fig. 4.6-3 (for fluids with $Z_C = 0.27$) and Eq. 4.6-8.[13]

The enthalpy change of a real fluid in going from $(T_0, P = 0)$ to (T, P) can then be computed using Fig. 4.6-4 as indicated here:

$$\begin{aligned}\underline{H}(T, P) - \underline{H}(T_0, P = 0) &= \underline{H}^{IG}(T, P) - \underline{H}(T_0, P = 0) + \{\underline{H}(T, P) - \underline{H}^{IG}(T, P)\} \\ &= \int_{T_0}^{T} C_P^* \, dT + T_C\left[\frac{\underline{H}(T, P) - \underline{H}^{IG}(T, P)}{T_C}\right]_{\text{(from Fig. 4.6-4)}}\end{aligned} \tag{4.6-9}$$

Similarly, the enthalpy change in going from any state (T_1, P_1) to state (T_2, P_2) can be computed from repeated application of Eq. 4.6-9, which yields

$$\begin{aligned}\underline{H}(T_2, P_2) - \underline{H}(T_1, P_1) = \int_{T_1}^{T_2} C_P^* \, dT + T_C\Bigg\{&\left[\frac{\underline{H}(T, P) - \underline{H}^{IG}(T, P)}{T_C}\right]_{T_{r_2}, P_{r_2}} \\ &- \left[\frac{\underline{H}(T, P) - \underline{H}^{IG}(T, P)}{T_C}\right]_{T_{r_1}, P_{r_1}}\Bigg\}_{\text{(from Fig. 4.6-4)}}\end{aligned} \tag{4.6-10}$$

The form of Eqs. 4.6-9 and 10 makes good physical sense in that each

[13]Note that Eqs. 4.6-8, 9, and 10 contain the term $(\underline{H} - \underline{H}^{IG})/T_C$, whereas Fig. 4.6-4 gives $(\underline{H}^{IG} - \underline{H})/T_C$.

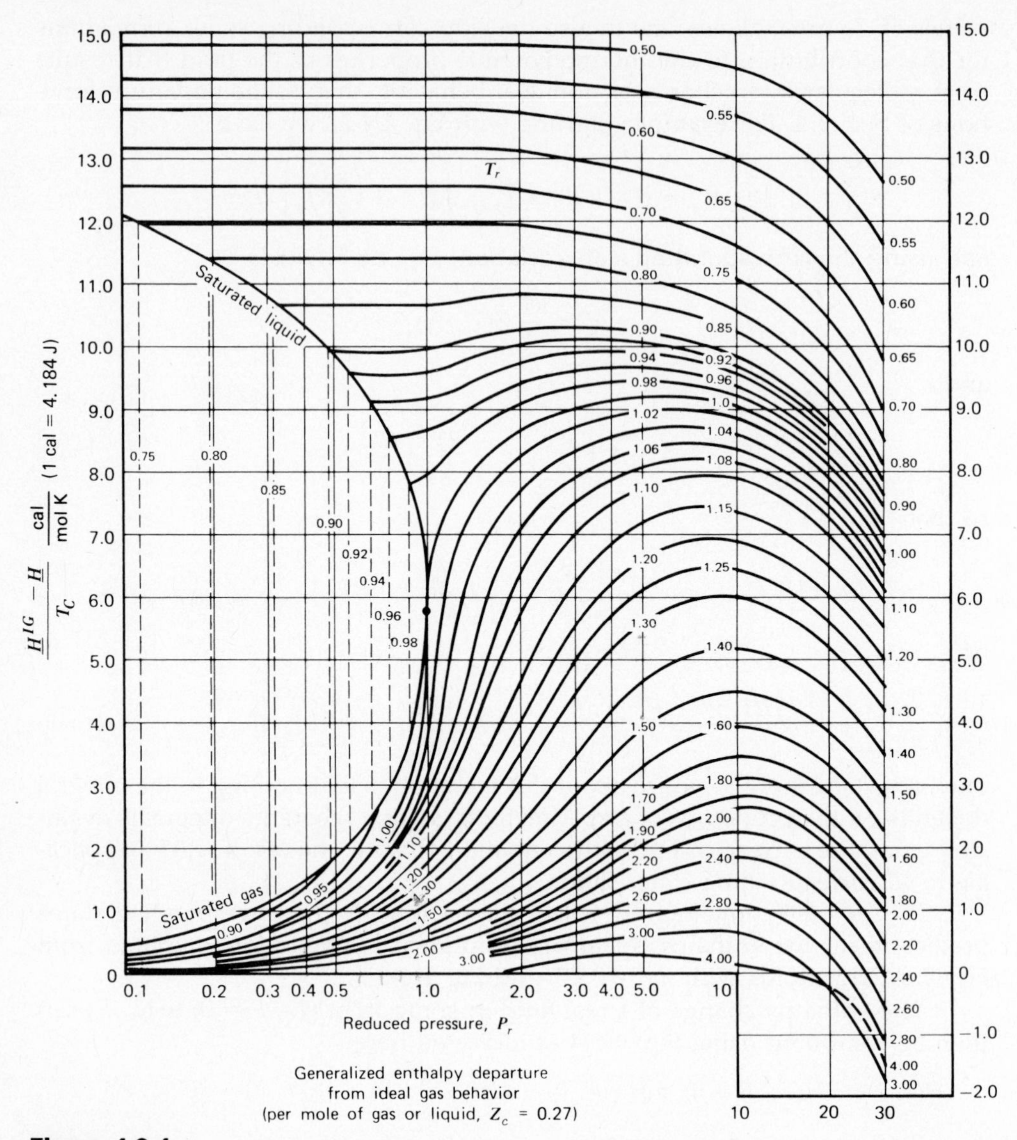

Figure 4.6-4
(Reprinted with permission from O. A. Hougen, K. M. Watson, and R. A. Ragatz, *Chemical Process Principles Charts*, 2nd ed., John Wiley & Sons, New York, 1960.)

consists of two terms with well-defined meanings. The first term depends only on the low-density heat capacity, which is a function of the molecular structure and is specific to the molecular species involved. The second term, on the other hand, represents the nonideal behavior of the fluid due to intermolecular interactions that do not exist in the ideal gas, but whose contribution can be estimated from the generalized correlation.

In a manner equivalent to that just used, it is also possible to show (Problem 4.6) that

$$\underline{S}(T, P) - \underline{S}^{IG}(T, P) = -R \int_{P_r=0,T_r}^{P_r,T_r} \left[\frac{Z-1}{P_r} + \frac{T_r}{P_r} \left(\frac{\partial Z}{\partial T_r} \right)_{P_r} \right] dP_r \tag{4.6-11}$$

This equation and Fig. 4.6-3 are the bases for the entropy departure plot given in Fig. 4.6-5.[14] The change in entropy between any two states (T_1, P_1) and (T_2, P_2) can then be computed from

$$\underline{S}(T_2, P_2) - \underline{S}(T_1, P_1) = \int_{T_1}^{T_2} \frac{C_P^*}{T} dT - R \int_{P_1}^{P_2} \frac{dP}{P}$$

$$+ \{[\underline{S} - \underline{S}^{IG}]_{T_{r_2}, P_{r_2}} - [\underline{S} - \underline{S}^{IG}]_{T_{r_1}, P_{r_1}}\}(\text{from Fig. 4.6-5}) \quad (4.6\text{-}12)$$

Similarly, corresponding states plots could be developed for the other thermodynamic properties, $\underline{U}$, $\underline{A}$, and $\underline{G}$, though these properties are usually computed from the relations

$$\underline{U} = \underline{H} - P\underline{V}$$

$$\underline{A} = \underline{U} - T\underline{S}$$

$$\underline{G} = \underline{H} - T\underline{S} \quad (4.6\text{-}13)$$

and the corresponding states figures already given.

ILLUSTRATION 4.6-2

Rework Illustration 4.5-1, assuming that nitrogen obeys the generalized correlations of Figs. 4.6-3, 4, and 5.

Solution

From Table 4.6-1 we have for nitrogen $T_C = 126.2$ K and $P_C = 33.94$ bar, and from the initial conditions of the problem

$$T_r = \frac{170}{126.2} = 1.347 \quad \text{and} \quad P_r = \frac{100}{33.94} = 2.946$$

From Fig. 4.6-3, $Z = 0.708$, so

$$\underline{V}(t = 0) = Z\frac{RT(t = 0)}{P(t = 0)} = Z\underline{V}^{IG}(t = 0) = 1.0007 \times 10^{-4} \text{ m}^3/\text{mol}$$

Therefore, following Illustration 4.5-1

$$N(t = 0) = 1499.0 \text{ mol}$$

$$N(t) = 1499.0 - 10t \text{ mol}$$

and

$$\underline{V}(t = 50 \text{ min}) = \frac{0.15 \text{ m}^3}{(1499.0 - 500) \text{ mol}} = 1.5015 \times 10^{-4} \text{ m}^3/\text{mol}$$

To compute the temperature and pressure at the end of 50 minutes we use

$$\underline{S}(t = 0) = \underline{S}(t = 50 \text{ min})$$

and recognize that

$$\underline{S}(t = 0) = \underline{S}^{IG}(T, \underline{V})_{t=0} + (\underline{S} - \underline{S}^{IG})_{t=0}$$

and

$$\underline{S}(t = 50 \text{ min}) = \underline{S}^{IG}(T, \underline{V})_{t=50 \text{ min}} + (\underline{S} - \underline{S}^{IG})_{t=50 \text{ min}}$$

[14]Note that Eqs. 4.6-11 and 12 contain the term $\underline{S} - \underline{S}^{IG}$, whereas Fig. 4.6-5 gives $\underline{S}^{IG} - \underline{S}$.

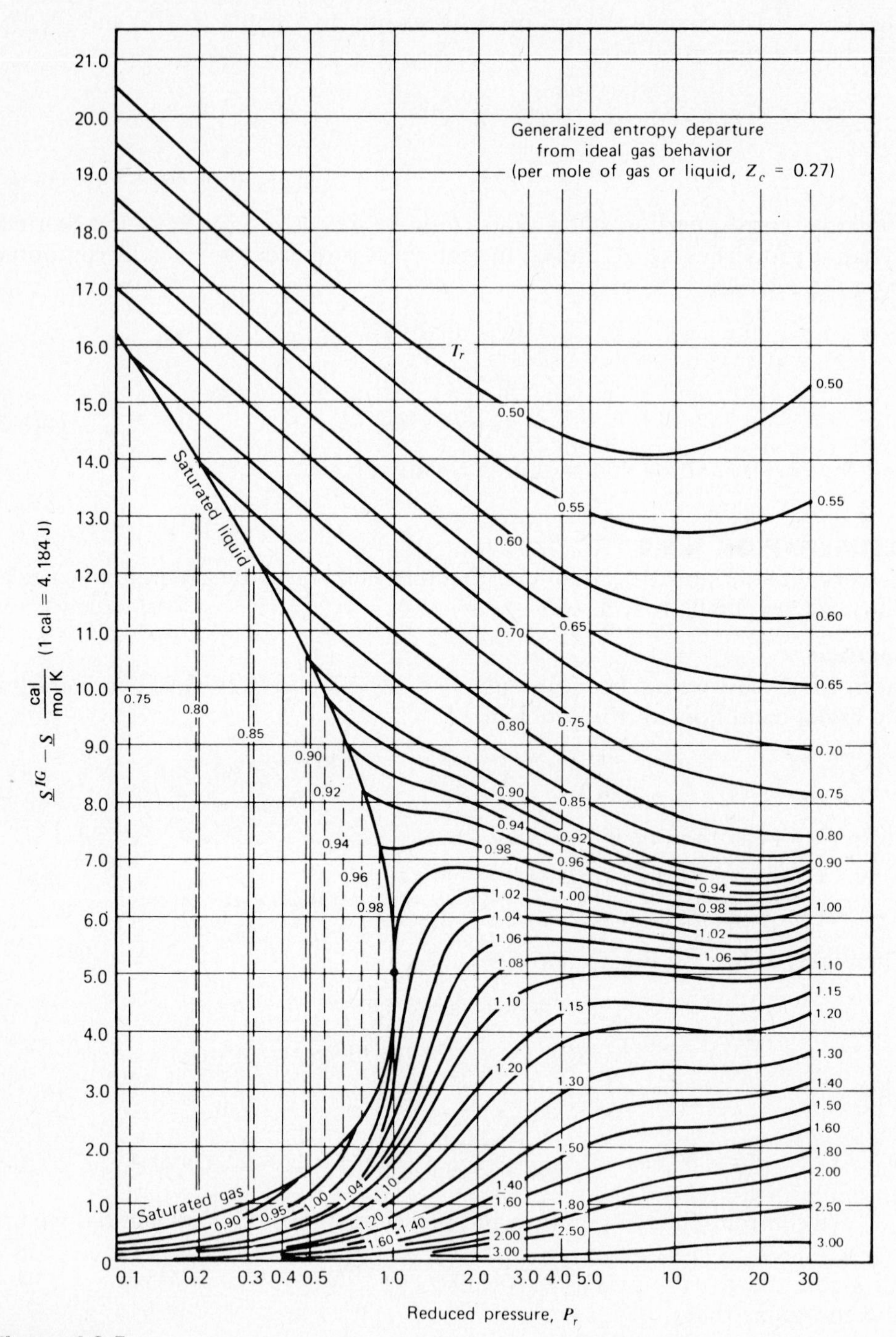

Figure 4.6-5
(Reprinted with permission from O. A. Hougen, K. M. Watson, and R. A. Ragatz, *Chemical Process Principles Charts,* 2nd ed., John Wiley & Sons, New York, 1960.)

so that

$$\underline{S}(t = 50 \text{ min}) - \underline{S}(t = 0) = \underline{S}^{IG}(T, \underline{V})_{t=50 \text{ min}} - \underline{S}^{IG}(T, \underline{V})_{t=0} + (\underline{S} - \underline{S}^{IG})_{t=50 \text{ min}} - (\underline{S} - \underline{S}^{IG})_{t=0}$$

where, from Illustration 4.5-1,

$$\underline{S}^{IG}(T, \underline{V})_{t=50 \text{ min}} - \underline{S}^{IG}(T, \underline{V})_{t=0} = 8.314 \ln \frac{\underline{V}(t = 50 \text{ min})}{\underline{V}(t = 0)} + 18.9 \ln \frac{T(t = 50 \text{ min})}{170} + 4.2 \times 10^{-3} [T(t = 50 \text{ min}) - 170]$$

Both the $(\underline{S} - \underline{S}^{IG})$ terms are gotten from the corresponding states charts. $(\underline{S} - \underline{S}^{IG})_{t=0}$ is easily evaluated, since the initial state is known; that is, $T_r = 1.347$ and $P_r = 2.946$, so that, from Fig. 4.6-5 $(\underline{S} - \underline{S}^{IG})_{t=0} = -2.23$ cal/mol K $= -9.31$ J/mol K. To compute $(\underline{S} - \underline{S}^{IG})$ at $t = 50$ minutes is more difficult because neither T_r nor P_r is known. The procedure to be followed is

1. $\underline{V}(t = 50 \text{ min})$ is known, so guess a value of $T(t = 50 \text{ min})$. (A reasonable first guess is obtained by assuming both $(\underline{S} - \underline{S}^{IG})$ terms are zero, and solving for T).
2. Use $\underline{V}(t = 50 \text{ min})$ and $T(t = 50 \text{ min})$ to compute, by trial and error, $P(t = 50 \text{ min})$ from

$$P = \frac{RT}{\underline{V}} Z\left(\frac{T}{T_C}, \frac{P}{P_C}\right) = \frac{RT}{\underline{V}} Z(T_r, P_r)$$

3. Use the values of P and T from steps 1 and 2 to compute $(\underline{S} - \underline{S}^{IG})_{t=50 \text{ min}}$.
4. Determine whether $\underline{S}(t = 50 \text{ min}) = \underline{S}(t = 0)$ is satisfied with the trial values of T and P. If not, guess another value of $T(t = 50)$ and go back to step 2.

Our solution after a number of trials, is

$$T(t = 50 \text{ min}) = 145 \text{ K}$$

$$P(t = 50 \text{ min}) = 49.2 \text{ bar} \quad \blacksquare$$

It should be pointed out that although the principle of corresponding states and Eqs. 4.6-7, 8, and 11 appear simple, the application of these equations can become tedious, as is evident from this illustration. Also, the use of generalized correlations will lead to results that are not as accurate as those obtained using tabulations of the thermodynamic properties for the fluid of interest. Therefore, the corresponding states principle is used in calculations only when reliable thermodynamic data are not available.

4.7 GENERALIZED EQUATIONS OF STATE

Although the discussion of the previous section focused on the van der Waals equation and corresponding states charts for both the compressibility factor Z and the thermodynamic departure functions, the modern application of the corresponding states idea is to use **generalized equations of state.** The concept is most easily demonstrated by again using the van der Waals equation of state. From Eqs. 4.2-33

$$P = \frac{RT}{\underline{V} - b} - \frac{a}{\underline{V}^2} \qquad (4.2\text{-}33b)$$

and the result of the inflection-point analysis of Sec. 4.6 we have that the constants a and b can be gotten from the fluid critical properties using

$$a = \frac{27R^2T_C^2}{64P_C} \quad \text{and} \quad b = \frac{RT_C}{8P_C} \tag{4.6-4a}$$

The combination of Eqs. 4.2-33b and 4.6-4a is an example of a generalized equation of state, since we now have an equation of state that is presumed to be valid for a class of fluids with parameters (a and b) that have not been fitted to a whole collection of experimental data, but rather are obtained only from the fluid critical properties. The important content of these equations is that they permit the calculation of the $P\underline{V}T$ behavior of a fluid knowing only its critical properties as was the case in corresponding states theory.

It must be emphasized that the van der Waals equation of state is *not very accurate* and has been used here merely for demonstration because of its simplicity. It is never used for engineering design predictions, though other cubic equations of state are used. To illustrate the use of generalized equations of state, we will consider only the Peng–Robinson equation, which is commonly used to represent hydrocarbons and inorganic gases such as nitrogen, oxygen, hydrogen sulfide, and others. The generalized form of the Peng–Robinson equation of state is

$$P = \frac{RT}{\underline{V} - b} - \frac{a(T)}{\underline{V}(\underline{V} + b) + b(\underline{V} - b)} \tag{4.4-2}$$

with

$$a(T) = 0.45724 \frac{R^2T_C^2}{P_C} \alpha(T) \tag{4.7-1}$$

$$b = 0.07780 \frac{RT_C}{P_C} \tag{4.7-2}$$

$$\sqrt{\alpha} = 1 + \kappa \left(1 - \sqrt{\frac{T}{T_C}}\right) \tag{4.7-3}$$

and

$$\kappa = 0.37464 + 1.54226\omega - 0.26992\omega^2 \tag{4.7-4}$$

where ω is the acentric factor defined earlier.

Equations 4.7-1 through 4 were obtained in the following manner. First, the critical-point restrictions of Eqs. 4.6-1 were used, which leads to (Problem 4.11)

$$a(T_C) = 0.45724 \frac{R^2T_C^2}{P_C} \quad \text{and} \quad b = 0.07780 \frac{RT_C}{P_C}$$

Next, to improve the predictions of the boiling pressure as a function of temperature, that is, the vapor pressure (which will be discussed in Sec. 5.5), Peng and Robinson added an additional temperature-dependent term into their equation by setting

$$a(T) = a(T_C)\alpha(T)$$

Note that to satisfy the critical-point restrictions, $\alpha(T = T_C)$ must equal unity, as does the form of Eq. 4.7-3. The specific form of α given by Eqs. 4.7-3 and 4 was chosen by fitting vapor pressure data for many fluids.

There are two points to be noticed when comparing the generalized van der Waals and Peng–Robinson equations of state. First, although the parameter a is a constant in the van der Waals equation, in the Peng–Robinson equation it is a function of temperature (actually reduced temperature $T_r = T/T_C$) through the temperature dependence of α. Second, the generalized parameters of the Peng–Robinson equation of state are functions of the critical temperature, the critical pressure, *and* the acentric factor ω of the fluid. Consequently, the Peng–Robinson equation of state, as generalized here, is said to be a three-parameter (T_C, P_C, ω) equation of state, whereas the van der Waals equation contains only the two parameters T_C and P_C.

This generalized form of the Peng–Robinson equation of state (or other equations of state) can be used to compute not only the compressibility, but also the departure functions for the other thermodynamic properties. This is done using Eqs. 4.4-29 and 30. In particular, to obtain numerical values for the enthalpy and/or entropy departure for a fluid that obeys the Peng–Robinson equation of state, one uses the following procedure:

1. Use the critical properties and acentric factor of the fluid to calculate b, κ, and the temperature-independent part of a using Eqs. 4.7-1, 2, and 4.
2. At the temperature of interest, compute numerical values for α and a using Eqs. 4.7-1 and 3.
3. Solve the equation of state, Eq. 4.4-2, for $\underline{V}$ and compute $Z = P\underline{V}/RT$. Alternatively, solve for Z directly from the equivalent equation

$$Z^3 - (1 - B)Z^2 + (A - 3B^2 - 2B)Z - (AB - B^2 - B^3) = 0 \tag{4.7-5}$$

   where $B = Pb/RT$ and $A = aP/R^2T^2$.
4. Use the computed value of Z and

$$\frac{da}{dT} = -0.45724\,\frac{R^2T_C^2}{P_C}\,\kappa\,\sqrt{\frac{\alpha}{TT_C}}$$

   to compute $[\underline{H}(T, P) - \underline{H}^{IG}(T, P)]$ and/or $[\underline{S}(T, P) - \underline{S}^{IG}(T, P)]$ as desired, using Eqs. 4.4-29 and 30.

The enthalpy and entropy departures from ideal gas behavior calculated in this way can be used to solve thermodynamic problems in the same manner as the similar functions obtained from corresponding states graphs were used in the previous section.

It is clear that the calculation outlined here using the Peng–Robinson equation of state is, when doing computations by hand, more tedious than merely calculating the reduced temperature and pressure and using the graphs in Section 4.6. However, the equations here have some important advantages when using digital computers or programmable calculators. First, this analytic computation avoids putting the three corresponding states graphs in a computer memory in numerical form. Second, the values of the compressibility factor and departures from ideal gas properties obtained in the present three-parameter calculation should be more accurate than those obtained from the simple two-parameter (T_C, P_C) corresponding states method of the previous section because of the additional fluid parameter (acentric factor) involved and the absence of interpolation errors. Finally, if, at some time in the future, it is decided to use a different equation of state, only a few lines of computer code need be changed, rather than drawing a new series of complicated graphs.

Before there was easy access to calculators and computers, it was common to apply the corresponding states principle by using tables and graphs as illustrated in the previous section. Now, however, the usual industrial practice is to directly use the corresponding states idea, that different fluids obey the same form of the equation of state, by using digital computer programs and generalized equations of state such as those discussed here.

ILLUSTRATION 4.7-1

Rework Illustration 4.5-1 assuming that nitrogen can be described using the Peng–Robinson equation of state.

Solution

The calculation of the thermodynamic properties needed here using the Peng–Robinson equation of state is done as follows:

1. Choose T and P.
2. Calculate a, b, A, and B using Eqs. 4.7-1 through 4 and $B = Pb/RT$ and $A = aP/R^2T^2$.
3. Find the vapor-phase compressibility Z using Eq. 4.7-5 by solving the cubic equation.
4. Using the calculated value of Z and Eq. 4.4-30 to calculate the entropy departure from ideal gas behavior, $\underline{S} - \underline{S}^{IG}$.

We have done these calculations using the BASIC program of Appendix A4.1. The calculational procedure to solve this problem is then as follows:

i. At the initial temperature and pressure compute Z, $\underline{V}$, and $\underline{S} - \underline{S}^{IG}$ using the Peng–Robinson equation of state and the critical properties and acentric factor of nitrogen. The results are

$$Z = 0.6769$$

$$\underline{V}(t = 0) = 0.9567 \times 10^{-4}\ \mathrm{m^3/mol}, \text{ and}$$

$$(\underline{S} - \underline{S}^{IG})_{t=0} = -9.18\ \mathrm{J/mol}$$

Thus,

$$N(t = 0) = 1567.9\ \mathrm{mol}$$

$$N(t = 50) = 1567.9 - 10 \times 50 = 1067.9\ \mathrm{mol}$$

$$\underline{V}(t = 50) = 0.15\ \mathrm{m^3}/1067.9\ \mathrm{mol} = 1.4046 \times 10^{-4}\ \mathrm{m^3/mol}$$

ii. Guess the final temperature of the expansion process (the ideal gas result is a good initial guess).

iii. Adjust the pressure until the correct value of $\underline{V}(t = 50)$ is obtained. Obtain $(\underline{S} - \underline{S}^{IG})_{t=50}$ for the guessed value of $T(t = 50)$.

iv. Determine whether the following equation is satisfied.

$$\underline{S}(t = 50) - \underline{S}(t = 0) = 0 = \underline{S}^{IG}(t = 50) - \underline{S}^{IG}(t = 0) + (\underline{S} - \underline{S}^{IG})_{t=50} - (\underline{S} - \underline{S}^{IG})_{t=0}$$

$$= 8.314 \ln \frac{\underline{V}(t = 50)}{\underline{V}(t = 0)} + 18.9 \ln \frac{T(t = 50)}{T(t = 0)} + 4.2 \times 10^{-3}[T(t = 50) - T(t = 0)]$$

$$+ (\underline{S} - \underline{S}^{IG})_{t=50} + 9.18 \overset{?}{=} 0$$

If it is, the correct solution has been obtained. If not, adjust guessed value of T (and P) and repeat the calculation. The result, after several iterations, is

$$T(t = 50 \text{ min}) = 143.6 \text{ K}$$

$$P(t = 50 \text{ min}) = 48.6 \text{ bar}$$

$$(\underline{S} - \underline{S}^{IG})_{t=50} = -9.09 \text{ J/mol K} \quad \blacksquare$$

4.8 MORE ABOUT THERMODYNAMIC PARTIAL DERIVATIVES (OPTIONAL)

In this section we first consider a general procedure for transforming the partial derivatives that occur in thermodynamics to derivatives or combinations of derivatives with different independent variables. We also consider the class of thermodynamic derivatives with respect to mole number or mass. These subjects are not central to the development elsewhere in this book.

There are numerous methods of making a transformation of variables in thermodynamic partial derivatives. Because of its generality, and the fact that it can be easily extended to more than two independent variables, the technique of **Jacobian transformations** will be used here.

If the variables $\underline{K}$ and $\underline{L}$ are functions of $\underline{X}$ and $\underline{Y}$, the Jacobian $\partial(\underline{K}, \underline{L})/\partial(\underline{X}, \underline{Y})$ is defined to be

$$\frac{\partial(\underline{K}, \underline{L})}{\partial(\underline{X}, \underline{Y})} = \left(\frac{\partial \underline{K}}{\partial \underline{X}}\right)_{\underline{Y}} \left(\frac{\partial \underline{L}}{\partial \underline{Y}}\right)_{\underline{X}} - \left(\frac{\partial \underline{K}}{\partial \underline{Y}}\right)_{\underline{X}} \left(\frac{\partial \underline{L}}{\partial \underline{X}}\right)_{\underline{Y}} = \begin{vmatrix} \left(\frac{\partial \underline{K}}{\partial \underline{X}}\right)_{\underline{Y}} & \left(\frac{\partial \underline{K}}{\partial \underline{Y}}\right)_{\underline{X}} \\ \left(\frac{\partial \underline{L}}{\partial \underline{X}}\right)_{\underline{Y}} & \left(\frac{\partial \underline{L}}{\partial \underline{Y}}\right)_{\underline{X}} \end{vmatrix} \tag{4.8-1}$$

where the last term on the right is the determinant of the matrix of derivatives. The property of the Jacobian especially useful in thermodynamics is that

$$\left(\frac{\partial \underline{K}}{\partial \underline{X}}\right)_{\underline{Y}} = \frac{\partial(\underline{K}, \underline{Y})}{\partial(\underline{X}, \underline{Y})} \tag{4.8-2}$$

as is evident from substituting $\underline{Y}$ for $\underline{L}$ in the previous equation. Other properties of Jacobians, which follow immediately from their definition and their relation to partial derivatives, are

$$\frac{\partial(\underline{K}, \underline{L})}{\partial(\underline{X}, \underline{Y})} = -\frac{\partial(\underline{L}, \underline{K})}{\partial(\underline{X}, \underline{Y})} \tag{4.8-3}$$

$$\frac{\partial(\underline{K}, \underline{L})}{\partial(\underline{X}, \underline{Y})} = \left[\frac{\partial(\underline{X}, \underline{Y})}{\partial(\underline{K}, \underline{L})}\right]^{-1} \quad \text{(see Eq. 4.1-2)} \tag{4.8-4}$$

and

$$\frac{\partial(\underline{K}, \underline{L})}{\partial(\underline{X}, \underline{Y})} = \frac{\partial(\underline{K}, \underline{L})}{\partial(\underline{B}, \underline{C})} \frac{\partial(\underline{B}, \underline{C})}{\partial(\underline{X}, \underline{Y})} = \frac{\partial(\underline{K}, \underline{L})}{\partial(\underline{B}, \underline{C})} \Big/ \frac{\partial(\underline{X}, \underline{Y})}{\partial(\underline{B}, \underline{C})} \tag{4.8-5}$$

(see Eq. 4.1-8)

where $\underline{B}$ and $\underline{C}$ are any two intensive variables.

The general procedure used in reducing any partial derivative of the form $(\partial \underline{X}/\partial \underline{Y})_{\underline{Z}}$ to combinations of only P, V, T, S, α, κ_T, and C_P or C_V is indicated here.

1. First, use, if possible, the entries of Table 4.2-1, together with the properties discussed in Sec. 4.1, to reduce the derivative. For example, to obtain the reduced form of $(\partial \underline{H}/\partial T)_{\underline{V}}$ we start with

$$d\underline{H} = C_P\,dT + \left[\underline{V} - T\left(\frac{\partial \underline{V}}{\partial T}\right)_P\right] dP$$

and use Eqs. 4.1-4, 6, and 7 to get

$$\left(\frac{\partial \underline{H}}{\partial T}\right)_{\underline{V}} = C_P + \left[\underline{V} - T\left(\frac{\partial \underline{V}}{\partial T}\right)_P\right]\left(\frac{\partial P}{\partial T}\right)_{\underline{V}}$$

$$= C_P + (\underline{V} - T\underline{V}\alpha)\,\frac{\partial P}{\kappa_T}$$

2. If Table 4.2-1 does not provide the information necessary for the reduction of the derivative, use one of the procedures suggested here, depending on the number of independent variables that appear in the derivative.
   a. Derivatives that contain the two desired independent variables (which here are presumed to be two variables from among P, T, and $\underline{V}$) should be written in Jacobian form and, using Eq. 4.8-5, the appropriate independent variables interposed to simplify the derivative. For example, in a Joule–Thomson expansion (Illustration 2.3-3) one can measure the change in temperature accompanying a differential change in pressure at constant enthalpy, that is, the derivative $(\partial T/\partial P)_{\underline{H}}$. To express this derivative in terms of simpler quantities we start by writing $(\partial T/\partial P)_{\underline{H}}$ in Jacobian form and interpose T and P as independent variables to get the desired expression. Thus

$$\left(\frac{\partial T}{\partial P}\right)_{\underline{H}} = \frac{\partial(T, \underline{H})}{\partial(P, \underline{H})} = \frac{\partial(T, \underline{H})}{\partial(T, P)} \times \frac{\partial(T, P)}{\partial(P, \underline{H})} = -\frac{\partial(\underline{H}, T)}{\partial(P, T)}\Big/\frac{\partial(\underline{H}, P)}{\partial(T, P)}$$

$$= -\left(\frac{\partial \underline{H}}{\partial P}\right)_T\Big/\left(\frac{\partial \underline{H}}{\partial T}\right)_P = -\frac{1}{C_P}\left[\underline{V} - T\left(\frac{\partial \underline{V}}{\partial T}\right)_P\right] = -(\underline{V}/C_P)(1 - T\alpha)$$

   b. Derivatives that contain only one of the desired independent variables can be reduced by first being rewritten in Jacobian form, and then using Eq. 4.8-5 to interpose the desired independent variables. The one resulting Jacobian that contains no variable common to both the numerator and the denominator is then expanded using Eq. 4.8-1 to obtain four derivatives that are easily reduced (see Illustration 4.8-1).
   c. Derivatives that do not contain any of the desired independent variables are the most difficult to reduce. Here, again, we start by rewriting the derivative in Jacobian form, interpose the desired independent variables, and then expand and reduce both the resulting Jacobians (neither of which will contain a common variable in the numerator and denominator). This procedure is demonstrated in Illustration 4.8-2.

ILLUSTRATION 4.8-1

Develop an expression for the derivative $(\partial \underline{S}/\partial T)_P$ in which T and $\underline{V}$ are the independent variables using the method of Jacobians.

Solution

This derivative explicitly contains only one (T) of the two desired independent variables (T, $\underline{V}$), therefore procedure 2b is to be used. Following this prescription we have

$$\left(\frac{\partial \underline{S}}{\partial T}\right)_P = \frac{\partial(\underline{S}, P)}{\partial(T, P)} = \frac{\partial(\underline{S}, P)}{\partial(T, \underline{V})} \bigg/ \frac{\partial(T, P)}{\partial(T, \underline{V})} = \left(\frac{\partial \underline{V}}{\partial P}\right)_T \frac{\partial(\underline{S}, P)}{\partial(T, \underline{V})}$$

Now expanding the remaining Jacobian we obtain

$$\frac{\partial(\underline{S}, P)}{\partial(T, \underline{V})} = \left(\frac{\partial \underline{S}}{\partial T}\right)_{\underline{V}} \left(\frac{\partial P}{\partial \underline{V}}\right)_T - \left(\frac{\partial \underline{S}}{\partial \underline{V}}\right)_T \left(\frac{\partial P}{\partial T}\right)_{\underline{V}} = \frac{C_V}{T}\left(\frac{\partial P}{\partial \underline{V}}\right)_T - \left(\frac{\partial P}{\partial T}\right)_{\underline{V}}^2$$

(from Eqs. 4.2-15 and 18)

so that

$$\left(\frac{\partial \underline{S}}{\partial T}\right)_P = \left(\frac{\partial \underline{V}}{\partial P}\right)_T \left[\frac{C_V}{T}\left(\frac{\partial P}{\partial \underline{V}}\right)_T - \left(\frac{\partial P}{\partial T}\right)_{\underline{V}}^2\right] = \frac{C_V}{T} - \left(\frac{\partial \underline{V}}{\partial P}\right)_T \left(\frac{\partial P}{\partial T}\right)_{\underline{V}}^2$$

As we saw in Illustration 4.2-2, this derivative can further be reduced to

$$\left(\frac{\partial \underline{S}}{\partial T}\right)_P = \frac{C_V}{T} + \frac{\underline{V}\alpha^2}{\kappa_T} \quad \blacksquare$$

ILLUSTRATION 4.8-2

In the Joule–Thomson experiment one measures the $\mu = (\partial T/\partial P)_{\underline{H}}$. Another quantity of interest, but not measurable, is the partial derivative $(\partial \underline{A}/\partial P)_{\underline{H}}$. Obtain an expression for this derivative in which T and $\underline{V}$ are the independent variables.

Solution

Following prescription 2c we have

$$\left(\frac{\partial \underline{A}}{\partial P}\right)_{\underline{H}} = \frac{\partial(\underline{A}, \underline{H})}{\partial(P, \underline{H})} = \frac{\partial(\underline{A}, \underline{H})}{\partial(T, \underline{V})} \bigg/ \frac{\partial(P, \underline{H})}{\partial(T, \underline{V})}$$

Next, each of the Jacobians is expanded

$$\frac{\partial(\underline{A}, \underline{H})}{\partial(T, \underline{V})} = \left(\frac{\partial \underline{A}}{\partial T}\right)_{\underline{V}} \left(\frac{\partial \underline{H}}{\partial \underline{V}}\right)_T - \left(\frac{\partial \underline{A}}{\partial \underline{V}}\right)_T \left(\frac{\partial \underline{H}}{\partial T}\right)_{\underline{V}} = -\underline{S}\left(\frac{\partial \underline{H}}{\partial \underline{V}}\right)_T + P\left(\frac{\partial \underline{H}}{\partial T}\right)_{\underline{V}}$$

and

$$\frac{\partial(P, \underline{H})}{\partial(T, \underline{V})} = \left(\frac{\partial P}{\partial T}\right)_{\underline{V}} \left(\frac{\partial \underline{H}}{\partial \underline{V}}\right)_T - \left(\frac{\partial \underline{H}}{\partial T}\right)_{\underline{V}} \left(\frac{\partial P}{\partial \underline{V}}\right)_T$$

Now

$$d\underline{H} = C_P\, dT + \left[\underline{V} - T\left(\frac{\partial \underline{V}}{\partial T}\right)_P\right] dP$$

so that

$$\left(\frac{\partial \underline{H}}{\partial P}\right)_T = \left[\underline{V} - T\left(\frac{\partial \underline{V}}{\partial T}\right)_P\right]$$

$$\left(\frac{\partial \underline{H}}{\partial \underline{V}}\right)_T = \left[\underline{V} - T\left(\frac{\partial \underline{V}}{\partial T}\right)_P\right]\left(\frac{\partial P}{\partial \underline{V}}\right)_T = \left(\frac{\partial \underline{H}}{\partial P}\right)_T \left(\frac{\partial P}{\partial \underline{V}}\right)_T$$

and

$$\left(\frac{\partial \underline{H}}{\partial T}\right)_{\underline{V}} = C_P + \left[\underline{V} - T\left(\frac{\partial \underline{V}}{\partial T}\right)_P\right]\left(\frac{\partial P}{\partial T}\right)_{\underline{V}} = C_P + \left(\frac{\partial \underline{H}}{\partial P}\right)_T \left(\frac{\partial P}{\partial T}\right)_{\underline{V}}$$

Thus

$$\left(\frac{\partial \underline{A}}{\partial P}\right)_{\underline{H}} = \frac{-\underline{S}\left(\frac{\partial \underline{H}}{\partial P}\right)_T\left(\frac{\partial P}{\partial \underline{V}}\right)_T + PC_P + P\left(\frac{\partial \underline{H}}{\partial P}\right)_T\left(\frac{\partial P}{\partial T}\right)_{\underline{V}}}{\left(\frac{\partial P}{\partial T}\right)_{\underline{V}}\left(\frac{\partial \underline{H}}{\partial P}\right)_T\left(\frac{\partial P}{\partial \underline{V}}\right)_T - C_P\left(\frac{\partial P}{\partial \underline{V}}\right)_T - \left(\frac{\partial \underline{H}}{\partial P}\right)_T\left(\frac{\partial P}{\partial T}\right)_{\underline{V}}\left(\frac{\partial P}{\partial \underline{V}}\right)_T}$$

$$\left(\frac{\partial \underline{A}}{\partial P}\right)_{\underline{H}} = \frac{-\underline{S}\left(\frac{\partial \underline{H}}{\partial P}\right)_T\left(\frac{\partial P}{\partial \underline{V}}\right)_T + PC_P + P\left(\frac{\partial \underline{H}}{\partial P}\right)_T\left(\frac{\partial P}{\partial T}\right)_{\underline{V}}}{-C_P\left(\frac{\partial P}{\partial \underline{V}}\right)_T}$$

$$= \frac{\underline{S}}{C_P}\left[\underline{V} - T\left(\frac{\partial \underline{V}}{\partial T}\right)_P\right] - P\left(\frac{\partial \underline{V}}{\partial P}\right)_T - \frac{P}{C_P}\left[\underline{V} - T\left(\frac{\partial \underline{V}}{\partial T}\right)_P\right]\left(\frac{\partial \underline{V}}{\partial P}\right)_T\left(\frac{\partial P}{\partial T}\right)_{\underline{V}}$$

Finally, by the triple product relation, Eq. 4.1-6

$$\left(\frac{\partial \underline{V}}{\partial P}\right)_T\left(\frac{\partial P}{\partial T}\right)_{\underline{V}}\left(\frac{\partial T}{\partial \underline{V}}\right)_P = -1 \quad \text{or} \quad \left(\frac{\partial \underline{V}}{\partial P}\right)_T\left(\frac{\partial P}{\partial T}\right)_{\underline{V}} = -\left(\frac{\partial \underline{V}}{\partial T}\right)_P$$

so that

$$\left(\frac{\partial \underline{A}}{\partial P}\right)_{\underline{H}} = \frac{\left(\underline{S} + P\left(\frac{\partial \underline{V}}{\partial T}\right)_P\right)}{C_P}\left[\underline{V} - T\left(\frac{\partial \underline{V}}{\partial T}\right)_P\right] - P\left(\frac{\partial \underline{V}}{\partial P}\right)_T$$

Note that for the ideal gas

$$\left[\underline{V} - T\left(\frac{\partial \underline{V}}{\partial T}\right)_P\right] = 0$$

and

$$\left(\frac{\partial \underline{V}}{\partial P}\right)_T = -\frac{\underline{V}}{P}$$

so

$$\left(\frac{\partial \underline{A}}{\partial P}\right)_{\underline{H}} = +\underline{V}$$

For a real (not ideal) gas the expression is much more complicated. ■

The evaluation of derivatives with respect to mole number or mass subject to various constraints (i.e., different variables held constant) will now be considered, as derivatives of this type are sometimes a source of confusion. For this discussion $\underline{X}$, $\underline{Y}$, and $\underline{Z}$ will be used to indicate the thermodynamic state variables (i.e., $\underline{U}$, $\underline{G}$, $\underline{H}$, $\underline{S}$, $\underline{A}$, $\underline{V}$, T, and P), X, Y, and Z to designate the extensive variables (U, G, H, S, A, and V), and N the number of moles in the system.[15]

We start the analysis by noting that since any intensive thermodynamic variable in a one-component system is a function of only two other state variables and not the number of moles

$$\left(\frac{\partial \underline{X}}{\partial N}\right)_{\underline{Y},\underline{Z}} = 0 \tag{4.8-6}$$

[15]Although the discussion here is in terms of moles and molar properties, we could equally well use mass M and properties per unit mass, $\hat{U}$, $\hat{V}$, and so forth.

as was stated in Sec. 4.1. Thus, for example

$$\left(\frac{\partial \underline{U}}{\partial N}\right)_{\underline{S},\underline{V}} = 0 \tag{4.8-7}$$

Similarly,

$$\left(\frac{\partial \hat{X}}{\partial M}\right)_{\hat{Y},\hat{Z}} = 0 \tag{4.8-8}$$

[Note, however, that the derivatives $(\partial \underline{X}/\partial N)_{\underline{Y},Z}$ and $(\partial \underline{X}/\partial N)_{Y,Z}$, which contain both intensive and extensive variables and are encountered only infrequently, are not equal to zero, as we will see.]

Next, we note that derivatives among only the extensive properties, $(\partial \underline{X}/\partial N)_{Y,Z}$, can be evaluated using the formulas and procedures discussed in this chapter. For example, from Eq. 4.2-9c, we know that

$$\left(\frac{\partial U}{\partial N}\right)_{S,V} = \underline{G} \tag{4.8-9}$$

Our interest here will mainly be in derivatives that contain both intensive and extensive variables.

Consider now the derivative of the extensive property X with respect to mole number along a path on which the intensive variables $\underline{Y}$ and $\underline{Z}$ are held constant. Since $X = N\underline{X}$, we have

$$\left(\frac{\partial X}{\partial N}\right)_{\underline{Y},\underline{Z}} = \left(\frac{\partial (N\underline{X})}{\partial N}\right)_{\underline{Y},\underline{Z}} = \underline{X} + N\left(\frac{\partial \underline{X}}{\partial N}\right)_{\underline{Y},\underline{Z}} = \underline{X} \tag{4.8-10}$$

so that

$$\left(\frac{\partial X}{\partial N}\right)_{\underline{Y},\underline{Z}} = \underline{X} \tag{4.8-11}$$

When $X = U$, $Y = S$, and $Z = V$ we have

$$\left(\frac{\partial U}{\partial N}\right)_{\underline{S},\underline{V}} = \underline{U} \tag{4.8-12}$$

which is to be compared with Eqs. 4.8-7 and 9. Note that the derivatives in each of these cases have different values and correspond to different physical situations. (Can you think of experiments to measure each of these derivatives?)

Consider next a derivative of the form $(\partial X/\partial N)_{\underline{Y},Z}$ where $\underline{Y}$ is a state variable other than T and P (i.e., a state variable such that $Y = N\underline{Y}$ is a meaningful extensive variable). To evaluate such a derivative we start from

$$dX = \left(\frac{\partial X}{\partial Y}\right)_{Z,N} dY + \left(\frac{\partial X}{\partial Z}\right)_{Y,N} dZ + \left(\frac{\partial X}{\partial N}\right)_{Y,Z} dN$$

Next we write Y as $N\underline{Y}$, $dY = d(N\underline{Y}) = N\,d\underline{Y} + \underline{Y}\,dN$, so that

$$dX = N\left(\frac{\partial X}{\partial Y}\right)_{Z,N} d\underline{Y} + \left(\frac{\partial X}{\partial Z}\right)_{Y,N} dZ + \left[\left(\frac{\partial X}{\partial N}\right)_{Y,Z} + \underline{Y}\left(\frac{\partial X}{\partial Y}\right)_{Z,N}\right] dN$$

and

$$\left(\frac{\partial X}{\partial N}\right)_{\underline{Y},Z} = \left(\frac{\partial X}{\partial N}\right)_{Y,Z} + \underline{Y}\left(\frac{\partial X}{\partial Y}\right)_{Z,N} \tag{4.8-13}$$

Thus, for example, we have

$$\left(\frac{\partial U}{\partial N}\right)_{\underline{S},V} = \left(\frac{\partial U}{\partial N}\right)_{S,V} + \underline{S}\left(\frac{\partial U}{\partial S}\right)_{V,N} = \underline{G} + \underline{S}T = \underline{H} \tag{4.8-14}$$

and

$$\left(\frac{\partial U}{\partial N}\right)_{S,\underline{V}} = \left(\frac{\partial U}{\partial N}\right)_{S,V} + \underline{V}\left(\frac{\partial U}{\partial V}\right)_{S,N} = \underline{G} + \underline{V}(-P) = \underline{A} \tag{4.8-15}$$

To evaluate a derivative of the form $(\partial \underline{X}/\partial N)_{Y,Z}$ we start from

$$\left(\frac{\partial X}{\partial N}\right)_{Y,Z} = \left(\frac{\partial (N\underline{X})}{\partial N}\right)_{Y,Z} = \underline{X} + N\left(\frac{\partial \underline{X}}{\partial N}\right)_{Y,Z} \tag{4.8-16}$$

and obtain

$$\left(\frac{\partial \underline{\underline{X}}}{\partial N}\right)_{Y,Z} = \frac{1}{N}\left[\left(\frac{\partial X}{\partial N}\right)_{Y,Z} - \underline{X}\right] \tag{4.8-17}$$

Using this formula we obtain

$$\left(\frac{\partial \underline{\underline{U}}}{\partial N}\right)_{S,V} = \frac{1}{N}\left[\left(\frac{\partial U}{\partial N}\right)_{S,V} - \underline{U}\right] = \frac{1}{N}(\underline{G} - \underline{U}) = \frac{P\underline{V} - T\underline{S}}{N}$$

Table 4.8-1
DERIVATIVES WITH RESPECT TO MOLE NUMBER

Derivative	Examples
$\left(\frac{\partial \underline{\underline{X}}}{\partial N}\right)_{\underline{Y},\underline{Z}} = 0$	$\left(\frac{\partial \underline{\underline{U}}}{\partial N}\right)_{\underline{S},\underline{V}} = 0$
$\left(\frac{\partial X}{\partial N}\right)_{Y,Z}$ (see Sec. 4.2)	$\left(\frac{\partial U}{\partial N}\right)_{S,V} = \underline{G}$
$\left(\frac{\partial X}{\partial N}\right)_{\underline{Y},\underline{Z}} = \underline{X}$	$\left(\frac{\partial U}{\partial N}\right)_{\underline{S},\underline{V}} = \underline{U}$
$\left(\frac{\partial X}{\partial N}\right)_{\underline{Y},Z} = \left(\frac{\partial X}{\partial N}\right)_{Y,Z} + \underline{Y}\left(\frac{\partial X}{\partial Y}\right)_{Z,N}$ (see note below)	$\left(\frac{\partial U}{\partial N}\right)_{\underline{S},V} = \underline{H}$
	$\left(\frac{\partial U}{\partial N}\right)_{S,\underline{V}} = \underline{A}$
$\left(\frac{\partial \underline{\underline{X}}}{\partial N}\right)_{Y,Z} = \frac{1}{N}\left[\left(\frac{\partial X}{\partial N}\right)_{Y,Z} - \underline{X}\right]$	$\left(\frac{\partial \underline{\underline{U}}}{\partial N}\right)_{S,V} = \frac{1}{N}(P\underline{V} - T\underline{S})$
$\left(\frac{\partial \underline{\underline{X}}}{\partial N}\right)_{\underline{Y},Z} = \frac{1}{N}\left[\left(\frac{\partial X}{\partial N}\right)_{Y,Z} + \underline{Y}\left(\frac{\partial X}{\partial Y}\right)_{Z,N} - \underline{X}\right]$ (see note below)	$\left(\frac{\partial \underline{\underline{U}}}{\partial N}\right)_{\underline{S},V} = \frac{P\underline{\underline{V}}}{N}$
	$\left(\frac{\partial \underline{\underline{U}}}{\partial N}\right)_{S,\underline{V}} = \frac{T\underline{\underline{S}}}{N}$

Note: In these two formulas $\underline{Y}$ is one of $\underline{U}$, $\underline{V}$, $\underline{G}$, $\underline{H}$, $\underline{S}$ or $\underline{A}$, but not T or P.

Finally, consider the derivative $(\partial \underline{X}/\partial N)_{\underline{Y},Z}$. Following the same procedure as above we have

$$\left(\frac{\partial \underline{\underline{X}}}{\partial N}\right)_{\underline{Y},Z} = \frac{1}{N}\left[\left(\frac{\partial X}{\partial N}\right)_{\underline{Y},Z} - \underline{X}\right]$$

and using Eq. 4.8-13 yields

$$\left(\frac{\partial \underline{\underline{X}}}{\partial N}\right)_{\underline{Y},Z} = \frac{1}{N}\left[\left(\frac{\partial X}{\partial N}\right)_{Y,Z} + \underline{Y}\left(\frac{\partial X}{\partial Y}\right)_{Z,N} - \underline{X}\right] \tag{4.8-18}$$

Two special cases of this equation are

$$\left(\frac{\partial \underline{\underline{U}}}{\partial N}\right)_{\underline{V},S} = \frac{1}{N}\left[\left(\frac{\partial U}{\partial N}\right)_{V,S} + \underline{V}\left(\frac{\partial U}{\partial V}\right)_{S,N} - \underline{U}\right] = \frac{1}{N}[\underline{G} - P\underline{V} - \underline{U}] = -\frac{T\underline{S}}{N}$$

$$\left(\frac{\partial \underline{\underline{U}}}{\partial N}\right)_{\underline{S},V} = \frac{1}{N}\left[\left(\frac{\partial U}{\partial N}\right)_{S,V} + \underline{S}\left(\frac{\partial U}{\partial S}\right)_{V,N} - \underline{U}\right] = \frac{1}{N}[\underline{G} - T\underline{S} - \underline{U}] = \frac{P\underline{V}}{N}$$

Table 4.8-1 summarizes the equations for the various derivatives considered here, and gives examples of their use. It is evident from this table that there are eight different values for the derivatives of internal energy with respect to mole number at fixed entropy and volume, depending on the particular combination of intensive and extensive variables involved. It is therefore not surprising that confusion sometimes arises in evaluating derivatives with respect to mole number (or mass).

APPENDIX **A4.1**

A BASIC Language Program for Thermodynamic Properties Calculations Using the Peng–Robinson Cubic Equation of State, PR1.BAS

This appendix describes a BASIC language program for pure fluid thermodynamic properties calculations using the Peng–Robinson equation of state. As written, the program runs using BASICA on an IBM PC or compatible computers; for more rapid execution, it may be compiled using a BASIC compiler. The program has been written in double precision to avoid roundoff errors. Given the critical temperature, critical pressure, acentric factor, and ideal gas heat capacity of the fluid (which the user must supply), the compressibility, specific volume, enthalpy and entropy departures from ideal gas behavior, the enthalpy and entropy with respect to a chosen reference state, and the fugacity can be calculated at a given state point. In addition, at temperatures below the critical temperature, the vapor pressure can be calculated using the procedure described in Chapter 5. To obtain a listing of the program, insert the disk accompanying this book into drive A of a PC compatible computer with a printer and type PRINT A:PR1.BAS.

The structure of the program is indicated here.

Statement Range	Function
10–410	Input routine for data entry and menu of calculation choices.
410–1280	Subroutine prfuga is used to calculate the compressibility, fugacity, enthalpy and entropy departures for given constants in equation of state and values of T and P.
450–540	Setting up constants in Eq.4.4-4.
550–890	Using Cardan's rule to solve cubic equation for real roots for the compressibility Z.
900–1020	Ordering Z roots and setting vapor root ($Z0$) to large compressibility and liquid root ($Z1$) to smallest.
1030–1040	Testing for erroneous root ($V < b$) and correcting if necessary.
1050–1170	If only compressibilities wanted (ICOMP = 0), then return; otherwise calculate fugacities of vapor and liquid phases.
1180–1280	If enthalpy and entropy not required (ICOMP = 1), then return; otherwise calculate enthalpy and entropy departures.
1290–1390	Subroutine prcons, calculates the constants a and b in the Peng–Robinson equation of state and temperature derivative of a for use in prfuga.
1400–1610	Subroutine comp is used for compressibility calculation. Requires T and P as input; output is vapor and liquid compressibilities and specific volumes.
1620–1870	Subroutine spfug is used to compute species fugacity. Requires T and P as input; output is compressibility, specific volume, fugacity, and fugacity coefficient of both the vapor and liquid phases.

1880–2240	Subroutine enthalpy calculates enthalpy and entropy departures, enthalpy and entropy, fugacity, compressibility, and specific volume of both the vapor and liquid phases. Requires T and P as input.
2250–2870	Subroutine vapor pressure requires T as input, output is vapor pressure and thermodynamic properties of the coexisting phases.
2320–2370	Generates initial guess for vapor pressure assuming $\ln P = A + B/T$ and using boiling point (P = 1 bar, $T = T_B$) and critical point ($P = P_C$, $T = T_C$) to find A and B.
2420	Call prfuga to calculate fugacities of coexisting phases.
2430–2460	Iteration on guessed pressure by successive substitution.
2500–2540	Testing to ensure that trivial solution of two identical phases (i.e., phases of equal compressibility) has not been obtained. If phases are identical, guessed pressure is corrected (increased if $Z^L > Z_C$ and decreased if $Z^L < Z_C$), and calculation is repeated.
2550–2600	Subroutine prfuga is called with correct pressure, this time with all thermodynamic properties being computed, and then entropy and enthalpy of both phases are calculated.
2880–3000	Subroutine ideal gas enthalpy and entropy computes those properties assuming

$$C_P^* = AAA + BBB \times T + CCC \times T^2 + DDD \times T^3$$

so that (with ideal gas at T_0 and P_0 as the reference state)

$$\begin{aligned} H^{IG}(T) &= \int_{T_0}^{T} C_P^* \, dT \\ &= AAA \times (T - T_0) + \frac{BBB}{2} \times (T^2 - T_0^2) + \frac{CCC}{3} \times (T^3 - T_0^3) \\ &\quad + \frac{DDD}{4} \times (T^4 - T_0^4) \end{aligned}$$

and

$$\begin{aligned} \underline{S}^{IG}(T, P) &= -R \ln (P/P_0) + \int_{T_0}^{T} C_P^* \, dT \\ &= -R \ln \left(\frac{P}{P_0}\right) + AAA \ln \left(\frac{T}{T_0}\right) + BBB \times (T - T_0) \\ &\quad + \frac{CCC}{2} \times (T^2 - T_0^2) + \frac{DDD}{3} \times (T^3 - T_0^2) \end{aligned}$$

Although the program on the disk is specific to the Peng–Robinson equation of state, only the following lines would have to be changed to use another cubic equation of state:

450–540, 1070–1260, and 1310–1380

PROBLEMS

4.1 Using the Mollier diagram, compute the Joule–Thomson coefficient $\mu = (\partial T/\partial P)_{\underline{H}}$ and the adiabatic compressibility $\kappa_S = (\partial T/\partial P)_{\underline{S}}$ for steam at 500°C and 10 MPa. Also, relate the ratio $(\partial \underline{H}/\partial \underline{S})_T/(\partial \underline{H}/\partial \underline{S})_P$ to μ and κ_S, and compute its value for steam at the same conditions.

4.2 Derive Eqns. 4.4-29 and 30.

4.3 Evaluate the difference

$$\left(\frac{\partial \underline{U}}{\partial T}\right)_P - \left(\frac{\partial \underline{U}}{\partial T}\right)_{\underline{V}}$$

for the ideal and van der Waals gases, and for a gas that obeys the virial equation of state.

4.4 One of the beauties of thermodynamics is that it provides interrelationships between various state variables and their derivatives so that information from one set of experiments can be used to predict the results of a completely different experiment. This is illustrated here.

a Show that

$$C_P = \frac{T^2}{\mu}\left(\frac{\partial(\underline{V}/T)}{\partial T}\right)_P$$

Thus, if the Joule–Thomson coefficient μ and the volumetric equation of state (in analytic or tabular form) are known for a fluid, C_P can be computed. Alternatively, if C_P and μ are known $(\partial(\underline{V}/T)/\partial T)_P$ can be calculated, or if C_P and $(\partial(\underline{V}/T)/\partial T)_P$ are known, μ can be calculated.

b Show that

$$\underline{V}(P, T_2) = \frac{T_2}{T_1}\underline{V}(P, T_1) + T_2\int_{P,T_1}^{P,T_2} \frac{\mu C_P}{T^2}\,dT$$

so that if μ and C_P are known as functions of temperature at the pressure P, and $\underline{V}$ is known at P and T_1, the specific volume at P and T_2 can be computed.

4.5 Derive Eqs. 4.6-2 and 4.6-3, and show that $Z_C|_{\text{van der Waals}} = 3/8$.

4.6 Derive Eq. 4.6-11.

4.7 100 m^3 of carbon dioxide initially at 150°C and 50 bar is to be isothermally compressed in a frictionless piston and cylinder device to a final pressure of 300 bar. Calculate:

i The volume of the compressed gas.
ii The work done to compress the gas.
iii The heat flow on compression.

Assuming carbon dioxide

a Is an ideal gas
b Obeys the principle of corresponding states
c Obeys the Peng–Robinson equation of state

4.8 By measuring the temperature change and the specific volume change accompanying a small pressure change in a reversible adiabatic process, one can evaluate the derivative

$$\left(\frac{\partial T}{\partial P}\right)_{\underline{S}}$$

and the adiabatic compressibility

$$\kappa_S = -\frac{1}{\underline{V}}\left(\frac{\partial \underline{V}}{\partial P}\right)_{\underline{S}}$$

Develop an expression for $(\partial T/\partial P)_{\underline{S}}$ in terms of T, $\underline{V}$, C_P, α, and κ_T, and show that

$$\frac{\kappa_S}{\kappa_T} = \frac{C_V}{C_P}$$

4.9 Prove that the following statements are true:

a $(\partial \underline{H}/\partial \underline{V})_T$ is equal to zero if $(\partial \underline{H}/\partial P)_T$ is equal to zero.

b The derivative $(\partial \underline{S}/\partial \underline{V})_P$ for a fluid has the same sign as its coefficient of thermal expansion α and is inversely proportional to it.

4.10 By measuring the temperature change accompanying a differential volume change in a free expansion across a valve and in a reversible adiabatic expansion, the two derivatives $(\partial T/\partial \underline{V})_{\underline{H}}$ and $(\partial T/\partial \underline{V})_{\underline{S}}$ can be experimentally evaluated.

a Develop expressions for these derivatives in terms of the more fundamental quantities.

b Evaluate these derivatives for a van der Waals fluid.

4.11 a Show for the Peng–Robinson equation of state (Eq. 4.4-2) that

$$a(T_C) = 0.45724\ R^2T_C^2/P_C$$

and

$$b = 0.07780\ RT_C/P_C$$

b Determine the critical compressibility of the Peng–Robinson equation of state.

4.12 Ethylene at 30 bar and 100°C passes through a heater-expander and emerges at 20 bar and 150°C. There is no flow of work into or out of the heater-expander, but heat is supplied. Assuming that ethylene obeys the Peng–Robinson equation of state, compute the flow of heat into the heater-expander per mole of ethylene.

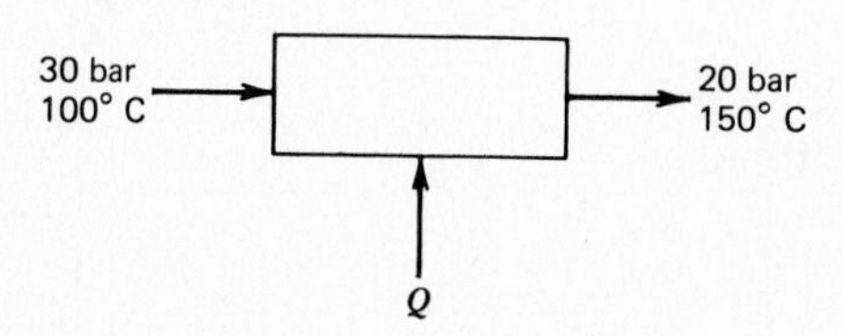

4.13 A natural gas stream (essentially pure methane) is available at 310 K and 14 bar. The gas is to be compressed to 345 bar before transmission by underground pipeline. If the compression is carried out adiabatically and reversibly, determine the compressor outlet temperature and the work of compression per mole of methane. You may assume that methane obeys the Peng–Robinson equation of state. For methane,

$$C_P^* = 14.72 + 0.987T\ \text{J/mol K}$$

for T in K.

4.14 Values of the virial coefficients B and C at a fixed temperature can be obtained from experimental PVT data by noting that

$$\lim_{\substack{P \to 0 \\ (\underline{V} \to \infty)}} \frac{P\underline{V}}{RT} = 1$$

$$\lim_{\substack{P \to 0 \\ (\underline{V} \to \infty)}} \underline{V}\left(\frac{P\underline{V}}{RT} - 1\right) = B$$

and

$$\lim_{\substack{P \to 0 \\ (\underline{V} \to \infty)}} \underline{V}^2\left(\frac{P\underline{V}}{RT} - 1 - \frac{B}{\underline{V}}\right) = C$$

a Using these formulas, show that the van der Waals equation leads to the following expressions for the virial coefficients

$$B = b - \frac{a}{RT}$$

$$C = b^2$$

b The temperature at which $\lim_{\substack{P \to 0 \\ (\text{or } \underline{V} \to \infty)}} \underline{V}\left(\frac{P\underline{V}}{RT} - 1\right) = B = 0$ is called the Boyle temperature. Show that for the van der Waals fluid

$$T_{\text{Boyle}} = 3.375\ T_C$$

where T_C is the critical temperature of the van der Waals fluid given by Eqs. 4.6-3. (For many real gases T_{Boyle} is approximately 2.5 T_C!)

4.15 From experimental data it is known that at moderate pressures the volumetric equation of state may be written as

$$P\underline{V} = RT + BP$$

where the second viral coefficient B is a function of temperature only. Data for nitrogen is given in the table.

T(K)	**B(cm^3/mol)**
75	−274
100	−160
125	−104
150	− 71.5
200	− 35.2
250	− 16.2
300	− 4.2
400	+ 9.0
500	+ 16.9
600	+ 21.3
700	+ 24.0

Source: J. H. Dymond and E. B. Smith, *The Viral Coefficients of Gases*, Clarendon Press, Oxford, 1969, p. 188.

a Identify the Boyle temperature (the temperature at which $B = 0$) and the inversion temperature [the temperature at which $\mu = (\partial T/\partial P)_{\underline{H}} = 0$] for gaseous nitrogen.

b Show, from the data in the table, that at temperatures above the inversion temperature the gas temperature increases in a Joule–Thomson expansion, whereas it decreases if the initial temperature is below the inversion temperature.

c Describe how you would find the inversion temperature as a function of pressure for nitrogen using Fig. 2.4-3 and for methane using Fig. 2.4-2.

4.16 Eighteen kilograms of freon-12 at 150°C is contained in a 0.03 m^3 tank. Compare the prediction for the pressure in the tank with that obtained using the figure in Problem 3.21.
Data for freon-12: see Table 4.6-1.

4.17 Calculate the molar volume, enthalpy, and entropy of carbon tetrachloride at 300°C and 35 bar using the Peng–Robinson equation of state and the principle of corresponding states. The following data are available:

$$\underline{H}(T = 16°\text{C, ideal gas at } P = 0.1 \text{ bar}) = 0$$

$$\underline{S}(T = 16°\text{C, ideal gas at } P = 0.1 \text{ bar}) = 0$$

$$T_C = 283.2°\text{C}$$

$$P_C = 45.6 \text{ bar}$$

$$Z_C = 0.272 \qquad \omega = 0.194$$

C_P^*: see Appendix II

4.18 The Clausius equation of state is

$$P(\underline{V} - b) = RT$$

a Show that for this volumetric equation of state

$$C_P(P, T) = C_V(P, T) + R$$

$$C_P(P, T) = C_P^*(T)$$

and

$$C_V(\underline{V}, T) = C_V^*(T)$$

b For a certain process the pressure of a gas must be reduced from an initial pressure P_1 to the final pressure P_2. The gas obeys the Clausius equation of state, and the pressure reduction is to be accomplished either by passing the gas through a flow constriction, such as a pressure-reducing valve, or by passing it through a small gas turbine (which we can assume to be both reversible and adiabatic). Obtain expressions for the final gas temperatures in each of these cases in terms of the initial state of the gas and the properties of the gas.

4.19 A tank is divided into two equal chambers by an internal diaphragm. One chamber contains methane at a pressure of 500 bar and a temperature of 20°C and the other chamber is evacuated. Suddenly, the diaphragm bursts. Compute the final temperature and pressure of the gas in the tank after sufficient time has passed for equilibrium to be attained. Assume that there is no heat transfer between the tank and the gas, and that methane

a Is an ideal gas.
b Obeys the principle of corresponding states.
c Is a van der Waals gas.
d Obeys the Peng–Robinson equation of state.

Data
For simplicity you may assume

$$C_P^* = 35.56 \text{ J/mol K}$$

4.20 The divided tank of the preceding problem is replaced with two interconnected tanks of equal volume; one tank is initially evacuated, and the other contains methane at 500 bar and 20°C. A valve connecting the two tanks is opened only long enough to allow the pressures in the tanks to equilibrate. If there is no heat transfer between the gas and the tanks, what is the temperature and pressure of the gas in each tank after the valve has been shut? Assume that methane
a Is an ideal gas.
b Obeys the principle of corresponding states.
c Obeys the Peng–Robinson equation of state.

4.21 Ammonia is to be isothermally compressed in a specially designed flow turbine from 1 bar and 100°C to 50 bar. If the compression is done reversibly, compute the heat and work flows needed per mole of ammonia if
a Ammonia obeys the principle of corresponding states.
b Ammonia satisfies the Clausius equation of state $P(\underline{V} - b) = RT$ with $b = 3.730 \times 10^{-2}$ m^3/kmol.
c Ammonia obeys the Peng–Robinson equation of state.

4.22 A tank containing carbon dioxide at 400 K and 50 bar is vented until the temperature in the tank falls to 300 K. Assuming there is no heat transfer between the gas and the tank, find the pressure in the tank at the end of the venting process and the fraction of the initial mass of gas remaining in the tank for each of the following cases.
a The equation of state of carbon doxide is

$$P(\underline{V} - b) = RT \qquad \text{with} \qquad b = 0.0441 \text{ m}^3/\text{kmol}$$

b Carbon dioxide obeys the law of corresponding states.
c Carbon dioxide obeys the Peng–Robinson equation of state.

In each case take the low pressure (ideal gas) heat capacity of CO_2 to be

$$C_P^* = 22.25 + 0.06T \text{ J/mol K } (T \text{ in K})$$

4.23 Derive the equations necessary to expand Illustration 4.4-1 to include the thermodynamic state variables internal energy, Gibbs free energy, and Helmholtz free energy.

4.24 Draw lines of constant Gibbs and Helmholtz free energy on the diagrams of Illustration 4.4-1.

4.25 The speed of propagation of a small pressure pulse or sound wave in a fluid, v_S, can be shown to be equal to

$$v_S = \sqrt{\left(\frac{\partial P}{\partial \rho}\right)_{\underline{S}}}$$

where ρ is the molar density.

a Show that an alternative expression for the sonic velocity is

$$v_S = \sqrt{\gamma \underline{V}^2 \left(\frac{\partial T}{\partial \underline{V}}\right)_P \left(\frac{\partial P}{\partial T}\right)_{\underline{V}}}$$

where $\gamma = C_P/C_V$.

b Show that $\gamma = 1 + R/C_V$ for both the ideal gas and a gas that obeys the Clausius equation of state

$$P(\underline{V} - b) = RT$$

and that γ is independent of specific volume for both gases.

c Develop expressions for v_S for both the ideal and the Clausius gases that do not contain any derivatives other than C_V and C_P.

4.26 The force required to maintain a polymeric fiber at a length L when its unstretched length is L_0 has been observed to be related to its temperature as follows

$$F = \gamma T(L - L_0)$$

where γ is a positive constant. The heat capacity of the fiber measured at the constant length L_0 is given by

$$C_L = \alpha + \beta T$$

where α and β are parameters that depend on the fiber length.

a Develop an equation that relates the change in entropy of the fiber to changes in its temperature and length, and evaluate the derivatives $(\partial S/\partial L)_T$ and $(\partial S/\partial T)_L$, which appear in this equation.

b Develop an equation that relates the change in internal energy of the fiber to changes in its temperature and length.

c Develop an equation that relates the entropy of the fiber at a temperature T_0 and an extension L_0 to its entropy at any other temperature T and extension L.

d If the fiber at $T = T_i$ and $L = L_i$ is stretched slowly and adiabatically until it attains a length L_f, what is the fiber temperature T_f?

e In polymer science it is common to attribute the force necessary to stretch a fiber to energetic and entropic effects. The energetic force (i.e., that part of the force that, on an isothermal extension of the fiber, increases its internal energy) is $F_U = (\partial U/\partial L)_T$, and the entropic force is $F_S = -T(\partial S/\partial L)_T$. Evaluate F_U and F_S for the fiber being considered here.

4.27 Derive the following Maxwell relations for open systems.

a Starting from Eq. 4.2-5a

$$\left(\frac{\partial T}{\partial V}\right)_{S,N} = -\left(\frac{\partial P}{\partial S}\right)_{V,N}; \quad \left(\frac{\partial T}{\partial N}\right)_{S,V} = \left(\frac{\partial \underline{\underline{G}}}{\partial S}\right)_{V,N}; \quad \text{and} \left(\frac{\partial P}{\partial N}\right)_{S,V} = -\left(\frac{\partial \underline{\underline{G}}}{\partial V}\right)_{S,N}$$

b Starting from Eq. 4.2-6a

$$\left(\frac{\partial T}{\partial P}\right)_{S,N} = \left(\frac{\partial V}{\partial S}\right)_{P,N}; \quad \left(\frac{\partial T}{\partial N}\right)_{S,P} = \left(\frac{\partial \underline{\underline{G}}}{\partial S}\right)_{P,N}; \quad \text{and} \left(\frac{\partial V}{\partial N}\right)_{S,P} = \left(\frac{\partial \underline{\underline{G}}}{\partial P}\right)_{S,N}$$

c Starting from Eq. 4.2-7a

$$\left(\frac{\partial S}{\partial V}\right)_{T,N} = \left(\frac{\partial P}{\partial T}\right)_{V,N}; \quad \left(\frac{\partial S}{\partial N}\right)_{T,V} = -\left(\frac{\partial \underline{\underline{G}}}{\partial T}\right)_{V,N}; \quad \text{and} \left(\frac{\partial P}{\partial N}\right)_{T,V} = -\left(\frac{\partial \underline{\underline{G}}}{\partial V}\right)_{T,N}$$

and

d Starting from Eq. 4.2-8a

$$\left(\frac{\partial S}{\partial P}\right)_{T,N} = -\left(\frac{\partial V}{\partial T}\right)_{P,N}; \qquad \left(\frac{\partial S}{\partial N}\right)_{T,P} = \left(\frac{\partial \underline{G}}{\partial T}\right)_{P,N}; \qquad \text{and} \left(\frac{\partial V}{\partial N}\right)_{T,P} = \left(\frac{\partial \underline{G}}{\partial P}\right)_{T,N}$$

4.28 For real gases the Joule–Thomson coefficient is greater than zero at low temperatures and less than zero at high temperatures. The temperature at which μ is equal to zero at a given pressure is called the inversion temperature.

a Show that the van der Waals equation of state exhibits this behavior, and develop an equation for the inversion temperature of this fluid as a function of its specific volume.

b Show that the van der Waals prediction for the inversion temperature can be written in the following corresponding states form

$$T_r^{\text{inv}} = \frac{3(3V_r - 1)^2}{4V_r^2}$$

c The following graph shows the inversion temperature of nitrogen as a function of pressure.[16] Plot, on this graph, the van der Waals prediction for inversion curve for nitrogen.

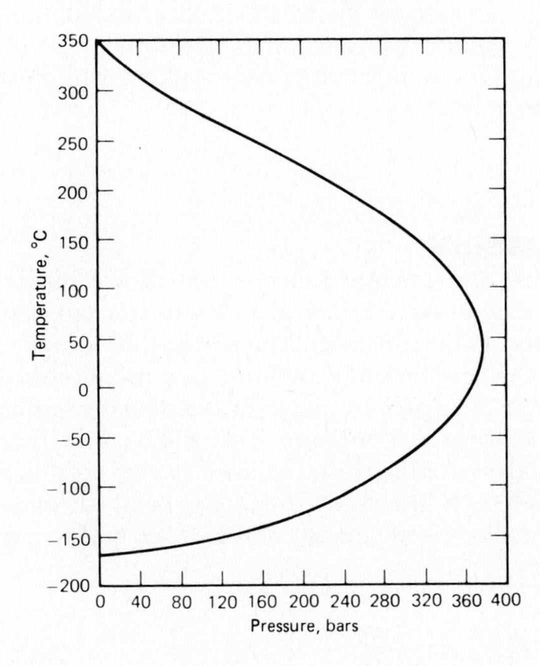

[16]From B. F. Dodge, *Chemical Engineering Thermodynamics*, Copyright © 1944 by McGraw–Hill, Inc. Used with permission of McGraw–Hill Book Company.

5

Equilibrium and Stability in One-Component Systems

Now that the basic principles of thermodynamics have been developed, and some computational details considered, we can study, in detail, a fundamental problem of thermodynamics: the prediction of the equilibrium state of a system. In this chapter we examine the conditions for the existence of a stable equilibrium state in a single-component system, with particular reference to the problem of phase equilibrium. Equilibrium in multicomponent systems will be considered in the following chapters.

5.1 THE CRITERIA FOR EQUILIBRIUM

In Chapter 3 we established that the entropy function provides a means of mathematically identifying the state of equilibrium in a closed, isolated system (i.e., a system in which M, U, and V are constant). The aim in this section is to develop a means of identifying the equilibrium state for closed thermodynamic systems subject to other constraints, especially those of constant temperature and volume, and constant temperature and pressure. This will be done by first reconsidering the equilibrium analysis for the closed isolated system used in Sec. 3.1, and then extending this analysis to the study of more general systems.

The starting point for the analysis is the energy and entropy balances for a closed system:

$$\frac{dU}{dt} = \dot{Q} - P\frac{dV}{dt} \tag{5.1-1}$$

and

$$\frac{dS}{dt} = \frac{\dot{Q}}{T} + \dot{S}_{\text{gen}} \tag{5.1-2}$$

with

$$\dot{S}_{\text{gen}} \geq 0 \quad \text{(the equality holding at equilibrium or for reversible processes)} \tag{5.1-3}$$

Here we have chosen the system in such a way that the only work term is that of the deformation of the system boundary. For a constant-volume system exchanging no heat with its surroundings, Eqs. 5.1-1 and 2 reduce to

$$\frac{dU}{dt} = 0$$

so that

$$U = \text{constant}$$

(or, equivalently, $\underline{U}$ = constant, since the total number of moles, or mass, is fixed in a closed, one-component system), and

$$\frac{dS}{dt} = \dot{S}_{\text{gen}} \geqslant 0 \tag{5.1-4}$$

Since the entropy function can only increase in value during the approach to equilibrium (because of the sign of $\dot{S}_{\text{gen}}$), the entropy must be a maximum at equilibrium. Thus, the equilibrium criterion for a closed, isolated system is

$$\left.\begin{array}{r} S = \text{maximum} \\ \text{or} \quad \underline{S} = \text{maximum} \end{array}\right\} \begin{array}{l} \text{at equilibrium in} \\ \text{a closed system at} \\ \text{constant } U \text{ and } V \end{array} \tag{5.1-5}$$

At this point you might ask how $\underline{S}$ can achieve a maximum value if $\underline{U}$ and $\underline{V}$ are fixed, since the specification of any two intensive variables completely fixes the values of all others. The answer to this question was given in Chapter 3, where it was pointed out that the specification of two intensive variables fixes the values of all the other state variables in the uniform equilibrium state of a single-component, single-phase system. Thus, the equilibrium criterion of Eq. 5.1-5 can be used for identifying the final equilibrium state in a closed isolated system that is initially nonuniform, or in which several phases or components are present.

To illustrate the use of this equilibrium criterion, consider the very simple, initially nonuniform system shown in Fig. 5.1-1. There a single-phase, single-component fluid in an adiabatic, constant-volume container has been divided into two subsystems by an imaginary boundary. Each of these subsystems will be assumed to contain the same molecular species of uniform thermodynamic properties. However, these subsystems are open to the flow of heat and mass across the boundary, and their temperature and pressure need not be the same. For the composite system consisting of the two subsystems, the total mass (though, in fact, we will use number of moles), internal energy, volume, and entropy, which are extensive variables, are the sums of these quantities for the

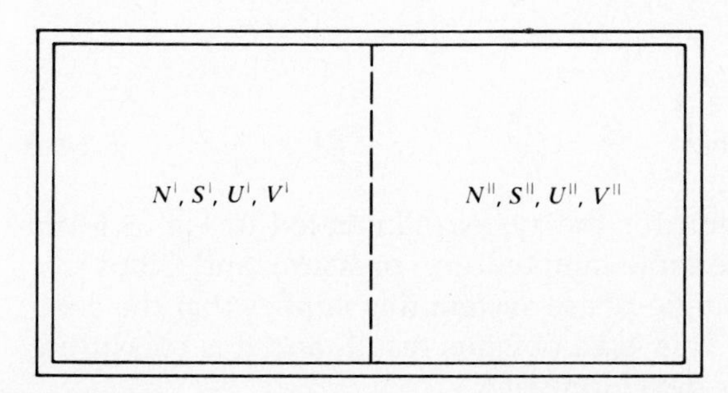

Figure 5.1-1
An isolated nonequilibrium system.

two subsystems, that is,

$$N = N^{\mathrm{I}} + N^{\mathrm{II}}$$
$$U = U^{\mathrm{I}} + U^{\mathrm{II}}$$
$$V = V^{\mathrm{I}} + V^{\mathrm{II}} \tag{5.1-6}$$

and

$$S = S^{\mathrm{I}} + S^{\mathrm{II}}$$

Now considering the entropy to be a function of internal energy, volume, and mole number, we can compute the change in the entropy of system I due to changes in N^{I}, U^{I}, and V^{I} from Eq. 4.2-5c

$$dS^{\mathrm{I}} = \left(\frac{\partial S^{\mathrm{I}}}{\partial U^{\mathrm{I}}}\right)_{V^{\mathrm{I}},N^{\mathrm{I}}} dU^{\mathrm{I}} + \left(\frac{\partial S^{\mathrm{I}}}{\partial V^{\mathrm{I}}}\right)_{U^{\mathrm{I}},N^{\mathrm{I}}} dV^{\mathrm{I}} + \left(\frac{\partial S^{\mathrm{I}}}{\partial N^{\mathrm{I}}}\right)_{U^{\mathrm{I}},V^{\mathrm{I}}} dN^{\mathrm{I}}$$
$$= \frac{1}{T^{\mathrm{I}}} dU^{\mathrm{I}} + \frac{P^{\mathrm{I}}}{T^{\mathrm{I}}} dV^{\mathrm{I}} - \frac{\underline{G}^{\mathrm{I}}}{T^{\mathrm{I}}} dN^{\mathrm{I}}$$

In a similar fashion dS^{II} can be computed. The entropy change of the composite system is

$$dS = dS^{\mathrm{I}} + dS^{\mathrm{II}} = \frac{1}{T^{\mathrm{I}}} dU^{\mathrm{I}} + \frac{1}{T^{\mathrm{II}}} dU^{\mathrm{II}} + \frac{P^{\mathrm{I}}}{T^{\mathrm{I}}} dV^{\mathrm{I}} + \frac{P^{\mathrm{II}}}{T^{\mathrm{II}}} dV^{\mathrm{II}} - \frac{\underline{G}^{\mathrm{I}}}{T^{\mathrm{I}}} dN^{\mathrm{I}} - \frac{\underline{G}^{\mathrm{II}}}{T^{\mathrm{II}}} dN^{\mathrm{II}}$$

However, the total number of moles, internal energy, and volume are constant by the constraints on the system, so that

$$dN = dN^{\mathrm{I}} + dN^{\mathrm{II}} = 0 \quad \text{or} \quad dN^{\mathrm{I}} = -dN^{\mathrm{II}}$$
$$dU = dU^{\mathrm{I}} + dU^{\mathrm{II}} = 0 \quad \text{or} \quad dU^{\mathrm{I}} = -dU^{\mathrm{II}}$$
$$dV = dV^{\mathrm{I}} + dV^{\mathrm{II}} = 0 \quad \text{or} \quad dV^{\mathrm{I}} = -dV^{\mathrm{II}} \tag{5.1-7}$$

Consequently,

$$dS = +\left(\frac{1}{T^{\mathrm{I}}} - \frac{1}{T^{\mathrm{II}}}\right) dU^{\mathrm{I}} + \left(\frac{P^{\mathrm{I}}}{T^{\mathrm{I}}} - \frac{P^{\mathrm{II}}}{T^{\mathrm{II}}}\right) dV^{\mathrm{I}} - \left(\frac{\underline{G}^{\mathrm{I}}}{T^{\mathrm{I}}} - \frac{\underline{G}^{\mathrm{II}}}{T^{\mathrm{II}}}\right) dN^{\mathrm{I}} \tag{5.1-8}$$

Now since $S =$ maximum or $dS = 0$ for all system variations at constant N, U, and V (here all variations of the independent variables dU^{I}, dV^{I}, and dN^{I} at constant total number of moles, total internal energy, and total volume), we conclude that

$$\left(\frac{\partial S}{\partial U^{\mathrm{I}}}\right)_{V^{\mathrm{I}},N^{\mathrm{I}}} = 0, \quad \text{so that} \quad \frac{1}{T^{\mathrm{I}}} = \frac{1}{T^{\mathrm{II}}}, \quad \text{or} \quad T^{\mathrm{I}} = T^{\mathrm{II}} \tag{5.1-9a}$$

$$\left(\frac{\partial S}{\partial V^{\mathrm{I}}}\right)_{U^{\mathrm{I}},N^{\mathrm{I}}} = 0, \quad \text{so that} \quad \frac{P^{\mathrm{I}}}{T^{\mathrm{I}}} = \frac{P^{\mathrm{II}}}{T^{\mathrm{II}}}, \quad \text{or} \quad P^{\mathrm{I}} = P^{\mathrm{II}} \tag{5.1-9b}$$

and

$$\left(\frac{\partial S}{\partial N^{\mathrm{I}}}\right)_{U^{\mathrm{I}},V^{\mathrm{I}}} = 0, \quad \text{so that} \quad \frac{\underline{G}^{\mathrm{I}}}{T^{\mathrm{I}}} = \frac{\underline{G}^{\mathrm{II}}}{T^{\mathrm{II}}}, \quad \text{or} \quad \underline{G}^{\mathrm{I}} = \underline{G}^{\mathrm{II}} \tag{5.1-9c}$$

Therefore, the equilibrium condition for the system illustrated in Fig. 5.1-1 is satisfied if the subsystems have the same temperature, pressure, and Gibbs free energy. For a single-component, single-phase system this implies that the composite system should be uniform. This is an obvious result, and it is reassuring that it arises so naturally from our development.

The foregoing discussion illustrates how the condition $dS = 0$ may be used to identify a possible equilibrium state of the system, that is, a state for which S = maximum. From calculus we know that $dS = 0$ is a necessary but not sufficient condition for S to achieve a maximum value. In particular, $dS = 0$ at a minimum value or an inflection point of S, as well as when S is a maximum. The condition $d^2S < 0$, when $dS = 0$, assures us that a maximum value of the entropy, and hence a true equilibrium state, has been identified rather than a metastable state (an inflection point) or an unstable state (minimum value of S). Thus, the sign of d^2S determines the stability of the state found from the condition that $dS = 0$. The implications of this stability condition are considered in the next section.

It is also possible to develop the equilibrium and stability conditions for systems subject to other constraints. For a closed system at constant temperature and volume, the energy and entropy balances are

$$\frac{dU}{dt} = \dot{Q}$$

and

$$\frac{dS}{dt} = \frac{\dot{Q}}{T} + \dot{S}_{\text{gen}}$$

Eliminating $\dot{Q}$ between these two equations, and using the fact that $T\,dS = d(TS)$, since T is constant, gives

$$\frac{d(U - TS)}{dt} = \frac{dA}{dt} = -T\dot{S}_{\text{gen}} \leqslant 0$$

Here we have also used the facts that $T \geqslant 0$ and $\dot{S}_{\text{gen}} \geqslant 0$, so that $(-T\dot{S}_{\text{gen}}) \leqslant 0$. Using the same argument that led from Eq. 5.1-4 to Eq. 5.1-5 here yields

$$\left.\begin{array}{l} A = \text{minimum} \\ \text{or } \underline{A} = \text{minimum} \end{array}\right\} \begin{array}{l} \text{for equilibrium in a closed} \\ \text{system at constant } T \text{ and } V \end{array} \qquad (5.1\text{-}10)$$

It should be noticed that this result is also a consequence of $\dot{S}_{\text{gen}} \geqslant 0$.

If we were to use Eq. 5.1-10 to identify the equilibrium state of an initially nonuniform, single-component system, such as that in Fig. 5.1-1, but now maintained a constant temperature and volume, we would find, following the previous analysis, that at equilibrium the pressures of the two subsystems are equal, as are the molar Gibbs free energies (cf. Eqs. 5.1-9 and Problem 5.4); since temperature is being maintained constant, it would, of course, be the same in both subsystems.

For a closed system maintained at constant temperature and pressure, we have

$$\frac{dU}{dt} = \dot{Q} - P\frac{dV}{dt} = \dot{Q} - \frac{d}{dt}(PV)$$

and

$$\frac{dS}{dt} = \frac{\dot{Q}}{T} + \dot{S}_{\text{gen}}$$

Again eliminating $\dot{Q}$ between these two equations gives

$$\frac{dU}{dt} + \frac{d}{dt}(PV) - \frac{d}{dt}(TS) = \frac{dG}{dt} = -T\dot{S}_{\text{gen}} \qquad (5.1\text{-}11)$$

so that the equilibrium criterion here is

$$\left.\begin{array}{r} G = \text{minimum} \\ \text{or} \quad \underline{G} = \text{minimum} \end{array}\right\} \begin{array}{l} \text{for equilibrium in a closed system} \\ \text{at constant } T \text{ and } P \end{array} \tag{5.1-12}$$

This equation leads to the equality of the molar Gibbs free energies (Eq. 5.1-9c) as a condition for equilibrium; of course, since both temperature and pressure are held constant, the system is also at uniform temperature and pressure at equilibrium.

Finally, the equilibrium criterion for a system consisting of an element of fluid moving with the velocity of the fluid around it is also of interest, as such a choice of system arises in the study of continuous processing equipment used in the chemical industry. The tubular chemical reactor discussed in Chapter 9 is perhaps the most common example. Since each fluid element is moving with the velocity of fluid surrounding it, there is no convected flow of mass into or out of this system. Therefore, each such element of mass in a pure fluid is a system closed to the flow of mass, and consequently is subject to precisely the same equilibrium criteria as the closed systems discussed previously (i.e., Eqs. 5.1-5, 10, or 12, depending on the constraints on the system).

The equilibrium and stability criteria for systems and constraints that will be of interest in this book are collected in Table 5.1-1. As we will see in Chapter 6, these equilibrium and stability criteria are also valid for multicomponent systems. Thus the entries in this table form the basis for the analysis of phase and chemical equilibrium problems to be considered during much of the remainder of this book.

Using the method of analysis indicated here, it is also possible to derive the equilibrium criteria for systems subject to other constraints. However, this task is left to you (Problem 5.2).

5.2 STABILITY OF THERMODYNAMIC SYSTEMS

In the previous section we used the result $dS = 0$ to identify the equilibrium state of an initially nonuniform system constrained to remain at constant mass, internal energy, and volume. In this section we explore the information content of

Table 5.1-1
EQUILIBRIUM AND STABILITY CRITERIA

System	Constraint	Equilibrium Criterion	Stability Criterion
Isolated, adiabatic fixed-boundary system	U = constant V = constant	S = maximum $dS = 0$	$d^2S < 0$
Isothermal closed system with fixed boundaries	T = constant V = constant	A = minimum $dA = 0$	$d^2A > 0$
Isothermal, isobaric closed system	T = constant P = constant	G = minimum $dG = 0$	$d^2G > 0$
Isothermal, isobaric open system moving with the fluid velocity	T = constant P = constant M = constant	G = minimum $dG = 0$	$d^2G > 0$

the stability criterion

$$d^2S < 0 \qquad \text{at constant } M,\ U, \text{ and } V$$

[The stability analysis for closed systems subject to other constraints (i.e., constant T and V or constant T and P) is similar to, and, in fact, somewhat simpler than the analysis here, and so is left to you (Problem 5.3.)]

By studying the sign of the second differential of the entropy we are really considering the following question: Suppose that a small fluctuation in a fluid property, say temperature or pressure, occurs in some region of a fluid that was initially at equilibrium. Is the character of the equilibrium state such that $d^2S < 0$, and the fluctuation will dissipate, or is $d^2S > 0$, in which case the fluctuation grows until the system evolves to a new equilibrium state of higher entropy?

In fact, since we know that fluids exist in thermodynamically stable states (Experimental Observation 7 of Sec. 1.7), we will take as an empirical fact that $d^2S < 0$ for all real fluids at equilibrium, and instead establish the restrictions placed on the equations of state of fluids by this stability condition. We first study the problem of the intrinsic stability of the equilibrium state in a pure, single-phase fluid and then the mutual stability of two interacting systems or phases.

We begin the discussion of intrinsic stability by considering further the example of the last section, equilibrium in a pure fluid at constant mass (actually we will use number of moles), internal energy, and volume. Using the (imaginary) subdivision of the system into two subsystems, and writing the extensive properties N, U, V, and S as sums of these properties for each subsystem, we were able to show, in Sec. 5.1, that the condition that

$$dS = dS^{\mathrm{I}} + dS^{\mathrm{II}} = 0 \tag{5.2-1}$$

for all system variations consistent with the constraints (i.e., all variations in dN^{I}, dV^{I}, and dU^{I}) led to the requirements that at equilibrium

$$T^{\mathrm{I}} = T^{\mathrm{II}}$$

$$P^{\mathrm{I}} = P^{\mathrm{II}}$$

and

$$\underline{G}^{\mathrm{I}} = \underline{G}^{\mathrm{II}}$$

Now, continuing, we write an expression for the stability requirement $d^2S < 0$ for this system, and obtain

$$\begin{aligned} d^2S = {} & S^{\mathrm{I}}_{\mathrm{UU}}(dU^{\mathrm{I}})^2 + 2S^{\mathrm{I}}_{\mathrm{UV}}(dU^{\mathrm{I}})(dV^{\mathrm{I}}) + S^{\mathrm{I}}_{\mathrm{VV}}(dV^{\mathrm{I}})^2 \\ & + 2S^{\mathrm{I}}_{\mathrm{UN}}(dU^{\mathrm{I}})(dN^{\mathrm{I}}) + 2S^{\mathrm{I}}_{\mathrm{VN}}(dV^{\mathrm{I}})(dN^{\mathrm{I}}) + S^{\mathrm{I}}_{\mathrm{NN}}(dN^{\mathrm{I}})^2 \\ & + S^{\mathrm{II}}_{\mathrm{UU}}(dU^{\mathrm{II}})^2 + 2S^{\mathrm{II}}_{\mathrm{UV}}(dU^{\mathrm{II}})(dV^{\mathrm{II}}) + S^{\mathrm{II}}_{\mathrm{VV}}(dV^{\mathrm{II}})^2 \\ & + 2S^{\mathrm{II}}_{\mathrm{UN}}(dU^{\mathrm{II}})(dN^{\mathrm{II}}) + 2S^{\mathrm{II}}_{\mathrm{VN}}(dV^{\mathrm{II}})(dN^{\mathrm{II}}) + S^{\mathrm{II}}_{\mathrm{NN}}(dN^{\mathrm{II}})^2 < 0 \end{aligned} \tag{5.2-2a}$$

In this equation we have used the abbreviated notation that

$$S^{\mathrm{I}}_{\mathrm{UU}} = \left(\frac{\partial^2 S}{\partial U^2}\right)^{\mathrm{I}}_{V,N}, \qquad S^{\mathrm{I}}_{\mathrm{UV}} = \left.\frac{\partial}{\partial U}\right|_{V,N}\left(\frac{\partial S}{\partial V}\right)^{\mathrm{I}}_{U,N}, \quad \ldots$$

Since the total number of moles, internal energy, and volume of the composite system are fixed, we have, as in Eqs. 5.1-7, that

$$dN^{\mathrm{I}} = -dN^{\mathrm{II}} \qquad dU^{\mathrm{I}} = -dU^{\mathrm{II}} \qquad \text{and} \qquad dV^{\mathrm{I}} = -dV^{\mathrm{II}}$$

and

$$\begin{aligned} d^2S = {} & (S_{UU}^{I} + S_{UU}^{II})(dU^{I})^2 + 2(S_{UV}^{I} + S_{UV}^{II})(dU^{I})(dV^{I}) \\ & + (S_{VV}^{I} + S_{VV}^{II})(dV^{I})^2 + 2(S_{UN}^{I} + S_{UN}^{II})(dU^{I})(dN^{I}) \\ & + 2(S_{VN}^{I} + S_{VN}^{II})(dV^{I})(dN^{I}) + (S_{NN}^{I} + S_{NN}^{II})(dN^{I})^2 \end{aligned} \tag{5.2-2b}$$

Furthermore, since the same fluid in the same state of aggregation is present in regions I and II, and since we have already established that the temperature, pressure, and molar Gibbs free energy each have the same value in both regions, the value of any state property must be the same in the two subsystems. It follows that any thermodynamic derivative that can be reduced to combinations of intensive variables must have the same value in both regions of the fluid. The second derivatives (Eq. 5.2-2b), as we will see shortly, are combinations of intensive and extensive variables. However, the quantities NS_{xy}, where x and y denote U, V, or N, are intensive variables. Therefore, it follows that

$$N^{I}S_{xy}^{I} = N^{II}S_{xy}^{II} \tag{5.2-3}$$

Using Eqs. 5.1-7 and 5.2-3 in Eq. 5.2-2 yields

$$\begin{aligned} d^2S = {} & \left(\frac{N^{I} + N^{II}}{N^{I}N^{II}}\right) [N^{I}S_{UU}^{I}(dU^{I})^2 + 2N^{I}S_{UV}^{I}(dU^{I})(dV^{I}) \\ & + N^{I}S_{VV}^{I}(dV^{I})^2 + 2N^{I}S_{UN}^{I}(dU^{I})(dN^{I}) \\ & + 2N^{I}S_{VN}^{I}(dV^{I})(dN^{I}) + N^{I}S_{NN}^{I}(dN^{I})^2] < 0 \end{aligned} \tag{5.2-4a}$$

The term $(N^{I} + N^{II})/N^{I}N^{II}$ must be greater than zero since mole numbers can only be positive. Also, we can eliminate the superscripts from the products NS_{xy}, as they are equal in both phases. Therefore, the inequality (Eq. 5.2-4a) can be rewritten as

$$\begin{aligned} & NS_{UU}(dU^{I})^2 + 2NS_{UV}(dU^{I})(dV^{I}) + NS_{VV}(dV^{I})^2 + 2NS_{UN}(dU^{I})(dN^{I}) \\ & \quad + 2NS_{VN}(dV^{I})(dN^{I}) + NS_{NN}(dN^{I})^2 < 0 \end{aligned} \tag{5.2-4b}$$

Equations 5.2-3 and 4 must be satisfied for all variations in N^{I}, U^{I}, and V^{I} if the fluid is to be stable. In particular, since Eq. 5.2-4b must be satisfied for all variations in U^{I} ($dU^{I} \neq 0$) at fixed values of N^{I} and V^{I} (i.e., $dN^{I} = 0$ and $dV^{I} = 0$), stable fluids must be such that

$$NS_{UU} < 0 \tag{5.2-5a}$$

Similarly, by considering variations in volume at fixed internal energy and mole number, and variations in mole number at fixed volume and internal energy, we obtain

$$NS_{VV} < 0 \tag{5.2-5b}$$

and

$$NS_{NN} < 0 \tag{5.2-5c}$$

as additional conditions for fluid stability.

More severe restrictions on the equation of state result from demanding that Eq. 5.2-4b be satisfied for all possible and simultaneous variations in internal energy, volume, and mole number and not merely for variations in one parameter, while the others are held fixed. Unfortunately, the present form of Eq. 5.2-4b is not well-suited for studying this more general situation since the cross-terms (i.e., $dU^I\, dV^I$, $dU^I\, dN^I$, and $dV^I\, dN^I$) may be positive or negative, depending on the sign of the variations dU^I, dV^I, and dN^I, so that little can be said about the coefficients of these terms.

By much algebraic manipulation (Problem 5.27) Eq. 5.2-4b can be written as

$$\theta_1(dX_1)^2 + \theta_2(dX_2)^2 + \theta_3(dX_3)^2 < 0 \tag{5.2-6}$$

where

$$\theta_1 = NS_{UU}$$

$$\theta_2 = (NS_{UU}NS_{VV} - N^2S_{UV}^2)/NS_{UU}$$

$$\theta_3 = \frac{(NS_{UU}NS_{NN} - N^2S_{UN}^2)}{NS_{UU}} - \frac{(NS_{UU}NS_{VN} - NS_{UV}NS_{UN})^2}{NS_{UU}(NS_{UU}NS_{VV} - N^2S_{UV}^2)}$$

$$dX_1 = dU^I + \frac{S_{UV}}{S_{UU}}\, dV^I + \frac{S_{UN}}{S_{UU}}\, dN^I$$

$$dX_2 = dV^I + \frac{(S_{UU}S_{VN} - S_{UV}S_{UN})}{(S_{UU}S_{VV} - S_{UV}^2)}\, dN^I$$

and

$$dX_3 = dN^I$$

The important feature of Eq. 5.2-6 is that it contains only square terms in the system variations. Thus, $(dX_1)^2$, $(dX_2)^2$, and $(dX_3)^2$ are greater than or equal to zero regardless of whether dU^I, dV^I, and dN^I individually are positive or negative. Consequently, if

$$\theta_1 = NS_{UU} \leq 0 \tag{5.2-7a}$$

$$\theta_2 = \frac{NS_{UU}NS_{VV} - N^2S_{UV}^2}{NS_{UU}} \leq 0 \tag{5.2-7b}$$

and

$$\theta_3 \leq 0 \tag{5.2-7c}$$

Eq. 5.2-6, and hence Eq. 5.2-4b, will be satisfied for all possible system variations. Equations 5.2-7 provide more restrictive conditions for fluid stability than Eqs. 5.2-5.

It now remains to evaluate the various entropy derivatives, so that the stability restrictions of Eqs. 5.2-7 can be put into more usable form. Starting from

$$NS_{UU} = N\frac{\partial}{\partial U}\bigg|_{N,V}\left(\frac{\partial S}{\partial U}\right)_{N,V} = N\left(\frac{\partial(1/T)}{\partial U}\right)_{N,V} = -\frac{N}{T^2}\left(\frac{\partial T}{\partial U}\right)_{N,V}$$

and using that for the open system

$$dU = NC_v\, dT + \left[T\left(\frac{\partial P}{\partial T}\right)_{\underline{V}} - P\right] dV + \underline{G}\, dN \tag{5.2-8}$$

leads to

$$\left(\frac{\partial U}{\partial T}\right)_{V,N} = NC_v$$

and

$$NS_{UU} = -\frac{1}{T^2C_v} < 0$$

Since T is positive, one condition for the existence of a stable equilibrium state of a fluid is that

$$C_v > 0 \tag{5.2-9}$$

That is, the constant-volume heat capacity must be positive, so that internal energy increases as the fluid temperature increases.

Next, we note that

$$NS_{UV} = N \frac{\partial}{\partial U}\bigg|_{V,N} \left(\frac{\partial S}{\partial V}\right)_{N,U} = N \frac{\partial}{\partial V}\bigg|_{U,N} \left(\frac{\partial S}{\partial U}\right)_{V,N} = N \frac{\partial}{\partial V}\bigg|_{U,N} \left(\frac{1}{T}\right) = -\frac{N}{T^2}\left(\frac{\partial T}{\partial V}\right)_{U,N}$$

and from Eq. 5.2-8 that

$$\left(\frac{\partial T}{\partial V}\right)_{U,N} = -\left[T\left(\frac{\partial P}{\partial T}\right)_{\underline{V}} - P\right]\Big/ NC_v$$

to obtain

$$NS_{UV} = \frac{\left[T\left(\frac{\partial P}{\partial T}\right)_{\underline{V}} - P\right]}{C_v T^2} \tag{5.2-10a^1}$$

Similarly, but with a great deal more algebra, we can show that

$$NS_{VV} = \frac{1}{T}\left(\frac{\partial P}{\partial \underline{V}}\right)_T - \frac{1}{C_v T^2}\left[T\left(\frac{\partial P}{\partial T}\right)_{\underline{V}} - P\right]^2 \tag{5.2-10b^1}$$

and

$$\theta_2 = \frac{NS_{UU}NS_{VV} - N^2S^2_{UV}}{NS_{UU}} = \frac{1}{T}\left(\frac{\partial P}{\partial \underline{V}}\right)_T$$

Thus, a further stability restriction on the equation of state, since T is positive, is that

$$\left(\frac{\partial P}{\partial \underline{V}}\right)_T < 0$$

or

$$\kappa_T = -\frac{1}{\underline{V}}\left(\frac{\partial \underline{V}}{\partial P}\right)_T > 0 \tag{5.2-11}$$

where κ_T is the isothermal compressibility of the fluid introduced in Sec. 4.2. This result indicates that if a fluid is to be stable, its volumetric equation of state must be such that the fluid volume decreases as the pressure increases at constant temperature. As we will see shortly, this restriction has important implications in the interpretation of phase behavior from the equation of state of a fluid.

Finally, and with a *great* deal more algebra (Problem 5.27), one can show that θ_3 is identically equal to zero, and thus does not provide any further restrictions on the equation of state.

The main conclusion from this exercise is that if a fluid is to exist in a stable equilibrium state, that is, an equilibrium state in which all small internal fluctuations will dissipate rather than grow, the fluid must be such that

$$C_v > 0 \tag{5.2-12}$$

[1]Note, from these equations, that NS_{UU}, NS_{UV}, and NS_{VV} are intensive variables as was suggested earlier, whereas S_{UU}, S_{UV}, and S_{VV} are proportional to N^{-1}.

and

$$\left(\frac{\partial P}{\partial \underline{V}}\right)_T < 0 \qquad \text{or} \qquad \kappa_T > 0 \tag{5.2-13}$$

Alternatively, since all real fluids exist in thermodynamically stable states, Eqs. 5.2-12 and 13 must be satisfied for real fluids. In fact, no real fluid state for which either $(\partial P/\partial \underline{V})_T > 0$ or $C_v < 0$ has yet been found.

Equations 5.2-12 and 13 may be thought of as part of the philosophical content of thermodynamics. In particular, thermodynamics alone does not give information on the heat capacity or the equation of state of any material; such information can be gotten only from statistical mechanics or experiment. However, thermodynamics does provide restrictions or consistency relations that must be satisfied by such data; Eqs. 5.2-12 and 13 are one example of this. (Several other consistency relations, for mixtures, will be discussed in later chapters.)

Next we consider the problems of identifying the equilibrium state for two interacting phases of the same molecular species in different states of aggregation, and of determining the requirements for the stability of this state. For generality, we again consider a composite system isolated from its environment, except here the boundary between the two subsystems is now the real interface between the phases. For this system we have

$$S = S^{\text{I}} + S^{\text{II}}$$

$$N = N^{\text{I}} + N^{\text{II}} = \text{constant}$$

$$V = V^{\text{I}} + V^{\text{II}} = \text{constant}$$

and

$$U = U^{\text{I}} + U^{\text{II}} = \text{constant} \tag{5.2-14}$$

Since N, U, and V are fixed, the equilibrium condition is that the entropy should attain a maximum value. Now, however, we allow for the fact that the states of aggregation in regions I and II are different, so that the fluids in these regions may follow different equations of state.

Using the analysis of Eqs. 5.1-6, 7, and 8, we find that at equilibrium (i.e., when $dS = 0$),

$$T^{\text{I}} = T^{\text{II}} \tag{5.2-15a}$$

$$P^{\text{I}} = P^{\text{II}} \tag{5.2-15b}$$

and

$$\underline{G}^{\text{I}} = \underline{G}^{\text{II}} \tag{5.2-15c}$$

Here Eqs. 5.2-15a and b provide the obvious conditions for equilibrium, and, since two different phases are present, Eq. 5.2-15c provides a less obvious condition for equilibrium.

Next, from the stability condition $d^2S < 0$ we obtain (following the analysis that led to Eq. 5.2-2b)

$$\begin{aligned} d^2S = {} & \{S^{\text{I}}_{\text{VV}} + S^{\text{II}}_{\text{VV}}\}(dV^{\text{I}})^2 + 2\{S^{\text{I}}_{\text{UV}} + S^{\text{II}}_{\text{UV}}\}(dU^{\text{I}})(dV^{\text{I}}) \\ & + \{S^{\text{I}}_{\text{UU}} + S^{\text{II}}_{\text{UU}}\}(dU^{\text{I}})^2 + 2\{S^{\text{I}}_{\text{VN}} + S^{\text{II}}_{\text{VN}}\}(dV^{\text{I}})(dN^{\text{I}}) \\ & + \{S^{\text{I}}_{\text{NN}} + S^{\text{II}}_{\text{NN}}\}(dN^{\text{I}})^2 + 2\{S^{\text{I}}_{\text{UN}} + S^{\text{II}}_{\text{UN}}\}(dU^{\text{I}})(dN^{\text{I}}) \end{aligned} \tag{5.2-16}$$

Here, however, the two partial derivatives within each of the bracketed terms need not be equal, since the two phases are in different states of aggregation, and thus obey different equations of state, or different roots of the same equation of state. It is clear from a comparison with Eq. 5.2-2b that a sufficient condition for Eq. 5.2-16 to be satisfied is for each phase to be intrinsically stable; that is, Eq. 5.2-16 is satisfied if, for *each* of the coexisting phases, the equations

$$C_V > 0 \quad \text{and} \quad \left(\frac{\partial P}{\partial \underline{V}}\right)_T < 0$$

are satisfied. Therefore, if two phases are each intrinsically stable, an equilibrium state involving the thermal and mechanical interaction of these phases is also thermodynamically stable. Stated differently, a condition for the mutual stability of two interacting subsystems is that each subsystem be intrinsically stable.

5.3 PHASE EQUILIBRIA: APPLICATION OF THE EQUILIBRIUM AND STABILITY CRITERIA TO THE EQUATION OF STATE

Figure 5.3-1 indicates the shape of various isotherms for a typical equation of state (for illustration we have used the van der Waals equation of state). In this figure the isotherms are labeled so that $T_5 > T_4 > T_3 > T_2 > T_1$. The isotherm T_3 has a single point, c, for which $(\partial P/\partial \underline{V})_T = 0$; at all other points on this isotherm

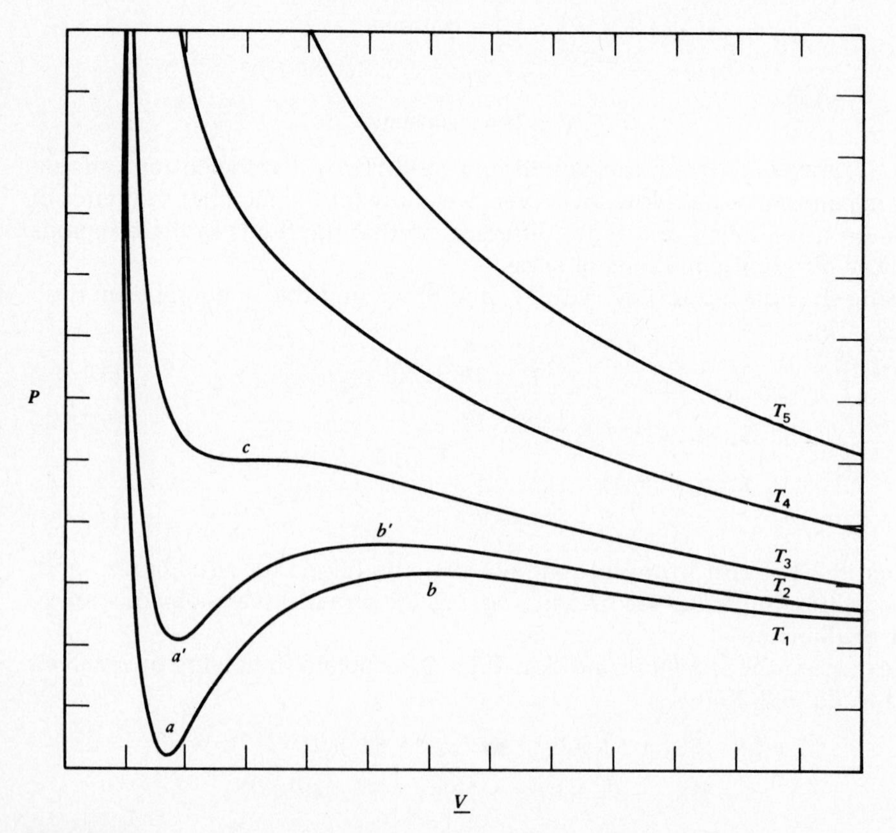

Figure 5.3-1
Isotherms of the van der Waals equation in the pressure–volume plane.

$(\partial P/\partial \underline{V})_T < 0$. On the isotherms T_4 and T_5, $(\partial P/\partial \underline{V})_T < 0$ everywhere, whereas on the isotherms T_1 and T_2, $(\partial P/\partial \underline{V})_T < 0$ in some regions and $(\partial P/\partial \underline{V})_T > 0$ in other regions (i.e., between the points a and b on isotherm T_1, and the points a' and b' on T_2). The criterion for fluid stability requires that $(\partial P/\partial \underline{V})_T < 0$, which is satisfied for the isotherms T_4 and T_5, but not in the aforementioned regions of the T_1 and T_2 isotherms. Thus we conclude that the regions a to b and a' to b' of the isotherms T_1 and T_2, respectively, are not physically realizable; that is, they will not be observed in any experiment.

This observation raises some question about the interpretation to be given to the T_1 and T_2 isotherms. We cannot simply attribute these oddities to a peculiarity of the van der Waals equation because many other, more accurate equations of state give essentially the same behavior. Some insight into the physical meaning of isotherms such as T_1 can be gained from Fig. 5.3-2, which shows this isotherm separately. If we look at any isobar (constant pressure line) between P_a and P_b in this figure, such as P_α, we see that it intersects the equation of state three times, corresponding to the fluid volumes $\underline{V}_\alpha$, $\underline{V}'_\alpha$, and $\underline{V}''_\alpha$. One of these, $\underline{V}''_\alpha$, is on the part of the isotherm that is unattainable by the stability criterion. However, there is no reason to think that either of the other two intersections at V_α and V'_α are physically unattainable. This suggests that at a given pressure and temperature the system can have two different volumes, a conclusion that apparently contradicts the experimental observation of Chapter 1 that two state variables completely determine the state of a single-component, single-phase system. However, this can occur if equilibrium can exist between two phases of the same species in different states of aggregation (and hence density). The equilibrium between liquid water and steam at 100°C and 101.325 kPa (1 atm) is one such example.

One experimental observation in phase equilibrium is that the two coexisting equilibrium phases must have the same temperature and pressure. Clearly, the arguments given in Secs. 5.1 and 5.2 establish this. Another experimental observation is that as the pressure is lowered along an isotherm on which a liquid–vapor phase transition occurs, the actual volume–pressure behavior is as

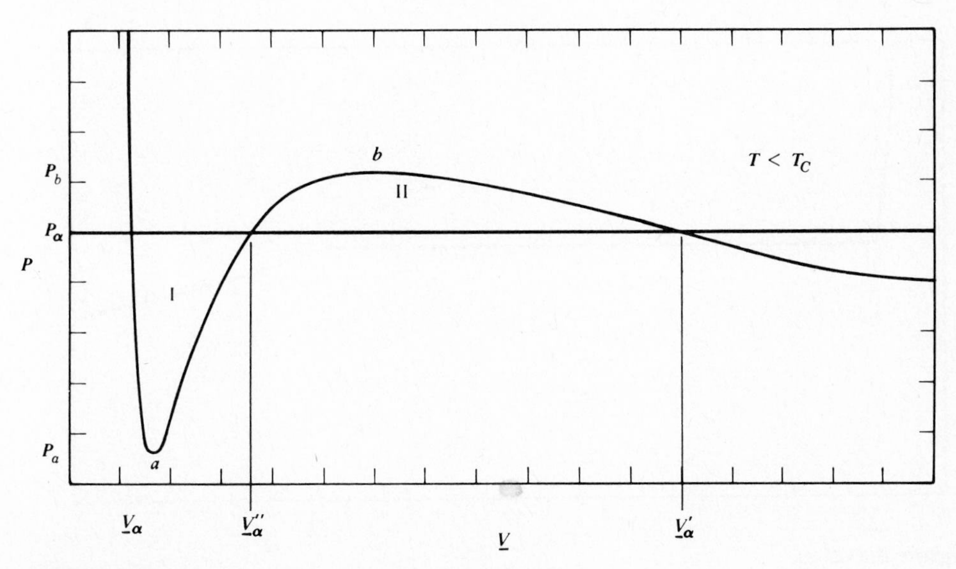

Figure 5.3-2
A low-temperature isotherm of the van der Waals equation.

shown in Fig. 5.3-3 and not as in Fig. 5.3-2. That is, there is a portion of the isotherm where the specific volume varies continuously at fixed temperature and pressure; this is the two-phase coexistence region. The continuous variation of specific volume in this region arises from the fact that since two phases of different density are present, here the vapor and liquid, the specific volume of the two-phase mixture ($\underline{V}$), is the sum of the product of the specific volume of each phase times the fraction of the total mass (or number of moles) in that phase (see Eq. 2.4-9). Clearly, the specific volume of both the vapor ($\underline{V}^{V}$) and liquid ($\underline{V}^{L}$) are constant through the coexistence region since two state variables, T and P, are fixed in each of these one-phase equilibrium subsystems. Letting x^{V} designate the fraction of vapor we can write

$$\underline{V} = x^{V}\underline{V}^{V} + (1 - x^{V})\underline{V}^{L} \tag{5.3-1a}$$

where x^{V} varies continuously between zero and one. Solving for x^{V} yields

$$x^{V} = \frac{\underline{V} - \underline{V}^{L}}{\underline{V}^{V} - \underline{V}^{L}} \tag{5.3-1b}$$

and

$$\frac{x^{V}}{1 - x^{V}} = \frac{\underline{V} - \underline{V}^{L}}{\underline{V}^{V} - \underline{V}} \tag{5.3-1c}$$

Equations analogous to those here also hold for the $\underline{H}$, $\underline{U}$, $\underline{G}$, $\underline{S}$, and $\underline{A}$. Equations of the form of Eq. 5.3-1c are called **Maxwell's rules** or **lever rules.**

The conclusion then is that an isotherm such as that shown in Fig. 5.3-2 is an approximate representation of the real phase behavior (shown in Fig. 5.3-3) by a relatively simple analytic equation of state. In fact, it is impossible to represent the discontinuities in the derivative $(\partial P/\partial \underline{V})_T$ that occur at $\underline{V}^{L}$ and $\underline{V}^{V}$ with *any* analytic equation of state. By its sigmoidal behavior in the two-phase region, the van der Waals equation of state is somewhat qualitatively and crudely exhibiting the essential features of vapor–liquid phase equilibrium; historically, it was the first equation of state to do so.

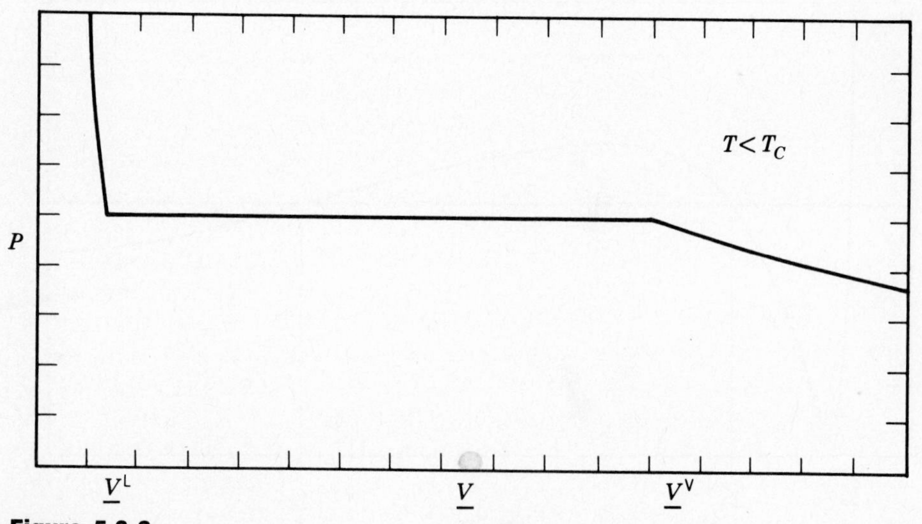

Figure 5.3-3
A low-temperature isotherm of a real fluid.

We can improve the representation of the two-phase region when using the van der Waals or other analytic equations of state by recognizing that all **van der Waals loops,** such as those shown in Fig. 5.3-2, should be replaced by horizontal lines (isobars), as shown in Fig. 5.3-3. This construction ensures that the equilibrium phases will have the same temperature and pressure (see Eqs. 5.1-9a and b). The question that remains is at which pressure should the isobar be drawn, since any pressure such that

$$P_a < P < P_b$$

will yield an isotherm like that in Fig. 5.3-3. The answer is that the pressure chosen must satisfy the last condition for equilibrium, that $\underline{G}^{\mathrm{I}} = \underline{G}^{\mathrm{II}}$.

To identify the equilibrium pressure we start from Eq. 4.2-8b

$$d\underline{G} = \underline{V}\,dP - \underline{S}\,dT$$

and recognize that for the integration between any two points along an isotherm of the equation of state we have

$$\Delta\underline{G} = \int_{P_1}^{P_2} \underline{V}\,dP$$

Thus, for a given equation of state we can identify the equilibrium pressure for each temperature by arbitrarily choosing pressures P_α along the van der Waals loop, until we find one for which

$$\underline{G}^{\mathrm{V}} - \underline{G}^{\mathrm{L}} = 0 = \int_{P_\alpha}^{P_a} \underline{V}\,dP + \int_{P_a}^{P_b} \underline{V}\,dP + \int_{P_b}^{P_\alpha} \underline{V}\,dP \qquad (5.3\text{-}2)$$

Here the specific volume in each of the integrations is to be computed from the equation of state for the appropriate part of the van der Waals loop. Alternatively, we can find the equilibrium pressure graphically by noting that Eq. 5.3-2 requires that areas I and II in Fig. 5.3-2 be equal at the pressure at which the vapor and liquid exist in equilibrium. This vapor–liquid coexistence pressure, which is a function of temperature, is called the **vapor pressure** of the liquid and will be denoted by $P^{\mathrm{vap}}(T)$.

We can continue in the manner described here to determine the phase behavior of the fluid for all temperatures and pressures. For the van der Waals fluid this result is shown in Fig. 5.3-4. An important feature of this figure is the dome-shaped, two-phase coexistence region. The inflection point C of Fig. 5.3-1 is the peak of this dome, and therefore is the highest temperature at which the condensed phase (the liquid) can exist; this point is the **critical point** of the liquid.

It is worthwhile retracing the steps followed in identifying the existence and location of the two-phase region in the P–V plane:

1. The stability condition $(\partial P/\partial \underline{V})_T < 0$ was used to identify the unstable region of an isotherm and thereby establish the existence of a two-phase region.
2. The conditions $T^{\mathrm{I}} = T^{\mathrm{II}}$ and $P^{\mathrm{I}} = P^{\mathrm{II}}$ were then used to establish the shape (but not the location) of the coexistence line in the P–V plane.
3. Finally, the equilibrium condition $\underline{G}^{\mathrm{I}} = \underline{G}^{\mathrm{II}}$ was used to locate the coexistence line.

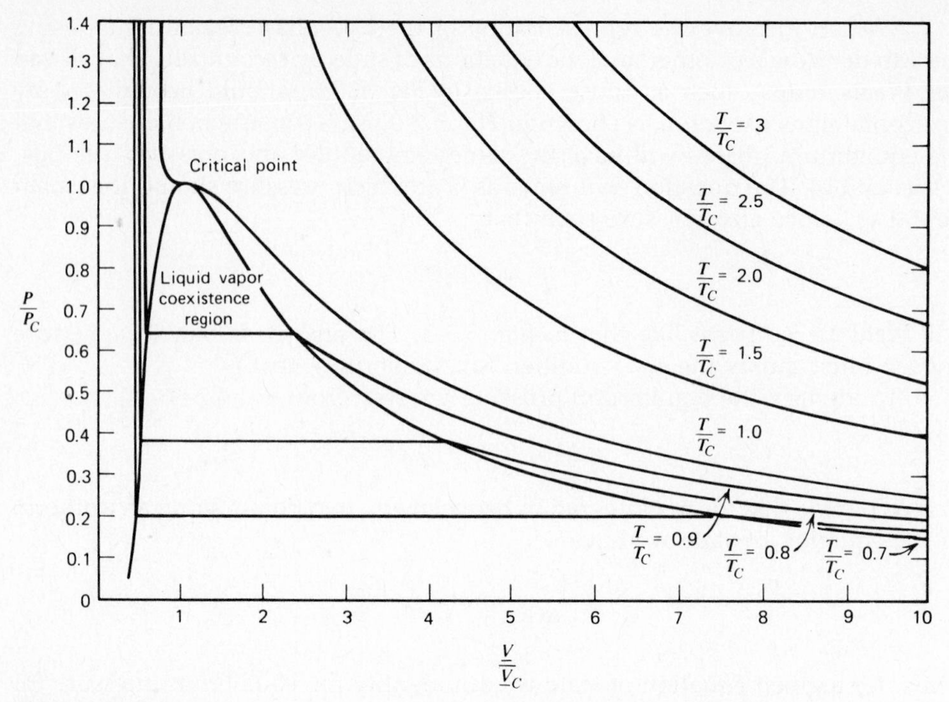

Figure 5.3-4
The van der Waals fluid with the vapor–liquid coexistence region identified.

A more detailed representation for phase equilibrium in a pure fluid, including the presence of a single solid phase,[2] is given in the isometric *PVT* phase diagram of Fig. 5.3-5. Actually, such complete phase diagrams are rarely available, although data may be available in the form of Fig. 5.3-4, which is a projection of the more complete diagram into the *P–V* plane, and Fig. 5.3-6, which is the projection into the *P–T* plane.

The concept of phase equilibrium and the critical point can also be considered from a somewhat different point of view. Presume it were possible to compute the Gibbs free energy as a function of temperature and pressure for any phase, either from an equation of state, experimental data, or statistical mechanics. Then, at fixed pressure, one could plot $\underline{G}$ as a function of T for each phase, as shown in Fig. 5.3-7 for the vapor and liquid phases. From the equilibrium condition that $\underline{G}$ be a minimum, one can conclude that the liquid is the equilibrium phase at temperatures below T_P, that the vapor is the equilibrium phase above T_P, and that both phases are present at the phase transition temperature T_P.

If such calculations are repeated for a wide range of temperatures and pressures, it is observed that the angle of intersection θ between the liquid and vapor free energy curves decreases as the pressure (and temperature) at which the intersection occurs increases (provided $P \leq P_c$). At the critical pressure, the two Gibbs free energy curves intersect with $\theta = 0$; that is, the two curves are

[2]If several solid phases occur corresponding to different crystal structures, as is frequently the case, the solid region is partitioned into several regions.

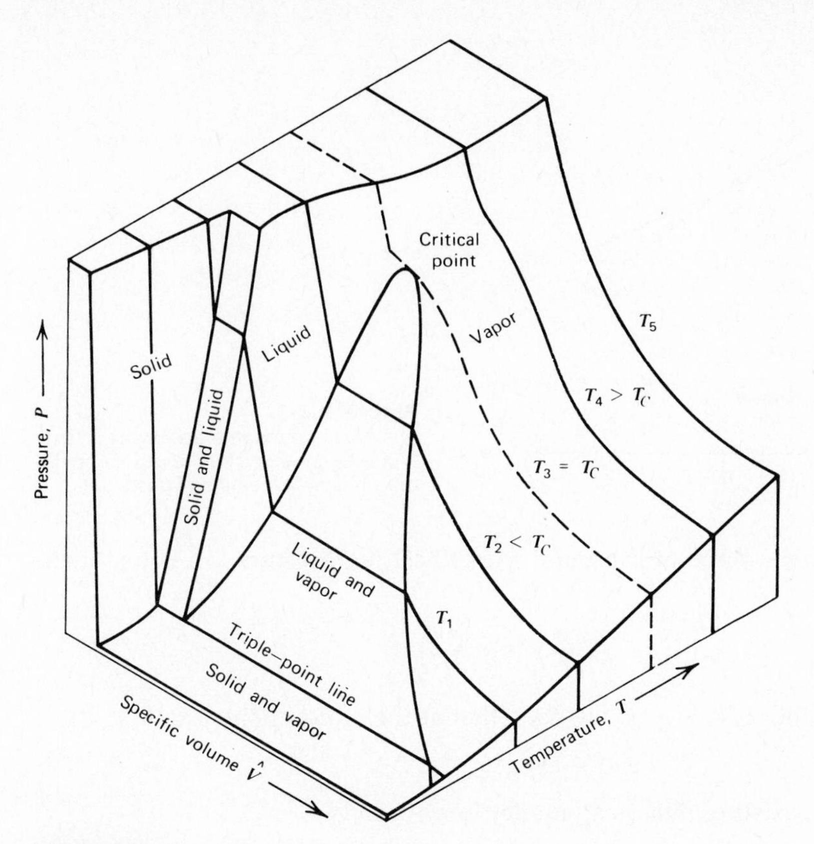

Figure 5.3-5
The *PVT* phase diagram for a substance with a single solid phase. (Adapted from J. Kestin, *A Course in Thermodynamics*, vol. 1. © 1966 by Blaisdell Publishing Co. (John Wiley & Sons, Inc.) Used with permission of John Wiley & Sons, Inc.)

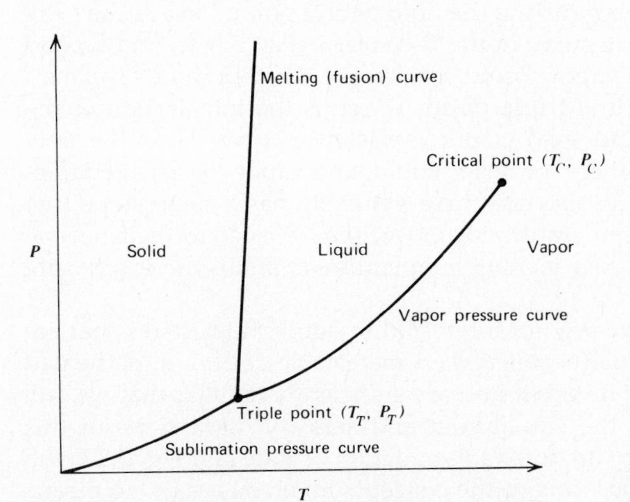

Figure 5.3-6
Phase diagram in the P–T plane.

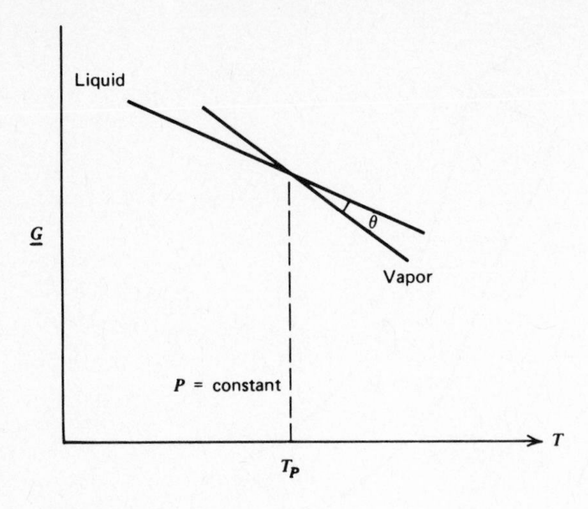

Figure 5.3-7
The molar Gibbs free energy as a function of temperature for the vapor and liquid phases of the same substance.

collinear for some range of T around the critical temperature T_c. Thus, at the critical point

$$\left(\frac{\partial \underline{G}^{\mathrm{L}}}{\partial T}\right)_P = \left(\frac{\partial \underline{G}^{\mathrm{V}}}{\partial T}\right)_P$$

Further, since $(\partial \underline{G}/\partial T)_P = -\underline{S}$, we have that at the critical point

$$\underline{S}^{\mathrm{L}}(T_C, P_C) = \underline{S}^{\mathrm{V}}(T_C, P_C)$$

Also, for the coexisting phases at equilibrium we have

$$\underline{G}^{\mathrm{L}}(T_C, P_C) = \underline{G}^{\mathrm{V}}(T_C, P_C)$$

by Eq. 5.2-15c. Since the molar Gibbs free energy, molar entropy, temperature, and pressure each have the same value in the vapor and the liquid, the values of all other state variables must be identical in the two equilibrium phases at the critical point. Consequently, the vapor and liquid phases should be indistinguishable at the critical point. This is exactly what is experimentally observed. At all temperatures higher than the critical temperature, regardless of the pressure, only the vapor phase exists. This is the reason for the abrupt terminus of the vapor–liquid coexistence line in the pressure–temperature plane (Fig. 5.3-6). [Thus, we have two ways of recognizing the fluid critical point. First as the peak in the vapor–liquid coexistence curve in the P–V plane (Fig. 5.3-4), and second as the abrupt terminus of the vapor–liquid coexistence curve in the P–T plane.]

Also interesting is the fluid **triple point,** which is the intersection of the solid–liquid, liquid–vapor, and solid–vapor coexistence curves. It is the only point on the phase diagram where the solid, liquid, and vapor coexist at equilibrium. Since the solid–liquid coexistence curve generally has a steep slope (see Fig. 5.3-6), the triple point temperature for most fluids is close to their normal melting temperature, that is, their melting temperature at atmospheric pressure (Problem 5.11).

Although, in general, we are not interested in equilibrium states that are unstable to large perturbations (usually called metastable states), superheated liquids and subcooled vapors do occur and are sufficiently familiar that we will briefly relate these states to the equilibrium and stability discussions of this chapter. For convenience, the van der Waals equation of state and Fig. 5.3-2 will be the basis for this discussion, though the concepts involved are by no means restricted to this equation. We start by noticing that although the liquid phase is thermodynamically stable along the isotherm T_1 in Fig. 5.3-2 down to a pressure P_a, the phase equilibrium analysis indicates that the vapor, and not the liquid, is the equilibrium phase at pressures below the vapor pressure $P_\alpha = P^{\text{vap}}(T_1)$. If care is taken to avoid vapor phase nucleation sites, such as by having only clean, smooth surfaces in contact with the liquid, it is possible to maintain a liquid at fixed temperature below its vapor pressure (but above its limit of stability, P_a), or at fixed pressure at a temperature higher than its boiling temperature, without having the liquid boil. Such a liquid is said to be **superheated.** The metastability of this state is illustrated by the fact that a superheated liquid, if slightly perturbed, may vaporize with explosive violence. (To avoid this occurrence "boiling stones" are used in chemistry laboratory experiments.) It is also possible, if no condensation nucleation sites, such as dust particles, are present, to prepare a vapor at a pressure higher than the liquid vapor pressure at the given temperature, but below its limit of stability (i.e., between $P_\alpha = P^{\text{vap}}(T_1)$ and P_b in Fig. 5.3-2) or, at lower than the liquid boiling temperature. Such a vapor is termed **subcooled** and is also metastable. (See Problem 5.8.)

At sufficiently low temperatures, the van der Waals equation predicts that the limit of stability of the liquid phase occurs at negative values of the pressure; that is, that a liquid could support a tensile force. In fact, such metastable behavior has been observed with water in capillary tubes and is thought to be important in the vascular system of plants.[3]

5.4 THE MOLAR GIBBS FREE ENERGY AND FUGACITY OF A PURE COMPONENT

In this section we consider how one uses an equation of state to identify the states of vapor–liquid equilibrium in a pure fluid. The starting point is the equality of molar Gibbs free energies in the coexisting phases

$$\underline{G}^{\text{L}}(T, P) = \underline{G}^{\text{V}}(T, P) \tag{5.1-9c}$$

To proceed, we note that from Eq. 4.2-8b

$$d\underline{G} = -\underline{S}\, dT + \underline{V}\, dP$$

so that

$$\left(\frac{\partial \underline{G}}{\partial T}\right)_P = -\underline{S} \tag{5.4-1}$$

and

$$\left(\frac{\partial \underline{G}}{\partial P}\right)_T = \underline{V} \tag{5.4-2}$$

Since our presumption here is that we have an equation of state from which we can compute $\underline{V}$ as a function of T and P, only Eq. 5.4-2 will be considered further.

Integration of Eq. 5.4-2 between any two pressures P_1 and P_2 (at constant temperature) yields

$$\underline{G}(T_1, P_2) - \underline{G}(T_1, P_1) = \int_{P_1}^{P_2} \underline{V}\, dP \tag{5.4-3}$$

[3]For a review of water under tension, especially in biological systems, see P. F. Scholander, *American Scientist* **60,** 584 (1972).

If the fluid under consideration were an ideal gas, then $\underline{V}^{IG} = RT/P$, so that

$$\underline{G}^{IG}(T_1, P_2) - \underline{G}^{IG}(T_1, P_1) = \int_{P_1}^{P_2} \frac{RT}{P}\, dP \tag{5.4-4}$$

Subtracting Eq. 5.4-4 from Eq. 5.4-3 gives

$$[\underline{G}(T_1, P_2) - \underline{G}^{IG}(T_1, P_2)] - [\underline{G}(T_1, P_1) - \underline{G}^{IG}(T_1, P_1)] = \int_{P_1}^{P_2} \left(\underline{V} - \frac{RT}{P}\right) dP \tag{5.4-5a}$$

Further, (1) setting P_1 equal to zero, (2) recognizing that at $P = 0$ all fluids are ideal gases so that $\underline{G}(T_1, P = 0) = \underline{G}^{IG}(T_1, P = 0)$, and (3) them omitting all subscripts yields

$$\underline{G}(T, P) - \underline{G}^{IG}(T, P) = \int_0^P \left(\underline{V} - \frac{RT}{P}\right) dP \tag{5.4-5b}$$

For convenience, we define a new thermodynamic function, the **fugacity**, by

$$f = P \exp\left\{\frac{\underline{G}(T, P) - \underline{G}^{IG}(T, P)}{RT}\right\} = P \exp\left\{\frac{1}{RT}\int_0^P \left(\underline{V} - \frac{RT}{P}\right) dP\right\} \tag{5.4-6a}$$

and the **fugacity coefficient** ϕ by

$$\phi = \frac{f}{P} = \exp\left\{\frac{\underline{G}(T, P) - \underline{G}^{IG}(T, P)}{RT}\right\} = \exp\left\{\frac{1}{RT}\int_0^P \left(\underline{V} - \frac{RT}{P}\right) dP\right\} \tag{5.4-6b}$$

From this definition it is clear that the fugacity has units of pressure, and that $f \rightarrow P$ as $P \rightarrow 0$; that is, the fugacity becomes equal to the pressure at pressures low enough that the fluid approaches the ideal gas state.[4] Similarly, the fugacity coefficient $\phi = f/P \rightarrow 1$ as $P \rightarrow 0$. Both the fugacity and the fugacity coefficient will be used extensively throughout this text.

The fugacity function has been introduced because its relation to the Gibbs free energy makes it useful in phase equilibrium calculations. The present criterion for equilibrium between two phases (Eq. 5.1-9c) is $\underline{G}^{\text{I}} = \underline{G}^{\text{II}}$, with the restriction that the temperature and pressure be constant and equal in both phases. Using this result in Eq. 5.4-6a written for each phase, and recognizing that the ideal gas molar Gibbs free energy is the same at fixed temperature and pressure regardless of which phase is considered, yields

$$f^{\text{I}}(T, P) = f^{\text{II}}(T, P) \tag{5.4-7}$$

as the condition for phase equilibrium. Since this equation follows directly from the equality of the molar Gibbs free energy in each phase at phase equilibrium, Eq. 5.4-7 can be used as the criterion for equilibrium; this will be done here.

Since the fugacity is, by Eq. 5.4-6, related to the equation of state, the equality of fugacities provides a direct way of doing phase equilibrium calculations using equations of state. In practice, however, Eq. 5.4-6 is somewhat difficult to use because although the molar volume $\underline{V}$ is needed as a function of T and P, it is clear that it is difficult to solve the equations of state considered in Sec. 4.4 explicitly for volume. In fact, all these equations of state are in a form in which pressure is an explicit function of volume and temperature. Therefore, it

[4]It is tempting to view the fugacity as a sort of corrected pressure; it is however, a well-defined function related to the exponential of the difference between the real and ideal gas Gibbs free energies.

is useful to have an equation relating the fugacity to an integral over volume (rather than pressure). We obtain such an equation by starting with Eq. 5.4-6b and using Eq. 4.4-25 at constant temperature in the form

$$dP = \frac{1}{\underline{V}} d(P\underline{V}) - \frac{P}{\underline{V}} d\underline{V} = \frac{P}{Z} dZ - \frac{P}{\underline{V}} d\underline{V}$$

to change the variable of integration to obtain (Problem 5.14)

$$\ln \frac{f(T, P)}{P} = \ln \phi = \frac{1}{RT} \int_{\underline{V}=\infty}^{\underline{V}} \left[\frac{RT}{\underline{V}} - P \right] d\underline{V} - \ln Z + (Z - 1) \tag{5.4-8}$$

where $Z = P\underline{V}/RT$ is the compressibility factor defined earlier. [Alternatively, Eq. 5.4-8 can be obtained from Eq. 5.4-6 using

$$\underline{G}(T, P) - \underline{G}^{IG}(T, P) = [\underline{H}(T, P)] - \underline{H}^{IG}(T, P)] - T[\underline{S}(T, P) - \underline{S}^{IG}(T, P)]$$

and Eqs. 4.4-28 and 29].

For later reference we note that from Eqs. 5.4-2 and 6 we have that

$$RT \left(\frac{\partial \ln f}{\partial P} \right)_T = \underline{V} = \left(\frac{\partial \underline{G}}{\partial P} \right)_T \tag{5.4-9a}$$

Also the temperature dependence of the fugacity is usually given as the temperature dependence of the logarithm of the fugacity coefficient (f/P), which is computed as:

$$\begin{aligned}
\frac{\partial}{\partial T} \left(\ln \frac{f}{P} \right)_P &= \frac{\partial}{\partial T} \bigg|_P \left\{ \frac{\underline{G}(T, P) - \underline{G}^{IG}(T, P)}{RT} \right\} \\
&= \frac{1}{RT} \frac{\partial}{\partial T} \{\underline{G}(T, P) - \underline{G}^{IG}(T, P)\} - \left\{ \frac{\underline{G}(T, P) - \underline{G}^{IG}(T, P)}{RT^2} \right\} \\
&= -\frac{1}{RT} \{\underline{S}(T, P) - \underline{S}^{IG}(T, P)\} - \left\{ \frac{\underline{G}(T, P) - \underline{G}^{IG}(T, P)}{RT^2} \right\} \\
&= -\frac{1}{RT^2} \{[\underline{G}(T, P) + T\underline{S}(T, P)] - [\underline{G}^{IG}(T, P) + T\underline{S}^{IG}(T, P)]\} \\
&= -\left\{ \frac{\underline{H}(T, P) - \underline{H}^{IG}(T, P)}{RT^2} \right\}
\end{aligned} \tag{5.4-9b}$$

In deriving this result we have used the relations $(\partial \underline{G}/\partial T)_P = -\underline{S}$ and $\underline{G} = \underline{H} - T\underline{S}$.

Since the fugacity function is of central importance in phase equilibrium calculations, we consider here the computation of the fugacity for pure gases, liquids, and solids.

a. Fugacity of a Pure Gaseous Species

To compute the fugacity of a pure gaseous species we will always use a volumetric equation of state and

$$\ln \frac{f^V(T, P)}{P} = \frac{1}{RT} \int_{\underline{V}=\infty}^{\underline{V}=Z^V RT/P} \left(\frac{RT}{\underline{V}} - P \right) d\underline{V} - \ln Z^V + (Z^V - 1) \tag{5.4-8}$$

where the superscript V has been used to designate the fugacity and compressibility of the vapor phase. Thus, given a volumetric equation of state of a gas applicable up to the pressure of interest, the fugacity of a pure gas can be computed by integration of Eq. 5.4-8. At very low pressures, where a gas can be described by the ideal gas equation of state

$$P\underline{V} = RT \qquad \text{or} \qquad Z^V = 1$$

we have that

$$\ln \frac{f^V(T, P)}{P} = 0 \qquad \text{or} \qquad f^V(T, P) = P \tag{5.4-10}$$

Thus, for a low pressure gas, the fugacity of a species is just equal to the total pressure.

At low to moderate pressures, the virial equation of state truncated after the second virial coefficient

$$\frac{P\underline{V}}{RT} = Z = 1 + \frac{B(T)}{\underline{V}} \tag{5.4-11}$$

may be used, if data for $B(T)$ are available. Using Eq. 5.4-11 in Eq. 5.4-8, we obtain (Problem 5.14)

$$\ln \frac{f^V(T, P)}{P} = \frac{2B(T)}{\underline{V}} - \ln Z = \frac{2PB(T)}{ZRT} - \ln Z \tag{5.4-12}$$

where

$$Z = 1 + \frac{B(T)}{\underline{V}} = \frac{1}{2}\left[1 + \sqrt{1 + \frac{4BP}{RT}}\right]$$

At higher pressures, a more complicated equation of state (or higher terms in the virial expansion) must be used. By using the (not very accurate) van der Waals equation, one obtains

$$\ln \frac{f^V}{P} = \ln \frac{\underline{V}}{\underline{V} - b} - \frac{a}{RT\underline{V}} + \left(\frac{P\underline{V}}{RT} - 1\right) - \ln\left(\frac{P\underline{V}}{RT}\right)$$

or

$$= (Z^V - 1) - \ln(Z^V - B) - \frac{A}{Z^V} \tag{5.4-13}$$

For hydrocarbons and simple gases, the Peng–Robinson equation (Eq. 4.4-2) provides a more accurate description. In this case we have

$$\ln \frac{f^V}{P} = (Z^V - 1) - \ln\left(Z^V - \frac{bP}{RT}\right) - \frac{a}{2\sqrt{2}bRT} \ln\left[\frac{Z^V + (1 + \sqrt{2})bP/RT}{Z^V + (1 - \sqrt{2})bP/RT}\right]$$

or

$$= (Z^V - 1) - \ln(Z^V - B) - \frac{A}{2\sqrt{2}B} \ln\left[\frac{Z^V + (1 + \sqrt{2})B}{Z^V + (1 - \sqrt{2})B}\right] \tag{5.4-14a}$$

where in Eqs. 5.4-13 and 14a, $A = aP/(RT)^2$ and $B = Pb/RT$. Of course, other equations of state could be used for the fugacity calculation starting from Eq. 5.4-8, though we will not consider such calculations here.

To use either the virial, van der Waals, Peng–Robinson, or other equations of state to calculate the fugacity of a gaseous species, the following procedure is used: (1) For a given value of T and P, use the chosen equation of state to calculate the molar volume $\underline{V}$ or, equivalently, the compressibility factor Z. When using cubic or more complicated equations of state, it is the low density (large $\underline{V}$ or Z) solution that is of interest. (2) This value of $\underline{V}$ or Z is then used in

Eq. 5.4-12, 13, or 14, as appropriate, to calculate the species fugacity coefficient, f/P, and thus the fugacity. The BASIC language program of Appendix A4.1 can be used for this calculation for the Peng–Robinson equation of state.

For hand calculations it is simpler, but less accurate, to compute the fugacity of a species using a specially prepared corresponding states fugacity chart. To do this, we note that since, for simple gases and hydrocarbons, the compressibility factor Z obeys a corresponding states relation (see Sec. 4.6), the fugacity coefficient, f/P, given by Eq. 5.4-6 can also be written in corresponding states form as follows:

$$\frac{f^V}{P} = \exp\left\{\frac{1}{RT}\int_0^P (\underline{V} - \underline{V}^{IG})\, dP\right\} = \exp\left\{\int_0^P \left(\frac{P\underline{V}}{RT} - 1\right)\frac{dP}{P}\right\}$$

$$= \exp\left\{\int_0^{P_r} [Z(T_r, P_r) - 1]\, d\ln P_r\right\} \tag{5.4-15a}$$

or

$$\ln\frac{f^V}{P} = \int_0^{P_r} [Z(T_r, P_r) - 1]\, d\ln P_r \tag{5.4-15b}$$

Consequently, the fugacity coefficient can be tabulated in the corresponding states manner. The corresponding states correlation for the fugacity coefficient of nonpolar gases and liquids given in Fig. 5.4-1 was obtained using Eq. 5.4-15b and the compressibility correlation (Fig. 4.6-3).

b. The Fugacity of a Pure Liquid

The fugacity of a pure liquid species can be computed in a number of ways, depending on the data available. If the equation of state for the liquid is known, we can again start from Eq. 5.4-8 now written as

$$\ln\frac{f^L(T, P)}{P} = \frac{1}{RT}\int_{\underline{V}=\infty}^{\underline{V}=Z^L RT/P}\left(\frac{RT}{\underline{V}} - P\right) d\underline{V} - \ln Z^L + (Z^L - 1) \tag{5.4-8}$$

where the superscript L has been used to indicate that the liquid phase compressibility (high density, small $\underline{V}$ and Z) is to be used, and it is the liquid-phase fugacity that is being calculated. Using, for example, the Peng–Robinson equation of state in Eq. 5.4-8 yields

$$\ln\frac{f^L}{P} = (Z^L - 1) - \ln\left(Z^L - \frac{bP}{RT}\right) - \frac{a}{2\sqrt{2}bRT}\ln\left[\frac{Z^L + (1+\sqrt{2})bP/RT}{Z^L + (1-\sqrt{2})bP/RT}\right]$$

$$= (Z^L - 1) - \ln(Z^L - B) - \frac{A}{2\sqrt{2}B}\ln\left[\frac{Z^L + (1+\sqrt{2})B}{Z^L + (1-\sqrt{2})B}\right] \tag{5.4-14b}$$

To use this equation, we first, at the specified value of T and P, solve the Peng–Robinson equation of state for the liquid compressibility, Z^L, and use this value to compute $f^L(T, P)$. Of course, other equations of state could be used in Eq. 5.4-8. The BASIC language computer program PR1.BAS of Appendix A4.1 can be used for this calculation for the Peng–Robinson equation of state.

If one has some data for a liquid, but not an equation of state, it is more convenient to start with Eq. 5.4-6, which can be rearranged to

$$RT\ln\frac{f^L}{P} = \int_0^P \left[\frac{\underline{V}}{RT} - \frac{1}{P}\right] dP \tag{5.4-16a}$$

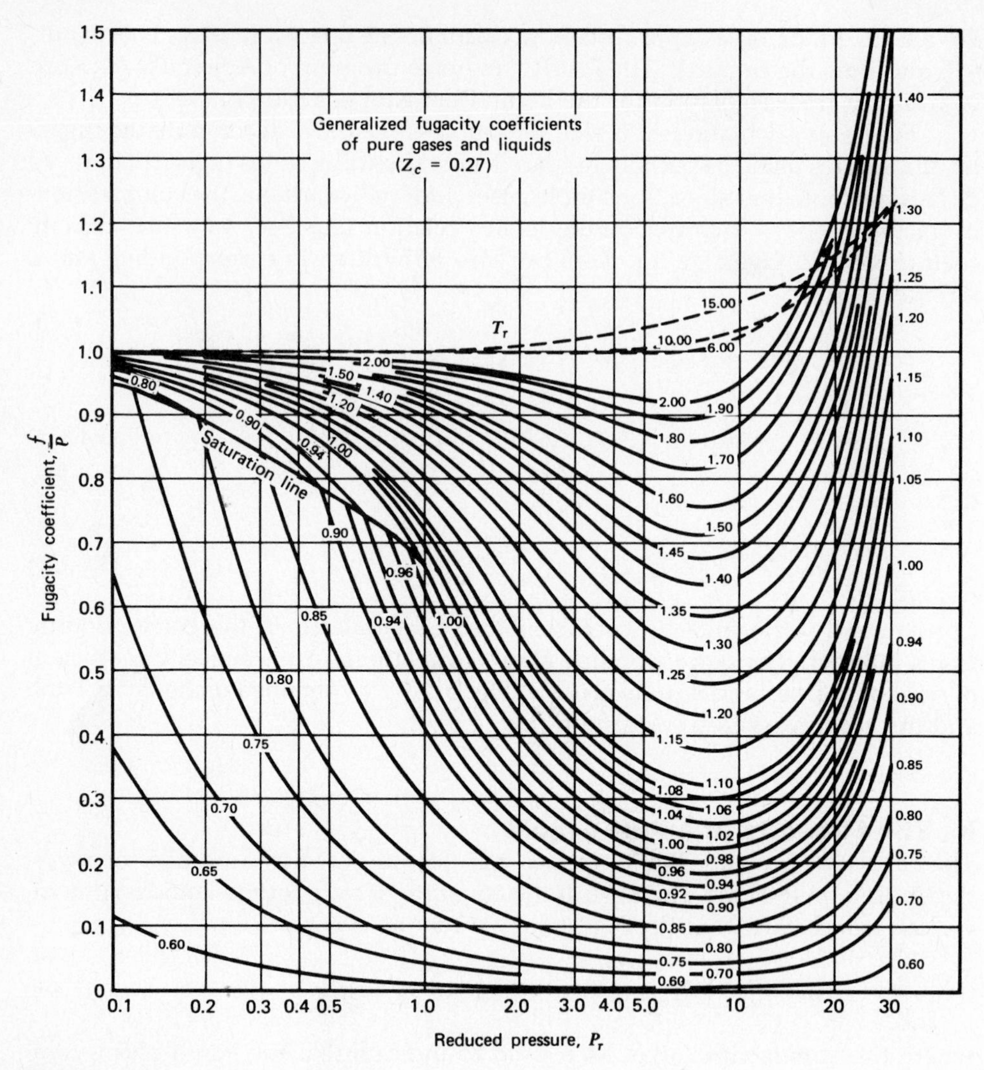

Figure 5.4-1
(From O. A. Hougen, K. M. Watson, and R. A. Ragatz, *Chemical Process Principles Charts*, 2nd ed. Copyright 1960 by John Wiley & Sons, Inc. Used with permission.)

and perform the integration. However, one has to recognize that a phase change from a vapor to a liquid occurs within the integration range, and that the molar volume of a fluid is discontinuous at this phase transition. Thus, the result of the integration is

$$RT \ln\left(\frac{f^{\mathrm{L}}}{P}\right) = \underline{G}(T, P) - \underline{G}^{IG}(T, P) = \int_{P=0}^{P^{\mathrm{vap}}(T)} \left(\underline{V} - \frac{RT}{P}\right) dP + RT\,\Delta\left(\ln\frac{f}{P}\right)_{\substack{\text{phase}\\ \text{change}}}$$
$$+ \int_{P^{\mathrm{vap}}(T)}^{P} \left(\underline{V} - \frac{RT}{P}\right) dP \tag{5.4-16b}$$

The first term on the right side of this equation is the difference between the real and ideal gas Gibbs free energy changes of compressing the vapor from zero pressure to the vapor pressure of the liquid at the temperature T. The second term allows for the free energy change at the phase transition. The last term is the difference between the liquid and ideal gas Gibbs free energy changes on compression of the liquid from its vapor pressure to the pressure of interest.

From Eq. 5.4-7 and the fact that the pressure is continuous at a phase change we have

$$\Delta \left(\ln \frac{f}{P}\right)_{\substack{\text{phase}\\ \text{change}}} = 0$$

and, from Eq. 5.4-6, we further have that

$$\int_0^{P^{\text{vap}}(T)} \left(\underline{V} - \frac{RT}{P}\right) dP = RT \ln \left(\frac{f}{P}\right)_{\text{sat},T} \tag{5.4-17}$$

where $(f/P)_{\text{sat},T}$ is the fugacity coefficient of the saturated fluid (either vapor or liquid, since these fugacities are equal) at the temperature T. Finally, the last term in Eq. 5.4-16 can be partially integrated as follows:

$$\int_{P^{\text{vap}}(T)}^{P} \left(\underline{V} - \frac{RT}{P}\right) dP = \int_{P^{\text{vap}}(T)}^{P} \underline{V}\, dP - RT \int_{P^{\text{vap}}(T)}^{P} \frac{1}{P}\, dP$$

$$= \int_{P^{\text{vap}}(T)}^{P} \underline{V}\, dP - RT \ln \frac{P}{P^{\text{vap}}(T)}$$

Combining these terms yields the following expressions for the fugacity of a pure liquid

$$f^{\text{L}}(T, P) = P^{\text{vap}}(T) \left(\frac{f}{P}\right)_{\text{sat},T} \exp \left[\frac{1}{RT} \int_{P^{\text{vap}}(T)}^{P} \underline{V}\, dP\right]$$

$$= f_{\text{sat}}(T) \exp \left[\frac{1}{RT} \int_{P^{\text{vap}}(T)}^{P} \underline{V}\, dP\right] \tag{5.4-18}$$

The exponential term in this equation, known as the **Poynting pressure correction,** accounts for the increase in fugacity due to the fact that the system pressure is higher than the vapor pressure of the liquid. Since the molar volume of a liquid is generally much less than that of a gas (so that $P\underline{V}^{\text{L}}/RT \ll 1$), the Poynting term is only important at high pressures. (An exception to this is for cryogenic systems where T is very low.)

Equation 5.4-18 lends itself to several levels of approximation. The simplest approximation is to neglect the Poynting and $(f/P)_{\text{sat}}$ terms and set the fugacity of the liquid equal to its vapor pressure at the same temperature, that is,

$$f^{\text{L}}(T, P) = P^{\text{vap}}(T) \tag{5.4-19}$$

This result is valid only when the vapor pressure is low and the liquid is at a low total pressure. While this equation applies to most fluids, it is not correct for fluids that associate, that is form dimers, and so on in the vapor phase, such as acetic acid. A more accurate approximation is

$$f^{\text{L}}(T, P) = f^{\text{L}}_{\text{sat}}(T) = f^{\text{V}}(T, P^{\text{vap}}) = P^{\text{vap}}(T) \left(\frac{f}{P}\right)_{\text{sat},T} \tag{5.4-20}$$

which states that the fugacity of a liquid at a given temperature and any pressure is just equal to its fugacity at its vapor pressure (as a vapor or liquid). This approximation is valid provided the system pressure is not greatly different from the species vapor pressure at the temperature of interest (so that the Poynting term is negligible). To evaluate $f^V(T, P^{vap})$, any of the methods considered in Sec. 5.4a may be used. For example, if second virial coefficient data are available, Eq. 5.4-13 can be used, or if an equation of state is available for the vapor, but not the liquid (as may be the case for nonhydrocarbon fluids), Eq. 5.4-8 (or its integrated form for the equations of state considered in Sec. 5.4a) can be used to estimate $f^V(T, P^{vap})$. Alternatively, but less accurately, $(f/P)_{sat,T}$ can be gotten from Fig. 5.4-1 using the saturation line and the reduced temperature of interest.

Another approximation that can be made in Eq. 5.4-18 is to take into account the Poynting pressure correction, but to assume that the liquid is incompressible (Fig. 5.3-4 for the van der Waals equation of state, for example, does indicate that at constant temperature the liquid volume is only slightly pressure dependent). In this case

$$f^L(T, P) = P^{vap}(T) \left(\frac{f}{P}\right)_{sat,T} \exp\left[\frac{\underline{V}(P - P^{vap})}{RT}\right] \qquad (5.4\text{-}21)$$

where the term on the right, but before the exponential, is the same as in Eq. 5.4-20. Alternatively, one can use

$$f^L(T, P) = P \left(\frac{f}{P}\right)_{T,P} \qquad (5.4\text{-}22)$$

where the fugacity coefficient is evaluated using the principle of corresponding states and Fig. 5.4-1.

c. Fugacity of a Pure Solid Phase

The extension of the previous discussion to solids is simple. In fact, if we recognize that a solid phase may undergo several phase transitions, and let $\underline{V}^J$ be the molar volume of the Jth phase, and P_J be the pressure above which this phase is stable at the temperature T, we have

$$f^S(T, P) = P^{sat}(T) \left(\frac{f}{P}\right)_{sat,T} \exp\left[\frac{1}{RT} \sum_{J=1} \int_{P_J}^{P_{J+1}} \underline{V}^J \, dP\right] \qquad (5.4\text{-}23)$$

Here P^{sat} is generally equal to sublimation pressure of the solid.[5] Since the sublimation (or vapor) pressure of a solid is generally small, so that the fugacity coefficient can be taken equal to unity, it is usually satisfactory to approximate Eq. 5.4-23 at low total pressures by

$$f^S(T, P) = P^{sat}(T) \qquad (5.4\text{-}24a)$$

or, for a solid subject to moderate or high total pressures, by

$$f^S(T, P) = P^{sat}(T) \exp\left[\frac{\underline{V}^S[P - P^{sat}(T)]}{RT}\right] \qquad (5.4\text{-}24b)$$

[5]Below the triple point temperature, as the pressure is reduced at constant temperature, a solid will sublimate directly to a vapor. However, near the triple point temperature some solids first melt to form liquids and then vaporize as the ambient pressure is reduced at constant temperature (see Fig. 5.3-6). In such cases $P^{sat}(T)$ in Eq. 5.4-23 is taken to be the vapor pressure of the liquid.

5.5 THE CALCULATION OF PURE FLUID PHASE EQUILIBRIUM. THE COMPUTATION OF VAPOR PRESSURE FROM AN EQUATION OF STATE

Now that the fugacity (or equivalently, the molar Gibbs free energy) of a pure fluid can be calculated, it is instructive to consider how one can compute the vapor-liquid equilibrium pressure of a pure fluid, that is, the vapor pressure, as a function of temperature, using a volumetric equation of state. Such calculations are straightforward in principle, but, because of the iterative nature of the computations involved, is best done on a computer or microcomputer. One starts by choosing a temperature between the melting point and critical point of the fluid of interest and guessing what the vapor pressure would be. Next, for these values of T and P, the equation of state is solved to find the liquid (smaller $\underline{V}$ and Z) and the vapor (larger $\underline{V}$ and Z) roots, which we denote by Z^{L} and Z^{V}, respectively. These values are then used in the fugacity coefficient expression appropriate to the equation of state to obtain f^{L} and f^{V}. Finally, these values are compared. If f^{L} is, within a specified tolerance, equal to f^{V}, the guessed pressure is the correct vapor pressure for the temperature of interest. If, however, the liquid phase fugacity is greater than the vapor phase fugacity (i.e., $f^{L} > f^{V}$), the guessed pressure is too low; it is too high if $f^{V} > f^{L}$. In either case, a new guess must be made for the pressure and the calculation repeated.

Figure 5.5-1 is a flow diagram for the calculation of the vapor pressure using the Peng–Robinson equation of state (this calculation is contained in the program that appears in Appendix A4.1); clearly other equations of state could have been used. Also, the algorithm could be modified slightly so that a pressure is chosen, and the boiling temperature at this pressure is found (in this case remember that, from Eq. 4.7-1, the a parameter in the equation of state is temperature dependent).

Figure 5.5-2 contains experimental vapor pressure versus temperature data for n-butane, together with vapor pressure predictions from (1) the van der Waals equation, (2) the Peng–Robinson equation but with $a = 0.45724\, R^2T_c^2/P_c$ rather than the correct expression of Eq. 4.7-1, and (3) the complete Peng–Robinson equation of state. From this figure we see that the vapor pressure predictions of the van der Waals equation are not very good, nor are the predictions of the simplified Peng–Robinson equation with the a parameter independent of temperature. However, the predictions with the complete Peng–Robinson equation are excellent. Indeed, the specific form of the temperature dependence of the $\alpha(T)$ term in the a parameter was chosen so that good vapor pressure predictions would be obtained at all temperatures, and so that $\alpha(T_c) = 1$ to ensure that the critical-point conditions are met.

Finally, it should be pointed out that in the calculation scheme suggested here, the initial guess for the vapor pressure at the chosen temperature (or temperature at fixed pressure) must be made with some care. In particular, the pressure (or temperature) must be within the range of the van der Waals loop of the equation of state, so that separate solutions for the vapor and liquid densities (or compressibilities) are obtained. If the guessed pressure is either too high so that only the high density root exists, or it is too low so that only the low density root exists, the presumed vapor and liquid phases will be identical, and the algorithm of Fig. 5.5-1 will accept the guessed pressure as the vapor pressure,

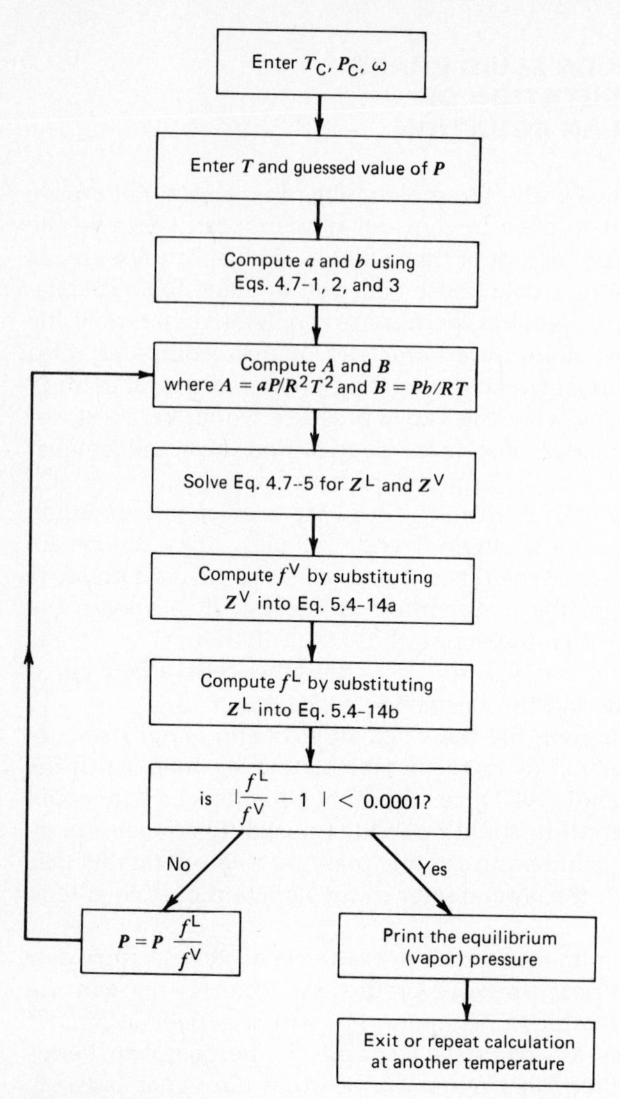

Figure 5.5-1
Flow sheet of a computer program for the calculation of the vapor pressure to a fluid using the Peng–Robinson equation of state.

even though it is not the correct solution to the problem. The "solution" so obtained is referred to as the **trivial solution** (in which both phases are identical) rather than the actual solution to the problem in which different vapor and liquid phases exist. The best method of avoiding this difficulty is to make a very good initial guess, though this is increasingly harder to do as one approaches the critical conditions of the fluid, and the van der Waals loop becomes very small (remember, it vanishes at the critical point).

Of course, using an equation of state, not only can the vapor pressure of a fluid be calculated, but so can other thermodynamic properties along the vapor–liquid phase boundary. This is demonstrated in the following illustration, which is a continuation of Illustration 4.4-1 dealing with the thermodynamic properties of oxygen.

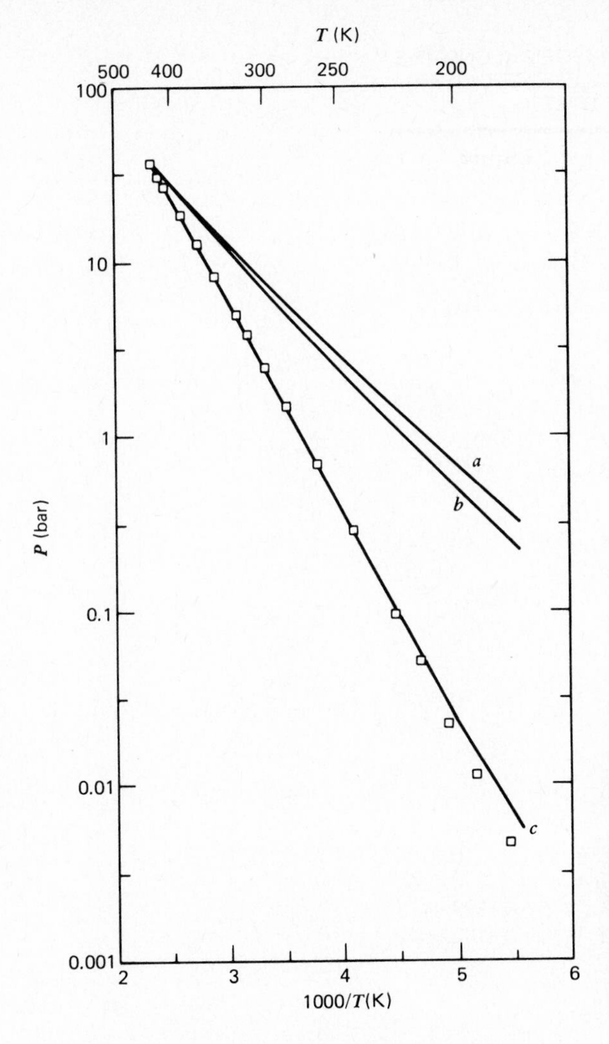

Figure 5.5-2
The vapor pressure of *n*-butane as a function of temperature. The points □ are the experimental data. Line *a* is the prediction of the van der Waals equation. Line *b* is the prediction of the Peng–Robinson equation with $\alpha = 1$, and line *c* is the prediction of the complete Peng–Robinson equation [i.e., $\alpha = \alpha(t)$].

ILLUSTRATION 5.5-1 (Illustration 4.4-1 continued)
Using the data in Illustration 4.4-1, and the same reference state, compute the vapor pressure of oxygen over the temperature range of −200°C to the critical temperature, and also compute the specific volume, enthalpy, and entropy along the vapor–liquid equilibrium phase envelope. Add these results to Figs. 4.4-3, 4, and 5.

Solution
Using the algorithm discussed in this section, and the computer program in Appendix A4.1, we obtain the results in Table 5.5-1. The vapor pressure as a function of temperature is plotted in Fig. 5.5-3. The specific volumes and molar enthalpies and entropies of the coexisting phases have been added as the two-phase envelopes in Figs. 4.4-3, 4, and 5. ■

ILLUSTRATION 5.5-2 (Illustration 4.4-1 concluded)
Complete the calculated thermodynamic properties chart for oxygen by considering temperatures between −100°C and −200°C.

Table 5.5-1
THERMODYNAMIC PROPERTIES OF OXYGEN ALONG THE VAPOR–LIQUID PHASE BOUNDARY CALCULATED USING THE PENG–ROBINSON EQUATION OF STATE

T(C)		Vapor	Liquid
	$P = 0.11$ bar		
	Z	0.9945	0.0004
−200	$\underline{V}$	53.3256	0.0232
	$\underline{H}$	−6311.60	−13501.04
	$\underline{S}$	−20.87	−119.14
	$P = 0.47$ bar		
	Z	0.9832	0.0016
−190	$\underline{V}$	14.5751	0.0241
	$\underline{H}$	−6064.97	−13013.02
	$\underline{S}$	−29.34	−112.90
	$P = 1.39$ bar		
	Z	0.9611	0.0045
−180	$\underline{V}$	5.3617	0.0252
	$\underline{H}$	−5842.71	−12517.98
	$\underline{S}$	−35.65	−107.30
	$P = 2.19$ bar		
	Z	0.9452	0.0070
−175	$\underline{V}$	3.5151	0.0259
	$\underline{H}$	−5744.89	−12261.51
	$\underline{S}$	−38.25	−104.64
	$P = 3.31$ bar		
	Z	0.9256	0.0103
−170	$\underline{V}$	2.3964	0.0266
	$\underline{H}$	−5658.53	−11997.44
	$\underline{S}$	−40.60	−102.05

Solution

The calculation here is much like that of Illustration 4.4-1 *except* that for some pressures at temperatures below the critical temperature, three solutions for the compressibility or specific volume may be obtained. To choose the correct solution in such cases, the vapor pressure calculated in the previous illustration is needed. If three solutions are obtained and the system pressure is above the vapor pressure, the smallest compressibility (or specific volume) is the correct solution, and it should be used in the calculation of all other thermodynamic properties. Conversely, if the system pressure is below the calculated vapor pressure, the largest compressibility or specific volume root is to be used.

Using this calculational procedure, the entries in Table 5.5-2 were obtained. These values also appear in Figs. 4.4-3, 4, and 5.

Note

This completes the calculation of the thermodynamic properties of oxygen. You should remember, however, that these results were obtained using only a simple, generalized three-parameter (T_C, P_C, ω) cubic equation of state. Therefore, although good estimates, the results are not of high accuracy as would be ob-

Table 5.5-1 *(Continued)*

T(C)		Vapor	Liquid
	P = 6.76 bar		
	Z	0.8746	0.0204
−160	$\underline{V}$	1.2175	0.0284
	$\underline{H}$	−5529.07	−11440.52
	$\underline{S}$	−44.75	−96.99
	P = 12.30 bar		
	Z	0.8062	0.0371
−150	$\underline{V}$	0.6713	0.0309
	$\underline{H}$	−5475.94	−10829.37
	$\underline{S}$	−48.49	−91.95
	P = 20.54 bar		
	Z	0.7170	0.0641
−140	$\underline{V}$	0.3864	0.0346
	$\underline{H}$	−5534.19	−10131.95
	$\underline{S}$	−52.20	−86.72
	P = 32.15 bar		
	Z	0.5983	0.1107
−130	$\underline{V}$	0.2215	0.0410
	$\underline{H}$	−5780.50	−9273.89
	$\underline{S}$	−56.43	−80.84
	P = 39.44 bar		
	Z	0.5191	0.1501
−125	$\underline{V}$	0.1621	0.0469
	$\underline{H}$	−6043.66	−8716.50
	$\underline{S}$	−59.19	−77.23
	P = 47.85 bar		
	Z	0.4000	0.2253
−120	$\underline{V}$	0.1065	0.0600
	$\underline{H}$	−6599.85	−7885.19
	$\underline{S}$	−63.62	−72.01

$\underline{V}$ [=] m^3/kmol $\underline{H}$ [=] J/mol = kJ/kmol
$\underline{S}$ [=] J/mol K = kJ/kmol K

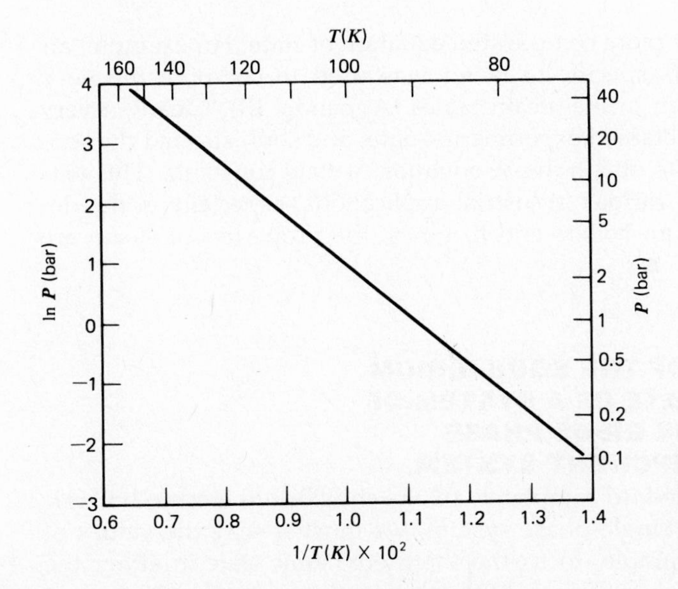

Figure 5.5-3
The vapor pressure of oxygen calculated using the Peng–Robinson equation of state.

Table 5.5-2
THE THERMODYNAMIC PROPERTIES OF OXYGEN IN THE LOW TEMPERATURE RANGE CALCULATED USING THE PENG–ROBINSON EQUATION OF STATE

T(C)	−125	−150	−175	−200
$P = 1$ bar				
Z	0.9915	0.9863	0.9756	0.0038
$\underline{V}$	12.2138	10.0988	7.9621	0.0232
$\underline{H}$	−4302.26	−4995.07	−5684.10	−13491.36
$\underline{S}$	−19.97	−25.09	−31.35	−119.04
$P = 2$ bar				
Z	0.9830	0.9723	0.9502	0.0076
$\underline{V}$	6.0543	4.9777	3.8775	0.0232
$\underline{H}$	−4330.39	−5031.67	−5734.76	−13489.39
$\underline{S}$	−25.85	−31.04	−37.42	−119.04
$P = 5$ bar				
Z	0.9569	0.9285	0.0158	0.0191
$\underline{V}$	2.3574	1.9014	0.0258	0.0232
$\underline{H}$	−4416.92	−5146.94	−12257.99	−13484.22
$\underline{S}$	−33.84	−39.24	−104.68	−119.07
$P = 10$ bar				
Z	0.9116	0.8478	0.0316	0.0382
$\underline{V}$	1.1229	0.8681	0.2580	0.0232
$\underline{H}$	−4569.30	−5362.75	−12251.78	−13475.52
$\underline{S}$	−40.27	−46.14	−104.75	−119.11
$P = 20$ bar				
Z	0.8119	0.0598	0.0630	0.0762
$\underline{V}$	0.5000	0.03063	0.0257	0.0232
$\underline{H}$	−4915.26	−10834.90	−12239.17	−13458.10
$\underline{S}$	−47.61	−92.19	−104.88	−119.19
$P = 30$ bar				
Z	0.6912	0.0889	0.0943	0.1142
$\underline{V}$	0.2838	0.0304	0.0256	0.0232
$\underline{H}$	−5356.44	−10840.08	−12226.32	−13440.62
$\underline{S}$	−53.14	−92.48	−105.01	−119.26

tained using a considerably more complicated equation of state. For example, an equation with 59 constants specific to water was used to compute the very accurate properties of steam in the steam tables (Appendix III). Clearly a very large amount of carefully obtained experimental data, and sophisticated numerical analysis, must be used to obtain the 59 equation of state constants. This was done for water because in various industrial applications, especially in the design and evaluation of steam boilers and turbines, the properties of steam are needed to high accuracy. ■

5.6 THE SPECIFICATION OF THE EQUILIBRIUM THERMODYNAMIC STATE OF A SYSTEM OF SEVERAL PHASES. THE GIBBS PHASE RULE FOR A ONE-COMPONENT SYSTEM.

As we have already indicated, to completely fix the equilibrium thermodynamic state of a one-component, single-phase system, we must specify the values of two state variables. For example, to fix the thermodynamic state in either the vapor, liquid, or solid regions of Fig. 5.3-6, both the temperature and pressure

Table 5.5-2 *(Continued)*

T(C)	−125	−150	−175	−200
P = 40 bar				
Z	0.1512	0.1176	0.1254	0.1521
$\underline{V}$	0.0466	0.0301	0.0256	0.0231
$\underline{H}$	−8734.00	−10843.32	−12213.23	−13423.11
$\underline{S}$	−77.37	−92.75	−105.14	−119.34
P = 50 bar				
Z	0.1740	0.1458	0.1563	0.1899
$\underline{V}$	0.0429	0.0299	0.0255	0.0231
$\underline{H}$	−8937.76	−10844.89	−12199.92	−13405.54
$\underline{S}$	−79.04	−93.01	−105.27	−119.42
P = 60 bar				
Z	0.1987	0.1737	0.1871	0.2277
$\underline{V}$	0.0408	0.0296	0.0256	0.0231
$\underline{H}$	−9054.31	−10845.02	−12186.40	−13387.93
$\underline{S}$	−80.11	−93.25	−105.39	−119.49
P = 70 bar				
Z	0.2235	0.2013	0.2178	0.2654
$\underline{V}$	0.0393	0.0294	0.0254	0.0231
$\underline{H}$	−9134.97	−10843.88	−12172.67	−13370.28
$\underline{S}$	−80.92	−93.48	−105.51	−119.57
P = 80 bar				
Z	0.2481	0.2285	0.2483	0.3030
$\underline{V}$	0.0382	0.0293	0.0253	0.0230
$\underline{H}$	−9195.27	−10841.61	−12158.77	−13352.59
$\underline{S}$	−81.59	−93.70	−105.62	−119.64
P = 90 bar				
Z	0.2724	0.2555	0.2787	0.3405
$\underline{V}$	0.0373	0.0291	0.0253	0.0230
$\underline{H}$	−9242.27	−10838.35	−12144.68	−13334.86
$\underline{S}$	−82.16	−93.91	−105.74	−119.71
P = 100 bar				
Z	0.2964	0.2823	0.3089	0.3780
$\underline{V}$	0.0365	0.0289	0.0252	0.0230
$\underline{H}$	−9279.82	−10834.20	−12130.43	−13317.09
$\underline{S}$	−82.67	−94.11	−105.85	−119.78

$\underline{V}$ [=] m³/kmol $\underline{H}$ [=] J/mol = kJ/kmol
$\underline{S}$ [=] J/mol K = kJ/kmol K

are needed. Thus, we say that a one-component, single-phase system has two **degrees of freedom**. In addition, to fix the total size or extent of the system we must also specify its mass or one of its extensive properties such as total volume or total energy from which the mass can be calculated.

In this section we are interested in determining the amount of information, and its type, that must be specified to completely fix the thermodynamic state of an equilibrium single-component, multiphase system. That is, we are interested in obtaining answers to the following questions:

1. How many state variables must be specified to completely fix the thermodynamic state of each phase when several phases are in equilibrium (i.e., how many degrees of freedom are there in a single-component multiphase system)?

2. How many additional variables need be specified, and what type of variable should they be, to fix the distribution of mass (or number of moles) between the phases, and thereby fix the overall molar properties of the composite, multiphase system?
3. What additional information is needed to fix the total size of the multiphase system?

To specify the thermodynamic state of any one phase of a single-component, multiphase system, two thermodynamic state variables of that phase must be specified; that is, each phase has two degrees of freedom. Thus, it might appear that if $\mathcal{P}$ phases are present, the system would have $2\mathcal{P}$ degrees of freedom. The actual number of degrees of freedom is considerably less, since the requirement that the phases be in equilibrium puts certain constraints on the values of the state variables in each phase. For example, from the analysis of Secs. 5.1 and 5.2 it is clear that at equilibrium the temperature in each phase must be the same. Thus, there are $\mathcal{P} - 1$ relations of the form

$$T^{\text{I}} = T^{\text{II}}$$
$$T^{\text{I}} = T^{\text{III}}$$
$$\text{etc.}$$

which must be satisfied. Similarly, at equilibrium the pressure in each phase must be the same, so that there are an additional $\mathcal{P} - 1$ restrictions on the state variables of the form

$$P^{\text{I}} = P^{\text{II}}$$
$$P^{\text{I}} = P^{\text{III}}$$
$$\text{etc.}$$

Finally, at equilibrium, the molar Gibbs free energies must be the same in each phase, so that

$$\underline{G}^{\text{I}}(T, P) = \underline{G}^{\text{II}}(T, P)$$
$$\underline{G}^{\text{I}}(T, P) = \underline{G}^{\text{III}}(T, P)$$
$$\text{etc.}$$

which provide an additional $\mathcal{P} - 1$ restrictions on the phase variables.

Since there are a total of $3(\mathcal{P} - 1)$ restrictions on the $2\mathcal{P}$ state variables needed to fix the thermodynamic state of each of the $\mathcal{P}$ phases, the number of degrees of freedom for the single-component multiphase system is

$$\begin{aligned}\mathcal{F} = \begin{array}{l}\text{number of degrees}\\ \text{of freedom}\end{array} &= \begin{pmatrix}\text{number of state}\\ \text{variables needed}\\ \text{to fix the state}\\ \text{of the } \mathcal{P} \text{ phases}\end{pmatrix} - \begin{pmatrix}\text{the restrictions on}\\ \text{these state variables}\\ \text{as a result of the}\\ \text{phases being in}\\ \text{equilibrium}\end{pmatrix} \\ &= 2\mathcal{P} - 3(\mathcal{P} - 1) \\ &= 3 - \mathcal{P} \end{aligned} \tag{5.6-1}$$

Thus, the specification of $3 - \mathcal{P}$ state variables of the individual phases is all that is needed, in principle, to completely fix the thermodynamic state of each of the phases in a one-component system. Of course, to fix the thermodynamic states of the phases in fact, we would need appropriate equation of state information.

It is easy to demonstrate that Eq. 5.6-1 is in agreement with our experience and with the phase diagram of Fig. 5.3-6. For example, we have already indicated that a single-phase ($\mathcal{P} = 1$) system has two degrees of freedom; this follows immediately from Eq. 5.6-1. To specify the thermodynamic state of each phase in a two-phase system (i.e., vapor–liquid, vapor–solid, or solid–liquid coexistence regions), it is clear from Fig. 5.3-6 that we need specify only the temperature *or* the pressure of the system; the value of the other variable can then be obtained from the appropriate coexistence curve. Setting $\mathcal{P}$ equal to 2 in Eq. 5.6-1 confirms that a two-phase system has only a single degree of freedom. Finally, since the solid, liquid, and vapor phases coexist at only a single point, the triple point, a single-component, three-phase system has no degrees of freedom. This also follows from Eq. 5.6-1 with $\mathcal{P}$ equal to 3.

The character of the variable to be specified as a degree of freedom is not completely arbitrary. To see this consider Eq. 5.3-1a, which gives the molar volume of a two-phase mixture as a function x^{V} and the two single-phase molar volumes. Clearly, a specification of either $\underline{V}^{\mathrm{V}}$ or $\underline{V}^{\mathrm{L}}$ is sufficient to fix the thermodynamic state of both phases because the two-phase system has only one degree of freedom. However, the specification of the two-phase molar volume $\underline{V}$ can be satisfied by a range of choices of temperatures along the coexistence curve by suitably adjusting x^{V}, so that $\underline{V}$ or any other molar property of the two phases combined is not suitable for a degree of freedom specification. Consequently, to fix the thermodynamic state of each of $\mathcal{P}$ phases in equilibrium, we must specify $3 - \mathcal{P}$ properties of the *individual phases*.

Next we want to consider how many variables, and of what type, must be specified to also fix the distribution of mass or number of moles between the phases, so that the molar thermodynamic properties of the system consisting of several phases in equilibrium can be determined. If there are $\mathcal{P}$ phases, there are $\mathcal{P}$ values of x^i, the mass fraction in phase i, which must be determined. Since the mass fractions must sum to unity, we have

$$x^{\mathrm{I}} + x^{\mathrm{II}} + \cdots = \sum_{i=1}^{\mathcal{P}} x^i = 1 \tag{5.6-2}$$

as one of the relations between the mass fractions. Thus, $\mathcal{P} - 1$ additional equations of the form

$$\sum_{i=1}^{\mathcal{P}} x^i \hat{V}^i = \hat{V}$$

$$\sum_{i=1}^{\mathcal{P}} x^i \hat{H}^i = \hat{H} \tag{5.6-3a}$$

or generally

$$\sum_{i=1}^{\mathcal{P}} x^i \hat{\theta}^i = \hat{\theta} \tag{5.6-3b}$$

are needed. In these equations, $\hat{\theta}^i$ is the property per unit mass in phase i, and $\hat{\theta}$ is the property per unit mass of the multiphase mixture.

From these equations, it is evident that to determine the mass distribution between the phases, we need to specify a sufficient number of variables of the individual phases to fix the thermodynamic state of each phase (i.e., the degrees of freedom $\mathcal{F}$) and $\mathcal{P} - 1$ thermodynamic properties of the multiphase system in the form of Eq. 5.6-3. For example, if we know that steam and water are in equilibrium at some temperature T (which fixes the single-degree freedom of this two-phase system), the equation of state or the steam tables can be used to obtain the equilibrium pressure, specific enthalpy, entropy, and volume of each of the phases, but not the mass distribution between the phases. If, in addition, the volume (or enthalpy or entropy, etc.) per unit mass of the two-phase mixture were known, this would be sufficient to determine the distribution of mass between the two phases, and then all the other overall thermodynamic properties.

Once the thermodynamic properties of all the phases are fixed (by the specification of the $\mathcal{F} = 3 - \mathcal{P}$ degrees of freedom), and the distribution of mass determined (by the specification of an additional $\mathcal{P} - 1$ specific properties of the multiphase system), the value of any one extensive variable (total volume, total enthalpy, etc.) of the multiphase system is sufficient to determine the total mass and all other extensive variables of the multiphase system.

Thus, to determine the thermodynamic properties per unit mass of a single-component, two-phase mixture we need specify the equivalent of one single-phase state variable (the one degree of freedom) and one variable that provides information on the mass distribution. The additional specification of one extensive property is needed to determine the total size of the system. Similarly, to fix the thermodynamic properties of a single-component, three-phase mixture, we need not specify any single state variable (since the triple point is unique), but two variables that provide information on the distribution of mass between the vapor, liquid, and solid phases and one extensive variable to determine the total mass of the multiphase system.

ILLUSTRATION 5.6-1

a. Show, using the steam tables, that fixing the equilibrium pressure of a steam–water mixture at 1.0135 bar is sufficient to completely fix the thermodynamic states of each phase. (This is an experimental verification of the fact that a one-component, two-phase system has only one degree of freedom).

b. Show that additionally fixing the specific volume of the two-phase system $\hat{V}$ at 1 m³/kg is sufficient to determine the distribution of mass between the two phases.

c. What is the total enthalpy of 3.2 m³ of this steam–water mixture?

Solution

a. Using the saturation steam tables of Appendix III, we see that fixing the pressure at 1.0135 bar is sufficient to determine the thermodynamic properties of each phase:

$$T = 100°\text{C}$$

$$\hat{V}^{\text{L}} = 0.001044 \text{ m}^3/\text{kg} \qquad \hat{V}^{\text{V}} = 1.6729 \text{ m}^3/\text{kg}$$

$$\hat{H}^{\text{L}} = 419.04 \text{ kJ/kg} \qquad \hat{H}^{\text{V}} = 2676.1 \text{ kJ/kg}$$

$$\hat{S}^{\text{L}} = 1.3069 \text{ kJ/kg K} \qquad \hat{S}^{\text{V}} = 7.3549 \text{ kJ/kg K}$$

Alternatively, specifying only the temperature, the specific volume, the specific enthalpy, or in fact any other intensive variable of one of the phases would be sufficient to fix the thermodynamic properties of both phases. However, a specification of only the system pressure, temperature, or any one-phase state variable is not sufficient to determine the relative amounts of the vapor and liquid phases.

b. To determine the relative amounts of each of the phases we need information from which x^V and x^L can be determined. Here the specific volume of the two-phase mixture is given, so we can use Eq. 5.3-1

$$\hat{V} = x^V \hat{V}^V + (1 - x^V)\hat{V}^L$$

and the relation

$$x^V + x^L = 1$$

to find the distribution of mass between the two phases. For the situation here

$$1 \text{ m}^3/\text{kg} = x^V (0.001044 \text{ m}^3/\text{kg}) + (1 - x^V)(1.6729 \text{ m}^3/\text{kg})$$

so that

$$x^V = 0.4025$$

and

$$x^L = 0.5975$$

c. Using the data in the problem statement, we have that the total mass of the steam–water mixture is

$$M = \frac{V}{\hat{V}} = \frac{3.2 \text{ m}^3}{1.0 \text{ m}^3/\text{kg}} = 3.2 \text{ kg}$$

From the results in parts (a) and (b) we can compute the enthalpy per unit mass of the two-phase mixture

$$\begin{aligned} \hat{H} &= x^L \hat{H}^L + x^V \hat{H}^V \\ &= 0.5975(419.04) + 0.4025(2676.1) = 1327.5 \text{ kJ/kg} \end{aligned}$$

Therefore

$$H = M\hat{H} = 3.2 \times 1327.5 = 4248 \text{ kJ} \quad \blacksquare$$

5.7 THERMODYNAMIC PROPERTIES OF PHASE TRANSITIONS

In this section several general properties of phase transitions are considered, as well as a phase transition classification system. The discussion and results of this section are applicable to all phase transitions (i.e., liquid–solid, solid–solid, vapor–solid, vapor–liquid, etc.), although special attention is given to vapor–liquid equilibrium.

One phase transition property important to chemists and engineers is the slope of the coexistence curves in the P–T plane; the slope of the vapor–liquid equilibrium line gives the rate of change of the vapor pressure of the liquid with temperature, the slope of the vapor–solid coexistence line is equal to the change with temperature of the vapor pressure of the solid (called the **sublimation pressure**), and the inverse of the slope of the liquid–solid coexistence line gives the change of the melting temperature of the solid with pressure. The slope of either of these phase equilibrium coexistence curves can be found by starting with Eq. 5.2-15c

$$\underline{G}^{\text{I}}(T, P) = \underline{G}^{\text{II}}(T, P) \tag{5.7-1}$$

where the superscripts label the phases. For a small change in the equilibrium temperature dT, the corresponding change of the coexistence pressure dP (i.e., the change in pressure following the coexistence curve) can be computed from the requirement that since Eq. 5.7-1 must be satisfied all along the coexistence curve, the changes in Gibbs free energies in the two phases corresponding to the temperature and pressure changes must be equal, that is

$$d\underline{G}^{\text{I}} = d\underline{G}^{\text{II}} \tag{5.7-2}$$

Using Eq. 4.2-8b in Eq. 5.7-2 gives

$$\underline{V}^{\text{I}}\, dP - \underline{S}^{\text{I}}\, dT = \underline{V}^{\text{II}}\, dP - \underline{S}^{\text{II}}\, dT$$

or, since P and T are the same in both phases for this equilibrium system (see Sec. 5.2)

$$\left(\frac{\partial P}{\partial T}\right)_{\underline{G}^{\text{I}}=\underline{G}^{\text{II}}} = \left(\frac{\underline{S}^{\text{I}} - \underline{S}^{\text{II}}}{\underline{V}^{\text{I}} - \underline{V}^{\text{II}}}\right) = \frac{\Delta \underline{S}}{\Delta \underline{V}} \tag{5.7-3}$$

From Eq. 5.7-1 we also have

$$\underline{G}^{\text{I}} = \underline{H}^{\text{I}} - T\underline{S}^{\text{I}} = \underline{G}^{\text{II}} = \underline{H}^{\text{II}} - T\underline{S}^{\text{II}}$$

or

$$\underline{S}^{\text{I}} - \underline{S}^{\text{II}} = \frac{\underline{H}^{\text{I}} - \underline{H}^{\text{II}}}{T}$$

so that Eq. 5.7-3 can be rewritten as

$$\left(\frac{\partial P^{\text{sat}}}{\partial T}\right)_{\underline{G}^{\text{I}}=\underline{G}^{\text{II}}} = \frac{\Delta \underline{S}}{\Delta \underline{V}} = \frac{\Delta \underline{H}}{T\Delta \underline{V}} \tag{5.7-4}$$

where $\Delta\underline{\theta} = \underline{\theta}^{\text{I}} - \underline{\theta}^{\text{II}}$, and we have used P^{sat} to denote the equilibrium coexistence pressure.[6] Equation 5.7-4 is the **Clapeyron equation**; it relates the slope of the coexistence curve to the enthalpy and volume changes at phase transition.

Figure 5.3-6 is, in many ways, a typical phase diagram. From this figure and Eqs. 5.7-3 and 4 one can make several observations about property changes at phase transitions. First, since none of the coexistence lines has zero slope, neither the entropy change nor the enthalpy change is equal to zero for solid–liquid, liquid–vapor, or solid–vapor phase transitions. Also, since the coexistence lines do not have infinite slope, $\Delta\underline{V}$ is not generally equal to zero. (In general, both the heat of fusion, $\Delta\underline{H}^{\text{fus}} = \underline{H}^{\text{L}} - \underline{H}^{\text{S}}$ and the volume change on melting $\Delta\underline{V}^{\text{fus}} = \underline{V}^{\text{L}} - \underline{V}^{\text{S}}$ are greater than zero for the liquid–solid transition, so that the liquid–solid coexistence line is as shown in the figure. One exception to this is water, for which $\Delta\underline{H}^{\text{fus}} > 0$ but $\Delta\underline{V}^{\text{fus}} < 0$, so the ice–water coexistence line has a negative slope.) From Sec. 5.3 we know that at the fluid critical point the coexisting phases are indistinguishable. Therefore, we can conclude that $\Delta\underline{H}$,

[6] P^{vap} will be used to denote the vapor pressure of the liquid, P^{sub} to denote the vapor pressure of the solid, and P^{sat} to designate a general equilibrium coexistence pressure; equations containing P^{sat} are applicable to both the vapor and sublimation pressures.

$\Delta \underline{V}$, and $\Delta \underline{S}$ are all nonzero away from the fluid critical point and approach zero as the critical point is approached.

Of particular interest is the application of Eq. 5.7-4 to the vapor–liquid coexistence curve because this gives the change in vapor pressure with temperature. At temperatures where the vapor pressure is not very high, it is found that $\underline{V}^{\text{V}} \gg \underline{V}^{\text{L}}$, and $\Delta \underline{V} \approx \underline{V}^{\text{V}}$. If, in addition, the vapor phase is ideal, we have $\Delta \underline{V}^{\text{vap}} = \underline{V}^{\text{V}} = RT/P$, so that

$$\frac{dP^{\text{vap}}}{dT} = \frac{P^{\text{vap}} \Delta \underline{H}^{\text{vap}}}{RT^2} \quad \text{or} \quad \frac{d \ln P^{\text{vap}}}{dT} = \frac{\Delta \underline{H}^{\text{vap}}}{RT^2} \tag{5.7-5a}$$

and

$$\ln \frac{P^{\text{vap}}(T_2)}{P^{\text{vap}}(T_1)} = \int_{T_1}^{T_2} \frac{\Delta \underline{H}^{\text{vap}}}{RT^2}\, dT \tag{5.7-5b}$$

which relates the fluid vapor pressures at two different temperatures to the heat of vaporization, $\Delta \underline{H}^{\text{vap}} = \underline{H}^{\text{V}} - \underline{H}^{\text{L}}$. Equation 5.7-5a is referred to as the **Clausius–Clapeyron equation**. If the heat of vaporization is assumed to be independent of temperature, Eq. 5.7-5 can be integrated to give

$$\ln \frac{P^{\text{vap}}(T_2)}{P^{\text{vap}}(T_1)} = -\frac{\Delta \underline{H}^{\text{vap}}}{R}\left(\frac{1}{T_2} - \frac{1}{T_1}\right) \tag{5.7-6}$$

a result that is also valid over small temperature ranges even when $\Delta \underline{H}^{\text{vap}}$ is temperature dependent. Equation 5.7-6 has been found to be fairly accurate for correlating the temperature dependence of the vapor pressure of liquids over limited temperature ranges.

ILLUSTRATION 5.7-1

The vapor pressure of liquid 2,2,4-trimethyl pentane at various temperatures is given here. Estimate the heat of vaporization of this compound at 25°C.

Vapor pressure (kPa)	0.667	1.333	2.666	5.333	8.000	13.33	26.66	53.33	101.32
Temperature (°C)	−15.0	−4.3	7.5	20.7	29.1	40.7	58.1	78.0	99.2

Solution

Over a relatively small range of temperature (say from 20.7 to 29.1°C), $\Delta \underline{H}^{\text{vap}}$ may be taken to be constant. Using Eq. 5.7-6 we obtain

$$\frac{\Delta \underline{H}^{\text{vap}}}{R} = \frac{-\ln[P^{\text{vap}}(T_2)/P^{\text{vap}}(T_1)]}{\dfrac{1}{T_2} - \dfrac{1}{T_1}} = \frac{-\ln(8.000/5.333)}{\dfrac{1}{302.25} - \dfrac{1}{293.85}} = 4287.8 \text{ K}$$

so that

$$\Delta \underline{H}^{\text{vap}} = 35.649 \text{ kJ/mol}$$

One can obtain an estimate of the temperature variation of the heat of vaporization by noting that the integration of Eq. 5.7-5 can be carried out as an indefinite rather than definite integral. In this case we obtain

$$\ln P^{\text{vap}} = -\frac{\Delta \underline{H}^{\text{vap}}}{RT} + C$$

where C is a constant. Therefore, if we were to plot $\ln P^{\text{vap}}$ versus $1/T$, we should get a straight line with a slope equal to $-\Delta \underline{H}^{\text{vap}}/R$ if the heat of vaporization is independent of temperature, and a curve if $\Delta \underline{H}^{\text{vap}}$ varies with temperature. Figure 5.7-1 is a vapor pressure–temperature plot for the 2,2,4-trimethyl pentane system. As is evident, $\Delta \underline{H}^{\text{vap}}$ is virtually constant over the whole temperature range. ■

This illustration is a nice example of the utility of thermodynamics in providing interrelationships between properties. In this case we see how data on the temperature dependence of the vapor pressure of a fluid can be used to determine its heat of vaporization.

The equation developed in the illustration may be rewritten as

$$\ln P^{\text{vap}} = A - \frac{B}{T} \tag{5.7-7}$$

with $B = \Delta \underline{H}^{\text{vap}}/R$ is reasonably good for estimating the vapor pressure over small temperature ranges. More commonly the **Antoine equation**

$$\ln P^{\text{vap}} = A - \frac{B}{T + C} \tag{5.7-8}$$

is used to correlate vapor pressures accurately over the range from 1 to 200 kPa. Antoine constants for many substances are given in the book by Reid, Prausnitz, and Poling.[7] Other commonly used vapor pressure correlations include the

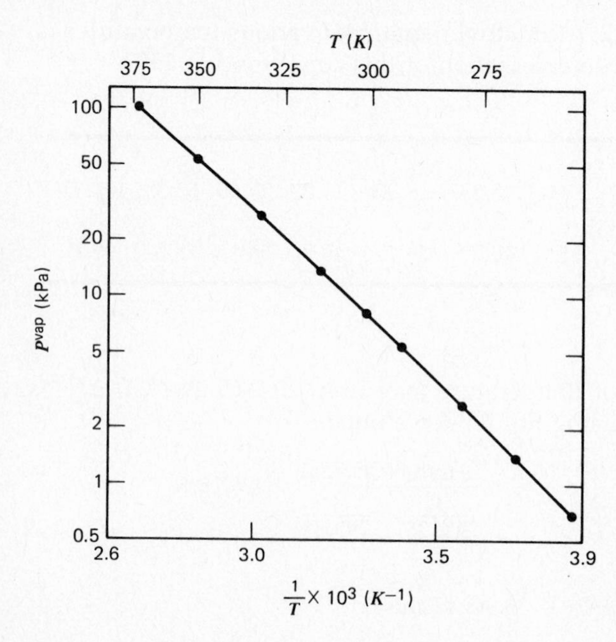

Figure 5.7-1 The vapor pressure of 2,2,4-trimethyl pentane as a function of temperature.

[7]R. C. Reid, J. M. Prausnitz, and B. E. Poling, *The Properties of Gases and Liquids*, 4th ed., McGraw-Hill, New York, 1987.

Riedel equation

$$\ln P^{\text{vap}} = A + \frac{B}{T} + C \ln T + DT^6 \tag{5.7-9}$$

and the **Harlecher–Braun equation**

$$\ln P^{\text{vap}} = A + \frac{B}{T} + C \ln T + \frac{D\, P^{\text{vap}}}{T^2} \tag{5.7-10}$$

which must be solved iteratively for the vapor pressure, but is reasonably accurate from low vapor pressure up to the critical pressure.

An important characteristic of the class of phase transitions we have been considering so far is that, except at the critical point, certain thermodynamic properties are discontinuous across the coexistence line; that is, these properties have different values in the two coexisting phases. For example, for the phase transitions indicated in Fig. 5.3-6 there is an enthalpy change, an entropy change, and a volume change on crossing the coexistence line. Also, the constant pressure heat capacity $C_P = (\partial \underline{H}/\partial T)_P$ becomes infinite at a phase transition for a pure component, because temperature is continuous and the enthalpy is discontinuous across the coexistence line. In contrast, the Gibbs free energy is, by Eq. 5.2-15c, continuous at a phase transition.

These observations may be summarized by noting that for the phase transitions considered here, the Gibbs free energy is continuous across the coexistence curve, but its first derivatives

$$\left(\frac{\partial \underline{G}}{\partial T}\right)_P = -\underline{S} \quad \text{and} \quad \left(\frac{\partial \underline{G}}{\partial P}\right)_T = \underline{V}$$

are discontinuous, as are all higher derivatives. This class of phase transition is called a **first-order phase transition.** The concept of higher-order phase transitions follows naturally. A second-order phase transition (at constant T and P) is one in which $\underline{G}$ and its first derivatives are continuous, but derivatives of second order and higher are discontinuous. Third-order phase transition is defined in a similar manner.

One example of a second-order phase change is the structural rearrangement of quartz, where $\underline{G}$, $\underline{S}$, and $\underline{V}$ are continuous across the coexistence line, but the constant-pressure heat capacity, which is related to the second temperature derivative of the Gibbs free energy

$$\left(\frac{\partial^2 \underline{G}}{\partial T^2}\right)_P = -\left(\frac{\partial \underline{S}}{\partial T}\right)_P = -\frac{C_P}{T}$$

is not only discontinuous, but has a singularity at the phase change (see Fig. 5.7-2 on page 242). No phase transitions higher than second order have been observed in the laboratory.

PROBLEMS

5.1 By doing some simple calculations and plotting several graphs, one can verify some of the statements made in this chapter concerning phase equilibrium and phase transitions. All the calculations should be done using the steam tables.

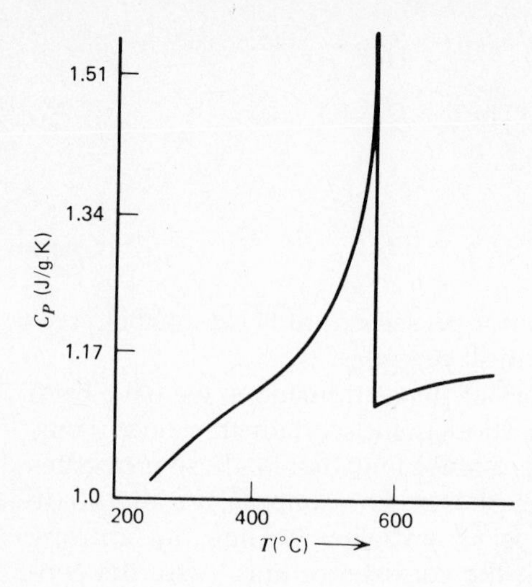

Figure 5.7-2
The specific heat of quartz near a second-order phase transition. (From H. B. Callen, *Thermodynamics*. © 1960 by John Wiley & Sons, Inc. Used with permission.)

a Establish, by direct calculation, that

$$\underline{G}^{L} = \underline{G}^{V}$$

for steam at 2.5 MPa and $T = 224°C$.

b Calculate $\underline{G}^{V}$ at $P = 2.5$ MPa for a collection of temperatures between 225 and 400°C and extrapolate this curve below 224°C.

c Find $\underline{G}^{L}$ at 160, 170, 180, 190, 200, and 210°C. Plot this result on the same graph as part b and extrapolate above 224°C. (*Hint*: For a liquid $\underline{H}$ and $\underline{S}$ can be taken to be independent of pressure. Therefore, the values of $\underline{H}$ and $\underline{S}$ for the liquid at any pressure can be gotten from the data for the saturated liquid at the same temperature). How does this graph compare with Fig. 5.3-7?

d Plot $\underline{V}$ versus T at $P = 2.5$ MPa for the temperature range of 150°C to 400°C, and show that $\underline{V}$ is discontinuous.

e Plot C_P versus T at $P = 2.5$ MPa over the temperature range of 150°C to 400°C; and thereby establish that C_P is discontinuous.

5.2 **a** Show that the condition for equilibrium in a closed system at constant entropy and volume is that the internal energy U achieve a minimum value subject to the constraints.

b Show that the condition for equilibrium in a closed system at constant entropy and pressure is that the enthalpy H achieve a minimum value subject to the constraints.

5.3 **a** Show that the intrinsic stability analysis for fluid equilibrium at constant temperature and volume leads to the single condition that

$$\left(\frac{\partial P}{\partial \underline{V}}\right)_T < 0$$

b Show that intrinsic stability analysis for fluid equilibrium at constant temperature and pressure does not lead to any restrictions on the equation of state.

5.4 **a** Show that the conditions for vapor–liquid equilibrium at constant N, T, and V are $\underline{G}^{V} = \underline{G}^{L}$ and $P^{V} = P^{L}$.

b Show that the condition for vapor–liquid equilibrium at constant N, T, and P is $\underline{G}^{V} = \underline{G}^{L}$.

5.5 Prove that $C_P \geqslant C_V$ for any fluid, and identify those conditions for which $C_P = C_V$.

5.6 Show that if the polymer fiber of Problem 4.26 is to be thermodynamically stable at all temperatures, the parameters α, β, and γ must be positive.

5.7 The entropy of a certain fluid has theoretically been found to be related to its internal energy and volume in the following way

$$\underline{S} = \underline{S}^{\circ} + \alpha \ln \frac{\underline{U}}{\underline{U}^{\circ}} + \beta \ln \frac{\underline{V}}{\underline{V}^{\circ}}$$

where $\underline{S}^{\circ}$, $\underline{U}^{\circ}$, and $\underline{V}^{\circ}$ are, respectively, the molar entropy, internal energy, and volume of the fluid in some appropriately chosen reference state, and α and β are positive constants.

a Develop an interrelationship between internal energy, temperature, and specific volume (the thermal equation of state) for this fluid.

b Develop an interrelationship between pressure, temperature, and volume (the volumetric equation of state) for this fluid.

c Show that this fluid does not have a first-order phase transition, by establishing that the fluid is stable in all thermodynamic states.

5.8 Figure 5.3-4 is the phase diagram for van der Waals fluid. Within the vapor–liquid coexistence envelope one can draw another envelope representing the limits of supercooling of the vapor and superheating of the liquid that can be observed in the laboratory; along each isotherm these are the points for which

$$\left(\frac{\partial P}{\partial \underline{V}}\right)_T = 0$$

Obtain this envelope for the van der Waals fluid. This is the spinodal curve.

The region between the coexistence curve and the curve just obtained is the metastable region of the fluid. Notice also that the critical point of the fluid is metastable.

5.9 Derive the following two independent equations for a second-order phase transition

$$\left.\frac{\partial P}{\partial T}\right|_{\substack{\text{along phase}\\ \text{transition}\\ \text{curve}}} = \frac{\underline{C}_P^{\mathrm{I}} - \underline{C}_P^{\mathrm{II}}}{T\left\{\left(\dfrac{\partial \underline{V}}{\partial T}\right)_P^{\mathrm{I}} - \left(\dfrac{\partial \underline{V}}{\partial T}\right)_P^{\mathrm{II}}\right\}}$$

and

$$\left.\frac{\partial P}{\partial T}\right|_{\substack{\text{along phase}\\ \text{transition}\\ \text{curve}}} = -\frac{\left\{\left(\dfrac{\partial \underline{V}}{\partial T}\right)_P^{\mathrm{I}} - \left(\dfrac{\partial \underline{V}}{\partial T}\right)_P^{\mathrm{II}}\right\}}{\left\{\left(\dfrac{\partial \underline{V}}{\partial P}\right)_T^{\mathrm{I}} - \left(\dfrac{\partial \underline{V}}{\partial P}\right)_T^{\mathrm{II}}\right\}} = \frac{\alpha^{\mathrm{I}} - \alpha^{\mathrm{II}}}{\kappa_T^{\mathrm{I}} - \kappa_T^{\mathrm{II}}}$$

These equations, which are the analogs of the Clausius–Clapeyron equation, are sometimes referred to as the Ehrenfest equations.

Also, show that these two equations can be derived by applying L'Hopitals' rule to the Clausius–Clapeyron equation for a first-order phase transition.

5.10 **a** The heat of fusion $\Delta \hat{H}^{fus}$ for the ice–water phase transition is 335 kJ/kg at 0°C and 1 bar. The density of water is 1000 kg/m^3 at these conditions, and that of ice is 915 kg/m^3. Develop an expression for the change of the melting temperature of ice with pressure.

b The heat of vaporization for the steam–water phase transition is 2255 kJ/kg at 100°C and 1 bar. Develop an expression for the change in the boiling temperature of water with pressure.

c Compute the freezing and boiling points of water in Denver, Colorado, where the mean atmosphere pressure is 84.6 kPa.

5.11 The triple point of iodine I_2 occurs at 112.9°C and 11.57 kPa. The heat of fusion at the triple point is 15.27 kJ/mol, and the following vapor pressure data are available for solid iodine:

Vapor pressure (kPa)	2.67	5.33	8.00
Temperature (°C)	84.7	97.5	105.4

Estimate the normal boiling temperature of molecular iodine.

5.12 The following data are available for water

$\ln P^{sub}(\text{ice}) = 28.8926 - 6140.1/T$	P in Pa
$\ln P^{vap}(\text{water}) = 26.3026 - 5432.8/T$	T in K

a Compute the triple-point temperature and pressure of water.

b Compute the heat of vaporization, the heat of sublimation, and the heat of fusion of water at its triple point.

5.13 **a** The following data have been reported for the vapor pressure of ethanol as a function of temperature.[8]

P^{vap}(Pa)	6.667	13.33	26.67	53.33	80.0	133.3
T(°C)	−12.0	−2.3	8.0	19.0	26.0	34.9

Use these data to calculate the heat of vaporization of ethanol at 17.33°C. (See Problem 6.18.)

b Ackermann and Rauh have measured the vapor pressure of liquid plutonium using a clever mass effusion technique.[9] Some of their results are given here

[8]Reference: R. H. Perry, D. W. Green and J. O. Maloney, eds., *The Chemical Engineers' Handbook*, 6th ed., McGraw–Hill, New York, 1984, pp. 3–55.

[9]Reference: R. J. Ackermann and E. G. Rauh, *J. Chem. Thermodynamics* **7**, 211 (1975).

T(K)	1343	1379	1424	1449
P^{vap}(bar)	7.336×10^{-9}	1.601×10^{-8}	3.840×10^{-8}	5.654×10^{-8}

Estimate the heat of vaporization of liquid plutonium at 1400 K.

5.14 **a** Derive Eq. 5.4-8.
b Derive Eq. 5.4-12.
c Obtain an expression for the fugacity of a pure species that obeys the van der Waals equation of state in terms of Z, $B = Pb/RT$ and $A = Pa/(RT)^2$ (i.e., derive Eq. 5.4-13).
d Repeat the derivation with the Peng–Robinson equation of state (i.e., derive Eq. 5.4-14a).

5.15 **a** Calculate the fugacity of liquid hydrogen sulfide in contact with its saturated vapor at 25.5°C and 20 bar.
b The vapor pressure of pure water at 310.6 K is 6.455 kPa. Compute the fugacity of pure liquid water at 310.6 K when it is under a pressure of 100 bar, 500 bar, and 1000 bar. Volume 7 of the *International Critical Tables* (McGraw–Hill, New York, 1929) gives values of 6.925, 9.175, and 12.966 kPa, respectively, for these conditions.

5.16 **a** Using only the steam tables, compute the fugacity of steam at 400°C and 2 MPa, and at 400°C and 50 MPa.
b Compute the fugacity of steam at 400°C and 2 MPa using the principle of corresponding states. Repeat the calculation at 400°C and 50 MPa.
c Repeat the calculations using the Peng–Robinson equation of state.
Comment on the causes of the differences among these predictions.

5.17 **a** Show that at moderately low pressures and densities the virial equation of state can be written as

$$\frac{PV}{RT} = 1 + B\left(\frac{P}{RT}\right) + (C - B^2)\left(\frac{P}{RT}\right)^2 + \cdots$$

b Prove that the fugacity coefficient for this form of the virial equation of state is

$$\frac{f}{p} = \exp\left\{B\left(\frac{P}{RT}\right) + \frac{(C - B^2)}{2}\left(\frac{P}{RT}\right)^2 + \cdots\right\}$$

The first two virial coefficients for methyl fluoride at 50°C are $B = -0.1663$ m³/kmol and $C = 0.01292$ (m³/kmol)². Plot the ratio (f/P) as a function of pressure at 50°C for pressures up to 150 bar. Compare the results with the corresponding states plot of (f/P) versus (P/P_C) given in the text and the results of the approximate calculation using the equation derived here.

5.18 The following data are available for carbon tetrachloride:

$T_C = 283.3$°C	$P_C = 4.56$ MPa	$Z_C = 0.272$	
Vapor pressure (MPa)	0.5065	1.013	2.026
T(°C)	141.7	178.0	222.0

a Compute the heat of vaporization of carbon tetrachloride at 200°C using only these data.

b Derive the following expression that can be used to compute the heat of vaporization from the principle of corresponding states:

$$\Delta \underline{H}^{\text{vap}} = T_C \left[\left(\frac{\underline{H} - \underline{H}^{IG}}{T_C} \right)_{\substack{\text{sat}\\\text{vap}}} - \left(\frac{\underline{H} - \underline{H}^{IG}}{T_C} \right)_{\substack{\text{sat}\\\text{liq}}} \right]$$

c Compute the heat of vaporization of carbon tetrachloride at 200°C using the principle of corresponding states.

d Comment on the reasons for the difference between the heats of vaporization computed in parts a and c and suggest a way to correct the results to improve the agreement.

5.19 An article in *Chemical and Engineering News* (Sept. 28, 1987) describes a hydrothermal autoclave. This device is of constant volume, is evacuated, and then filled with water so that a fraction x of the total volume is filled with liquid water and the remainder is filled with water vapor. The autoclave is then heated, so that the temperature and pressure in the sealed vessel increases. It is observed that if x is greater than a "critical fill" value, x_C, the liquid volume fraction increases as the temperature increases and the vessel becomes completely filled with liquid at temperatures below the critical temperature. On the other hand, if $x < x_C$, the liquid evaporates as temperature is increased, and the autoclave becomes completely filled with vapor below the water critical temperature. If, however, $x = x_C$, the volume fraction of water in the autoclave remains constant as the temperature increases, and the temperature–pressure trajectory passes right through the water critical point. Assuming the hydrothermal autoclave is to be loaded at 25°C, calculate the critical fill x_C.

a Using the steam tables.

b Assuming the water obeys the Peng–Robinson equation of state.

5.20 **a** Quantitatively explain why ice skates slide along the surface of ice. Can it get too cold to ice skate?

b Is it possible to ice skate on other materials such as frozen carbon dioxide?

c What is the approximate lowest temperature at which a good snowball can be made? Why can't a snowball be made if the temperature is too low?

5.21 A thermally insulated (adiabatic) constant-volume bomb has been very carefully prepared so that half its volume is filled with water vapor and half with subcooled liquid water, both at −10°C and 0.2876 kPa (the saturation pressure of the subcooled liquid). Find the temperature, pressure, and fraction of water in each phase after equilibrium has been established in the bomb. What is the entropy change for the process?

For simplicity, neglect the heat capacity of the bomb, assume the vapor phase is ideal, and, for the limited temperature range of interest here, that each of the quantities below are independent of temperature.

Data

$$\Delta \hat{H}(\text{solid} \rightarrow \text{liquid}) = 335 \text{ J/g}$$

$$\Delta \hat{H}(\text{liquid} \rightarrow \text{vapor}, T \sim 0°\text{C}) = 2530 \text{ J/g}$$

$C_P(\text{liquid}) = 4.22$ J/g °C; $\hat{V}(\text{liquid}) \cong 1 \times 10^{-3}$ m³/kg

$C_P(\text{solid}) = 2.1$ J/g °C; $\hat{V}(\text{solid}) = 1.11 \times 10^{-3}$ m³/kg

$C_P(\text{vapor}) = 2.03$ J/g °C

5.22 Estimate the triple point temperature and pressure of benzene. The following data are available:

Vapor Pressure				
T(°C)	−36.7	−19.6	−11.5	−2.6
P^{vap}(Pa)	1.333	6.667	13.33	26.67

Melting point at atmospheric pressure = 5.49°C
Heat of fusion at 5.49°C = 127 J/g
Liquid volume at 5.49°C = 0.901×10^{-3} m³/kg
Volume change on melting = 0.1317×10^{-3} m³/kg

5.23 Many thermodynamic and statistical mechanical theories of fluids lead to predictions of the Helmholtz free energy $\underline{A}$ with T and $\underline{V}$ as the independent variables; that is, the result of the theory is an expression of the form $\underline{A} = \underline{A}(T, \underline{V})$. The following figure is a plot of $\underline{A}$ for one molecular species as a function of specific volume at constant temperature. The curve on the left has been calculated assuming the species is present as a liquid and the curve on the right assuming the species is present as a gas.

Prove that for the situation indicated in the figure, the vapor and liquid can coexist at equilibrium, that the specific volumes of the two coexisting phases are given by the points of tangency of the Helmholtz free energy curves with the line that is tangent to both curves, and that the slope of this tangent line is equal to the negative of the equilibrium (vapor) pressure.

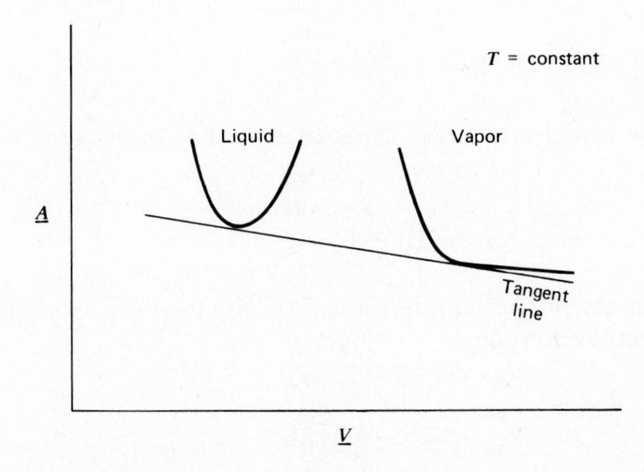

5.24 The principle of corresponding states has far greater applicability in thermodynamics than was indicated in the discussion of Sec. 4.6. For example, it is possible to construct corresponding states relations for both the vapor

and liquid densities along the coexistence curve, and for the vapor pressure and enthalpy change on vaporization ($\Delta\underline{H}^{\text{vap}}$) as a function of reduced temperature. This will be demonstrated here using the van der Waals equation as a model equation of state.

a Show that at vapor–liquid equilibrium

$$\int_{\underline{V}^{\text{L}}}^{\underline{V}^{\text{V}}} \underline{V}\left(\frac{\partial P}{\partial \underline{V}}\right)_T d\underline{V} = 0$$

b Show that the solution to this equation for the van der Waals fluid is

$$\ln\left[\frac{3V_r^{\text{L}} - 1}{3V_r^{\text{V}} - 1}\right] + (3V_r^{\text{V}} - 1)^{-1} + (3V_r^{\text{L}} - 1)^{-1} + \frac{9}{4T_r V_r^{\text{L}}}\left(1 - \frac{V_r^{\text{L}}}{V_r^{\text{V}}}\right) = 0$$

where $V_r^{\text{L}} = \underline{V}^{\text{L}}/\underline{V}_{\text{C}}$ and $V_r^{\text{V}} = \underline{V}^{\text{V}}/\underline{V}_{\text{C}}$.

c Construct a corresponding states curve for the reduced vapor pressure P_r^{vap} of the van der Waals fluid as a function of its reduced temperature.

d Describe how a corresponding states curve for the enthalpy change on vaporization as a function of the reduced temperature can be obtained for the van der Waals fluid.

5.25 **a** Show that it is not possible for a pure fluid to have a quaternary point where the vapor, liquid, and two solid phases are all in equilibrium.

b Show that the specification of the overall molar volume of a one-component, two-phase mixture is not sufficient to identify the thermodynamic state of either phase, but that the specification of both the molar volume and the molar enthalpy (or any two other molar properties) is sufficient to determine the thermodynamic states of each of the phases and the distribution of mass between the phases.

5.26 From Eqs. 4.2-18 and 20, we have the following as definitions of the heat capacities at constant volume and constant pressure

$$C_{\text{V}} = T\left(\frac{\partial \underline{S}}{\partial T}\right)_{\underline{V}}$$

and

$$C_{\text{P}} = T\left(\frac{\partial \underline{S}}{\partial T}\right)_P$$

More generally, we can define a heat capacity subject to some other constraint X by

$$C_X = T\left(\frac{\partial \underline{S}}{\partial T}\right)_X$$

One such heat capacity of special interest is C_{vap}, the heat capacity along the vapor–liquid equilibrium line.

a Show that

$$C_{\text{vap}}^i = C_{\text{P}}^i - \alpha^i \underline{V}^i \frac{\Delta\underline{H}^{\text{vap}}}{\Delta\underline{V}^{\text{vap}}}$$

where α^i is the coefficient of thermal expansion for phase i (liquid or vapor), C_{P}^i is its constant-pressure heat capacity, and $\underline{V}^i$ its molar volume. $\Delta\underline{H}^{\text{vap}}$ and $\Delta\underline{V}^{\text{vap}}$ are the molar enthalpy and volume changes on vaporization.

b Show that
(i) $C^{L}_{vap} \approx C^{L}_{P}$
and that
(ii) C^{V}_{vap} may be negative by considering saturated steam at its normal boiling point and at 370°C.

5.27 **a** By multiplying out the various terms, show that Eqs. 5.2-4 and 6 are equivalent. Also show that

b $$S_{UV} = \frac{\partial}{\partial U}\bigg|_{N,V}\left(\frac{\partial S}{\partial N}\right)_{U,V} = -\frac{1}{T^2}\left(\frac{\partial T}{\partial N}\right)_{U,V} = \frac{-\underline{V}}{NC_VT}\left(\frac{\partial P}{\partial T}\right)_{\underline{V}} + \frac{\underline{H}}{NT^2C_V}$$

c $$S_{VN} = \frac{1}{T}\left(\frac{\partial P}{\partial N}\right)_{U,V} - \frac{P}{T^2}\left(\frac{\partial T}{\partial N}\right)_{U,V}$$

$$= -\frac{1}{NC_VT^2}\left\{\underline{H} - \underline{V}T\left(\frac{\partial P}{\partial T}\right)_{\underline{V}}\right\}\left\{T\left(\frac{\partial P}{\partial T}\right)_{\underline{V}} - P\right\} - \frac{\underline{V}}{NT}\left(\frac{\partial P}{\partial \underline{V}}\right)_T$$

d $$S_{NN} = \frac{2\underline{H}\underline{V}}{NC_VT}\left(\frac{\partial P}{\partial T}\right)_{\underline{V}} - \frac{\underline{H}^2}{NC_VT^2} - \frac{\underline{V}^2}{NC_V}\left(\frac{\partial P}{\partial T}\right)^2_V + \frac{\underline{V}^2}{NT}\left(\frac{\partial P}{\partial \underline{V}}\right)_T$$

and that

e $$\theta_3 = 0$$

5.28 Assuming that nitrogen obeys the Peng–Robinson equation of state, develop tables and charts of its thermodynamic properties such as those in Illustrations 4.4-1, 5.5-1, and 5.5-2. Compare your results with those in Figure 2.4-3 and comment on the differences.

5.29 Assuming that water obeys the Peng–Robinson equation of state, develop tables and charts of its thermodynamic properties such as those in Illustrations 4.4-1, 5.5-1, and 5.5-2. Compare your results with those in Figures 2.4-1 and Appendix III.

5.30 A well-insulated gas cylinder, containing ethylene at 85 bar and 25°C, is exhausted until the pressure drops to 10 bar. This process occurs fast enough so that no heat transfer occurs between the gas and the cylinder walls, but not so rapidly as to produce large velocity or temperature gradients within the cylinder.
a Calculate the final temperature of the ethylene in the cylinder.
b What fraction of the ethylene remaining in the cylinder is present as a liquid?

5.31 The following vapor–liquid equilibrium data are available for methyl ethyl ketone:

Heat of vaporization at 75°C: 31,602 J/mol

Molar volume of saturated liquid at 75°C: 9.65×10^{-2} m^3/kmol

$$\ln P^{vap} = 43.552 - \frac{5622.7}{T} - 4.70504 \ln T$$

where P^{vap} is the vapor pressure in bar and T is in K. Assuming the saturated vapor obeys the volume-explicit form of the virial equation,

$$\underline{V} = \frac{RT}{P} + B$$

Calculate the second virial coefficient, B, for methyl ethyl ketone at 75°C.

5.32 The freezing point of *n*-hexadecane is approximately 18.5°C, where its vapor pressure is very low. By how much would the freezing point of *n*-hexadecane be depressed if *n*-hexadecane were under a 200 bar pressure? For simplicity in this calculation you may assume that the changes in volume and heat capacity on fusion of *n*-hexadecane are zero, and that its heat of fusion is 48.702 kJ/mol.

5.33 Ten grams of liquid water at 95°C are contained in the insulated container shown here.

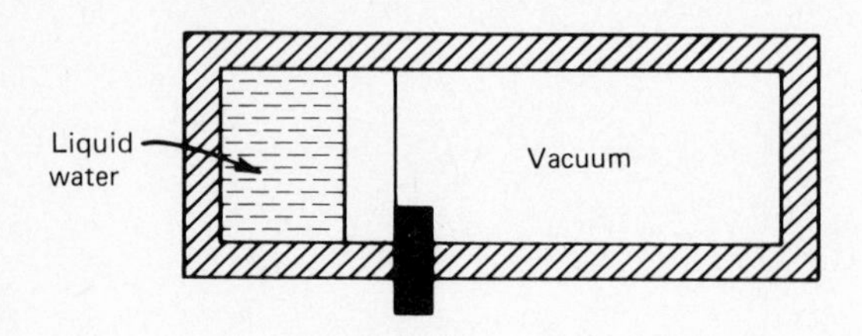

The pin holding the frictionless piston in place breaks, and the volume available to the water increases to 1×10^{-3} m^3. During the expansion some of the water evaporates, but no heat is transferred to the cylinder.

Find the temperature, pressure, and amounts of vapor, liquid, and solid water present after the expansion.

5.34 At a subcritical temperature, the branch of the Peng–Robinson equation of state for $\underline{V} > b$ exhibits the familiar van der Waals loop. However, there is also interesting behavior of the equation in the ranges $\underline{V} < b$ and $\underline{V} < 0$, though these regions do not have any physical meaning. At supercritical temperatures the van der Waals loop disappears, but much of the structure in the $V < 0$ region remains. Establish this by drawing a P–$\underline{V}$ plot for *n*-butane at temperatures of 0.6 T_C, 0.8 T_C, 1.0 T_C, 1.5 T_C, and 5 T_C for all values of $\underline{V}$.

6

The Thermodynamics of Multicomponent Mixtures

In this chapter the study of thermodynamics, which was restricted to pure fluids in the first part of this book, is extended to mixtures. First, the problem of relating the thermodynamic properties of a mixture, such as the enthalpy, free energies, or entropy, to its temperature, pressure, and composition is discussed; this is the problem of the thermodynamic description of mixtures. Next, the equations of change for mixtures are developed. Using the equations of change and the positive nature of the entropy generation term, the general equilibrium criteria for mixtures are considered and shown to be formally identical to the pure fluid equilibrium criteria. These criteria are then used to establish the conditions of phase and chemical equilibrium in mixtures; results that are of central importance in the remainder of this book.

6.1 THE THERMODYNAMIC DESCRIPTION OF MIXTURES

In Chapter 1 we pointed out that the thermodynamic state of a pure, single-phase system was completely specified by fixing the values of two of its intensive state variables (e.g., T and P or T and $\underline{V}$), and that its size and thermodynamic state would be fixed by a specification of its mass or number of moles and two of its state variables (e.g., N, T, and P or N, T, and $\underline{V}$). For a single-phase mixture of $\mathscr{C}$ components, one intuitively expects that the thermodynamic state will be fixed by specifying the values of two of the intensive variables of the system *and* all but one of the species mole fractions (e.g., $T, P, x_1, x_2, \ldots, x_{\mathscr{C}-1}$); the remaining mole fraction being calculable from the fact that the mole fractions must sum to one. (Though we shall largely use mole fraction in the discussion, other composition scales such as mass fraction, volume fraction, molality, etc. could be used. Indeed, there are applications in which each is used.) Similarly, one anticipates that the size and state of the system will be fixed by specifying the values of two state variables and the mole numbers of all species present. That is, for a $\mathscr{C}$ component mixture we expect equations of state of the form:

$$U = U(T, P, N_1, N_2, \ldots, N_{\mathscr{C}}) \quad \text{or} \quad \underline{U} = \underline{U}(T, P, x_1, x_2, \ldots, x_{\mathscr{C}-1})$$

$$V = V(T, P, N_1, N_2, \ldots, N_{\mathscr{C}}) \quad \text{or} \quad \underline{V} = \underline{V}(T, P, x_1, x_2, \ldots, x_{\mathscr{C}-1})$$

etc.

Here x_i is the mole fraction of species i, that is, $x_i = N_i/N$, where N_i is the number of moles of species i and N is the total number of moles, and $\underline{U}$ and $\underline{V}$ are the internal energy and volume per mole of the mixture.[1] Our main interest here will be in determining how the thermodynamic properties of the mixture depend on species concentrations. That is, we would like to know the concentration-dependence of the mixture equations of state.

A naive expectation is that each thermodynamic mixture property is a sum of the analogous property for the pure components at the same temperature and pressure weighted with their fractional compositions. That is, if $\underline{U}_i$ is the internal energy per mole of pure species i at temperature T and pressure P, then it would be convenient if, for a $\mathscr{C}$-component mixture

$$\underline{U}(T, P, x_1, \ldots, x_{\mathscr{C}-1}) = \sum_{i=1}^{\mathscr{C}} x_i \underline{U}_i(T, P) \tag{6.1-1}$$

Alternatively, if $\hat{U}_i$ is the internal energy per unit mass of pure species i, then the development of the mixture equation of state would be simple if

$$\hat{U}(T, P, w_1, \ldots, w_{\mathscr{C}-1}) = \sum_{i=1}^{\mathscr{C}} w_i \hat{U}_i(T, P) \tag{6.1-2}$$

where w_i is the mass fraction of species i.

Unfortunately, relations as simple as Eq. 6.1-1 and 2 are generally not valid. This is evident from Fig. 6.1-1, which shows the enthalpy of a sulfuric acid–water mixture at various temperatures. If equations of the form of Eq. 6.1-2 were valid, then

$$\hat{H} = w_1 \hat{H}_1 + w_2 \hat{H}_2 = w_1 \hat{H}_1 + (1 - w_1) \hat{H}_2$$

and the enthalpy of the mixture would be a linear combination of the two pure component enthalpies; this is indicated by the dashed line connecting the pure component enthalpies at 0°C. The actual thermodynamic behavior of the mixture, given by the solid lines in the figure, is quite different. In Fig. 6.1-2 are plotted the **volume change on mixing**, defined to be[2]

$$\Delta \underline{V}_{\text{mix}} = \underline{V}(T, P, x_i) - x_1 \underline{V}_1(T, P) - x_2 \underline{V}_2(T, P)$$

and the **enthalpy change on mixing**

$$\Delta \underline{H}_{\text{mix}} = \underline{H}(T, P, x_i) - x_1 \underline{H}_1(T, P) - x_2 \underline{H}_2(T, P)$$

for several mixtures. (The origin of such data is discussed in Sec. 6.6.) If equations like Eq. 6.1-1 were valid for these mixtures, then both the $\Delta \underline{V}_{\text{mix}}$ and $\Delta \underline{H}_{\text{mix}}$ would be equal to zero; the data in the figures clearly show that this is not the case.

There is a simple reason why such elementary mixture equations as Eqs. 6.1-1 and 2 fail for real mixtures. We have already pointed out that one contribution to the internal energy (and other thermodynamic properties) of a fluid is the interactions between its constituent molecules. For a pure fluid this contribution

[1]In writing these equations we have chosen T and P as the independent state variables; other choices could have been made. However, since temperature and pressure are easily measured and controlled, they are the most practical choice of independent variables for processes of interest to chemists and engineers.

[2]In these equations, and most of those that follow, the abbreviated notation $\underline{\theta}(T, P, x_1, x_2, \ldots, x_{\mathscr{C}-1}) = \underline{\theta}(T, P, x_i)$ will be used.

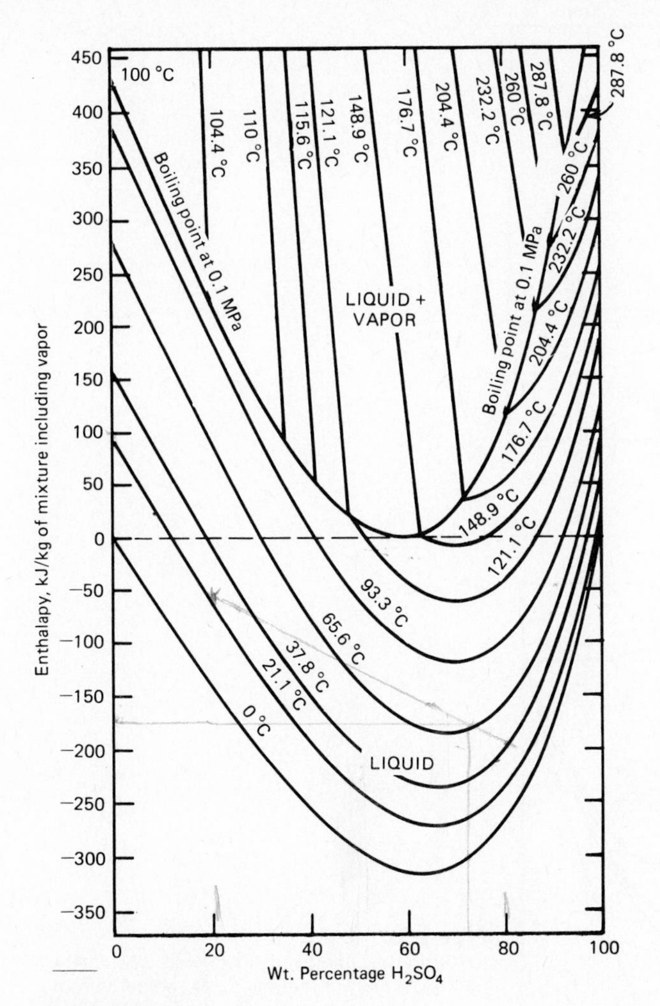

Figure 6.1-1
Enthalpy-concentration diagram for aqueous sulfuric acid at 0.1 mPa. The sulfuric acid percentage is by weight (which, in the two-phase region, includes the vapor). Reference states: the enthalpies of the pure liquids at 0°C and their vapor pressures are zero. (Based on Fig 81, p 325 in O. A. Hougen, K. M. Watson, and R. A. Ragatz, *Chemical Process Principles. I. Material and Energy Balances*, 2nd ed. Copyright 1954 by John Wiley & Sons, Inc. Used with permission.)

to the internal energy is the result of molecules interacting only with other molecules of the same species. However, in a binary fluid mixture of species 1 and 2, the total interaction energy is a result of 1–1, 2–2, and 1–2 interactions. Although the 1–1 and 2–2 molecular interactions that occur in the mixture are approximately taken into account in Eqs. 6.1-1 and 2 through the pure component properties, the effects of the 1–2 interactions are not. Therefore, it is not surprising that these equations are poor approximations for the properties of most real fluid mixtures.

What we must do, instead of making simple guesses such as Eqs. 6.1-1 and 2, is develop a general framework for relating the thermodynamic properties of a mixture to the composition variables. Since the interaction energy of a mixture is largely determined by the number of each type of molecular interaction, the mole fractions or, equivalently, the mole numbers, are the most logical composi-

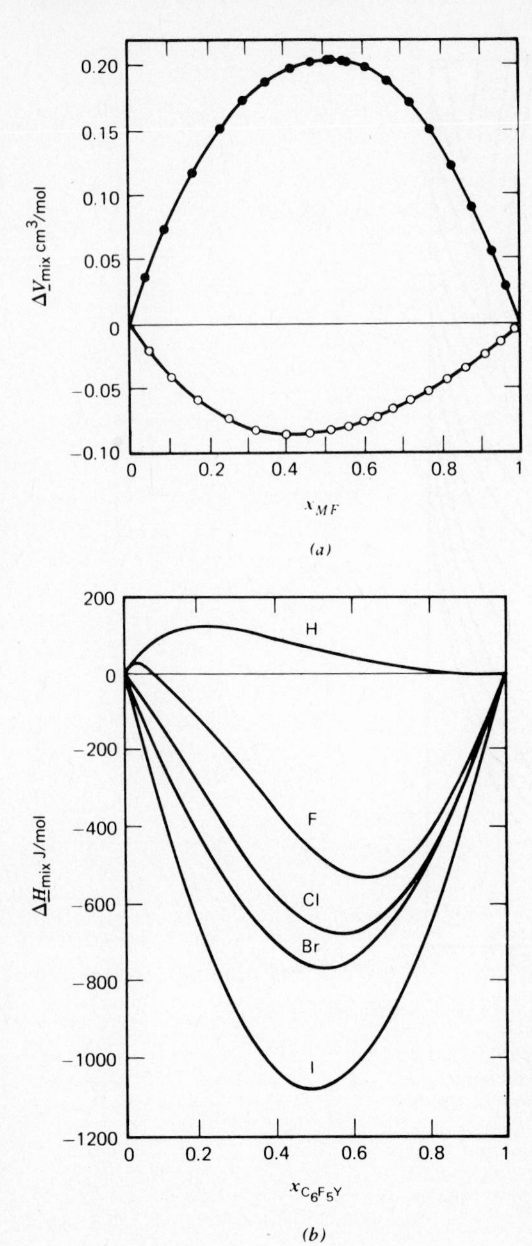

Figure 6.1-2
(*a*) Volume change on mixing at 298.15K; ○ methyl formate + methanol. ● methyl formate + ethanol. [From J. Polak and B. C.-Y. Lu, *J. Chem. Thermodynamics* **4,** 469 (1972). Reprinted with permission of Academic Press, Inc.] (*b*) Enthalpy change on mixing at 298.15 K for mixtures of benzene (C_6H_6) and aromatic fluorocarbons C_6F_5Y, with Y = H, F, Cl, Br, and I. [From D. V. Fenby and S. Ruenkrairergsa, *J. Chem. Thermodynamics* **5,** 227 (1973). Reprinted with permission of Academic Press, Inc.]

tion variables for the analysis. We will use the symbol $\underline{\theta}$ to be any molar property (molar volume, molar enthalpy, etc.) of a mixture consisting of N_1 moles of species 1, N_2 moles of species 2, and so forth; $N = \Sigma_i^{\mathscr{C}} N_i$ being the total number of moles in the system. For the present our main interest is in the composition variation of $\underline{\theta}$ at fixed T and P, so that we write

$$N\underline{\theta} = \phi(N_1, N_2, \ldots, N_{\mathscr{C}}) \qquad \text{at constant } T \text{ and } P \tag{6.1-3}$$

to indicate that $N\underline{\theta}$ is a function of all the mole numbers.[3]

[3]The total property $\theta = N\underline{\theta}$ is, of course, a function of T, P, and the mole numbers; that is,

$$N\underline{\theta} = N\underline{\theta}(T, P, x_1, x_2, \ldots) = \theta(T, P, N_1, N_2, \ldots)$$

If the number of moles of each species in the mixture under consideration were doubled, so that $N_1 \rightarrow 2N_1$, $N_2 \rightarrow 2N_2$, . . . , and $N \rightarrow 2N$ (thereby keeping the relative amounts of each species, and the relative numbers of each type of molecular interaction in the mixture, unchanged), the value of θ *per mole of mixture*, $\underline{\theta}$, would remain unchanged; that is, $\underline{\theta}(T, P, N_1, N_2, \ldots) = \underline{\theta}(T, P, 2N_1, 2N_2, \ldots)$. Rewriting Eq. 6.1-3 for this situation gives

$$2N\underline{\theta} = \phi(2N_1, 2N_2, \ldots, 2N_{\mathscr{C}}) \tag{6.1-4}$$

while multiplying Eq. 6.1-3 by the factor 2 yields

$$2N\underline{\theta} = 2\phi(N_1, N_2, \ldots, N_{\mathscr{C}}) \tag{6.1-5}$$

so that

$$\phi(2N_1, 2N_2, \ldots, 2N_{\mathscr{C}}) = 2\phi(N_1, N_2, \ldots, N_{\mathscr{C}})$$

This relation indicates that as long as the relative amounts of each species are kept constant, ϕ is a linear function of the total number of moles. This suggests ϕ should really be considered to be a function of the mole ratios. Thus, a more general way of indicating the concentration dependence of θ is as follows

$$N\underline{\theta} = N_1\phi_1\left(\frac{N_2}{N_1}, \frac{N_3}{N_1}, \ldots, \frac{N_c}{N_1}\right) \tag{6.1-6}$$

where we have introduced a new function, ϕ_1, of only $\mathscr{C} - 1$ mole ratios in a $\mathscr{C}$-component mixture (at constant T and P). In this equation the multiplicative factor N_1 establishes the extent or total mass of the mixture, and the function ϕ_1 is dependent only on the relative amounts of the various species. (Of course, any species other than species 1 could have been singled out in writing Eq. 6.1-6.)

To determine how the property $N\underline{\theta}$ varies with the number of moles of each species present, we look at the derivatives of $N\underline{\theta}$ with respect to mole number. The change in the value of $N\underline{\theta}$ as the number of moles of species 1 changes (all other mole numbers, T, and P being held constant) is

$$\left.\frac{\partial}{\partial N_1}(N\underline{\theta})\right|_{T,P,N_2,N_3\ldots} = \phi_1\left(\frac{N_2}{N_1}, \ldots\right) + N_1 \sum_{i=2}^{\mathscr{C}} \frac{\partial \phi_1}{\partial\left(\frac{N_i}{N_1}\right)} \frac{\partial\left(\frac{N_i}{N_1}\right)}{\partial N_1} \tag{6.1-7}$$

where the last term arises from the chain rule of partial differentiation and the fact that ϕ_1 is a function of the mole ratios. Now multiplying by N_1, and using the fact that

$$\left.\frac{\partial\left(\frac{N_i}{N_1}\right)}{\partial N_1}\right|_{i\neq 1} = -\frac{N_i}{N_1^2}$$

gives

$$\left.N_1\frac{\partial}{\partial N_1}(N\underline{\theta})\right|_{T,P,N_{j\neq i}} = N_1\left\{\phi_1\left(\frac{N_2}{N_1}, \ldots\right) - \frac{1}{N_1}\sum_{i=2}^{\mathscr{C}} N_i \frac{\partial \phi_1}{\partial\left(\frac{N_i}{N_1}\right)}\right\}$$

$$= N\underline{\theta} - \sum_{i=2}^{\mathscr{C}} N_i \frac{\partial \phi_1}{\partial\left(\frac{N_i}{N_1}\right)} \tag{6.1-8a}$$

or

$$N\underline{\theta} = N_1 \frac{\partial}{\partial N_1}(N\underline{\theta})\bigg|_{T,P,N_{j\neq 1}} + \sum_{i=2}^{\mathscr{C}} \frac{\partial \phi_1}{\partial\left(\dfrac{N_i}{N_1}\right)} \tag{6.1-8b}$$

Here we have introduced the notation $N_{j\neq 1}$ in the partial derivative to indicate that all mole numbers except N_1 are being held constant. Similarly, for the derivative with respect to N_i, $i \neq 1$ we have

$$\frac{\partial(N\underline{\theta})}{\partial N_i}\bigg|_{T,P,N_{j\neq i}} = N_1 \sum_{j=2}^{\mathscr{C}} \frac{\partial \phi_1}{\partial\left(\dfrac{N_j}{N_1}\right)} \frac{\partial\left(\dfrac{N_j}{N_1}\right)}{\partial N_i} = \frac{\partial \phi_1}{\partial\left(\dfrac{N_i}{N_1}\right)} \tag{6.1-9}$$

since $(\partial N_j/\partial N_i)$ is equal to one for $i = j$, and zero for $i \neq j$. Using Eq. 6.1-9 in Eq. 6.1-8b yields

$$N\underline{\theta} = N_1 \frac{\partial}{\partial N_1}(N\underline{\theta})\bigg|_{T,P,N_{j\neq 1}} + \sum_{i=2}^{\mathscr{C}} N_i \frac{\partial(N\underline{\theta})}{\partial N_i}\bigg|_{T,P,N_{j\neq i}}$$

or

$$N\underline{\theta} = \sum_{i=1}^{\mathscr{C}} N_i \frac{\partial(N\underline{\theta})}{\partial N_i}\bigg|_{T,P,N_{j\neq i}} \tag{6.1-10}$$

and dividing by $N = \Sigma N_i$

$$\underline{\theta} = \sum_{i=1}^{\mathscr{C}} x_i \frac{\partial(N\underline{\theta})}{\partial N_i}\bigg|_{T,P,N_{j\neq i}} \tag{6.1-11}$$

Common thermodynamic notation is to define a **partial molar thermodynamic property** $\bar{\theta}_i$ as[4]

$$\bar{\theta}_i = \bar{\theta}_i(T, P, x_i) = \frac{\partial(N\underline{\theta})}{\partial N_i}\bigg|_{T,P,N_{j\neq i}} \tag{6.1-12}$$

so that Eq. 6.1-11 can be written as

$$\underline{\theta} = \sum_{i=1}^{\mathscr{C}} x_i \bar{\theta}_i(T, P, x_i) \tag{6.1-13}$$

Although this equation is similar in form to the naive assumption of Eq. 6.1-1, there is the very important difference that $\bar{\theta}_i$ which appears here is a *true mixture property*, to be evaluated experimentally for each mixture (see Sec. 6.6), whereas $\underline{\theta}_i$, which appears in Eq. 6.1-1, is a pure component property. It should be emphasized that generally $\bar{\theta}_i \neq \underline{\theta}_i$; that is, *the partial molar and pure component thermodynamic properties are not equal*! Some common partial molar thermodynamic properties are listed in Table 6.1-1. An important problem in the thermodynamics of mixtures is the measurement or estimation of these partial molar properties; this will be the focus of part of this chapter and all of Chapter 7.

[4]Note that the temperature and pressure (as well as certain mole numbers) are being held constant in the partial molar derivative. Later in this chapter we will be concerned with similar derivatives in which T and/or P are not held constant, such derivatives are not partial molar quantities!

Table 6.1-1

Partial Molar Property		Molar Property of the Mixture
$\bar{U}_i =$ partial molar internal energy	$= \left(\frac{\partial(N\underline{U})}{\partial N_i}\right)_{T,P,N_{j\neq i}}$	$\underline{U} = \sum x_i \bar{U}_i$
$\bar{V}_i =$ partial molar volume	$= \left(\frac{\partial(N\underline{V})}{\partial N_i}\right)_{T,P,N_{j\neq i}}$	$\underline{V} = \sum x_i \bar{V}_i$
$\bar{H}_i =$ partial molar enthalpy	$= \left(\frac{\partial(N\underline{H})}{\partial N_i}\right)_{T,P,N_{j\neq i}}$	$\underline{H} = \sum x_i \bar{H}_i$
$\bar{S}_i =$ partial molar entropy	$= \left(\frac{\partial(N\underline{S})}{\partial N_i}\right)_{T,P,N_{j\neq i}}$	$\underline{S} = \sum x_i \bar{S}_i$
$\bar{G}_i =$ partial molar Gibbs free energy	$= \left(\frac{\partial(N\underline{G})}{\partial N_i}\right)_{T,P,N_{j\neq i}}$	$\underline{G} = \sum x_i \bar{G}_i$
$\bar{A}_i =$ partial molar Helmholtz free energy	$= \left(\frac{\partial(N\underline{A})}{\partial N_i}\right)_{T,P,N_{j\neq i}}$	$\underline{A} = \sum x_i \bar{A}_i$

The fact that the partial molar properties of each species in a mixture are different from the pure component molar properties gives rise to changes in thermodynamic properties during the process of forming a mixture from the pure components. For example, consider the formation of a mixture from N_1 moles of species 1, N_2 moles of species 2, and so on. The total volume and enthalpy of the *unmixed* pure components are

$$V = \sum_i^{\mathscr{C}} N_i \underline{V}_i(T, P)$$

and

$$H = \sum_i^{\mathscr{C}} N_i \underline{H}_i(T, P)$$

whereas the volume and enthalpy of the mixture at the same temperature and pressure are, from Eq. 6.1-13

$$V(T, P, N_1, N_2, \ldots) = N\underline{V}(T, P, x_1, x_2, \ldots) = \sum_i^{\mathscr{C}} N_i \bar{V}_i(T, P, x_1, x_2, \ldots)$$

and

$$H(T, P, N_1, N_2, \ldots) = N\underline{H}(T, P, x_1, x_2, \ldots) = \sum_i^{\mathscr{C}} N_i \bar{H}_i(T, P, x_1, x_2, \ldots)$$

Therefore, the isothermal volume change on mixing, ΔV_{mix} and the isothermal enthalpy change on mixing, or the **heat of mixing**, ΔH_{mix}, are

$$\Delta V_{\text{mix}}(T, P, N_1, N_2, \ldots) = V(T, P, N_1, N_2, \ldots) - \sum_{i=1}^{\mathscr{C}} N_i \underline{V}_i(T, P)$$

$$= \sum_{i=1}^{\mathscr{C}} N_i[\bar{V}_i(T, P, x_i) - \underline{V}_i(T, P)] \qquad (6.1\text{-}14)$$

and

$$\Delta H_{\text{mix}}(T, P, N_1, N_2, \ldots) = H(T, P, N_1, N_2, \ldots) - \sum_{i=1}^{\mathscr{C}} N_i \underline{H}_i(T, P)$$
$$= \sum_{i=1}^{\mathscr{C}} N_i[\overline{H}_i(T, P, x_i) - \underline{H}_i(T, P)] \qquad (6.1\text{-}15)$$

Other thermodynamic property changes on mixing can be defined similarly.

In Chapter 3 the Helmholtz and Gibbs free energies of pure components were introduced by the relations

$$\underline{A} = \underline{U} - T\underline{S} \qquad (6.1\text{-}16a)$$

$$\underline{G} = \underline{H} - T\underline{S} \qquad (6.1\text{-}16b)$$

These definitions are also valid for mixtures, provided the values of $\underline{U}$, $\underline{H}$, and $\underline{S}$ used are those for the mixture. That is, the Helmholtz and Gibbs free energies of a mixture bear the same relation to the mixture internal energy, enthalpy, and entropy as do the pure component free energies to the pure component internal energy, enthalpy, and entropy.

Equations like Eqs. 6.1-16a and b are also satisfied by the partial molar properties. To see this we multiply Eq. 6.1-16a by the total number of moles N and take the derivative with respect to N_i at constant T, P, and $N_{j\neq i}$ to obtain

$$\frac{\partial}{\partial N_i}(N\underline{A})\bigg|_{T,P,N_{j\neq i}} = \frac{\partial}{\partial N_i}(N\underline{U})\bigg|_{T,P,N_{j\neq i}} - T\frac{\partial}{\partial N_i}(N\underline{S})\bigg|_{T,P,N_{j\neq i}}$$

or

$$\overline{A}_i = \overline{U}_i - T\overline{S}_i$$

Similarly, starting with Eq. 6.1-16b, one can easily show that $\overline{G}_i = \overline{H}_i - T\overline{S}_i$.

The constant-pressure heat capacity of a mixture is given by

$$C_P = \left(\frac{\partial \underline{H}}{\partial T}\right)_{P,N_j}$$

where $\underline{H}$ is the mixture enthalpy, and the subscript N_j indicates that the derivative is to be taken at constant number of moles of all species present; the partial molar heat capacity for species i in a mixture is defined to be

$$\overline{C}_{P,i} = \frac{\partial}{\partial N_i}(NC_P)\bigg|_{T,P,N_{j\neq i}}$$

Now

$$NC_P = \frac{\partial(N\underline{H})}{\partial T}\bigg|_{N_j,P}$$

and

$$\overline{C}_{P,i} = \frac{\partial(NC_P)}{\partial N_i}\bigg|_{T,P,N_{j\neq i}} = \frac{\partial}{\partial N_i}\bigg|_{T,P,N_{j\neq i}} \frac{\partial}{\partial T}(N\underline{H})\bigg|_{P,N_j}$$
$$= \frac{\partial}{\partial T}\bigg|_{P,N_j} \frac{\partial(N\underline{H})}{\partial N_i}\bigg|_{T,P,N_{j\neq i}} = \left(\frac{\partial \overline{H}_i}{\partial T}\right)_{P,N_j}$$

so that $\overline{C}_{P,i} = (\partial \overline{H}_i/\partial T)_{P,N_j}$ for species i in a mixture, just as $C_{P,i} = (\partial \underline{H}_i/\partial T)_{P,N}$ for the pure species i.

In a similar fashion a large collection of relations among the partial molar quantities can be developed. For example, since $(\partial \underline{G}_i/\partial T)_{P,N} = -\underline{S}_i$ for a pure fluid, one can easily show that $(\partial \bar{G}_i/\partial T)_{P,N_i} = -\bar{S}_i$ for a mixture. In fact, by continuing this argument to other mixture properties, one finds that for each relationship among the thermodynamic variables in a pure fluid, there exists an identical relationship for the partial molar thermodynamic properties in a mixture!

6.2 THE PARTIAL MOLAR GIBBS FREE ENERGY AND THE GENERALIZED GIBBS–DUHEM EQUATION

Since the Gibbs free energy of a multicomponent mixture is a function of temperature, pressure, and species mole numbers, the total differential of the Gibbs function can be written as

$$dG = \left(\frac{\partial G}{\partial T}\right)_{P,N_j} dT + \left(\frac{\partial G}{\partial P}\right)_{T,N_j} dP + \sum_{i=1}^{\mathscr{C}} \left(\frac{\partial G}{\partial N_i}\right)_{T,P,N_{j\neq i}} dN_i$$

$$= -S\,dT + V\,dP + \sum_{i=1}^{\mathscr{C}} \bar{G}_i\,dN_i \tag{6.2-1}$$

Here the first two derivatives follow from Eqs. 4.2-12 for the pure fluid, and the last from the definition of the partial molar Gibbs free energy. Historically, the partial molar Gibbs free energy has been called the **chemical potential,** and designated by the symbol μ_i; this notation will not be used here.

Since the enthalpy can be written as a function of entropy and pressure (see Eq. 4.2-6), we have that

$$dH = \left(\frac{\partial H}{\partial P}\right)_{S,N_j} dP + \left(\frac{\partial H}{\partial S}\right)_{P,N_j} dS + \sum_{i=1}^{\mathscr{C}} \left(\frac{\partial H}{\partial N_i}\right)_{P,S,N_{j\neq i}} dN_i$$

$$= V\,dP + T\,dS + \sum_{i=1}^{\mathscr{C}} \left(\frac{\partial H}{\partial N_i}\right)_{P,S,N_{j\neq i}} dN_i \tag{6.2-2}$$

In this equation it should be noted that $(\partial H/\partial N_i)_{P,S,N_{j\neq i}}$ is not equal to the partial molar enthalpy which is $\bar{H}_i = (\partial H/\partial N_i)_{T,P,N_{j\neq i}}$ (see Problem 6.1).

However, from $H = G + TS$ and Eq. 6.2-1 we have[5]

$$dH = dG + T\,dS + S\,dT = -S\,dT + V\,dP + \sum_i \bar{G}_i\,dN_i + T\,dS + S\,dT$$

$$= V\,dP + T\,dS + \sum_i \bar{G}_i\,dN_i \tag{6.2-3}$$

Comparing Eqs. 6.2-2 and 3 establishes that

$$\left(\frac{\partial H}{\partial N_i}\right)_{P,S,N_{j\neq i}} = \bar{G}_i = \left(\frac{\partial G}{\partial N_i}\right)_{T,P,N_{j\neq i}}$$

so that the derivative $(\partial H/\partial N_i)_{P,S,N_{j\neq i}}$ is equal to the partial molar Gibbs free energy.

[5]You should compare Eqs. 6.2-3, 4, and 5 with Eqs. 4.2-6a, 5a, and 7a, respectively.

Using the procedure established here, it is also easily shown (Problem 6.1) that

$$dU = T\,dS - P\,dV + \sum_{i}^{\mathscr{C}} \bar{G}_i\,dN_i \tag{6.2-4}$$

$$dA = -P\,dV - S\,dT + \sum_{i}^{\mathscr{C}} \bar{G}_i\,dN_i \tag{6.2-5}$$

with

$$\bar{G}_i = \left(\frac{\partial U}{\partial N_i}\right)_{S,V,N_{j\neq i}} = \left(\frac{\partial A}{\partial N_i}\right)_{T,V,N_{j\neq i}}$$

From these equations we see that the partial molar Gibbs free energy assumes special importance in mixtures, as did the molar Gibbs free energy in pure fluids.

Based on the discussion of partial molar properties in the previous section, any thermodynamic function θ can be written as

$$N\underline{\theta} = \sum N_i \bar{\theta}_i$$

so that

$$d(N\underline{\theta}) = \sum N_i\,d\bar{\theta}_i + \sum \bar{\theta}_i\,dN_i \tag{6.2-6}$$

by the product rule of differentiation. However, $N\underline{\theta}$ can also be considered to be a function of T, P, and all the mole numbers, in which case we have, following Eq. 6.2-1

$$\begin{aligned} d(N\underline{\theta}) &= \left.\frac{\partial(N\underline{\theta})}{\partial T}\right|_{P,N_j} dT + \left.\frac{\partial(N\underline{\theta})}{\partial P}\right|_{T,N_j} dP + \sum_{i=1}^{\mathscr{C}} \left.\frac{\partial(N\underline{\theta})}{\partial N_i}\right|_{T,P,N_{j\neq i}} dN_i \\ &= N\left(\frac{\partial\underline{\theta}}{\partial T}\right)_{P,N_j} dT + N\left(\frac{\partial\underline{\theta}}{\partial P}\right)_{T,N_j} dP + \sum_{i=1}^{\mathscr{C}} \bar{\theta}_i\,dN_i \end{aligned} \tag{6.2-7}$$

Subtracting Eq. 6.2-7 from Eq. 6.2-6 gives the following relation

$$-N\left(\frac{\partial\underline{\theta}}{\partial T}\right)_{P,N_j} dT - N\left(\frac{\partial\underline{\theta}}{\partial P}\right)_{T,N_j} dP + \sum_{i=1}^{\mathscr{C}} N_i\,d\bar{\theta}_i = 0 \tag{6.2-8a}$$

and dividing by the total number of moles N yields

$$-\left(\frac{\partial\underline{\theta}}{\partial T}\right)_{P,N_j} dT - \left(\frac{\partial\underline{\theta}}{\partial P}\right)_{T,N_j} dP + \sum_{i=1}^{\mathscr{C}} x_i\,d\bar{\theta}_i = 0 \tag{6.2-8b}$$

These results are forms of the **generalized Gibbs–Duhem equation**.

For changes at constant temperature and pressure the Gibbs–Duhem equation reduces to

$$\sum_{i=1}^{\mathscr{C}} N_i\,d\bar{\theta}_i|_{T,P} = 0 \tag{6.2-9a}$$

and

$$\sum_{i=1}^{\mathscr{C}} x_i\,d\bar{\theta}_i|_{T,P} = 0 \tag{6.2-9b}$$

Finally, for a change in any property Y (except mole fraction[6]) at constant temperature and pressure, Eq. 6.2-9a can be rewritten as

$$\sum_{i}^{\mathscr{C}} N_i \left(\frac{\partial \overline{\theta}_i}{\partial Y}\right)_{T,P} = 0 \tag{6.2-10}$$

whereas for a change in the number of moles of species j at constant temperature, pressure, and all other mole numbers, we have

$$\sum_{i}^{\mathscr{C}} N_i \left(\frac{\partial \overline{\theta}_i}{\partial N_j}\right)_{T,P,N_{k\neq j}} = 0 \tag{6.2-11}$$

Since in much of the remainder of this book we are concerned with equilibrium at constant temperature and pressure, the Gibbs free energy will be of central interest. The Gibbs–Duhem equations for the Gibbs free energy, obtained by setting $\theta = G$ in Eqs. 6.2-8, are

$$0 = S\,dT - V\,dP + \sum_{i=1}^{\mathscr{C}} N_i\,d\overline{G}_i \tag{6.2-12a}$$

$$0 = \underline{S}\,dT - \underline{V}\,dP + \sum_{i=1}^{\mathscr{C}} x_i\,d\overline{G}_i \tag{6.2-12b}$$

and the relations analogous to Eqs. 6.2-9, 10, and 11 are

$$\sum_{i=1}^{\mathscr{C}} N_i\,d\overline{G}_i|_{T,P} = 0 \tag{6.2-13a}$$

$$\sum_{i=1}^{\mathscr{C}} x_i\,d\overline{G}_i|_{T,P} = 0 \tag{6.2-13b}$$

$$\sum_{i=1}^{\mathscr{C}} N_i \left(\frac{\partial \overline{G}_i}{\partial Y}\right)_{T,P} = 0 \tag{6.2-14}$$

and

$$\sum_{i=1}^{\mathscr{C}} N_i \left(\frac{\partial \overline{G}_i}{\partial N_j}\right)_{T,P,N_{k\neq j}} = 0 \tag{6.2-15}$$

Each of these equations will be used later.

The Gibbs–Duhem equation is a thermodynamic consistency relation which expresses the fact that among the set of $\mathscr{C} + 2$ state variables T, P, and $\mathscr{C}$ partial molar Gibbs free energies in a $\mathscr{C}$-component system, only $\mathscr{C} + 1$ of these variables are independent. Thus, for example, although temperature, pressure, $\overline{G}_1, \overline{G}_2, \ldots,$ and $\overline{G}_{\mathscr{C}-1}$ can be independently varied, $\overline{G}_{\mathscr{C}}$ cannot also be changed at will; instead, its change $d\overline{G}_{\mathscr{C}}$ is related to the changes $dT, dP, d\overline{G}_1, \ldots, d\overline{G}_{\mathscr{C}-1}$ by Eq. 6.2-12a, so that

$$d\overline{G}_{\mathscr{C}} = \frac{1}{N_{\mathscr{C}}}\left\{-S\,dT + V\,dP - \sum_{i=1}^{\mathscr{C}-1} N_i\,d\overline{G}_i\right\}$$

[6]Since mole fractions can be varied in several ways (i.e., by varying the number of moles of only one species holding all other mole numbers fixed, by varying the mole numbers of only two species, etc.), a more careful derivation than that given here must be used to obtain the mole fraction analog of Eq. 6.2-10. The result of such an analysis is Eq. 6.2-18.

Thus, the interrelationships provided by Eqs. 6.2-8 through 15 are really restrictions on the mixture equation of state. As such, these equations are important in minimizing the amount of experimental data necessary in evaluating the thermodynamic properties of mixtures, in simplifying the description of multicomponent systems, and in testing the consistency of certain types of experimental data (see Chapter 8). Later in this chapter we show how the equations of change for mixtures and the Gibbs–Duhem equations provide a basis for the experimental determination of partial molar properties.

Although Eqs. 6.2-8 through 11 are well-suited for calculations in which temperature, pressure, and the partial molar properties are the independent variables, it is usually more convenient to have T, P, and the mole fractions x_i as the independent variables. A change of variables is accomplished by realizing that for a $\mathscr{C}$-component mixture there are only $\mathscr{C} - 1$ independent mole fractions (since $\Sigma_{i=1}^{\mathscr{C}} x_i = 1$). Thus we write

$$d\bar{\theta}_i = \left(\frac{\partial \bar{\theta}_i}{\partial T}\right)_{P,x_j} dT + \left(\frac{\partial \bar{\theta}_i}{\partial P}\right)_{T,x_j} dP + \sum_{j=1}^{\mathscr{C}-1} \left(\frac{\partial \bar{\theta}_i}{\partial x_j}\right)_{T,P} dx_j \tag{6.2-16}$$

where $T, P, x_1, \ldots, x_{\mathscr{C}-1}$ have been chosen as the independent variables. Substituting this expansion in Eq. 6.2-8b gives

$$\begin{aligned} 0 = &- \left(\frac{\partial \underline{\theta}}{\partial T}\right)_{P,N_j} dT - \left(\frac{\partial \underline{\theta}}{\partial P}\right)_{T,N_j} dP \\ &+ \sum_{i=1}^{\mathscr{C}} x_i \left[\left(\frac{\partial \bar{\theta}_i}{\partial T}\right)_{P,x_j} dT + \left(\frac{\partial \bar{\theta}_i}{\partial P}\right)_{T,x_j} dP + \sum_{j=1}^{\mathscr{C}-1} \left(\frac{\partial \bar{\theta}_i}{\partial x_j}\right)_{T,P} dx_j\right] \end{aligned} \tag{6.2-17}$$

Since

$$\sum_{i=1}^{\mathscr{C}} x_i \left(\frac{\partial \bar{\theta}_i}{\partial T}\right)_{P,x_j} dT = \frac{\partial}{\partial T}\bigg|_{P,x_j} \left(\sum_{i=1}^{\mathscr{C}} x_i \bar{\theta}_i\right) dT = \left(\frac{\partial \underline{\theta}}{\partial T}\right)_{P,x_j} dT$$

and since holding all the mole numbers constant is equivalent to keeping all the mole fractions fixed, the first and third terms in Eq. 6.2-17 cancel. Similarly, the second and fourth terms cancel. Thus we are left with

$$\sum_{i=1}^{\mathscr{C}} x_i \sum_{j=1}^{\mathscr{C}-1} \left(\frac{\partial \bar{\theta}_i}{\partial x_j}\right)_{T,P} dx_j = 0 \tag{6.2-18}$$

which for a binary mixture reduces to

$$\sum_{i=1}^{2} x_i \left(\frac{\partial \bar{\theta}_i}{\partial x_1}\right)_{T,P} dx_j = 0 \tag{6.2-19a}$$

or, equivalently

$$x_1 \left(\frac{\partial \bar{\theta}_1}{\partial x_1}\right)_{T,P} + x_2 \left(\frac{\partial \bar{\theta}_2}{\partial x_1}\right)_{T,P} = 0 \tag{6.2-19b}$$

For the special case in which θ is equal to the Gibbs free energy we have

$$x_1 \left(\frac{\partial \bar{G}_1}{\partial x_1}\right)_{T,P} + x_2 \left(\frac{\partial \bar{G}_2}{\partial x_1}\right)_{T,P} = 0 \tag{6.2-20}$$

Finally, note that several different forms of Eqs. 6.2-19 and 20 can be obtained by using $x_2 = 1 - x_1$ and $dx_2 = -dx_1$.

6.3 A NOTATION FOR CHEMICAL REACTIONS

Since our interest in this chapter is with the equations of change and the equilibrium criteria for general thermodynamic systems, it is useful to introduce a convenient notation for the description of chemical reactions. Throughout this book the chemical reaction

$$\alpha A + \beta B + \cdots \rightleftarrows \rho R + \cdots$$

where α, β, . . . are the molar stoichiometric coefficients, will be written as

$$\rho R + \cdots - \alpha A - \beta B \cdots = 0$$

or

$$\sum_i \nu_i I = 0 \tag{6.3-1}$$

Here ν_i is the **stoichiometric coefficient** of species I, so defined that ν_i is positive for reaction products, negative for reactants, and equal to zero for inert species. In this notation the reaction $H_2 + \frac{1}{2}O_2 = H_2O$ is written as $H_2O - H_2 - \frac{1}{2}O_2 = 0$, so that $\nu_{H_2O} = +1$, $\nu_{H_2} = -1$ and $\nu_{O_2} = -\frac{1}{2}$.

We will use N_i to represent the number of moles of species i in the system at any time t, and $N_{i,0}$ for the initial number of moles of species i. N_i and $N_{i,0}$ are related through the reaction variable X, called the **molar extent of reaction**, and the stoichiometric coefficient ν_i by the equation

$$N_i = N_{i,0} + \nu_i X \tag{6.3-2a}$$

or

$$X = \frac{N_i - N_{i,0}}{\nu_i} \tag{6.3-2b}$$

An important characteristic of the reaction variable X defined in this way is that it has the same value for each molecular species involved in a reaction; this is illustrated in the following example. Thus, given the initial mole numbers of all species and X (or the number of moles of one species from which X can be calculated) at time t, one can easily compute all other mole numbers in the system. In this way the complete progress of a chemical reaction (i.e., the change in mole numbers of all the species involved in the reaction) is given by the value of the single variable X.

ILLUSTRATION 6.3-1

The electrolytic decomposition of water to form hydrogen and oxygen occurs as follows: $H_2O \rightarrow H_2 + \frac{1}{2}O_2$. Initially, only 3.0 mol of water are present in a closed system. At some later time it is found that 1.2 mol of H_2 and 1.8 mol of H_2O are present.

a. Show that the molar extent of reaction based on H_2 and H_2O are equal.

b. Compute the number of moles of O_2 in the system.

Solution

a. The reaction $H_2O \rightarrow H_2 + \frac{1}{2}O_2$ is rewritten as

$$H_2 + \tfrac{1}{2}O_2 - H_2O = 0$$

so that

$$\nu_{H_2O} = -1 \qquad \nu_{H_2} = +1 \qquad \text{and} \qquad \nu_{O_2} = +\tfrac{1}{2}$$

From the H_2 data

$$X = \frac{1.2 - 0.0}{+1} = +1.2 \text{ mol}$$

From the H_2O data

$$X = \frac{1.8 - 3.0}{-1} = +1.2 \text{ mol}$$

b. Starting from $N_i = N_{i,0} + \nu_i X$, we have that

$$N_{O_2} = 0 + (+\tfrac{1}{2})(1.2) = 0.6 \text{ mol}$$

Comment

Note that the molar extent of reaction is *not* a fractional conversion variable, and, therefore, its value is *not* restricted to lie between 0 and 1. As defined here X, which has units of number of moles, is the number of moles of a species that has reacted divided by the stoichiometric coefficient for the species. In fact, X may be negative if the reaction proceeds in the reverse direction to that indicated (e.g., if hydrogen and oxygen react to form water). ■

The rate of change of the number of the moles of species i resulting from chemical reaction is obtained by taking the derivative with respect to time of Eq. 6.3-2

$$\left(\frac{dN_i}{dt}\right)_{rxn} = \nu_i \left(\frac{dX}{dt}\right) \tag{6.3-3}$$

where the subscript rxn is meant to indicate that this is the rate of change of species i attributable to chemical reaction alone.[7] The total number of moles in a closed system at any time (i.e., for any extent of reaction) is

$$N = \sum_{i=1}^{\mathscr{C}} N_i = \sum_{i=1}^{\mathscr{C}} (N_{i,0} + \nu_i X) = \sum_{i}^{\mathscr{C}} N_{i,0} + X \sum_{i}^{\mathscr{C}} \nu_i \tag{6.3-4}$$

The notion of the extent of reaction variable can easily be extended to multiple chemical reactions. Before doing this, however, it is necessary to introduce the concept of **independent chemical reactions**. The term independent reactions is used here to designate the smallest collection of reactions that, on forming various linear combinations, includes all possible chemical reactions among the species present. Since we have used the adjective "smallest" in defining a set of independent reactions, it follows that no reaction in the set can itself be a linear combination of the others. For example, one can write the following three reactions between carbon and oxygen

$$C + O_2 = CO_2$$

$$2C + O_2 = 2CO$$

$$2CO + O_2 = 2CO_2$$

[7]For an open system the number of moles of species i may also change due to mass flows into and out of the system.

However, if we add the second and third of these equations, we get twice the first, so that these three reactions are not independent. In this case, any two of the three reactions form an independent set.

As we will see in Chapter 9, to describe a chemically reacting system it is not necessary to consider all the chemical reactions that can occur between the reactant species, only the independent reactions. Furthermore, the molar extent of reaction for any chemical reaction among the species can be computed from an appropriate linear combination of the known extents of reaction for the set of independent chemical reactions (see Sec. 9.3). Thus, in the carbon–oxygen reaction system just considered, only two reaction variables (and the initial mole numbers) need be specified to completely fix the mole numbers of each species; the specification of a third reaction variable would be redundant.

To avoid unnecessary complications in the analysis of multiple reactions, we restrict the following discussion to sets of independent chemical reactions. This means, of course, that we need a method of identifying a set of independent reactions from a larger collection of reactions. Where only a few reactions are involved, as in the foregoing example, this can be done by inspection. If many reactions occur, the methods of matrix algebra[8] may be used to determine a set of independent reactions, though we will employ a simpler procedure developed by Denbigh.[9] In the Denbigh procedure one first writes the stoichiometric equations for the formation, from its constituent atoms, of each of the molecular species present in the chemical reaction system. One of the equations that contains an atomic species not actually present in the atomic state at the reaction conditions is then used, by appropriate addition and subtraction, to eliminate that atomic species from the remaining equations. In this way the number of stoichiometric equations is reduced by one. The procedure is repeated until all atomic species not present have been eliminated. The equations that remain form a set of independent chemical reactions; the molar extents of reaction for these reactions are the variables to be used for the description of the multiple reaction system and to follow the composition changes in the mixture.[10]

As an example of this method consider again the oxidation of carbon. We start by writing the following equations for the formation of each of the compounds

$$2O = O_2$$

$$C + O = CO$$

$$C + 2O = CO_2$$

Now since a free oxygen atom does not occur, the first equation is used to eliminate O from the other two equations to obtain

$$2C + O_2 = 2CO$$

$$C + O_2 = CO_2$$

[8] N. R. Amundson, *Mathematical Methods in Chemical Engineering,* Prentice–Hall, Englewood Cliffs, N.J., 1966, p. 50.

[9] K. Denbigh, *Principles of Chemical Equilibrium,* 4th ed., Cambridge University Press, Cambridge, 1981, pp. 169–172.

[10] When considering chemical reactions involving isomers, for example, ortho-, meta-, and paraxylene, one proceeds as described here, treating each isomer as a separate chemical species.

All the remaining atomic species in these equations (here only carbon) are present in the reaction system, so no further reduction of the equations is possible, and these two equations form a set of independent reactions.

For the multiple reaction case, X_j is defined to be the molar extent of reaction (or simply the reaction variable) for the jth independent reaction, and ν_{ij} the stoichiometric coefficient for species i in the jth reaction. The number of moles of species i present at any time (in a closed system) is

$$N_i = N_{i,0} + \sum_{j=1}^{\mathcal{M}} \nu_{ij} X_j \tag{6.3-5}$$

where the summation is over the set of $\mathcal{M}$ independent chemical reactions. The instantaneous rate of change of the number of moles of species i due only to chemical reaction is

$$\left(\frac{dN_i}{dt}\right)_{rxn} = \sum_{j=1}^{\mathcal{M}} \nu_{ij} \frac{dX_j}{dt} \tag{6.3-6}$$

Finally, defining $\hat{X}_j = X_j/V$ to be the molar extent of reaction per unit volume, Eq. 6.3-6 can be written as

$$\left(\frac{dN_i}{dt}\right)_{rxn} = \sum_{j=1}^{\mathcal{M}} \nu_{ij} \frac{d}{dt}(V\hat{X}_j) = \frac{dV}{dt}\sum_{j=1}^{\mathcal{M}} \nu_{ij}\hat{X}_j + V\sum_{j=1}^{\mathcal{M}} \nu_{ij}\frac{d\hat{X}_j}{dt} \tag{6.3-7}$$

Note that $d\hat{X}_j/dt$, the rate of reaction per unit volume, is the reaction variable most frequently used by chemists and chemical engineers in chemical reactor analysis.

6.4 THE EQUATIONS OF CHANGE FOR A MULTICOMPONENT SYSTEM

The next step in the development of the thermodynamics of multicomponent systems is the formulation of the equations of change. These equations can, in completely general form, be considerably more complicated than the analogous pure component equations since (1) the mass or number of moles of each species may not be conserved due to chemical reactions, and (2) the diffusion of one species relative to the others may occur if concentration gradients are present. Furthermore, there is the computational difficulty that each thermodynamic property depends, in a complicated fashion, on the temperature, pressure, and composition of the mixture.

To simplify the development of the equations, we will neglect all diffusional processes, since diffusion has very little effect on the thermodynamic state of the system. (This assumption is equivalent to setting the average velocity of each species equal to the mass average velocity of the fluid.) In Chapter 2 the kinetic and potential energy terms were found to make a small contribution to the pure component energy balance; the relative importance of these terms to the energy balance for a reacting mixture is even less, due to the large energy changes that accompany chemical transformations. Therefore, we will neglect the potential and kinetic energy terms in the formulation of the mixture energy balance. This omission will cause serious errors only in the analysis of rocket engines and similar high-speed devices.

With these assumptions, the formulation of the equations of change for a multicomponent reacting mixture is not nearly so formidable a task as it might first appear. In fact, merely by making the proper identifications, the equations of change for a mixture can be written as simple generalizations of the equations of change for a single-component system. The starting point is Eq. 2.1-4, which is rewritten as

$$\frac{d\theta}{dt} = \begin{pmatrix}\text{rate of change of} \\ \theta \text{ in the system}\end{pmatrix} = \begin{pmatrix}\text{rate at which } \theta \text{ enters the system across} \\ \text{system boundaries}\end{pmatrix} - \begin{pmatrix}\text{rate at which } \theta \text{ leaves the system across} \\ \text{system boundaries}\end{pmatrix} + \begin{pmatrix}\text{rate at which } \theta \text{ is generated within the} \\ \text{system}\end{pmatrix} \tag{2.1-4}$$

The balance equation for species i is gotten by setting θ equal to N_i, the number of moles of species i, by letting $(\dot{N}_i)_k$ be equal to the molar flow rate of species i into the system at the kth port, and by recognizing that species i may be generated within the system by chemical reaction, to obtain

$$\begin{aligned}\frac{dN_i}{dt} &= \sum_{k=1}^{K} (\dot{N}_i)_k + \begin{pmatrix}\text{rate at which species } i \text{ is being} \\ \text{produced by chemical reaction}\end{pmatrix} \\ &= \sum_{k=1}^{K} (\dot{N}_i)_k + \left(\frac{dN_i}{dt}\right)_{rxn} \\ &= \sum_{k=1}^{K} (\dot{N}_i)_k + \sum_{j=1}^{\mathcal{M}} \nu_{ij} \frac{dX_j}{dt}\end{aligned} \tag{6.4-1a}$$

We can obtain a balance equation on the total number of moles in the system by summing Eq. 6.4-1a over all species i, recognizing that $\Sigma_{i=1}^{\mathscr{C}} N_i = N$ is the total number of moles, and that

$$\sum_{i=1}^{\mathscr{C}} \sum_{k=1}^{K} (\dot{N}_i)_k = \sum_{k=1}^{K} \sum_{i=1}^{\mathscr{C}} (\dot{N}_i)_k = \sum_{k=1}^{K} (\dot{N})_k$$

where $(\dot{N})_k = \Sigma_{i=1}^{\mathscr{C}} (\dot{N}_i)_k$ is the total molar flow rate at the kth entry port, so that

$$\frac{dN}{dt} = \sum_{k=1}^{K} (\dot{N})_k + \sum_{i=1}^{\mathscr{C}} \sum_{j=1}^{\mathcal{M}} \nu_{ij} \frac{dX_j}{dt} \tag{6.4-1b}$$

Since neither the number of moles of species i nor the total number of moles are conserved quantities, we need information on the rates at which all chemical reactions occur to use Eqs. 6.4-1. That is, we need detailed information on the reaction processes internal to the black-box system (unless, of course, no reactions occur, in which case $dX_j/dt = 0$ for all j).

Although we will mainly be interested in the equations of change on a molar basis, for completeness and for several illustrations that follow, certain of the equations of change will also be given on a mass basis. To obtain a balance equation for the mass of species i we need only multiply Eq. 6.4-1a by the molecular weight of species i, m_i, and use the notation $M_i = m_i N_i$ and $(\dot{M}_i)_k = m_i(\dot{N}_i)_k$ to get

$$\frac{dM_i}{dt} = \sum_{k=1}^{K} (\dot{M}_i)_k + \sum_{j=1}^{\mathcal{M}} m_i \nu_{ij} \frac{dX_j}{dt} \tag{6.4-2a}$$

Also, summing the equation over species i, we get an overall mass balance equation

$$\frac{dM}{dt} = \sum_{k=1}^{K} (\dot{M})_k \tag{6.4-2b}$$

where $(\dot{M})_k = \Sigma_{i=1}^{\mathscr{C}} (\dot{M}_i)_k$ is the total mass flow at the kth entry port. Here the chemical reaction term vanishes since total mass is a conserved quantity (Problem 6.5).

If θ in Eq. 2.1-4 is now taken to be the total energy of the system (really only the internal energy, since we are neglecting the kinetic and potential energy terms), and the same energy transfer mechanisms identified in Section 2.3 are used here, we obtain

$$\frac{dU}{dt} = \sum_{k=1}^{K} (\dot{N}\underline{H})_k + \dot{Q} + \dot{W}_s - P\frac{dV}{dt} \qquad \text{(molar basis)} \tag{6.4-3}$$

and

$$\frac{dU}{dt} = \sum_{k=1}^{K} (\dot{M}\hat{H})_k + \dot{Q} + \dot{W}_s - P\frac{dV}{dt} \qquad \text{(mass basis)}$$

Here $(\dot{N}\underline{H})_k$ is the product of the molar flow rate and the molar enthalpy of the fluid, and $(\dot{M}\hat{H})_k$ the product of the mass flow rate and enthalpy per unit mass, both at the kth entry port. Since some or all of the flow streams may be mixtures, to evaluate a term such as $(\dot{N}\underline{H})_k$ the relation

$$(\dot{N}\underline{H})_k = \sum_{i=1}^{\mathscr{C}} (\dot{N}_i\bar{H}_i)_k \tag{6.4-4}$$

must be used, where $(\dot{N}_i)_k$ is the molar flow rate of species i in the kth flow stream, and $(\bar{H}_i)_k$ is its partial molar enthalpy. Similarly, the internal energy of the system must be gotten from

$$U = \sum_{i=1}^{\mathscr{C}} N_i\bar{U}_i$$

where N_i is the number of moles of species i and $\bar{U}_i$ is its partial molar internal energy.

At this point you might be concerned that the heat of reaction (i.e., the energy released or absorbed on reaction since the total energy of the chemical bonds in the reactant and product molecules are not equal) does not explicitly appear in the energy balance equation. Since the only assumptions made in deriving this equation were to neglect the potential and kinetic energy terms and diffusion, neither of which involves chemical reaction, the heat of reaction is in fact contained in the energy balance, though it appears implicitly rather than explicitly. Although energy balances on chemical reactors will be studied in detail in Chapter 9, to demonstrate that the heat of reaction is contained in Eq. 6.4-3, we will consider its application to the well-stirred, steady-state chemical reactor in Fig. 6.4-1. Here by steady-state we mean that the rate at which each species i leaves the reactor just balances the rate at which species i enters the reactor and is produced within it, so that N_i does not vary with time; that is

$$\frac{dN_i}{dt} = 0 = (\dot{N}_i)_{\text{in}} - (\dot{N}_i)_{\text{out}} + \nu_i\frac{dX}{dt} \tag{6.4-5a}$$

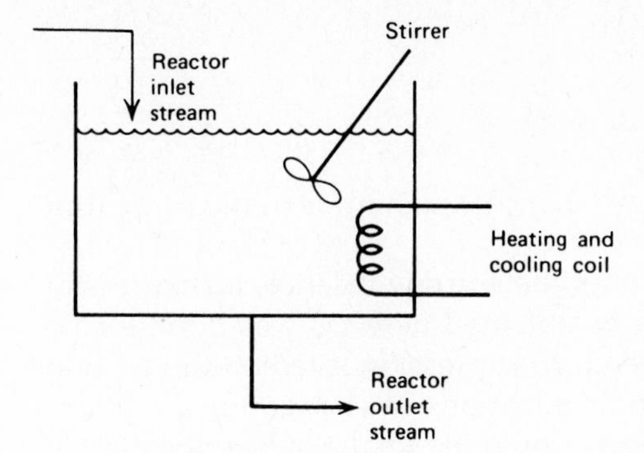

Figure 6.4-1
A simple stirred-tank reactor.

and, further, that the rates of flow of energy into and out of the reactor just balance, so that the internal energy of the contents of the reactor is time independent

$$\frac{dU}{dt} = 0 = \sum_{i=1}^{\mathscr{C}} (\dot{N}_i\overline{H}_i)_{\text{in}} - \sum_{i=1}^{\mathscr{C}} (\dot{N}_i\overline{H}_i)_{\text{out}} + \dot{Q} \tag{6.4-5b}$$

For the very simple case in which the inlet and outlet streams, and the reactor contents are all at temperature T, and with the assumption that the partial molar enthalpy of each species is just equal to its pure component enthalpy, we obtain

$$\dot{Q} = \sum_{i=1}^{\mathscr{C}} [(\dot{N}_i)_{\text{out}} - (\dot{N}_i)_{\text{in}}]\underline{H}_i$$

Now if there were no chemical reaction $(\dot{N}_i)_{\text{out}} = (\dot{N}_i)_{\text{in}}$ and the heat flow rate $\dot{Q}$ should be equal to zero to maintain the constant temperature T. However, when a chemical reaction occurs, $(\dot{N}_i)_{\text{out}} \neq (\dot{N}_i)_{\text{in}}$, and the steady heat flow $\dot{Q}$ required to keep the reactor at constant temperature is

$$\dot{Q} = \sum_{i=1}^{\mathscr{C}} [(\dot{N}_i)_{\text{out}} - (\dot{N}_i)_{\text{in}}]\underline{H}_i = \sum_{i=1}^{\mathscr{C}} \frac{dX}{dt} \nu_i\underline{H}_i$$

$$= \frac{dX}{dt} \sum_{i=1}^{\mathscr{C}} \nu_i\underline{H}_i = \Delta H_{rxn} \frac{dX}{dt}$$

so that the heat flow rate is equal to the product of $\Delta H_{rxn} = \sum \nu_i\underline{H}_i$, the **isothermal heat of reaction**, and dX/dt, the reaction rate. Thus, we see that the heat of reaction is implicitly contained in the energy balance equation through the difference in species mole numbers in the inlet and outlet flow streams.

For nonideal mixtures (i.e., mixtures for which $\overline{H}_i \neq \underline{H}_i$) and nonisothermal reactors, the expression for the heat of reaction is buried somewhat deeper in the energy balance equations; this will be discussed in Chapter 9.

To derive the macroscopic entropy balance equation for mixtures we now set $\theta = S$ in Eq. 2.1-4 and make the same identification for the entropy flow terms as was made in Sec. 3.1, to obtain

$$\frac{dS}{dt} = \sum_{k=1}^{K} (\dot{N}\underline{S})_k + \frac{\dot{Q}}{T} + \dot{S}_{\text{gen}} \tag{6.4-6}$$

where

$$(\dot{N}\underline{S})_k = \sum_{i=1}^{\mathscr{C}} (\dot{N}_i\bar{S}_i)_k$$

Thus the entropy balance, like the energy balance equation, is of the same form for the pure fluid and for mixtures.

As the final step in the development of the entropy balance, an expression for the entropy generation term should be obtained here, as was done for the pure fluid in Sec. 3.6. The derivation of such an expression is tedious and would require us to develop detailed microscopic equations of change for a mixture. Rather than doing this we will merely write down the final result, referring you to deGroot and Mazur for the details.[11] Their result, with the slight modification of writing the chemical reaction term on a molar rather than mass basis, is

$$\dot{S}_{gen} = \int \dot{\sigma}_s \, dV$$

where

$$\dot{\sigma}_s = \frac{\lambda}{T^2}(\nabla T)^2 + \frac{\mu}{T}\phi^2 - \frac{1}{T}\sum_{j=1}^{\mathscr{M}}\sum_{i=1}^{\mathscr{C}} \nu_{ij}\bar{G}_i \frac{dX_j}{dt} \tag{6.4-7}$$

(see Sec. 2.7 for the definitions of λ, ϕ, and the gradient operator ∇). The last term, which is new, represents the contribution of chemical reactions to the entropy generation rate. One can establish that this term makes a positive contribution to the entropy generation term. It is of interest to note that had diffusional processes also been considered, there would be an additional contribution to $\dot{S}_{gen}$ due to diffusion. That term would be in the form of a diffusional flux times a driving force for diffusion and would also be greater than or equal to zero.

In Chapter 3 we defined a reversible process to be a process for which $\dot{S}_{gen} = 0$. This led to the conclusion that in a reversible process in a one-component system only infinitesimal temperature and velocity gradients could be present. Clearly, on the basis of Eq. 6.4-7, a reversible process in a multicomponent system is one in which only infinitesimal gradients in temperature, velocity, and concentration may be present, and in which all chemical reactions proceed at only infinitesimal rates.

The differential form of the multicomponent mass, energy, and entropy balances can be integrated over time to get difference forms of the balance equations, as was done in Chapters 2 and 3 for the pure component equations. The differential and difference forms of the balance equations are listed in Tables 6.4-1 and 2, respectively. It is left to you to work out the various simplifications of these equations that arise for special cases of closed systems, adiabatic processes, and so forth.

With the equations of change for mixtures, and given mixture thermodynamic data, such as the enthalpy data for sulfuric acid–water mixtures in Fig. 6.1-1, it is possible to solve many thermodynamic energy flow problems for mixtures. One example is given in the following illustration.

[11] S. R. de Groot and P. Mazur, *Non-equilibrium Thermodynamics*, North-Holland, Amsterdam, 1962, Chapter 3.

Table 6.4-1
THE DIFFERENTIAL FORM OF THE EQUATIONS OF CHANGE FOR A MULTICOMPONENT SYSTEM

Species balance

$$\frac{dN_i}{dt} = \sum_{k=1}^{K} (\dot{N}_i)_k + \sum_{j=1}^{\mathcal{M}} \nu_{ij} \frac{dX_j}{dt} \qquad \text{(molar basis)}$$

$$\frac{dM_i}{dt} = \sum_{k=1}^{K} (\dot{M}_i)_k + \sum_{j=1}^{\mathcal{M}} m_i \nu_{ij} \frac{dX_j}{dt} \qquad \text{(mass basis)}$$

Overall mass balance

$$\frac{dN}{dt} = \sum_{k=1}^{K} (\dot{N})_k + \sum_{i=1}^{\mathcal{C}} \sum_{j=1}^{\mathcal{M}} \nu_{ij} \frac{dX_j}{dt} \qquad \text{(molar basis)}$$

$$\frac{dM}{dt} = \sum_{k=1}^{K} (\dot{M})_k \qquad \text{(mass basis)}$$

Energy balance

$$\frac{dU}{dt} = \sum_{k=1}^{K} (\dot{N}\underline{H})_k + \dot{Q} + \dot{W}_s - P\frac{dV}{dt} \qquad \text{(molar basis)}$$

$$\frac{dU}{dt} = \sum_{k=1}^{K} (\dot{M}\hat{H})_k + \dot{Q} + \dot{W}_s - P\frac{dV}{dt} \qquad \text{(mass basis)}$$

Entropy balance

$$\frac{dS}{dt} = \sum_{k=1}^{K} (\dot{N}\underline{S})_k + \frac{\dot{Q}}{T} + \dot{S}_{\text{gen}} \qquad \text{(molar basis)}$$

$$\frac{dS}{dt} = \sum_{k=1}^{K} (\dot{M}\hat{S})_k + \frac{\dot{Q}}{T} + \dot{S}_{\text{gen}} \qquad \text{(mass basis)}$$

where

$$(\dot{N})_k = \sum_{i=1}^{\mathcal{C}} (\dot{N}_i)_k \qquad (\dot{N}\underline{H})_k = \sum_{i=1}^{\mathcal{C}} (\dot{N}_i\overline{H}_i)_k \qquad (\dot{N}\underline{S})_k = \sum_{i=1}^{\mathcal{C}} (\dot{N}_i\overline{S}_i)_k$$

and

$$(\dot{M})_k = \sum_{i=1}^{\mathcal{C}} (\dot{M}_i)_k$$

Table 6.4-2
THE DIFFERENCE FORM OF THE EQUATIONS OF CHANGE FOR A MULTICOMPONENT MIXTURE

Species balance

$$N_i(t_2) - N_i(t_1) = \sum_{k=1}^{K} \int_{t_1}^{t_2} (\dot{N}_i)_k \, dt + \sum_{j=1}^{\mathcal{M}} \nu_{ij} X_j$$

$$M_i(t_2) - M_i(t_1) = \sum_{k=1}^{K} \int_{t_1}^{t_2} (\dot{M}_i)_k \, dt + \sum_{j=1}^{\mathcal{M}} \nu_{ij} m_i X_j$$

Overall mass balance

$$N(t_2) - N(t_1) = \sum_{k=1}^{K} \int_{t_1}^{t_2} (\dot{N})_k \, dt + \sum_{i=1}^{\mathcal{C}} \sum_{j=1}^{\mathcal{M}} \nu_{ij} X_j$$

$$M(t_2) - M(t_1) = \sum_{k=1}^{K} \int_{t_1}^{t_2} (\dot{M})_k \, dt$$

Energy balance

$$U(t_2) - U(t_1) = \sum_{k=1}^{K} \int_{t_1}^{t_2} (\dot{N}\underline{H})_k \, dt + Q + W_s - \int P \, dV$$

$$U(t_2) - U(t_1) = \sum_{k=1}^{K} \int_{t_1}^{t_2} (\dot{M}\hat{H})_k \, dt + Q + W_s - \int P \, dV$$

Entropy balance

$$S(t_2) - S(t_1) = \sum_{k=1}^{K} \int_{t_1}^{t_2} (\dot{N}\underline{S})_k \, dt + \int_{t_1}^{t_2} \frac{\dot{Q}}{T} \, dt + S_{\text{gen}}$$

$$S(t_2) - S(t_1) = \sum_{k=1}^{K} \int_{t_1}^{t_2} (\dot{M}\hat{S})_k \, dt + \int_{t_1}^{t_2} \frac{\dot{Q}}{T} \, dt + S_{\text{gen}}$$

where $(\dot{N})_k$, $(\dot{N}\underline{H})_k$, $(\dot{N}\underline{S})_k$, and $(\dot{M})_k$ are defined in Table 6.4-1

ILLUSTRATION 6.4-1

A continuous flow steam-heated mixing kettle will be used to produce a 20 wt% sulfuric acid solution at 65.56°C from a solution of 90 wt% sulfuric acid at 0°C and pure water at 21.1°C.
Estimate

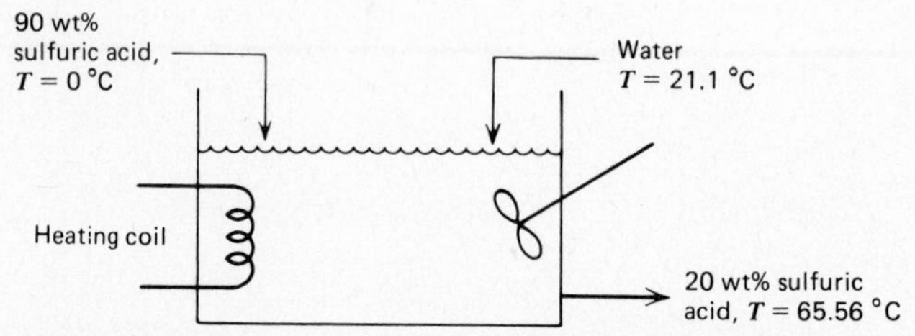

a. The kilograms of pure water needed per kilogram of initial sulfuric acid solution to produce a mixture of the desired concentration.

b. The amount of heat needed per kilogram of initial sulfuric acid solution to heat the mixture to the desired temperature.

c. The temperature of the kettle effluent if the mixing process were carried out adiabatically.

Solution

We choose the contents of the mixing kettle as the system. The difference form of the equations of change will be used for a time interval in which 1 kg of concentrated sulfuric acid enters the kettle.

a. Since there is no chemical reaction, and the mixing tank operates continuously, the total and species mass balances reduce to

$$0 = \sum_{k=1}^{3} (M)_k \qquad \text{and} \qquad \sum_{k=1}^{3} (M_{H_2SO_4})_k = 0$$

Denoting the 90 wt% acid stream by the subscript 1 and its mass flow by $(M)_1$, the water stream by the subscript 2, and the dilute acid stream by subscript 3, we have, from the total mass balance,

$$0 = (M)_1 + (M)_2 + (M)_3 = (M)_1 + Z(M)_1 + (M)_3$$

or

$$(M)_3 = -(1 + Z)(M)_1$$

where Z is equal to the kilograms of water used per kilogram of the 90 wt% acid. Also, from the mass balance on sulfuric acid, we have

$$0 = 0.90(M)_1 + 0(M)_2 + 0.20(M)_3 = 0.90(M)_1 - 0.20(1 + Z)(M)_1$$

Therefore,

$$1 + Z = \frac{0.90}{0.20} = 4.5 \qquad Z = 3.5$$

so that 3.5 kg of water must be added to each 1 kg of 90 wt% acid solution to produce a 20 wt% solution.

b. The steady-state energy balance is

$$0 = \sum (M\hat{H})_k + Q$$

since $W_s = 0$ and $\int P\,dV = 0$. From the mass balance of part a

$$(M)_2 = 3.5(M)_1$$

$$(M)_3 = -4.5(M)_1$$

From the enthalpy-concentration chart, Fig. 6.1-1, we have

$$\hat{H}_1 = \hat{H}(90 \text{ wt\% } H_2SO_4,\ T = 0°\text{C}) = -183 \text{ kJ/kg}$$

$$\hat{H}_2 = \hat{H}(\text{pure } H_2O,\ T = 21.1°\text{C}) = 91 \text{ kJ/kg}$$

$$\hat{H}_3 = \hat{H}(20 \text{ wt\% } H_2SO_4,\ T = 65.56°\text{C}) = 87 \text{ kJ/kg}$$

so that

$$\begin{aligned} Q &= (4.5 \times 87 - 3.5 \times 91 - 1 \times (-183)) \\ &= (391.5 - 318.5 + 183) = 256 \text{ kJ/kg of initial acid solution} \end{aligned}$$

c. For adiabatic operation, the energy balance is

$$0 = \sum_k (M\hat{H})_k$$

or

$$0 = 4.5\hat{H}_3 - 3.5(91) - 1(-183)$$

$$4.5\hat{H}_3 = 135.5 \qquad \hat{H}_3 = 58.3 \text{ kJ/kg}$$

Referring to the enthalpy-concentration diagram, we find that $T \sim 50°C$. ■

6.5 THE HEAT OF REACTION AND A CONVENTION FOR THE THERMODYNAMIC PROPERTIES OF REACTING MIXTURES

In the first part of this book we were interested in the change in the internal energy, enthalpy, and entropy accompanying a change in the thermodynamic state of a pure substance. For convenience in such calculations, one state of each substance was chosen as the reference state with zero values for both enthalpy and entropy, and the values of the thermodynamic properties of the substance in other states were given relative to this reference state. Thus in the steam tables in Appendix III the enthalpy and entropy are zero for liquid water at the triple point, whereas the zero values of these thermodynamic properties for methane (Fig. 2.4-2) and nitrogen (Fig. 2.4-3) correspond to different conditions of temperature and pressure. The reference states for sulfuric acid and water in the enthalpy-concentration diagram of Fig. 6.1-1 were also chosen on the basis of convenience. Such arbitrary choices for the reference state were satisfactory because our interest was only with the changes in thermodynamic properties in going from one thermodynamic state to another in a nonreacting system. However, when chemical reactions occur, the reference state for the thermodynamic properties of each species must be chosen with greater care. In particular, the thermodynamic properties for each species must be such that the differences between the reactant and product species in any chemical transformation will be equal to that which would be measured in the appropriate experiment. For example, in the ideal gas-phase reaction

$$H_2 + \tfrac{1}{2}O_2 \rightarrow H_2O$$

it is observed that 241.83 kJ are liberated for each mole of water vapor produced when this reaction is run in an isothermal, constant-pressure calorimeter at 25°C, 1.013 bar = 1 atm, and with all species in the vapor phase. Clearly, then, the enthalpies of the reacting molecules must be related as follows:

$$\begin{aligned}\Delta H(T = 25°C, P = 1.013 \text{ bar}) &= \underline{H}_{H_2O}(\text{vapor}, T = 25°C, P = 1.013 \text{ bar}) \\ &\quad - \underline{H}_{H_2}(\text{gas}, T = 25°C, P = 1.013 \text{ bar}) \\ &\quad - \tfrac{1}{2}\underline{H}_{O_2}(\text{gas}, T = 25°C, P = 1.013 \text{ bar}) \\ &= -241.83 \frac{\text{kJ}}{\text{mol } H_2O \text{ produced}}\end{aligned}$$

so that we are not free to choose the value of enthalpy of hydrogen, oxygen, and water vapor all arbitrarily. A similar argument applies to the other thermodynamic properties such as the entropy and Gibbs free energy.

By considering a large collection of chemical reactions, one finds that, to be consistent with the heat of reaction data for all possible reactions, the zero value of internal energy or enthalpy can be set *only once for each element*. Thus, one could set the enthalpy of each element to be zero at 25°C and 1.013 bar (or some other reference state) and then determine the enthalpy of every compound by measuring the heat liberated or absorbed during its production, by isothermal chemical reaction, from its elements.

Although such a procedure is a conceptually pleasing method of devising a thermodynamic enthalpy scale, it is experimentally difficult. For example, to determine the enthalpy of water vapor, one would have to measure the heat liberated on its formation from hydrogen and oxygen atoms. However, since hydrogen and oxygen molecules, and not their atoms, are the thermodynamically stable species at 25°C, one would first have to dissociate the oxygen and hydrogen molecules into their constituent atoms and then accurately measure the heat evolved when the atoms combined to form a water molecule—a very difficult task. Instead, the thermodynamic energy scale that is most frequently used is based on choosing as the zero state of both enthalpy and Gibbs free energy[12] of each atomic species its simplest thermodynamically stable state at 25°C and 1 atm = 1.013 bar. Thus, the reference state for argon and helium is as the atomic gases; for oxygen, nitrogen, and hydrogen, it is as molecular gases; for mercury and bromine, it is the atomic and molecular liquids, respectively; whereas for iron and carbon, it is as the α-crystalline and graphite solids, respectively, all at 25°C and 1.013 bar.

Starting from this basis, and using the data from a large number of heats of reaction, heats of mixing, and chemical equilibrium measurements, the enthalpy and Gibbs free energy content of all other molecular species relative to their constituent atoms in their reference states can be determined; these quantities are called the **enthalpy of formation** $\Delta \underline{H}_f^\circ$ and the **Gibbs free energy of formation,** $\Delta \underline{G}_f^\circ$. Thus by definition

$$
\begin{aligned}
\Delta \underline{H}^\circ_{f,\mathrm{H_2O}}(\text{vapor, 25°C, 1.013 bar}) &= \underline{H}_{\mathrm{H_2O}}(\text{vapor, 25°C, 1.013 bar}) \\
&\quad - \underline{H}_{\mathrm{H_2}}(\text{gas, 25°C, 1.013 bar}) \\
&\quad - \tfrac{1}{2}\underline{H}_{\mathrm{O_2}}(\text{gas, 25°C, 1.013 bar})
\end{aligned}
$$

Appendix IV contains a listing of $\Delta \underline{H}_f^\circ$ and $\Delta \underline{G}_f^\circ$ for a large collection of substances in their normal states of aggregation at 25°C and 1.013 bar.[13]

Isothermal heats (enthalpies) and Gibbs free energies of formation of species may be summed to compute the enthalpy change and Gibbs free energy change that would occur if the molecular species at 25°C, 1.013 bar, and the state of aggregation listed in Appendix IV reacted to form products at 25°C, 1.013 bar, and the listed state of aggregation. We will denote these changes by ΔH°_{rxn} (25°C, 1.013 bar) and ΔG°_{rxn} (25°C, 1.013 bar), respectively. For example, for the gas-phase reaction

$$3\mathrm{NO_2} + \mathrm{H_2O} = 2\mathrm{HNO_3} + \mathrm{NO}$$

we have

$$
\begin{aligned}
\Delta H^\circ_{rxn}(\text{25°C, 1 atm}) &= 2\underline{H}_{\mathrm{HNO_3}}(\text{25°C, 1.013 bar}) + \underline{H}_{\mathrm{NO}}(\text{25°C, 1.013 bar}) \\
&\quad -3\underline{H}_{\mathrm{NO_2}}(\text{25°C, 1.013 bar}) - \underline{H}_{\mathrm{H_2O}}(\text{25°C, 1.013 bar})
\end{aligned}
$$

[12]For chemical equilibrium the Gibbs free energy, rather than the entropy, is of central importance. Therefore, generally the enthalpy and Gibbs free energy are set equal to zero in the reference state.

[13]Unfortunately, the table in Appendix IV is not in SI units in that pressure is in atmospheres and energy is in calories.

$$= 2[\Delta \underline{H}^\circ_{f,HNO_3} + \tfrac{3}{2}\underline{H}_{O_2} + \tfrac{1}{2}\underline{H}_{H_2} + \tfrac{1}{2}\underline{H}_{N_2}]_{25^\circ C, 1.013\ bar}$$
$$+ [\Delta \underline{H}^\circ_{f,NO} + \tfrac{1}{2}\underline{H}_{O_2} + \tfrac{1}{2}\underline{H}_{N_2}]_{25^\circ C, 1.013\ bar}$$
$$- 3[\Delta \underline{H}^\circ_{f,NO_2} + \underline{H}_{O_2} + \tfrac{1}{2}\underline{H}_{N_2}]_{25^\circ C, 1.013\ bar}$$
$$- [\Delta \underline{H}^\circ_{f,H_2O} + \underline{H}_{H_2} + \tfrac{1}{2}\underline{H}_{O_2}]_{25^\circ C, 1.013\ bar}$$
$$= [2\Delta \underline{H}^\circ_{f,HNO_3} + \Delta \underline{H}^\circ_{f,NO} - 3\Delta \underline{H}^\circ_{f,NO_2} - \Delta \underline{H}^\circ_{f,H_2O}]_{25^\circ C, 1.013\ bar}$$
$$= \sum \nu_i \Delta \underline{H}^\circ_{f,i}(25^\circ C,\ 1.013\ \text{bar})$$

Notice that the enthalpies of the reference state atomic species cancel; by stoichiometry this will always occur. Thus, we have, as general results, that

$$\Delta H^\circ_{rxn}(25^\circ C,\ 1.013\ \text{bar}) = \sum_i \nu_i \Delta \underline{H}^\circ_{f,i}(25^\circ C,\ 1.013\ \text{bar}) \tag{6.5-1}$$

and

$$\Delta G^\circ_{rxn}(25^\circ C,\ 1.013\ \text{bar}) = \sum_i \nu_i \Delta \underline{G}^\circ_{f,i}(25^\circ C,\ 1.013\ \text{bar}) \tag{6.5-2}$$

In some cases, notably with salts that dissociate when dissolved in aqueous solution, it is convenient to have enthalpy and Gibbs free energy of formation data for other than the pure component state. Consequently, Appendix IV contains entries for several substances in aqueous solution. This is designated by the notation aq, N, where N denotes the number of moles of water per mole of solute; $N = \infty$ corresponds to the limit of infinite dilution of the solute. In these cases the enthalpy of formation is for the dilution indicated, but the Gibbs free energy of formation is that for the species in an ideal 1 molal solution (see Sec. 7.8). To compute the enthalpy or Gibbs free energy of a solute in solution at temperatures, pressures, and dilutions other than those listed in Appendix IV, data on the properties of the mixture, especially heats of mixing, partial molar heat capacities, and partial molar volumes, are needed.

In this book we will use the term **standard state** of a substance to indicate the state of aggregation given in Appendix IV (or, in certain cases, the reference states discussed in Chapter 7), at the temperature T and a pressure of 1 atm = 1.013 bar. The **standard heat of reaction** at temperature T, $\Delta H^\circ_{rxn}(T)$, is defined as the change in enthalpy that results when a stoichiometric amount of the reactants, each in their standard states at the temperature T, chemically react to form the reaction products in their standard states at the temperature T. Thus, in analogy with Eqs. 6.5-1 and 2

$$\Delta H^\circ_{rxn}(T,\ P = 1.013\ \text{bar}) = \sum_i \nu_i \Delta \underline{H}^\circ_{f,i}(T,\ P = 1.013\ \text{bar}) \tag{6.5-3}$$

and

$$\Delta G^\circ_{rxn}(T,\ P = 1.013\ \text{bar}) = \sum_i \nu_i \Delta \underline{G}^\circ_{f,i}(T,\ P = 1.013\ \text{bar}) \tag{6.5-4}$$

where the subscript i indicates the species and the superscript $^\circ$ denotes the standard state. From

$$\underline{H}^\circ_i(T, P) = \underline{H}^\circ_i(T_0, P) + \int_{T_0}^{T} C^\circ_{P,i}\, dT$$

and the fact that the enthalpies of the standard state cancel in Eq. 6.5-1 we have that

$$\Delta H^\circ_{rxn}(T, 1.013 \text{ bar}) = \sum_i \nu_i \Delta \underline{H}^\circ_{f,i}(25°\text{C}, 1.013 \text{ bar}) + \sum_i \nu_i \int_{T=25°\text{C}}^{T} C^\circ_{\text{P},i}\, dT$$

$$= \Delta H^\circ_{rxn}(25°\text{C}, 1.013 \text{ bar}) + \sum_i \nu_i \int_{T=25°\text{C}}^{T} C^\circ_{\text{P},i}\, dT \qquad (6.5\text{-}5)$$

where $C^\circ_{\text{P},i}$ is the heat capacity of species i in its standard state.

In certain instances, such as in the study of large organic compounds that require a complicated synthesis procedure, it is not possible to measure the heat of formation directly. A substitute procedure in these cases is to measure the energy change for some other reaction, usually the heat of combustion; that is, the energy liberated when the compound is completely oxidized (all the carbon is oxidized to carbon dioxide, all the hydrogen to water, etc.) The **standard heat of combustion** $\Delta \underline{H}^\circ_{\text{C}}(T)$ is defined to be the heat of combustion with both the reactants and the products in their standard states, at the temperature T. The standard heat of combustion at 25°C and 1.013 bar is listed in Appendix V for a number of compounds. (Note that there are two entries in this table; one corresponding to the liquid phase and the other to the vapor phase being the standard state for water.) Given the standard heat of combustion, the standard heat of reaction is computed as indicated in Illustration 6.5-1.

ILLUSTRATION 6.5-1

Compute the standard heat of reaction for the hydrogenation of benzene to cyclohexane

$$C_6H_6 + 3H_2 \rightarrow C_6H_{12}$$

from the standard heat of combustion data.

Solution

The standard heat of reaction can, in principle, be computed from Eq. 6.5-3; however, the enthalpy of formation of cyclohexane is unavailable. From the standard heat of combustion data in Appendix V we have that $\Delta \underline{H}^\circ_{\text{C}} = -936{,}880$ cal/mol of cyclohexane for the following reaction

$$C_6H_{12}(l) + 9O_2 \rightarrow 6CO_2 + 6H_2O(l)$$

Thus

$$\Delta \underline{H}^\circ_{\text{C},C_6H_{12}} = 6\underline{H}_{CO_2} + 6\underline{H}_{H_2O} - \underline{H}_{C_6H_{12}} - 9\underline{H}_{O_2} = -936{,}880 \text{ cal/mol}$$

or

$$\underline{H}_{C_6H_{12}} = -\Delta \underline{H}^\circ_{\text{C},C_6H_{12}} - 9\underline{H}_{O_2} + 6\underline{H}_{CO_2} + 6\underline{H}_{H_2O}$$

Similarly,

$$\underline{H}_{C_6H_6} = -\Delta \underline{H}^\circ_{\text{C},C_6H_6} - 7\tfrac{1}{2}\underline{H}_{O_2} + 6\underline{H}_{CO_2} + 3\underline{H}_{H_2O}$$

and

$$3\underline{H}_{H_2} = -3\Delta \underline{H}^\circ_{\text{C},H_2} - 1\tfrac{1}{2}\underline{H}_{O_2} + 3\underline{H}_{H_2O}$$

Therefore,

$$\begin{aligned}\Delta H^\circ_{rxn} = & -\Delta \underline{H}^\circ_{\text{C},C_6H_{12}} - 9\underline{H}_{O_2} + 6\underline{H}_{CO_2} + 6\underline{H}_{H_2O} \\ & + \Delta \underline{H}^\circ_{\text{C},C_6H_6} + 7\tfrac{1}{2}\underline{H}_{O_2} - 6\underline{H}_{CO_2} - 3\underline{H}_{H_2O} \\ & + 3\Delta \underline{H}^\circ_{\text{C},H_2} + 1\tfrac{1}{2}\underline{H}_{O_2} \qquad - 3\underline{H}_{H_2O}\end{aligned}$$

$$= -\Delta \underline{H}^\circ_{C,C_6H_{12}} + \Delta \underline{H}^\circ_{C,C_6H_6} + 3\Delta \underline{H}^\circ_{C,H_2} = -\sum_i \nu_i \Delta \underline{H}^\circ_{C,i}$$

$$= +936{,}880 - 780{,}980 - 3 \times 68{,}317.4 = -49{,}052 \frac{\text{cal}}{\text{mol benzene}}$$

$$= -205.23 \frac{\text{kJ}}{\text{mol benzene}}$$

Comment

Note that in the final equation, $\Delta H^\circ_{rxn} = -\Sigma \nu_i \Delta \underline{H}^\circ_{C,i}$, the enthalpies of the reference state atomic species cancel as they must owing to conservation of atomic species on chemical reaction. ■

The equation developed in this illustration,

$$\Delta H^\circ_{rxn}(T = 25°\text{C}, P = 1.013 \text{ bar}) = -\sum_i \nu_i \Delta \underline{H}^\circ_{C,i}(T = 25°\text{C}, P = 1.013 \text{ bar}) \quad (6.5\text{-}6)$$

is always valid and provides a way of computing the standard heat of reaction from standard heat of combustion data.

ILLUSTRATION 6.5-2

Compute the standard state heat of reaction for the gas-phase reaction $N_2O_4 = 2NO_2$ over the temperature range of 200 to 600 K.[14]

Data

See Appendixes II and IV.

Solution

The heat of reaction at a temperature T can be computed from

$$\Delta H^\circ_{rxn}(T) = \sum_i \nu_i \Delta \underline{H}^\circ_{f,i}(T)$$

At $T = 25°\text{C}$ we have, from the data in Appendix IV, that for each mole of N_2O_4 reacted

$$\begin{aligned}\Delta H^\circ_{rxn}(T = 25°\text{C}) &= [2 \times 7.96 - 2.23] \text{ kcal/mol} \\ &= 13{,}690 \text{ cal/mol } N_2O_4 \text{ reacted} \\ &= 13.69 \text{ kcal/mol} = 57.28 \text{ kJ/mol}\end{aligned}$$

To compute the heat of reaction at any temperature T, we start from Eq. 6.5-5 and note that since the standard state for each species is a low-pressure gas, $C^\circ_{P,i} = C^*_{P,i}$. Therefore,

$$\Delta H^\circ_{rxn}(T) = \Delta H^\circ_{rxn}(T = 25°\text{C}) + \int_{T=298.2\text{ K}}^{T} \sum_i \nu_i C^*_{P,i}\, dT$$

For the case here we have, from Appendix II, that

$$\begin{aligned}\sum \nu_i C^*_{P,i} &= 2C^*_{P,NO_2} - C^*_{P,N_2O_4} \\ &= 3.06 - 1.73 \times 10^{-2}T + 1.028 \times 10^{-5}T^2 + 3.76 \times 10^{-9}T^3 \text{ cal/mol K}\end{aligned}$$

[14]Since, as we will see, $\overline{H}_i = \underline{H}_i$ for an ideal gas mixture, the standard state heat of reaction and the actual heat of reaction are identical in this case.

Thus

$$\Delta H^\circ_{rxn}(T) = 13{,}690 + \int_{298.2}^{T} (3.06 - 1.73 \times 10^{-2}T + 1.028 \times 10^{-5}T^2 + 3.76 \times 10^{-9}T^3)\, dT$$

$$= 13{,}690 + 3.06(T - 298.2) - \frac{1.73}{2} \times 10^{-2}(T^2 - 298.2^2)$$

$$+ \frac{1.028}{3} \times 10^{-5}(T^3 - 298.2^3) + \frac{3.76}{4} \times 10^{-9}(T^4 - 298.2^4)$$

$$= 13{,}449 + 3.06T - 0.865 \times 10^{-2}T^2 + 0.343 \times 10^{-5}T^3$$

$$+ 0.940 \times 10^{-9}T^4 \text{ cal/mol } N_2O_4$$

Values of ΔH°_{rxn} for various values of T are given in the table.

T(K)	200	300	400	500	600
ΔH°_{rxn}(cal/mol N_2O_4)	13,744	13,689	13,533	13,304	13,034
ΔH°_{rxn}(kJ/mol N_2O_4)	57.505	57.275	56.622	55.664	54.534

Comment

The heat of reaction can be, and in fact usually is, a much stronger function of temperature than is the case here. ■

6.6 THE EXPERIMENTAL DETERMINATION OF THE PARTIAL MOLAR VOLUME AND ENTHALPY

Experimental values for some of the partial molar quantities introduced in Sec. 6.1 can be obtained from laboratory measurements on mixtures. In particular, mixture density measurements can be used to obtain partial molar volumes, and heat of mixing data yields information on partial molar enthalpies. Both these experiments will be considered here. In Chapter 8 equilibrium experiments that provide information on the partial molar Gibbs free energy of a component in a mixture will be discussed. Once the partial molar enthalpy and partial molar Gibbs free energy are known at the same temperature, the partial molar entropy can be computed from the relation $\bar{S}_i = (\bar{G}_i - \bar{H}_i)/T$.

Table 6.6-1 contains data on the mixture density, at constant temperature and pressure, for the water(1)–methanol(2) system. Column 4 of this table contains calculated values for the molar volume change on mixing (i.e., $\Delta \underline{V}_{\text{mix}} = \underline{V} - x_1\underline{V}_1 - x_2\underline{V}_2$) at various compositions; these data have also been plotted in Fig. 6.6-1. The slope of the $\Delta \underline{V}_{\text{mix}}$ versus mole fraction curve at any mole fraction is related to the partial molar volumes. To obtain this relationship we start from

$$\Delta \underline{V}_{\text{mix}} = (x_1\bar{V}_1 + x_2\bar{V}_2) - x_1\underline{V}_1 - x_2\underline{V}_2 = x_1(\bar{V}_1 - \underline{V}_1) + x_2(\bar{V}_2 - \underline{V}_2) \quad (6.6\text{-}1)$$

so that

$$\left.\frac{\partial(\Delta \underline{V}_{\text{mix}})}{\partial x_1}\right|_{T,P} = (\bar{V}_1 - \underline{V}_1) + x_1\left(\frac{\partial \bar{V}_1}{\partial x_1}\right)_{T,P} - (\bar{V}_2 - \underline{V}_2) + x_2\left(\frac{\partial \bar{V}_2}{\partial x_1}\right)_{T,P} \quad (6.6\text{-}2)$$

Table 6.6-1
DENSITY DATA FOR THE WATER(1)–METHANOL(2) SYSTEM AT $T = 298.15$ K

x_1	ρ(kg/m³)	$\underline{V}$(m³/mol) × 10⁶	$\Delta\underline{V}_{mix}$(m³/mol) × 10⁶
0.0	786.846	40.7221	0
0.1162	806.655	37.7015	−0.3883
0.2221	825.959	35.0219	−0.6688
0.2841	837.504	33.5007	−0.7855
0.3729	855.031	31.3572	−0.9174
0.4186	864.245	30.2812	−0.9581
0.5266	887.222	27.7895	−1.0032
0.6119	905.376	25.9108	−0.9496
0.7220	929.537	23.5759	−0.7904
0.8509	957.522	20.9986	−0.4476
0.9489	981.906	19.0772	−0.1490
1.0	997.047	18.0686	0

*Note that the notation $\underline{V}(\text{m}^3/\text{mol}) \times 10^6$ means that the entries in the table have been multiplied by the factor 10^6. Therefore, for example, the volume of pure methanol $\underline{V}(x_1 = 0)$ is 40.7221×10^{-6} m³/mol = 40.7221 cm³/mol.

Here we have used the facts that the pure component molar volumes are independent of mixture composition (i.e., $(\partial \underline{V}_i/\partial x_1)_{T,P} = 0$), and that for a binary mixture $x_2 = 1 - x_1$, so $(\partial x_2/\partial x_1) = -1$.

From the Gibbs–Duhem equation, Eq. 6.2-19b with $\bar{\theta}_i = \bar{V}_i$, we have

$$x_1 \left(\frac{\partial \bar{V}_1}{\partial x_1}\right)_{T,P} + x_2 \left(\frac{\partial \bar{V}_2}{\partial x_1}\right)_{T,P} = 0$$

so that Eq. 6.6-2 becomes

$$\left.\frac{\partial(\Delta \underline{V}_{mix})}{\partial x_1}\right|_{T,P} = (\bar{V}_1 - \underline{V}_1) - (\bar{V}_2 - \underline{V}_2) \tag{6.6-3}$$

Now multiplying this equation by x_1 and subtracting the result from Eq. 6.6-1 gives

$$\Delta \underline{V}_{mix} - x_1 \left.\frac{\partial(\Delta \underline{V}_{mix})}{\partial x_1}\right|_{T,P} = (\bar{V}_2 - \underline{V}_2) \tag{6.6-4a}$$

Similarly,

$$\begin{aligned}\Delta \underline{V}_{mix} - x_2 \left.\frac{\partial(\Delta \underline{V}_{mix})}{\partial x_2}\right|_{T,P} &= \Delta \underline{V}_{mix} + x_2 \left.\frac{\partial(\Delta \underline{V}_{mix})}{\partial x_1}\right|_{T,P} \\ &= (\bar{V}_1 - \underline{V}_1)\end{aligned} \tag{6.6-4b}$$

Therefore, given data for the volume change on mixing as a function of concentration, so that $\Delta \underline{V}_{mix}$ and the derivative $\partial(\Delta \underline{V}_{mix})/\partial x_1$ can be evaluated at x_1, we can immediately compute $(\bar{V}_1 - \underline{V}_1)$ and $(\bar{V}_2 - \underline{V}_2)$ at this composition. A knowledge of the pure component molar volumes, then, is all that is necessary to compute $\bar{V}_1$ and $\bar{V}_2$ at the specified composition x_1. By repeating the calculation at other values of the mole fraction, the complete partial molar volume versus

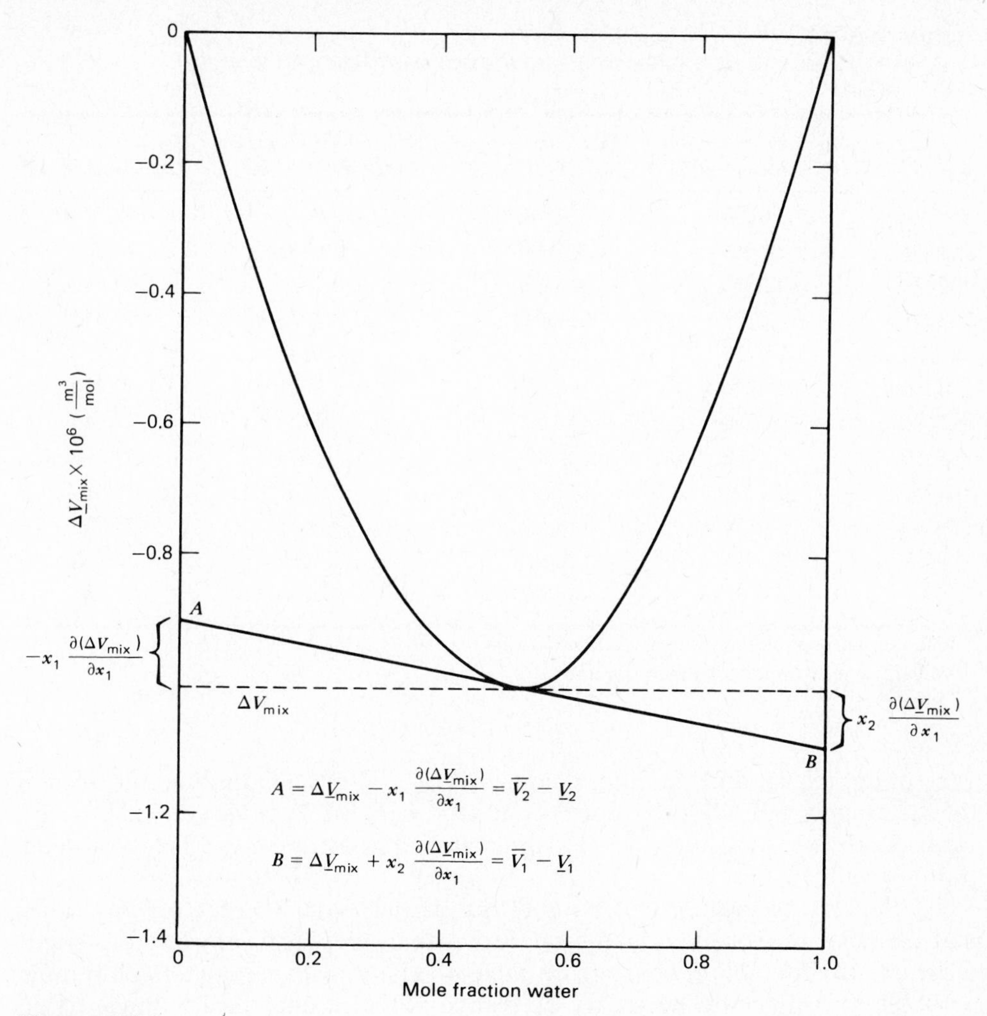

Figure 6.6-1
The isothermal volume change on mixing for the water–methanol system.

composition curve can be obtained. The results of this computation are given in Table 6.6-2.

It is also possible to evaluate $(\overline{V}_1 - \underline{V}_1)$ and $(\overline{V}_2 - \underline{V}_2)$ in a more direct, graphical manner. At a given composition, say x_1^*, a tangent line to the $\Delta\underline{V}_{mix}$ curve is drawn; the intersections of this tangent line with the ordinate at $x_1 = 0$ and $x_1 = 1$ are designated by the symbols A and B in Fig. 6.6-1. The slope of this tangent line is

$$\left.\frac{\partial(\Delta\underline{V}_{mix})}{\partial x_1}\right|_{x_1^*}$$

and

$$-x_1^* \left.\frac{\partial(\Delta\underline{V}_{mix})}{\partial x_1}\right|_{x_1^*}$$

represents the distance indicated in Fig. 6.6-1. Referring to the figure, it is evident that the numerical value on the ordinate at point A is equal to the left

Table 6.6-2
THE PARTIAL MOLAR VOLUMES OF THE WATER(1)–METHANOL(2) SYSTEM AT T = 298.15 K

x_1	$\overline{V}_1 - \underline{V}_1$ (m³/mol) × 10⁶	$\overline{V}_1$ (m³/mol) × 10⁶	$\overline{V}_2 - \underline{V}_2$ (m³/mol) × 10⁶	$\overline{V}_2$ (m³/mol) × 10⁶
0	−3.8893	14.180*	0	40.722†
0.1162	−2.9741	15.095	−0.0530	40.669
0.2221	−2.3833	15.686	−0.1727	40.549
0.2841	−2.0751	15.994	−0.2773	40.445
0.3729	−1.6452	16.424	−0.4884	40.234
0.4186	−1.4260	16.643	−0.6321	40.090
0.5266	−0.9260	17.143	−1.0822	39.640
0.6119	−0.5752	17.494	−1.5464	39.176
0.7220	−0.2294	17.840	−2.2363	38.486
0.8509	−0.0254	18.044	−2.9631	37.759
0.9489	−0.0026	18.072	−3.1689	37.553
1.0	0.	18.069†	−3.0348	37.687*

*Value of partial molar volume at infinite dilution.
†Value of pure component molar volume.

side of Eq. 6.6-4a, and, therefore, equal to $(\overline{V}_2 - \underline{V}_2)$. Similarly, the intersection of the tangent line with the ordinate at $x_1 = 1$ (point B) gives $(\overline{V}_1 - \underline{V}_1)$ at x_1^*. Thus, both $(\overline{V}_1 - \underline{V}_1)$ and $(\overline{V}_2 - \underline{V}_2)$ are obtained by a simple graphical construction.

For very accurate calculations of the partial molar volume (or any other partial molar property) an analytical, rather than graphical, procedure is used. First one fits the volume change on mixing, $\Delta\underline{V}_{\text{mix}}$, with a polynomial in mole fraction, and then the necessary derivative is found analytically. Since $\Delta\underline{V}_{\text{mix}}$ must equal zero at $x_1 = 0$ and $x_1 = 1$ $(x_2 = 0)$, it is usually fit with a polynomial of the **Redlich–Kister** form

$$\Delta\underline{V}_{\text{mix}} = x_1 x_2 \sum_{i=0}^{n} a_i (x_1 - x_2)^i \tag{6.6-5a}$$

(Similar expansions are also used for $\Delta\underline{H}_{\text{mix}}$, $\Delta\underline{U}_{\text{mix}}$, and the excess properties to be defined in Chapter 7.) Then, rewriting Eq. 6.6-5a we have

$$\Delta\underline{V}_{\text{mix}} = x_1(1 - x_1) \sum_{i=0}^{n} a_i (2x_1 - 1)^i$$

$$\left.\frac{\partial \Delta\underline{V}_{\text{mix}}}{\partial x_1}\right|_{T,P} = -\sum_{i=0}^{n} a_i (2x_1 - 1)^{i+1} + 2x_1(1 - x_1) \sum_{i=0}^{n} a_i i (2x_1 - 1)^{i-1} \tag{6.6-5b}$$

and

$$\Delta\underline{V}_{\text{mix}} - x_1 \left.\frac{\partial(\Delta\underline{V}_{\text{mix}})}{\partial x_1}\right|_{T,P} = \overline{V}_2 - \underline{V}_2$$

$$= x_1^2 \sum_i a_i [(x_1 - x_2)^i - 2ix_2(x_1 - x_2)^{i-1}] \tag{6.6-6a}$$

Also

$$\overline{V}_1 - \underline{V}_1 = \Delta\underline{V}_{\text{mix}} + x_2 \left.\frac{\partial(\Delta\underline{V}_{\text{mix}})}{\partial x_1}\right|_{T,P}$$

$$= x_2^2 \sum_i a_i[(x_1 - x_2)^i + 2ix_1(x_1 - x_2)^{i-1}] \tag{6.6-6b}$$

An accurate representation of the water–methanol data has been obtained using Eq. 6.6-5 with

$$a_0 = -4.0034 \times 10^{-6} \text{ m}^3/\text{mol}$$

$$a_1 = -0.17756 \times 10^{-6}$$

$$a_2 = 0.54139 \times 10^{-6}$$

$$a_3 = 0.60481 \times 10^{-6}$$

and the partial molar volumes in Table 6.6-2 have been computed using these constants and Eqs. 6.6-6.

Finally, we note that for the water–methanol system the volume change on mixing was negative, as were $(\overline{V}_1 - \underline{V}_1)$ and $(\overline{V}_2 - \underline{V}_2)$. This is not a general characteristic in that, depending on the system, these three quantities can be positive, negative, or even positive over part of the composition range and negative over the rest.

The partial molar enthalpy of a species in a binary mixture can be obtained by a similar analysis, but using enthalpy change on mixing (or heat of mixing) data. Such measurements are frequently made using the steady-state flow calorimeter schematically indicated in Fig. 6.6-2. Two streams, one of pure fluid 1 and the second of pure fluid 2, both at a temperature T and a pressure P enter this steady-state mixing device, and a single mixed stream, also at T and P, leaves. Taking the contents of the calorimeter to be the system, the mass and energy balances (from Eqs. 6.4-1 and 6.4-3) are

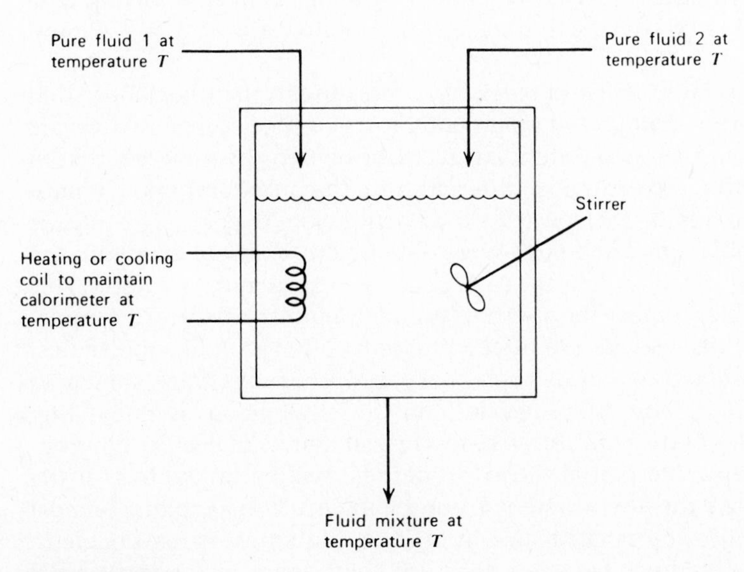

Figure 6.6-2
An isothermal flow calorimeter.

$$0 = (\dot{N})_1 + (\dot{N})_2 - (\dot{N})_3 \tag{6.6-7}$$

and

$$0 = (\dot{N})_1\underline{H}_1 + (\dot{N})_2\underline{H}_2 - (\dot{N})_3\underline{H}_{\text{mix}} + \dot{Q} \tag{6.6-8}$$

Thus

$$\dot{Q} = [(\dot{N})_1 + (\dot{N})_2]\underline{H}_{\text{mix}} - (\dot{N})_1\underline{H}_1 - (\dot{N})_2\underline{H}_2 = [(\dot{N})_1 + (\dot{N})_2]\Delta\underline{H}_{\text{mix}}$$

and

$$\Delta\underline{H}_{\text{mix}} = \dot{Q}/[(\dot{N})_1 + (\dot{N})_2]$$

where $\Delta\underline{H}_{\text{mix}} = \underline{H}_{\text{mix}} - x_1\underline{H}_1 - x_2\underline{H}_2$. Therefore, by monitoring $(\dot{N})_1$ and $(\dot{N})_2$ and the heat flow rate $\dot{Q}$ necessary to maintain constant temperature, the heat of mixing $\Delta\underline{H}_{\text{mix}} = \dot{Q}/[(\dot{N})_1 + (\dot{N})_2]$ can be determined at the composition $x_1 = (\dot{N})_1/[(\dot{N})_1 + (\dot{N})_2]$. Measurements at a collection of values of the ratio $\dot{N}_1/\dot{N}_2$ give the complete heat of mixing versus composition curve at fixed T and P.

Once the composition dependence of the heat of mixing is known, $\bar{H}_1$ and $\bar{H}_2$ may be computed in a manner completely analogous to the procedure used for the partial molar volumes. In particular, it is easily established that

$$\Delta\underline{H}_{\text{mix}} - x_1 \left.\frac{\partial(\Delta\underline{H}_{\text{mix}})}{\partial x_1}\right|_{T,P} = (\bar{H}_2 - \underline{H}_2) \tag{6.6-9a}$$

and

$$\Delta\underline{H}_{\text{mix}} - x_2 \left.\frac{\partial(\Delta\underline{H}_{\text{mix}})}{\partial x_2}\right|_{T,P} = \Delta\underline{H}_{\text{mix}} + x_2 \left.\frac{\partial(\Delta\underline{H}_{\text{mix}})}{\partial x_1}\right|_{T,P} = (\bar{H}_1 - \underline{H}_1) \tag{6.6-9b}$$

so that either the computational or graphical technique may also be used to calculate the partial molar enthalpy.

Table 6.6-3 and Fig. 6.6-3 contain the heat of mixing data for the water-methanol system. These data have been used to compute the partial molar enthalpies given in Table 6.6-4. Note that one feature of the heat of mixing data of Fig. 6.6-3 is that it is skewed, with the largest absolute value at $x_1 = 0.73$ (and not $x_1 = 0.5$).

The entries in Tables 6.6-2 and 6.6-4 are interesting in that they show that the partial molar volume and partial molar enthalpy of a species in a mixture are very similar to the pure component molar quantities when the mole fraction of that species is near unity and are most different from the pure component values at infinite dilution; that is, as the species mole fraction goes to zero. (The infinite dilution values in Tables 6.6-2 and 6.6-4 were obtained by extrapolating both the $\bar{V}$ and $\bar{H}$ versus mole fraction data for each species to zero mole fraction.) This behavior is reasonable because in a strongly nonequimolar mixture the molecules of the concentrated species are interacting most often with like molecules, so that their environment and thus their molar properties are very similar to those of the pure fluid. The dilute species, on the other hand, is interacting mostly with molecules of the concentrated species, so that its molecular environment, and consequently its partial molar properties, will be unlike that of the pure component. Since the environment around a molecule in a mixture is most dissimilar from its pure component state at infinite dilution, it is reasonable to expect the greatest difference between the pure component molar and partial molar properties to occur in this limit.

Table 6.6-3
HEAT OF MIXING DATA FOR THE WATER(1)–METHANOL(2) SYSTEM AT T = 19.69°C

x_1	Q^+ kJ/mol B	Q kJ/mol = $\Delta \underline{H}_{mix}$
0.05	−0.134	−0.127
0.10	−0.272	−0.245
0.15	−0.419	−0.356
0.20	−0.569	−0.455
0.25	−0.716	−0.537
0.30	−0.862	−0.603
0.35	−1.017	−0.661
0.40	−1.197	−0.718
0.45	−1.398	−0.769
0.50	−1.632	−0.816
0.55	−1.896	−0.853
0.60	−2.218	−0.887
0.65	−2.591	−0.907
0.70	−3.055	−0.917
0.75	−3.666	−0.917
0.80	−4.357	−0.871
0.85	−5.114	−0.767
0.90	−5.989	−0.599
0.95	−6.838	−0.342

Source: International Critical Tables, vol. 5, McGraw–Hill, New York, 1929, p. 159.
$Q = (1 - x_1)Q^+$

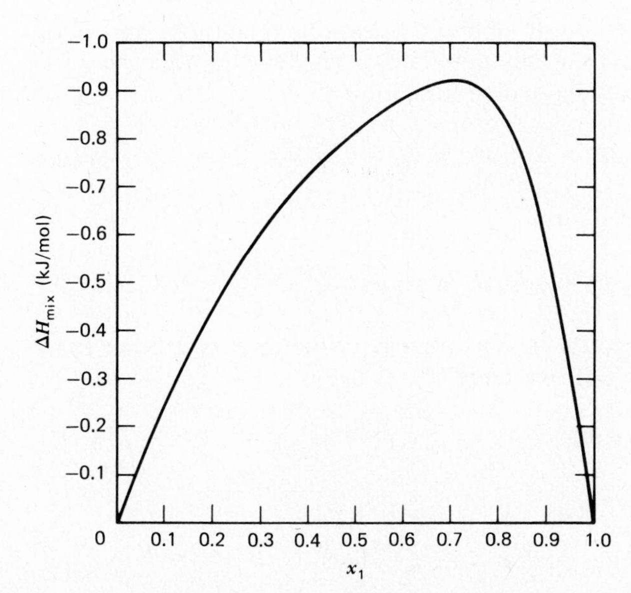

Figure 6.6-3
Heat of mixing data for the water (1)–methanol (2) system at T = 19.69°C.

Table 6.6-4
THE DIFFERENCE BETWEEN THE PARTIAL MOLAR AND PURE COMPONENT ENTHALPIES FOR THE WATER(1)–METHANOL(2) SYSTEM AT T = 16.69°C

x_1	$\bar{H}_1 - \underline{H}_1$ (kJ/mol)	$\bar{H}_2 - \underline{H}_2$ (kJ/mol)
0	−2.703*	0
0.05	−2.482	−0.006
0.10	−2.251	−0.025
0.15	−2.032	−0.056
0.20	−1.838	−0.097
0.25	−1.678	−0.143
0.30	−1.551	−0.191
0.35	−1.456	−0.237
0.40	−1.383	−0.280
0.45	−1.325	−0.323
0.50	−1.270	−0.373
0.55	−1.209	−0.441
0.60	−1.131	−0.548
0.65	−1.028	−0.719
0.70	−0.898	−0.992
0.75	−0.740	−1.412
0.80	−0.560	−2.036
0.85	−0.371	−2.935
0.90	−0.193	−4.192
0.95	−0.056	−5.905
1.00	0	−8.188*

*Indicates value at infinite dilution.

Finally, the analyses used here to obtain expressions relating $\bar{V}_i$ and $\bar{H}_i$ to $\Delta\underline{V}_{\text{mix}}$ and $\Delta\underline{H}_{\text{mix}}$, respectively, are easily generalized yielding the following for the partial molar property of any extensive function θ

$$\bar{\theta}_1(T, P, x_1) - \underline{\theta}_1(T, P) = \Delta\underline{\theta}_{\text{mix}}(T, P, x_1) - x_2 \left.\frac{\partial(\Delta\underline{\theta}_{\text{mix}})}{\partial x_2}\right|_{T,P} \tag{6.6-10a}$$

and

$$\bar{\theta}_2(T, P, x_2) - \underline{\theta}_2(T, P) = \Delta\underline{\theta}_{\text{mix}}(T, P, x_2) - x_1 \left.\frac{\partial(\Delta\underline{\theta}_{\text{mix}})}{\partial x_1}\right|_{T,P} \tag{6.6-10b}$$

One can also show that if $\underline{\theta}_{\text{mix}}$ is any molar property of the mixture (rather than change on mixing, which is $\Delta\underline{\theta}_{\text{mix}}$), we have

$$\bar{\theta}_1(T, P, x_1) = \underline{\theta}_{\text{mix}}(T, P, x_1) - x_2 \left.\frac{\partial(\underline{\theta}_{\text{mix}})}{\partial x_2}\right|_{T,P} \tag{6.6-11a}$$

and

$$\bar{\theta}_2(T, P, x_2) = \underline{\theta}_{\text{mix}}(T, P, x_2) - x_1 \left.\frac{\partial(\underline{\theta}_{\text{mix}})}{\partial x_1}\right|_{T,P} \tag{6.6-11b}$$

6.7 CRITERIA FOR PHASE EQUILIBRIUM IN MULTICOMPONENT SYSTEMS

An important observation of this chapter is that the equations of change for a multicomponent mixture are identical, in form, to those for a pure fluid. The difference between the two is that the pure fluid equations contain thermodynamic properties ($\underline{U}$, $\underline{H}$, $\underline{S}$, etc.) that can be computed from pure fluid equations of state and heat capacity data, whereas in the multicomponent case these thermodynamic properties can only be computed if the appropriate mixture equation of state and heat capacity data or enthalpy-concentration and entropy-concentration data are given, or if we otherwise have enough information to evaluate the necessary concentration-dependent partial molar quantities at all temperatures, pressures, and compositions of interest. Although this represents an important computational difference between the two sets of equations, it has no effect on their fundamental structure. Consequently, for a closed system, we have for both the pure component and multicomponent cases that

$$\frac{dU}{dt} = \dot{Q} + \dot{W}_s - P\frac{dV}{dt} \tag{6.7-1}$$

and

$$\frac{dS}{dt} = \frac{\dot{Q}}{T} + \dot{S}_{\text{gen}} \tag{6.7-2}$$

where

$$U = N\underline{U}(T, P) \qquad S = N\underline{S}(T, P) \qquad \left\{\begin{array}{l}\text{for the pure component}\\ \text{system (molar basis)}\end{array}\right.$$

and

$$U = \sum_{i=1}^{\mathscr{C}} N_i\bar{U}_i(T, P, x_1, \ldots x_{\mathscr{C}-1}) \qquad S = \sum_{i=1}^{\mathscr{C}} N_i\bar{S}_i(T, P, x_1, \ldots x_{\mathscr{C}-1}) \qquad \left\{\begin{array}{l}\text{for a multicomponent}\\ \text{system (molar basis)}\end{array}\right. \tag{6.7-3}$$

Since the form of the balance equations are unchanged, we can use, without modification, the analysis of the last chapter to establish that the equilibrium criteria for a closed multicomponent mixture are (Problem 6.20)

$$\begin{aligned} S &= \text{maximum for equilibrium at constant } M, U, \text{ and } V \\ A &= \text{minimum for equilibrium at constant } M, T, \text{ and } V \\ G &= \text{minimum for equilibrium at constant } M, T, \text{ and } P \end{aligned} \tag{6.7-4}$$

Thus, although it may be computationally more difficult to identify the equilibrium state in a multicomponent mixture than is the case for a pure fluid, the basic criteria used in this identification are the same.

As the first application of these criteria, consider the problem of identifying the state of equilibrium in a closed, nonreacting multicomponent system at constant internal energy and volume. To be specific, suppose N_1 moles of species 1, N_2 moles of species 2, and so on are put into an adiabatic container that will be maintained at constant volume, and that these species are only partially soluble in one another and do not chemically react. What we would like to be able to do is to predict the composition of each of the phases present at equilibrium. (A more difficult, but solvable problem is to also predict the number of phases that will be present. This problem is briefly considered in Chapter 8.) In the analysis that follows, we develop the equations that will be used in Chapter 8 to compute the equilibrium compositions.

The starting point for solving this problem is the general equilibrium criteria of Eqs. 6.7-4. In particular, the equilibrium criterion for a closed, adiabatic, constant-volume system is

$$S = \text{maximum}$$

subject to the constraints of constant U, V, and total number of moles of each species N_i. For the two-phase system each extensive property (e.g., N_i, S, U, V, etc.) is the sum of the properties for the individual phases, for example

$$N_i = N_i^{\mathrm{I}} + N_i^{\mathrm{II}}$$

where the superscripts I and II refer to the phase. In general, the problem of finding the extreme value of a function subject to constraints is not a straightforward task, as will become evident later. However, here this can be done easily. We start by setting the differential of the entropy for the two-phase system equal to zero

$$dS = dS^{\mathrm{I}} + dS^{\mathrm{II}} = 0 \tag{6.7-5}$$

and then use Eq. 6.2-4, rearranged as follows,

$$dS = \frac{1}{T}\,dU + \frac{P}{T}\,dV - \frac{1}{T}\sum_{i=1}^{\mathscr{C}} \overline{G}_i\,dN_i$$

for each phase. Now recognizing that since the total internal energy, total volume, and the number of moles of each species are fixed, we have

$$dU^{\mathrm{II}} = -dU^{\mathrm{I}}$$
$$dV^{\mathrm{II}} = -dV^{\mathrm{I}}$$
$$dN_i^{\mathrm{II}} = -dN_i^{\mathrm{I}}$$

which can be used in Eq. 6.7-5 to obtain

$$dS = \left(\frac{1}{T^{\mathrm{I}}} - \frac{1}{T^{\mathrm{II}}}\right) dU^{\mathrm{I}} + \left(\frac{P^{\mathrm{I}}}{T^{\mathrm{I}}} - \frac{P^{\mathrm{II}}}{T^{\mathrm{II}}}\right) dV^{\mathrm{I}}$$
$$- \sum_{i=1}^{\mathscr{C}} \left(\frac{\overline{G}_i^{\mathrm{I}}}{T^{\mathrm{I}}} - \frac{\overline{G}_i^{\mathrm{II}}}{T^{\mathrm{II}}}\right) dN_i^{\mathrm{I}} = 0 \tag{6.7-6}$$

The condition for equilibrium is that the differential of the entropy be zero with respect to all variations of the independent and unconstrained variables, here dU^{I}, dV^{I}, and dN_i^{I}. In order for Eq. 6.7-6 to be satisfied, we must have (1) that

$$\left(\frac{\partial S}{\partial U^{\mathrm{I}}}\right)_{V^{\mathrm{I}},N_i^{\mathrm{I}}} = 0$$

which implies that

$$\frac{1}{T^{\mathrm{I}}} = \frac{1}{T^{\mathrm{II}}} \qquad \text{or simply} \qquad T^{\mathrm{I}} = T^{\mathrm{II}} \tag{6.7-7a}$$

(2) that

$$\left(\frac{\partial S}{\partial V^{\mathrm{I}}}\right)_{U^{\mathrm{I}}, N_i^{\mathrm{I}}} = 0$$

which implies

$$\frac{P^{\mathrm{I}}}{T^{\mathrm{I}}} = \frac{P^{\mathrm{II}}}{T^{\mathrm{II}}}$$

or, in view of Eq. 6.7-7a that

$$P^{\mathrm{I}} = P^{\mathrm{II}} \tag{6.7-7b}$$

and (3) that $(\partial S/\partial N_i^{\mathrm{I}})_{U^{\mathrm{I}}, V^{\mathrm{I}}, N_{j \neq i}^{\mathrm{I}}} = 0$ for each species i, which implies that

$$\overline{G}_i^{\mathrm{I}} = \overline{G}_i^{\mathrm{II}} \qquad \text{for each species } i \tag{6.7-7c}$$

since $T^{\mathrm{I}} = T^{\mathrm{II}}$.

Thus, for phase equilibrium to exist in a closed, nonreacting multicomponent system at constant energy and volume, the pressure must be the same in both phases (so that mechanical equilibrium exists), the temperature must be the same in both phases (so that thermal equilibrium exists), and the partial molar Gibbs free energy of each species must be the same in each phase (so that equilibrium with respect to species diffusion exists[15]). Note that with the replacement of the partial molar Gibbs free energy by the pure component Gibbs free energy, Eqs. 6.7-7 become identical with the conditions for phase equilibrium in a one-component system derived in Sec. 5.1 (see Eqs. 5.1-9). In principle, we could now continue to follow the development of Chapter 5 and derive the conditions for stability of the equilibrium state. However, this task is algebraically complicated and will not be considered here.[16]

To derive the conditions for phase equilibrium in a closed system at constant (and, of course, uniform) temperature and pressure, we start from the equilibrium criterion that G should be a minimum and set the differential of G for the two-phase system equal to zero, that is,

$$dG = dG^{\mathrm{I}} + dG^{\mathrm{II}} = 0 \tag{6.7-8}$$

Now recognizing that at constant T and P (from Eq. 6.2-1)

$$dG|_{T,P} = \sum_{i=1}^{\mathscr{C}} \overline{G}_i \, dN_i$$

[15]Clearly, from these results, it is a species partial molar Gibbs free energy difference between phases, rather than a concentration difference, that is the driving force for interphase mass transfer in the approach to equilibrium.

[16]See, for example, Chap. 9 of M. Modell and R. C. Reid, *Thermodynamics and Its Applications*, 2nd edition, Prentice–Hall, Englewood Cliffs, N.J., 1983.

and that the total number of moles of each species are fixed, so that $N_i = N_i^{\mathrm{I}} + N_i^{\mathrm{II}}$ or $dN_i^{\mathrm{II}} = -dN_i^{\mathrm{I}}$, we obtain

$$dG = \sum_{i=1}^{\mathscr{C}} \overline{G}_i^{\mathrm{I}}\, dN_i^{\mathrm{I}} + \sum_{i=1}^{\mathscr{C}} \overline{G}_i^{\mathrm{II}}\, dN_i^{\mathrm{II}} = \sum_{i=1}^{\mathscr{C}} (\overline{G}_i^{\mathrm{I}} - \overline{G}_i^{\mathrm{II}})\, dN_i^{\mathrm{I}} = 0$$

Setting the derivative of the Gibbs free energy with respect to each of its independent variables (here the mole numbers N_i^{I}) equal to zero yields

$$\left(\frac{\partial G}{\partial N_i^{\mathrm{I}}}\right)_{N_{j\neq i}^{\mathrm{I}}} = 0 = \overline{G}_i^{\mathrm{I}} - \overline{G}_i^{\mathrm{II}} \qquad \text{or} \qquad \overline{G}_i^{\mathrm{I}} = \overline{G}_i^{\mathrm{II}} \tag{6.7-9}$$

so that here, as in Chapter 5, we find that the equality of Gibbs free energies is a necessary condition for the existence of phase equilibrium for systems subject to a variety of constraints (see Problem 6.3).

Although we will not do so here, it is easy to show that these analyses for two-phase equilibrium are easily generalized to multiphase equilibrium and yield

$$\overline{G}_i^{\mathrm{I}} = \overline{G}_i^{\mathrm{II}} = \overline{G}_i^{\mathrm{III}} = \cdots \tag{6.7-10}$$

6.8 THE CRITERIA FOR CHEMICAL EQUILIBRIUM, AND COMBINED CHEMICAL AND PHASE EQUILIBRIUM

Equations 6.7-4 also provide a means of identifying the equilibrium state when chemical reactions occur. To see this consider the case of a single chemical reaction occurring in a single phase (both these restrictions will be removed shortly) in a closed system at constant temperature and pressure.[17] The total Gibbs free energy for this system, using the reaction variable notation introduced in Sec. 6.3, is

$$G = \sum_{i=1}^{\mathscr{C}} N_i \overline{G}_i = \sum_{i=1}^{\mathscr{C}} (N_{i,0} + \nu_i X)\overline{G}_i$$

Since the only variation possible in a one-phase closed system at constant temperature and pressure is in the extent of reaction X, the equilibrium criterion is

$$\left(\frac{\partial G}{\partial X}\right)_{T,P} = 0$$

which yields

$$0 = \sum_{i=1}^{\mathscr{C}} \nu_i \overline{G}_i \tag{6.8-1}$$

(Note that $\Sigma_i^{\mathscr{C}} N_i(\partial \overline{G}_i/\partial X)_{T,P}$ is equal to zero by the Gibbs–Duhem equation, Eq. 6.2-14 with $Y = X$.)

It is possible to show that the criterion for chemical equilibrium developed here is also applicable to systems subject to constraints other than constant

[17]Since chemists and chemical engineers are usually interested in chemical and phase equilibria at constant temperature and pressure, the discussions that follow largely concern equilibrium under these constraints.

temperature and pressure (Problem 6.4). In fact, Eq. 6.8-1, like the phase equilibrium criterion of Eq. 6.7-9, is of general applicability. Of course, the difficulty that arises in using either of these equations is in translating their simple form into a useful prescription for equilibrium calculations by relating the partial molar Gibbs free energies to quantities of more direct interest, such as temperature, pressure, and mole fractions. This problem will be the focus of much of the rest of this book.

The first generalization of the analysis given here is to the case of multiple chemical reactions in a closed single-phase constant temperature and pressure system. Using the notation of Sec. 6.3, the number of moles of species i present at any time is

$$N_i = N_{i,0} + \sum_{j=1}^{\mathcal{M}} \nu_{ij} X_j \tag{6.3-5}$$

where the summation is over the *independent* reactions. The total Gibbs free energy of the system is

$$\begin{aligned} \underline{G} &= \sum_{i=1}^{\mathcal{C}} N_i \overline{G}_i = \sum_{i=1}^{\mathcal{C}} \left(N_{i,0} + \sum_{j=1}^{\mathcal{M}} \nu_{ij} X_j \right) \overline{G}_i \\ &= \sum_{i=1}^{\mathcal{C}} N_{i,0} \overline{G}_i + \sum_{i=1}^{\mathcal{C}} \sum_{j=1}^{\mathcal{M}} \nu_{ij} X_j \overline{G}_i \end{aligned} \tag{6.8-2}$$

The condition for chemical equilibrium in this multireaction system is that $G =$ minimum or $dG = 0$ for all variations consistent with the stoichiometry at constant temperature, pressure, and total mass. For the present case this implies

$$\left(\frac{\partial G}{\partial X_j} \right)_{T,P,X_{i \neq j}} = 0 \qquad j = 1, 2, \ldots, \mathcal{M} \tag{6.8-3}$$

so that

$$\left(\frac{\partial G}{\partial X_j} \right)_{T,P,X_{i \neq j}} = 0 = \sum_{i=1}^{\mathcal{C}} \nu_{ij} \overline{G}_i + \sum_{i=1}^{\mathcal{C}} N_i \left(\frac{\partial \overline{G}_i}{\partial X_j} \right)_{T,P,X_{k \neq j}} \qquad \begin{array}{l} \text{for all independent} \\ \text{reactions} \\ \quad j = 1, 2, \ldots, \mathcal{M} \end{array}$$

Since the sum $\sum_{i=1}^{\mathcal{C}} N_i (\partial \overline{G}_i / \partial X_j)_{T,P,X_{k \neq j}}$ vanishes by the Gibbs–Duhem equation, the equilibrium criterion is

$$\sum_{i=1}^{\mathcal{C}} \nu_{ij} \overline{G}_i = 0 \qquad j = 1, 2, \ldots, \mathcal{M} \tag{6.8-4}$$

This equation is analogous to Eq. 6.8-1 for the single reaction case. The interpretation of Eq. 6.8-4 is clear; in a system in which several chemical reactions occur, chemical equilibrium is achieved only when each reaction is itself in equilibrium.

The final case to be considered is that of combined phase and chemical equilibrium in a closed system at constant temperature and pressure. At this point you can probably guess the final result: if both phase and chemical transformations are possible, equilibrium occurs only when each possible transformation is itself in equilibrium. Thus, Eqs. 6.7-7 and 6.8-4 must be simultaneously satisfied for all species and all reactions in all phases.

To prove this assertion, it is first useful to consider the mathematical technique of Lagrange multipliers, a method used to extremalize a function subject to constraints. Rather than develop the method in complete generality, we merely introduce it by application to the problem just considered, equilibrium in a single-phase multiple chemical reaction system.

We identified the equilibrium state for several chemical reactions occurring in a single-phase system at constant temperature and pressure by finding the state for which $G = \sum_{i=1}^{\mathscr{C}} N_i\overline{G}_i$ was equal to a minimum subject to the stoichiometric constraints

$$N_i = N_{i,0} + \sum_{j=1}^{\mathscr{M}} \nu_{ij}X_j \qquad \text{for all species } i$$

The procedure used was to incorporate the constraints directly into the Gibbs free energy function and then extremalize the resulting unconstrained equation (Eq. 6.8-2) by setting each of the $\mathscr{M}$ derivatives $(\partial G/\partial X_j)_{T,P,X_{i\neq j}}$ equal to zero. Unfortunately, this direct substitution technique can be very cumbersome when the constraints are complicated, as is the case in the problem of combined chemical and phase equilibrium.

An alternative method of obtaining a solution to the multiple-reaction, single-phase equilibrium problem is to use the method of Lagrange multipliers.[18] Here one first rewrites the constraints as

$$N_i - N_{i,0} - \sum_{j=1}^{\mathscr{M}} \nu_{ij}X_j = 0 \qquad i = 1, 2, \ldots, \mathscr{C} \tag{6.8-5}$$

and then creates a new function $\mathscr{G}$ by adding the constraints, each with a multiplying parameter α_i, to the original Gibbs function,

$$\mathscr{G} = \sum_{i=1}^{\mathscr{C}} N_i\overline{G}_i + \sum_{i=1}^{\mathscr{C}} \alpha_i \left[N_i - N_{i,0} - \sum_{j=1}^{\mathscr{M}} \nu_{ij}X_j\right] \tag{6.8-6}$$

The independent variables of this new function are $N_1, N_2, \ldots, N_{\mathscr{C}}, X_1, X_2, \ldots, X_{\mathscr{M}}$, and $\alpha_1, \alpha_2, \ldots, \alpha_{\mathscr{C}}$. To determine the state for which the Gibbs free energy is a minimum subject to the stoichiometric constraints of Eq. 6.8-5, the partial derivatives of this new unconstrained function $\mathscr{G}$ with respect to each of its independent variables are set equal to zero. From this procedure we obtain the following sequence of simultaneous equations to be solved:

$$\left(\frac{\partial \mathscr{G}}{\partial N_k}\right)_{T,P,X_j,N_{j\neq k}} = 0 = \overline{G}_k + \alpha_k + \underbrace{\sum_{i=1}^{\mathscr{C}} N_i \left(\frac{\partial \overline{G}_i}{\partial N_k}\right)_{T,P,X_j,N_{j\neq k}}}_{\to 0 \text{ by Gibbs–Duhem equation}}$$

or

$$\alpha_k = -\overline{G}_k \qquad k = 1, 2, \ldots, \mathscr{C} \tag{6.8-7}$$

$$\left(\frac{\partial \mathscr{G}}{\partial X_k}\right)_{T,P,N_j,X_{j\neq k}} = 0 = -\sum_{i=1}^{\mathscr{C}} \nu_{ik}\alpha_i \qquad k = 1, 2, \ldots, \mathscr{M} \tag{6.8-8}$$

[18] A more complete discussion of Lagrange multipliers may be found in M. H. Protter and C. B. Morrey, *College Calculus with Analytic Geometry*, Addison–Wesley, Reading, Mass., 1964, pp. 708–715 and V. G. Jenson and G. V. Jeffreys, *Mathematical Methods in Chemical Engineering*, Academic Press, New York, 1963, pp. 482–483.

or, using Eq. 6.8-7

$$\sum_{i=1}^{\mathscr{C}} \nu_{ik}\overline{G}_i = 0 \qquad k = 1, 2, \ldots, \mathscr{M} \tag{6.8-9}$$

and, finally

$$\left(\frac{\partial \mathscr{G}}{\partial \alpha_i}\right)_{T,P,X_j,N_k} = 0 = N_i - N_{i,0} - \sum_{j=1}^{\mathscr{M}} \nu_{ij}X_j \qquad i = 1, 2, \ldots, \tag{6.8-10}$$

Clearly, Eq. 6.8-9 gives the same equilibrium requirement as before (see Eq. 6.8-4), whereas Eqs. 6.8-10 ensure that the stoichiometric constraints are satisfied in solving the problem. Thus the Lagrange multiplier method yields the same results as the direct or brute force approach. Although the Lagrange multiplier method appears awkward when applied to the very simple problem here, its real utility is for complicated problems in which the number of constraints is large or the constraints are nonlinear in the independent variables, so that direct substitution is very difficult or impossible.

To derive the criteria for combined chemical and phase equilibrium, the following notation will be used: N_i^k and $\overline{G}_i^k$ are, respectively, the number of moles and partial molar Gibbs free energy of species i in the kth phase; $N_{i,0}$ is the initial number of moles of species i in the closed system; and X_j is overall molar extent of reaction (reaction variable) for the jth independent reaction, regardless of which phase or in how many different phases the reaction occurs. Thus

$$\begin{pmatrix}\text{total number of}\\ \text{moles of species}\\ i \text{ in all } \mathscr{P} \text{ phases}\end{pmatrix} = \sum_{k=1}^{\mathscr{P}} N_i^k = N_{i,0} + \sum_{j=1}^{\mathscr{M}} \nu_{ij}X_j \tag{6.8-11}$$

and

$$\begin{pmatrix}\text{total Gibbs free}\\ \text{energy of system}\end{pmatrix} = G = \sum_{k=1}^{\mathscr{P}} \sum_{i=1}^{\mathscr{C}} N_i^k\overline{G}_i^k \tag{6.8-12}$$

where $\mathscr{P}$ is the number of phases, $\mathscr{C}$ is the number of components, and $\mathscr{M}$ is the number of independent reactions.

The equilibrium state at constant temperature and pressure is that state for which the Gibbs free energy G achieves a minimum value from among all states consistent with the reaction stoichiometry. The identification of the equilibrium state is then a problem of minimizing the Gibbs free energy subject to the stoichiometric constraints of Eq. 6.8-11. Since the easiest way of solving this problem is to use the method of Lagrange multipliers, we define a set of Lagrange multipliers, $\alpha_1, \alpha_2, \ldots, \alpha_{\mathscr{C}}$ and construct the augmented function

$$\mathscr{G} = \sum_{k=1}^{\mathscr{P}} \sum_{i=1}^{\mathscr{C}} N_i^k\overline{G}_i^k + \sum_{i=1}^{\mathscr{C}} \alpha_i \left\{\sum_{k=1}^{\mathscr{P}} N_i^k - N_{i,0} - \sum_{j=1}^{\mathscr{M}} \nu_{ij}X_j\right\} \tag{6.8-13}$$

whose minimum we wish to find for all variations of the independent variables $X_1, X_2, \ldots, X_{\mathscr{M}}, N_1^{\mathrm{I}}, N_1^{\mathrm{II}}, \ldots, N_{\mathscr{C}}^{\mathscr{P}}$, and $\alpha_1, \alpha_2, \ldots, \alpha_{\mathscr{C}}$.

Setting the partial derivatives of $\mathscr{G}$ with respect to $N_1^{\mathrm{I}}, N_1^{\mathrm{II}}, \ldots$ equal to zero, remembering that the N's, X's and α's are now to be treated as independent variables, yields[19]

$$\left(\frac{\partial \mathcal{G}}{\partial N_1^{\mathrm{I}}}\right) = \overline{G}_1^{\mathrm{I}} + \sum_{k=1}^{\mathcal{P}} \sum_{i=1}^{\mathcal{C}} N_i^k \left(\frac{\partial \overline{G}_i^k}{\partial N_1^{\mathrm{I}}}\right) + \alpha_1 = \overline{G}_1^{\mathrm{I}} + \alpha_1 = 0$$

$$\left(\frac{\partial \mathcal{G}}{\partial N_1^{\mathrm{II}}}\right) = \overline{G}_1^{\mathrm{II}} + \sum_{k=1}^{\mathcal{P}} \sum_{i=1}^{\mathcal{C}} N_i^k \left(\frac{\partial \overline{G}_i^k}{\partial N_1^{\mathrm{II}}}\right) + \alpha_1 = \overline{G}_1^{\mathrm{II}} + \alpha_1 = 0 \tag{6.8-14}$$

$$\vdots$$

In each case the double summation term vanishes by application of the Gibbs–Duhem equation (Eq. 6.2-15) to each phase. The net information content of Eqs. 6.8-14 is

$$\overline{G}_1^{\mathrm{I}} = \overline{G}_1^{\mathrm{II}} = \cdots = \overline{G}_1^{\mathcal{P}} = -\alpha_1$$

and by generalization

$$\overline{G}_i^{\mathrm{I}} = \overline{G}_i^{\mathrm{II}} = \cdots = \overline{G}_i^{\mathcal{P}} = -\alpha_i \tag{6.8-15}$$

These equations establish that one of the equilibrium conditions in a multiple reaction, multiphase system is that phase equilibrium must be established for each of the species among the phases in which the species is present.

Another set of equilibrium criteria is obtained by minimizing $\mathcal{G}$ with respect to each of the reaction variables X_j ($j = 1$ to $\mathcal{M}$). Thus[20]

$$\left(\frac{\partial \mathcal{G}}{\partial X_1}\right) = 0 = + \sum_{k=1}^{\mathcal{P}} \sum_{i=1}^{\mathcal{C}} N_i^k \left(\frac{\partial \overline{G}_i^k}{\partial X_1}\right) - \sum_{i=1}^{\mathcal{C}} \alpha_i \nu_{i1} = 0 \tag{6.8-16}$$

The first term of the right side of this equation vanishes by the Gibbs–Duhem equation, and, from Eq. 6.8-15, we can set $\overline{G}_i^{\mathrm{I}} = \overline{G}_i^{\mathrm{II}} = \cdots = \overline{G}_i^k = \cdots = -\alpha_i$, and so obtain

$$\left(\frac{\partial \mathcal{G}}{\partial X_1}\right) = 0 = \sum_{i=1}^{\mathcal{C}} \nu_{i1} \overline{G}_i^k \tag{6.8-17}$$

Similarly, from $\partial \mathcal{G}/\partial X_2 = 0$, $\partial \mathcal{G}/\partial X_3 = 0, \cdots$, we obtain

$$\sum_{i=1}^{\mathcal{C}} \nu_{ij} \overline{G}_i^k = 0 \qquad \begin{array}{l}\text{for all phases } k = \mathrm{I}, \mathrm{II}, \ldots, \mathcal{P} \\ \text{and all reactions } j = 1, 2, \ldots, \mathcal{M}\end{array} \tag{6.8-18}$$

which establishes that a further condition for equilibrium in a multiphase, multireaction system is that each reaction must be in equilibrium in every phase. In fact, since at equilibrium the partial molar Gibbs free energy of each species is the same in every phase (see Eq. 6.8-15), if Eq. 6.8-18 is satisfied in any phase, it is satisfied in all phases.

Finally, setting the partial derivatives of $\mathcal{G}$ with respect to the parameters α_i equal to zero yields the stoichiometric constraints of Eq. 6.8-11. The equilibrium state is that state for which Eqs. 6.8-11, 15, and 18 are simultaneously satisfied.

[19]Though we have not listed the variables being held constant, you should recognize that all variables in the set T, P, N_i^k ($i = 1, \ldots, \mathcal{C}$; $k = 1, \ldots, \mathcal{P}$), X_j ($j = 1, \ldots, \mathcal{M}$) and α_i ($i = 1, \ldots, \mathcal{C}$), except the one being varied in the derivative, have been held constant.

[20]See footnote 19.

Thus, we have proved the assertion that in the case of combined chemical and phase equilibrium the conditions of phase equilibrium must be satisfied for all the species in each of the phases and, furthermore, that chemical equilibrium must exist for each reaction in each phase. (The fact that each reaction must be in chemical equilibrium in each phase does not imply that each mole fraction will be the same in each phase. This point is demonstrated in Chapter 9.)

6.9 THE SPECIFICATION OF THE EQUILIBRIUM THERMODYNAMIC STATE OF A MULTICOMPONENT, MULTIPHASE SYSTEM; THE GIBBS PHASE RULE

As has been mentioned several times, the equilibrium state of a single-phase, one-component system is completely fixed by the specification of two independent, intensive variables. From this observation we were able, in Sec. 5.6, to establish a simple relation for determining the number of degrees of freedom for a single-component multiphase system. Here an analogous equation is developed for determining the number of degrees of freedom in a reacting multicomponent, multiphase system; this relationship is called the **Gibbs phase rule**.

The starting point for the present analysis is the observation of Sec. 6.1 that the equilibrium thermodynamic state of a single-phase $\mathscr{C}$-component system can be fixed by specifying the values of two intensive variables and $\mathscr{C} - 1$ mole fractions. Alternatively, the specification of any $\mathscr{C} + 1$ independent state variables could be used to fix the state of this system.[21] Thus, we can say that a $\mathscr{C}$ component single-phase system has $\mathscr{C} + 1$ degrees of freedom, that is, we are free to adjust $\mathscr{C} + 1$ independent intensive thermodynamic properties of this system; however, once this is done, all the other intensive thermodynamic properties are fixed. This is equivalent to saying that if $T, P, x_1, x_2, \ldots, x_{\mathscr{C}-1}$ are taken as the independent variables, there exist equations of state in nature of the form

$$\underline{V} = \underline{V}(T, P, x_1, \ldots, x_{\mathscr{C}-1})$$

$$\underline{S} = \underline{S}(T, P, x_1, \ldots, x_{\mathscr{C}-1})$$

$$\underline{G} = \underline{G}(T, P, x_1, \ldots, x_{\mathscr{C}-1})$$

etc.

though we may not have been clever enough in our experiments to have determined the functional relationship between the variables.

Our interest here is in determining the number of degrees of freedom in a general multicomponent, multiphase chemically reacting system consisting of $\mathscr{C}$ components distributed among $\mathscr{P}$ phases and in which $\mathscr{M}$ independent chemical reactions occur. Since $\mathscr{C} + 1$ variables are required to completely specify the state of each phase, and there are $\mathscr{P}$ phases present, it would appear that a total of $\mathscr{P}(\mathscr{C} + 1)$ variables must be specified to fix the state of each of the phases.

[21] By $\mathscr{C} + 1$ independent state variables we mean $\mathscr{C} + 1$ nonredundant pieces of information about the thermodynamic state of the system. For example, temperature, pressure, and $\mathscr{C} - 1$ mole fractions form a set of $\mathscr{C} + 1$ independent variables; temperature and $\mathscr{C}$ mole fractions are not independent, however, since $\sum_i^{\mathscr{C}} x_i = 1$, so that only $\mathscr{C} - 1$ mole fractions are independent. Similarly, for a gas mixture composed of ideal gases, the enthalpy or internal energy, temperature, and $\mathscr{C} - 1$ mole fractions do not form an independent set of variables, since $\underline{H}$ and $\underline{U}$ are calculable from the mole fractions and the temperature. However, $\underline{H}$, P, and $\mathscr{C} - 1$ mole fractions are independent.

Actually, the number of variables that must be specified is considerably less than this, since the requirement that equilibrium exists provides a number of interrelationships between the state variables in each of the phases. In particular, the fact that the temperature must be the same in all phases

$$T^{\mathrm{I}} = T^{\mathrm{II}} = T^{\mathrm{III}} = \cdots = T^{\mathcal{P}} \tag{6.9-1}$$

results in $(\mathcal{P} - 1)$ restrictions on the values of the state variables of the phases. Similarly, the requirement that the pressure be the same in all phases

$$P^{\mathrm{I}} = P^{\mathrm{II}} = P^{\mathrm{III}} = \cdots = P^{\mathcal{P}} \tag{6.9-2}$$

provides an additional $(\mathcal{P} - 1)$ restrictions. Since each of the partial molar Gibbs free energies are calculable from equations of state of the form

$$\overline{G}_i^j = \overline{G}_i^j(T, P, x_1^j, x_2^j, \ldots, x_{\mathcal{C}-1}^j) \tag{6.9-3}$$

the condition for phase equilibrium

$$\overline{G}_i^{\mathrm{I}} = \overline{G}_i^{\mathrm{II}} = \cdots = \overline{G}_i^{\mathcal{P}} \qquad (i = 1, 2, \ldots, \mathcal{C}) \tag{6.9-4}$$

provides $\mathcal{C}(\mathcal{P} - 1)$ additional relationships among the variables without introducing any new unknowns.

Finally, if $\mathcal{M}$ *independent* chemical reactions occur, there are $\mathcal{M}$ additional relations of the form

$$\sum_{i=1}^{\mathcal{C}} \nu_{ij}\overline{G}_i = 0 \qquad j = 1, 2, \ldots, \mathcal{M} \tag{6.9-5}$$

(where we have omitted the superscript indicating the phase since, by Eq. 6.9-4, the partial molar Gibbs free energy for each species is the same in all phases).

Now designating the number of degrees of freedom by the symbol $\mathcal{F}$ we have

$$\begin{aligned} \mathcal{F} &= \begin{pmatrix} \text{number of unknown} \\ \text{thermodynamic} \\ \text{parameters} \end{pmatrix} - \begin{pmatrix} \text{number of independent relations} \\ \text{among the unknown parameters} \end{pmatrix} \\ &= \mathcal{P}(\mathcal{C} + 1) \qquad\qquad - [2(\mathcal{P} - 1) + \mathcal{C}(\mathcal{P} - 1) + \mathcal{M}] \\ &= \mathcal{C} - \mathcal{M} - \mathcal{P} + 2 \end{aligned} \tag{6.9-6}$$

Therefore, in a $\mathcal{C}$ component, $\mathcal{P}$ phase system in which $\mathcal{M}$ independent chemical reactions occur, the specification of $\mathcal{C} - \mathcal{M} - \mathcal{P} + 2$ state variables of the individual phases completely fixes the thermodynamic state of each of the phases. This result is known as the Gibbs phase rule.

In practice, temperature, pressure, and phase compositions are most commonly used to fix the thermodynamic state of multicomponent-multiphase systems, though any other information about the thermodynamic state of the individual phases could be used as well. However, thermodynamic information about the composite multiphase system is not useful in fixing the state of the system. That is, we could use the specific volume of any one of the phases as one of the $\mathcal{C} - \mathcal{M} - \mathcal{P} + 2$ degrees of freedom, but not the molar volume of the multiphase system. Finally we note that for a pure fluid $\mathcal{C} = 1$ and $\mathcal{M} = 0$, so that Eq. 6.9-6 reduces to

$$\mathcal{F} = 3 - \mathcal{P} \tag{6.9-7}$$

the result found in Sec. 5.6.

ILLUSTRATION 6.9-1

In Chapter 9 we consider the reaction equilibrium when styrene is hydrogenated to form ethylbenzene. Depending on the temperature and pressure of system, this reaction may take place in the vapor phase, or in a vapor–liquid mixture. Show that the system has three degrees of freedom if a single phase exists, but only two degrees of freedom if the reactants and products form a two-phase mixture.

Solution

The styrene–hydrogen–ethylbenzene system is a three-component ($\mathscr{C} = 3$), single-reaction ($\mathscr{M} = 1$) system. Thus

$$\mathscr{F} = \mathscr{C} - \mathscr{M} - \mathscr{P} + 2 = 4 - \mathscr{P}$$

Clearly, if only the vapor phase exists ($\mathscr{P} = 1$), there are three degrees of freedom; if, however, both the vapor and liquid are present ($\mathscr{P} = 2$), the system has only two degrees of freedom. ■

It is also of interest to determine the amount and nature of the additional information needed to fix the relative amounts of each of the phases in equilibrium, once their thermodynamic states are known. We can obtain this from an analysis that equates the number of variables to the number of restrictions on these variables. It is convenient for this discussion to write the specific thermodynamic properties of the multiphase system in terms of the distribution of mass between the phases. The argument could be based on a distribution of numbers of moles; however, it is somewhat more straight forward on a mass basis because total mass, and not total moles, is a conserved quantity. Thus, we will use x^i to represent the mass fraction of the ith phase; clearly the x^i must satisfy the equation

$$1 = x^{\mathrm{I}} + x^{\mathrm{II}} + x^{\mathrm{III}} + \cdots + x^{\mathscr{P}} \tag{6.9-8}$$

The total volume per unit mass $\hat{V}$, the total entropy per unit mass $\hat{S}$, and so on, are related to the analogous quantities in each of the phases by the equations

$$\begin{aligned} \hat{V} &= x^{\mathrm{I}}\hat{V}^{\mathrm{I}} + x^{\mathrm{II}}\hat{V}^{\mathrm{II}} + \cdots + x^{\mathscr{P}}\hat{V}^{\mathscr{P}} \\ \hat{S} &= x^{\mathrm{I}}\hat{S}^{\mathrm{I}} + x^{\mathrm{II}}\hat{S}^{\mathrm{II}} + \cdots + x^{\mathscr{P}}\hat{S}^{\mathscr{P}} \\ &\text{etc.} \end{aligned} \tag{6.9-9}$$

In writing these equations we are presuming that the specific volumes, entropies, and so forth for each phase (denoted by the superscript) are known from the equations of state or experimental data for the individual phases and a previous specification of the $\mathscr{C} - \mathscr{M} - \mathscr{P} + 2$ degrees of freedom.

Since there are $\mathscr{P}$ unknown mass distribution variables, the x^i's, it is evident that we need $\mathscr{P}$ equations to determine the relative amounts of each of the phases. Therefore, Eq. 6.9-8, together with the specification of $\mathscr{P} - 1$ intensive thermodynamic variables for the multiphase system (which can be written in the form of Eq. 6.9-9), are needed. This is in addition to the $\mathscr{C} - \mathscr{M} - \mathscr{P} + 2$ intensive variables of the individual phases that must be specified to completely fix the thermodynamic state of all the phases. (You should convince yourself that this conclusion is in agreement with Illustration 5.6-1.)

Thus far we have not considered the fact that the initial composition of a chemical or phase equilibrium system may be known. Such information can be used in the formulation of species mass balances and the energy balance, which lead to additional equations relating the phase variables. Depending on the extent of initial information available and the number of phases present, the initial state information may or may not reduce the degrees of freedom of the system. This point is most easily demonstrated by reference to specific examples, so the effect of initial state information will be considered in the illustrations of Chapters 8 and 9.

We should point out that the Gibbs phase rule is of use in deciding whether or not an equilibrium problem is "well-posed," that is, whether enough information has been given for the problem to be solvable, but it is not of use in actually solving for the equilibrium state. This too will be demonstrated by examples in Chapters 8 and 9. The Gibbs phase rule, being general in its scope and application, is regarded as another part of the philosophical content of thermodynamics.

6.10 SOME CONCLUDING REMARKS

The discussion in this chapter essentially concludes our development of thermodynamic theory. The remainder of this book will largely be concerned with how this theory is used to solve problems of interest to the chemical process industry. Since the partial molar Gibbs free energy has emerged as the central function in equilibrium computations, Chapter 7 will be concerned with the techniques used for estimating this quantity in gaseous, liquid, and solid mixtures. The final two chapters are then devoted to the use of thermodynamics in explaining and predicting the great diversity of physical and chemical equilibria that occurs in mixtures.

PROBLEMS

6.1 Prove that

a.
$$\left(\frac{\partial H}{\partial N_i}\right)_{P,S,N_{j\neq i}} = \overline{H}_i - T\overline{S}_i = \overline{G}_i$$

b.
$$\overline{G}_i = \left(\frac{\partial U}{\partial N_i}\right)_{S,V,N_{j\neq i}} = \left(\frac{\partial A}{\partial N_i}\right)_{V,T,N_{j\neq i}}$$

6.2 Derive the analogs of the Gibbs–Duhem equations (Eqs. 6.2-8 and 9) for the constraints of

a. Constant temperature and volume.
b. Constant internal energy and volume.
c. Constant entropy and volume.

6.3 In Sec. 6.7 we established that the condition for equilibrium between two phases is

$$\overline{G}_i^{\mathrm{I}} = \overline{G}_i^{\mathrm{II}} \quad \text{(for all species present in both phases)}$$

for closed systems either at constant temperature and pressure or at constant internal energy and volume. Show that this equilibrium condition must also be satisfied for closed systems at

a. Constant temperature and volume.
b. Constant entropy and volume.

6.4 Show that the criterion for chemical equilibrium developed in the text

$$\sum_{i}^{\mathscr{C}} \nu_i \bar{G}_i = 0$$

for a closed system at constant temperature and pressure is also the equilibrium condition to be satisfied for closed systems subject to the following constraints:

a. Constant temperature and volume.
b. Constant internal energy and volume.

6.5 Prove that since total mass is conserved during a chemical reaction, that

$$\sum_{i=1}^{\mathscr{C}} \nu_i m_i = 0 \qquad \text{for a single-reaction system}$$

and

$$\sum_{i=1}^{\mathscr{C}} \nu_{ij} m_i = 0 \qquad j = 1, 2, \ldots, \mathscr{M} \quad \text{for a multiple-reaction system}$$

where m_i is equal to the molecular weight of species i. Also show, by direct substitution, that the first of these equations is satisfied for the reaction

$$H_2O = H_2 + \tfrac{1}{2}O_2$$

6.6 Show that the partial molar volumes computed from Eqs. 6.6-4 and the partial molar enthalpies computed from Eqs. 6.6-9 must satisfy the Gibbs–Duhem equation.

6.7 Compute the partial molar volumes of methyl formate in methanol–methyl formate and ethanol–methyl formate mixtures at 298.15 K for various compositions using the experimental data in Fig. 6.1-2*a* and the following pure component data:

$$\underline{V}_{MF} = 0.06728 \text{ m}^3/\text{kmol}$$

$$\underline{V}_M = 0.0473$$

$$\underline{V}_E = 0.05868$$

6.8 Compute the difference between the pure component and partial molar enthalpies for both components at 298.15 K and various compositions in each of the following mixtures using the data in Fig. 6.1-2*b*.

a. benzene-C_6F_5H
b. benzene-C_6F_6
c. benzene-C_6F_5Cl
d. benzene-C_6F_5Br
e. benzene-C_6F_5I

6.9 **a.** In vapor–liquid equilibrium in a binary mixture, both components are generally present in both phases. How many degrees of freedom are there for this system?

b. The reaction between nitrogen and hydrogen to form ammonia occurs in the gas phase. How many degrees of freedom are there for this system?

c. Steam and coal react at high temperatures to form hydrogen, carbon monoxide, carbon dioxide, and methane. The following reactions have been suggested as being involved in the chemical transformation

$$C + 2H_2O = CO_2 + 2H_2$$

$$C + H_2O = CO + H_2$$

$$C + CO_2 = 2CO$$

$$C + 2H_2 = CH_4$$

$$CO + H_2O = CO_2 + H_2$$

$$CO + 3H_2 = CH_4 + H_2O$$

How many degrees of freedom are there for this system? [*Hint*: (1) How many independent chemical reactions are there in this sequence? (2) How many phase equilibrium equations are there?]

6.10 **a.** In vapor–liquid equilibrium mixtures sometimes occur in which the compositions of the coexisting vapor and liquid phases are the same. Such mixtures are called azeotropes. Show that a binary azeotropic mixture has only one degree of freedom.

b. In osmotic equilibrium, two mixtures at different pressures and separated by a rigid membrane permeable to only one of the species present attain a state of equilibrium in which the two phases have different compositions. How many degrees of freedom are there for osmotic equilibrium in a binary mixture?

c. The phase equilibrium behavior of furfural ($C_5H_4O_2$)–water mixtures is complicated because furfural and water are only partially soluble in the liquid phase.

(i) How many degrees of freedom are there for the vapor–liquid mixture if only a single liquid phase is present?

(ii) How many degrees of freedom are there for the vapor–liquid mixture if two liquid phases are present?

6.11 The temperature achieved when two fluid streams of differing temperature and/or composition are adiabatically mixed is termed the adiabatic mixing temperature. Compute the adiabatic mixing temperature for the following two cases:

a. Equal weights of aqueous solutions containing 10 wt% sulfuric acid at 20°C and 90 wt% sulfuric acid at 70°C are mixed.

b. Equal weights of aqueous solutions containing 10 wt% sulfuric acid at 20°C and 60 wt% sulfuric acid at 0°C are mixed.

Explain why the adiabatic mixing temperature is greater than that of either of the initial solutions in one of these cases, and intermediate to those of the initial solutions in the other case.

6.12 The molar integral heat of solution $\Delta \underline{H}_s$ is defined to be the change in enthalpy that results when 1 mole of solute (component 1) is isothermally mixed with N_2 moles of solvent (component 2)

$$\Delta \underline{H}_s = (1 + N_2)\underline{H}_{mix} - \underline{H}_1 - N_2\underline{H}_2 = \bar{H}_1 + N_2\bar{H}_2 - \underline{H}_1 - N_2\underline{H}_2$$

$\Delta \underline{H}_s$ is easily measured in an isothermal calorimeter by monitoring the heat evolved or absorbed on successive additions of solvent to a given amount of solute. The table gives the integral heat of solution data for 1 mole sulfuric acid in water at 25°C (the negative sign indicates that heat is evolved in the dilution process).

N_2 (moles of water)	0.25	1.0	1.5	2.33	4.0	5.44	9.0	10.1	19.0	20.0
$-\Delta \underline{H}_s$ (J)	8242	28200	34980	44690	54440	58370	62800	64850	70710	71970

a. Calculate the heat evolved when 100 g of pure sulfuric acid is added isothermally to 100 g of water.

b. Calculate the heat evolved when the solution prepared in part (a) is diluted with an additional 100 g of water.

c. Calculate the heat evolved when 100 g of a 60 wt% solution of sulfuric acid is mixed with 75 g of a 25 wt% sulfuric acid solution.

d. Relate $(\overline{H}_1 - \underline{H}_1)$ and $(\overline{H}_2 - \underline{H}_2)$ to only N_1, N_2, $\Delta \underline{H}_s$ and the derivatives of $\Delta \underline{H}_s$ with respect to the ratio N_2/N_1.

e. Compute the numerical values of $(\overline{H}_1 - \underline{H}_1)$ and $(\overline{H}_2 - \underline{H}_2)$ in a 50 wt% sulfuric acid solution.

6.13 The following data have been reported for the constant-pressure heat capacity of a benzene-carbon tetrachloride mixture at 20°C.[22]

Wt% CCl_4	C_P(J/g °C)	Wt% CCl_4	C_P
0	1.7655	60	1.004
10	1.630	70	0.927
20	1.493	80	0.858
30	1.358	90	0.816
40	1.222	100	0.807
50	1.100		

On a single graph plot the constant pressure partial molar heat capacity for both benzene and carbon tetrachloride as a function of composition.

6.14 A 20 wt% solution of sulfuric acid in water is to be enriched to a 60 wt% sulfuric acid solution by adding pure sulfuric acid.

a. How much pure sulfuric acid should be added?

b. If the 20 wt% solution is available at 5°C, and the pure sulfuric acid at 50°C, how much heat will have to be removed to produce the 60 wt% solution at 70°C? How much heat will have to be added or removed to produce the 60 wt% solution at its boiling point?

6.15 Develop a procedure for determining the partial molar properties for each constituent in a three-component (ternary) mixture. In particular, what data would you want, and what would you do with the data? Based on your analysis, do you suppose there is much partial molar property data available for ternary and quartenary mixtures?

6.16 The partial molar enthalpies of species in simple binary mixtures can sometimes be approximated by the following expressions:

$$\overline{H}_1 = a_1 + b_1 x_2^2$$

[22] Data reference: *International Critical Tables*, vol. 5, McGraw–Hill, New York, 1929.

and

$$\bar{H}_2 = a_2 + b_2 x_1^2$$

a. For these expressions show that b_1 must equal b_2.
b. Making use of the fact that

$$\lim_{x_i \to 1} \bar{\theta}_i = \underline{\theta}_i$$

for any thermodynamic property θ, show that

$$a_1 = \underline{H}_1 \qquad a_2 = \underline{H}_2 \qquad \text{and} \qquad \Delta \underline{H}_{\text{mix}} = b_1 x_1 x_2$$

6.17 A partial molar property of a component in a mixture may be either greater than or less than the corresponding pure component molar property. Furthermore, the partial molar property may vary with composition in a complicated way. Show this to be the case by computing (a) the partial molar volumes and (b) the partial molar enthalpies of ethanol and water in an ethanol–water mixture. (The data that follow are from Volumes 3 and 5 of the *International Critical Tables*, McGraw–Hill, New York, 1929).

Alcohol wt%	Density at 20°C (kg m^{-3}) × 10^{-3}	Mole% Water	Heat Evolved on Mixing at 17.33°C (kJ/mol of Ethanol)
0	0.9982		
5	0.9894	5	0.042
10	0.9819	10	0.092
15	0.9751	15	0.167
20	0.9686	20	0.251
25	0.9617	25	0.335
30	0.9538	30	0.423
35	0.9494	35	0.519
40	0.9352	40	0.636
45	0.9247	45	0.757
50	0.9138	50	0.946
55	0.9026	55	1.201
60	0.8911	60	1.507
65	0.8795	65	1.925
70	0.8677	70	2.478
75	0.8556	75	3.218
80	0.8434	80	4.269
85	0.8310	85	5.821
90	0.8180	90	7.801
95	0.8042	95	9.818
100	0.7893		

6.18 Using the information in Problems 5.13 and 6.17, estimate the heat of vaporization for the first bit of ethanol from ethanol–water solutions containing 25, 50, and 75 mole% ethanol and from a solution infinitely dilute in ethanol. How do these heats of vaporization compare with that for pure ethanol computed in Problem 5.13? Why is there a difference between the various heats of vaporization?

6.19 The volume of a binary mixture has been reported in the following polynomial form

$$\underline{V}(T, P, x_1, x_2) = x_1 b_1 + x_2 b_2 + x_1 x_2 \sum_{i=0}^{n} a_i (x_1 - x_2)^i$$

a. What values should be used for b_1 and b_2?
b. Derive, from the equation here, expressions for $\overline{V}_1$, $\overline{V}_2$, $\overline{V}_1^{ex} = \overline{V}_1 - \underline{V}_1$ and $\overline{V}_2^{ex} = \overline{V}_2 - \underline{V}_2$.
c. Derive, from the equation here, expressions for the partial molar excess volumes of each species at infinite dilution, that is, $\overline{V}_1^{ex}$ $(T, P, x_1 \to 0)$ and $\overline{V}_2^{ex}$ $(T, P, x_2 \to 0)$.

6.20 Prove the validity of Eqs. 6.7-4.

6.21 The definition of a partial molar property is

$$\overline{M}_i = \left(\frac{\partial(N\underline{M})}{\partial N_i}\right)_{T,P,N_{j\neq i}}$$

It is tempting, but incorrect, to assume that this equation can be written as

$$\overline{M}_i = \left(\frac{\partial(\underline{M})}{\partial x_i}\right)_{T,P}$$

Prove that the correct result is

$$\overline{M}_i = \underline{M} + \left(\frac{\partial(\underline{M})}{\partial x_i}\right)_{T,P,x_{\kappa\neq i}} - \sum x_j \left(\frac{\partial(\underline{M})}{\partial x_j}\right)_{T,P,x_{\kappa\neq j}}$$

6.22 In some cases if pure liquid A and pure liquid B are mixed at constant temperature and pressure, two liquid phases are formed at equilibrium, one rich in species A and the other in species B. We know that the equilibrium state at constant T and P is a state of minimum Gibbs free energy and the Gibbs free energy of a two-phase mixture is the sum of the number of moles times the molar Gibbs free energy for each phase. What would the molar Gibbs free energy versus mole fraction curve look like for this system if we could prevent phase separation from occurring? Identify the equilibrium compositions of the two phases on this diagram. The limit of stability of a single phase at constant temperature and pressure can be found from $d^2G = 0$ or

$$\left(\frac{\partial^2 \underline{G}}{\partial x_1^2}\right)_{T,P} = 0$$

Identify the limits of single-phase stability on the Gibbs free energy versus mole fraction curve.

6.23 Mattingley and Fenby [*J. Chem. Thermo.* **7**, 307 (1975)] have reported that the enthalpies of triethylamine–benzene solutions at 298.15 K are given by

$$\underline{H}_{mix} - [x_B\underline{H}_B + (1 - x_B)\underline{H}_{EA}] =$$
$$x_B(1 - x_B)\{1418 - 482.4(1 - 2x_B) + 187.4(1 - 2x_B)^3\}$$

where x_B is the mole fraction of benzene and $\underline{H}_{mix}$, $\underline{H}_B$, and $\underline{H}_{EA}$ are the molar enthalpies of the mixture, pure benzene, and pure triethylamine, respectively, units of J/mol.
a. Develop expressions for $(\overline{H}_B - \underline{H}_B)$ and $(\overline{H}_{EA} - \underline{H}_{EA})$.
b. Compute values for $(\overline{H}_B - \underline{H}_B)$ and $(\overline{H}_{EA} - \underline{H}_{EA})$ at $x_B = 0.5$.
c. One gram mole of a 25 mole% benzene mixture is to be mixed with one gram mole of a 75 mole% benzene mixture at 298.15 K. How much heat must be added or removed for the process to be isothermal?
[*Note:* $\underline{H}_{mix} - x_B\underline{H}_B - (1 - x_B)\underline{H}_{EA}$ is the enthalpy change on mixing defined in Sec. 6.1.]

6.24 When water and n-propanol are isothermally mixed, heat may be either absorbed ($Q > 0$) or evolved ($Q < 0$), depending on the final composition of the mixture. Volume 5 of the *International Critical Tables* (McGraw–Hill, New York, 1929) gives the following data:

Mole% Water	Q, kJ/mol of n-propanol
5	+0.042
10	+0.084
15	+0.121
20	+0.159
25	+0.197
30	+0.230
35	+0.243
40	+0.243
45	+0.209
50	+0.167
55	+0.084
60	−0.038
65	−0.201
70	−0.431
75	−0.778
80	−1.335
85	−2.264
90	−4.110
95	−7.985

Plot $(\bar{H}_W - \underline{H}_W)$ and $(\bar{H}_{NP} - \underline{H}_{NP})$ over the whole composition range.

6.25 The heat of mixing data of Featherstone and Dickinson [*J. Chem. Thermo.* **9**, 75 (1977)] for the n-octanol + n-decane system is approximately fit by

$$\Delta \underline{H}_{\text{mix}} = x_1 x_2 \left(A + B(x_1 - x_2)\right) \qquad \text{J/mol}$$

where

$$A = -12{,}974 + 51.505T$$

and

$$B = +8782.8 - 34.129T$$

with T in K and x_1 being the n-octanol mole fraction.

a. Compute the difference between the partial molar and pure component enthalpies of n-octanol and n-decane at $x_1 = 0.5$ and $T = 300$ K.

b. Compute the difference between the partial molar and pure component heat capacities of n-octanol and n-decane at $x_1 = 0.5$ and $T = 300$ K.

c. An $x_1 = 0.2$ solution and an $x_1 = 0.9$ solution are to flow continuously into an isothermal mixer in the mole ratio 2 : 1 at 300 K. Will heat have to be added or removed to keep the temperature of the solution leaving the mixer at 300 K? What will be the heat flow per mole of solution leaving the mixer?

7

The Estimation of the Gibbs Free Energy and Fugacity of a Component in a Mixture

The most important ingredient in the thermodynamic analysis of mixtures is information about the partial molar properties of each species in the mixture. The partial molar Gibbs free energy is of special interest since it is needed in the study of phase and chemical equilibria, which will be considered in great detail in the following two chapters. For many mixtures the experimental partial molar property information needed for equilibrium calculations is not available. Consequently, in this chapter we consider methods for estimating the partial molar Gibbs free energy and its equivalent, the fugacity. Before proceeding with this detailed study, we will consider two very simple cases, a mixture of ideal gases (Sec. 7.1) and the ideal mixture (Sec. 7.3), for which the partial molar properties are simply related to the pure component properties.

7.1 THE IDEAL GAS MIXTURE

As defined in Chapter 2, the ideal gas is a gas whose volumetric equation of state at all temperatures, pressures, and densities is

$$PV = NRT \qquad \text{or} \qquad P\underline{V} = RT \tag{7.1-1}$$

and whose internal energy is a function of temperature only. By the methods of statistical mechanics, one can show that such behavior occurs when a gas is sufficiently dilute that interactions between the molecules make a negligible contribution to the total energy of the system. That is, a gas is ideal when each molecule in the gas is (energetically) unaware of the presence of other molecules.

An **ideal gas mixture** is a gas mixture with a density so low that its molecules do not appreciably interact. In this case the volumetric equation of state of the gas mixture will also be of the form of Eq. 7.1-1, and its internal energy will merely be the sum of the internal energies of each of the constituent ideal gases, and thus a function of temperature and mole number only. That is,

$$PV^{IGM} = (N_1 + N_2 + \cdots)RT = \left(\sum_j^{\mathscr{C}} N_j\right) RT \tag{7.1-2}$$

and

$$U^{IGM}(T, N_j) = \sum_{j}^{\mathscr{C}} N_j \underline{U}_j^{IG}(T) \tag{7.1-3}$$

Here we have used the superscripts *IG* and *IGM* to indicate properties of the ideal gas and the ideal gas mixture, respectively, and taken pressure and temperature to be the independent variables. From Eq. 6.1-12 it then follows that for the ideal gas mixture

$$\bar{U}_i^{IGM}(T, x_i) = \left.\frac{\partial U^{IGM}(T, N_i)}{\partial N_i}\right|_{T,P,N_{j\neq i}} = \left.\frac{\partial}{\partial N_i}\right|_{T,P,N_{j\neq i}} \sum_{j=1}^{\mathscr{C}} N_j \underline{U}_j^{IG}(T) = \underline{U}_i^{IG}(T) \tag{7.1-4}$$

and

$$\bar{V}_i^{IGM}(T, P, x_i) = \left.\frac{\partial V^{IGM}(T, P, N_i)}{\partial N_i}\right|_{T,P,N_{j\neq i}} = \left.\frac{\partial}{\partial N_i}\right|_{T,P,N_{j\neq i}} \sum_{j}^{\mathscr{C}} N_j \frac{RT}{P}$$

$$= \frac{RT}{P} = \underline{V}_i^{IG}(T, P) \tag{7.1-5}$$

Equation 7.1-4 indicates that the partial molar internal energy of species i in a mixture at a given temperature is equal to the pure component molar internal energy of that component at the same temperature. Similarly, Eq. 7.1-5 establishes that the partial molar volume of species i in an ideal gas mixture at a given temperature and pressure is identical with the pure component molar volume at that temperature and pressure.

Consider now the process of forming an ideal gas mixture at temperature T and pressure P from a collection of pure ideal gases, all at that temperature and pressure. From the discussion here it is clear that for each species $\bar{V}_i(T, P, x_i) = \underline{V}_i(T, P)$, and $\bar{U}_i(T, x_i) = \underline{U}_i(T)$. It then follows immediately from equations like Eqs. 6.1-14 and 15 that $\Delta V_{\text{mix}} = 0$ and $\Delta U_{\text{mix}} = 0$ for this process. Also $\Delta H_{\text{mix}} \equiv \Delta U_{\text{mix}} + P\,\Delta V_{\text{mix}} = 0$.

The **partial pressure** of species i in a gas mixture, denoted by P_i, is defined for both ideal and nonideal gas mixtures to be the product of the mole fraction of species i and total pressure P, that is,

$$P_i = x_i P \tag{7.1-6}$$

For the ideal gas mixture

$$P_i^{IGM}(N, V, T, x_i) = \frac{N_i}{\sum_{j=1}^{\mathscr{C}} N_j} P = \frac{N_i}{\sum_{j=1}^{\mathscr{C}} N_j} \left\{ \sum_{j=1}^{\mathscr{C}} N_j \frac{RT}{V} \right\} = \frac{N_i RT}{V} = P^{IG}(N_i, V, T)$$

Thus, for the ideal gas mixture, the partial pressure of species i is equal to the pressure that would be exerted if the same number of moles of that species, N_i, alone were contained in the same volume V and maintained at the same temperature T as the mixture.

Since there is no energy of interaction in an ideal gas mixture, the effect on each species of forming an ideal gas mixture at constant temperature and total pressure is equivalent to reducing the pressure from P to its partial pressure in the mixture P_i. Alternatively, the effect is equivalent to expanding each gas from its initial volume $V_i = N_i RT/P$ to the volume of the mixture $V = \Sigma_i N_i RT/P$. Thus, from Eqs. 3.4-2 and 3, we have

$$\bar{S}_i^{IGM}(T, P, x_i) - \underline{S}_i^{IG}(T, P) = -R \ln \frac{P_i}{P} = -R \ln x_i \tag{7.1-7}$$

or

$$\bar{S}_i^{IGM}(T, V, x_i) - \underline{S}_i^{IG}(T, V_i) = +R \ln \frac{V}{V_i} = R \ln \frac{\sum N_i RT/P}{N_i RT/P} = -R \ln x_i$$

Consequently,

$$\Delta S_{\text{mix}}^{IGM} = \sum_{i=1}^{\mathscr{C}} N_i[\bar{S}_i^{IGM}(T, P, x_i) - \underline{S}_i^{IG}(T, P)] = -R \sum_{i=1}^{\mathscr{C}} N_i \ln x_i$$

and

$$\Delta \underline{S}_{\text{mix}}^{IGM} = \frac{\Delta S_{\text{mix}}^{IGM}}{N} = -R \sum_{i=1}^{\mathscr{C}} x_i \ln x_i \tag{7.1-8}$$

The statistical mechanical interpretation of Eq. 7.1-8 is that an ideal gas mixture is a completely mixed or random mixture. This is discussed in Appendix A7.1.

Using the energy, volume, and entropy changes on mixing given here, one can easily compute the other thermodynamic properties of an ideal gas mixture (Problem 7.1). The results are given in Table 7.1-1. Of particular interest are the expressions for $\bar{G}_i^{IGM}(T, P, x_i)$ and $\Delta \underline{G}_{\text{mix}}^{IGM}$

$$\begin{aligned} \bar{G}_i^{IGM}(T, P, x_i) &= \bar{H}_i^{IGM}(T, P, x_i) - T\bar{S}_i^{IGM}(T, P, x_i) \\ &= \underline{H}_i^{IG}(T, P) - T(\underline{S}_i^{IG}(T, P) - R \ln x_i) \\ &= \underline{G}_i^{IG}(T, P) + RT \ln x_i \end{aligned} \tag{7.1-9}$$

and

$$\begin{aligned} \Delta \underline{G}_{\text{mix}}^{IGM} &= \sum_{i=1}^{\mathscr{C}} x_i \{\bar{G}_i^{IGM}(T, P, x_i) - \underline{G}_i^{IG}(T, P)\} \\ &= RT \sum x_i \ln x_i \end{aligned} \tag{7.1-10}$$

Table 7.1-1
PROPERTIES OF AN IDEAL GAS MIXTURE (MIXING AT CONSTANT T AND P)

Internal energy	$\bar{U}_i^{IGM}(T, x_i) = \underline{U}_i^{IG}(T)$	$\Delta \underline{U}_{\text{mix}}^{IGM} = 0$
Enthalpy	$\bar{H}_i^{IGM}(T, x_i) = \underline{H}_i^{IG}(T)$	$\Delta \underline{H}_{\text{mix}}^{IGM} = 0$
Volume	$\bar{V}_i^{IGM}(T, P, x_i) = \underline{V}^{IG}(T, P)$	$\Delta \underline{V}_{\text{mix}}^{IGM} = 0$
Entropy	$\bar{S}_i^{IGM}(T, P, x_i) = \underline{S}_i^{IG}(T, P) - R \ln x_i$	$\Delta \underline{S}_{\text{mix}}^{IGM} = -R \sum_{i=1}^{\mathscr{C}} x_i \ln x_i$
Gibbs free energy	$\bar{G}_i^{IGM}(T, P, x_i) = \underline{G}_i^{IG}(T, P) + RT \ln x_i$	$\Delta \underline{G}_{\text{mix}}^{IGM} = RT \sum_{i=1}^{\mathscr{C}} x_i \ln x_i$
Helmholtz free energy	$\bar{A}_i^{IGM}(T, P, x_i) = \underline{A}_i^{IG}(T, P) + RT \ln x_i$	$\Delta \underline{A}_{\text{mix}}^{IGM} = RT \sum_{i=1}^{\mathscr{C}} x_i \ln x_i$

7.2 THE PARTIAL MOLAR GIBBS FREE ENERGY AND FUGACITY

Unfortunately, very few mixtures are ideal gas mixtures, so that general methods must be developed for estimating the thermodynamic properties of real mixtures. As we saw in the discussion of phase equilibrium in a pure fluid of Sec. 5.4, the fugacity function was especially useful; the same is true for mixtures. Therefore, in an analogous fashion to the derivation in Sec. 5.4, we start from

$$dG = -S\,dT + V\,dP + \sum_{i=1}^{\mathscr{C}} \overline{G}_i\,dN_i \tag{6.2-1}$$

and, using the commutative property of second derivatives of the thermodynamic functions (c.f. Eq. 4.1-3),

$$\frac{\partial}{\partial N_i}\bigg|_{T,P,N_{j\neq i}} \left(\frac{\partial G}{\partial T}\right)_{P,N_j} = \frac{\partial}{\partial T}\bigg|_{P,N_j} \left(\frac{\partial G}{\partial N_i}\right)_{T,P,N_{j\neq i}}$$

and

$$\frac{\partial}{\partial N_i}\bigg|_{T,P,N_{j\neq i}} \left(\frac{\partial G}{\partial P}\right)_{T,N_j} = \frac{\partial}{\partial P}\bigg|_{T,N_j} \left(\frac{\partial G}{\partial N_i}\right)_{T,P,N_{j\neq i}}$$

obtain the two equations

$$\overline{S}_i = -\left(\frac{\partial \overline{G}_i}{\partial T}\right)_{P,N_j} \tag{7.2-1}$$

and

$$\overline{V}_i = \left(\frac{\partial \overline{G}_i}{\partial P}\right)_{T,N_j} \tag{7.2-2}$$

As in the pure component case, the second of these equations is more useful than the first, and leads to the relation

$$\overline{G}_i(T_1, P_2) - \overline{G}_i(T_1, P_1) = \int_{P_1}^{P_2} \overline{V}_i\,dP$$

In analogy with Eq. 5.4-6, the fugacity of species i in a mixture, denoted by $\bar{f}_i$, is defined with reference to the ideal gas mixture as follows:

$$\bar{f}_i(T, P, x_i) = x_i P \exp\left\{\frac{\overline{G}_i(T, P, x_i) - \overline{G}_i^{IGM}(T, P, x_i)}{RT}\right\}$$

$$= x_i P \exp\left\{\frac{1}{RT}\int_0^P (\overline{V}_i - \overline{V}_i^{IGM})\,dP\right\} \tag{7.2-3a}$$

so that $\bar{f}_i \rightarrow x_i P \equiv P_i$ as $P \rightarrow 0$. Here P_i is the partial pressure of species i, and the superscript IGM indicates an ideal gas mixture property. The fact that as the pressure goes to zero all mixtures become ideal gas mixtures (just as all pure fluids become ideal gases) is embedded in this definition. Also, the fugacity coefficient for a component in a mixture, $\bar{\phi}_i$, is defined as

$$\bar{\phi}_i = \frac{\bar{f}_i}{x_i P} = \exp\left\{\frac{\overline{G}_i(T, P, x_i) - \overline{G}_i^{IGM}(T, P, x_i)}{RT}\right\}$$

$$= \exp\left\{\frac{1}{RT}\int_0^P (\overline{V}_i - \overline{V}_i^{IGM})\,dP\right\} \tag{7.2-3b}$$

The multicomponent analog of Eq. 5.4-9a, obtained by differentiating $\ln \bar{f}_i$ with respect to pressure at constant temperature and composition, is

$$RT\left(\frac{\partial \ln \bar{f}_i}{\partial P}\right)_{T,x_i} = \left(\frac{\partial \bar{G}_i}{\partial P}\right)_{T,x_i} = \bar{V}_i \tag{7.2-4}$$

To relate the fugacity of pure component i to the fugacity of component i in a mixture, we first subtract Eq. 5.4-9a from Eq. 7.2-4, and then integrate between $P = 0$ and the pressure of interest P, to obtain

$$RT \ln\left\{\frac{\bar{f}_i(T, P, x_i)}{\bar{f}_i(T, P \to 0, x_i)}\right\} - RT \ln\left\{\frac{f_i(T, P)}{f_i(T, P \to 0)}\right\} = \int_{P\to 0}^{P} (\bar{V}_i - \underline{V}_i)\, dP \tag{7.2-5}$$

We now use the fact that as $P \to 0$, $\bar{f}_i \to x_i P$ and $f_i \to P$, to obtain

$$RT \ln\left\{\frac{\bar{f}_i(T, P, x_i)}{x_i f_i(T, P)}\right\} = \int_0^P (\bar{V}_i - \underline{V}_i)\, dP \tag{7.2-6}$$

Therefore, for a mixture in which $\bar{V}_i = \underline{V}_i$ *under all conditions*, the fugacity of each species in the mixture is equal to its mole fraction times its pure component fugacity evaluated at the same temperature and pressure as the mixture. If $\bar{V}_i \neq \underline{V}_i$, then $\bar{f}_i$ and f_i are related through the integral over all pressures of the difference between the species partial molar and pure component molar volumes.

The temperature dependence of the fugacity $\bar{f}_i$ (actually the fugacity coefficient $\phi_i = \bar{f}_i / x_i P$) can be gotten by differentiating Eq. 7.2-3 with respect to temperature at constant pressure and composition

$$\left[\frac{\partial \ln (\bar{f}_i / x_i P)}{\partial T}\right]_{P,x_i} = -\frac{(\bar{G}_i - \bar{G}_i^{IGM})}{RT^2} + \frac{1}{RT}\left[\frac{\partial(\bar{G}_i - \bar{G}_i^{IGM})}{\partial T}\right]_{P,x_i} \tag{7.2-7}$$

and then using $d\underline{G} = \underline{V}\, dP - \underline{S}\, dT$ and $\underline{G} = \underline{H} - T\underline{S}$, to obtain

$$\left[\frac{\partial \ln (\bar{f}_i / x_i P)}{\partial T}\right]_{P,x_i} = -\frac{(\bar{H}_i - \bar{H}_i^{IGM})}{RT^2} \tag{7.2-8}$$

It is useful to have an expression for the change in partial molar Gibbs free energy of a species between two states of the same temperature and pressure, but of differing composition. To derive such an equation we start by writing

$$\Delta \bar{G}_i = \bar{G}_i(T, P, x_i^{\text{II}}) - \bar{G}_i(T, P, x_i^{\text{I}})$$

where the superscripts I and II denote states of different composition. Now substituting the logarithm of Eq. 7.2-3

$$\bar{G}_i(T, P, x_i) = \bar{G}_i^{IGM}(T, P, x_i) + RT \ln\left(\frac{\bar{f}_i(T, P, x_i)}{x_i P}\right) \tag{7.2-9}$$

and using Eq. 7.1-9

$$\bar{G}_i^{IGM}(T, P, x_i) = \underline{G}_i^{IG}(T, P) + RT \ln x_i$$

yields

$$\begin{aligned}
\bar{G}_i(T, P, x_i^{\text{II}}) - \bar{G}_i(T, P, x_i^{\text{I}}) &= \bar{G}_i^{IGM}(T, P, x_i^{\text{II}}) + RT \ln\left\{\frac{\bar{f}_i(T, P, x_i^{\text{II}})}{x_i^{\text{II}} P}\right\} \\
&\quad - \bar{G}_i^{IGM}(T, P, x_i^{\text{I}}) - RT \ln\left\{\frac{\bar{f}_i(T, P, x_i^{\text{I}})}{x_i^{\text{I}} P}\right\} \\
&= \underline{G}_i^{IG}(T, P) + RT \ln x_i^{\text{II}} + RT \ln\left\{\frac{\bar{f}_i(T, P, x_i^{\text{II}})}{x_i^{\text{II}} P}\right\} \\
&\quad - \underline{G}_i^{IG}(T, P) - RT \ln x_i^{\text{I}} - RT \ln\left\{\frac{\bar{f}_i(T, P, x_i^{\text{I}})}{x_i^{\text{I}} P}\right\}
\end{aligned}$$

or, finally

$$\bar{G}_i(T, P, x_i^{\text{II}}) - \bar{G}_i(T, P, x_i^{\text{I}}) = RT \ln \left\{ \frac{\bar{f}_i(T, P, x_i^{\text{II}})}{\bar{f}_i(T, P, x_i^{\text{I}})} \right\} \tag{7.2-10}$$

The fugacity function has been introduced because its relation to the Gibbs free energy function makes it useful in phase equilibrium calculations. The present criterion for equilibrium between two phases is that $\bar{G}_i^{\text{I}} = \bar{G}_i^{\text{II}}$ for all species i, with the restriction that the temperature and pressure be constant and, of course, equal in both phases. Using Eq. 7.2-10 and the equality of partial molar Gibbs free energies yields

$$\bar{f}_i^{\text{I}} = \bar{f}_i(T, P, x_i^{\text{I}}) = \bar{f}_i(T, P, x_i^{\text{II}}) = \bar{f}_i^{\text{II}} \tag{7.2-11}$$

Therefore, at equilibrium, the fugacity of each species must be the same in both phases. Since this result follows directly from Eq. 6.7-10, it may be substituted for it. Furthermore, since we can make estimates for the fugacity of a species in a mixture in a more direct fashion than for partial molar Gibbs free energies, it is more convenient to use Eq. 7.2-11 as the basis for phase equilibrium calculations.

In analogy with Eqs. 5.4-6 and 7.2-3, the overall fugacity of a mixture $\bar{f}$ is defined by the relation

$$\bar{f} = P \exp \left\{ \frac{\underline{G}(T, P, x_i) - \underline{G}^{\text{IGM}}(T, P, x_i)}{RT} \right\} \tag{7.2-12}$$

Note that the fugacity of species i in a mixture, $\bar{f}_i$, as defined by Eq. 7.2-3 is not a partial molar fugacity, that is,

$$\bar{f}_i \neq \left(\frac{\partial (N\bar{f})}{\partial N_i} \right)_{T,P,N_{j\neq i}}$$

Finally, you should note that Eq. 5.4-6 for computing the fugacity of a pure component and Eq. 7.2-3 for computing the fugacity of a species in a mixture both require that the volumetric equation of state be solved explicitly for the volume in terms of the temperature and pressure. However, as was discussed in Sec. 5.4, equations of state are usually pressure-explicit (i.e., easily solved for pressure as a function of temperature and volume and not vice versa; see Eqs. 4.4-1 through 3), so that calculations based on Eqs. 5.4-6 and 7.2-3 can be difficult. Starting from Eq. 7.2-3, using Eq. 4.4-25 in the form

$$dP = \frac{1}{\underline{V}} d(P\underline{V}) - \frac{P}{\underline{V}} d\underline{V} = \frac{P}{Z} dZ - \frac{P}{\underline{V}} d\underline{V}$$

and the triple product rule (Eq. 4.1-6a)

$$\left(\frac{\partial V}{\partial N_i} \right)_{T,P,N_{j\neq i}} \left(\frac{\partial P}{\partial V} \right)_{T,N_j} \left(\frac{\partial N_i}{\partial P} \right)_{T,V,N_{j\neq i}} = -1$$

in the form

$$\left(\frac{\partial V}{\partial N_i} \right)_{T,P,N_{j\neq i}} dP = - \left(\frac{\partial P}{\partial N_i} \right)_{T,V,N_{j\neq i}} dV$$

we obtain for the fugacity (actually the fugacity coefficient) of a species in a mixture

$$\ln \phi_i = \ln \frac{\bar{f}_i(T, P, x_i)}{x_i P} = \frac{1}{RT} \int_{\underline{V}=\infty}^{\underline{V}=ZRT/P} \left[\frac{RT}{\underline{V}} - N \left(\frac{\partial P}{\partial N_i} \right)_{T,V,N_{j\neq i}} \right] d\underline{V} - \ln Z \tag{7.2-13}$$

(see Problem 7.20) as we had previously obtained

$$\ln \phi = \ln \frac{f(T, P)}{P} = \frac{1}{RT} \int_{\underline{V}=\infty}^{\underline{V}=ZRT/P} \left(\frac{RT}{\underline{V}} - P \right) d\underline{V} - \ln Z + (Z - 1) \tag{5.4-8}$$

for the fugacity of a pure component. Equation 7.2-13 is especially useful for computing the fugacity of a species in a mixture from a pressure-explicit equation of state, as we will see in Sec. 7.4.

7.3 THE IDEAL MIXTURE AND EXCESS MIXTURE PROPERTIES

The estimation of the thermodynamic properties of a real fluid or fluid mixture in the absence of direct experimental data is a very complicated problem involving detailed spectroscopic, structural, and interaction potential data and the use of statistical mechanics. Such a calculation is beyond the scope of this book and would not be of interest to most engineers. Instead, we use procedures for estimating mixture thermodynamic properties that are far simpler than starting from "first principles." In particular, we will use either (1) equations of state, (2) the principle of corresponding states as extended to mixtures (Sec. 7.7), or, as in this section, (3) choose a state for each system in which the thermodynamic properties are reasonably well known and then try to estimate how the departure of the real system from the chosen reference state affects the system properties. This last procedure is philosophically similar to the method used in Chapter 4, where the properties of real fluids were computed as the sum of an ideal gas contribution plus the departure from ideal gas behavior.

Clearly, the accuracy of a property estimation technique based on such a procedure increases as the difference between the reference state and actual state of the system diminishes. Therefore, choosing the reference state to be an ideal gas mixture at the same temperature and composition as the mixture under consideration is not very satisfactory, because the reference state and actual state, particularly for liquids, may be too dissimilar. It is in this context that we introduce the concept of an ideal mixture.

An **ideal mixture**, which may be either a gaseous or liquid mixture, is defined to be a mixture in which

$$\bar{H}_i^{IM}(T, P, x_i) = \underline{H}_i(T, P) \tag{7.3-1}$$

and

$$\bar{V}_i^{IM}(T, P, x_i) = \underline{V}_i(T, P) \tag{7.3-2}$$

for *all* temperatures, pressures, and compositions (the superscript *IM* indicates an ideal mixture property). From Eqs. 6.1-14 and 15 it is evident that for such a mixture

$$\Delta V_{\text{mix}}^{IM} = \sum_{i=1}^{\mathscr{C}} N_i(\bar{V}_i^{IM} - \underline{V}_i) = 0$$

and

$$\Delta H_{\text{mix}}^{IM} = \sum_{i=1}^{\mathscr{C}} N_i(\bar{H}_i^{IM} - \underline{H}_i) = 0 \tag{7.3-3}$$

so that there are no volume or enthalpy changes on the formation of an ideal mixture from its pure components at the same temperature and pressure. Also, since $\overline{V}_i^{IM} = \underline{V}_i$ at all temperatures and pressures we have, from Eq. 7.2-6,

$$\bar{f}_i^{IM}(T, P, x_i) = x_i f_i(T, P) \tag{7.3-4}$$

Next, using Eq. 7.3-1 and 2 it is easily established that

$$\overline{U}_i^{IM}(T, P, x_i) = \underline{U}_i(T, P) \tag{7.3-5a}$$

and from Eq. 7.2-10 written as

$$\overline{G}_i^{IM}(T, P, x_i) - \underline{G}_i(T, P) = RT \ln \left\{\frac{\bar{f}_i^{IM}(T, P, x_1)}{f_i(T, P)}\right\} = RT \ln \left\{\frac{x_i f_i(T, P)}{f_i(T, P)}\right\}$$

$$= RT \ln x_i$$

we get

$$\overline{G}_i^{IM}(T, P, x_i) = \underline{G}_i(T, P) + RT \ln x_i$$

$$\overline{A}_i^{IM}(T, P, x_i) = \underline{A}_i(T, P) + RT \ln x_i \tag{7.3-5b}$$

and

$$\overline{S}_i^{IM}(T, P, x_i) = \underline{S}_i(T, P) - R \ln x_i$$

These equations resemble those obtained in Sec. 7.1 for the ideal gas mixture. There is an important difference, however. In the present case we are considering an ideal mixture of fluids that are not ideal gases, so that each of the pure component properties here will not be an ideal gas property, but rather a real fluid property that must either be measured or computed using the techniques described in Chapter 4. Thus, the molar volume $\underline{V}_i$ is not equal to RT/P, and the fugacity of each species is not equal to the pressure.

Table 7.3-1 lists some of the properties of ideal mixtures. One additional property, which follows directly from Eq. 7.3-4, is

$$\left[\frac{\partial \ln (\bar{f}_i^{IM}/x_i P)}{\partial T}\right]_{P,x_i} = \left[\frac{\partial \ln (f_i/P)}{\partial T}\right]_P \tag{7.3-6}$$

Note that an ideal mixture identically satisfies the Gibbs–Duhem equation (Eq. 6.2-12b),

$$\begin{aligned} &\underline{S}^{IM}\, dT - \underline{V}^{IM}\, dP + \sum x_i\, d\overline{G}_i^{IM} \\ &= \sum x_i \overline{S}_i^{IM}\, dT - \sum x_i \overline{V}_i^{IM}\, dP + \sum x_i d(\underline{G}_i + RT \ln x_i) \\ &= \sum x_i \underline{S}_i\, dT - R \sum x_i \ln x_i\, dT - \sum x_i \underline{V}_i\, dP + \sum x_i\, d\underline{G}_i \\ &\qquad + R \sum x_i \ln x_i\, dT + RT \sum x_i d \ln x_i \\ &= \sum x_i(\underline{S}_i\, dT - \underline{V}_i\, dP + d\underline{G}_i) + RT \sum dx_i \\ &\equiv 0 \end{aligned} \tag{7.3-7}$$

since $d\underline{G} = \underline{V}\, dP - \underline{S}\, dT$ for a pure component, and $\Sigma\, x_i = 1$, so that $d\, \Sigma\, x_i = \Sigma\, dx_i = d(1) = 0$.

Any property change on mixing $\Delta\underline{\theta}_{\text{mix}}$ of a real mixture can be written in terms of the analogous property of an ideal mixture as follows:

$$\Delta\underline{\theta}_{\text{mix}}(T, P, x_i) = \Delta\underline{\theta}_{\text{mix}}^{IM}(T, P, x_i) + [\Delta\underline{\theta}_{\text{mix}}(T, P, x_i) - \Delta\underline{\theta}_{\text{mix}}^{IM}(T, P, x_i)]$$

$$= \Delta\underline{\theta}_{\text{mix}}^{IM} + \underline{\theta}^{ex}$$

Table 7.3-1
MIXTURE THERMODYNAMIC PROPERTIES

Property	Ideal Mixtures	Real Mixtures
Volume	$\bar{V}_i^{IM} = \underline{V}_i$	$\bar{V}_i^{ex} = \bar{V}_i - \underline{V}_i$
	$\Delta \underline{V}_{\text{mix}}^{IM} = 0$	$\underline{V}^{ex} = \Delta \underline{V}_{\text{mix}}$
Internal energy	$\bar{U}_i^{IM} = \underline{U}_i$	$\bar{U}_i^{ex} = \bar{U}_i - \underline{U}_i$
	$\Delta \underline{U}_{\text{mix}}^{IM} = 0$	$\underline{U}^{ex} = \Delta \underline{U}_{\text{mix}}$
Enthalpy	$\bar{H}_i^{IM} = \underline{H}_i$	$\bar{H}_i^{ex} = \bar{H}_i - \underline{H}_i$
	$\Delta \underline{H}_{\text{mix}}^{IM} = 0$	$\underline{H}^{ex} = \Delta \underline{H}_{\text{mix}}$
Entropy	$\bar{S}_i^{IM} = \underline{S}_i - R \ln x_i$	$\bar{S}_i^{ex} = \bar{S}_i - \underline{S}_i + R \ln x_i$
	$\Delta \underline{S}_{\text{mix}}^{IM} = -R \sum_{i=1}^{\mathscr{C}} x_i \ln x_i$	$\underline{S}^{ex} = \Delta \underline{S}_{\text{mix}} + R \sum_{i=1}^{\mathscr{C}} x_i \ln x_i$
Gibbs free energy	$\bar{G}_i^{IM} = \underline{G}_i + RT \ln x_i$	$\bar{G}_i^{ex} = \bar{G}_i - \underline{G}_i - RT \ln x_i$
	$\Delta \underline{G}_{\text{mix}}^{IM} = RT \sum_{i=1}^{\mathscr{C}} x_i \ln x_i$	$\underline{G}^{ex} = \Delta \underline{G}_{\text{mix}} - RT \sum_{i=1}^{\mathscr{C}} x_i \ln x_i$
Helmholtz free energy	$\bar{A}_i^{IM} = \underline{A}_i + RT \ln x_i$	$\bar{A}_i^{ex} = \bar{A}_i - \underline{A}_i - RT \ln x_i$
	$\Delta \underline{A}_{\text{mix}}^{IM} = RT \sum_{i=1}^{\mathscr{C}} x_i \ln x_i$	$\underline{A}^{ex} = \Delta \underline{A}_{\text{mix}} - RT \sum_{i=1}^{\mathscr{C}} x_i \ln x_i$

where

$$\underline{\theta}^{ex} = \sum_i x_i \bar{\theta}_i - \sum_i x_i \bar{\theta}_i^{IM} = \sum_i x_i (\bar{\theta}_i - \bar{\theta}_i^{IM}) \tag{7.3-8}$$

is the **excess mixing property**; that is, the change in $\underline{\theta}$ that occurs on mixing at constant temperature and pressure in addition to that which would occur if an ideal mixture were formed. Excess properties $\underline{\theta}^{ex}$ are generally complicated, nonlinear functions of the composition, temperature, and pressure, and usually must be gotten from experiment. The hope is, however, that the excess properties will be small (particularly when $\underline{\theta}$ is the Gibbs free energy or entropy) compared to $\Delta \underline{\theta}_{\text{mix}}^{IM}$ so that even an approximate theory for $\underline{\theta}^{ex}$ may be sufficient to compute $\Delta \underline{\theta}_{\text{mix}}$ with reasonable accuracy. Table 7.3-1 also contains a list of excess thermodynamic properties that will be of interest in this book.

Finally, in analogy with Eq. (6.1-13), we define a partial molar excess quantity by the relation

$$\bar{\theta}_i^{ex} = \left.\frac{\partial (N \underline{\theta}^{ex})}{\partial N_i}\right|_{T,P,N_{j \neq i}} = \left.\frac{\partial}{\partial N_i}\right|_{T,P,N_{j \neq i}} \sum_k N_k (\bar{\theta}_k - \bar{\theta}_k^{IM})$$

which reduces to

$$\bar{\theta}_i^{ex} = \bar{\theta}_i - \bar{\theta}_i^{IM} \tag{7.3-9}$$

since

$$\sum_k N_k \frac{\partial}{\partial N_i} (\bar{\theta}_k - \bar{\theta}_k^{IM})|_{T,P,N_{j\neq i}} = 0$$

by the Gibbs–Duhem equation (Eq. 6.2-11). For the case in which $\bar{\theta}_i$ is equal to the partial molar Gibbs free energy, we have

$$\begin{aligned}
\bar{G}_i^{ex} &= (\bar{G}_i - \bar{G}_i^{IM}) = (\bar{G}_i - \bar{G}_i^{IGM}) + (\bar{G}_i^{IGM} - \bar{G}_i^{IM}) \\
&= (\bar{G}_i - \bar{G}_i^{IGM}) + (\underline{G}_i^{IG} + RT \ln x_i - \underline{G}_i - RT \ln x_i) \\
&= (\bar{G}_i - \bar{G}_i^{IGM}) + (\underline{G}_i^{IG} - \underline{G}_i) \\
&= RT \ln \frac{\bar{f}_i}{x_i P} - RT \ln \frac{f_i}{P} = RT \ln \left(\frac{\bar{f}_i}{x_i f_i}\right) \\
&= \int_0^P [\bar{V}_i - \underline{V}_i]\, dP
\end{aligned} \tag{7.3-10}$$

Common thermodynamic notation for liquid mixtures not describable by an equation of state is to define an **activity coefficient** $\gamma_i(T, P, x_i)$, which is a function of temperature, pressure, and composition, by the equation

$$\bar{f}_i^{\mathrm{L}}(T, P, x_i) = x_i \gamma_i(T, P, x_i) f_i^{\mathrm{L}}(T, P) \tag{7.3-11}$$

where the superscript L has been used to denote the liquid phase. Alternatively,

$$RT \ln \gamma_i(T, P, x_i) = \bar{G}_i^{ex} = \frac{\partial (N\underline{G}^{ex})}{\partial N_i} \tag{7.3-12}$$

[Note that the fugacity of the pure liquid, $f_i^{\mathrm{L}}(T, P)$, in Eq. 7.3-11 can be found from the methods of Sec. 5.4b]. As will be seen in Chapter 8, the calculation of the activity coefficients for each species in a mixture is an important step in many phase equilibrium calculations. Therefore, much of this chapter and the following one deals with models (equations) for $\underline{G}^{ex}$ and activity coefficients.

From Eq. 7.3-4, the activity coefficient is equal to unity for species in ideal mixtures and

$$\gamma_i(T, P, x_i) = \exp\left(\frac{\bar{G}_i^{ex}}{RT}\right) = \exp\left(\frac{1}{RT} \int_0^P [\bar{V}_i(T, P, x_i) - \underline{V}_i(T, P)]\, dP\right) \tag{7.3-13}$$

for nonideal mixtures. Since both real and ideal mixtures satisfy the Gibbs–Duhem equation of Sec. 6.2

$$0 = \underline{S}\, dT - \underline{V}\, dP + \sum_{i=1}^{\mathscr{C}} x_i\, d\bar{G}_i \tag{6.2-12b}$$

and

$$0 = \underline{S}^{IM}\, dT - \underline{V}^{IM}\, dP + \sum_{i=1}^{\mathscr{C}} x_i\, d\bar{G}_i^{IM} \tag{7.3-7}$$

we can subtract these two equations to obtain a form of the Gibbs–Duhem equation applicable to excess thermodynamic properties

$$0 = \underline{S}^{ex}\, dT - \underline{V}^{ex}\, dP + \sum_{i=1}^{\mathscr{C}} x_i\, d\bar{G}_i^{ex} \tag{7.3-14}$$

Now using Eq. 7.3-13, in the form of $\overline{G}_i^{ex} = RT \ln \gamma_i$, and that

$$\underline{G}^{ex} = \sum x_i \overline{G}_i^{ex} = \underline{H}^{ex} - T\underline{S}^{ex}$$

we obtain

$$0 = \frac{\underline{H}^{ex}}{T} dT - \underline{V}^{ex} dP + RT \sum_{i=1}^{\mathscr{C}} x_i \, d \ln \gamma_i \tag{7.3-15}$$

Thus, for changes in composition at constant temperature and pressure we have

$$0 = \sum_{i=1}^{\mathscr{C}} x_i \, d \ln \gamma_i \big|_{T,P} \tag{7.3-16}$$

which is a special case of Eqs. 6.2-9b and 13b. For a binary mixture we have, from Eq. 6.2-20,

$$x_1 \left(\frac{\partial \ln \gamma_1}{\partial x_1} \right)_{T,P} + x_2 \left(\frac{\partial \ln \gamma_2}{\partial x_1} \right)_{T,P} = 0 \tag{7.3-17}$$

To determine the dependence of the activity coefficient γ_i on temperature and pressure, we rewrite Eq. 7.3-13 as follows

$$\ln \gamma_i(T, P, x_i) = \frac{\overline{G}_i^{ex}(T, P, x_i)}{RT} = \frac{1}{RT} \int_0^P [\overline{V}_i(T, P, x_i) - \underline{V}_i(T, P)] \, dP \tag{7.3-18}$$

Taking the derivative with respect to pressure at constant temperature and composition we obtain

$$\left(\frac{\partial \ln \gamma_i(T, P, x_i)}{\partial P} \right)_{T,x_i} = \frac{\partial}{\partial P} \left(\frac{1}{RT} \int_0^P [\overline{V}_i(T, P, x_i) - \underline{V}_i(T, P)] \, dP \right)_{T,x_i}$$

$$= \frac{\overline{V}_i(T, P, x_i) - \underline{V}_i(T, P)}{RT} = \frac{\overline{V}_i^{ex}(T, P, x_i)}{RT} \tag{7.3-19}$$

Therefore,

$$\gamma_i(T, P_2, x_i) = \gamma_i(T, P_1, x_i) \exp \left[\int_{P_1}^{P_2} \frac{\overline{V}_i^{ex}(T, P, x_i)}{RT} \, dP \right]$$

$$\cong \gamma_i(T, P_1, x_i) \exp \left[\frac{\overline{V}_i^{ex}(T, x_i)(P_2 - P_1)}{RT} \right] \tag{7.3-20}$$

since the excess partial molar volume is approximately pressure independent. (Note that although Eqs. 7.3-18 and 19 are correct, they are difficult to use in practice since the activity coefficient description is applied to fluid mixtures not well described by an equation of state.) Next, taking the temperature derivative of Eq. 7.3-18 at constant pressure and composition, we obtain

$$\left(\frac{\partial \ln \gamma_i(T, P, x_i)}{\partial T} \right)_{P,x_i} = \frac{\partial}{\partial T} \left(\frac{\overline{G}_i^{ex}}{RT} \right)_{P,x_i} = -\frac{\overline{H}_i^{ex}(T, P, x_i)}{RT^2} \tag{7.3-21}$$

so that

$$\gamma_i(T_2, P, x_i) = \gamma_i(T_1, P, x_i) \exp \left[-\int_{T_1}^{T_2} \frac{\overline{H}_i^{ex}(T, P, x_i)}{RT^2} \, dT \right] \tag{7.3-22}$$

For small temperature changes, or if the excess partial molar enthalpy is independent of temperature, we have

$$\gamma_i(T_2, P, x_i) = \gamma_i(T_1, P, x_i) \exp\left[\frac{\overline{H}_i^{ex}(x_i)}{R}\left(\frac{1}{T_2} - \frac{1}{T_1}\right)\right] \tag{7.3-23}$$

These equations, especially Eqs. 7.3-20, 22, and 23, are useful when one wants to correct the numerical values of activity coefficients obtained at one temperature and pressure for use at other temperatures or pressures. For example, Eq. 7.3-22 or 23 would be needed in the prediction of low-temperature liquid–liquid phase equilibrium (Sec. 8.4) if the only activity coefficient data available were those obtained from higher temperature vapor–liquid equilibrium measurements (Sec. 8.1). Alternatively, if activity coefficients have been measured at different pressures, Eq. 7.3-20 can be used to calculate partial molar excess volumes, or if activity coefficients have been measured at different temperatures, Eq. 7.2-23 can be used to calculate partial molar excess enthalpies (and heats of mixing).

7.4 THE FUGACITY OF SPECIES IN GASEOUS, LIQUID, AND SOLID MIXTURES

The fugacity function is central to the calculation of phase equilibrium. This should be apparent from the earlier discussion of this chapter, and from the calculations of Sec. 5.5, which established that once we had the pure fluid fugacity, vapor–liquid equilibrium could be predicted. Consequently, for the remainder of this chapter we will be concerned with estimating the fugacity of species in gaseous, liquid, and solid mixtures.

a. Gaseous Mixtures[1]

Several methods are commonly used for estimating the fugacity of a species in a gaseous mixture. The most approximate method is based on the observation that some gaseous mixtures follow Amagat's law; that is,

$$V(T, P, N_1, N_2, \ldots) = \sum_{i=1}^{\mathscr{C}} N_i \underline{V}_i(T, P) \tag{7.4-1}$$

which, on multiplying by P/NRT, can be rewritten as

$$Z(T, P, y_1, y_2, \ldots) = \sum_{i=1}^{\mathscr{C}} y_i Z_i(T, P) \tag{7.4-2}$$

Here $Z = PV/NRT$ is the compressibility of the mixture, Z_i is the compressibility of pure component i, and $y_i = N_i/N$ is the mole fraction of species i. (Hereafter y_i will be used to indicate gas-phase mole fractions and x_i to denote mole fractions in the liquid phase and in equations that are applicable to both gases and liquids.)

The data in Fig. 7.4-1 for the nitrogen–butane system show that the mixture compressibility is nearly a linear function of mole fraction at both low and high pressures, so that Eq. 7.4-2 is approximately satisfied at these conditions, but fails in the intermediate pressure range. If we, nonetheless, accept Eq. 7.4-2 as being a reasonable approximation over the whole pressure range, we

[1]See also Secs. 7.7 and 7.8.

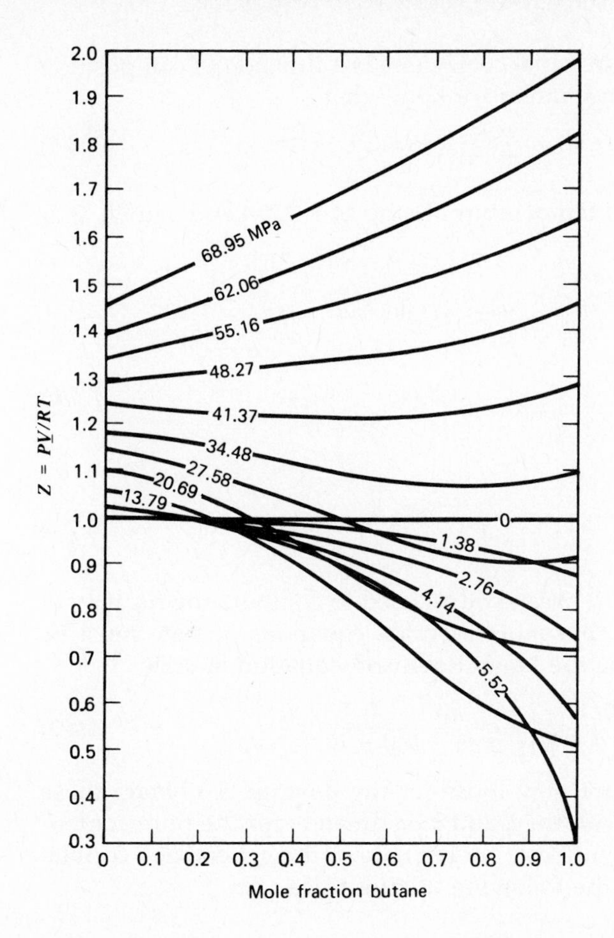

Figure 7.4-1
Compressibility factors for nitrogen–butane mixtures at 444.3 K. [Data of R. B. Evans and C. M. Watson, *Chem. Eng. Data Series* **1,** 67 (1956). Based on figure in J. M. Prausnitz, *Molecular Thermodynamics of Fluid-Phase Equilibria*, 1969. Reprinted by permission of Prentice–Hall, Inc., Englewood Cliffs, N.J.]

then have, from Eq. 7.4-1 and the definition of the partial molar volume, that $\bar{V}_i(T, P, x_i) = \underline{V}_i(T, P)$ and, from Eq. 7.2-6

$$\bar{f}_i^V(T, P, y_i) = y_i f_i^V(T, P) \tag{7.4-3}$$

This result, which relates the fugacity of a species in a gaseous mixture only to its mole fraction and the fugacity of the pure *gaseous* component at the same temperature and pressure, is known as the **Lewis–Randall rule.**

A more accurate way to estimate the fugacity of a species in a gaseous mixture is to start with Eq. 7.2-13

$$\ln \frac{\bar{f}_i^V(T, P, y_i)}{y_i P} = \frac{1}{RT} \int_{\underline{V}=\infty}^{\underline{V}=Z^V RT/P} \left[\frac{RT}{\underline{V}} - N \left(\frac{\partial P}{\partial N_i} \right)_{T,V,N_{j\neq i}} \right] d\underline{V} - \ln Z^V \tag{7.2-13}$$

and use an appropriate equation of state. At low pressures, the truncated virial equation of state

$$\frac{P\underline{V}}{RT} = 1 + \frac{B_{\text{mix}}(T, y_i)}{\underline{V}} = Z_{\text{mix}} \tag{7.4-4}$$

can be used if data for the mixture virial coefficient as a function of composition are available.[2] From statistical mechanics we know that

$$B_{\text{mix}}(T, y_i) = \sum_i \sum_j y_i y_j B_{ij}(T) \tag{7.4-5}$$

where each $B_{ij}(T)$ is a function of temperature. Using Eqs. 7.4-4 and 5 in Eq. 7.2-13 yields (see Problem 7.6).

$$\ln \frac{\bar{f}_i^V(T, P, y_i)}{y_i P} = \frac{2}{\underline{V}} \sum_j y_j B_{ij}(T) - \ln Z_{\text{mix}}$$

$$= \frac{2P}{Z_{\text{mix}} RT} \sum_j y_j B_{ij}(T) - \ln Z_{\text{mix}} \tag{7.4-6}$$

where

$$Z_{\text{mix}} = \frac{1}{2}\left(1 + \sqrt{1 + \frac{4B_{\text{mix}}P}{RT}}\right) \tag{7.4-7}$$

At higher pressures, Eq. 7.2-13 can still be used to compute the fugacity of a species in a gaseous mixture, but more accurate equations of state must be used. One can, for example, use the Peng–Robinson equation of state

$$P = \frac{RT}{\underline{V} - b} - \frac{a(T)}{\underline{V}(\underline{V} + b) + b(\underline{V} - b)} \tag{4.4-2}$$

where the parameters a and b are now those for the mixture. To obtain these mixture parameters one starts with the a and b parameters for the pure components gotten from either fitting pure component data or the generalized correlations of Sec. 4.7 and then uses the following **mixing rules**

$$a = \sum_{i=1}^{\mathscr{C}} \sum_{j=1}^{\mathscr{C}} y_i y_j a_{ij}$$

$$b = \sum_{i=1}^{\mathscr{C}} y_i b_i \tag{7.4-8}$$

where a_{ii} and b_i and the parameters for pure component i, and

$$a_{ij} = \sqrt{a_{ii} a_{jj}}\,(1 - k_{ij}) = a_{ji} \tag{7.4-9}$$

Here a new parameter k_{ij}, known as the **binary interaction parameter**, has been introduced to obtain better agreement in mixture equation of state calculations. This parameter is found by fitting the equation of state to mixture data (usually vapor–liquid equilibrium data, as will be discussed in Chapter 8). Values of the interaction parameter k_{ij} that have been reported for a number of binary mixtures appear in Table 7.4-1.

As a result of the mole fraction dependence of the equation of state parameters, the pressure is a function of mole fraction or, alternatively, the number of moles of each species present. Evaluating the derivative $(\partial P/\partial N_i)_{T, V, N_{j \neq i}}$, which

[2]The book, *The Virial Coefficients of Pure Gases and Mixtures. A Critical Compilation*, by J. H. Dymond and E. B. Smith, Clarendon Press, Oxford, 1980, contains second virial coefficient data for more than one-thousand mixtures.

Table 7.4-1

BINARY INTERACTION PARAMETERS k_{12} FOR THE PENG–ROBINSON EQUATION OF STATE*

	C_2H_4	C_2H_6	C_3H_6	C_3H_8	$i\text{-}C_4H_{10}$	$n\text{-}C_4H_{10}$	$i\text{-}C_5H_{12}$	$n\text{-}C_5H_{12}$	$n\text{-}C_6H_{14}$	C_6H_6	$c\text{-}C_6H_{12}$	$n\text{-}C_7H_{16}$	$n\text{-}C_8H_{18}$	$n\text{-}C_{10}H_{22}$	N_2	CO	CO_2	SO_2	H_2S
CH_4	0.022	−0.003	0.033	0.016	0.026	0.019	−0.006	0.026	0.040	0.055	0.039	0.035	0.050	0.049	0.030	0.030	0.09	0.136	0.08
C_2H_4		0.010				0.092				0.031		0.014		0.025	0.086	−0.022	0.056		
C_2H_6			0.089	0.001	−0.007	0.010		0.008	−0.04	0.042	0.018	0.007	0.019	0.014	0.044	0.026	0.130		0.086
C_3H_6				0.007	−0.014										0.09	0.026	0.093		0.08
C_3H_8					−0.007	0.003	0.011	0.027	0.001	0.023		0.006	0.0	0.0	0.078	0.03	0.12		0.08
$i\text{-}C_4H_{10}$						0.0									0.10	0.04	0.13		0.047
$n\text{-}C_4H_{10}$								0.017	−0.006			0.003	0.007	0.008	0.087	0.04	0.135		0.07
$i\text{-}C_5H_{12}$								0.06							0.092	0.04	0.121		0.06
$n\text{-}C_5H_{12}$										0.018	0.004	0.007	0.0		0.10	0.04	0.125		0.063
$n\text{-}C_6H_{14}$										0.010	−0.004	−0.008			0.15	0.04	0.11		0.06
C_6H_6											0.013	0.001	0.003	0.1	0.164	0.11	0.077	0.015	
$c\text{-}C_6H_{12}$															0.14	0.10	0.105		
$n\text{-}C_7H_{16}$													0.0		0.1	0.04	0.10		0.06
$n\text{-}C_8H_{18}$															0.1	0.04	0.12		0.06
$n\text{-}C_{10}H_{22}$															0.11	0.04	0.114		0.033
N_2																0.012	−0.02	0.08	0.17
CO																	0.03		0.054
CO_2																		0.136	0.097
SO_2																			
H_2S																			

* Obtained from data in "Vapor-Liquid Equilibria for Mixtures of Low Boiling Substances", by H. Knapp, R. Döring, L. Oellrich, U. Plöcker and J. M. Prausnitz, DECHEMA Chemistry Data Series Vol. VI, Frankfurt/Main, 1982, and other sources. Blanks indicate no data are available from which the k_{12} could be evaluated. In such case use estimates from mixtures of similar compounds.

appears in Eq. 7.2-13, using the Peng–Robinson equation of state yields

$$\ln \frac{\bar{f}_i^{\mathrm{V}}(T, P, y_i)}{y_i P} = \frac{b_i}{b}(Z^{\mathrm{V}} - 1) - \ln\left(Z^{\mathrm{V}} - \frac{bP}{RT}\right) - \frac{a}{2\sqrt{2}\, bRT}\left[\frac{2\sum_j y_j a_{ij}}{a} - \frac{b_i}{b}\right] \ln\left[\frac{Z^{\mathrm{V}} + (\sqrt{2} + 1)\frac{bP}{RT}}{Z^{\mathrm{V}} - (\sqrt{2} - 1)\frac{bP}{RT}}\right]$$

$$= \frac{B_i}{B}(Z^{\mathrm{V}} - 1) - \ln(Z^{\mathrm{V}} - B) - \frac{A}{2\sqrt{2}\, B}\left[\frac{2\sum_j y_i A_{ij}}{A} - \frac{B_i}{B}\right] \ln\left[\frac{Z^{\mathrm{V}} + (\sqrt{2} + 1) B}{Z^{\mathrm{V}} - (\sqrt{2} - 1) B}\right] \tag{7.4-10}$$

where, again, $A = aP/(RT)^2$, $B = bP/RT$, and the superscript V has been used to remind us that the vapor-phase compressibility factor is to be used.

To calculate the fugacity of each species in a gaseous mixture using Eq. 7.4-10 at specified values of T, P, and mole fractions of all components $y_1, y_2, \ldots y_{\mathscr{C}}$, the following procedure is used:

1. Obtain the parameters a_{ii} and b_i for each component of the mixture either from the generalized correlations given in Sec. 4.7 or, if necessary, from pure component data.
2. Compute a and b parameters for the mixture using the mixing rules of Eq. 7.4-8. (If the value for any binary interaction parameter, k_{ij}, is not available, either use a value for similar mixtures or assume it is zero.)
3. Solve the cubic equation of state for the vapor compressibility Z^{V}.
4. Use this value of Z^{V} to compute the vapor fugacity for each species using Eq. 7.4-10 repeatedly for $i = 1, 2, \ldots \mathscr{C}$, where $\mathscr{C}$ is the number of components.

A BASIC language computer program to do this calculation is discussed in Appendix A7.2. The program is on the disk accompanying this book.

b. Liquid Mixtures

The estimation of species fugacities in liquid mixtures is done in two different ways, depending on the data available and the components in the mixture. For liquid mixtures involving only hydrocarbons and dissolved gases such as nitrogen, hydrogen sulfide, and carbon dioxide, simple equations of state may be used to describe liquid state behavior. For example, if the approximate Peng–Robinson equation of state is used, the fugacity of each species in the mixture is, following the same development as for gaseous mixtures, given by

$$\ln \frac{\bar{f}_i^{\mathrm{L}}(T, P, x_i)}{x_i P} = \frac{b_i}{b}(Z^{\mathrm{L}} - 1) - \ln\left(Z^{\mathrm{L}} - \frac{bP}{RT}\right) - \frac{a}{2\sqrt{2}\, bRT}\left[\frac{2\sum_j x_j a_{ij}}{a} - \frac{b_i}{b}\right] \ln\left[\frac{Z^{\mathrm{L}} + (\sqrt{2} + 1)\frac{bP}{RT}}{Z^{\mathrm{L}} - (\sqrt{2} - 1)\frac{bP}{RT}}\right]$$

$$= \frac{B_i}{B}(Z^{\mathrm{L}} - 1) - \ln(Z^{\mathrm{L}} - B)$$

$$- \frac{A}{2\sqrt{2}\,B}\left[\frac{2\sum_j x_i A_{ij}}{A} - \frac{B_i}{B}\right] \ln \left[\frac{Z^{L} + (\sqrt{2}+1)\,B}{Z^{L} - (\sqrt{2}-1)\,B}\right] \tag{7.4-11}$$

where A and B are as before, and Z^L is the liquid (high density or small Z) root for the compressibility factor. Consequently, the calculational scheme for species fugacities in liquid mixtures that can be described by equations of state is similar to that for gaseous mixtures, except that the liquid phase, rather than vapor phase, compressibility factor is used in the calculations. The computer program of Appendix A7.2 can be used for this calculation as well.

For liquid mixtures in which one or more of the components cannot presently be described by an equation of state in the liquid phase, for example, mixtures containing alcohols, bases, organic or inorganic acids, and electrolytes, another procedure for estimating species fugacities must be used. The most common starting point is Eq. 7.3-11

$$\bar{f}_i(T, P, x_i) = x_i \gamma_i(T, P, x_i) f_i(T, P) \tag{7.3-11}$$

where γ_i is the activity coefficient of species i, and $f_i(T, P)$ is the fugacity of pure species i as a *liquid* at the temperature and pressure of the mixture.[3] However, since volumetric equation of state data are not available, the activity coefficient is not obtained from integration of Eq. 7.3-13, but rather from

$$RT \ln \gamma_i(T, P, x_i) = \bar{G}_i^{ex} = \frac{\partial(N\underline{G}^{ex})}{\partial N_i} \tag{7.3-12}$$

and approximate models for the excess Gibbs free energy $\underline{G}^{ex}$ of the liquid mixture. Several such liquid solution models are considered in the next section.

c. Solid Mixtures

The molecules in a crystalline solid are arranged in an ordered lattice structure consisting of repetitions of a unit cell. To form a solid solution, it is necessary to replace some molecules in the lattice with molecules of another component. If the two species are different in size, shape, or in the nature of their intermolecular forces, a solid solution can only be obtained by greatly distorting the crystalline structure. Such a distortion requires a great deal of energy, and, therefore, is energetically unfavorable. That is, the total energy (and in this case the Gibbs free energy) of the distorted crystalline solid solution is greater than the energy of the two pure solids (each in its undistorted crystalline state). Consequently, with the exception of metals and metal alloys, the energetically preferred state for many solid mixtures is as heterogeneous mixtures consisting of regions of pure single species in their usual crystalline state. If these regions are large enough that macroscopic thermodynamics can be applied without making corrections for the thermodynamic state of the molecules at the interface between two solid regions, the solid mixture can be treated as an agglomeration of pure species, each having its own *pure component* solid fugacity. Thus

$$\bar{f}_i^{S}(T, P, x_i) = f_i^{S}(T, P) \tag{7.4-13}$$

[3]If the pure species does not exist as a liquid at the temperature and pressure of interest, as, for example, in the case of a gas or solid dissolved in a liquid, other, more appropriate starting points are used, as discussed in Sec. 7.8.

Since alloys and solutions of metals will not be considered in this book, Eq. 7.4-13 will be used for computing the species fugacity in a solid mixture.

The point that the discrete pure component regions in a solid mixture should not be too small is worth dwelling on. A collection of molecules can be thought of as being composed of molecules at the surface of the region and molecules in the interior. The interior molecules experience interactions only with similar molecules and therefore are in the same energy state as molecules in a macroscopically large mass of the pure substance. The surface or interfacial molecules, however, interact with both like and unlike molecules, and hence their properties are representative of molecules in a mixture. The assumption being made in Eq. 7.4-13 is that the number of interfacial molecules is small compared to the number of interior molecules, so that even if the properties of the interfacial molecules are poorly represented, the effect on the total thermodynamic properties of the system is small.

7.5 SEVERAL CORRELATIVE LIQUID MIXTURE (ACTIVITY COEFFICIENT) MODELS

In this section we are interested in mixtures for which equation of state models are inapplicable. In such cases attempts are made to estimate $\underline{G}^{ex}$ or $\Delta\underline{G}_{\mathrm{mix}}$ directly, either empirically or semitheoretically. When making such estimates it is useful to distinguish between simple and nonsimple liquid mixtures. We define a **simple liquid mixture** to be one that is formed by mixing pure fluids, each of which is a liquid at the temperature and pressure of the mixture. A **nonsimple liquid mixture**, on the other hand, is formed by mixing pure species, at least one of which is not a liquid at the temperature and pressure of the mixture. Examples of nonsimple liquid mixtures are solutions of dissolved solids in liquids and dissolved gases in liquids. In this section we are concerned with liquid mixture models that are applicable to simple solutions. Nonsimple liquid solutions will be considered in Sec. 7.8 and elsewhere in this book. Also, for simplicity, most of the excess Gibbs free energy models we will consider here are for two-component (binary) mixtures. Multicomponent mixture models are considered in Appendix A7.3.

From the definition of a simple liquid mixture and excess mixing properties, Eq. 7.3-8, we have that the Gibbs free energy change on mixing is

$$\Delta\underline{G}_{\mathrm{mix}} = \Delta\underline{G}^{IM}_{\mathrm{mix}} + \underline{G}^{ex} = RT\sum_{i=1}^{\mathscr{C}} x_i \ln x_i + \underline{G}^{ex}$$

The excess Gibbs free energy of mixing for the benzene–2,2,4 trimethyl pentane system (a simple liquid mixture) is shown in Fig. 7.5-1 (the origin of these data will be discussed in Sec. 8.1); data for the excess Gibbs free energy for several other mixtures are given in Fig. 7.5-2. Each of the curves in these figures is of simple form. Thus, one approach to estimating the excess Gibbs free energy of mixing has been merely to try to fit results such as those given in Figs. 7.5-1 and 2 with polynomials in the composition. The hope is that given a limited amount of experimental data, one can determine the parameters in the appropriate polynomial expansion, and then predict the excess Gibbs free energy and liquid phase activity coefficients over the whole composition range. Of course, any expression chosen for the excess Gibbs free energy must satisfy the Gibbs–Duhem equation (i.e., Eqs. 7.3-14 through 17), and, like the data in the figures, go to zero as $x_1 \rightarrow 0$ and $x_1 \rightarrow 1$.

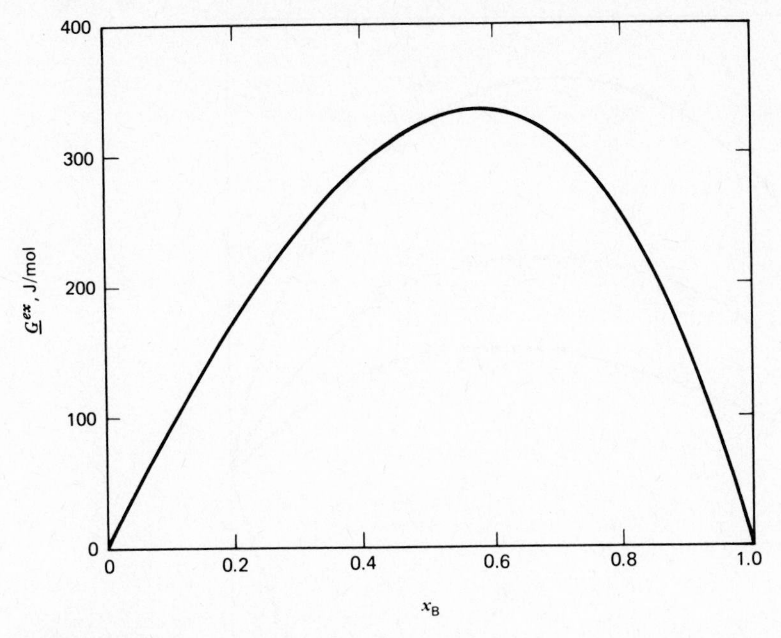

Figure 7.5-1
The excess Gibbs free energy for the benzene–2,2,4-trimethyl pentane system at 55°C.

The simplest polynomial representation of $\underline{G}^{ex}$ satisfying these criteria is

$$\underline{G}^{ex} = Ax_1x_2 \tag{7.5-1}$$

for which

$$\begin{aligned}\bar{G}_1^{ex} &= \left.\frac{\partial(N\underline{G}^{ex})}{\partial N_1}\right|_{T,P,N_2} = \frac{\partial}{\partial N_1}\left(\frac{AN_1N_2}{N_1+N_2}\right) \\ &= A\left\{\frac{N_2}{N_1+N_2} - \frac{N_1N_2}{(N_1+N_2)^2}\right\} = Ax_2^2\end{aligned} \tag{7.5-2}$$

so that

$$\gamma_1 = \exp\left\{\frac{\bar{G}_1^{ex}}{RT}\right\} = \exp\left\{\frac{Ax_2^2}{RT}\right\}$$

and

$$\gamma_2 = \exp\left\{\frac{Ax_1^2}{RT}\right\} \tag{7.5-3}$$

Consequently, for this solution model we have

$$\bar{f}_i^{L}(T, P, x_i) = x_i\gamma_i f_i^{L}(T, P) \tag{7.5-4}$$

where

$$\begin{aligned}RT \ln \gamma_1 &= Ax_2^2 \\ RT \ln \gamma_2 &= Ax_1^2\end{aligned} \tag{7.5-5}$$

These relations, known as the **one-constant Margules equations**, are plotted in Fig. 7.5-3. Two interesting features of the one-constant Margules equations are

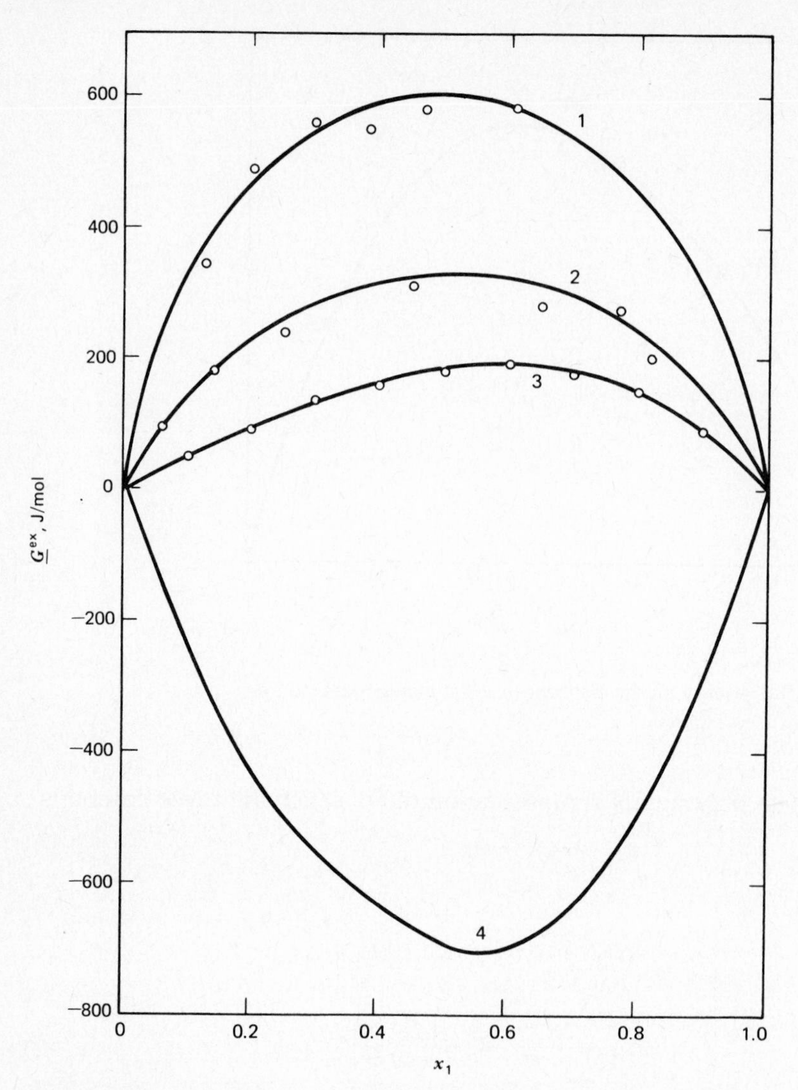

Figure 7.5-2
The excess Gibbs free energy for several mixtures:
Curve 1: Trimethyl methane (1) and benzene at 0°C. [Data from V. Mathot and A. Desmyter, *J. Chem. Phys.* **21,** 782 (1953).]
Curve 2: Trimethyl methane (1) and carbon tetrachloride at 0°C. [Data from *op. cit.*)]
Curve 3: Methane (1) and propane at 100 K. [Data from A. J. B. Cutler and J. A. Morrison, *Trans. Farad. Soc.* **61,** 429 (1965).]
Curve 4: Water (1) and hydrogen peroxide at 75°C. [From the smoothed data of G. Scatchard, G. M. Kavanagh, and L. B. Ticknor, *J. Am. Chem. Soc.* **74,** 3715 (1952).]

apparent from this figure. First, the two species activity coefficients are mirror images of each other as a function of the composition. This is not a general result, but follows from the choice of a symmetric function in the compositions for $\underline{G}^{ex}$. Second, $\gamma_i \to 1$ as $x_i \to 1$, so that $\bar{f}_i \to f_i$ in this limit. This makes good physical sense since one expects the fugacity of a component in a mixture to tend toward that of the pure liquid as the mixture becomes concentrated in that component. Conversely, γ_i is larger the greater the dilution of species i, as

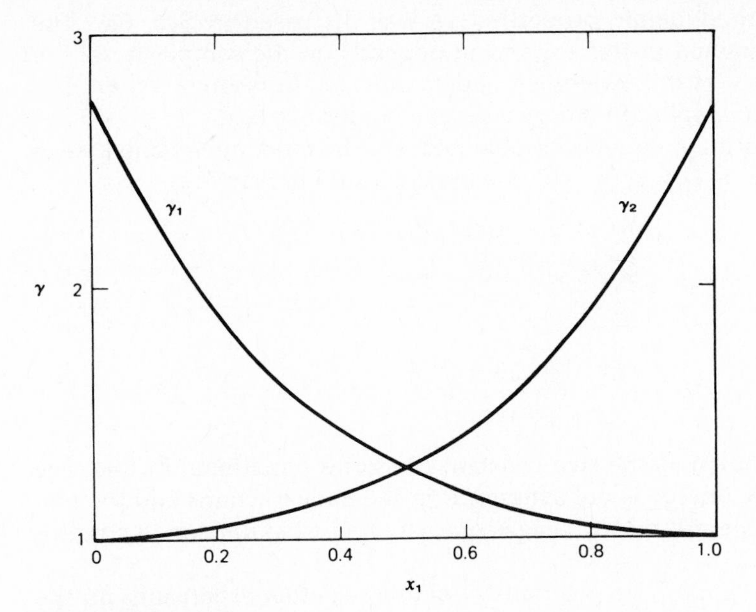

Figure 7.5-3
The activity coefficients for the one-constant Margules equation with $(A/RT) = 1.0$.

would be expected from the discussion in Chapter 6. Note that the parameter A can be either positive or negative, so that Eq. 7.5-5 can lead to activity coefficients which are greater than one ($A > 0$) or less than one ($A < 0$).

It is interesting to note that in the ideal solution model (i.e., $\bar{f}_i = x_i f_i$), the only role of components other than the one of interest is as diluents, so that they affect $\bar{f}_i$ only through x_i. No account is taken of the fact that the energy of a species 1–species 2 interaction can be different from that of species 1–species 1 or species 2–species 2 interactions. For the nonideal solution of Eq. 7.5-1 the parameter A is dependent on the species, or more precisely, on the differences in species interaction energies, involved. This results in the behavior of species 1 being influenced by both the nature and composition of species 2, and vice versa. The parameter A is a complicated function of macroscopic and molecular properties and so is difficult to estimate a priori; its value may be either positive or negative and generally is a function of temperature. Over small temperature ranges A may be assumed to be a constant, so that its value found from experiments at one temperature can be used at neighboring temperatures.

The one-constant Margules equation provides a satisfactory representation for activity coefficient behavior only for liquid mixtures containing constituents of similar size, shape, and chemical nature. For more complicated systems, particularly mixtures of dissimilar molecules, simple relations such as Eqs. 7.5-1 or 5 should not be expected to be valid. In particular, the excess Gibbs free energy of a general mixture is not likely to be a symmetric function of the mole fraction and the activity coefficients of the two species in a mixture should not be expected to be mirror images of each other. One possible generalization of Eq. 7.5-1 to such cases is to set

$$\underline{G}^{ex} = x_1 x_2 \{A + B(x_1 - x_2) + C(x_1 - x_2)^2 + \cdots\} \tag{7.5-6}$$

where $A, B, C, \ldots$ are temperature-dependent parameters. This expression for $\underline{G}^{ex}$ is another example of the Redlich–Kister expansion used for the representation of excess thermodynamic properties, as was discussed in Sec. 6.6. The number of terms retained in this expansion depends on the complexity of the mixture, the accuracy of the experimental data, and the fit desired. When $A = B = C = \cdots = 0$, ideal solution theory is recovered; for $A \neq 0$, $B = C = \cdots = 0$ the one-constant Margules equation is obtained. For the more interesting case of $A \neq 0$, $B \neq 0$, but $C = D = \cdots = 0$ one obtains (see Problem 7.7)

$$RT \ln \gamma_1 = \alpha_1 x_2^2 + \beta_1 x_2^3$$
$$RT \ln \gamma_2 = \alpha_2 x_1^2 + \beta_2 x_1^3 \tag{7.5-7}$$

where

$$\alpha_i = A + 3(-1)^{i+1} B$$
$$\beta_i = 4(-1)^i B$$

These results are known as the **two-constant Margules equations**. In this case the excess Gibbs free energy is not symmetric in the mole fractions and the two activity coefficients are not mirror images of each other as a function of concentration.

The expansion of Eq. 7.5-6 is certainly not unique; other expansions for the excess Gibbs free energy could also be used. Another expansion is that of Wohl[4]

$$\frac{\underline{G}^{ex}}{RT(x_1 q_1 + x_2 q_2)} = 2a_{12} z_1 z_2 + 3a_{112} z_1^2 z_2 + 3a_{122} z_1 z_2^2 + 4a_{1112} z_1^3 z_2 + 4a_{1222} z_1 z_2^3 + 6a_{1122} z_1^2 z_2^2 + \cdots \tag{7.5-8}$$

where q_i is some measure of the volume of molecule i (e.g., its liquid molar volume or van der Waals b parameter) and the a's are interaction parameters for unlike molecule interactions. The z_i's in Eq. 7.5-8 are, essentially, volume fractions defined by

$$z_i = \frac{x_i q_i}{x_1 q_1 + x_2 q_2}$$

Equation 7.5-8 is modeled after the virial expansion for gaseous mixtures, and, in fact, the constants in the expansion (2, 3, 3, 4, 4, 6, etc.) are those that arise in that equation.

The liquid-phase activity coefficients for the Wohl expansion can be obtained from Eq. 7.5-8 by taking the appropriate derivatives

$$\ln \gamma_i = \frac{\bar{G}_i^{ex}}{RT} = \frac{\partial}{\partial N_i}\left(\frac{N\underline{G}^{ex}}{RT}\right)_{T,P,N_{j\neq i}}$$

In particular, for the case in which we assume $a_{112} = a_{122} = \cdots = 0$, we have that

$$\frac{\underline{G}^{ex}}{RT} = (x_1 q_1 + x_2 q_2) 2a_{12} z_1 z_2 = \frac{2a_{12} x_1 q_1 x_2 q_2}{x_1 q_1 + x_2 q_2}$$

[4]K. Wohl, *Trans. A.I.Ch.E.* **42,** 215 (1946).

and (see Problem 7.7)

$$\ln \gamma_1 = \frac{\alpha}{\left[1 + \dfrac{\alpha}{\beta}\dfrac{x_1}{x_2}\right]^2} \quad \text{and} \quad \ln \gamma_2 = \frac{\beta}{\left[1 + \dfrac{\beta}{\alpha}\dfrac{x_2}{x_1}\right]^2} \tag{7.5-9}$$

where $\alpha = 2q_1a_{12}$ and $\beta = 2q_2a_{12}$. Equations 7.5-9 are known as the **van Laar equations**[5]; they are frequently used to correlate activity coefficient data. Other, more complicated, activity coefficient equations can be derived from Eq. 7.5-8, though this will not be done here.

The values of the parameters in the activity coefficient equations are usually found by fitting these equations to experimental activity coefficient data (see Problem 7.18). This is particularly simple in the case of the van Laar equations, which can be written as

$$\alpha = \left(1 + \frac{x_2 \ln \gamma_2}{x_1 \ln \gamma_1}\right)^2 \ln \gamma_1$$

and

$$\beta = \left(1 + \frac{x_1 \ln \gamma_1}{x_2 \ln \gamma_2}\right)^2 \ln \gamma_2 \tag{7.5-10}$$

so that data for γ_1 and γ_2 at only a single mole fraction can be used to evaluate the two van Laar constants, and hence to compute the activity coefficients at other compositions. This procedure will be used in Chapter 8. Alternatively, if activity coefficient data are available at several compositions, a regression procedure can be used to obtain the "best" fit values for α and β. Table 7.5-1 contains values that have been reported for these parameters for a number of binary mixtures.

ILLUSTRATION 7.5-1

The points in the following figures represent smoothed values of the activity coefficients for both species in a benzene–2,2,4 trimethyl pentane mixture at 55°C gotten from the vapor–liquid equilibrium measurements of Weissman and Wood (see Illustration 8.1-4). Test the accuracy of the one-constant Margules equation and the van Laar equations in correlating these data.

Solution

a. The one-constant Margules Equation

From the data presented in Fig. 7.5-4 it is clear that the activity coefficient for benzene is not the mirror image of that for trimethyl pentane. Therefore, the one-constant Margules equation cannot be made to fit both sets of activity coefficients simultaneously. (It is interesting to note that the Margules form, $RT \ln \gamma_i = A_i x_j^2$, will fit these data well if A_1 and A_2 are separately chosen. In practice, A_1 can be expected to be equal to A_2 only if the molar volumes of the two species are approximately equal. However, this suggestion does not satisfy the Gibbs–Duhem equation! Can you prove this?)

b. The van Laar Equation

One can use Eqs. 7.5-10 and a single activity coefficient-composition data point (or a least squares analysis of all the data points) to find values for the van Laar

[5] J. J. van Laar, *Z. Physik. Chem.* **72,** 723 (1910): **83,** 599 (1913).

Table 7.5-1
THE VAN LAAR CONSTANTS FOR SOME BINARY MIXTURES

Component 1–Component 2	Temperature Range (°C)	α	β
Acetaldehyde–water	19.8–100	1.59	1.80
Acetone–benzene	56.1–80.1	0.405	0.405
Acetone–methanol	56.1–64.6	0.58	0.56
Acetone–water	25°C	1.89	1.66
	56.1–100	2.05	1.50
Benzene–isopropanol	71.9–82.3	1.36	1.95
Carbon disulfide–acetone	39.5–56.1	1.28	1.79
Carbon disulfide–carbon tetrachloride	46.3–76.7	0.23	0.16
Carbon tetrachloride–benzene	76.4–80.2	0.12	0.11
Ethanol–benzene	67.0–80.1	1.946	1.610
Ethanol–cyclohexane	66.3–80.8	2.102	1.729
Ethanol–toluene	76.4–110.7	1.757	1.757
Ethanol–water	25	1.54	0.97
Ethyl acetate–benzene	71.1–80.2	1.15	0.92
Ethyl acetate–ethanol	71.7–78.3	0.896	0.896
Ethyl acetate–toluene	77.2–110.7	0.09	0.58
Ethyl ether–acetone	34.6–56.1	0.741	0.741
Ethyl ether–ethanol	34.6–78.3	0.97	1.27
n-Hexane–ethanol	59.3–78.3	1.57	2.58
Isobutane–furfural	37.8	2.62	3.02
	51.7	2.51	2.83
Isopropanol–water	82.3–100	2.40	1.13
Methanol–benzene	55.5–64.6	0.56	0.56
Methanol–ethyl acetate	62.1–77.1	1.16	1.16
Methanol–water	25	0.58	0.46
	64.6–100	0.83	0.51
Methyl acetate–methanol	53.7–64.6	1.06	1.06
Methyl acetate–water	57.0–100	2.99	1.89
n-Propanol–water	88.0–100	2.53	1.13
Water–phenol	100–181	0.83	3.22

Source: This table is an adaptation of one given in J. H. Perry, ed., *Chemical Engineers' Handbook*, 4th ed., McGraw–Hill, New York, 1963, p. 13-7.

Note: When $\alpha = \beta$, the van Laar equation $\ln \gamma_1 = \dfrac{\alpha}{[1 + (\alpha x_1/\beta x_2)]^2}$ reduces to the Margules form $\ln \gamma_1 = \alpha x_2^2$.

parameters. Using the data at $x_1 = 0.6$ we find that $\alpha = 0.415$ and $\beta = 0.706$. The activity coefficient predictions based on these values of the van Laar parameters are shown in Fig. 7.5-5. The agreement between the correlation and the experimental data is excellent. ■

The molecular level assumption underlying the Redlich–Kister expansion is that completely random mixtures are formed; that is, that the ratio of species 1

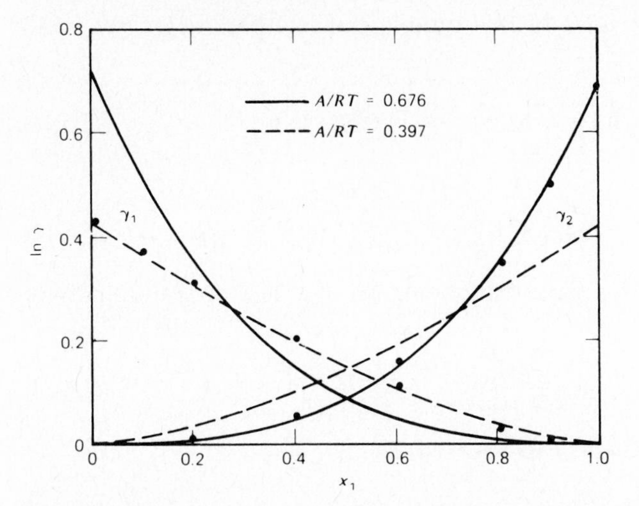

Figure 7.5-4
Experimental activity coefficient data for the benzene–2,2,4-trimethyl pentane mixture and the correlation of these data obtained using the one-constant Margules equation.

to species 2 molecules in the vicinity of any molecule is, on the average, the same as the ratio of the mole fractions. A different class of excess Gibbs free energy models can be formulated by assuming that the ratio of species 1 to species 2 molecules surrounding any molecule also depends on the differences in size and energies of interaction of the chosen molecule with species 1 and species 2. Thus, around each molecule there is a local composition that is different from the bulk composition. From this picture, the several binary mixture models have been developed.

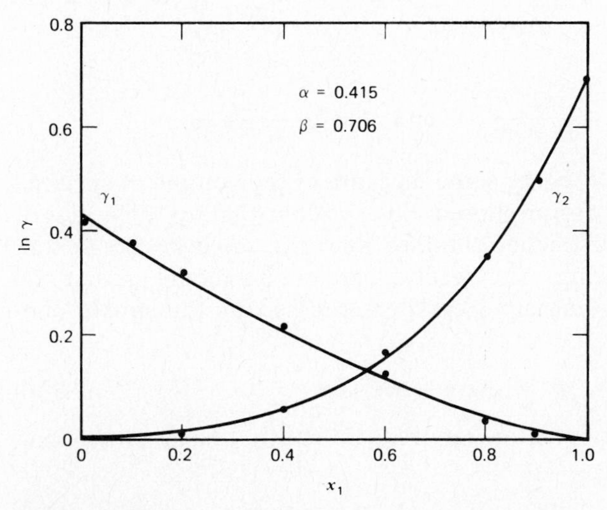

Figure 7.5-5
Experimental activity coefficient data for the benzene–2,2,4-trimethyl pentane mixture and the correlation of these data obtained using the van Laar equation.

The first model we consider of this type is the two-parameter (Λ_{12}, Λ_{21}) Wilson equation[6]

$$\frac{\underline{G}^{ex}}{RT} = -x_1 \ln (x_1 + \Lambda_{12}x_2) - x_2 \ln (x_2 + \Lambda_{21}x_1) \quad (7.5\text{-}11)$$

for which

$$\ln \gamma_1 = -\ln (x_1 + \Lambda_{12}x_2) + x_2 \left[\frac{\Lambda_{12}}{x_1 + \Lambda_{12}x_2} - \frac{\Lambda_{21}}{\Lambda_{21}x_1 + x_2}\right] \quad (7.5\text{-}12)$$

The second model is the three-parameter (α, τ_{12}, τ_{21}) nonrandom two-liquid (NRTL) equation[7]

$$\frac{\underline{G}^{ex}}{RT} = x_1 x_2 \left(\frac{\tau_{21}G_{21}}{x_1 + x_2 G_{21}} + \frac{\tau_{12}G_{12}}{x_2 + x_1 G_{12}}\right) \quad (7.5\text{-}13)$$

with $\ln G_{12} = -\alpha\tau_{12}$ and $\ln G_{21} = -\alpha\tau_{21}$ for which

$$\ln \gamma_1 = x_2^2 \left[\tau_{21}\left(\frac{G_{21}}{x_1 + x_2 G_{21}}\right)^2 + \frac{\tau_{12}G_{12}}{(x_2 + x_1 G_{12})^2}\right] \quad (7.5\text{-}14)$$

In both models the expressions for $\ln \gamma_2$ are obtained from the equations for $\ln \gamma_1$ by interchanging the subscripts 1 and 2. Also note that in these models there are different weightings of the mole fractions of the species due to the parameters (Λ_{ij} and τ_{ij}), the values of which depend on differences in size and energy of the molecules in the mixture. (The multicomponent forms of the Wilson and NRTL models are given in Appendix A7.3.)

Another model, that of Flory and Huggins, is meant to apply to mixtures of molecules of different size, including solutions of polymers. This solution model contains two parts. The first is an expression for the entropy of mixing per mole

$$\Delta \underline{S}_{\text{mix}} = -R(x_1 \ln \phi_1 + x_2 \ln \phi_2) \quad (7.5\text{-}15a)$$

or

$$\underline{S}^{ex} = \Delta \underline{S}_{\text{mix}} - \Delta \underline{S}^{IM}_{\text{mix}} = -R\left(x_1 \ln \frac{\phi_1}{x_1} + x_2 \ln \frac{\phi_2}{x_2}\right) \quad (7.5\text{-}15b)$$

Here

$$\phi_1 = \frac{x_1 v_1}{x_1 v_1 + x_2 v_2} = \frac{x_1}{x_1 + m x_2} \quad \text{and} \quad \phi_2 = \frac{m x_2}{x_1 + m x_2}$$

are the volume fractions, with v_i being some measure of the volume of species i molecules, and $m = v_2/v_1$. The assumption in Eqs. 7.5-15 is that for molecules of different size it is the volume fractions, rather than the mole fractions, that determine the entropy of mixing. The second part of the model is that the enthalpy of mixing, or excess enthalpy, can be expressed by the simple one-constant term

$$\Delta \underline{H}_{\text{mix}} = \underline{H}^{ex} = \chi RT(x_1 + m x_2)\phi_1\phi_2 \quad (7.5\text{-}16)$$

where χ is an adjustable parameter referred to as the Flory interaction parameter.

[6]G. M. Wilson, *J. Am. Chem. Soc.* **86,** 127 (1964).

[7]H. Renon and J. M. Prausnitz, *A.I.Ch.E.J.* **14,** 135 (1968).

Combining Eqs. 7.5-15 and 16 gives

$$\frac{\underline{G}^{ex}}{RT} = \frac{\underline{H}^{ex} - T\underline{S}^{ex}}{RT} = \left[x_1 \ln \frac{\phi_1}{x_1} + x_2 \ln \frac{\phi_2}{x_2}\right] + \chi(x_1 + mx_2)\phi_1\phi_2 \qquad (7.5\text{-}17)$$

which is the **Flory–Huggins** model. The first term on the right side of this equation is the entropic contribution to the excess Gibbs free energy, and the second term is the enthalpic contribution. These two terms are also referred to as the combinatorial and residual terms, respectively. The activity coefficient expressions for this model (Problem 7.24) are

$$\ln \gamma_1 = \ln \frac{\phi_1}{x_1} + \left(1 - \frac{1}{m}\right)\phi_2 + \chi\phi_2^2 \qquad (7.5\text{-}18)$$

and

$$\ln \gamma_2 = \ln \frac{\phi_2}{x_2} + (m - 1)\phi_1 + \chi\phi_1^2$$

We will consider only one additional activity coefficient equation here, the UNIQUAC (universal quasichemical) model of Abrams and Prausnitz.[8] This model, based on statistical mechanical theory, allows local compositions to result from both the size and energy differences between the molecules in the mixture. The result is the expression

$$\frac{\underline{G}^{ex}}{RT} = \frac{\underline{G}^{ex}\text{ (combinatorial)}}{RT} + \frac{\underline{G}^{ex}\text{ (residual)}}{RT} \qquad (7.5\text{-}19)$$

where the first term accounts for molecular size and shape differences, and the second term accounts largely for energy differences. These terms, in multicomponent form, are given by

$$\frac{\underline{G}^{ex}\text{ (combinatorial)}}{RT} = \sum_i x_i \ln \frac{\phi_i}{x_i} + \frac{z}{2}\sum_i x_i q_i \ln \frac{\theta_i}{\phi_i} \qquad (7.5\text{-}20)$$

and

$$\frac{\underline{G}^{ex}\text{ (residual)}}{RT} = -\sum q_i x_i \ln\left(\sum \theta_j \tau_{ji}\right) \qquad (7.5\text{-}21)$$

where

r_i = volume parameter for species i

q_i = surface area parameter for species i

θ_i = area fraction of species i $= x_i q_i \Big/ \sum x_j q_j$

ϕ_i = segment or volume fraction of species i $= x_i r_i \Big/ \sum x_j r_j$

and

$$\ln \tau_{ij} = -\frac{(u_{ij} - u_{jj})}{RT}$$

[8] D. S. Abrams and J. M. Prausnitz, *A.I.Ch.E.J.* **21,** 116 (1975).

with u_{ij} being the average interaction energy for a species i–species j interaction and z being the average coordination number, usually taken to be 10. Combining Eqs. 7.5-19, 20, and 21 gives

$$\ln \gamma_i = \ln \gamma_i \text{ (combinatorial)} + \ln \gamma_i \text{ (residual)} \tag{7.5-22}$$

$$\ln \gamma_i \text{ (combinatorial)} = \ln \frac{\phi_i}{x_i} + \frac{z}{2} q_i \ln \frac{\theta_i}{\phi_i} + l_i - \frac{\phi_i}{x_i} \sum_j x_j l_j \tag{7.5-23a}$$

and

$$\ln \gamma_i \text{ (residual)} = -q_i \left[1 - \ln \left(\sum_j \theta_j \tau_{ji} \right) - \sum_j \frac{\theta_j \tau_{ij}}{\sum_k \theta_k \tau_{kj}} \right] \tag{7.5-23b}$$

with $l_i = (r_i - q_i)z/2 - (r_i - 1)$.

Since the size and surface area parameters r_i and q_i can be evaluated from molecular structure information as will be discussed, the UNIQUAC equation contains only two adjustable parameters, τ_{12} and τ_{21} (or, equivalently, u_{12}–u_{22} and u_{21}–u_{11}) for each binary pair. Thus, like the van Laar or Wilson equations, it is a two-parameter activity coefficient model. It does have a better theoretical basis than these models, and it is somewhat more complicated.

Instead of listing the volume (r) and surface area (q) parameters for each molecular species for use in the UNIQUAC model, these parameters are evaluated by a **group contribution** method. The underlying idea is that a molecule can be considered to be a collection of **functional groups**, and that volume R_i and surface area Q_i of group i will be approximately the same in any molecule in which that group occurs. For example, we expect the contribution to the total volume and surface area of a molecule from a methyl (CH_3–) group to be the same independent of whether the methyl group is at the end of an ethane, propane, or dodecane molecule. Thus, the volume and surface area parameters r and q of a molecule will be gotten from a sum over its functional groups of the R and Q parameters. The advantage of this group contribution approach is that from a relatively small number of functional groups, the properties of the millions upon millions of different molecules can be obtained. [This idea is analogous to the alphabet; from only 26 letters (each of which can be considered to be a functional group), all the words in the English language can be made.]

Table 7.5-2 contains the R and Q parameters for 85 functional groups referred to as subgroups in the table (the main group, subgroup terminology will be explained in the following section). All the values that appear in the table have been normalized to the properties of a methylene group in polymethylene, and therefore are unitless. There are several things to note about the entries in this table. First, several molecules, such as water and furfural, have such unique properties that they have been treated as functional groups. Second, similar groups, such as –CH_3 and –C– may have different R and Q parameters. This is because a –CH_3 group, being at the end of a molecule, increases the molecular volume and surface area, whereas the –C– group, which is at the interior of a molecule, makes a small contribution to molecular volume and none to surface area. Finally, some groups, such as dimethylformamide (DMF) are considered to be both a molecule (DMF-1) and a functional group (DMF-2) in a larger molecule such as diethylformamide.

Table 7.5-2
THE GROUP VOLUME AND SURFACE AREA PARAMETERS, R AND Q, FOR USE WITH THE UNIQUAC AND UNIFAC MODELS

Main Group	Subgroup	R	Q	Example Assignments
CH_2	CH_3	0.9011	0.8480	
	CH_2	0.6744	0.5400	n-Hexane: 4 CH_2, 2 CH_3
	CH	0.4469	0.2280	Isobutane: 1 CH, 3 CH_3
	C	0.2195	0.0000	Neopentane: 1 C, 4 CH_3
C=C	CH_2=CH	1.3454	1.1760	1-Hexene: 1 CH_2=CH, 3 CH_2, 1 CH_3
	CH=CH	1.1167	0.8670	2-Hexene: 1 CH=CH, 2 CH_3, 2 CH_2
	CH_2=C	1.1173	0.9880	
	CH=C	0.8886	0.6760	
	C=C	0.6605	0.4850	
ACH	ACH	0.5313	0.4000	Benzene: 6 ACH
	AC	0.3652	0.1200	
$ACCH_2$	$ACCH_3$	1.2663	0.9680	Toluene: 5 ACH, 1 $ACCH_3$
	$ACCH_2$	1.0396	0.6600	Ethylbenzene: 5 ACH, 1 $ACCH_2$, 1 CH_3
	ACCH	0.8121	0.3480	
OH	OH	1.0000	1.2000	n-Propanol: 1 OH, 1 CH_3, 2 CH_2
CH_3OH	CH_3OH	1.4311	1.4320	Methanol
Water	H_2O	0.9200	1.400	Water
ACOH	ACOH	0.8952	0.6800	Phenol: 1 ACOH, 5 ACH
CH_2CO	CH_3CO	1.6724	1.4880	Dimethylketone: 1 CH_3CO, 1 CH_3
	CH_2CO	1.4457	1.1800	Diethylketone: 1 CH_2CO, 2 CH_3, 1 CH_2
CHO	CHO	0.9980	0.9480	Ethanal: 1 CHO, 1 CH_3
CCOO	CH_3COO	1.9031	1.7280	Methyl acetate: 1 CH_3COO, 1 CH_3
	CH_2COO	1.6764	1.4200	Methyl propanate: 1 CH_2COO, 2 CH_3
HCOO	HCOO	1.2420	1.1880	Methyl formate: 1 HCOO, 1 CH_3
CH_2O	CH_3O	1.1450	1.0880	
	CH_2O	0.9183	0.7800	Ethyl ether: 1 CH_2O, 1 CH_3, 1 CH_2
	CH-O	0.6908	0.4680	
	FCH_2O	0.9183	1.1000	Tetrahydrofuran: 1 FCH_2O, 3CH_2
CNH_2	CH_3NH_2	1.5959	1.5440	
	CH_2NH_2	1.3692	1.2360	Propyl amine: 1 CH_2NH_2, 1 CH_3, 1 CH_2
	$CHNH_2$	1.1417	0.9240	
CNH	CH_3NH	1.4337	1.2440	
	CH_2NH	1.2070	0.9360	Diethyl amine: 1 CH_2NH, 2 CH_3, 1 CH_2
	CHNH	0.9795	0.6240	
$(C)_3N$	CH_3N	1.1865	0.9400	
	CH_2N	0.9597	0.6320	Triethylamine: 1 CH_2N, 2 CH_2, 3 CH_3
$ACNH_2$	$ACNH_2$	1.0600	0.8160	Aniline: 1 $ACNH_2$, 5 ACH
Pyridine	C_5H_5N	2.9993	2.1130	
	C_5H_4N	2.8332	1.8330	Methyl pyridine: 1 C_5H_4N, 1 CH_3
	C_5H_3N	2.6670	1.5530	
CCN	CH_3CN	1.8701	1.7240	
	CH_2CN	1.6434	1.4160	Propionnitrile: 1 CH_2CN, 1 CH_3

Table 7.5-2 *(Continued)*

Main Group	Subgroup	R	Q	Example Assignments
COOH	COOH	1.3013	1.2240	Acetic acid: 1 COOH, 1 CH_3
	HCOOH	1.5280	1.5320	Formic acid
CCl	CH_2Cl	1.4654	1.2640	Chloroethane: 1 CH_2Cl, 1 CH_3
	CHCl	1.2380	0.9520	
	CCl	1.0060	0.7240	
CCl_2	CH_2Cl_2	2.2564	1.9880	
	$CHCl_2$	2.0606	1.6840	1,1-Dichloroethane: 1 $CHCl_2$, 1 CH_3
	CCl_2	1.8016	1.4480	
CCl_3	$CHCl_3$	2.8700	2.4100	
	CCl_3	2.6401	2.1840	1,1,1-Trichloroethane: 1 CCl_3, 1 CH_3
CCl_4	CCl_4	3.3900	2.9100	Chloroform
ACCl	ACCl	1.1562	0.8440	Chlorbenzene: 1 ACCl, 5 ACH
CNO_2	CH_3NO_2	2.0086	1.8680	Nitromethane
	CH_2NO_2	1.7818	1.5600	Nitroethane: 1 CH_2NO_2, 1 CH_3
	$CHNO_2$	1.5544	1.2480	
$ACNO_2$	$ACNO_2$	1.4199	1.1040	Nitrobenzene: 1 $ACNO_2$, 5 ACH
CS_2	CS_2	2.0570	1.6500	Carbon disulfide
CH_3SH	CH_3SH	1.8770	1.6760	Methanethiol
	CH_2SH	1.6510	1.3680	Ethanethiol: 1 CH_2SH, 1 CH_3
Furfural	Furfural	3.1680	2.4810	Furfural
DOH	$(CH_2OH)_2$	2.4088	2.2480	Ethylene glycol
I	I	1.2640	0.9920	Iodomethane: 1 I, 1 CH_3
Br	Br	0.9492	0.8320	Bromomethane: 1 Br, 1 CH_3
C≡C	CH≡C	1.2920	1.0880	1-Hexyne: 1 CH≡C, 1 CH_3, 3 CH_2
	C≡C	1.0613	0.7840	2-Hexyne: 1 C≡C, 2 CH_3, 2 CH_2
Me_2SO	Me_2SO	2.8266	2.4720	Dimethyl sulfoxide
ACRY	ACRY	2.3144	2.0520	Acrylonitrile
ClCC	Cl(C=C)	0.7910	0.7240	Trichloroethylene: 3 Cl(C=C), 1 CH=C
ACF	ACF	0.6948	0.5240	Hexafluorobenzene: 6 ACF
DMF	DMF-1	3.0856	2.7360	Dimethylformamide
	DMF-2	2.6322	2.1200	Diethylformamide: 1 DMF-2, 2 CH_3
CF_2	CF_3	1.4060	1.3800	
	CF_2	1.0105	0.9200	Perfluorohexane: 4 CF_2, 2 CF_3
	CF	0.6150	0.4600	
COO	COO	1.3800	1.2000	Butylacetate: 1 COO, 2 CH_3, 3 CH_2
SiH_2	SiH_3	1.6035	1.2632	Methylsilane: 1 SiH_3, 1 CH_3
	SiH_2	1.4443	1.0063	
	SiH	1.2853	0.7494	
	Si	1.0470	0.4099	Hexamethyldisiloxane: 1 Si, 1 SiO, 6 CH_3
SiO	SiH_2O	1.4838	1.0621	
	SiHO	1.3030	0.7639	
	SiO	1.1044	0.4657	Hexamethyldisiloxane: 1 Si, 1 SiO, 6 CH_3
NMP	NMP	3.9810	3.2000	N-methylpyrrolidone

ILLUSTRATION 7.5-2

One mole each of benzene and 2,2,4-trimethyl pentane are mixed together. Using the data in Table 7.5-2, compute the volume fraction and surface area fractions of benzene and 2,2,4-trimethyl pentane.

Solution

Benzene consists of six aromatic CH (ACH) groups. Therefore,

$$r_B = 6 \times 0.5313 = 3.1878$$

$$q_B = 6 \times 0.4000 = 2.4000$$

The structure of 2,2,4-trimethyl pentane is

$$\begin{array}{c} \quad CH_3 \qquad\quad CH_3 \\ \quad | \qquad\qquad\;\; | \\ CH_3{-}C{-}CH_2{-}CH{-}CH_3 \\ | \qquad\qquad\qquad \\ CH_3 \qquad\qquad\quad \end{array}$$

This molecule consists of 5 CH_3 groups, 1 CH_2 group, 1 CH group, and 1 C group. Thus

$$r_{TMP} = 5 \times 0.9011 + 0.6744 + 0.4469 + 0.2195 = 5.8463$$

and

$$q_{TMP} = 5 \times 0.8480 + 0.5400 + 0.2280 + 0.0 = 5.0080$$

Consequently,

$$\theta_B = \frac{0.5 \times 2.4000}{0.5 \times 2.4000 + 0.5 \times 5.0080} = 0.324, \qquad \theta_{TMP} = 0.676$$

$$\phi_B = \frac{0.5 \times 3.1878}{0.5 \times 3.1878 + 0.5 \times 5.8463} = 0.353, \qquad \phi_{TMP} = 0.647 \quad \blacksquare$$

Finally, it is useful to note that the Gibbs–Duhem equation can be used to get information about the activity coefficient behavior of one component in a binary mixture if the concentration dependence of the other species is known. This is demonstrated in the next illustration.

ILLUSTRATION 7.5-3

The activity coefficient for species 1 in a binary mixture can be represented by

$$\ln \gamma_1 = ax_2^2 + bx_2^3 + cx_2^4$$

where a, b, and c are concentration-independent parameters. What is the expression for $\ln \gamma_2$ in terms of these same parameters?

Solution

For a binary mixture at constant temperature and pressure we have, from Eqs. 6.2-20 or 7.3-17, that

$$x_1 \frac{\partial \ln \gamma_1}{\partial x_2} + x_2 \frac{\partial \ln \gamma_2}{\partial x_2} = 0$$

Since $x_1 = 1 - x_2$, and $\ln \gamma_1 = ax_2^2 + bx_2^3 + cx_2^4$, we have

$$\frac{\partial \ln \gamma_2}{\partial x_2} = -\frac{x_1}{x_2}\frac{\partial \ln \gamma_1}{\partial x_2} = -\frac{(1-x_2)}{x_2}(2ax_2 + 3bx_2^2 + 4cx_2^3)$$

$$= -2a + (2a - 3b)x_2 + (3b - 4c)x_2^2 + 4cx_2^3$$

Now, by definition, $\gamma_2(x_2 = 1) = 1$, and $\ln \gamma_2(x_2 = 1) = 0$. Therefore,

$$\int_{x_2=1}^{x_2} \frac{d \ln \gamma_2}{\partial x_2} dx_2 = \ln \gamma_2(x_2) - \ln \gamma_2(x_2 = 1) = \ln \gamma_2(x_2)$$

$$= \int_{x_2=1}^{x_2} [-2a + (2a - 3b)x_2 + (3b - 4c)x_2^2 + 4cx_2^3]\, dx_2$$

$$= -2a(x_2 - 1) + \frac{(2a - 3b)}{2}(x_2^2 - 1) + \frac{(3b - 4c)}{3}(x_2^3 - 1) + \frac{4c}{4}(x_2^4 - 1)$$

Again using $x_1 + x_2 = 1$, or $x_2 = 1 - x_1$, yields

$$\begin{aligned}\ln \gamma_2 = {} & -2a(1 - x_1 - 1) + \tfrac{1}{2}(2a - 3b)(1 - 2x_1 + x_1^2 - 1) \\ & + \tfrac{1}{3}(3b - 4c)(1 - 3x_1 + 3x_1^2 - x_1^3 - 1) \\ & + c(1 - 4x_1 + 6x_1^2 - 4x_1^3 + x_1^4 - 1)\end{aligned}$$

$$\ln \gamma_2 = \left(a + \frac{3b}{2} + 2c\right) x_1^2 - (b + \tfrac{8}{3}c)x_1^3 + cx_1^4 \quad \blacksquare$$

7.6 TWO PREDICTIVE ACTIVITY COEFFICIENT MODELS

As the variety of organic compounds of interest in chemical processing is very large, and the number of possible binary, ternary, etc., mixtures is essentially uncountable, situations frequently arise in which engineers need to make activity coefficient predictions for systems, or at conditions, for which experimental data are not available. Although the models discussed in the previous section are useful for correlating experimental data, or for making predictions given a limited amount of experimental information, they are of little value in making predictions when no experimental data are available. This is because little physical significance has been attributed to the parameters in these equations.

Most of the recent theories of liquid solution behavior have been based on well-defined thermodynamic or statistical mechanical assumptions, so that the parameters that appear can be related to the molecular properties of the species in the mixture and the resulting models have some predictive ability. Although a detailed study of the more fundamental approaches to liquid solution theory is beyond the scope of this book, we will consider two examples here: the theory of van Laar,[9] which leads to regular solution theory, and the UNIFAC group contribution model, which arises from the UNIQUAC model introduced in the previous section. Both regular solution theory and the UNIFAC model are useful for estimating solution behavior in the absence of experimental data. However, neither one is considered sufficiently accurate for the design of a chemical plant.

The theory of van Laar (i.e., the argument that originally led to Eqs. 7.5-9) is based on the assumptions that (1) the binary mixture is composed of two

[9]See Footnote 5.

species of similar size and energies of interaction, and (2) the van der Waals equation of state applies to both the pure fluids and the binary mixture.[10] The implication of assumption 1 is that the molecules of each species will be uniformly distributed throughout the mixture (see Appendix A7.1) and the intermolecular spacing will be similar to that in the pure fluids. Consequently, it is not unreasonable to expect that at a given temperature and pressure

$$\Delta \underline{V}_{\text{mix}} = 0 \qquad \text{or} \qquad \underline{V}^{ex} = 0$$

and

$$\Delta \underline{S}_{\text{mix}} = -R \sum_{i=1}^{2} x_i \ln x_i \qquad \text{or} \qquad \underline{S}^{ex} = 0 \tag{7.6-1}$$

so that

$$\underline{G}^{ex} = \underline{U}^{ex} + P\underline{V}^{ex} - T\underline{S}^{ex} = \underline{U}^{ex}$$

for such liquid mixtures. Thus to obtain the excess Gibbs free energy change, and thereby the activity coefficients for this liquid mixture, we need only compute the excess internal energy change on mixing.

Since the internal energy, and therefore the excess internal energy, is a state function, $\underline{U}^{ex}$ may be computed along any convenient path leading from the pure components to the mixture. The path we will use is:

1. Start with x_1 moles of pure liquid 1 and x_2 moles of pure liquid 2 (where $x_1 + x_2 = 1$) at the temperature and pressure of the mixture, and, at constant temperature, lower the pressure so that each of the pure liquids vaporizes to an ideal gas.
2. At constant temperature and (very low) pressure, mix the ideal gases to form an ideal gas mixture.
3. Now compress the gas mixture, at constant temperature, to a liquid mixture at the final pressure P.

The total internal energy change for this process (which is just the excess internal energy change since $\Delta \underline{U}^{IM}_{\text{mix}} = 0$) is the sum of the internal energy changes of each of the steps,

$$\underline{G}^{ex} = \underline{U}^{ex} = \Delta \underline{U}^{\text{I}} + \Delta \underline{U}^{\text{II}} + \Delta \underline{U}^{\text{III}} \tag{7.6-2}$$

Noting that

$$\left(\frac{\partial \underline{U}}{\partial \underline{V}}\right)_T = T\left(\frac{\partial P}{\partial T}\right)_{\underline{V}} - P$$

and using the facts that $\underline{V} \rightarrow \infty$ as $P \rightarrow 0$, and that $\Delta \underline{U}^{\text{II}} = \Delta \underline{U}^{IGM}_{\text{mix}} = 0$, one obtains

$$\underline{G}^{ex} = \Delta \underline{U} = x_1 \left[\int_{\underline{V}_1}^{\infty} \left\{T\left(\frac{\partial P}{\partial T}\right)_{\underline{V}} - P\right\} d\underline{V}\right]_{\substack{\text{pure}\\\text{fluid}\\1}} + x_2 \left[\int_{\underline{V}_2}^{\infty} \left\{T\left(\frac{\partial P}{\partial T}\right)_{\underline{V}} - P\right\} d\underline{V}\right]_{\substack{\text{pure}\\\text{fluid}\\2}}$$

$$- \left[\int_{\underline{V}_{\text{mix}}}^{\infty} \left\{T\left(\frac{\partial P}{\partial T}\right)_{\underline{V}} - P\right\} d\underline{V}\right]_{\text{mixture}} \tag{7.6-3}$$

[10] van Laar was a student of van der Waals.

Each of the bracketed terms represents the internal energy change on going from a liquid to an ideal gas and is equal to the internal energy change on vaporization.

Next we use assumption 2, that the van der Waals equation of state is applicable to both pure fluids and the mixture. From

$$P = \frac{RT}{(\underline{V} - b)} - \frac{a}{\underline{V}^2} \tag{7.6-4}$$

it is easily established that

$$\left(\frac{\partial \underline{U}}{\partial \underline{V}}\right)_T = T\left(\frac{\partial P}{\partial T}\right)_{\underline{V}} - P = \frac{a}{\underline{V}^2}$$

so that Eq. 7.6-3 becomes

$$\begin{aligned} \underline{G}^{ex} = \Delta \underline{U} &= x_1 \int_{\underline{V}_1}^{\infty} \frac{a_1}{\underline{V}^2} d\underline{V} + x_2 \int_{\underline{V}_2}^{\infty} \frac{a_2}{\underline{V}^2} d\underline{V} - \int_{\underline{V}_{\text{mix}}}^{\infty} \frac{a_{\text{mix}}}{\underline{V}^2} d\underline{V} \\ &= x_1 \frac{a_1}{\underline{V}_1} + x_2 \frac{a_2}{\underline{V}_2} - \frac{a_{\text{mix}}}{\underline{V}_{\text{mix}}} \end{aligned} \tag{7.6-5}$$

where the molar volumes appearing in this equation are those of the liquid. Since the molecules of a liquid are closely packed, liquids are relatively incompressible; that is, enormous pressures are required to produce relatively small changes in the molar volume. Such behavior is predicted by Eq. 7.6-4 if $\underline{V} \approx b$. Making this substitution in Eq. 7.6-5 gives

$$\underline{G}^{ex} = \Delta \underline{U} = x_1 \frac{a_1}{b_1} + x_2 \frac{a_2}{b_2} - \frac{a_{\text{mix}}}{b_{\text{mix}}}$$

Now using the mixing rules of Eq. 7.4-8 (with $k_{ij} = 0$), we obtain the following expression for the excess internal energy of a van Laar mixture:

$$\underline{U}^{ex} = \underline{G}^{ex} = \Delta \underline{U} = \frac{x_1 x_2 b_1 b_2}{x_1 b_1 + x_2 b_2} \left(\frac{\sqrt{a_1}}{b_1} - \frac{\sqrt{a_2}}{b_2}\right)^2$$

Finally, differentiating this expression with respect to composition yields $\overline{G}_i^{ex}$ and, from Eq. 7.3-12, one obtains the van Laar equations for the activity coefficients

$$\ln \gamma_1 = \frac{\alpha}{\left[1 + \dfrac{\alpha x_1}{\beta x_2}\right]^2} \quad \text{and} \quad \ln \gamma_2 = \frac{\beta}{\left[1 + \dfrac{\beta x_2}{\alpha x_1}\right]^2}$$

with

$$\alpha = \frac{b_1}{RT}\left[\frac{\sqrt{a_1}}{b_1} - \frac{\sqrt{a_2}}{b_2}\right]^2 \quad \text{and} \quad \beta = \frac{b_2}{RT}\left[\frac{\sqrt{a_1}}{b_1} - \frac{\sqrt{a_2}}{b_2}\right]^2 \tag{7.6-6}$$

This development provides both a justification for the van Laar equations and a method of estimating the van Laar parameters for liquid-phase activity coefficients from the parameters in the van der Waals equation of state. Since we know the van der Waals equation is not very accurate, it is not surprising that, if α and β are treated as adjustable parameters, the correlative value of the van Laar equations is greater than when α and β are determined from Eqs. 7.6-6 (Problem 7.8).

Scatchard,[11] based on the observations of Hildebrand,[12] concluded that although $\underline{V}^{ex}$ and $\underline{S}^{ex}$ were approximately equal to zero for many solutions (such mixtures are called **regular solutions**), few obeyed the van der Waals equation of state. Therefore, he suggested that instead of using an equation of state to predict the internal energy change on vaporization as in the van Laar theory, the experimental internal energy change on vaporization (usually at 25°C) be used. It was also suggested that the internal energy change of vaporization for the mixture be estimated from the following approximate mixing rule

$$\Delta\underline{U}_{\text{mix}}^{\text{vap}} = \left(x_1\sqrt{\frac{\underline{V}_1\Delta\underline{U}_1^{\text{vap}}}{\underline{V}_{\text{mix}}}} + x_2\sqrt{\frac{\underline{V}_2\Delta\underline{U}_2^{\text{vap}}}{\underline{V}_{\text{mix}}}}\right)^2 \tag{7.6-7}$$

since experimental data on the heat of vaporization of mixtures are rarely available. In this equation all molar volumes are those of the liquids, and $\underline{V}_{\text{mix}} = x_1\underline{V}_1 + x_2\underline{V}_2$, since $\Delta\underline{V}_{\text{mix}} = 0$ by the first van Laar assumption. Defining the volume fraction Φ_i and **solubility parameter** δ_i of species i by

$$\Phi_i \equiv \frac{x_i\underline{V}_i}{\underline{V}_{\text{mix}}} \tag{7.6-8}$$

and

$$\delta_i \equiv \left(\frac{\Delta\underline{U}_i^{\text{vap}}}{\underline{V}_i}\right)^{1/2} \tag{7.6-9}$$

one obtains (see Eq. 7.6-3)

$$\begin{aligned}\underline{G}^{ex} = \underline{U}^{ex} &= x_1\Delta\underline{U}_1^{\text{vap}} + x_2\Delta\underline{U}_2^{\text{vap}} - \Delta\underline{U}_{\text{mix}}^{\text{vap}} \\ &= (x_1\underline{V}_1 + x_2\underline{V}_2)\Phi_1\Phi_2[\delta_1 - \delta_2]^2\end{aligned}$$

which, on differentiation, yields

$$\begin{aligned}RT\ln\gamma_1 &= \underline{V}_1\Phi_2^2[\delta_1 - \delta_2]^2 \\ RT\ln\gamma_2 &= \underline{V}_2\Phi_1^2[\delta_1 - \delta_2]^2\end{aligned} \tag{7.6-10}$$

Thus, we have a recipe for estimating the activity coefficients of each species in a binary liquid mixture from a knowledge of the pure component molar volumes, the mole (or volume) fractions, and the solubility parameters (or internal energy changes on vaporization) of each species. Table 7.6-1 gives a list of the solubility parameters and molar volumes for a number of nonpolar liquids. Note, however, that the assumptions contained in regular solution theory (i.e., $\underline{V}^{ex} = \underline{S}^{ex} = 0$ and Eq. 7.6-7) are not generally applicable to polar substances. Therefore, regular solution theory should be used only for the components listed in Table 7.6-1 and similar compounds.

Regular solution theory is functionally equivalent to the van Laar theory since, with the substitutions

$$\alpha = \frac{\underline{V}_1}{RT}(\delta_1 - \delta_2)^2 \qquad \text{and} \qquad \beta = \frac{\underline{V}_2}{RT}(\delta_1 - \delta_2)^2$$

Eqs. 7.5-9 and 7.6-10 are identical. The important advantage of regular solution theory, however, is that its parameters are calculable without resorting to activ-

[11]G. Scatchard, *Chem. Rev.* **8,** 321 (1931).

[12]J. H. Hildebrand, *J. Am. Chem. Soc.* **41,** 1067 (1919). This work is discussed in J. H. Hildebrand, J. M. Prausnitz, and R. L. Scott, *Regular and Related Solutions,* Van Nostrand–Reinhold, Princeton, N.J., 1970.

Table 7.6-1
MOLAR LIQUID VOLUMES AND SOLUBILITY PARAMETERS OF SOME NONPOLAR LIQUIDS

Liquefied Gases at 90 K	$\underline{V}^L$ (cc/mol)	δ (cal/cc)$^{1/2}$
Nitrogen	38.1	5.3
Carbon monoxide	37.1	5.7
Argon	29.0	6.8
Oxygen	28.0	7.2
Methane	35.3	7.4
Carbon tetrafluoride	46.0	8.3
Ethane	45.7	9.5
Liquid Solvents at 25°C		
Perfluoro-*n*-heptane	226	6.0
Neopentane	122	6.2
Isopentane	117	6.8
n-Pentane	116	7.1
n-Hexane	132	7.3
1-Hexene	126	7.3
n-Octane	164	7.5
n-Hexadecane	294	8.0
Cyclohexane	109	8.2
Carbon tetrachloride	97	8.6
Ethyl benzene	123	8.8
Toluene	107	8.9
Benzene	89	9.2
Styrene	116	9.3
Tetrachloroethylene	103	9.3
Carbon disulfide	61	10.0
Bromine	51	11.5

Source: From J. M. Prausnitz, *Molecular Thermodynamics of Fluid-Phase Equilibria,* 1969. Reprinted by permission of Prentice–Hall, Inc., Englewood Cliffs, N.J.

Note: In regular solution theory the solubility parameter has traditionally been given in the units shown. For this reason the traditional units, rather than SI units, appear in this table.

ity coefficient measurements. Unfortunately, the parameters gotten in this way are not as accurate as those fitted to experimental data.

ILLUSTRATION 7.6-1

Compare the regular solution theory predictions for the activity coefficients of the benzene–2,2,4-trimethyl pentane mixture with the experimental data given in Illustration 7.5-1.

Solution

From Table 7.6-1 we have $\underline{V}^L = 89$ cc/mol, and $\delta = 9.2$ (cal/cc)$^{1/2}$ for benzene. The parameters for 2,2,4-trimethyl pentane are not given. However, the molar volume of this compound is, approximately, 165 cc/mol, and the solubility parameter can be estimated from

$$\delta = \left(\frac{\Delta \underline{U}^{\text{vap}}}{\underline{V}^{\text{L}}}\right)^{1/2} = \left(\frac{\Delta \underline{H}^{\text{vap}} - RT}{\underline{V}^{\text{L}}}\right)^{1/2}$$

where, in the numerator, we have neglected the liquid molar volume with respect to that of the vapor, and further assumed ideal vapor-phase behavior. Using the value of $\Delta \underline{H}^{\text{vap}}$ found in Illustration 5.7-1, we obtain $\delta = 6.93\ (\text{cal/cc})^{1/2}$.

Our predictions for the activity coefficients, together with the experimental data, are plotted in Fig. 7.6-1. It is evident that although the regular solution theory prediction leads to activity coefficient behavior that is qualitatively correct, the quantitative agreement in this case is, in fact, poor. This example should serve as a warning that theoretical predictions may not always be accurate and that experimental data are to be preferred. ■

The extension of regular solution theory to multicomponent mixtures is an algebraically messy task; the final result is:

$$RT \ln \gamma_i = \underline{V}_i(\delta_i - \bar{\delta})^2 \tag{7.6-11}$$

where

$$\bar{\delta} = \begin{pmatrix} \text{volume fraction} \\ \text{averaged solubility} \\ \text{parameter} \end{pmatrix} = \sum_j \Phi_j \delta_j$$

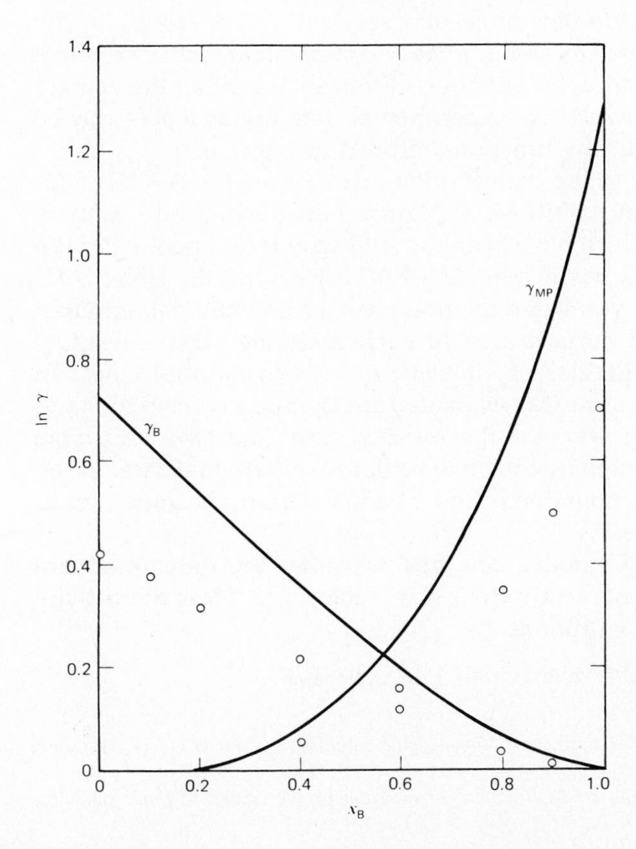

Figure 7.6-1
The regular solution theory predictions for the activity coefficients of the benzene–2,2,4-trimethyl pentane mixture. The points indicated by ○ are the experimental data.

and

$$\Phi_j = \frac{x_j \underline{V}_j}{\sum_l x_l \underline{V}_l}$$

where each of the sums are over all the species in the mixture.

Several other characteristics of regular solution theory are worth noting. The first is that the theory leads only to positive deviations from ideal solution behavior in the sense that $\gamma_i \geqslant 1$. This result can be traced back to the assumption of Eq. 7.6-7, which requires that $\Delta \underline{U}^{\text{vap}}_{\text{mix}}$ always be intermediate to that of the two pure components. Also, the solubility parameters are clearly functions of temperature since $\delta \to 0$ as $T \to T_C$ for each species; however, $(\delta_1 - \delta_2)$ is often nearly temperature independent, at least over limited temperature ranges. Therefore, the solubility parameters listed in Table 7.6-1 may be used at temperatures other than the one at which they were obtained (see Eq. 7.3-22, however). Also, referring to Eq. 7.6-10, it is evident that liquids with very different solubility parameters, such as neopentane and carbon disulfide, can be expected to exhibit highly nonideal solution behavior (i.e., $\gamma_i > 1$), whereas adjacent liquids in Table 7.6-1 will form nearly ideal solutions.

The most successful recent activity coefficient prediction methods are based on the idea of group contributions in which each molecule is considered to be a collection of basic building blocks, the functional groups discussed in the last section. Thus, a mixture of molecules is considered to be a mixture of functional groups, and solution properties result from functional group interactions. As the number of different types of functional groups is very much smaller than the number of different molecular species, it is possible, by the regression of experimental data, to obtain a fairly compact table of parameters for the interaction of each group with all others. From such a table, the activity coefficients in a mixture for which no experimental data are available can be estimated from a knowledge of the functional groups present.

The two most developed group contribution methods are the ASOG (*A*nalytical *S*olution *O*f *G*roups) and UNIFAC (*UNI*quac *F*unctional-group *A*ctivity *C*oefficient) models, both of which have been the subject of recent books.[13,14] We will consider only the UNIFAC model here. UNIFAC is based on the UNIQUAC model of Sec. 7.5. This model, you will remember, had a combinatorial term that depends on the volume and surface area of each molecule, and a residual term that is a result of the energies of interaction between the molecules. In UNIQUAC the combinatorial term was evaluated using group contributions to compute the size parameters, whereas the residual term had two adjustable parameters for each binary system that were to be fit to experimental data. In the UNIFAC model, both the combinatorial and residual terms are gotten from group contribution methods.

When using the UNIFAC model, one first identifies the functional subgroups present in each molecule using the list in Table 7.5-2. Next the activity coefficient for each species is written as

$$\ln \gamma_i = \ln \gamma_i(\text{combinatorial}) + \ln \gamma_i(\text{residual})$$

[13] A. Fredenslund, J. Gmehling, and P. Rasmussen, *Vapor-Liquid Equilibrium Using UNIFAC*, Elsevier, Amsterdam, 1977.

[14] K. Kojima and T. Tochigi, *Prediction of Vapor-Liquid Equilibrium by the ASOG Method*, Elsevier, Amsterdam, 1979.

The combinatorial term is evaluated as before using Eq. 7.5-23a. Here, however, the residual term is also evaluated by a group contribution method, so that the mixture is envisioned as being a mixture of functional groups, rather than of molecules. The residual contribution to the logarithm of the activity coefficient of group k in the mixture, $\ln \Gamma_k$, is computed from the group contribution analog of Eq. 7.5-23b, which we write as

$$\ln \Gamma_k = Q_k \left[1 - \ln \left(\sum_m \Theta_m \Psi_{mk} \right) - \sum_m \frac{\Theta_m \Psi_{km}}{\sum_n \Theta_n \Psi_{nm}} \right] \tag{7.6-12}$$

where

$$\Theta_m = \begin{pmatrix} \text{surface area} \\ \text{fraction of} \\ \text{group } m \end{pmatrix} = \frac{X_m Q_m}{\sum_n X_n Q_n}$$

$$X_m = \text{mole fraction of group } m \text{ in mixture} \tag{7.6-13a}$$

and

$$\Psi_{mn} = \exp\left[\frac{-(u_{mn} - u_{nn})}{kT}\right] = \exp\left[\frac{-a_{mn}}{T}\right] \tag{7.6-13b}$$

where u_{mn} is a measure of the interaction energy between groups m and n, and the sums are over all groups in the mixture.

Finally, the residual contribution to the residual part of the activity coefficient of species i is computed from

$$\ln \gamma_i(\text{residual}) = \sum_k \upsilon_k^{(i)} [\ln \Gamma_k - \ln \Gamma_k^{(i)}] \tag{7.6-14}$$

Here $\upsilon_i^{(i)}$ is the number of k groups present in species i and $\ln \Gamma_k^{(i)}$ is the residual contribution to the activity coefficient of group k in a pure fluid of species i molecules. This last term has been included to ensure that in the limit of pure species i, which is still a mixture of groups (unless species i molecules consist of only a single functional group), $\ln \gamma_i$(residual) goes to zero.

The combination of Eqs. 7.5-23a and 7.6-12 through 14 is the UNIFAC model. Since the volume (R_i) and surface (Q_i) parameters are known (Table 7.5-2), the only unknowns are binary parameters, a_{nm} and a_{mn}, for each pair of functional groups. Continuing with the group contribution idea, it is next assumed that any pair of functional groups m and n, will interact in the same manner, that is, have the same value of a_{mn} and a_{nm}, independent of the mixtures in which these two groups occur. Thus, for example, it is assumed that the interaction between an alcohol (–OH) group and a methyl (–CH_3) group will be the same, regardless of whether these groups occur in ethanol–n-pentane, isopropanol–decane, or 2-octanol–2,2-4-trimethyl pentane mixtures.

Consequently, by a regression analysis of great quantities of activity coefficient (or as we will see in Sec. 8.1, vapor–liquid equilibrium) data, the binary parameters a_{nm} and a_{mn} for many group–group interactions can be determined. These parameters can then be used to predict the activity coefficients in mixtures (binary or multicomponent) for which no experimental data are available.

In the course of such an analysis it was found that (1) experimental data existed to determine many of, but not all, the binary group parameters a_{nm} and a_{mn}, and (2) that some very similar groups, such as the –CH_3, –CH_2, –CH, and –C groups, interact with other groups in approximately the same way and, therefore, have the same interaction parameters with other groups. Such very similar subgroups are considered to belong to the same main group. Consequently, in Table 7.5-2 there are 44 main groups among the 85 subgroups. All the subgroups within a main group (i.e., the CH_3, CH_2, CH, and C subgroups within the CH_2 main group) have the same values of the binary parameters for interactions with other main groups and zero values for the interactions with other subgroups in their own main group.

Appendix A7.4 discusses a BASIC language program to use the UNIFAC model for activity coefficient predictions. This program is on the disk accompanying this book.

ILLUSTRATION 7.6-2

Compare the UNIFAC predictions for the activity coefficients of the benzene–2,2,4-trimethyl pentane mixture with the experimental data given in Illustration 7.5-1.

Solution

Benzene consists of 6 aromatic CH groups (i.e., subgroup 10 of Table 7.5-2) whereas 2,2,4-trimethyl pentane contains 5 CH_3 groups (subgroup 1), 1 CH_2 group (subgroup 2), 2 CH groups (subgroup 3), and 1 C group (subgroup 4). Using the UNIFAC program of Appendix A7.4, we obtain the results plotted in Fig. 7.6-2.

It is clear from this figure that, for this simple system, the UNIFAC predictions are good; in particular, they are much better than the regular solution predictions of Fig. 7.6-1. Although the UNIFAC predictions for all systems are not always as good as for the benzene–2,2,4-trimethyl pentane system, UNIFAC is the best activity coefficient prediction method currently available. ■

Table 7.6-3 summarizes both the prescriptions so far developed and those that are discussed elsewhere in this book for the estimation of species fugacities. The discussion in this text is certainly not a comprehensive one; for more general fugacity estimation techniques, refer to the excellent book by Prausnitz et al.[15]

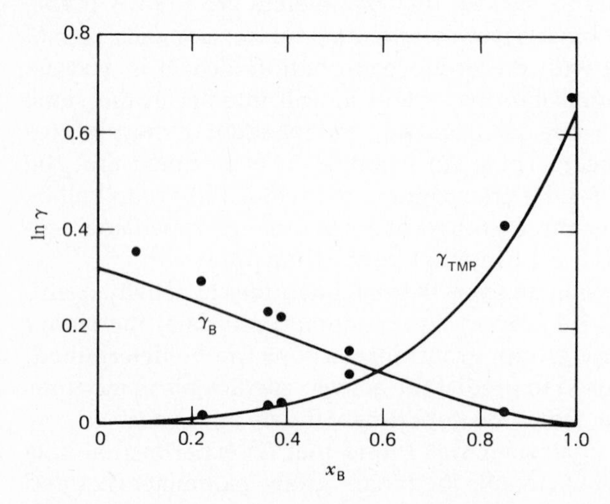

Figure 7.6-2
The UNIFAC predictions for the activity coefficients of the benzene-2,2,4-trimethyl pentane system. The points are activity coefficients derived from experimental data.

[15] J. M. Prausnitz, R. N. Lichtenthaler, and E. G. de Azevedo, *Molecular Thermodynamics of Fluid-phase Equilibria*, 2nd ed., Prentice–Hall, Englewood Cliffs, N.J., 1986.

Table 7.6-3
FUGACITY ESTIMATION PROCEDURES

A. Pure Fluids

State	Method	Comments
Pure vapor	**a.** Virial equation of state See Eqs. 5.4-11 and 12	Applicable only at low pressures
	b. Equation of state See, for example, Eq. 5.4-14a	
	c. Corresponding states $f^V(T, P) = \left(\frac{f}{P}\right) P$	Least accurate method; (f/P) gotten from Fig. 5.4-1
Pure liquid	**a.** Equation of state See, for example, Eq. 5.4-14b	Applicable to hydrocarbons and a few other liquids
	b. Corresponding states $f^L(T, P) = P_i^{vap}(T) \left(\frac{f}{P}\right)_{sat} \exp\left[\int_{P_i^{vap}}^{P} \frac{V_i}{RT} dP\right]$ $\cong P_i^{vap}(T) \left(\frac{f}{P}\right)_{sat}$	Poynting correction is frequently negligible, and (f/P) is gotten from a vapor-phase equation of state or Fig. 5.4-1
Pure solid	$f^S(T, P) = P_i^{sat}(T)$	Poynting and (f/P) corrections are usually unimportant

B. Mixtures

State	Method	Comments
Vapor mixtures	**a.** Virial equation of state See Eqs. 7.4-4 through 7	Applicable only at low pressures
	b. Equation of state See, for example, Eqs. 7.4-8 through 10	
	c. Lewis–Randall rule $\bar{f}_i^V(T, P) = y_i f_i^V(T, P)$ $= y_i P \left(\frac{f}{P}\right)_i$	Least accurate. (f/P) gotten from a pure vapor equation of state or Fig. 5.4-1
Simple liquid mixtures	**a.** Equations of state See, for example, Eqs. 7.4-8 through 10	Applicable to mixtures of hydrocarbons and some other liquids
	b. Activity coefficient method $\bar{f}_i^L(T, P, x_i) = x_i \gamma_i(T, P, x_i) f_i^L(T, P)$	Activity coefficient γ_i gotten from experimental data (Sec. 7.5) or prediction methods (Sec. 7.6)
	c. Corresponding states theory Sec. 7.7	Least accurate procedure
Nonsimple liquid mixtures	**a.** Equations of state See, for example, Eqs. 7.4-8 through 10	Applicable to mixtures of hydrocarbons and a few other fluids
	b. Mixture models of Secs. 7.8, 8.2, and 8.4	
Solid mixtures	$\bar{f}_i^S(T, P, x_i) = f_i^S(T, P) = P_i^{sat}(T)$	Applicable unless an alloy or intermetallic compound is formed

7.7 A CORRESPONDING STATES PRINCIPLE FOR MIXTURES; THE PSEUDOCRITICAL CONSTANT METHOD

When one has access to a digital computer, the easiest, reasonably accurate way to compute the fugacity of each species in a mixture consisting of hydrocarbons (or hydrocarbons and light gases) is by use of equations of state (see Secs. 5.4 and 7.4). If a digital computer is not available, the principle of corresponding states may be used, though the results will be less accurate. In Sec. 4.6 we showed how to estimate the compressibility and the enthalpy and entropy departures from ideal gas behavior for a real fluid using its critical constants and several generalized property correlations. In this section we consider the extension of the corresponding states principle to mixtures. Here, as in Sec. 4.6, we consider only the simplest form of the corresponding states principle in that the reduced temperature and reduced pressure will be the only parameters used.

The basis for the extension of the corresponding states principle to mixtures is the observation that the dependence of the compressibility factor of a mixture on temperature and pressure *at fixed composition* is not very different from the dependence of Z on T and P for a pure substance. This suggests that the generalized property correlations developed for pure substances could also be used for mixtures if we knew what values to use for the effective mixture critical constants. Since these reducing parameters for the corresponding states theory of mixtures play the same role as the critical constants in the pure fluid corresponding states theory, they are called **pseudocritical constants**. However, the pseudocritical constants are those parameters that lead to the most accurate corresponding states predictions and, as we will see in Chapter 8, they are not related to actual critical phenomena in mixtures.

A number of rules have been proposed for estimating the pseudocritical temperature $T_{C,m}$ and pseudocritical pressure $P_{C,m}$ of a mixture from the critical properties of the pure species. We will consider here only the simple rules proposed by Kay

$$T_{C,m} = \sum_{i=1}^{\mathscr{C}} x_i T_{C,i} \tag{7.7-1}$$

$$P_{C,m} = \sum_{i=1}^{\mathscr{C}} x_i P_{C,i} \tag{7.7-2}$$

and the slightly more complicated, modified Prausnitz–Gunn rule, in which $T_{C,m}$ is as given in Eq. 7.7-1, and

$$P_{C,m} = R\left(\sum_{i=1}^{\mathscr{C}} x_i Z_{C,i}\right)\left(\sum_{i=1}^{\mathscr{C}} x_i T_{C,i}\right) \Big/ \left(\sum_{i=1}^{\mathscr{C}} x_i \underline{V}_{C,i}\right) \tag{7.7-3}$$

In these equations $T_{C,i}$, $P_{C,i}$, $\underline{V}_{C,i}$, and $Z_{C,i}$ are the critical temperature, pressure, volume, and compressibility of species i, respectively, and x_i is its mole fraction in the vapor or liquid.

With the pseudocritical constants calculated here, the compressibility and the enthalpy and entropy departures from *ideal gas mixture* behavior can be calculated using the generalized property correlations given in Chapter 4. In particular,

$$Z_m = Z\left(\frac{T}{T_{C,m}}, \frac{P}{P_{C,m}}\right) = Z(T_{r,m}, P_{r,m}) \tag{7.7-4}$$

where Z is given as a function of T_r and P_r in Fig. 4.6-3,

$$\frac{\underline{H}(T, P, x_i) - \underline{H}^{IGM}(T, P, x_i)}{T_{C,m}} = \left[\frac{\underline{H} - \underline{H}^{IG}}{T_C}\right]_{T_r, P_r} \quad (7.7\text{-}5)$$

where the term on the right is given as a function of the reduced temperature and pressure in Fig. 4.6-4, and

$$\underline{S}(T, P, x_i) - \underline{S}^{IGM}(T, P, x_i) = [\underline{S} - \underline{S}^{IG}]_{T_r, P_r} \quad (7.7\text{-}6)$$

where the function $[\underline{S} - \underline{S}^{IG}]$ is given as a function T_r and P_r in Fig. 4.6-5. With these corresponding states correlations, and the ideal gas mixture heat capacity

$$C^*_{P,m} = \sum_{i=1}^{\mathscr{C}} x_i C^*_{P,i} \quad (7.7\text{-}7)$$

the change in the thermodynamic properties of a mixture between any two states of the same composition can be computed using Eqs. 4.6-10, 12, and 13. Similarly, using the corresponding states correlation of Fig. 5.4-1, a value for the mixture fugacity coefficient $(\bar{f}/P)$ can be obtained.

ILLUSTRATION 7.7-1

Compute Z, $(\underline{H} - \underline{H}^{IGM})$, $(\underline{S} - \underline{S}^{IGM})$ and $\bar{f}/P$ for an equimolar mixture of carbon dioxide and methane at 500 K and 500 bar using the pseudocritical constants determined from (a) Kay's rule and (b) the modified Prausnitz–Gunn rule.

Solution

From Table 4.6-1 we have

	T_C(K)	P_C(bar)	$\underline{V}_C$(m^3/kmol)	Z_C
CO_2	304.2	73.76	0.094	0.274
CH_4	190.6	46.00	0.099	0.288

a. Kay's Rule

$$T_{C,m} = \sum x_i T_{C,i} = 247.4 \text{ K} \quad \text{and} \quad T_{r,m} = 2.02$$

$$P_{C,m} = \sum x_i P_{C,i} = 59.88 \text{ bar} \quad \text{and} \quad P_{r,m} = 8.35$$

From Fig. 4.6-3

$$Z = 1.09$$

From Fig. 4.6-4

$$\frac{\underline{H}^{IGM} - \underline{H}}{T_{C,m}} = 2.55 \text{ cal/mol K} = 10.7 \text{ J/mol K}$$

so that

$$(\underline{H} - \underline{H}^{IGM}) = -10.7 \frac{\text{J}}{\text{mol K}} \times 247.4 \text{ K} = -2.647 \text{ kJ/mol}$$

from Fig. 4.6-5

$$\underline{S}^{IGM} - \underline{S} = 1.13 \text{ cal/mol K} = 4.73 \text{ J/mol K}$$

and from Fig. 5.4-1

$$\frac{\bar{f}}{P} = 0.93$$

b. The Modified Prausnitz–Gunn Rule

$$\sum x_i Z_{C,i} = 0.281, \quad \sum x_i \underline{V}_{C,i} = 0.0965 \text{ m}^3/\text{kmol}$$

$$\begin{aligned} P_{C,m} &= R\left(\sum x_i Z_{C,i}\right)\left(\sum x_i T_{C,i}\right)\Big/\left(\sum x_i \underline{V}_{C,i}\right) \\ &= 8.314 \times 10^{-5} \frac{\text{bar m}^3}{\text{mol K}} \times \frac{0.281 \times 247.4 \text{ K}}{0.0965 \text{ m}^3/\text{kmol}} \times \frac{10^3 \text{ mol}}{\text{kmol}} \\ &= 59.89 \text{ bar} \end{aligned}$$

Since the pseudocritical constants here are so close to those computed in part a, each of the corresponding states thermodynamic properties will, essentially, be the same as in part a. ■

The corresponding states principle and the rules for computing the pseudocritical constants may also be used to calculate the fugacity of species i in a mixture, $\bar{f}_i(T, P, x_i)$, although this is a more difficult computation. To establish the relationship of $\bar{f}_i(T, P, x_i)$ to corresponding states theory, we first consider the following expression for the difference between the Gibbs free energy of a real fluid mixture and an ideal gas mixture

$$\Delta G = N[\underline{G}(T, P, x_i) - \underline{G}^{IGM}(T, P, x_i)] = \sum_i N_i[\bar{G}_i(T, P, x_i) - \bar{G}_i^{IGM}(T, P, x_i)] \quad (7.7\text{-}8)$$

Now taking the derivative of this Gibbs free energy difference with respect to the number of moles of species j, holding T, P, and the mole numbers of all other species constant, and using the Gibbs–Duhem equation, we obtain

$$\begin{aligned} \bar{G}_j(T, P, x_i) - \bar{G}_j^{IGM}(T, P, x_i) &= \underline{G}(T, P, x_i) - \underline{G}^{IGM}(T, P, x_i) \\ &\quad + N\frac{\partial}{\partial N_j}[\underline{G}(T, P, x_i) - \underline{G}^{IGM}(T, P, x_i)] \end{aligned} \quad (7.7\text{-}9)$$

Using the definition of fugacity, we further obtain

$$\ln\left(\frac{\bar{f}_j}{x_j P}\right) = \ln\left(\frac{\bar{f}}{P}\right) + N\frac{\partial}{\partial N_j}\left[\frac{\underline{G}(T, P, x_i) - \underline{G}^{IGM}(T, P, x_i)}{RT}\right]_{T,P,N_{i\neq j}}$$

To evaluate the derivative in this equation, we first note that since $\underline{H} - \underline{H}^{IGM}$ and $\underline{S} - \underline{S}^{IGM}$ satisfy corresponding states relations (Eqs. 7.7-5 and 6), so does $\underline{G} - \underline{G}^{IGM}$, that is

$$\frac{\underline{G}(T, P, x_i) - \underline{G}^{IGM}(T, P, x_i)}{T_{C,m}} = \left[\frac{\underline{H} - \underline{H}^{IG}}{T_C}\right]_{\text{Fig. 4.6-4}} - T_{r,m}[\underline{S} - \underline{S}^{IG}]_{\text{Fig. 4.6-5}} \quad (7.7\text{-}10)$$

Thus the effect of a change in the number of moles of species j on the term $\underline{G} - \underline{G}^{IGM}$, which, by the corresponding states principle, is only a function of reduced temperature and pressure, is a result of the variation of $T_{r,m}$ and $P_{r,m}$ with species mole number at constant temperature and pressure; that is,

$$\begin{aligned} N\frac{\partial}{\partial N_j}\left(\frac{\underline{G} - \underline{G}^{IGM}}{RT}\right) &= N\frac{\partial}{\partial T_r}\left(\frac{\underline{G} - \underline{G}^{IGM}}{RT}\right)\left(\frac{\partial T_{r,m}}{\partial N_j}\right)_{T,P,N_{i\neq j}} \\ &\quad + N\frac{\partial}{\partial P_r}\left(\frac{\underline{G} - \underline{G}^{IGM}}{RT}\right)\left(\frac{\partial P_{r,m}}{\partial N_j}\right)_{T,P,N_{i\neq j}} \end{aligned} \quad (7.7\text{-}11)$$

It is left to you (Problem 7.14) to show that

$$\frac{\partial}{\partial T_r}\left(\frac{\underline{G} - \underline{G}^{IGM}}{RT}\right)_{P_r,N_j} = -\frac{T_{C,m}(\underline{H} - \underline{H}^{IGM})}{RT^2} \tag{7.7-12}$$

and

$$\frac{\partial}{\partial P_r}\left(\frac{\underline{G} - \underline{G}^{IGM}}{RT}\right)_{T_r,N_j} = \frac{Z_m - 1}{P_r} \tag{7.7-13}$$

so that

$$\ln\left(\frac{\bar{f}_j}{x_j P}\right) = \ln\left(\frac{\bar{f}}{P}\right) - T_{r,m}^{-2}\frac{(\underline{H} - \underline{H}^{IGM})}{RT_{C,m}}\psi_1^j + \frac{(Z_m - 1)}{P_{r,m}}\psi_2^j \tag{7.7-14a}$$

where

$$\psi_1^j = N\left(\frac{\partial T_{r,m}}{\partial N_j}\right)_{T,P,N_{i\neq j}} \quad \text{and} \quad \psi_2^j = N\left(\frac{\partial P_{r,m}}{\partial N_j}\right)_{T,P,N_{i\neq j}} \tag{7.7-14b}$$

In using Eqs. 7.7-14, the terms $(\bar{f}/P)$, $(\underline{H} - \underline{H}^{IGM})$, and Z_m are to be computed from the corresponding states correlation, whereas the terms ψ_1^j and ψ_2^j are computed from the critical properties once the pseudocritical constant rules are specified. It is left to you to show (Problem 7.14) that[16]

$$\psi_1^j(K) = \psi_1^j(P \,\&\, G) = -\frac{T}{T_{C,m}^2}(T_{C,j} - T_{C,m})$$

$$\psi_2^j(K) = -P(P_{C,j} - {}_{C,m})/P_{C,m}^2 \tag{7.7-15}$$

and

$$\psi_2^j(P \,\&\, G) = \frac{P}{P_{C,m}}\left\{\frac{(\underline{V}_{C,j} - \underline{V}_{C,m})}{\underline{V}_{C,m}} - \frac{(T_{C,j} - T_{C,m})}{T_{C,m}} - \frac{(Z_{C,j} - Z_{C,m})}{Z_{C,m}}\right\}$$

where

$$\underline{V}_{C,m} = \sum_i x_i \underline{V}_{C,i} \quad \text{and} \quad Z_{C,m} = \sum_i x_i Z_{C,i}$$

ILLUSTRATION 7.7-2

Compute the fugacity of both carbon dioxide and methane in an equimolar mixture at 500 K and 500 bar using (a) the Lewis–Randall rule, (b) Eqs. 7.7-14 with the Prausnitz–Gunn pseudocritical constants, and (c) the Peng–Robinson equation of state.

Solution

a. The Lewis–Randall rule is

$$\bar{f}_i(T, P, y_i) = y_i f_i(T, P) = y_i\left(\frac{f}{P}\right)_i P$$

[16]Here K and P & G are used to indicate the functions ψ that arise from the use of Kay's rule and the modified Prausnitz–Gunn rule for the pseudocritical constants.

From the data in Illustration 7.7-1 and Fig. 5.4-1 we have

	T_r	P_r	f/P
CO_2	1.644	6.78	~0.77
CH_4	2.623	10.87	~1.01[17]

Thus

$$\bar{f}_{CO_2} = 0.5 \times 0.77 \times 500 \text{ bar} = 192.5 \text{ bar}$$

and

$$\bar{f}_{CH_4} = 0.5 \times 1.01 \times 500 \text{ bar} = 252.5 \text{ bar}$$

b. Using the Prausnitz–Gunn pseudocritical rules, and noting that for a equimolar binary mixture we have $\psi_1^i = -\psi_1^j$ and $\psi_2^i = -\psi_2^j$, we obtain

$$\psi_1^{CO_2} = -\frac{T}{T_{C,m}^2}(T_{C,CO_2} - T_{C,m}) = -\frac{500(304.2 - 247.4)}{(247.4)^2} = -0.4640 = -\psi_1^{CH_4}$$

and

$$\psi_2^{CO_2} = \frac{P}{P_{C,m}}\left\{\frac{(\underline{V}_{C,CO_2} - \underline{V}_{C,m})}{\underline{V}_{C,m}} - \frac{(T_{C,CO_2} - T_{C,m})}{T_{C,m}} - \frac{(Z_{C,CO_2} - Z_{C,m})}{Z_{C,m}}\right\}$$

$$= \frac{500}{59.89}\left\{\frac{(0.094 - 0.0965)}{0.0965} - \frac{(304.2 - 247.4)}{247.4} - \frac{(0.274 - 0.281)}{0.281}\right\}$$

$$= -1.925 = -\psi_2^{CH_4}$$

From the previous illustration we have

$$\frac{\bar{f}}{P} = 0.93 \quad \text{or} \quad \ln\frac{\bar{f}}{P} = -0.0726$$

$$\frac{\underline{H} - \underline{H}^{IGM}}{T_{C,m}} = -10.7 \text{ J/mol K} \quad \text{and} \quad Z_m = 1.09$$

Therefore

$$\ln\left(\frac{\bar{f}_{CO_2}}{x_{CO_2}P}\right) = \ln\left(\frac{\bar{f}}{P}\right) - T_{r,m}^{-2}\frac{(\underline{H} - \underline{H}^{IGM})}{RT_{C,m}}\psi_1^{CO_2} + \frac{(Z_m - 1)}{P_{r,m}}\psi_2^{CO_2}$$

$$= -0.0726 - (-10.7)(-0.4640)/8.314(2.02)^2 + (0.09)(-1.925)/8.35$$

$$= -0.2397$$

so that

$$\frac{\bar{f}_{CO_2}}{x_{CO_2}P} = 0.787 \quad \text{and} \quad \bar{f}_{CO_2} = 0.787 \times 0.5 \times 500 \text{ bar} = 196.75 \text{ bar}$$

[17]Since it is very difficult to read the fugacity coefficient curve in the range of the reduced temperature and pressure for methane in this problem, this entry was obtained from the tables in O. A. Hougen, K. M. Watson, and R. A. Ragatz, *Chemical Process Principles, Part II, Thermodynamics*, 2nd ed., John Wiley & Sons, New York, 1959.

Also,

$$\ln \frac{\bar{f}_{CH_4}}{x_{CH_4}P} = -0.0726 - (-10.7)(+0.4640)/8.314(2.02)^2 + (0.09)(+1.925)/8.35 = +0.0945$$

$$\frac{\bar{f}_{CH_4}}{x_{CH_4}P} = 1.099 \quad \text{and} \quad \bar{f}_{CH_4} = 1.099 \times 0.5 \times 500 \text{ bar} = 274.8 \text{ bar}$$

c. Using the Peng–Robinson equation of state (Eqs. 4.4-2 and 7.4-8 through 10) and the computer program discussed in Appendix A7.2 we find, for $k_{12} = 0.0$,

$$\bar{f}_{CO_2} = 208.71 \text{ bar and } \bar{f}_{CH_4} = 264.72 \text{ bar}$$

while using $k_{12} = 0.09$ (from Table 7.4-1)

$$\bar{f}_{CO_2} = 212.81 \text{ bar and } \bar{f}_{CH_4} = 269.35 \text{ bar}$$

This last value should be the most accurate in this illustration. ■

From this example we see that the Lewis–Randall rule leads to the simplest calculations and the least accurate results. Using the corresponding states method with the Prausnitz–Gunn pseudocritical rules leads to somewhat more accurate results from a slightly more tedious calculation. The use of a cubic equation of state, such as the Peng–Robinson equation, leads to reasonably accurate values for the species fugacities, but requires the use of a computer.

7.8 THE FUGACITY OF SPECIES IN NONSIMPLE MIXTURES

To estimate the fugacity of a species in a gaseous mixture using the Lewis–Randall rule

$$\bar{f}_i^V(T, P, y_i) = y_i f_i^V(T, P) \tag{7.8-1}$$

we need the fugacity of pure species i as a *vapor* at the temperature and pressure of the mixture. In a nonsimple gaseous mixture at least one of the pure components does not exist as a vapor at the mixture temperature and pressure, so that $f_i^V(T, P)$ for that species cannot be computed without some sort of approximation or assumption. To estimate the fugacity of a species in a liquid mixture from an activity coefficient model, one uses

$$\bar{f}_i^L(T, P, x_i) = x_i \gamma_i(T, P, x_i) f_i^L(T, P) \tag{7.8-2}$$

and needs the fugacity of pure species i as a *liquid* at the temperature and pressure of the mixture. In a nonsimple liquid mixture, at least one of the components, if pure, would be a solid or a vapor at the temperature and pressure of the mixture, so that it is not evident how to compute $f_i^L(T, P)$ in this case, either.

The study of vapor–liquid equilibria (Sec. 8.1), the solubility of gases in liquids (Sec. 8.3), and the solubility of solids in liquids (Sec. 8.5) all involve nonsimple mixtures. To see why this occurs, consider the criterion for vapor–liquid equilibrium

$$\bar{f}_i^L(T, P, x_i) = \bar{f}_i^V(T, P, y_i)$$

Using Eqs. 7.8-1 and 2, we have

$$x_i \gamma_i(T, P, x_i) f_i^L(T, P) = y_i f_i^V(T, P) \tag{7.8-3}$$

To use this equation we must estimate the pure component fugacity of each species as both a liquid and a vapor at the temperature and pressure of the mixture. However, at this temperature and pressure either the liquid or the vapor will be the stable phase for each species, generally not both.[18] Consequently, at least one of the phases will be a nonsimple mixture. In most vapor–liquid problems both phases will be nonsimple mixtures, in that species with pure component vapor pressures above the system pressure appear in the liquid phase and others with vapor pressures below the system pressure appear in the vapor phase.

For these situations one can proceed in several ways. The simplest, and most accurate when it is applicable, is to use equations of state to compute species fugacities in mixtures, and thereby avoid the use of Eqs. 7.8-1 and 2 and the necessity of computing a pure component fugacity in a thermodynamic state that does not occur. Thus, when an equation of state (virial, cubic, etc.) can be used for the vapor mixture, a nonsimple gaseous mixture can be treated using the methods of Sec. 7.5. Similarly, if a nonsimple liquid mixture can be described by an equation of state, which is likely to be the case only for hydrocarbons and perhaps hydrocarbons with dissolved gases, it also can be treated by the methods of Sec. 7.5. Alternatively, but less accurately, one can use the principle of corresponding states (Section 7.7 and especially Eqs. 7.7-14) and also avoid the main problem that arises in nonsimple mixtures, that of computing the pure component fugacity in a state that does not occur.

If an equation of state or the principle of corresponding states cannot be used, one can, in principle, proceed in either of two ways. The first procedure is to write the Gibbs free energy of mixing as

$$\Delta \underline{G}_{\text{mix}} = \Delta \underline{G}' + \Delta \underline{G}^{IM}_{\text{mix}} + \underline{G}^{ex}$$

where $\Delta \underline{G}^{IM}_{\text{mix}}$ and $\underline{G}^{ex}$ have their usual meanings, and $\Delta \underline{G}'$ is the Gibbs free energy change of converting to the same phase as the mixture those species that exist in other phases as pure substances. For example, $\Delta \underline{G}'$ might be the Gibbs free energy change of producing pure liquids from either gases or solids before forming a liquid mixture. The partial molar Gibbs free energy and fugacity of each species in the mixture would then be computed directly from $\Delta \underline{G}_{\text{mix}}$. In practice, however, the calculation of $\Delta \underline{G}'$ can be difficult, since it may involve the estimation of the Gibbs free energy of substances in hypothetical states (i.e., a liquid above its critical point).

A second, somewhat more straightforward procedure is to use Eqs. 7.8-1 and 2 and the models for γ_i considered in Secs. 7.5 and 6, but with estimates for the pure component fugacities of the nonexistent gases and liquids obtained by simple extrapolation procedures. Of course, such extrapolation schemes have no real physical or chemical basis; they are, instead, calculational methods that have been found empirically to lead to satisfactory predictions, provided the extrapolation is not too great. The first extrapolation scheme to be considered is for the liquid-phase fugacity of a species that would be a vapor as a pure component at the temperature and pressure of the mixture. The starting point here is the observation that the fugacity of a pure liquid is equal to its vapor pressure if the vapor pressure is not too high (or the product of P^{vap} and the fugacity coefficient at higher vapor pressures). Consequently, we will take the fugacity of

[18]Unless, fortuitously, the system pressure is equal to the vapor pressure of that species.

the hypothetical liquid at the temperature T and the pressure P to be equal to the vapor pressure of the real liquid at the same temperature [even though $P^{vap}(T) > P$]. Thus, in Illustration 8.1-2 where we consider vapor–liquid equilibrium for an n-pentane, n-hexane, and n-heptane mixture at 69°C and 1 bar, the fugacity of "liquid" n-pentane, for use in Eq. 7.8-2, will be taken equal to its vapor pressure at 69°C, 2.729 bar.

To calculate the fugacity of a pure vapor that, at the conditions of the mixture, exists only as a liquid, we will use Eq. 7.8-1 with $f_i^{V}(T, P)$ equal to the total pressure, if the pressure is low enough, or

$$f_i^{V}(T, P) = P\left(\frac{f}{P}\right)$$

at higher pressures. Here, however, the fugacity coefficient is not obtained from Fig. 5.4-1, but rather from Fig. 7.8-1, which is a corresponding states correlation in which the fugacity coefficient for gases has been extrapolated into the liquid region. (You should compare Figs. 5.4-1 and 7.8-1.)

Extrapolation schemes may also be used in some circumstances where the desired phase does not exist at any pressure for the temperature of interest, for example, to estimate the fugacity of a "liquid" not too far above its critical temperature, or not too far below its triple point temperature. As an example of the methods used, consider the estimation of the liquid-phase fugacity for the substance whose pure component phase diagram is given in Fig. 7.8-2*a* at either temperatures below the triple point (so that the solid is the stable phase) or at temperatures above the critical temperature (where the vapor is the stable phase). In either case the first step in the procedure is to extend the vapor pressure curve, either analytically (using the Clausius–Clapeyron equation) or graphically as indicated in Fig. 7.8-2*b*, to obtain the vapor pressure of the hypothetical liquid.[19] In the case of the subcooled liquid, which involves an extrapolation into the solid region, the vapor pressure is usually so low that the fugacity coefficient is close to unity, and the fugacity of this hypothetical liquid is equal to the extrapolated vapor pressure. For the superheated liquid, however, the extrapolation is above the critical temperature of the liquid and yields very high vapor pressures, so that the fugacity of this hypothetical liquid is equal to the product of the extrapolated vapor pressure and the fugacity coefficient (which is gotten from the corresponding states tabulation of Fig. 7.8-1).

In a similar fashion, the fugacity of a hypothetical superheated solid can be estimated by extrapolating the sublimation pressure line into the liquid region of the phase diagram. This is indicated in Fig. 7.8-2*c*. The fugacity coefficient correction may be small in this case also.

For species whose thermodynamic properties are needed in hypothetical states far removed from their stable states, such as a liquid well above its critical point, the extrapolation procedures discussed here are usually inaccurate. In some cases special correlations or prescriptions are used; one such correlation will be discussed in Chapter 8. In other cases different procedures, such as those discussed next, are used.

The fugacity of a very dilute species in a liquid mixture (e.g., a dissolved gas or solid of limited solubility) is experimentally found to be linearly proportional to its mole fraction at low mole fractions, that is,

[19] For accurate extrapolations $\ln P^{vap}$ should be plotted versus $1/T$.

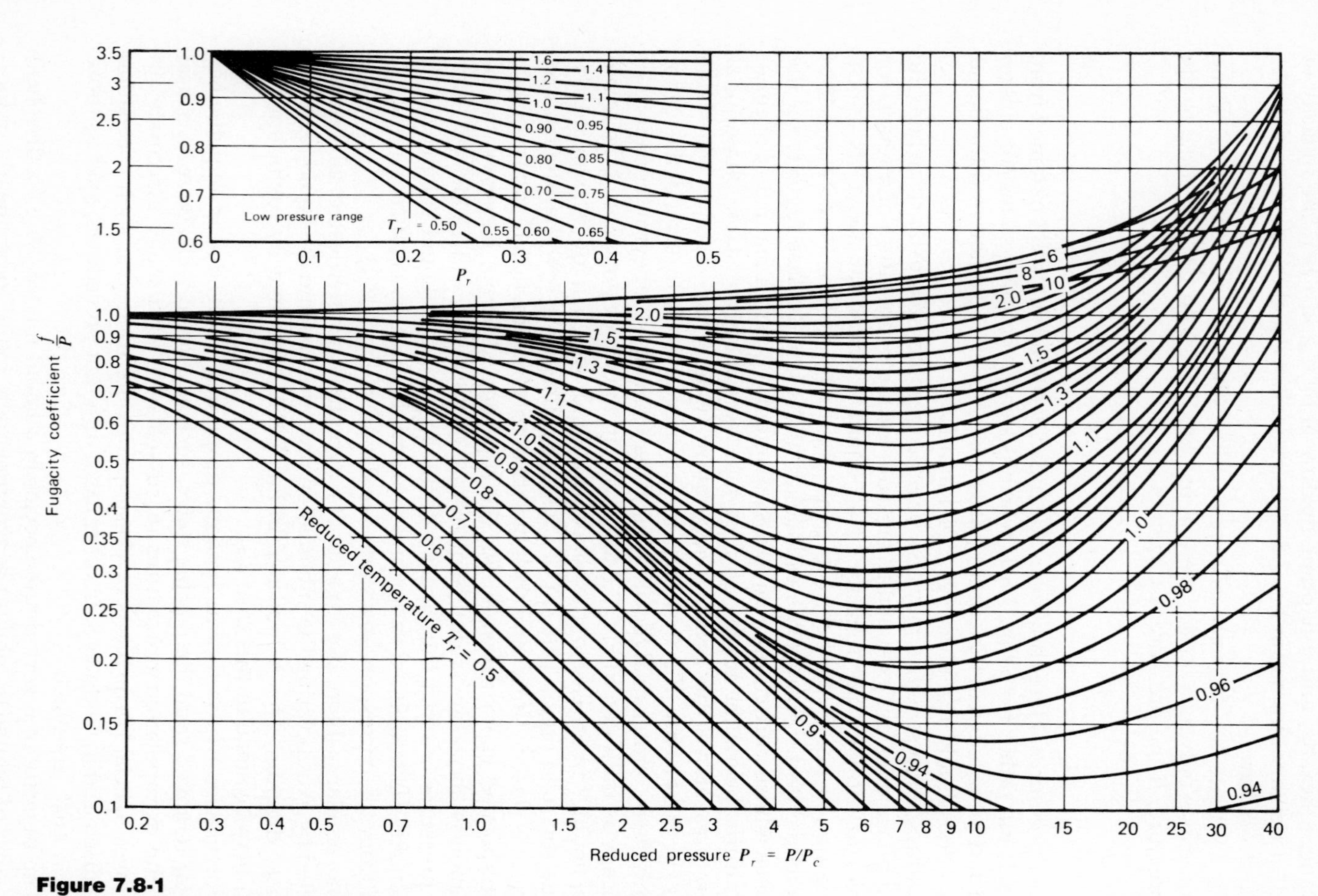

Figure 7.8-1
Fugacity coefficients of gases and vapors. (From O. A. Hougen and K. M. Watson, *Chemical Process Principle Charts*. Copyright 1946 by John Wiley Sons, Inc., New York. Used with permission.)

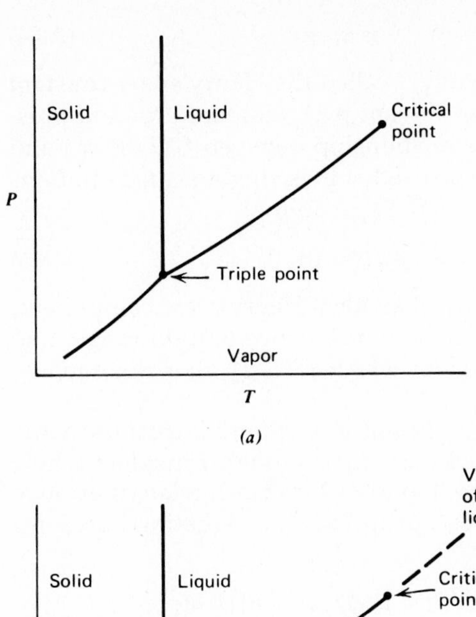

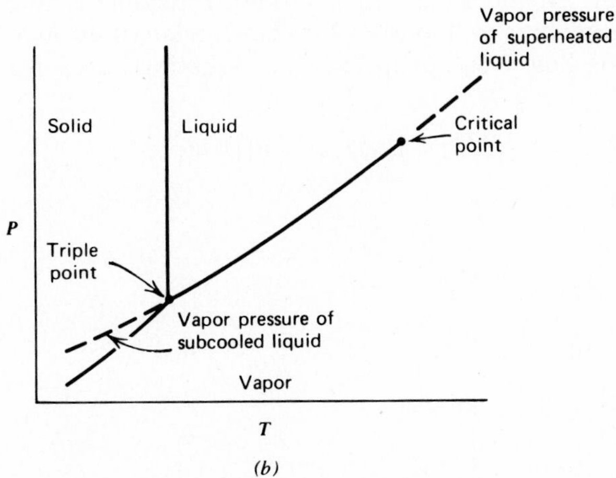

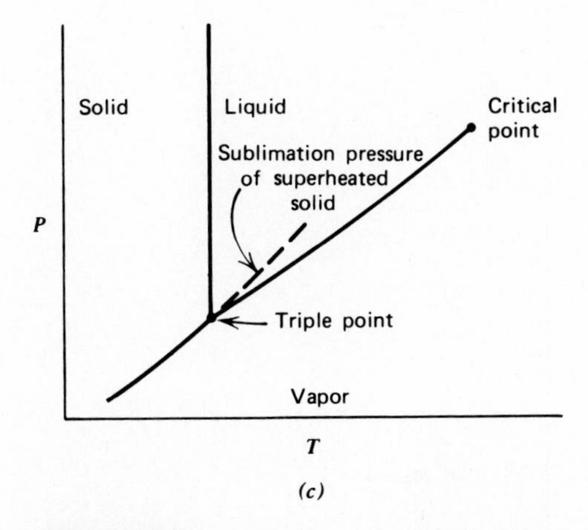

Figure 7.8-2
(a) $P-T$ Phase diagram for a typical substance. (b) $P-T$ phase diagram with dashed lines indicating extrapolation of the liquid-phase vapor pressure into the solid and vapor regions. (c) $P-T$ phase diagram with a dashed line indicating extrapolation of the sublimation pressure of the solid into the liquid region.

$$\bar{f}_i^L(T, P, x_i) = x_i H_i(T, P) \qquad \text{as } x_i \to 0 \tag{7.8-4}$$

The value of the "constant of proportionality," called the **Henry's law constant** (see Sec. 8.3), is dependent on the solute–solvent pair, temperature, and pressure. At higher concentrations the linear relationship between $\bar{f}_i^L(T, P, x_i)$ and mole fraction fails; a form of Eq. 7.8-4 can be used at these higher concentrations by introducing a new activity coefficient $\gamma_i^*(T, P, x_i)$ so that

$$\bar{f}_i^L(T, P, x_i) = x_i \gamma_i^*(T, P, x_i) H_i(T, P) \tag{7.8-5}$$

The dashed line in Fig. 7.8-3*a* is the fugacity of an ideal Henry's law component, that is, a species that obeys Eq. 7.8-4 over the whole concentration range, and the solid lines represent two real solutions for which $\gamma_i^*(T, P, x_i)$ is not equal to unity at all concentrations.

Note that the Henry's law activity coefficient γ_i^* is different than the activity coefficient γ_i defined earlier. In particular, in solutions considered here $\gamma_i^* \to 1$ as $x_i \to 0$, whereas previously $\gamma_i \to 1$ as $x_i \to 1$. We can relate these two activity coefficients by comparing Eqs. 7.8-2, 4, and 5. First, equating Eqs. 7.8-2 and 5 gives

$$\bar{f}_i^L(T, P, x_i) = x_i \gamma_i^*(T, P, x_i) H_i(T, P) = x_i \gamma_i(T, P, x_i) f_i^L(T, P)$$

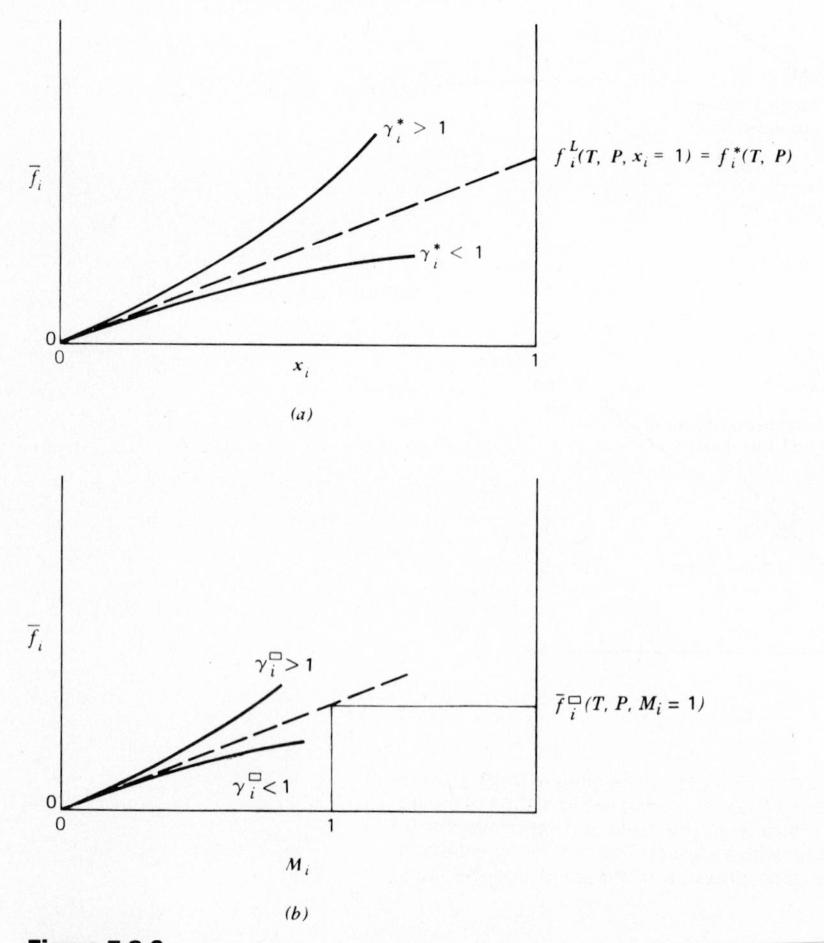

Figure 7.8-3
Solute fugacity in real and ideal Henry's law solutions: (*a*) solute fugacity versus mole fraction; (*b*) solute fugacity versus molality.

or

$$\gamma_i^*(T, P, x_i)H_i(T, P) = \gamma_i(T, P, x_i)f_i^{\mathrm{L}}(T, P) \tag{7.8-6}$$

and taking the limit as $x_i \to 0$ [remembering that $\gamma_i^*(T, P, x_i = 0) = 1$] yields

$$H_i(T, P) = \gamma_i(T, P, x_i = 0)f_i^{\mathrm{L}}(T, P) \tag{7.8-7}$$

Using this relation in Eq. 7.8-6 gives

$$\gamma_i^*(T, P, x_i)\gamma_i(T, P, x_i = 0)f_i^{\mathrm{L}}(T, P) = \gamma_i(T, P, x_i)f_i^{\mathrm{L}}(T, P)$$

or simply

$$\gamma_i^*(T, P, x_i) = \frac{\gamma_i(T, P, x_i)}{\gamma_i(T, P, x_i = 0)} \tag{7.8-8}$$

so that the activity coefficient γ_i^* is equal to the ratio of the activity coefficient γ_i to its value at infinite dilution.

If a solute–solvent pair were ideal in the Henry's law sense, Eq. 7.8-4 would be satisfied at all mole fractions; in particular, at $x_i = 1$

$$\bar{f}_i^{\mathrm{L}}(T, P, x_i = 1) = H_i(T, P)$$

Thus, the Henry's law constant is the hypothetical fugacity of a solute species as a pure liquid extrapolated from its infinite dilution behavior; we will denote this by $f_i^*(T, P)$ (see Fig. 7.8-3*a*). Thus

$$\bar{f}_i^{\mathrm{L}}(T, P, x_i) = x_i\gamma_i^*(T, P, x_i)H_i(T, P) = x_i\gamma_i^*(T, P, x_i)f_i^*(T, P) \tag{7.8-9}$$

Using Eq. 7.2-10, we can also write

$$\bar{G}_i(T, P, x_i) - \underline{G}_i^*(T, P) = RT \ln \left\{\frac{\bar{f}_i^{\mathrm{L}}(T, P, x_i)}{f_i^*(T, P)}\right\} = RT \ln \{x_i\gamma_i^*(T, P, x_i)\} \tag{7.8-10}$$

where $\underline{G}_i^*(T, P)$ is the (hypothetical) Gibbs free energy of the solute species as a pure liquid obtained from extrapolation of its dilute solution behavior.

The fugacity of a very dilute species can also be written as

$$\bar{f}_i^{\mathrm{L}}(T, P, M_i) = M_i\mathcal{H}_i(T, P) \qquad \text{as } M_i \to 0 \tag{7.8-11}$$

Here M_i is the **molality** of species i, that is, the number of moles of this species per 1000 g of solvent,[20] and $\mathcal{H}_i$ is the Henry's law constant for molality; its value depends on the solute–solvent pair, temperature, and pressure. For real solutions the activity coefficient $\gamma_i^\square(T, P, M_i)$ is introduced, so that

$$\bar{f}_i^{\mathrm{L}}(T, P, M_i) = M_i\gamma_i^\square(T, P, M_i)\mathcal{H}_i(T, P) \tag{7.8-12}$$

clearly $\gamma_i^\square(T, P, M_i) \to 1$ as $M_i \to 0$. The behavior of real and ideal solutions are indicated by dashed and solid lines in Fig. 7.8-3*b*.

[20]The molality of a solution consisting of n_i moles of solute in n_s moles of a solvent of molecular weight m_s is

$$M_i = \frac{n_i\, 1000}{m_s n_s}$$

whereas the mole fraction of solute i is

$$x_i = \frac{n_i}{n_s + \Sigma\, n_j}$$

where the summation is over all solutes. At low-solute concentration $n_s \gg \Sigma_j n_j$, and these equations reduce to $x_i \simeq n_i/n_s$, and $M_i \simeq x_i\, 1000/m_s$, so that M_i and x_i are linearly related. Therefore, it is not surprising that both Eqs. 7.8-4 and 11 are satisfied. Furthermore, $H_i = 1000\, \mathcal{H}_i/m_s$.

From Eq. 7.8-12 we have that the molal Henry's law constant is equal to the fugacity of the solute species at unit molality in an *ideal Henry's law* solution; that is

$$\mathcal{H}_i(T, P) = \bar{f}_i^{\square}(T, P, M_i = 1) \tag{7.8-13}$$

where the ideal solution, unit molality fugacity $\bar{f}_i^{\square}(T, P, M_i = 1)$ is obtained by extrapolation of dilute solution behavior, as indicated in Fig. 7.8-3*b*. Using the analysis which led to Eq. 7.8-8, one can show that

$$\gamma_i^{\square}(M_i) = \frac{x_i\, 1000}{m_s M_i} \frac{\gamma_i(x_i)}{\gamma_i(x_i = 0)} = \frac{x_i\, 1000}{m_s M_i} \gamma_i^*(x_i) \tag{7.8-14}$$

and that

$$\bar{G}_i(T, P, M_i) = \bar{G}_i^{\square}(T, P, M_i = 1) + RT \ln \left(\frac{M_i \gamma_i^{\square}(M_i)}{M_i = 1} \right) \tag{7.8-15}$$

The value of the partial molar Gibbs free energy of species i in the (hypothetical) ideal solution $G_i^{\square}(T, P, M_i = 1)$ is obtained by assuming ideal solution behavior and extrapolating to one molal the behavior of $\bar{G}_i(T, P, M_i)$ in very low molality solutions. The value of $\bar{G}_i^{\square}$ (and $\underline{G}^*$, as well) obtained in this way depends on temperature, pressure, and the solute–solvent pair.

It is useful to identify the physical significance of the quantities used here and to relate them to the analogous quantities for simple mixtures. In a simple liquid mixture, the properties of the pure components dominate the partial molar properties, and we have

$$\bar{f}_i^{L}(T, P, x_i) = x_i \gamma_i(T, P, x_i) f_i^{L}(T, P)$$

where $f_i^{L}(T, P)$ is the pure component fugacity (i.e., the fugacity of species i when it interacts only with other molecules of the same species), and the explicit mole fraction accounts for its dilution. The activity coefficient γ_i arises because the nature of the interaction between the solute species i and the solvent is different than that between species i molecules, so that γ_i accounts for the effect of replacing solute–solute interactions with solute–solvent interactions. By using a Henry's law description for a nonsimple mixture we recognize that, for the solute species, the liquid mixture and pure component states are very different. The implication of using

$$\bar{f}_i^{L}(T, P, x_i) = x_i \gamma_i^*(T, P, x_i) H_i(T, P)$$

or

$$\bar{f}_i^{L}(T, P, M_i) = M_i \gamma_i^{\square}(T, P, M_i) \mathcal{H}_i(T, P)$$

is that the properties of the solute species in solution are largely determined by solute molecules interacting only with solvent molecules, which are taken into account by the Henry's law constants H_i and $\mathcal{H}_i$. In this case, the activity coefficients γ_i^* and $\gamma_i^{\square}$ account for the effect of replacing solute–solvent interactions with solute–solute interactions.

We will have occasion to use the Henry's law descriptions (on both a mole fraction and molality basis) and the associated activity coefficients several times in this book. The immediate disadvantage of these choices is that $\bar{f}_i^*(T, P, x_i = 1)$ and $\bar{f}_i^{\square}(T, P, M_i = 1)$ can only be gotten by extrapolation of experimental information on very dilute solutions. However, this information may be easier to obtain and more accurate than that gotten by estimating the liquid-phase fugacity of a species whose equilibrium state is a supercritical gas or solid below the triple point temperature.

It is left as an exercise for you to relate the regular solution and UNIFAC model predictions for γ_i^* and $\gamma_i^{\square}$ to those already obtained for γ_i (Problem 7.9).

7.9 ELECTROLYTE SOLUTIONS

So far in this chapter we have considered mixtures of electrically neutral molecules. However, liquid solutions containing ionic species, especially aqueous solutions of acids, bases, and salts, occur frequently in chemical and biological processes. Charged particles interact with coulombic forces at small separations and, because of the formation of ion clouds around each ion, damped coulombic forces at larger separation distances. These forces are much longer range than those involved in the interactions of neutral molecules, so that solute ions in solution interact at very low concentrations. Consequently, electrolyte solutions are very nonideal in the sense that the electrolyte Henry's law activity coefficient $\gamma_i^{\square}$ of Eq. 7.8-12 is significantly different from unity at very low electrolyte concentrations; also the greater the charge on the ions, the stronger their interaction and the more nonideal the solution. Since the solution models discussed in the preceding sections do not allow for the formation of ion clouds, they do not apply to electrolyte solutions.

In this section we discuss certain characteristics of electrolyte solutions and present equations for the prediction or correlation of electrolyte activity coefficients in solution. Since the derivation of these equations are complicated, and beyond the scope of this book, the derivations will not be given.

Our interest is with an electrically neutral electrolyte, designated by $A_{\nu_+}B_{\nu_-}$ which, in solution, dissociates as follows:

$$A_{\nu_+}B_{\nu_-} = \nu_+A^{z+} + \nu_-B^{z-} \tag{7.9-1}$$

Here ν_+ and ν_- are the numbers of positive ions (cations) and negative ions (anions) obtained from the dissociation of one electrolyte molecule, and z_+ and z_- are the charges of the ions in units of charge of a proton (i.e., z_+ and z_- are the valences of the ions). For an electrically neutral salt, ν_+, ν_-, z_+, and z_- are related by the charge conservation (or electrical neutrality) condition that

$$\nu_+z_+ + \nu_-z_- = 0 \tag{7.9-2}$$

An important consideration in the study of electrolytes is that the concentration of any one ionic species is not independently variable, because the electrical neutrality of the solution must be maintained. Thus, if N_A and N_B are the number of moles of the A^{z+} and B^{z-} ions, respectively, which result from the dissolution of $A_{\nu_+}B_{\nu_-}$, it follows that N_A and N_B are related by

$$z_+N_A + z_-N_B = 0 \tag{7.9-3}$$

This restriction has an important implication with regard to the description of electrolyte solutions, as will be evident shortly.

A solution of a single electrolyte in a solvent contains four identifiable species: the solvent, undissociated electrolyte, anions, and cations. Therefore, it might seem appropriate, following Eqs. 6.1-12 and 13, to write the Gibbs free energy of the solution as

$$G = N_S\overline{G}_S + N_{AB}\overline{G}_{AB} + N_A\overline{G}_A + N_B\overline{G}_B \tag{7.9-4}$$

where N_S and N_{AB} are the mole numbers of solvent and undissociated electrolyte and $\overline{G}_i$ is the partial molar Gibbs free energy of species i, that is,

$$\overline{G}_i = \left(\frac{\partial G}{\partial N_i}\right)_{T,P,N_{j\neq i}} \tag{7.9-5}$$

Since solutions with low electrolyte concentrations are of most interest, the solute activity coefficients in electrolyte solutions could, in principle, be defined, following Eq. 7.8-15, by

$$\overline{G}_i(T, P, M_i) = \overline{G}_i^{\square} + RT \ln (\gamma_i^{\square} M_i/(M_i = 1)) \tag{7.9-6}$$

where M_i is the molality of species i, $\overline{G}_i^{\square}$ is its Gibbs free energy in an ideal solution of unit molality, and $\gamma_i^{\square}$ is the activity coefficient defined so that $\gamma_i^{\square}$ approaches unity as M_i approaches zero. Thus, we have for the undissociated electrolyte, and, in principle, for each of the ions, that

$$\overline{G}_{AB}(T, P, M_{AB}) = \overline{G}_{AB}^{\square} + RT \ln (\gamma_{AB}^{\square} M_{AB}/(M_{AB} = 1))$$

$$\overline{G}_A(T, P, M_A) = \overline{G}_A^{\square} + RT \ln (\gamma_A^{\square} M_A/(M_A = 1)) \tag{7.9-7}$$

and

$$\overline{G}_B(T, P, M_B) = \overline{G}_B^{\square} + RT \ln (\gamma_B^{\square} M_B/(M_B = 1))$$

The difficulty with this description is that $\overline{G}_A$ and $\overline{G}_B$ are not separately measurable, because, by Eq. 7.9-3, it is not possible to vary the number of moles of cations holding the number of moles of anions fixed, or vice versa. (Even in mixed electrolyte solutions, that is, solutions of several electrolytes, the condition of overall electrical neutrality makes it impossible to vary the number of only one ionic species.) To maintain the present thermodynamic description of mixtures, and, in particular, the concept of the partial molar Gibbs free energy, we instead consider a single electrolyte solution to be a three-component system: solvent, undissociated electrolyte, and dissociated electrolyte. Letting $N_{AB,D}$ be the moles of dissociated electrolyte, we then have

$$G = N_S\overline{G}_S + N_{AB}\overline{G}_{AB} + N_{AB,D}\overline{G}_{AB,D} \tag{7.9-8}$$

where $\overline{G}_{AB,D}$, the partial molar Gibbs free energy of the dissociated electrolyte, $\overline{G}_S$, and $\overline{G}_{AB}$ are all measurable.

Comparing Eqs. 7.9-4 and 8 yields

$$N_{AB,D}\overline{G}_{AB,D} = N_A\overline{G}_A + N_B\overline{G}_B$$

or

$$\overline{G}_{AB,D} = \nu_+\overline{G}_A + \nu_-\overline{G}_B \tag{7.9-9}$$

so that

$$\overline{G}_{AB,D} = \nu_+[\overline{G}_A^{\square} + RT \ln (\gamma_A^{\square} M_A/M_A = 1)] + \nu_-[\overline{G}_B^{\square} + RT \ln (\gamma_B^{\square} M_B/M_B = 1)]$$

$$= [\nu_+\overline{G}_A^{\square} + \nu_-\overline{G}_B^{\square}] + RT \ln \left\{\frac{(\gamma_A^{\square} M_A)^{\nu_+}(\gamma_B^{\square} M_B)^{\nu_-}}{(M_A = 1)^{\nu_+}(M_B = 1)^{\nu_-}}\right\} \tag{7.9-10}$$

Finally, we define a **mean ionic activity coefficient**, $\gamma_\pm$, by

$$\gamma_\pm^{\nu} = (\gamma_A^{\square})^{\nu_+}(\gamma_B^{\square})^{\nu_-} \tag{7.9-11}$$

a **mean ionic molality, $M_\pm$**,

$$M_\pm^{\nu} = M_A^{\nu_+}M_B^{\nu_-} \tag{7.9-12}$$

and

$$\overline{G}_{AB,D}^{\square} = \nu_+ \overline{G}_A^{\square} + \nu_- \overline{G}_B^{\square} \qquad (7.9\text{-}13)$$

to obtain

$$\overline{G}_{AB,D} = \overline{G}_{AB,D}^{\square} + RT \ln [(M_\pm \gamma_\pm)^\nu/(M = 1)^\nu] \qquad (7.9\text{-}14)$$

where $\nu = \nu_+ + \nu_-$.

Since only the mean activity coefficient, and not the activity coefficients of the individual ions, is measurable, our interest in the remainder of this section is in formulas for $\gamma_\pm$. Also, since we will be concerned mostly with low electrolyte concentrations, in the application of these formulas the distinction between molality and concentration in moles/liter will sometimes be ignored.

P. Debye and E. Hückel,[21] using a statistical mechanical model to obtain the average ion distribution around ions in solution, derived the following expression for the dependence of $\gamma_\pm$ on electrolyte concentration

$$\ln \gamma_\pm = -\alpha |z_+ z_-| \sqrt{\mu} \qquad (7.9\text{-}15)$$

The bracketed term in this equation is the absolute value of the product of ion valences, α is a parameter that depends on the solvent and the temperature (see Table 7.9-1 for the values of water), and μ is the **ionic strength** defined by

$$\mu = \frac{1}{2} \sum_i z_i^2 M_i \qquad (7.9\text{-}16)$$

where the summation is over all ions in solution.

Table 7.9-1
VALUES OF THE PARAMETERS IN THE EQUATIONS FOR $\gamma_\pm$ FOR AQUEOUS SOLUTIONS[22]

T(°C)	α(mol/l)$^{-1/2}$	β[(mol/l)$^{1/2}$ Å]$^{-1}$
0	1.132	0.3248
5	1.140	0.3256
10	1.149	0.3264
15	1.158	0.3273
20	1.167	0.3282
25	1.178	0.3291
30	1.188	0.3301
40	1.212	0.3323
50	1.237	0.3346
60	1.265	0.3371
70	1.295	0.3397
80	1.328	0.3426
90	1.363	0.3456
100	1.401	0.3488

Source: Adapted from a table in R. A. Robinson and R. H. Stokes, *Electrolyte Solutions*, 2nd ed., Butterworths, London, 1959.

[21] P. Debye and E. Hückel, *Phys. Z.* **24,** 185 (1923).
[22] 1 Å = 1 Angstrom unit = 10^{-10} m.
l = liter = 10^{-3} m^3.

Equation 7.9-15 is exact at very low ionic strengths and is usually referred to as the **Debye–Hückel limiting law**. Unfortunately, significant deviations from the limiting law expression are observed at ionic strengths as low as 0.01 molal (see Fig. 7.9-1). For higher electrolyte concentrations, the following empirical and semitheoretical modifications of Eq. 7.9-15 have been proposed:

$$\ln \gamma_{\pm} = \frac{\alpha|z_+z_-|\sqrt{\mu}}{1 + \beta a\sqrt{\mu}} \tag{7.9-17}$$

and

$$\ln \gamma_{\pm} = \frac{\alpha|z_+z_-|\sqrt{\mu}}{1 + \beta a\sqrt{\mu}} + \delta\mu \tag{7.9-18}$$

In these equations β is the parameter given in Table 7.9-1 and a is a constant related to the average hydrated radius of ions, usually about 4 Å. In practice, however, the product βa is sometimes set equal to unity or treated as an adjust-

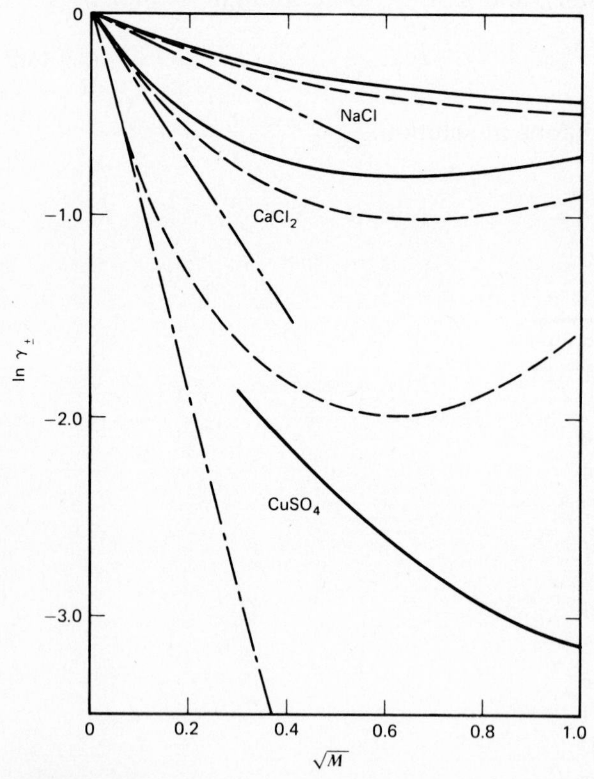

Figure 7.9-1
The activity coefficients of various salts in aqueous solution at 25°C as a function of the salt molarity M. The solid line is the experimental data (from R. A. Robinson and R. H. Stokes, *Electrolyte Solutions, 2 ed., Buttersworth, London, 1959), the line* ——–—— *is the result of the Debye–Hückel limiting law, Eq. 7.9-15, and the dashed line is the prediction of Eq. 7.9-18, with* $\beta\alpha = 1$ and $\delta = 0.1\ |z_1z_-|$. Note that for NaCl $\mu = M$, for $CaCl_2$ $\mu = 3M$, and for $CuSO_4$ $\mu = 4\ M$.

able parameter. Similarly, δ is sometimes set equal to $0.1|z_+z_-|$ and sometimes taken to be an additional adjustable parameter.

Equations 7.9-17 and 18 are frequently satisfactory for solutions of 0.1 molal and somewhat higher (see Fig. 7.9-1). In general, the accuracy of the equations is best for a $z_+ = 1$, $z_- = -1$ electrolyte (termed a 1:1 electrolyte) and becomes progressively less satisfactory for 1:2, 2:2, etc. electrolytes, which are increasingly more nonideal.

We should point out that these equations are valid for solutions of several electrolytes, or mixed electrolyte solutions, as well as single electrolytes. In such cases there are several mean activity coefficients, one for each electrolyte, and the equations here are used to compute the value of each. In this calculation the ionic strength μ is that computed by summing over all ions in solution, and, consequently, is the same for each electrolyte in the solution.

There are few solution models valid for moderate and high concentrations of electrolytes. Perhaps the most successful is the recent model of Pitzer:[23]

$$\frac{G^{ex}}{n_w RT} = f(\mu) + \sum_i \sum_j \lambda_{ij}(\mu) M_i M_j + \sum_i \sum_j \sum_k \delta_{ijk} M_i M_j M_k \qquad (7.9\text{-}19)$$

where G^{ex}/n_w is the excess Gibbs free energy per kilogram of solvent, M_i is the molality of each ion or neutral solute present, and $f(\mu)$ is the Debye–Hückel term. Finally, λ_{ij} and δ_{ijk} are the second and third virial coefficients among the species present, with λ_{ij} having some dependence on ionic strength μ.

Although Eq. 7.9-19 has a (very) large number of constants, especially in mixed electrolyte solutions, it has been useful in representing the thermodynamic behavior of electrolyte solutions all the way from dilute solutions to molten salts.

7.10 CONCLUDING REMARKS

The objective of this chapter has been to develop methods of estimating species fugacities in mixtures, which are so important in phase equilibrium calculations, as will be seen in the following chapter. Because of the variety of methods discussed, there may be some confusion as to which fugacity estimation technique applies in a given situation. The comments that follow may be helpful in choosing among the three main methods discussed in this chapter:

1. Equations of state
2. Activity coefficient (or excess Gibbs free energy) models, including those based on the Henry's law standard state

and

3. Corresponding states

(Electrolyte solutions are rather special and can only be treated by the methods considered in Sec. 7.9. Therefore, electrolyte solutions will not be considered in this discussion.)

First, it should be noted that the principle of corresponding states and equations of state have approximately the same range of applicability; though

[23]See, for example, K. S. Pitzer, "Thermodynamics of Aqueous Systems with Industrial Applications," *ACS Symposium Series*, **133**, 451 (1980).

predictions based on equations of state will be more accurate and computationally more difficult. With the availability of digital computers, equations of state calculations are to be preferred over predictions based on the principle of corresponding states. Consequently, in the following discussion we will refer only to equation of state and activity coefficient models; you should keep in mind that corresponding states predictions can be used whenever reference is made to equation of state calculations.

In low to moderate density vapors, mixture nonidealities are not very large, and, therefore, equations of state of the type discussed in this text can generally be used for the prediction of vapor-phase fugacities of all species. [However, mixtures containing species that associate (i.e., form dimers, trimers, etc.) in the vapor phase, such as acetic acid, must be treated using more complicated equations of state and mixing rules.] At very low densities, the vapor-phase fugacity of species in all mixtures can be estimated using the virial equation of state with experimentally determined virial coefficients. The Lewis–Randall rule should only be used for approximate calculations.

At liquid densities, solution nonidealities can be large. In this case, equation of state predictions will be reasonably accurate only for mixtures composed of relatively simple molecules that are similar. Thus, equation of state (and corresponding states) methods will be satisfactory only for mixtures involving light hydrocarbons including with some dissolved inorganic gases (CO, CO_2, H_2S, N_2, etc.) Present equation of state methods will not be accurate if complicated molecular phenomena are involved, for example, hydrogen bonding as occurs in mixtures containing water, alcohols, and organic acids, or if chemically dissimilar molecular species are involved. In these cases, activity coefficient (or Gibbs free energy) models must be used for estimating the liquid-phase species fugacities, even though an equation of state model may be used for the vapor phase. The disadvantage of presently available equations of state is their inapplicability to other than hydrocarbon mixtures at liquid densities.

Consequently, in the petroleum and natural gas industries, where mostly light and moderate molecular weight hydrocarbons (with only small concentrations of aromatic and inorganic compounds) are involved, equation of state methods predominate for the prediction of both liquid- and vapor-phase properties. In the chemicals industry, on the other hand, where oxygen, nitrogen, sulfur, or halogen-containing organic compounds and inorganic (electrolyte) compounds are involved, activity coefficient or Gibbs free energy models are used for the prediction of liquid-phase properties, with equation of state models being used for the vapor phase if the pressure is above ambient.

The critical point of a mixture, to be discussed in the next chapter, is similar to the critical point of a pure component in that it is the temperature and pressure at which the vapor and liquid phases of a mixture of given composition become identical. Although we have not yet discussed phase equilibrium in multicomponent mixtures (Chapter 8), we can guess that if we are to accurately predict the point at which two phases become identical (the critical point), the models we use must give identical results for all phase properties at this point. This will occur if we use the same equation of state model to describe both the vapor and liquid phases, but clearly cannot, in general, be expected to occur if we use an activity coefficient model for the liquid phase and an equation of state for the vapor phase. This ability to predict critical phenomena is an important advantage of using the same equation of state model for both phases. Another advantage of the equation of state description is that the concept of standard states, and especially hypothetical or extrapolated standard states, never arises.

Finally, since a number of different activity coefficient (or excess Gibbs free energy) models have been discussed, it is useful to consider their range of application. The most important observation is that none of the completely predictive methods, such as regular solution theory, UNIFAC, or ASOG, can be regarded as highly accurate. Therefore, these methods should only be used when no experimental data are available for the system of interest. Of the predictive methods, UNIFAC is the best developed and regular solution theory is the least accurate.

For moderately nonideal systems, all the models (van Laar, two-constant Margules, Wilson, UNIQUAC, and NRTL) perform comparably; in such cases the van Laar, two-constant Margules, UNIQUAC, and Wilson equations, which have only two parameters for a binary mixture, are preferred over the three-constant NRTL equation. For mixtures of very different species, such as polar or associating compounds (e.g., alcohols and other oxy-hydrocarbons) in nonpolar solvents (e.g., hydrocarbons), the Wilson, UNIQUAC, and NRTL models are superior to the van Laar and two-constant Margules equations. For species that are so dissimilar that they are only partially soluble and form two liquid phases (See Sec. 8.3), the NRTL and UNIQUAC equations are frequently useful, whereas the Wilson equation is inapplicable (see Problem 8.4-11).

APPENDIX A7.1

A Statistical Mechanical Interpretation of the Entropy of Mixing in an Ideal Mixture

The entropy change of mixing for an ideal binary gaseous or liquid mixture

$$\Delta \underline{S}^{IM} = -R \sum x_i \ln x_i$$
$$= -R(x_1 \ln x_1 + x_2 \ln x_2) \tag{A7.1-1}$$

(cf. Eq. 7.1-8 and the appropriate entry in Table 7.3-1) has a very simple statistical mechanical interpretation.

To demonstrate this we will consider a collection of N_1 molecules of species 1 and N_2 molecules of species 2 maintained at constant total internal energy U and constant volume V. Furthermore, it will be assumed that either the molecules do not interact (this is the case for an ideal gas mixture) or that the molecules do interact, but that the potential energy function is the same for all species, that is, the species 1–species 1, species 2–species 2, and species 1–species 2 interactions are all alike (this is the assumption that underlies the ideal mixture model).

In courses on advanced physical chemistry and statistical mechanics it is shown that the entropy function for a binary mixture at constant internal energy, volume, and number of particles is

$$S = N_1 S_{1,\text{intra}} + N_2 S_{2,\text{intra}} + S_{\text{conf}} \tag{A7.1-2}$$

Here $S_{i,\text{intra}}$ is the entropy of one molecule of species i due to its intramolecular structure (which can be calculated from detailed structural and spectroscopic information on the species), and S_{conf} is the entropy contribution due to the configuration of the system, that is, the way species 1 and 2 are distributed in the mixture. The statistical mechanical expression for the configurational entropy for a specific macroscopic state of the mixtures being considered here is

$$S_{\text{conf}} = k \ln \Omega \tag{A7.1-3}$$

where Ω is the number of different arrangements of the molecules that results in the desired macroscopic state and k is the Boltzmann constant. Our interest here is in computing the value of S_{conf} for different distributions of the two species within the system.

To compute S_{conf} we will imagine that within the volume V there is a three-dimensional lattice with $N = N_1 + N_2$ equally spaced lattice points. Different macroscopic states for this model system correspond to different arrangements of the N_1 molecules of species 1 and the N_2 molecules of species 2 among the N lattice points. By the ideal mixture assumption, each distribution of molecules among the lattice points has the same energy as any other.

The number of ways of distributing the N_1 identical molecules of species 1 and the N_2 identical molecules of species 2 among the $N_1 + N_2$ lattice sites in a completely random fashion is, from simple probability theory, equal to

$$\frac{(N_1 + N_2)!}{N_1! N_2!}$$

Therefore, the configurational entropy of the completely random mixture is

$$S_{\text{conf}}\begin{pmatrix}\text{random}\\ \text{mixture}\end{pmatrix} = k \ln \frac{(N_1 + N_2)!}{N_1! N_2!} \tag{A7.1-4}$$

This equation can be simplified by using Stirling's approximation for the logarithm of a factorial number

$$\ln M! = M \ln M - M \tag{A7.1-5}$$

which is valid for large M. Using Eq. A7.1-5 in Eq. A7.1-4 yields

$$\begin{aligned} S_{\text{conf}}\begin{pmatrix}\text{random}\\ \text{mixture}\end{pmatrix} &= k[\ln (N_1 + N_2)! - \ln N_1! - \ln N_2!] \\ &= k[(N_1 + N_2) \ln (N_1 + N_2) - (N_1 + N_2) - N_1 \ln N_1 + N_1 - N_2 \ln N_2 + N_2] \\ &= -k\left[N_1 \ln \frac{N_1}{N_1 + N_2} + N_2 \ln \frac{N_2}{N_1 + N_2}\right] \\ &= -k(N_1 + N_2)[x_1 \ln x_1 + x_2 \ln x_2] \end{aligned} \tag{A7.1-6}$$

Now dividing by the sum of N_1 and N_2 and multiplying by Avogadro's number $\tilde{N}(= 6.022 \times 10^{23})$, we obtain an expression for the configurational entropy per mole of mixture

$$\begin{aligned} \underline{S}_{\text{conf}}\begin{pmatrix}\text{random}\\ \text{mixture}\end{pmatrix} &= -\tilde{N}k[x_1 \ln x_1 + x_2 \ln x_2] \\ &= -R[x_1 \ln x_1 + x_2 \ln x_2] \end{aligned} \tag{A7.1-7}$$

where $R = \tilde{N}k$ is the gas constant of Table 1.4-1.

Now consider the completely ordered configuration in which the molecules of species 1 are restricted to the first N_1 lattice sites, and the molecules of species 2 to the remaining N_1 lattice sites; that is, the two species are not mixed. The number of different ways that this can be accomplished is

$$\frac{N_1!N_2!}{N_1!N_2!} = 1 \tag{A7.1-8}$$

so that

$$\underline{S}_{\text{conf}}\begin{pmatrix}\text{completely}\\ \text{ordered}\end{pmatrix} = R \ln 1 = 0 \tag{A7.1-9}$$

Therefore, the entropy change on going from the completely ordered (unmixed) state to the randomly ordered (completely mixed) state is

$$\begin{aligned} \Delta\underline{S}_{\text{mix}} &= \underline{S}_{\text{conf}}\begin{pmatrix}\text{random}\\ \text{mixture}\end{pmatrix} - \underline{S}_{\text{conf}}\begin{pmatrix}\text{completely}\\ \text{ordered}\end{pmatrix} \\ &= -R[x_1 \ln x_1 + x_2 \ln x_2] \end{aligned} \tag{A7.1-10}$$

which is in agreement with Eq. A7.1-1.

One can easily establish, though we will not do so here, that any partially ordered state will have a molar configurational entropy intermediate to the randomly mixed and completely ordered states. Therefore, the randomly mixed state is the state of highest entropy. Since the criterion for equilibrium at constant internal energy and volume is that the entropy of the system achieve a maximum, the randomly mixed or completely disordered state is the equilibrium state in an ideal mixture.

It is tempting to try to generalize this result by suggesting that the completely disordered state is always the equilibrium state. However, *this is not correct*! In a mixture in which not all interactions are alike, different distributions of molecules among the lattice sites will result in different total energies of the system. In this case, energetic, as well as entropic, effects are important in determining the equilibrium state. Thus, in a system at constant internal energy and volume, the equilibrium state will be the state of maximum entropy (or maximum randomness) among *only those states* that have the required internal energy. For systems at constant temperature and volume, the equilibrium state is a state of minimum Helmholtz free energy. Since $A = U - TS$, it is evident that increasing the entropy (or disorder) in the system decreases the Helmholtz free energy only if it does not increase the internal energy of the system. For example, consider the lattice model used here, but now letting u_{ii} be the interaction energy of a species i–species i interaction, where $u_{11} \neq u_{12} \neq u_{22}$ so the mixture is not ideal. Clearly, if u_{12} is greater than the arithmetic average of u_{11} and u_{22}, that is, if

$$u_{12} > \tfrac{1}{2}(u_{11} + u_{22})$$

increasing the randomness of the mixture (and the number of 1–2 interactions at the expense of 1–1 and 2–2 interactions), increases both the entropy and the internal energy of the system. Therefore, the equilibrium state for this system will be that compromise between energetic and entropic effects for which the Helmholtz free energy is a minimum.

In real mixtures this balance between energetic and entropic effects is illustrated, for example, in liquid–liquid phase equilibrium (to be discussed in Chapter 8), in which the equilibrium state of some liquid mixtures is as two coexisting phases of different composition, a distinctly ordered, rather than random state. The equilibrium state is not a single phase of uniform composition, since the increase in randomness (and hence decrease in $-TS$) of producing such a state from the two-phase mixture would be less than the increase in the internal energy of the system. Consequently, the random single-phase mixture would have a higher Helmholtz free energy than the more ordered two-phase mixture, and therefore would not be a stable equilibrium state.

A BASIC Language Program for Multicomponent Phase Equilibrium Calculations Using the Peng–Robinson Equation of State, VLMU.BAS

This appendix describes a BASIC language program for species fugacity and vapor–liquid equilibrium calculations using the Peng–Robinson equation of state. Given the critical temperatures, critical pressures, acentric factors, and the binary interaction parameters for the species in the mixture, which must be supplied by the user, a species fugacity, bubble-point temperature or pressure, dew-point temperature or pressure, or isothermal vapor–liquid flash calculation may be performed. As written, the program runs using BASICA on an IBM PC or compatible computers. The program has been written in double precision to avoid roundoff errors. For faster execution it may be compiled using a BASIC compiler. To obtain a listing of this program, insert the disk accompanying this book into drive A of a PC compatible computer with a printer and type PRINT A: VLMU.BAS.

The structure of the program is indicated here.

Statement Range	Function
10–720	Input routine for data entry and menu of calculation choices.
720–1740	Subroutine prfuga is similar to the subroutine of the same name in program PR1.BAS. It is used to calculate the compressibility and species fugacity for given constants in equation of state, and values of T, P and mole fraction.
800–1040	Setting up constants in Eq. 4.4-4.
1050–1420	Using Cardan's rule to solve cubic equation for real roots for the compressibility Z.
1430–1520	Ordering Z roots and setting vapor root (Z0) to largest compressibility and liquid root (Z1) to smallest.
1530–1540	Testing for erroneous root ($V < b$) and correcting if necessary.
1550–1740	If only compressibilities wanted (ICOMP = 0), then return; otherwise calculate species fugacities in the vapor or liquid phase, as appropriate.
1760–1930	Subroutine prcons, calculates the constants a and b in the Peng–Robinson equation of state for each species for use in prfuga.
1950–2240	Subroutine comp is used for compressibility calculation. Requires T, P, and mole fraction as input; output is vapor and liquid compressibilities and specific volumes.
2260–3280	Subroutine flash calculates the compositions of the coexiting vapor and liquid phases. The required input is temperature (T), pressure (P), feed composition (z_i), the output is the composition (x_i and y_i) and compressibilities (ZZ1 and ZZ2) of the coexiting phases, the vapor–liquid split (XLF) and the values of $K_i = y_i/x_i = XK1(\text{i})$.
2260–2420	Data input.
2430	Call prcons to calculate a_i and b_i.
2440–2560	Generate initial guesses for K_i and XLF.

Statement Range	Function
2600–2830	Iterate to obtain the phase compositions and split for guessed K values.
2870–2940	Call prfuga to calculate fugacities of species in each phase at compositions calculated above.
2950–3040	Check if $\bar{f}_i^L = \bar{f}_i^V$. If this is not true for any species, obtain new K_i value for this species from $$K_i^{new} = K_i^{old} \times \bar{f}_i^L/\bar{f}_i^V$$
3060	Check that fraction of liquid lies between zero and one.
3070–3280	Output routine.
3300–3620	Subroutine spfug is used to compute species fugacity. This subroutine requests T, P, and feed mole fractions as input, calls prcons and prfuga, and provides the fugacities and fugacity coefficients as output.
3630–4940	Subroutine bubptt calculates the bubble-point temperature of a liquid feed at a specified pressure and the composition of the first bubble of vapor that forms.
3630–3770	Data input.
3780–3860	Initial guess procedure for bubble-point temperature.
3870–3960	Initial guess procedure for $K_i = y_i/x_i$ and y_i.
3990–4070	Calculation of species fugacities at liquid (feed) composition and estimated bubble-point temperature.
4080–4160	Normalization of vapor-phase mole fractions.
4170–4220	Calculation of species fugacities in vapor at estimated vapor composition.
4230–4270	Recalculation of vapor-phase compositions to ensure equality of fugacities.
4280–4380	Check for convergence of vapor-phase compositions Store Σy_i [as $S(1)$].
4390–4450	Repeat calculation at $T - 0.005$ Store Σy_i [as $S(2)$].
4460	Evaluate $d(\Sigma y_i)/dT$ using $S(1)$ and $S(2)$.
4470–4570	Use bounded Newton–Raphson procedure (with numerical derivative calculated above) to generate new bubble-point temperature estimate.
4590–4750	Test if calculation converged to a nontrivial solution (trivial solution occurs when liquid and vapor phases are identical and equal to feed). Also check whether a real vapor solution (i.e., $Z > 0.307$) has been obtained.
4760–4940	Output routine.
4950–6240	Subroutine dewptt calculates the dew point temperature of a vapor at a specified pressure and the composition of the first drop of vapor that forms. The structure of this subroutine is very similar to that of bubptt except that dew point temperature is the Newton–Raphson iteration variable.
6250–7590	Subroutine bubptp calculates the bubble-point pressure of a liquid at a specified temperature and the composition of the first bubble of vapor that forms. The structure of this subroutine is similar to that of bubptt, except that the Newton–Raphson iteration variable is bubble-point pressure.
7600–8870	Subroutine dewptp calculates the dew-point pressure of a vapor at a specified pressure and the composition of the first drop of liquid that forms. The structure of this subroutine is similar to that of bubptt except that dew-point pressure is the iteration variable.

Although the program is specific to the Peng–Robinson equation of state, only the following lines would have to be changed to use another cubic equation of state:

750–770, 1020–1040, 1600–1700, and 1780–1840

Multicomponent Excess Gibbs Free Energy (Activity Coefficient) Models

The excess Gibbs free energy models for binary mixtures discussed in Sec. 7.5 can be extended to multicomponent mixtures. For example, the Wohl expansion of Eq. 7.5-8 can be extended to ternary mixtures

$$\begin{aligned}\frac{G^{ex}}{RT(x_1q_1 + x_2q_2 + x_3q_3)} &= 2a_{12}z_1z_2 + 2a_{13}z_1z_3 + 2a_{23}z_2z_3 \\ &+ 3a_{112}z_1^2z_2 + 3a_{122}z_1z_2^2 + 3a_{113}z_1^2z_3 \\ &+ 3a_{133}z_1z_3^2 + 3a_{223}z_2^2z_3 + 3a_{233}z_2z_3^2 \\ &+ 6a_{123}z_1z_2z_3 + \cdots \end{aligned} \tag{A7.3-1}$$

which by neglecting all terms of third and higher order in the volume fractions, yields for species 1

$$\ln \gamma_1 = \frac{\left\{x_2^2\alpha_{12}\left(\frac{\beta_{12}}{\alpha_{12}}\right)^2 + x_3^2\alpha_{13}\left(\frac{\beta_{13}}{\alpha_{13}}\right)^2 + x_2x_3\frac{\beta_{12}}{\alpha_{12}}\frac{\beta_{13}}{\alpha_{13}}\left(\alpha_{12} + \alpha_{13} - \alpha_{23}\frac{\alpha_{12}}{\beta_{12}}\right)\right\}}{\left[x_1 + x_2\left(\frac{\beta_{12}}{\alpha_{12}}\right) + x_3\left(\frac{\beta_{13}}{\alpha_{13}}\right)\right]^2} \tag{A7.3-2}$$

where $\alpha_{ij} = 2q_ia_{ij}$, $\beta_{ij} = 2q_ja_{ij}$, and $a_{ij} = a_{ji}$. (Thus, $\alpha_{ji} = \beta_{ij}$ and $\beta_{ji} = \alpha_{ij}$.) The expression for $\ln \gamma_2$ is obtained by interchanging the subscripts 1 and 2 in Eq. A7.3-2 and for $\ln \gamma_3$ by interchanging the subscripts 1 and 3. The collection of equations obtained in this way are the van Laar equations for a ternary mixture. Note that there are two parameters, α_{ij} and β_{ij}, for each pair of components in the mixture.

The multicomponent form of the van Laar model is obtained by starting from

$$\frac{G^{ex}}{RT\sum_{i=1}^{\mathscr{C}} x_iq_i} = \sum_{i=1}^{\mathscr{C}}\sum_{j=1}^{\mathscr{C}} \alpha_{ij}z_iz_j + \sum_{i=1}^{\mathscr{C}}\sum_{j=1}^{\mathscr{C}}\sum_{k=1}^{\mathscr{C}} \alpha_{ijk}z_iz_jz_k + \cdots \tag{A7.3-3}$$

and neglecting third and higher order terms in the volume fraction. Thus, $\alpha_{ij} = \alpha_{ji}$, $\alpha_{ii} = 0$, and $\alpha_{ijk} = \alpha_{ikj} = \alpha_{kij} = \alpha_{iii} = 0$.

The multicomponent form of the Wilson equation is

$$\frac{G^{ex}}{RT} = -\sum_{i=1}^{\mathscr{C}} x_i \ln\left(\sum_{j=1}^{\mathscr{C}} x_j\Lambda_{ij}\right) \tag{A7.3-4}$$

for which

$$\ln \gamma_i = 1 - \ln\left(\sum_{j=1}^{\mathscr{C}} x_j\Lambda_{ij}\right) - \sum_{j=1}^{\mathscr{C}} \frac{x_j\Lambda_{ji}}{\sum_{k=1}^{\mathscr{C}} x_k\Lambda_{jk}} \tag{A7.3-5}$$

Since $\Lambda_{ii} = 1$, there are also two parameters Λ_{ij} and Λ_{ji} for each binary pair of components in this multicomponent mixture model.

The multicomponent NRTL equation is

$$\frac{\underline{G}^{ex}}{RT} = \sum_{i=1}^{\mathscr{C}} x_i \frac{\sum_{j=1}^{\mathscr{C}} \tau_{ji} G_{ji} x_j}{\sum_{j=1}^{\mathscr{C}} G_{ji} x_j} \tag{A7.3-6}$$

with $\ln G_{ij} = -\alpha_{ij}\tau_{ij}$, $\alpha_{ij} = \alpha_{ji}$ and $\tau_{ii} = 0$ for which

$$\ln \gamma_i = \frac{\sum_{j=1}^{\mathscr{C}} \tau_{ji} G_{ji} x_j}{\sum_{j=1}^{\mathscr{C}} G_{ji} x_j} + \sum_{j=1}^{\mathscr{C}} \frac{x_j G_{ij}}{\sum_{j=1}^{\mathscr{C}} x_k G_{kj}} \left(\tau_{ij} - \frac{\sum_{k=1}^{\mathscr{C}} x_k \tau_{kj} G_{kj}}{\sum_{k=1}^{\mathscr{C}} x_k G_{kj}} \right) \tag{A7.3-7}$$

This equation has three parameters, τ_{ij}, τ_{ji}, and α_{ij}, for each pair of components in the multicomponent mixture.

The important feature of each of the equations discussed here is that all the parameters that appear can be determined from activity coefficient data for binary mixtures. That is, by correlating activity coefficient data for the species 1–species 2 mixture using any of the models, the 1–2 parameters can be determined. Similarly, from data for species 2–species 3 and species 1–species 3 binary mixtures, the 2–3 and 1–3 parameters can be found. These coefficients can then be used to estimate the activity coefficients for the ternary 1–2–3 mixture without any experimental data for the three-component system.

One should keep in mind that this ability to predict multicomponent behavior from data on binary mixtures is not an exact result, but rather arises from the assumptions made or the models used. This is most clearly seen in going from Eq. A7.3-1 to Eq. A7.3-2. Had the term $\alpha_{123}z_1z_2z_3$ been retained in the Wohl expansion, Eq. A7.3-2 would contain this α_{123} term, which could only be obtained from experimental data for the ternary mixture. Thus, if this more complete expansion were used, binary data *and* some ternary data would be needed to determine the activity coefficient model parameters for the ternary mixture.

APPENDIX A7.4

A BASIC Language Program for the Prediction of Activity Coefficients and Low Pressure Vapor–Liquid Equilibrium Using the UNIFAC Model, UNIFAC.BAS

This appendix describes a BASIC language program for the prediction of activity coefficients in multicomponent systems using the fourth revision of the UNIFAC method [D. Tiegs, J. Gmehling, P. Rasmussen, and A. Fredenslund, *Ind. Eng. Chem. Res.* **26,** 159 (1987)]. In addition, for binary systems the low pressure vapor–liquid equilibrium is calculated (see Sec. 8.1) on the assumption that all fugacity coefficients are unity. If a graphics monitor is used, an *x-y* diagram can be plotted. This program will only run in advanced BASIC (BASICA) on IBM PC or compatible computers. For faster execution it may be compiled using a BASIC compiler. This program uses the data files UFNRQM.DTA and UNFA44.DTA.

The user first is asked to supply the number of groups in each molecule from the menu supplied. If *x-y* values or an *x-y* diagram is desired, the user must supply the pure component vapor pressures at the temperature of interest or the constants in the Antoine vapor pressure equation. The flow of the program is controlled by the function keys F1 through F10. At each point in the program only those function keys that are active are displayed at the bottom of the screen. To obtain a listing of this program, insert the disk accompanying this book into drive A of an IBM-PC compatible computer with a printer and type PRINT A:UNIFAC.BAS. Similarly, typing PRINT A:UFNRQM.DTA or PRINT A:UNFA44.DTA will print the data files containing the group R and Q parameters, and the a_{mn} parameters, respectively.

The structure of the program is indicated here.

Statement Range	Function
1000–1440	Initiation routine, including reading R and Q parameters and the interaction parameters a_{ij} from data files UFNRQM.DTA and UNFA44.DTA, respectively.
1450–1700	Menu of choices available using functions keys. Note, however, that **1.** On start-up, program immediately goes to component entry section. **2.** Activity coefficient calculation option will only appear after T and x are specified. **3.** Activity coefficient list calculation option will only appear after T is specified. **4.** x-y list option will only appear after γ-list calculation. **5.** x-y graph option will only appear if graphics monitor is used and γ-list calculation has been completed. If Antoine constants have not been previously entered (using AntCon option), user will be asked to provide pure component vapor pressures.
1710–2170	Component entry routine
1720–1880	Enter component name and subgroup frequency
1820	Call subgroup entry routine
1890–2060	Print list of components, subgroups, and subgroup frequencies
2070–2170	Check for availability of interaction parameters by calling UNIFAC program with flag IAPFLG=0

Statement Range	Function
2180–2470	Concentration entry routine. Will normalize mole fractions that do not sum to unity and will avoid division by zero errors by setting a 0 mole fraction to 0.000001
2480–2560	Temperature entry
2570–3820	UNIFAC calculation of activity coefficients
2610–2820	Calculation of combinatorial contribution to the activity coefficient
2830–3010	Setting up terms for residual calculation
3020–3200	Calculation of residual contribution to the activity coefficient of each group in each molecule
3210–3410	Calculation of residual contribution to the activity coefficient of each group in the mixture of molecules
3340–3370	Checks for availability of interaction parameters between the groups after component entry routine (i.e., when IAPFLG=0)
3420–3530	Calculates activity coefficient for each molecule
3540–3680	Prints list of pairs of groups in the mixture for which interaction coefficients are not available.
3690–3820	Either returns or prints activity coefficients, depending on way the subprogram was called.
3830–3850	Exit routine
3860–4050	Calculates group mole fractions in molecule (ANM) and in mixture (ANMX)
4060–4570	Routine that allows entry of frequency of groups in each molecule using menu of groups and cursor keys
4580–4760	Routine to prepare list of activity coefficients in a binary mixture over the range of 0 to 1 mole fraction in increments of 0.05 mole fraction
4770–5000	Routine to prepare list of x_1-y_1-y_2-P after γ's have been calculated above. If Antoine constants have been entered, pure component vapor pressures are calculated; if not, the user is asked to supply the vapor pressures.
5010–5440	Routine to graph x_1 versus y_1 using the data prepared above
5450–5540	Entry routine for Antoine constants

PROBLEMS

7.1 Show that for mixing of ideal gases at constant temperature and pressure to form an ideal gas mixture

$$\Delta \underline{U}_{\text{mix}} = \Delta \underline{H}_{\text{mix}} = \Delta \underline{V}_{\text{mix}} = 0$$

and

$$\Delta \underline{G}_{\text{mix}} = \Delta \underline{A}_{\text{mix}} = RT \sum_i x_i \ln x_i$$

7.2 In Section 7.1 we considered the changes in thermodynamic properties on forming an ideal gas mixture from a collection of ideal gases at the same temperature and pressure. A second, less common way of forming an ideal gas mixture is to start with a collection of pure ideal gases, each at the temperature T and volume V, and mix and compress the mixture to produce an ideal gas mixture at temperature T and volume V.

a Show that the mixing process described here is mixing at constant partial pressure of each component.

b Derive each of the entries in the following table:

Ideal Gas Mixing Properties* at Constant Temperature and Partial Pressure of Each Species

Internal energy	$\bar{U}_i^{IGM}(T, x_i) = \underline{U}_i^{IG}(T)$	$\Delta\underline{U}_{mix}^{IGM} = 0$
Volume†	$\bar{V}_i^{IGM}(T, P, x_i) = x_i\underline{V}_i^{IG}(T, P_i)$	$\Delta\underline{V}_{mix}^{IGM} = (1 - \mathscr{C})V \Big/ \sum_{i=1}^{\mathscr{C}} N_i$
Enthalpy	$\bar{H}_i^{IGM}(T, x_i) = \underline{H}_i^{IG}(T)$	$\Delta\underline{H}_{mix}^{IGM} = 0$
Entropy	$\bar{S}_i^{IGM}(T, P, x_i) = \underline{S}_i^{IG}(T, P_i)$	$\Delta\underline{S}_{mix}^{IGM} = 0$
Helmholtz free energy	$\bar{A}_i^{IGM}(T, P, x_i) = \underline{A}_i^{IG}(T, P_i)$	$\Delta\underline{A}_{mix}^{IGM} = 0$
Gibbs free energy	$\bar{G}_i^{IGM}(T, P, x_i) = \underline{G}_i^{IG}(T, P_i)$	$\Delta\underline{G}_{mix}^{IGM} = 0$

*For mixing at constant temperature and partial pressure of each species we have the following

$$\Delta\underline{\theta}_{mix}^{IGM} = \sum_i x_i[\bar{\theta}_i(T, P, x_i) - \underline{\theta}_i(T, P_i)]$$

†$\mathscr{C}$ = number of components.

7.3 Repeat the derivations of the previous problem for a mixing process in which both pure fluids, initially at a temperature T and pressure P, are mixed at constant temperature and then the pressure adjusted so that the final volume of the mixture is equal to the sum of the initial volumes of the pure components (i.e., there is no volume change on mixing).

7.4 Assuming that two pure fluids and their mixture can be described by the van der Waals equation of state

$$P = \frac{RT}{\underline{V} - b} - \frac{a}{\underline{V}^2}$$

and that for the mixture the van der Waals one-fluid mixing rules apply

$$a = \sum_i \sum_j x_i x_j a_{ij} \quad \text{and} \quad b = \sum x_i b_i$$

a Show that the fugacity coefficient for species i in the mixture is

$$\ln \phi_i = \ln \frac{\bar{f}_i}{x_i P} = \frac{B_i}{Z - B} - \ln (Z - B) - \frac{2 \sum_j x_j a_{ij}}{RT\underline{V}}$$

where $B = Pb/RT$.

b Derive an expression for the activity coefficient of each species.

7.5 Assuming that the van der Waals equation of state

$$P = \frac{RT}{\underline{V} - b} - \frac{a}{\underline{V}^2}$$

is satisfied by two pure fluids and by their mixture, and that the van der Waals one-fluid rules

$$a = \sum_i \sum_j x_i x_j a_{ij}$$

and

$$b = \sum_i \sum_j x_i x_j b_{ij}$$

apply to the mixture, derive expressions for the

a Excess volume change on mixing at constant T and P.

b Excess enthalpy and internal energy changes on mixing at constant T and P.

c Excess entropy change on mixing at constant T and P.

d Excess Helmholtz and Gibbs free energy changes on mixing at constant T and P.

7.6 The virial equation for a binary mixture is

$$\frac{P\underline{V}}{RT} = 1 + \frac{B_{\text{mix}}}{\underline{V}} + \cdots$$

with

$$B_{\text{mix}} = y_1^2 B_{11} + y_2^2 B_{22} + 2y_1 y_2 B_{12}$$

Here B_{11} and B_{22} are the second virial coefficients for pure species 1 and pure species 2, respectively, and B_{12} is the cross second virial coefficient. For a binary mixture

a Obtain an expression for the fugacity coefficient of a species (Eq. 7.4-6).

b Show that the activity coefficient for species 1 is

$$\ln \gamma_1 = \delta_{12} y_2^2 P/RT$$

where $\delta_{12} = 2B_{12} - B_{11} - B_{22}$.

c Generalize the results in part b to a multicomponent mixture.

7.7 **a** Derive the two-constant Margules equations for the activity coefficients of a binary mixture (Eqs. 7.5-7).

b Derive Eqs. 7.5-9.

c Use the results of part a to derive van Laar expressions for the activity coefficients of a ternary mixture (Eqs. A7.3-2).

7.8 Using the van Laar theory, estimate the activity coefficients for the benzene–2,2,4-trimethyl pentane system at 55°C. Compare the predictions with the results in Illustrations 7.5-1 and 2.

Data

	Benzene	**2,2,4-Trimethyl Pentane**
Critical temperature	562.1 K	554 K
Critical pressure	4.894 MPa	2.482 MPa
Critical density	301 kg/m^3	235 kg/m^3

7.9 Develop expressions for γ_1^* and $\gamma_1^\square$ using the one-constant and two-constant Margules equations, the van Laar equation, regular solution theory, and the UNIFAC model.

7.10 Use the lattice model discussed in Appendix A7.1 to show that the state of maximum entropy for an ideal gas at constant temperature (energy) and contained in a volume V is the state of uniform density.

7.11 Calculate the fugacity for each species in the following gases at 290 K and 800 bar:

a Pure oxygen

b Pure nitrogen

c Oxygen and nitrogen in a 30 mole % O_2, 70 mole % N_2 mixture using the Lewis–Randall rule

d Oxygen and nitrogen in the mixture in part c using Kay's rule for the pseudocritical constants.

e Oxygen and nitrogen in the mixture in part c using the Prausnitz–Gunn rules for the pseudocritical constants.

f Oxygen and nitrogen in the mixture in part c using the Peng–Robinson equation of state.

7.12 Chemically similar compounds (e.g., ethanol and water or benzene and toluene) generally form mixtures that are close to ideal as evidenced by activity coefficients that are near unity and small excess Gibbs free energies of mixing. On the other hand, chemically dissimilar species (e.g., benzene and water or toluene and ethanol) form strongly nonideal mixtures. Show, by considering the binary mixtures that can be formed from ethanol, water, benzene, and toluene, that the UNIFAC model predicts such behavior.

7.13 Repeat the calculations of the previous problem with the regular solution model. Compare the two results.

7.14 Derive Eqs. 7.7-12 through 15.

7.15 **a** Show that the minimum amount of work, W_s^{min}, necessary to separate 1 mole of a binary mixture into its pure components at constant temperature and pressure is

$$W_s^{min} = x_1 RT \ln \frac{f_1(T, P)}{\bar{f}_1(T, P, x_1)} + x_2 RT \ln \frac{f_2(T, P)}{\bar{f}_2(T, P, x_2)}$$

b Show that this expression reduces to

$$W_s^{min} = -x_1 RT \ln x_1 - x_2 RT \ln x_2$$

for (i) an ideal liquid mixture or (ii) a gaseous mixture for which the Lewis–Randall rule is obeyed.

7.16 Experimentally it is observed that

$$\lim_{x_i \to 1} \left(\frac{\partial \ln \gamma_i}{\partial x_i} \right)_{T,P} = 0 \qquad \text{for any species } i$$

This equation implies that the activity coefficient γ_i (or its logarithm) is weakly dependent on mole fraction near the pure component limit. Since we also know that $\gamma_i \to 1$ as $x_i \to 1$, this equation further implies that the activity coefficient is near unity for an almost pure substance.

a Show that this equation is satisfied by all the liquid solution models discussed in Sec. 7.5.

b Show that since this equation is satisfied for any substance, we also have that

$$\lim_{x_i \to 0} x_i \left(\frac{\partial \ln \gamma_i}{\partial x_i} \right)_{T,P} = 0$$

7.17 The following data are available for mean activity coefficients of single electrolytes in water at 25°C.[24]

	$\gamma_{\pm}$		
Molality	**KCl**	$\mathbf{CrCl_3}$	$\mathbf{Cr_2(SO_4)_3}$
0.1	0.770	0.331	0.0458
0.2	0.718	0.298	0.0300
0.3	0.688	0.294	0.0238
0.5	0.649	0.314	0.0190
0.6	0.637	0.335	0.0182
0.8	0.618	0.397	0.0185
1.0	0.604	0.481	0.0208

Compare these data with the predictions of the Debye–Hückel limiting law, Eq. 7.9-15, and Eq. 7.9-18 with $\beta a = 1$ and $\delta = 0.1|z_+z_-|$.

7.18 **a** Given experimental data either for the excess Gibbs free energy $\underline{G}^{ex}$, or for species activity coefficients from which $\underline{G}^{ex}$ can be computed, it is sometimes difficult to decide whether to fit the data to the two-constant Margules or van Laar expressions for $\underline{G}^{ex}$ and γ_i. One method of making this decision is to plot $\underline{G}^{ex}/x_1x_2$ versus x_1 and $x_1x_2/\underline{G}^{ex}$ versus x_1 and determine which of the two plots is most nearly linear. If it is the first, the data are fit with the two-constant Margules expression, if the second, the van Laar expression is used. Justify this procedure, and suggest how these plots can be used to obtain the parameters in the activity coefficient equations.

b The following data have been obtained for the benzene–2,2,4-trimethyl pentane mixture (Illustration 8.1-4). Using the procedure in part a, decide which of the two solution methods is likely to give the best fit of the data.

x_B	0.0819	0.2192	0.3584	0.3831	0.5256	0.8478	0.9872
$\underline{G}^{ex}$(J/mol)	83.7	203.8	294.1	302.5	351.9	223.8	23.8

7.19 The excess Gibbs free energies for liquid argon–methane mixtures have been measured at several temperatures.[25] The results are

$$\frac{\underline{G}^{ex}}{RT} = x_{\text{Ar}}(1 - x_{\text{Ar}})\{A - B(1 - 2x_{\text{Ar}})\}$$

[24]Reference: Appendix 8.10 in R. A. Robinson and R. H. Stokes, *Electrolyte Solutions*, 2nd ed., Butterworths, London, 1959.

[25]Reference: A. G. Duncan and M. J. Hiza; *I.E.C. Fund.* **11**, 38 (1972).

where numerical values for the parameters are given as:

T(°K)	A	B
109.0	0.3036	−0.0169
112.0	0.2944	+0.0118
115.74	0.2804	+0.0546

Compute the following:

a The activity coefficients of argon and methane at 112.0 K and $x_{Ar} = 0.5$.

b The molar isothermal enthalpy change on producing an $x_{Ar} = 0.5$ mixture from its pure components at 112.0 K.

c The molar isothermal entropy change on producing an $x_{Ar} = 0.5$ mixture from its pure components at 112.0 K.

7.20 Derive Eq. 7.2-13.

7.21 Wilson[26] has proposed that the excess Gibbs free energy of a multicomponent system be given by

$$\underline{G}^{ex} = -RT \sum_{i=1}^{\mathscr{C}} x_i \ln \left[\sum_{j=1}^{\mathscr{C}} x_j \Lambda_{ij} \right]$$

where

$$\Lambda_{ij} = \frac{\underline{V}_j^{\mathrm{L}}}{\underline{V}_i^{\mathrm{L}}} \exp \left[-\frac{(\lambda_{ij} - \lambda_{ii})}{RT} \right]$$

Note that this equation contains only the interaction parameters Λ_{ij} for binary mixtures. Also, the parameters $(\lambda_{ij} - \lambda_{ii})$ appear to be insensitive to temperature.

Holmes and van Winkle[27] have tested this equation and found it to be accurate for the prediction of binary and ternary vapor–liquid equilibria. They also report values of the parameters $\underline{V}_i^{\mathrm{L}}$ and $(\lambda_{ij} - \lambda_{ii})$ for many binary mixtures.

Use the Wilson equation to:

a Obtain the following expression for the activity coefficient of species 1 in a binary mixture

$$\ln \gamma_1 = -\ln (x_1 + x_2\Lambda_{12}) + x_2 \left[\frac{\Lambda_{12}}{x_1 + x_2\Lambda_{12}} - \frac{\Lambda_{21}}{x_1\Lambda_{21} + x_2} \right]$$

b Obtain the following expression for the activity coefficient of species 1 in a multicomponent mixture

$$\ln \gamma_1 = 1 - \ln \left[\sum_{j=1}^{\mathscr{C}} x_j \Lambda_{1j} \right] - \sum_{i=1}^{\mathscr{C}} \frac{x_i \Lambda_{i1}}{\sum_j x_j \Lambda_{ij}}$$

[26] G. M. Wilson, *J. Am. Chem. Soc.* **86,** 127 (1964).

[27] M. J. Holmes and M. van Winkle, *Ind. Eng. Chem.* **62,** 21 (1970).

7.22 The fugacity of a species in a mixture can have a peculiar dependence on composition at fixed temperature and pressure, especially if there is a change of phase with composition. Show this by developing plots of the fugacity of isobutane and of carbon dioxide in their binary mixture as a function of isobutane composition using the Peng–Robinson equation of state for each of the following conditions

a $T = 377.6$ K and P in the range from 20 to 80 bar

b $T = 300$ K and P in the range from 7 to 35 bar

7.23 It has been suggested that since the one-parameter Margules expansion is not flexible enough to fit most activity coefficient data, it should be expanded by adding additional constants. In particular, the following have been suggested

2 parameter models:	$\ln \gamma_1 = Ax_2^2$	$\ln \gamma_2 = Bx_1^2$
or	$\ln \gamma_1 = Ax_2^n$	$\ln \gamma_2 = Ax_1^n$
3 parameter model:	$\ln \gamma_1 = Ax_2^n$	$\ln \gamma_2 = Bx_1^n$

In each case the reference states are the pure components at the temperature and pressure of the mixture.

a Which of these models are reasonable?

b What are the allowable values for the parameters A, B, and n in each of the models?

7.24 Derive Eqs. 7.5-18

8

Phase Equilibrium in Mixtures

The objective of this chapter is to illustrate how the principles of Chapter 6 and the calculational procedures of Chapter 7 can be used to study many different types of phase equilibria. In particular, we consider here:

1. Vapor–liquid equilibria
2. The solubility of gas in a liquid
3. The solubility of a liquid in a liquid
4. Liquid–liquid–vapor equilibria
5. The solubility of a solid in a liquid
6. The distribution of a solute among two liquid phases
7. The freezing-point depression of a solvent due to the presence of a solute
8. Osmotic equilibrium and osmotic pressure

Our interest in phase equilibria will be twofold: to make predictions about the equilibrium state for the types of phase equilibria listed above using activity coefficient models and/or equations of state; and to use experimental phase equilibrium data to obtain activity coefficient and other partial molar property information.

8.1 VAPOR–LIQUID EQUILIBRIA USING ACTIVITY COEFFICIENT MODELS

In the analysis of distillation and other vapor–liquid separations processes one must estimate the compositions of vapor and liquid mixtures in equilibrium. This topic is discussed here with particular reference to the preparation of mixture vapor–liquid equilibrium phase diagrams, partial vaporization and condensation calculations, and the use of vapor–liquid equilibrium measurements to get information on the partial molar properties of mixtures.

The starting point for all vapor–liquid calculations is the equilibrium criterion

$$\bar{f}_i^{\mathrm{L}}(T, P, x_i) = \bar{f}_i^{\mathrm{V}}(T, P, y_i)$$

where the superscripts L and V refer to the liquid and vapor phases, respectively. From the entries in Table 7.6-3, it is clear that to compute the fugacity of a species in a vapor, $\bar{f}_i^V(T, P, y_i)$ we must use either an equation of state, less accurately the principle of corresponding states, or, still less accurately a simplifying assumption such as the Lewis–Randall rule or the ideal gas mixture model. All these methods have the same essential basis, the equation of state of the gas mixture, and represent different assumptions in the implementation. For the fugacity of a species in a liquid, $\bar{f}_i^L(T, P, x_i)$, we have, however, two different ways of proceeding; one based on activity coefficient or excess Gibbs free energy models, and another based on the equation of state description of the liquid phase.

These two descriptions represent different methods of analysis of the equilibrium problem, and hence we will consider them separately. The activity coefficient description will be considered here, and the equation of state description in Sec. 8.2. However, before proceeding with the discussion it is again useful to consider the range of validity of the two models. The activity coefficient (or excess Gibbs free energy) models can be used for liquid mixtures of all species. However, this description generally does not contain density, and therefore will not give a good description of an expanded liquid, as occurs near the vapor–liquid critical point of a mixture. Also, when two different models are used, such as an activity coefficient model for the liquid phase and an equation of state model for the vapor phase, the properties of the two phases cannot become identical, so that usually the vapor–liquid critical region behavior is predicted incorrectly.

When using an identical equation of state description for both phases, good phase equilibrium predictions can be made over a wide range of temperatures and pressures, including near the critical region. Furthermore, not only can the compositions of the coexisting phases be predicted from an equation of state, but so can other properties such as their densities and enthalpies. Unfortunately, at present only mixtures of hydrocarbons, inorganic gases, and a few other substances can be described using the simple equations of state we have considered in this book. Consequently, the present state is that reasonably accurate predictions or correlations of vapor–liquid equilibrium can be made for hydrocarbons and inorganic gases at all conditions using equations of state and for other fluids at low to moderate pressures (i.e., away from the mixture critical point) using an activity coefficient model for the liquid and an equation of state for the vapor.

At low pressures we can use the activity coefficient description and the Lewis–Randall rule to obtain

$$x_i\gamma_i(T, P, x_i)P_i^{\text{vap}}(T)\left(\frac{f}{P}\right)_{\text{sat},i} = y_iP\left(\frac{f}{P}\right)_i \tag{8.1-1a}$$

This equation provides a relation between the compositions in the two coexisting phases. Summing Eq. 8.1-1 over all species yields

$$\sum_{i=1}^{\mathscr{C}} x_i\gamma_i(T, P, x_i)P_i^{\text{vap}}(T)\left(\frac{f}{P}\right)_{\text{sat},i} = P\sum_{i=1}^{\mathscr{C}} y_i\left(\frac{f}{P}\right)_i \tag{8.1-2a}$$

There are several simplifications that can be made in these equations. First, if the total pressure and the vapor pressures of the species are sufficiently low that all fugacity coefficient corrections are negligible, we have

$$x_i \gamma_i(T, P, x_i) P_i^{\text{vap}}(T) = y_i P \tag{8.1-1b}$$

and, since $\Sigma y_i = 1$,

$$\sum x_i \gamma_i(T, P, x_i) P_i^{\text{vap}}(T) = P \tag{8.1-2b}$$

Further, if the liquid phase forms an ideal mixture (i.e., $\gamma_i = 1$ for all species), these equations further reduce to

$$x_i P_i^{\text{vap}}(T) = y_i P = P_i \tag{8.1-3}$$

and

$$\sum x_i P_i^{\text{vap}}(T) = \sum P_i = P \tag{8.1-4}$$

where $P_i = y_i P$ is the partial pressure of species i in the vapor phase. Equation 8.1-3, which is known as **Raoult's law,** indicates that the partial pressure of a component in an ideal solution is equal to the product of the species mole fraction and its pure component vapor pressure.

Equations 8.1-1 and 2 or their simplifications here, together with the restrictions that

$$\sum_i x_i = 1 \tag{8.1-5}$$

and

$$\sum_i y_i = 1 \tag{8.1-6}$$

and the mass and energy balance equations, are the basic relations for all vapor–liquid equilibrium calculations.

As the first illustration of the use of these equations consider vapor–liquid equilibrium in the hexane–triethylamine system at 60°C. These species form an essentially ideal mixture. The vapor pressures of hexane and triethylamine at this temperature are 0.7583 bar and 0.3843 bar, respectively, so that the fugacity coefficients at saturation and for the vapor phase can be neglected. Consequently, Eqs. 8.1-3 and 4 should be applicable to this system. The three solid lines in Fig. 8.1-1a represent the two species partial pressures and the total

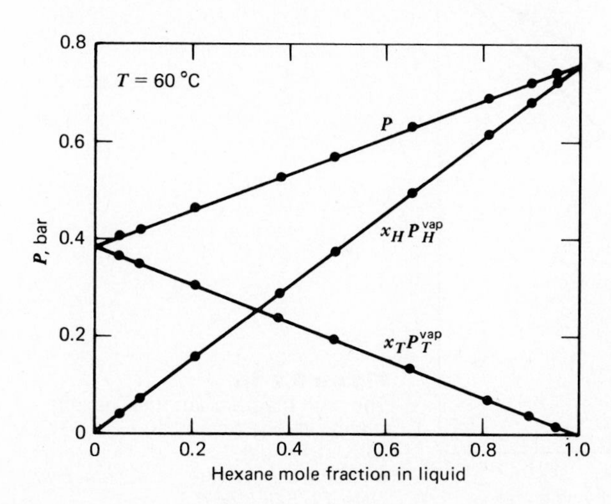

Figure 8.1-1a
Equilibrium total pressure and species liquid phase fugacities ($x_i P_i^{\text{vap}}$) versus mole fraction for the essentially ideal system to hexane–triethylamine at 60°C. [Based on data of J. L. Humphrey and M. Van Winkle, *J. Chem. Eng. Data* **12,** 526 (1967).]

pressure as a function of liquid phase composition using these equations; the dots are the experimental results. The close agreement between the computations and the laboratory data indicates that the hexane–triethylamine system is ideal.

One interesting characteristic of Fig. 8.1-1*a* is that the equilibrium pressure (at fixed temperature) is a linear function of the liquid–phase mole fraction. This is true only for ideal mixtures.

Once the equilibrium total pressure has been computed for a given liquid composition using Eqs. 8.1-2 or 4, the equilibrium composition of the vapor can be calculated using Eqs. 8.1-1 or 3, as appropriate. Indeed, we can prepare a complete vapor–liquid equilibrium composition diagram, or x-y diagram, at constant temperature, by choosing a collection of values for the composition of one of the phases, say x_i, and then using the vapor pressure data (and, if the mixture is nonideal, activity coefficient data) to compute the total pressure and value of y_i corresponding to each x_i. For example, again consider the hexane–triethylamine system just discussed. To calculate the composition of the vapor in equilibrium with a 50 mole % hexane mixture at T = 60°C, Eq. 8.1-4 is first used to compute the equilibrium pressure

$$P = \sum x_i P_i^{\text{vap}} = x_H P_H^{\text{vap}} + x_T P_T^{\text{vap}}$$

$$= 0.5 \times 0.7583 + 0.5 \times 0.3843 = 0.5713 \text{ bar}$$

and then Eq. 8.1-3 is used to calculate the vapor phase mole fraction

$$y_H = \frac{x_H P_H^{\text{vap}}}{P} = \frac{0.5 \times 0.7583}{0.5713} = 0.6637$$

By choosing other liquid compositions and repeating the calculation, the complete vapor–liquid equilibrium composition diagram or x-y diagram can be constructed. The results are shown in Fig. 8.1-1*b* along with points representing the experimental data. The second line in this figure is the line $x = y$; the greater the difference between the x-y curve of the mixture and the $x = y$ line, the larger the difference in composition between the liquid and vapor phases and the easier it

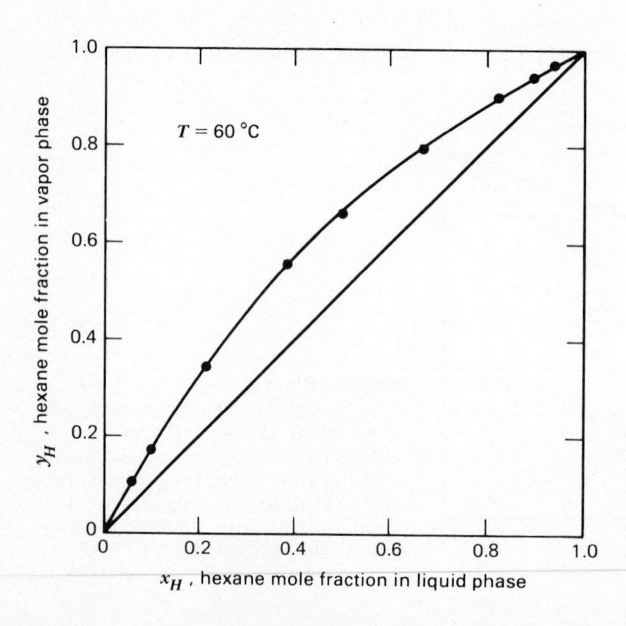

Figure 8.1-1*b* The x-y diagram for the hexane–triethylamine system at T = 60°C. [Based on data of J. L. Humphrey and M. Van Winkle, *J. Chem Eng Data* **12,** 526 (1967).]

is to separate the two components by distillation. (Since x-y diagrams are most often used in the study of distillation, it is common practice to include the $x = y$ line.)

An alternate way of presenting vapor–liquid equilibrium data is to plot, on a single figure, the equilibrium pressure and the compositions for both phases at fixed temperature. This has been done for the hexane–triethylamine system in Fig. 8.1-1*c*. In this figure the equilibrium compositions of the vapor and liquid at a given pressure are given by the curves labeled vapor and liquid, respectively; the compositions of the two coexisting phases at a given pressure are given by the intersection of a horizontal line (i.e., a line of constant pressure) with the vapor and liquid curves. The term **tie line** will be used here, and generally in this chapter, to indicate a line connecting the equilibrium compositions in two coexisting phases. The tie line drawn in Fig. 8.1-1*c* shows that at 60°C, a liquid containing 50 mole % hexane is in equilibrium with a vapor containing 66.37 mole % hexane at 0.5713 bar.

So far the discussion has been specific to systems at constant temperature; equivalently, pressure could be fixed and temperature and phase composition be taken as the variables. Although much experimental vapor–liquid equilibrium data are obtained in constant-temperature experiments, distillation columns and other vapor–liquid separations equipment in the chemical process industry are operated more nearly at constant pressure. Therefore, it is important that chemical engineers be familiar with both types of calculations.

The vapor–liquid equilibrium temperature for fixed pressure and composition is found as the solution to Eqs. 8.1-2 or, if the system is ideal, as the solution to Eq. 8.1-4. However, since the temperature appears only implicitly in these equations through the species vapor pressures,[1] and since there is a nonlinear relationship between the vapor pressure and temperature (cf., the Clausius–

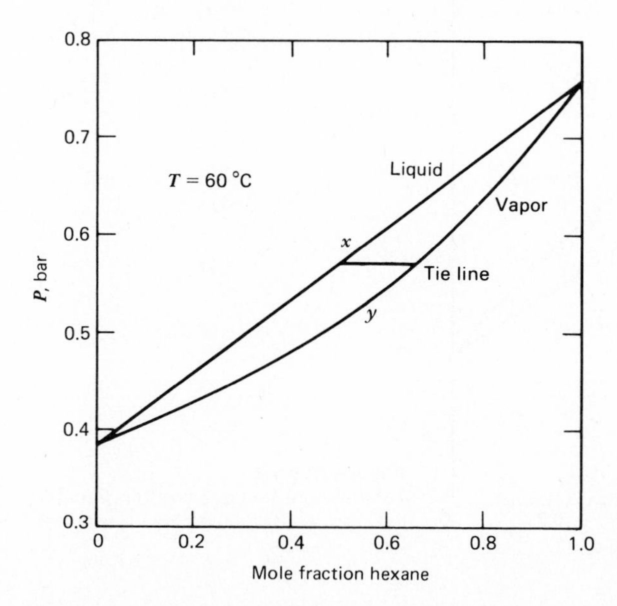

Figure 8.1-1c
Pressure-composition diagram for the hexane–triethylamine system at fixed temperature.

[1]In fact, the species activity coefficients also depend on temperature; see Eq. 7.3-22. However, since this temperature dependence is usually small compared to the temperature variation of the vapor pressure, it is neglected here.

Clapeyron equation, Eq. 5.7-5*a*), these equations are usually solved by iteration. That is, one guesses a value of the equilibrium temperature, computes the value of the vapor pressure of each species at this temperature, and then tests whether the pressure computed from Eqs. 8.1-2 (or Eq. 8.1-4 if the system is ideal) is in agreement with the fixed pressure. If the two agree, the guessed equilibrium temperature is correct, and the vapor phase mole fractions can be computed from Eq. 8.1-1 (or, if the system is ideal, from Eq. 8.1-3.)[2] If the two pressures do not agree, a new trial temperature is chosen, and the calculation repeated. Figure 8.1-1d is a plot, on a single graph, of the equilibrium temperature and mole fractions for the hexane–triethylamine system at 0.7 bar calculated in this way, and Fig. 8.1-1*e* is the x-y diagram for this system. Note that tie lines drawn on Fig. 8.1-1*d* are again horizontal lines.

Figures 8.1-1*c* and *d* are two-dimensional sections of the three-dimensional phase diagram of Figure 8.1-1*f*. The intersections of this three-dimensional equilibrium surface with planes of constant temperature (the vertical, unshaded planes) produce two-dimensional figures such as Figure 8.1-1*c*, whereas the intersection of a plane of constant pressure (horizontal, shaded plane) results in a diagram such as Figure 8.1-1*d*.

Since few liquid mixtures are ideal, vapor–liquid equilibrium calculations can be considerably more complicated than is the case for the hexane–triethylamine system, and the system phase diagrams can be more structured than Fig. 8.1-1*f*. These complications arise from the (nonlinear) composition dependence of the species activity coefficients. For example, as a result of the composition dependence of γ_i, the equilibrium pressure in a fixed temperature experiment

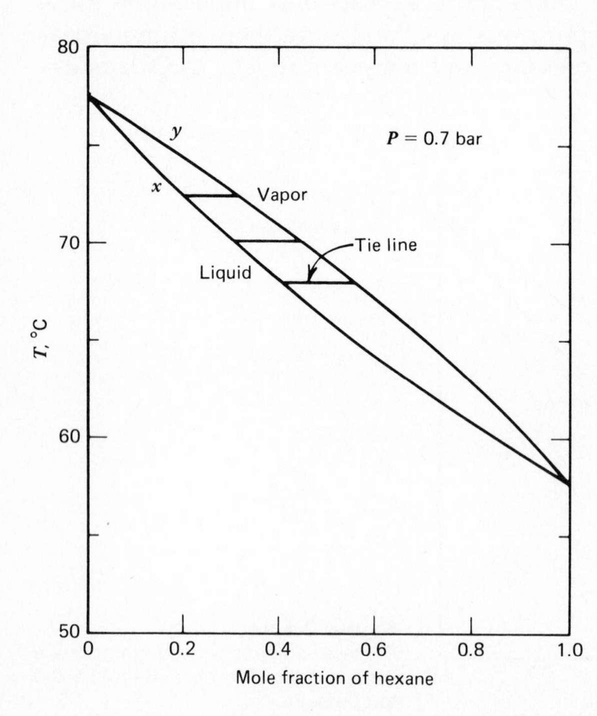

Figure 8.1-1*d*
Temperature-composition diagram for the hexane–triethylamine system at fixed pressure.

[2]If the vapor phase mole fractions calculated in this way do not sum to 1 (i.e., $\Sigma\, y_i \neq 1$) then either (a) only a single phase, vapor or liquid, is present at equilibrium or (b) both phases are present, but the activity coefficient data used in the calculation are inaccurate.

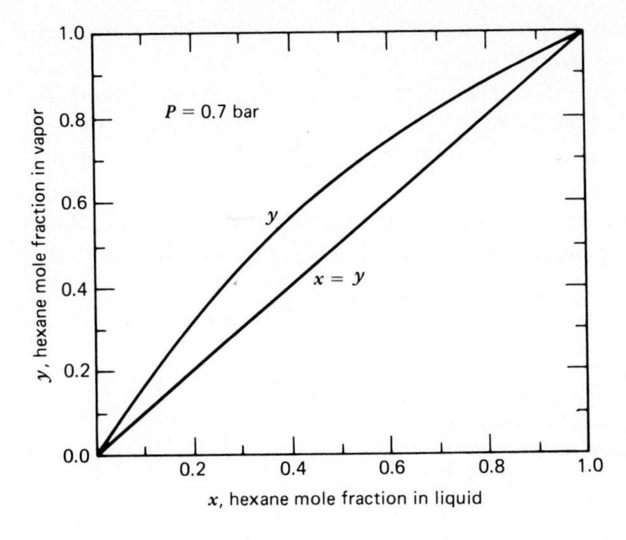

Figure 8.1-1e
The *x-y* diagram for the hexane–triethylamine system at a constant pressure of 0.7 bar.

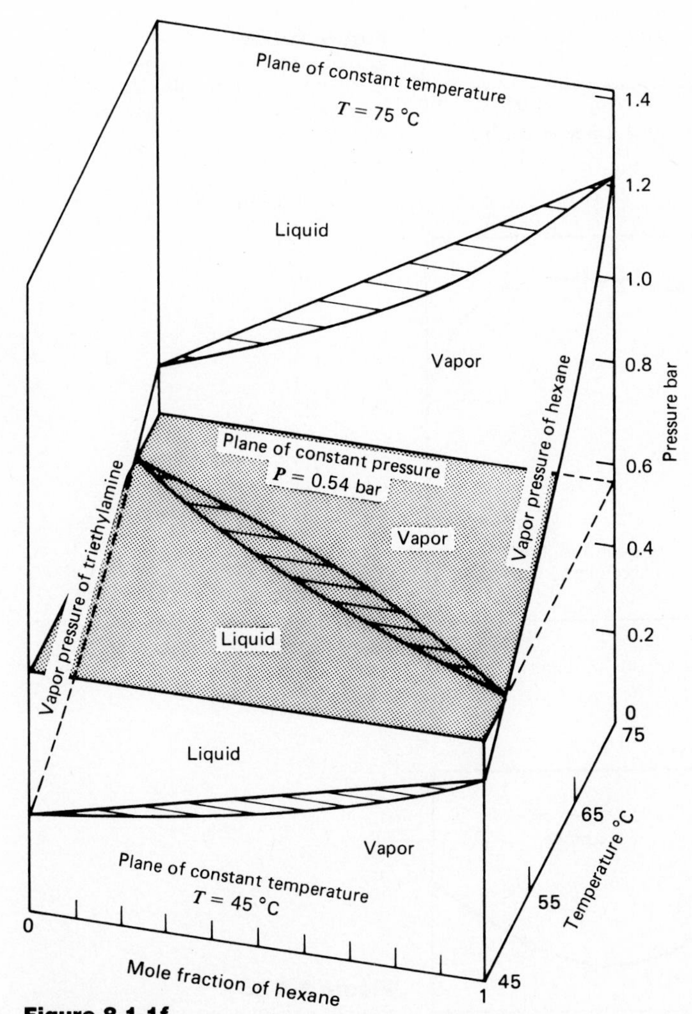

Figure 8.1-1f
Vapor–liquid equilibria to hexane + triethylamine.

will no longer be a linear function of mole fraction (see Eq. 8.1-1). Thus nonideal solutions exhibit deviations from Raoult's law.

Several examples of experimental data for nonideal solutions are given in Fig. 8.1-2. It is easy to establish that if

$$P > \sum x_i P_i^{\text{vap}}$$

(see Figs. 8.1-2*a* and *b*), $\gamma_i > 1$ for at least one of the species in the mixture, that is, $P_i > x_i P_i^{\text{vap}}$ (so that positive deviations from Raoult's law occur). Similarly, in real solutions

$$P < \sum x_i P_i^{\text{vap}}$$

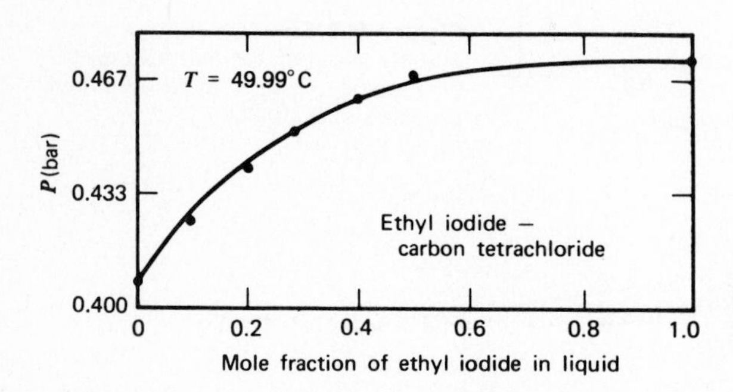

Figure 8.1-2*a* Equilibrium total pressure versus composition for the ethyl iodine–carbon tetrachloride system at T = 49.99°C.

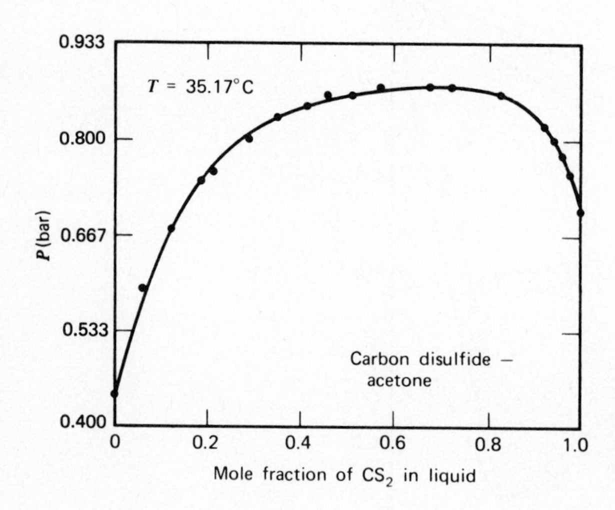

Figure 8.1-2*b* Equilibrium total pressure versus composition for the carbon disulfide–acetone system at T = 35.17°C.

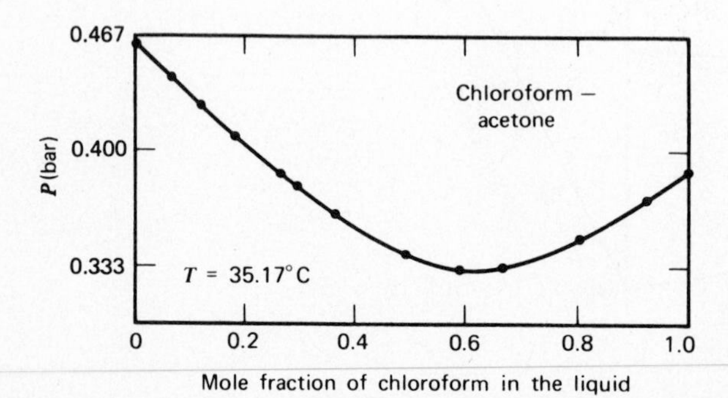

Figure 8.1-2c Equilibrium total pressure versus composition for the chloroform–acetone system at T = 35.17°C.

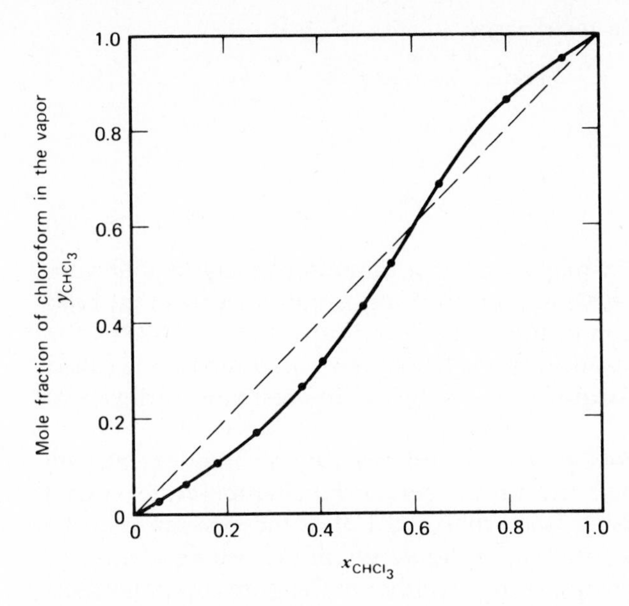

Figure 8.1-2d
The *x-y* diagram for the chloroform–acetone system at T = 35.17°C.

(see Fig. 8.1-2*c*) occurs when the activity coefficients for one or more of the species in the mixture are less than one, that is, when there are negative deviations from Raoult's law.

Of special interest are binary mixtures in which the total pressure versus liquid composition curve exhibits an extremum (either a maximum or a minimum), as illustrated in Figs. 8.1-2*b* and *c*. To locate this extremum we set $(\partial P/\partial x_1)_T = 0$, keeping in mind that $x_2 = 1 - x_1$, to get

$$\gamma_1 P_1^{\text{vap}}\left(1 + x_1 \frac{\partial \ln \gamma_1}{\partial x_1}\right) - \gamma_2 P_2^{\text{vap}}\left(1 + x_2 \frac{\partial \ln \gamma_2}{\partial x_2}\right) = 0 \tag{8.1-7}$$

To proceed further we note that the Gibbs–Duhem equation given in Eq. 7.3-15

$$0 = \frac{\underline{H}^{ex}}{T}\, dT - \underline{V}^{ex}\, dP + RT \sum_{i=1}^{\mathscr{C}} x_i\, d \ln \gamma_i$$

can be rewritten as

$$0 = \frac{\underline{H}^{ex}}{T}\left(\frac{\partial T}{\partial x_1}\right)_T - \underline{V}^{ex}\left(\frac{\partial P}{\partial x_1}\right)_T + RT\, x_1 \left(\frac{\partial \ln \gamma_1}{\partial x_1}\right)_T + RT\, x_2 \left(\frac{\partial \ln \gamma_2}{\partial x_1}\right)_T \tag{8.1-8}$$

Next, using $dx_1 = -dx_2$, and observing that $(\partial T/\partial x_1)_T = 0$ since temperature is fixed, and that $(\partial P/\partial x_1)_T = 0$ at an extreme value of the pressure, yields

$$x_1 \left(\frac{\partial \ln \gamma_1}{\partial x_1}\right)_T + x_2 \left(\frac{\partial \ln \gamma_2}{\partial x_1}\right)_T = 0 \tag{8.1-9}$$

Finally, using this expression in Eq. 8.1-7 and recognizing that $(1 + x_1\, \partial \ln \gamma_1/\partial x_1)_T \neq 0$ (the results of Problem 7.16 provide a verification of this fact), establishes that when $(\partial P/\partial x_1)_T = 0$,

$$\gamma_1 P_1^{\text{vap}} - \gamma_2 P_2^{\text{vap}} = 0 \tag{8.1-10}$$

or, from Eq. 8.1-1b, that

$$\frac{y_1}{x_1} = \frac{\gamma_1 P_1^{\text{vap}}}{P} = \frac{\gamma_2 P_2^{\text{vap}}}{P} = \frac{y_2}{x_2} \tag{8.1-11}$$

This equation, together with the restrictions

$$x_1 + x_2 = 1 \quad \text{and} \quad y_1 + y_2 = 1$$

has the solution

$$x_1 = y_1$$
$$x_2 = y_2 \tag{8.1-12}$$

Thus, the occurrence of either a maximum or a minimum in the equilibrium pressure versus mole fraction curve at a given composition indicates that both the vapor and liquid are of this composition (this is indicated in Fig. 8.1-2*d* by the intersection of the x-y and $x = y$ curves). Such a vapor–liquid mixture is called an **azeotrope** or an **azetropic mixture,** and is of special interest (and annoyance) in distillation processes.

The occurrence of an azeotrope is also indicated by an interior extreme value of the equilibrium temperature on an equilibrium temperature versus composition curve at fixed pressure (see Problem 8.1-3). If the extreme value of temperature is a maximum (i.e., the vapor–liquid equilibrium temperature of the mixture is greater than the boiling points of either of the pure components at the chosen pressure), then the mixture is said to be a **maximum-boiling** azeotrope; a maximum boiling azeotrope occurs when there are negative deviations from Raoult's law; that is, when the activity coefficient of one or more species is less than one. Similarly, if the azeotropic temperature is below the boiling points of the pure components, the mixture is a **minimum-boiling** azeotrope, the activity coefficient of at least one of the species in the mixture is greater than unity, and there are positive deviations from Raoult's law.

At a low pressure azeotropic point the liquid phase activity coefficients can be calculated from the relation

$$\gamma_i = \frac{P}{P_i^{\text{vap}}} \tag{8.1-13}$$

which results from Eqs. 8.1-11 and 12. Because of the simplicity of this result, azeotropic data are frequently used to evaluate liquid-phase activity coefficients. In particular, given the azeotropic composition, temperature, and pressure, one can calculate the liquid-phase activity coefficients at the azeotropic composition. This information can be used to obtain the values of the parameters in a liquid solution model (e.g., the van Laar model), which, in turn, can be used to calculate the complete pressure-composition and x-y diagrams for the system. This procedure is illustrated next.

ILLUSTRATION 8.1-1

Benzene and cyclohexane form an azeotrope at 0.525 mole fraction benzene at a temperature of 77.6°C and a total pressure of 1.013 bar. At this temperature the vapor pressure of pure benzene is 0.993 bar, and that of pure cyclohexane is 0.980 bar.

a. Using the van Laar model, estimate the activity coefficients of benzene and cyclohexane over the whole composition range. Use this activity coefficient information to compute the equilibrium pressure versus liquid composition and equilibrium vapor composition versus liquid composition curves at 77.6°C.

b. Make predictions for the activity coefficients of benzene and cyclohexane using regular solution theory, and compare these with the results obtained in part a.

Solution

a. The van Laar Model
The starting point is the equilibrium relation

$$\bar{f}_i^{\mathrm{L}}(T, P, x_i) = \bar{f}_i^{\mathrm{V}}(T, P, y_i)$$

which, at the pressures here, reduces to

$$x_i \gamma_i P_i^{\mathrm{vap}} = y_i P$$

Since $x_i = y_i$ at an azeotropic point, we have $\gamma_i = P/P_i^{\mathrm{vap}}$, so that

$$\gamma_{\mathrm{B}} = \frac{1.013}{0.993} = 1.020$$

$$\gamma_{\mathrm{C}} = \frac{1.013}{0.980} = 1.034$$

and, using Eqs. 7.5-10, we obtain $\alpha = 0.126$ and $\beta = 0.0916$. Therefore,

$$\ln \gamma_{\mathrm{B}} = \frac{0.126}{\left[1 + 1.376\left(\frac{x_{\mathrm{B}}}{x_{\mathrm{C}}}\right)\right]^2} \quad \text{and} \quad \ln \gamma_{\mathrm{C}} = \frac{0.0916}{\left[1 + 0.727\left(\frac{x_{\mathrm{C}}}{x_{\mathrm{B}}}\right)\right]^2} \tag{i}$$

The values of the activity coefficients for benzene (B) and cyclohexane (C) calculated from these equations are given in the following table and figure.

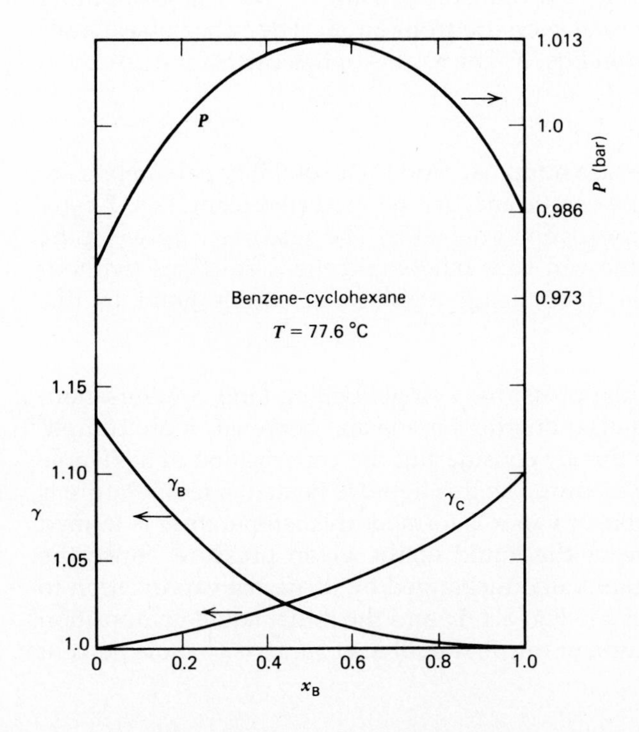

	van Laar				Regular Solution	
x_B	γ_B	γ_C	y_B	P(bar)	γ_B	γ_C
0	1.13	1.00	0	0.980	1.14	1.00
0.1	1.10	1.00	0.110	0.992	1.11	1.00
0.2	1.07	1.01	0.212	1.001	1.09	1.00
0.3	1.05	1.01	0.311	1.006	1.07	1.01
0.4	1.03	1.02	0.406	1.012	1.06	1.02
0.5	1.02	1.03	0.501	1.013	1.04	1.03
0.525	1.02	1.03	0.525	1.013	1.04	1.04
0.6	1.01	1.04	0.596	1.010	1.03	1.05
0.7	1.01	1.05	0.693	1.006	1.02	1.07
0.8	1.00	1.07	0.792	1.005	1.01	1.10
0.9	1.00	1.08	0.894	1.000	1.01	1.13
1.0	1.00	1.10	1.0	0.993	1.00	1.17

To compute the composition of the vapor in equilibrium with the liquid we use Eqs. 8.1-1b

$$x_B \gamma_B P_B^{vap} = y_B P \qquad \text{and} \qquad x_C \gamma_C P_C^{vap} = y_C P \tag{ii}$$

and Eq. 8.1-2b

$$x_B \gamma_B P_B^{vap} + x_C \gamma_C P_C^{vap} = P \tag{iii}$$

In these equations the vapor composition, y_B and y_C, and the equilibrium pressure P are unknown (the equilibrium pressure is 1.013 bar only at $x_B = 0.525$). The solution is obtained by choosing a value of x_B, using $x_C = 1 - x_B$, computing γ_B and γ_C from Eqs. i, and the total pressure from Eq. iii. The vapor phase mole fractions are then computed from Eqs. ii. The results of this calculation are given in the table.

b. Regular Solution Theory
Since benzene and cyclohexane are nonpolar, and their solubility parameters are given in Table 7.6-1, the activity coefficients can be predicted using Eqs. 7.6-10. The results of this calculation are given in the table. The agreement between the correlation of the data using the van Laar model and the predictions (without reference to the experimental data) using regular solution is good in this case. ■

A pure fluid has, at a given pressure, a single boiling (and condensation) temperature; boiling and condensation phenomena are, however, more complicated in mixtures. We can see this by considering the vaporization of a 50 mole percent hexane–triethylamine mixture. As this liquid is heated, a temperature is reached at which the first bubble of vapor is formed; this temperature is termed the **bubble-point temperature** of the liquid at the given pressure. Since the composition of the liquid is essentially unchanged by its partial vaporization to form one small bubble, we can use Fig. 8.1-1*d* and the initial liquid composition to determine that the composition of this first bubble of vapor is 66 mole percent hexane and the bubble-point temperature is 66.04°C. Since the vapor is richer in hexane than the liquid mixture, the liquid will be depleted in hexane as the boiling proceeds. Thus, as more and more liquid vaporizes, the liquid will become increasingly more dilute in hexane and its boiling temperature will increase.

Conversely, we can consider the problem of condensing a 50 mole percent hexane–triethylamine vapor mixture. As the vapor temperature decreases at fixed pressure, a temperature, called the **dew-point temperature,** is reached at which the first drop of vapor condenses. Since the condensation to form a single small drop of liquid leaves the vapor virtually unchanged, we can use Fig. 8.1-1*d* to determine that, at 0.7 bar, the first drop of condensate will appear at about 69.26°C, and its composition will be about 34.2 mole percent hexane. Clearly, as the condensation process continues, the vapor will become increasingly richer in hexane, and the equilibrium condensation temperature will decrease.

Thus, boiling and condensation phenomena at fixed pressure, which occur at a single temperature in a pure fluid, take place over the range of temperatures between the dew point and bubble point in a mixture. Generally, the dew-point and bubble-point temperatures differ by many degrees (see Illustration 8.1-2); the two are close for the hexane–triethylamine system because the species vapor pressures are close and the components form an ideal mixture.

If constant-pressure vapor–liquid equilibrium diagrams, such as Fig. 8.1-1d, have been previously prepared, dew-point and bubble-point temperatures can easily be found. Generally, such information is not available, and the trial and error procedure indicated in Illustration 8.1-2 is used to estimate these temperatures. (The calculation of the dew-point and bubble-points *pressures* at fixed temperature is considered in Problem 8.9-13).

ILLUSTRATION 8.1-2

Estimate the bubble point and dew point temperatures of a 25 mole percent *n*-pentane, 45 mole percent *n*-hexane, and 30 mole percent *n*-heptane mixture at 1.013 bar (1 atm).

Data

$$\ln P_5^{\text{vap}} = 10.442 - \frac{26799}{RT} \qquad \delta_5 = 7.02\ (\text{cal/cc})^{1/2}$$

$$\ln P_6^{\text{vap}} = 10.456 - \frac{29676}{RT} \qquad \delta_6 = 7.27\ (\text{cal/cc})^{1/2}$$

$$\ln P_7^{\text{vap}} = 11.431 - \frac{35200}{RT} \qquad \delta_7 = 7.43\ (\text{cal/cc})^{1/2}$$

for P in bar, T in K, and $R = 8.314$ J/mol K. The subscripts 5, 6, and 7 designate pentane, hexane, and heptane, respectively.

Solution

Before solving this problem it is useful to check whether this problem is, in fact, solvable. We can get this information from the Gibbs phase rule. At the dew point or bubble point there are 3 components, 2 phases, and no chemical reactions, so that there are

$$\mathcal{F} = 3 - 0 - 2 + 2 = 3$$

degrees of freedom. Since the pressure and two independent mole fractions of one phase have been fixed, the problem is well-posed and, in principle, solvable.

Since the solubility parameters for these hydrocarbons are sufficiently close, we will assume the liquid mixture is ideal; that is, $\bar{f}_i^L = x_i P_i^{vap}$. (This assumption should be reasonably accurate here, and greatly simplifies the calculations.) Finally, since the pressure is so low, we will assume the vapor phase is ideal, so that $\bar{f}_i^V = y_i P$.

a. At the bubble point of the liquid mixture Eqs. 8.1-3, 4, and 6 must be satisfied. Therefore, the procedure is to

i. Pick a trial value of the bubble-point temperature.

ii. Compute the values of the y_i from

$$y_i = \frac{P_i^{vap}}{P} x_i$$

iii. If $\Sigma y_i = 1$, the trial value of T is the bubble-point temperature. If $\Sigma y_i > 1$, T is too high, if $\Sigma y_i < 1$, T is too low; in either case adjust the value of T and go back to step ii.

Following this calculational procedure, we find

$$T \text{ (bubble point)} = 334.6 \text{ K}$$

$$y_5 = 0.554$$

$$y_6 = 0.359$$

$$y_7 = 0.087$$

b. To find the dew point of the vapor mixture, we

i. Pick a trial value for the dew-point temperature.

ii. Compute the values of the liquid phase composition from

$$x_i = y_i \frac{P}{P_i^{vap}}$$

iii. If $\Sigma x_i = 1$, the trial value of T is the dew-point temperature. If $\Sigma x_i > 1$, T is too low, if $\Sigma x_i < 1$, T is too high, in either case adjust the value of T and go back to step ii.

In this case we find

$$T \text{ (dew point)} = 350.5 \text{ K}$$

$$x_5 = 0.073$$

$$x_6 = 0.347$$

$$x_7 = 0.580$$

Comment

Note that the dew point and bubble point differ by 16 K for this mixture. ■

Another type of vapor–liquid equilibrium problem is the calculation of two-phase equilibrium state when either a liquid of known composition is partially vaporized or a vapor is partially condensed as a result of a change in temperature and/or pressure. Such problems are generically referred to as flash problems. To solve such problems one uses the equilibrium criterion, Eq. 8.1-1, the restrictions of Eqs. 8.1-5 and 6, the state variable constraints, the species mass balance equations, and, if the process is not at constant temperature, the energy balance equation.

For a process in which 1 mole of a mixture with species mole fractions x_1°, $x_2^\circ, \ldots, x_\mathscr{C}^\circ$ is, by partial vaporization or condensation, separated into L moles of liquid of composition $x_1, x_2, \ldots, x_\mathscr{C}$, and V moles of vapor of composition $y_1, y_2, \ldots, y_\mathscr{C}$, the species mass balance yields

$$x_i L + y_i V = x_i^\circ \qquad i = 1, 2, \ldots, \mathscr{C} \tag{8.1-14}$$

since no chemical reactions occur. From the total mass balance we also have

$$L + V = 1 \tag{8.1-15}$$

though this is not an *independent* equation, since it can be obtained by summing Eq. 8.1-14 over all species, and using Eqs. 8.1-6.

Equations 8.1-14 and 15, together with the equilibrium relations, can be used to solve problems involving partial vaporization and condensation processes at constant temperature. For partial vaporization and condensation processes that occur adiabatically, as is frequently the case in the chemical industry, the final temperature of the vapor–liquid mixture is also unknown and must be found as part of the solution. This is done by including the energy balance among the equations to be solved. Since the isothermal partial vaporization or isothermal flash calculation is already tedious (see Illustration 8.1-3), the adiabatic partial vaporization (or adiabatic flash) problem will not be considered here.[3]

ILLUSTRATION 8.1-3

A liquid mixture of 25 mole percent *n*-pentane, 45 mole percent *n*-hexane, and 30 mole percent *n*-heptane, initially at 69°C and a high pressure, is partially vaporized by isothermally lowering the pressure to 1.013 bar (1 atm). Find the relative amounts of vapor and liquid in equilibrium and their compositions.

Solution

From the Antoine equation data in the preceding problem, we have

$$P_5^{\text{vap}} = 2.735 \text{ bar} \qquad P_6^{\text{vap}} = 1.013 \text{ bar} \qquad \text{and} \qquad P_7^{\text{vap}} = 0.387 \text{ bar}$$

Also, using the simplifications for this system introduced in the previous illustration, the equilibrium relation $\bar{f}_i^{\text{L}} = \bar{f}_i^{\text{V}}$ reduces to $x_i P_i^{\text{vap}} = y_i P$, or

$$\frac{y_i}{x_i} = \frac{P_i^{\text{vap}}}{P}$$

For convenience, we will use the **K-factor** defined by the relation $y_i = K_i x_i$ in the calculations; here $K_i = P_i^{\text{vap}}(T)/P$. Thus we obtain the following three equations:

$$y_5 = x_5 K_5 \qquad \text{where } K_5 = 2.70 \tag{1}$$

$$y_6 = x_6 K_6 \qquad K_6 = 1.00 \tag{2}$$

$$y_7 = x_7 K_7 \qquad K_7 = 0.382 \tag{3}$$

[3]Flash vaporization processes are usually considered in books on mass transfer processes and stagewise operations. See, for example, R. E. Treybal, *Mass Transfer Operations* 3rd ed., McGraw–Hill, N.Y., 1980, p. 363 e seq., and C. J. King, *Separation Processes*, 2nd ed. McGraw–Hill, N.Y., 1980, pp. 68–90

We also have, from Eqs. 8.1-5, 6, 14, and 15, that

$$x_5 + x_6 + x_7 = 1 \tag{4}$$

$$y_5 + y_6 + y_7 = 1 \tag{5}$$

and

$$x_5L + y_5V = 0.25 \tag{6}$$

$$x_6L + y_6V = 0.45 \tag{7}$$

$$x_7L + y_7V = 0.30 \tag{8}$$

$$L + V = 1.0 \tag{9}$$

Thus we have eight independent equations for eight unknowns (x_5, x_6, x_7, y_5, y_6, y_7, L, and V), and any numerical procedure for solving linear algebraic equations may be used to solve this set of equations. One method is to use Eqs. 1, 2, and 3 to eliminate the vapor-phase mole fractions and the overall mass balance, Eq. 9, to eliminate the total amount of vapor. In this way the eight algebraic equations are reduced to five linear algebraic equations

$$x_5 + x_6 + x_7 = 1 \tag{4}$$

$$K_5x_5 + K_6x_6 + K_7x_7 = 1 \tag{5'}$$

$$x_5[L + K_5(1 - L)] = x_5[L(1 - K_5) + K_5] = 0.25 \tag{6'}$$

$$x_6[L(1 - K_6) + K_6] = 0.45 \tag{7'}$$

$$x_7[L(1 - K_7) + K_7] = 0.30 \tag{8'}$$

These equations are most easily solved by trial and error. In particular, a value of L is guessed, used in Eqs. 6′–8′ to compute x_5, x_6, and x_7. These trial values of the liquid-phase mole fractions are then tested in Eqs. 4 and 5′. If those equations are satisfied, the guessed value of L and the computed values of the x_i's are correct, and the vapor-phase mole fractions can be computed from Eqs. 1–3. If Eqs. 4 and 5′ are not satisfied, a new guess for L is made, and the procedure repeated.

Our solution for this problem is

$$L = 0.585 \qquad V = 0.415$$

$$x_5 = 0.147 \qquad y_5 = 0.396$$

$$x_6 = 0.450 \qquad y_6 = 0.450$$

$$x_7 = 0.403 \qquad y_7 = 0.154$$

Comments

1. The K-factor formulation introduced in this calculation is frequently useful in solving vapor–liquid equilibrium problems. The procedure is easily generalized to nonideal liquid and vapor phases as follows:

   $$\frac{y_i}{x_i} \equiv K_i = \frac{\gamma_i P_i^{\text{vap}}}{\phi_i P}$$

   In this case K_i is a nonlinear function of the liquid phase mole fraction through the activity coefficient γ_i, and also a function of the vapor phase composition through the fugacity coefficient ϕ_i.

2. It was not necessary to assume ideal solution behavior to solve this problem. One could, for example, assume that the solution was regular, in

which case γ_i (and K_i) would be a transcendental function of the mole fractions. The calculation of the vapor- and liquid-phase mole fractions is then considerably more complicated than was the case here; the basis of the calculation, the equality of the fugacity of each species in both phases, remains unchanged, however.

3. In some cases it may not be possible to find an *acceptable* solution to the algebraic equations. Here by acceptable we mean a solution such that each mole fraction, L, and V are all greater than zero and less than one. This difficulty occurs when the vapor and liquid phases cannot coexist in the equilibrium state. For example, if the flash vaporization temperature were sufficiently high or the total pressure so low that all the K_i's were greater than one, there would be no set of mole fractions that satisfies both Eqs. 4 and 5′. In this case only the vapor is present. Similarly, if all the K_i's are less than one (which occurs at low temperatures or high pressures), only the liquid is present, and again there is no acceptable solution to the equations.
4. For this system $\mathscr{C} = 3$, $\mathscr{P} = 2$, and $\mathscr{M} = 0$, so that, from the Gibbs phase rule, the number of degrees of freedom is

   $\mathscr{F} = 3 - 0 - 2 + 2 = 3$

   Since the equilibrium temperature and pressure were specified, one degree of freedom remains. If no further information about the system were given, that is, if one were asked to determine the equilibrium compositions of vapor and liquid for a pentane–hexane–heptane mixture at 69°C and 1.013 bar with no other restrictions, many different vapor and liquid compositions would be solutions to the problem. A problem that does not have a unique solution is said to be ill-posed. With the initial liquid composition given, however, the species mass balances (Eqs. 6, 7, and 8) provide the additional equations that must be satisfied to ensure that there will be no more than one solution to the problem. ■

Although the emphasis so far has been on the prediction of vapor–liquid equilibria, one can instead use accurate equilibrium measurements to compute liquid-phase activity coefficients and excess Gibbs free energies. This is demonstrated in the following illustration.

ILLUSTRATION 8.1-4

Weissman and Wood[4] have made very accurate measurements of vapor–liquid equilibria in benzene–2,2,4-trimethyl pentane mixtures over a range of temperatures. Their data for the vapor and liquid compositions and equilibrium total pressures at 55°C are given in the following table:

x_B	y_B	P (bar)
0.0819	0.1869	0.26892
0.2192	0.4065	0.31573
0.3584	0.5509	0.35463
0.3831	0.5748	0.36088
0.5256	0.6786	0.39105
0.8478	0.8741	0.43277
0.9872	0.9863	0.43641

[4]S. Weissman and S. E. Wood, *J. Chem. Phys.* **32,** 1153 (1960).

The vapor pressure of pure benzene at 55°C is 0.43596 bar, and that of 2,2,4-trimethyl pentane is 0.23738 bar.

a. Calculate the activity coefficients of benzene and 2,2,4-trimethyl pentane and $\underline{G}^{ex}$ at each of the experimental points.

b. Obtain smoothed values for the excess Gibbs free energy and activity coefficients for this system as a function of composition using the following procedure:

i. Assume an expansion like Eq. 7.5-6 or 8 for $\underline{G}^{ex}$.

ii. Use the data from part a to compute the expansion coefficients in this expression.

iii. Use the expansion coefficients to compute smoothed values of $\underline{G}^{ex}$

iv. Derive an expression for the activity coefficients for the assumed form of $\underline{G}^{ex}$ and compute smoothed values of the activity coefficients.

Solution

a. The condition for vapor–liquid equilibrium is

$$\bar{f}_i^L(T, P, x_i) = \bar{f}_i^V(T, P, y_i)$$

For the conditions here we can assume an ideal gas-phase mixture and neglect all fugacity and Poynting corrections (though Weissman and Wood included these in their analysis of the data), so that

$$x_i \gamma_i P_i^{vap} = y_i P$$

or

$$\gamma_i = y_i P / x_i P_i^{vap}$$

The activity coefficients calculated in this manner are given in the following table. Values of $\underline{G}^{ex}$ computed from

$$\underline{G}^{ex} = x_1 \bar{G}_1^{ex} + x_2 \bar{G}_2^{ex} = RT(x_1 \ln \gamma_1 + x_2 \ln \gamma_2)$$

are also given.

b. Weissman and Wood used the Redlich–Kister type expansion for the excess Gibbs free energy in the form

$$\underline{G}^{ex} = x_1 x_2 [a + b(x_1 - x_2) + c(x_1 - x_2)^2] \tag{1}$$

Since this equation is linear in the unknown parameters a, b, and c, it is easily fitted, by least squares analysis, to the experimental data. The results, in J/mol are

$$a = 1339.0, \qquad b = 419.45, \qquad \text{and} \qquad c = 109.83$$

Now using $x_i = N_i/(N_1 + N_2)$ we obtain

$$G^{ex} = N\underline{G}^{ex} = \frac{N_1 N_2}{(N_1 + N_2)} \left[a + b\left(\frac{N_1 - N_2}{N_1 + N_2}\right) + c\left(\frac{N_1 - N_2}{N_1 + N_2}\right)^2 \right]$$

and, from Eqs. 7.3-9 and 12,

$$\begin{aligned} \bar{G}_1^{ex} = RT \ln \gamma_1 &= \left(\frac{\partial G^{ex}}{\partial N_1}\right)_{T,P,N_2} \\ &= x_2^2[a + b(x_1 - x_2) + c(x_1 - x_2)^2] \\ &\quad + 2x_1 x_2^2[b + 2c(x_1 - x_2)] \end{aligned} \tag{2}$$

In a similar fashion one finds

$$RT \ln \gamma_2 = x_1^2[a + b(x_1 - x_2) + c(x_1 - x_2)^2]$$
$$-2x_1^2 x_2[b + 2c(x_1 - x_2)] \tag{3}$$

Equations 2 and 3, with the parameter values given above, were used to compute the smoothed activity coefficient values and the calculated values for $\underline{G}^{ex}$ given in the table and plotted in Fig. 1. In Fig. 2 the partial pressures of each species (i.e., $P_B = y_B P$ and $P_{TMP} = y_{TMP} P$) are plotted as a function of the liquid-phase composition. The dashed lines indicate the behavior to be expected if the solution were ideal (i.e., if Raoult's law were obeyed).

	Experimental Data			Smoothed Data			
x_B	γ_B	γ_{TMP}	$\underline{G}^{ex}$	γ_B	γ_{TMP}	$\underline{G}^{ex}$	$\Delta\underline{G}_{mix}$
0.0819	1.408	1.003	84.68	1.428	1.002	83.85	− 689.40
0.2192	1.343	1.011	199.74	1.342	1.013	203.34	−1231.56
0.3584	1.250	1.046	296.73	1.261	1.039	294.09	−1486.24
0.3831	1.242	1.048	305.22	1.246	1.046	306.52	−1509.50
0.5256	1.158	1.116	352.63	1.166	1.107	351.75	−1535.95
0.8478	1.023	1.508	224.30	1.023	1.508	223.72	− 940.02
0.9872	1.000	1.968	23.97	1.000	1.961	24.02	− 162.88

Note: All Gibbs free energy data in units of J/mol. ■

Several further aspects of vapor–liquid equilibria need to be considered. The first is the additional information that can be obtained if vapor–liquid equilibrium measurements are made at a collection of temperatures. Weissman and

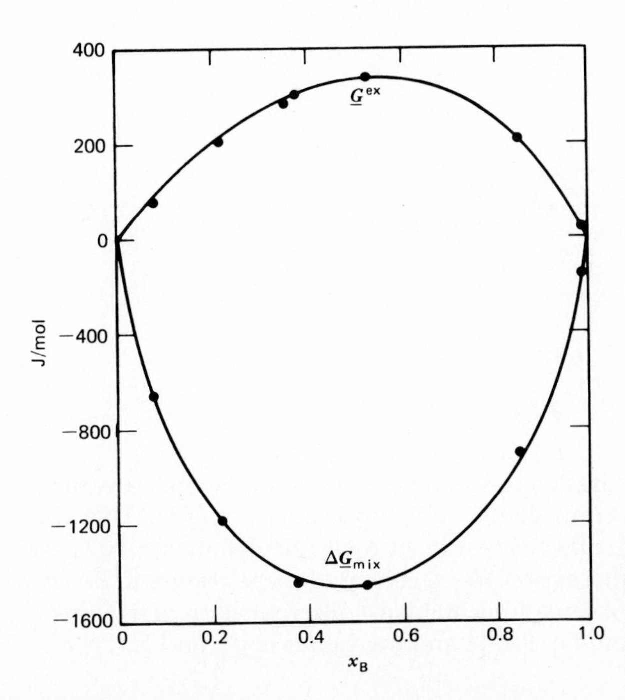

Figure 1 The excess Gibbs free energy and Gibbs free energy change on mixing for the benzene–2,2,4-trimethyl pentane system at 55°C.

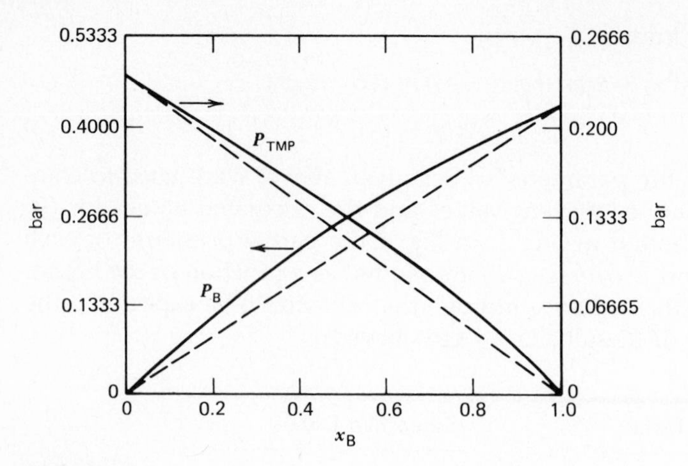

Figure 2
The equilibrium benzene and 2,2,4-trimethyl pentane partial pressures as a function of benzene mole fraction at 55°C.

Wood carried out their experiments at 35, 45, 55, 65, and 75°C and obtained data like that given in Illustration 8.1-4 at all these temperatures. From $\underline{G} = \underline{H} - T\underline{S}$, and the discussion of Chapter 6, we have

$$\Delta\underline{G}_{\text{mix}} = \Delta\underline{H}_{\text{mix}} - T\Delta\underline{S}_{\text{mix}}$$

$$\underline{G}^{ex} = \underline{H}^{ex} - T\underline{S}^{ex} \tag{8.1-16}$$

and

$$\frac{\partial}{\partial T}\left(\frac{\Delta\underline{G}_{\text{mix}}}{T}\right)_{P,x_i} = \frac{1}{T}\left(\frac{\partial\Delta\underline{G}_{\text{mix}}}{\partial T}\right)_{P,x_i} - \frac{1}{T^2}(\Delta\underline{G}_{\text{mix}}) \tag{8.1-17}$$

Using the Maxwell relation

$$\left(\frac{\partial\underline{G}}{\partial T}\right)_P = -\underline{S}$$

which for mixtures becomes

$$\left(\frac{\partial\Delta\underline{G}_{\text{mix}}}{\partial T}\right)_{P,x_i} = -\Delta\underline{S}_{\text{mix}}; \qquad \left(\frac{\partial\underline{G}^{ex}}{\partial T}\right)_{P,x_i} = -\underline{S}^{ex} \tag{8.1-18}$$

and using Eq. 8.1-16 in Eq. 8.1-17 gives

$$\frac{\partial}{\partial T}\left(\frac{\Delta\underline{G}_{\text{mix}}}{T}\right)_{P,x_i} = -\frac{1}{T}\Delta\underline{S}_{\text{mix}} - \frac{1}{T^2}(\Delta\underline{H}_{\text{mix}} - T\Delta\underline{S}_{\text{mix}})$$

$$= -\frac{\Delta\underline{H}_{\text{mix}}}{T^2} = -\frac{\underline{H}^{ex}}{T^2} \tag{8.1-19}$$

Therefore, given $\Delta\underline{G}_{\text{mix}}$ or $\underline{G}^{ex}$ data at a collection of temperatures, one can obtain information about $\Delta\underline{H}_{\text{mix}}$ (or, equivalently, $\underline{H}^{ex}$, since $\Delta\underline{H}_{\text{mix}} = \underline{H}^{ex}$). This was done by Weissman and Wood, and the results of their calculations for 40°C are plotted in Fig. 8.1-3. One would expect the accuracy of these results to be less than that of $\Delta\underline{G}_{\text{mix}}$, since these calculations involve a differentiation of the experimental data. $\underline{S}^{ex}$, calculated from Eq. 8.1-16 and the values of $\underline{G}^{ex}$ and $\underline{H}^{ex}$, is also plotted in Fig. 8.1-3.

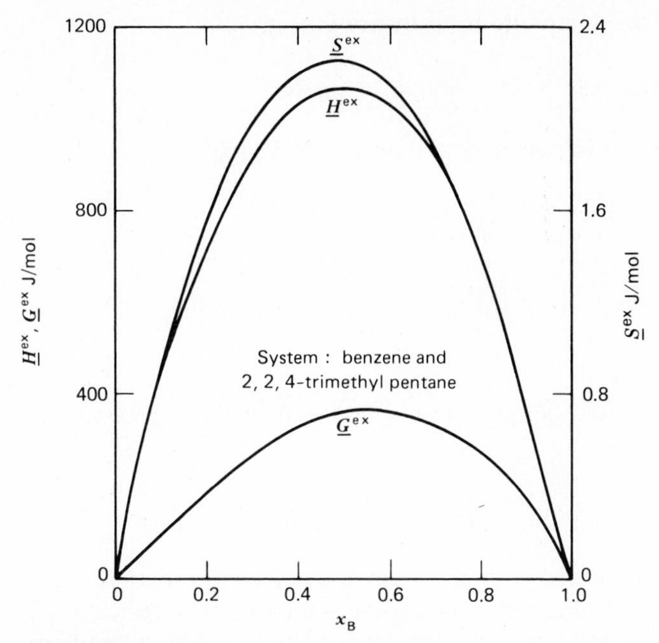

Figure 8.1-3
Excess Gibbs free energy, enthalpy, and entropy as a function of mole fraction for the benzene–2,2,4-trimethyl pentane system.

Next, one frequently would like to be able to make some assessment of the accuracy of a set of experimental vapor–liquid (or more correctly, activity coefficient) measurements. Thermodynamics (as opposed to solution modelling) provides no means of predicting the values of liquid-phase activity coefficients to which the experimental results could be compared. Also, since the liquid solution models discussed in Chapter 7 only approximate real solution behavior, any discrepancy between these theoretical models and experiment is undoubtedly more a reflection of the inadequacy of the theory than a test of the experimental results.

In Chapter 5 we found that although thermodynamics could not be used to predict the equation of state of a real fluid, it did provide certain consistency tests (i.e., Eqs. 5.2-12 and 13) that had to be satisfied by any equation of state. The situation is much the same here, in that starting from

$$\underline{G}^{ex} = \sum_{i=1}^{\mathscr{C}} x_i \overline{G}_i^{ex} = RT \sum_{i=1}^{\mathscr{C}} x_i \ln \gamma_i \tag{8.1-20}$$

and using the Gibbs–Duhem equation, we can develop a consistency test that must be satisfied by activity coefficient data and thus can be used to accept or reject experimental data. In particular, for a binary mixture, we have, from Eq. 8.1-20, that

$$d\left(\frac{\underline{G}^{ex}}{RT}\right) = x_1\, d \ln \gamma_1 + \ln \gamma_1\, dx_1 + x_2\, d \ln \gamma_2 - \ln \gamma_2\, dx_1$$

since $dx_2 = -dx_1$. Also, from the Gibbs–Duhem equation, Eq. 7.3-15, we have that

$$0 = \frac{\underline{H}^{ex}}{RT^2}\, dT - \frac{\underline{V}^{ex}}{RT}\, dP + x_1\, d \ln \gamma_1 + x_2\, d \ln \gamma_2$$

Subtracting the second of these equations from the first gives

$$d\left(\frac{\underline{G}^{ex}}{RT}\right) = \ln\frac{\gamma_1}{\gamma_2}\,dx_1 - \frac{\underline{H}^{ex}}{RT^2}\,dT + \frac{\underline{V}^{ex}}{RT}\,dP$$

Now integrating this equation between $x_1 = 0$ and $x_1 = 1$ yields

$$\int_{x_1=0}^{x_1=1} d\left(\frac{\underline{G}^{ex}}{T}\right) = \left.\frac{\underline{G}^{ex}}{T}\right|_{x_1=1} - \left.\frac{\underline{G}^{ex}}{T}\right|_{x_1=0}$$

$$= +R\int_{x_1=0}^{x_1=1}\ln\frac{\gamma_1}{\gamma_2}\,dx_1 + \int_{P(x_1=0)}^{P(x_1=1)}\frac{\underline{V}^{ex}}{T}\,dP - \int_{T(x_1=0)}^{T(x_1=1)}\frac{\underline{H}^{ex}}{T^2}\,dT$$

$$= 0$$

where we have used the fact that $\underline{G}^{ex}(x_1 = 1) = \underline{G}^{ex}(x_1 = 0) = 0$ (cf. Section 7.5). Therefore

$$\int_{x_1=0}^{x_1=1}\ln\frac{\gamma_2}{\gamma_1}\,dx_1 = +\int_{P(x_1=0)}^{P(x_1=1)}\frac{\underline{V}^{ex}}{RT}\,dP - \int_{T(x_1=0)}^{T(x_1=1)}\frac{\underline{H}^{ex}}{RT^2}\,dT \qquad (8.1\text{-}21)$$

This equation provides a thermodynamic consistency test for experimental activity coefficient data. As an illustration of its use, consider its application to the Weissman–Wood experiments, which were carried out at constant temperature but varying total pressure. In this case Eq. 8.1-21 reduces to

$$\int_{x_1=0}^{x_1=1}\ln\frac{\gamma_2}{\gamma_1}\,dx_1 = +\int_{P(x_1=0)}^{P(x_1=1)}\frac{\underline{V}^{ex}}{RT}\,dP$$

Now since the total pressure variation in the experiments was small, and $\underline{V}^{ex}$ is usually very small for liquid mixtures, the integral on the right side of this equation can be neglected. Thus to test the thermodynamic consistency of the Weissman–Wood activity coefficient data we have

$$\int_{x_1=0}^{x_1=1}\ln\frac{\gamma_2}{\gamma_1}\,dx_1 = 0 \qquad (8.1\text{-}22)$$

Figure 8.1-4 is a plot of ln (γ_2/γ_1) versus mole fraction using the activity coefficient data of Illustration 8.1-4. The two areas, I and II, between the curve and the ln $(\gamma_2/\gamma_1) = 0$ line are virtually equal in size but opposite in sign, so that Eq. 8.1-22 can be considered satisfied. Of course, as a result of experimental errors, this equation is never satisfied exactly. One usually considers vapor–liquid (or activity coefficient) data to be thermodynamically consistent if the two areas are such that

$$-0.02 \leqslant \frac{|\text{area I}| - |\text{area II}|}{|\text{area I}| + |\text{area II}|} \leqslant 0.02$$

where the vertical bars indicate that absolute values of the areas are to be used.

For activity coefficient data obtained from measurements at constant total pressure, but varying temperature, the appropriate consistency relation is

$$\int_{x_1=0}^{x_1=1}\ln\frac{\gamma_2}{\gamma_1}\,dx_1 = -\int_{T(x_1=0)}^{T(x_1=1)}\frac{\underline{H}^{ex}}{RT^2}\,dT$$

Depending on the system, and especially the magnitude of the excess enthalpy and the temperature range of the experiments, the integral on the right side of this equation may or may not be negligible.

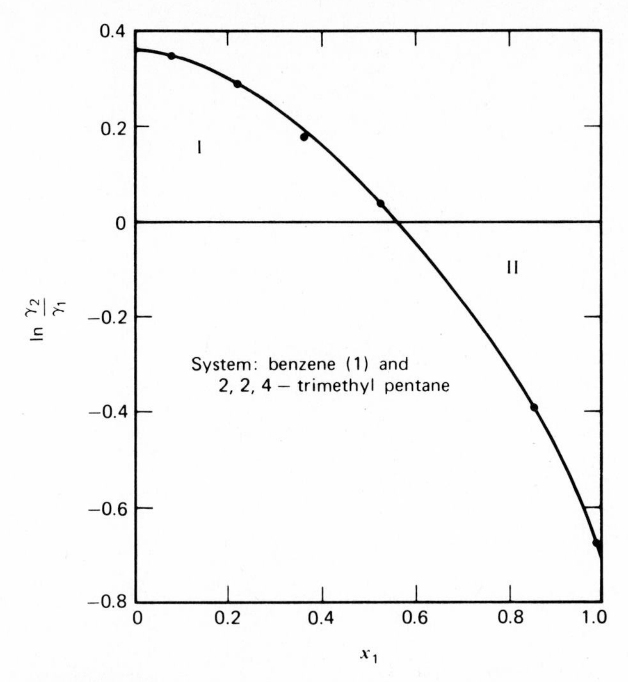

Figure 8.1-4
Thermodynamic consistency test for the activity coefficients of the benzene–2,2,4-trimethyl pentane system.

It is possible, as a result of cancellation, to satisfy the integral test of Eq. 8.1-21 while violating the differential form Gibbs–Duhem equation, Eq. 7.3-15, on which Eq. 8.1-21 is based, at some or all data points. In this case the experimental data should be rejected as being thermodynamically inconsistent. Thus, the integral consistency test is a necessary, but not sufficient, condition for accepting experimental data.

As another example of low pressure vapor–liquid equilibrium, we consider the *n*-pentane–propionaldehyde mixture at 40.0°C. Eng and Sandler[5] took data for this system using a small dynamic still. In such equipment the liquid mixture is boiled in a still pot, the resulting liquid and vapor are allowed to equilibrate, and then are separated. The vapor is condensed, and the liquid and condensed vapor are sampled and analyzed before being mixed and returned to the still pot. In this way, the *x-y-P-T* data in Table 8.1-1 and Figs. 8.1-5*a* and *b* were obtained. (Such data can be tested for thermodynamic consistency, Problem 8.1-12.) As is evident, this system is nonideal and has an azeotrope at about 0.656 mole fraction pentane and 1.3640 bar. We will use these data to test the UNIFAC prediction method, as well as to examine the effect of making predictions based on a less complete data set than appears in Table 8.1-1.

First, we use the UNIFAC program, discussed in Appendix A7.4 to compute the activity coefficients in the *n*-pentane–propionaldehyde mixture over the complete liquid concentration range, and then, using Eqs. 8.1-1b and 2b, we compute the vapor compositions and equilibrium pressures. The results also appear in Fig. 8.1-5*a* as a *P-x-y* diagram and in Fig. 8.1-5*b* as an *x-y* diagram. The

[5]R. Eng and S. I. Sandler, *J. Chem. Eng. Data* **29**, 156 (1984).

Table 8.1-1
VAPOR–LIQUID EQUILIBRIUM DATA FOR THE *n*-PENTANE (1)–PROPIONALDEHYDE (2) SYSTEM AT 40.0°C

x_1	y_1	P(bar)
0	0	0.7609
0.0503	0.2121	0.9398
0.1014	0.3452	1.0643
0.1647	0.4288	1.1622
0.2212	0.4685	1.2173
0.3019	0.5281	1.2756
0.3476	0.5539	1.2949
0.4082	0.5686	1.3197
0.4463	0.5877	1.3354
0.5031	0.6146	1.3494
0.5610	0.6311	1.3568
0.6812	0.6827	1.3636
0.7597	0.7293	1.3567
0.8333	0.7669	1.3353
0.9180	0.8452	1.2814
1.0	1.0	1.1541

UNIFAC predictions are in very good agreement with the experimental data, including a reasonably accurate prediction of the azeotropic point. Clearly, an engineer needing information on the *n*-pentane–propionaldehyde system, but having no experimental data, would be better to assume the UNIFAC model applies to this mixture than to assume that the system were ideal. Since propionaldehyde is strongly polar, the regular solution model could not be used for this mixture.

The dynamic still method of obtaining vapor–liquid equilibrium data has several disadvantages. First, it is a slow process. Second, the compositions of the vapor and liquid must be analyzed (usually by gas chromatography), which is less precise and direct than measuring temperature, pressure, or weight. Consequently, alternative methods of measuring partial vapor–liquid equilibrium data have been developed that do not require chemical analysis.

One method is to use a static still, which consists of a small vessel that is evacuated and almost completely filled with a gravimetrically prepared liquid binary mixture. This vessel is placed in a temperature bath, and the pressure inside the vessel is measured after equilibrium has been reached. As weighing can be done very accurately, and since only a small amount (a bubble) of vapor is formed, the liquid composition barely changes during the vaporization process and therefore is known very accurately. By repeating this process with a number of prepared solutions, one obtains a set of P-T-x data. Such data can be studied in two ways. First is to use the Gibbs–Duhem equation and numerical integration methods to calculate the vapor-phase mole fractions, as considered in Problem 8.1-6.

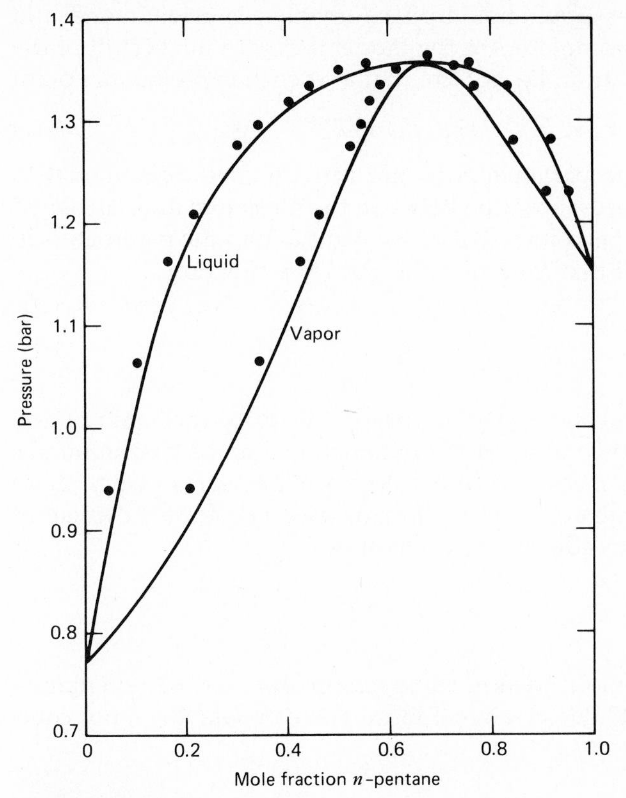

Figure 8.1-5a
P-x-y diagram for the n-pentane–propionaldehyde system at 40°C. The lines are the UNIFAC predictions and the points are the experimental data of Eng and Sandler.

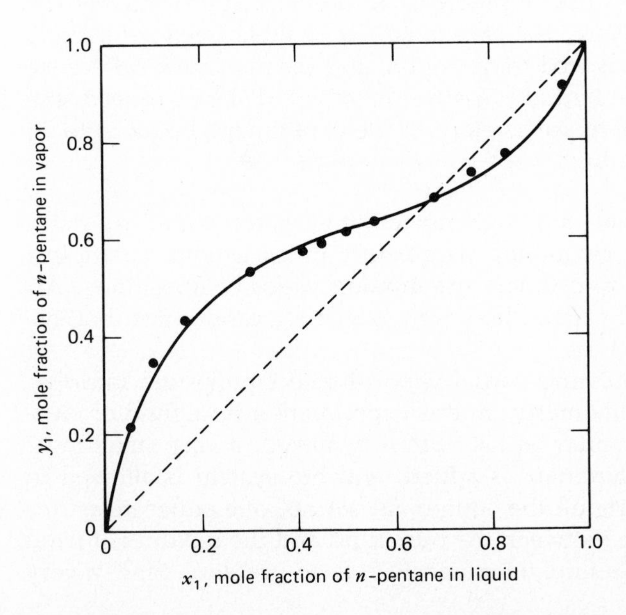

Figure 8.1-5b
x-y diagram for the n-pentane–propionaldehyde system at 40°C. The solid line is the UNIFAC prediction, and the points are the experimental data of Eng and Sandler.

A second method is to assume a liquid-phase activity coefficient model and determine the values of the parameters in the model that give the best fit of the experimental data. We have, from Eq. 8.1-2b, that at the jth experimental point

$$P_j = x_1^j \gamma_1^j P_1^{\text{vap}} + (1 - x_1^j) \gamma_2^j P_2^{\text{vap}} \tag{8.1-23}$$

We now want to choose the parameters in the activity coefficient model to minimize the sum of the squares deviation between the measured and calculated pressures over all experimental points; that is, we want to find the parameters in the activity coefficient model that minimize the objective function

$$\text{Min} \sum_{\substack{\text{exp}\\ \text{pts}\\ j}} [P_j^{\text{exp}} - P_j^{\text{calc}}]^2 = \text{Min} \sum_{\substack{\text{exp}\\ \text{pts}\\ j}} [P_j^{\text{exp}} - x_1^j \gamma_1^j P_1^{\text{vap}} - (1 - x_1^j) \gamma_2^j P_2^{\text{vap}}]^2 \tag{8.1-24}$$

Thus, for example, if the van Laar equation is used to describe the liquid phase, then we want to determine the values of the parameters α and β that minimize the deviations between the measured and calculated pressures. Once these parameters have been determined, they can then be used to calculate the vapor-phase compositions. This procedure is next illustrated.

ILLUSTRATION 8.1-5

Using only the liquid-phase mole fraction and pressure data for the n-pentane–propionaldehyde system at 40.0°C given in Table 8.1-1, estimate the vapor compositions.

Solution

Using the method just described, the van Laar equation, and a parameter estimation computer program with the objective function of Eq. 8.1-24, we find

$$\alpha = 1.4106 \quad \text{and} \quad \beta = 1.3438$$

With these parameter values, we obtain the calculated pressures and vapor mole fractions given in Table 8.1-2. It is clear, from this table, that the predictions are reasonably accurate. The azeotrope is predicted to occur at approximately the correct composition and pressure, the calculated vapor mole fractions usually agree within ± 0.015 of the measured composition, and the maximum difference between the calculated and measured pressures is only 0.0174 bar. Indeed, the calculated results for this system are so close to the experimental data as to be almost indistinguishable from them on x-y or p-x-y plots. ■

These results suggest that, although not quite as good as P-T-x-y data, P-T-x data can be useful for estimating parameters in an activity coefficient model that can then be used to estimate the missing vapor compositions. An important disadvantage of P-T-x data, however, is that we cannot test its thermodynamic consistency.

A second method of obtaining partial vapor–liquid equilibrium information is by infinite dilution ebulliometry. In this experiment a pure fluid of measured weight is boiled. Then, after equilibrium is achieved, a very small measured weight of a second component is added, and the system is allowed to re-equilibrate. Then, depending on the equipment set up, one either measures the change in boiling pressure (between the pure fluid and the mixture) at fixed temperature or the change in boiling temperature at fixed pressure. Since a very

Table 8.1-2
COMPARISON OF MEASURED VAPOR-PHASE MOLE FRACTIONS FOR THE *n*-PENTANE–PROPIONALDEHYDE SYSTEM AT 40°C WITH VALUES PREDICTED FROM *P-T-x* DATA

x_1	y_1^{exp}	y_1^{calc}	P_1^{exp}	P_1^{calc}
0	0	0	0.7609	0.7609
0.0503	0.2121	0.2211	0.9398	0.9312
0.1014	0.3452	0.3424	1.0643	1.0555
0.1647	0.4288	0.4309	1.1622	1.1618
0.2212	0.4685	0.4810	1.2173	1.2257
0.3019	0.5281	0.5286	1.2756	1.2846
0.3476	0.5539	0.5484	1.2949	1.3067
0.4082	0.5686	0.5702	1.3197	1.3208
0.4463	0.5877	0.5824	1.3354	1.3379
0.5031	0.6146	0.5996	1.3494	1.3490
0.5610	0.6311	0.6173	1.3568	1.3566
0.6812	0.6827	0.6609	1.3636	1.3604
0.7597	0.7293	0.7005	1.3567	1.3500
0.8333	0.7669	0.7529	1.3353	1.3244
0.9180	0.8452	0.8455	1.2814	1.2640
1.0	1.0	1.0	1.1541	1.1541

small amount of the second component has been added, and the weights are known, so that the mole fractions can be determined, one has measured either

$$\left(\frac{\partial P}{\partial x_2}\right)_T \quad \text{or} \quad \left(\frac{\partial T}{\partial x_2}\right)_P$$

depending on the apparatus. Furthermore, if the amount of the added second component is small, these quantities have been determined in the limit of $x_2 \to 0$.

To analyze the data from such an experiment, assuming an ideal vapor phase, we start from

$$P = x_1 \gamma_1 P_1^{\text{vap}} + x_2 \gamma_2 P_2^{\text{vap}}$$

For the constant temperature experiment (noting that the pure component vapor pressures depend only on temperature, which is being held fixed) we have

$$\left(\frac{\partial P}{\partial x_2}\right)_T = -\gamma_1 P_1^{\text{vap}} + x_1 \left(\frac{\partial \gamma_1}{\partial x_2}\right)_T P_1^{\text{vap}} + \gamma_2 P_2^{\text{vap}} + x_2 \left(\frac{\partial \gamma_2}{\partial x_2}\right)_T P_2^{\text{vap}}$$

Now in the limit of $x_2 \to 0$, we have that $\gamma_1 \to 1$ and $(\partial \gamma_1 / \partial x_2) = 0$ as γ_1 is proportional to higher than a linear power of x_2 (i.e., see Eqs. 7.5-5, 7, etc.). Thus

$$\left(\frac{\partial P}{\partial x_2}\right)_{T, x_2 \to 0} = \gamma_2(x_2 \to 0) P_2^{\text{vap}} - P_1^{\text{vap}}$$

or

$$\gamma_2(x_2 \rightarrow 0) = \gamma_2^{\infty} = \frac{P_1^{\text{vap}} + \left(\dfrac{\partial P}{\partial x_2}\right)_{T, x_2 \rightarrow 0}}{P_2^{\text{vap}}} \tag{8.1-25}$$

In a similar fashion, for the constant-pressure ebulliometer, we have (Problem 8.1-13)

$$\gamma_2(x_2 \rightarrow 0) = \gamma_2^{\infty} = \frac{P_1^{\text{vap}} - \left(\dfrac{dP_1^{\text{vap}}}{dT}\right)\left(\dfrac{\partial T}{\partial x_2}\right)_{P, x_2 \rightarrow 0}}{P_2^{\text{vap}}} \tag{8.1-26}$$

It is, of course, possible to derive equations analogous to Eqs. 8.1-25 and 26 for a nonideal vapor phase.

Thus, from the ebulliometric experiment, one obtains the infinite dilution activity coefficient directly. Now repeating the experiment by starting with pure component 2 and adding an infinitesimal amount of component 1, $\gamma_1(x_1 \rightarrow 0) = \gamma_1^{\infty}$ can be obtained. These two data points can then be used to determine the parameters in a two-constant activity coefficient model. For example, from the van Laar model of Eqs. 7.5-9, we have

$$\ln \gamma_1^{\infty} = \alpha \qquad \text{and} \qquad \ln \gamma_2^{\infty} = \beta \tag{8.1-27}$$

Thus once the infinite dilution activity coefficients have been measured, and the parameters in an activity coefficient model determined, the complete *P-T-x-y* behavior of the system can be estimated.

ILLUSTRATION 8.1-6

A recent ebulliometric study of the *n*-pentane–propionaldehyde at 40°C has shown that $\gamma_1^{\infty} = 3.848$ and $\gamma_2^{\infty} = 3.979$. Use this information to compute the *P-x-y* diagram for this system at 40°C.

Solution

The van Laar activity coefficient model will be used. From Eqs. 8.1-27 we have

$$\alpha = \ln \gamma_1^{\infty} = \ln (3.848) = 1.3476$$

and

$$\beta = \ln \gamma_2^{\infty} = \ln (3.979) = 1.3810$$

These values are in reasonable agreement with, but slightly different from, those found in the previous illustration. Using the values for the van Laar parameters, we obtain the y and P values in Table 8.1-3. Clearly the agreement is excellent. ■

The previous two illustrations demonstrate the utility of both *P-T-x* and ebulliometric γ^{∞} data in determining values in activity coefficient models, and then using these parameters to compute the missing data. It should be remembered, however, that in the analysis of both the static still and ebulliometric measurements, an activity coefficient model that satisfies the Gibbs–Duhem equation has been used. Therefore, the calculated results must satisfy the thermodynamic consistency test. Consequently, there is no independent test of the quality of the results as when complete *P-T-x-y* data have been measured. However, both static still and ebulliometric measurements provide valuable data and

Table 8.1-3
COMPARISON OF MEASURED PRESSURES AND VAPOR-PHASE MOLE FRACTIONS FOR THE *n*-PENTANE–PROPIONALDEHYDE SYSTEM AT 40°C WITH VALUES CALCULATED USING γ^{∞} DATA

x_1	y_1^{exp}	y_1^{calc}	P_1^{exp}	P_1^{calc}
0	0	0	0.7609	0.7609
0.0503	0.2121	0.2131	0.9398	0.9214
0.1014	0.3452	0.3352	1.0643	1.0424
0.1647	0.4288	0.4267	1.1622	1.1493
0.2212	0.4685	0.4798	1.2173	1.2157
0.3019	0.5281	0.5307	1.2756	1.2784
0.3476	0.5539	0.5520	1.2949	1.3024
0.4082	0.5686	0.5752	1.3197	1.3255
0.4463	0.5877	0.5879	1.3354	1.3363
0.5031	0.6146	0.6056	1.3494	1.3483
0.5610	0.6311	0.6232	1.3568	1.3566
0.6812	0.6827	0.6652	1.3636	1.3616
0.7597	0.7293	0.7028	1.3567	1.3524
0.8333	0.7669	0.7528	1.3353	1.3283
0.9180	0.8452	0.8432	1.2814	1.2684
1.0	1.0	1.0	1.1541	1.1541

can be done quickly, which may be important for components that chemically react or decompose.

The experimental data for the hexafluorobenzene–benzene system in Table 8.1-4 and Fig. 8.1-6 show a rarely encountered degree of complexity in low pressure vapor–liquid equilibrium. This system exhibits both minimum and maximum boiling azeotropes. This occurs because the excess Gibbs free energy for this system, though small, is first positive and then negative as the concentration of hexafluorobenzene is increased. However, as the vapor pressures of hexafluorobenzene and benzene are almost identical, the solution nonidealities produce the double azeotrope.

Finally, although the discussion and illustrations of this section have been concerned only with low pressure vapor–liquid equilibria, phase equilibrium at somewhat higher pressures could have been considered also. The most important change in the analysis is that the gas phase can no longer be considered ideal or described by the Lewis–Randall rule; rather, an equation of state (the virial equation at low to moderate pressures and more complicated equations at higher pressures) would have to be used. Also, the Poynting pressure correction of Eq. 5.4-18 may have to be used in the calculation of the pure liquid-phase fugacities. Both these changes add some complexity to the calculations, but improve their accuracy. For simplicity, these factors will not be considered here. We do, however, consider high pressure vapor–liquid equilibria in the next section; a situation in which an equation of state must be used.

Table 8.1-4
VAPOR–LIQUID EQUILIBRIUM FOR THE SYSTEM HEXAFLUOROBENZENE(1)–BENZENE(2) AT 60°C

x_1	y_1	P (bar)	$\underline{G}^{ex}$ (J/mol)
0.0000	0.0000	0.52160	0
0.0941	0.0970	0.52570	32
0.1849	0.1788	0.52568	40
0.2741	0.2567	0.52287	33
0.3648	0.3383	0.51818	16
0.4538	0.4237	0.50989	−4
0.5266	0.4982	0.50773	−21
0.6013	0.5783	0.50350	−35
0.6894	0.6760	0.49974	−44
0.7852	0.7824	0.49757	−45
0.8960	0.8996	0.49794	−30
1.0000	1.0000	0.50155	0

Data of W. J. Gaw and F. L. Swinton, *Trans. Faraday Soc.* **64,** 2023 (1968).

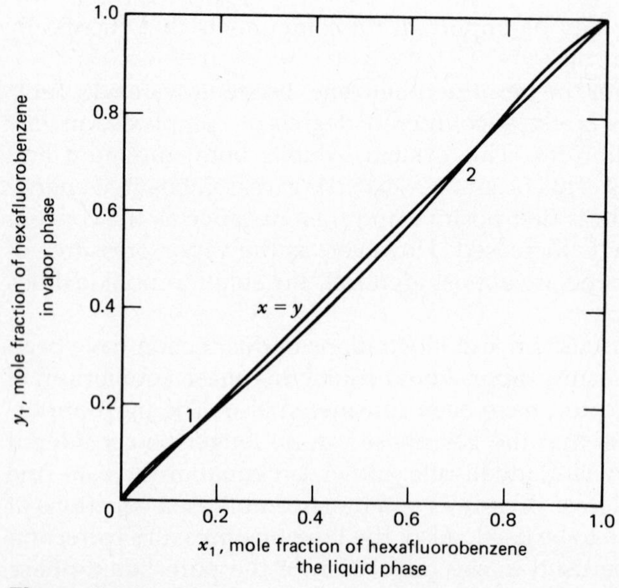

Figure 8.1-6
The *x-y* diagram for the hexafluorobenzene–benzene system at 60°C based on the data of Gaw and Swinton. [*Trans. Faraday Soc.* **64,** 2023 (1968)].

PROBLEMS FOR SECTION 8.1

8.1-1 For a separations process it is necessary to determine the vapor–liquid equilibrium compositions for a mixture of ethyl bromide and n-heptane at 30°C. At this temperature the vapor pressure of pure ethyl bromide is 0.7569 bar, and the vapor pressure of pure n-heptane is 0.0773 bar.

a Calculate the composition of the vapor in equilibrium with a liquid containing 47.23 mole percent ethyl bromide at a total pressure of 0.4537 bar at $T = 30°\text{C}$ assuming the solution is ideal.

b Recalculate the vapor composition in part a assuming the solution is regular. The regular solution parameters are:

	$\underline{V}^L$ (cc/mol)	δ (cal/cc)$^{1/2}$
Ethyl bromide	75	8.9
n-heptane	148	7.4

c Recalculate the vapor composition of part a using the UNIFAC model.

d Recalculate the vapor composition of part a given that a vapor of composition 81.5 mole percent ethyl bromide is in equilibrium with 28.43 mole percent liquid ethyl bromide solution at a total pressure of 0.3197 bar at $T = 30°\text{C}$.

8.1-2 A vapor–liquid mixture of furfural ($C_5H_4O_2$) and water is maintained at 1.013 bar and 109.5°C. It is observed that at equilibrium the water content of the liquid is 10 mole percent and that of the vapor is 81 mole percent. The temperature of the mixture is changed to 100.6°C, and some (but not all) of the vapor condenses. Assuming the vapor phase is ideal, and that the liquid-phase activity coefficients are independent of temperature, but dependent on concentration, compute the equilibrium vapor and liquid compositions at the new temperature.

Data

$$P^{\text{vap}}_{\text{H}_2\text{O}}(T = 109.5°\text{C}) = 1.4088 \text{ bar}$$

$$P^{\text{vap}}_{\text{H}_2\text{O}}(T = 100.6°\text{C}) = 1.0352 \text{ bar}$$

$$P^{\text{vap}}_{\text{FURF}}(T = 109.5°\text{C}) = 0.1690 \text{ bar}$$

$$P^{\text{vap}}_{\text{FURF}}(T = 100.6°\text{C}) = 0.1193 \text{ bar}$$

8.1-3 In this section it was shown that if the equilibrium pressure versus mole fraction curve for a binary mixture has an interior extreme value at some liquid phase mole fraction, (i.e., if $(\partial P/\partial x)_T = 0$ for $0 < x < 1$), an azeotrope was formed at that composition. Show that if, at constant pressure, the equilibrium temperature versus liquid-phase mole fraction has an interior extreme value (i.e., if $(\partial T/\partial x)_P = 0$ for $0 < x < 1$), the mixture forms an azeotrope.

8.1-4 Benzene and ethanol form azeotropic mixtures. Consequently, benzene is sometimes added to solvent grades of ethanol to prevent industrious chemical engineering students from purifying solvent-grade ethanol by distillation for use at an after-finals party. Prepare an x-y and a P-x diagram for the benzene–ethanol system at 45°C assuming, separately,

a The mixture is ideal.
b The mixture is regular.
c The mixture is described by the UNIFAC model.
d The activity coefficients for this system obey the van Laar equation and the datum point at $x_{EA} = 0.6155$ is used to obtain the van Laar parameters.

Compare the results obtained in parts a–d with the experimental data in the following table.

x_{EA}	y_{EA}	P(bar)	x_{EA}	y_{EA}	P(bar)
0	0	0.2939	0.5284	0.4101	0.4093
0.0374	0.1965	0.3613	0.6155	0.4343	0.4028
0.0972	0.2895	0.3953	0.7087	0.4751	0.3891
0.2183	0.3370	0.4088	0.8102	0.5456	0.3615
0.3141	0.3625	0.4124	0.9193	0.7078	0.3036
0.4150	0.3842	0.4128	0.9591	0.8201	0.2711
0.5199	0.4065	0.4100	1.00	1.00	0.2321

Source: I. Brown and F. Smith, *Austral. J. Chem.* **7**, 264 (1954).

Also compare the computed van Laar coefficients with those given in Table 7.5-1.

8.1-5 The system toluene–acetic acid forms an azeotrope containing 62.7 mole percent toluene and having a minimum boiling point of 105.4°C at 1.013 bar. The following vapor pressure data are available:

	P^{vap}(bar)	
T°C	Toluene	Acetic Acid
70	0.2699	0.1813
80	0.3863	0.2697
90	0.5395	0.3916
100	0.7429	0.5561
110	—	0.7744
110.7	1.0133	—
118.5	—	1.0133
120	—	1.0586

a Calculate the van Laar constants for this system and plot $\ln \gamma_T$ and $\ln \gamma_A$ versus x_T.
b Assuming that the activity coefficients (and van Laar constants) are independent of temperature over the limited range of temperatures involved, develop an x-y diagram for this system at $P = 1$ bar. Com-

pare this with the x-y diagram that would be obtained if the system were ideal.

8.1-6 The illustrations of this section were meant to demonstrate how one can determine activity coefficients from measurements of temperature, pressure and the mole fractions in both phases of a vapor–liquid equilibrium system. An alternative procedure is, at constant temperature, to measure the total equilibrium pressure above liquid mixtures of known (or measured) composition. This replaces time-consuming measurements of vapor phase compositions with a more detailed analysis of the experimental data and more complicated calculations.

a Starting with the Gibbs–Duhem equation show, that at constant temperature,

$$RT \sum x_i \, d \ln (x_i \gamma_i) = \underline{V}^{ex} \, dP$$

and, for a binary mixture, that

$$\left(\frac{x_1}{y_1} - \frac{x_2}{y_2} \right) dy_1 = \left(\frac{P\underline{V}^{ex}}{RT} - 1 \right) d \ln P$$

which can also be rewritten as

$$\frac{(y_1 - x_1)}{y_1(1 - y_1)} \frac{dy_1}{dx_1} = \frac{d \ln P}{dx_1}$$

since $P\underline{V}^{ex}/RT \ll 1$.

b The equilibrium pressure above various mixtures of carbon tetrachloride and n-heptane at 50°C are given in the following table.

Mole Percent of CCl_4 in the Liquid	Pressure (bar)
0.0	0.1873
3.32	0.1956
9.83	0.2131
17.14	0.2320
30.24	0.2649
35.14	0.2765
43.24	0.2943
50.12	0.3097
57.00	0.3263
64.96	0.3425
73.23	0.3616
81.26	0.3765
89.92	0.3939
96.49	0.4055
100.0	0.4113

Source: C. P. Smith and E. W. Engel, *J. Am. Chem. Soc.* **51,** 2646 (1929).

Develop the x-y diagram for this system, and compute the liquid-phase activity coefficients of carbon tetrachloride and n-heptane.

c The equilibrium pressure above various mixtures of ethylene bromide and 1-nitropropane at 75°C are given in the following table.

Mole Percent of Ethylene Bromide in Liquid	Pressure (bar)
0.0	0.1533
2.98	0.1556
3.52	0.1568
5.75	0.1597
15.80	0.1659
26.98	0.1719
39.95	0.1760
50.85	0.1773
65.62	0.1773
76.48	0.1745
88.01	0.1699
94.31	0.1652
100.0	0.1596

Source: J. R. Lacher, W. B. Buck, and W. H. Parry, *J. Am. Chem. Soc.* **66,** 2422 (1941).

Develop the x-y diagram for this system, and compute the liquid-phase activity coefficients of ethylene bromide and 1-nitropentane. Compute the activity coefficients at the azeotropic point, and use the van Laar model to predict activity coefficients at other compositions. How do the two sets of activity coefficients compare?

8.1-7 The following mixture of hydrocarbons is obtained as one stream in a petroleum refinery

Component	Mole%	A	B
Ethane	5	817.08	4.402229
Propane	10	1051.38	4.517190
n-Butane	40	1267.56	4.617679
2-Methyl propane	45	1183.44	4.474013

The parameters A and B in this table are the constants in the equation

$$\log_{10} P^{\text{vap}} = -\frac{A}{T} + B$$

where P^{vap} is the vapor pressure in bar and T is the temperature in K. These paraffinic hydrocarbons form an ideal mixture.

a Compute the bubble point of the mixture at 5 bar.

b Compute the dew point of the mixture at 5 bar.

c Find the amounts and compositions of the vapor and liquid phases that would result if this mixture were isothermally flash vaporized at 30°C from a high pressure to 5 bar.

d Set up the equations to be used, and the information needed, to compute the amounts, compositions, and temperature of the vapor and liquid phases that would result if this mixture were *adiabatically* flash vaporized from a high pressure and 50°C to 5 bar.

8.1-8 The following vapor–liquid equilibrium data for the *n*-pentane–acetone system at 1.013 bar were obtained by Lo, Bieker, and Karr [*J. Chem. Eng. Data* **7**, 327 (1962)].

x_P	y_P	T(°C)	P_P^{vap} (bar)	P_A^{vap}
0.021	0.108	49.15	1.560	0.803
0.061	0.307	45.76	1.397	0.703
0.134	0.475	39.58	1.146	0.551
0.210	0.550	36.67	1.036	0.493
0.292	0.614	34.35	0.960	0.453
0.405	0.664	32.85	0.913	0.425
0.503	0.678	33.35	0.903	0.421
0.611	0.711	31.97	0.887	0.413
0.728	0.739	31.93	0.880	0.410
0.869	0.810	32.27	0.896	0.419
0.953	0.906	33.89	0.954	0.445

a Are these data thermodynamically consistent?

b Determine which of the activity coefficient models discussed in Chapter 7 best fits these data.

8.1-9 Use the UNIFAC model to predict the vapor–liquid behavior of the system in the previous problem, and compare the results with the experimental data.

8.1-10 Estimate, as best you can, the vapor–liquid equilibrium coexistence pressure and the composition of the vapor in equilibrium with a liquid containing 20 mole percent ethanol, 40 mole percent benzene, and 40 mole percent ethyl acetate at 78°C.

8.1-11 The experimental data for the hexafluorobenzene–benzene mixture in Table 8.1-4 and Fig. 8.1-6 show a double azeotrope. Test the ability of common thermodynamic models, such as the equations of Wilson and van Laar, and the Redlich–Kister expansion to fit these data.

8.1-12 Determine whether the data in Table 8.1-1 satisfy the Gibbs–Duhem integral consistency test.

8.1-13 Derive Eq. 8.1-26 for the constant pressure ebulliometer.

8.1-14 Using the following data, estimate the total pressure and composition of the vapor in equilibrium with a 20 mole percent ethanol (1) solution in water(2) at 78.15°C. Data (at 78.15°C):

$$\text{Vapor pressure of ethanol (1)} = 1.006 \text{ bar}$$

$$\text{Vapor pressure of water(2)} = 0.439 \text{ bar}$$

$$\lim_{x_1 \to 0} \gamma_1 = \gamma_1^\infty = 1.6931$$

$$\lim_{x_2 \to 0} \gamma_2 = \gamma_2^\infty = 1.9523$$

8.1-15 In vapor–liquid equilibrium the **relative volatility** α_{ij} is defined to be the ratio of the separation or K factor for species i to that for species j, that is,

$$\alpha_{ij} = \frac{K_i}{K_j} = \frac{y_i/x_i}{y_j/x_j}$$

In approximate distillation column calculations the relative volatility is assumed to be a constant (independent of composition, temperature, and pressure). Test this assumption for the ethanol–ethyl acetate system using the following data:

$$RT \ln \gamma_i = 8.163 x_j^2 \text{ kJ/mol}$$

$$\ln P_{\text{EOH}}^{\text{vap}} = \frac{-4728.98}{T} + 13.4643 \qquad P\,[=]\,\text{bar};\ T\,[=]\,\text{K}$$

$$\ln P_{\text{EAC}}^{\text{vap}} = \frac{-3570.58}{T} + 10.4575$$

8.1-16 The binary mixture of benzene and ethylene chloride forms an ideal solution (i.e., one that obeys Raoult's law) at 49.99°C as shown by the data of J. von Zawidzki [*Z. Phys. Chem.* **35,** 129 (1900)]. At this temperature pure benzene boils at 0.357 bar, and pure ethylene chloride boils at 0.315 bar. Develop the analogs of Figures 8.1-1*a*, *b*, and *c* for this system.

8.2 VAPOR–LIQUID EQUILIBRIA USING EQUATIONS OF STATE

The discussion of the previous section was concerned with low pressure vapor–liquid equilibria and involved the use of activity coefficient models. Here we are interested in high pressure phase equilibrium in fluids describable by equations of state. One example of the type of data we are interested in describing (or predicting) is that given in Fig. 8.2-1 for the ethane–propylene system. There we see the liquid (bubble-point) and vapor (dew-point) curves for this system at three different isotherms. The coexisting vapor and liquid phases have the same pressure, and, thus, are joined by horizontal tie lines, only one of which has been drawn. The intersections of these tie lines with the bubble and dew curves give the compositions of the coexisting equilibrium liquid and vapor phases, respectively.

Figure 8.2-1 shows some of the variety of phase behavior that occurs in hydrocarbon mixtures at high pressure. The lowest isotherm, 261 K, is below the critical temperatures of both ethane ($T_C = 305.4$ K) and propylene ($T_C = 365.0$ K). Consequently, both vapor and liquid exist at all compositions, and the isotherm is qualitatively similar to the low pressure isotherm for the n-hexane–triethylamine system of Fig. 8.1-1c. The next isotherm, at $T = 311$ K, is slightly above the

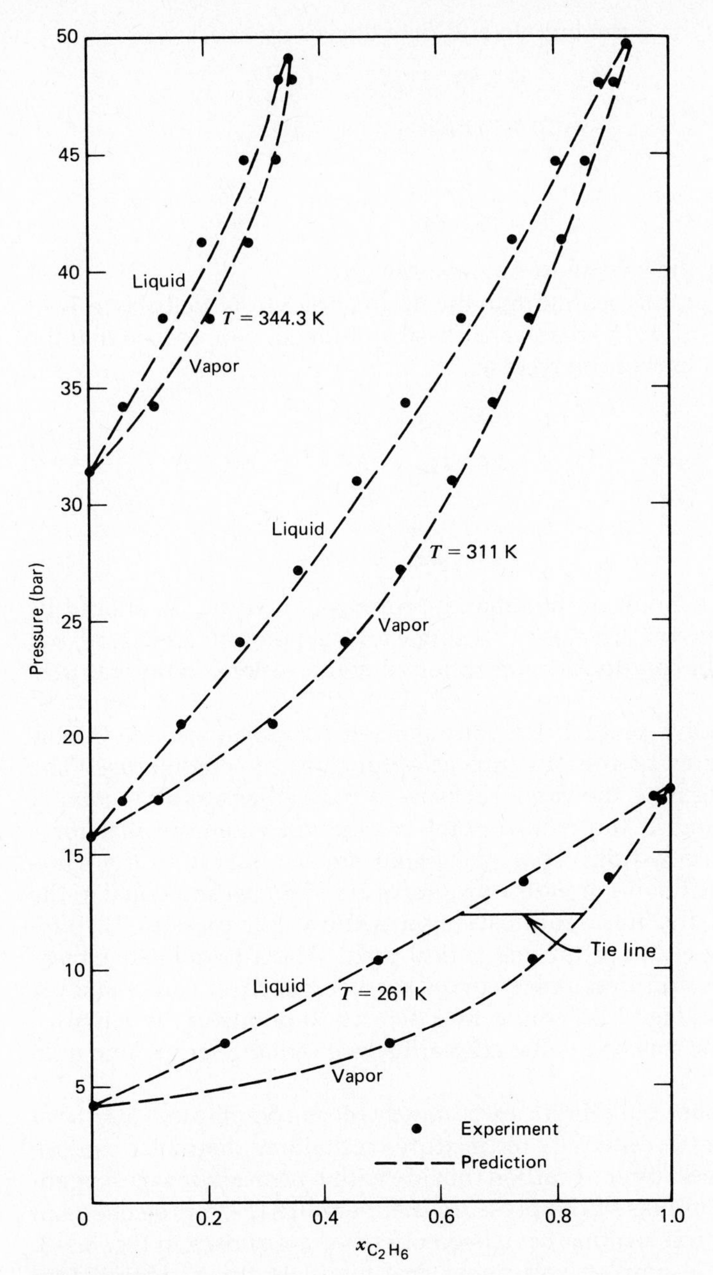

Figure 8.2-1
Constant temperature vapor–liquid equilibrium data for the ethane–propylene system. [R.A. McKay, H. H. Reamer, B. H. Sage, and W. N. Lacey, *Ind. Eng. Chem.* **43,** 2112 (1951).]

critical temperature of ethane, but below T_C for propylene. Thus, pure ethane cannot exist as a liquid at this temperature, nor can mixtures very rich in ethane. As a result, we see the bubble-point and dew-point curves join, not at pure ethane (as with the $T = 261$ K isotherm), but rather at an ethane mole fraction of 0.93. That is, at $T = 311$ K, both vapor and liquid can exist for ethane mole fractions in the range of 0 to 0.93, but above this composition only a vapor is present. The point at which the bubble and dew curves intersect

$$x_{C_2H_6} = 0.93$$

$$T = 311 \text{ K}$$

and

$$P = 49.8 \text{ bar}$$

is one critical point for the ethane–propylene mixture.

The highest temperature isotherm in the figure, 344.3 K, is well above T_C of ethane, so only liquids dilute in ethane are possible. Indeed, here we see that the bubble- and dew-point curves intersect at

$$x_{C_2H_6} = 0.35$$

$$T = 344.2 \text{ K}$$

and

$$P = 48.6 \text{ bar}$$

which is another critical point of the ethane–propylene mixture. As should be evident from this discussion, there is not a single critical point for a mixture, but rather a range of critical conditions with different compositions, temperatures, and pressures.

In Fig. 8.2-2 we have plotted, for various fixed compositions, the bubble and dew point pressures of this mixture as a function of temperature. The leftmost curve in this figure is the vapor pressure of pure ethane as a function of temperature, terminating in the critical point of ethane (remember that for a pure component, the coexisting vapor and liquid are necessarily of the same composition, so that the bubble and dew pressures are identical and equal to the vapor pressure). Similarly, the rightmost curve is the vapor pressure of pure propylene, terminating at the propylene critical point. The intermediate curves (loops) are the bubble- and dew-point curves for various fixed compositions. Finally, there is a line in Fig. 8.2-2 connecting the critical points of the mixtures of various compositions; this line is the **critical locus** of ethane–propylene mixtures.

High pressure phase equilibria can be much more complicated than that shown in Figs. 8.2-1 and 2, especially for mixtures containing dissimilar components, for example, either water or carbon dioxide with hydrocarbons or oxygenated hydrocarbons. Examples of the pressure–temperature (*P-T*) projections of the various types of critical loci that have been observed are shown in Fig. 8.2-3. When looking at these examples, remember that there are three independent variables (temperature, pressure, and the composition of one of the species), so that these figures are two-dimensional projections of surfaces in a three-dimensional figure. Thus, the mixture composition, which is not shown, varies along each of these critical lines.

The critical locus in Figure 8.2-3*a* is very much like that for the ethane–propylene system in Fig. 8.2-2; this is referred to as category I phase behavior. For such systems the critical line starts at the critical point of pure component one (C_1) and, as the mixture becomes richer in the second component the critical line goes smoothly to the critical point C_2.

Category II behavior is slightly more complicated in that at low temperatures and high pressures there is a region of liquid–liquid equilibrium (LLE)

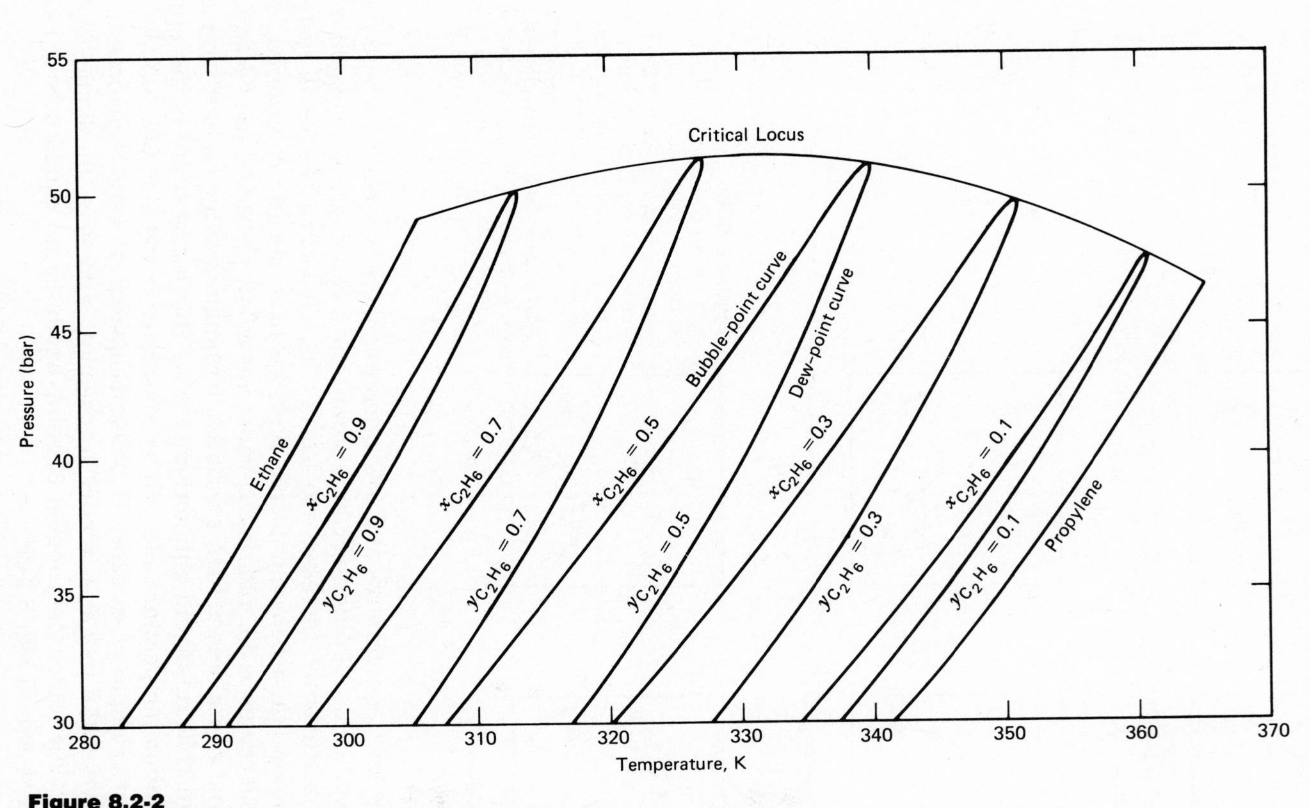

Figure 8.2-2
The bubble point, dew point, and critical locus for the ethane–propylene system.

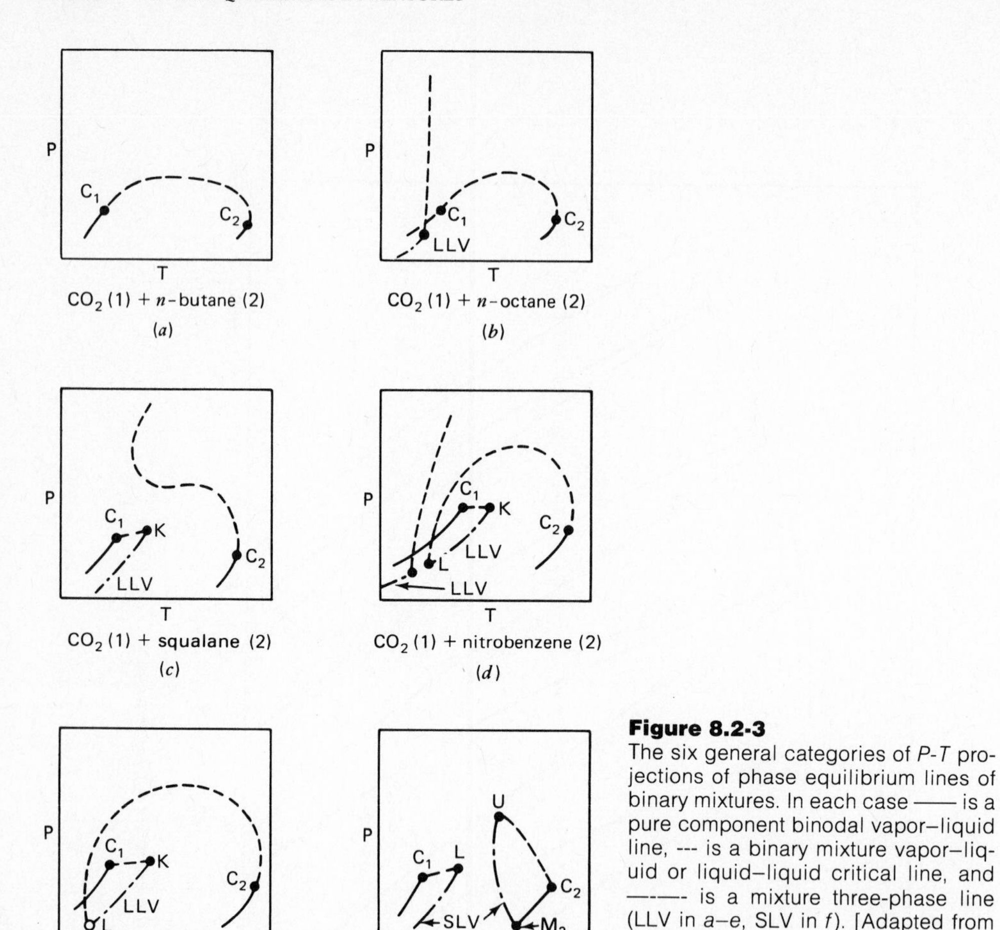

Figure 8.2-3
The six general categories of *P-T* projections of phase equilibrium lines of binary mixtures. In each case —— is a pure component binodal vapor–liquid line, --- is a binary mixture vapor–liquid or liquid–liquid critical line, and —-—- is a mixture three-phase line (LLV in *a–e*, SLV in *f*). [Adapted from M. E. Paulaitis, V. J. Krukonis, R. T. Kurnick, and R. C. Reid, *Rev. Chem. Eng.* **1,** 179 (1983). Used with permission.]

whereas at low temperatures and low pressures there is a region where three phases, two liquids (of differing composition) and a vapor, are all in equilibrium (LLVE). Consequently, there are two sets of critical lines in a category II system; one for vapor–liquid behavior of the type we have already considered, and another that begins as a liquid–liquid-vapor line at low pressures and becomes a liquid–liquid critical line at high pressures, terminating at the highest temperature at which liquid–liquid equilibrium exists. This temperature is the upper **critical solution temperature** and will be considered in detail in Sec. 8.3. Figure 8.2-3*b* is an example of category II phase behavior. At very high pressures, usually beyond the range of interest to chemical engineers, the liquid–liquid equilibrium line intersects a region of solid–liquid–liquid equilibrium (SLLE). This is not shown in Fig. 8.2-3*b*.

Category III phase behavior, shown in FIg. 8.2-3*c*, is similar to category II behavior, except that the region of liquid–liquid–vapor behavior occurs at higher temperatures and, as the composition varies, intersects the vapor–liquid critical curve at the **critical end point** K. Thus, there is one vapor–liquid critical line originating at component 1 (C_1) and terminating at the critical end point, and a second starting at C_2 that merges smoothly into the liquid–liquid equilibrium line at high pressures.

Category V phase behavior, shown in Fig. 8.2-3*e*, is similar to category III behavior, except that the critical line starting at component 2 intersects the liquid–liquid–vapor three-phase region at a **lower critical end point** L. Note that there is no region of liquid–liquid equilibrium (LLE) in a category V system and two critical end points (K and L) of the three-phase region.

Category IV phase behavior, shown in Fig. 8.2-3*d*, has two regions of liquid–liquid–vapor equilibrium. The low temperature LLVE region exhibits category II behavior, whereas the higher temperature LLVE region behaves like a category V system.

Category VI phase behavior, shown in Fig. 8.2-3*f*, occurs with components that are so dissimilar that component 2 has a melting or triple point (M_2) that is well above the critical temperature of component 1. In this case there are two regions of solid–liquid–vapor equilibrium (SLVE). One starts at the triple point of pure component 2 (M_2) and intersects the liquid–vapor critical line at the **upper critical end point** U. The second solid–liquid–vapor critical line starts below the melting point M_2 and intersects the vapor–liquid critical line starting at component 1 at the lower critical end point L. Between the lower and upper critical points only solid–vapor (or solid–fluid) equilibria exists.

A detailed analysis of the complicated phase behavior of Fig. 8.2-3 is beyond the scope of this textbook, but can be found in the book by Rowlinson and Swinton.[6] It is useful to note that all the types of phase behavior discussed here can be predicted using equations of state, though we will restrict our attention to category I systems.

We now turn from the qualitative description of high pressure phase equilibria to its quantitative description, that is, to the correlation and/or prediction of vapor–liquid equilibrium for hydrocarbon (and light gas) systems, of which the ethane–propylene system is merely one example. Our interest will be only in systems describable by a single equation of state for both the vapor and liquid phases, as the case in which the liquid is described by an activity coefficient model was considered in the previous section.

The starting point for any phase equilibrium calculation is, of course, the equality of fugacities of each species in each phase, that is,

$$\bar{f}_i^{\mathrm{L}}(T, P, x_i) = \bar{f}_i^{\mathrm{V}}(T, P, y_i) \tag{8.2-1}$$

Here, however, we will use an equation of state to calculate species fugacities in both phases. For example, when using the Peng–Robinson equation of state, Eq. 7.4-10 is used to compute the fugacity of a species in the vapor phase and Eq. 7.4-11 for the liquid phase (of course, with the composition and compressibility appropriate to each phase).

Phase equilibrium calculations with equations of state are iterative and sufficiently complicated to best be done on a digital computer. Consider, for example, the calculation of the bubble-point pressure and vapor composition for a liquid of known composition at a temperature T. One would need to make an initial guess for the bubble-point pressure, P_B, and the vapor mole fractions (or, perhaps more easily, for the values of $K_i = y_i/x_i$), and then check to see if $\Sigma\, y_i = 1$ and the equality of species fugacities (Eq. 8.2-1) were satisfied for each species with the fugacities calculated from the equation of state. If these restrictions are not satisfied, the pressure and K_i values must be adjusted and the calculation

[6] J. S. Rowlinson and F. L. Swinton, *Liquids and Liquid Mixtures*, 3rd ed., Butterworths, London, 1982, Chapter 6.

repeated. A flow diagram for one algorithm for solving this problem is given in Fig. 8.2-4.

The initial guesses for the bubble-point pressure P_B and for the $K_i = y_i/x_i$ values for all species in the mixture do not affect the final solution to the problem, but may influence the number of iterations required to obtain the solution. One possible set of initial guesses is obtained by assuming ideal liquid and vapor mixtures so that

$$P_B = \sum x_i P_i^{\text{vap}}(T) \tag{8.2-2}$$

and

$$K_i = \frac{y_i}{x_i} = \frac{P_i^{\text{vap}}(T)}{P_B} \tag{8.2-3}$$

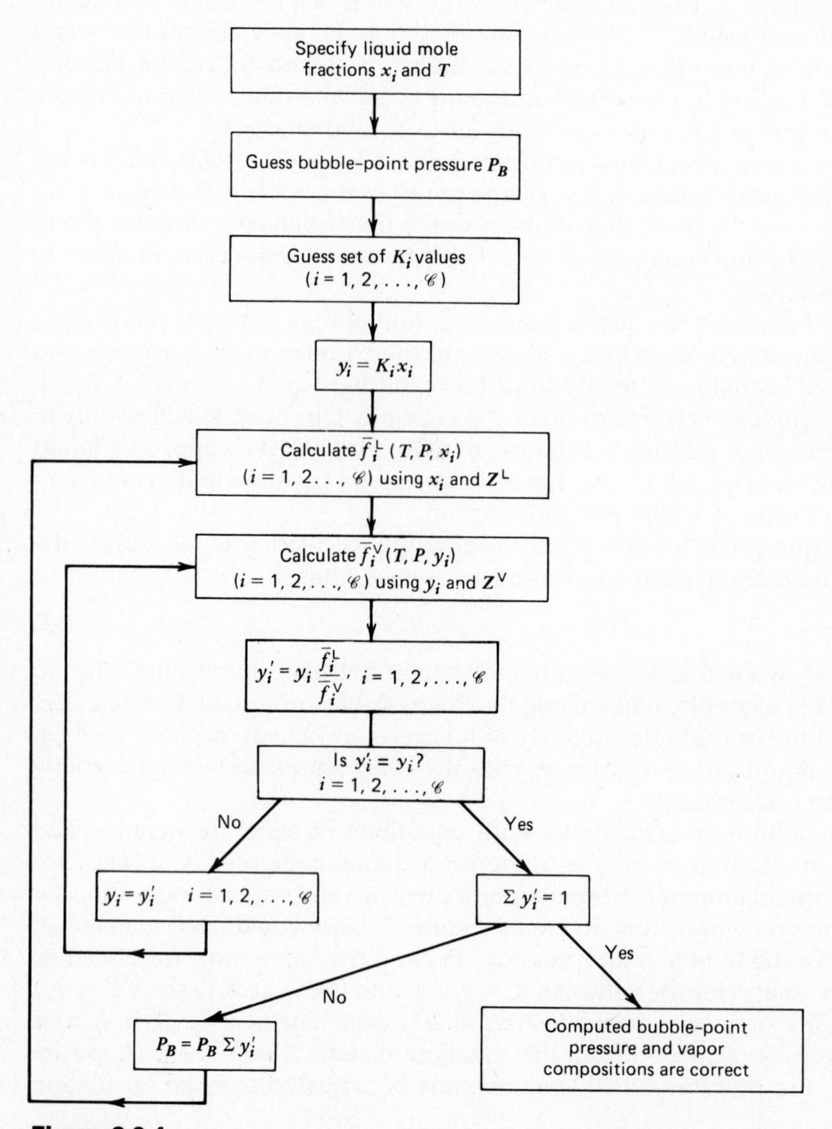

Figure 8.2-4
An algorithm for solving the bubble-point pressure problem using an equation of state.

where the pure component vapor pressure can be estimated using the Antoine equation, Eq. 5.7-8, with parameters for the fluid of interest, or by using the equation of state as described in Sec. 5.5.

Although the algorithm described in Fig. 8.2-4 is specific to the computation of the bubble-point pressure, slight changes make it applicable to other phase equilibrium calculations. For example, by specifying P, and replacing $P_B = P_B \sum y_i'$ by $T_B = T_B / \sum y_i'$, an algorithm for the bubble-point temperature calculation at fixed pressure is obtained. It is only slightly more difficult to change the calculational procedure so that dew-point pressure or temperature calculations can be made; this is left to you (Problem 8.2-5).

We will consider only one additional type of phase equilibrium calculation here, the isothermal flash calculation mentioned at the end of Sec. 8.1. In this calculation one needs to satisfy the equality of species fugacities relation (Eq. 8.2-1) as in other phase equilibrium calculations and also the mass balances (based on 1 mole of feed of mole fractions x_i°) discussed earlier

$$x_i L + y_i V = x_i^\circ \qquad i = 1, 2, \ldots, \mathscr{C} \tag{8.1-14}$$

$$L + V = 1 \tag{8.1-15}$$

and the summation conditions

$$\sum x_i = 1 \quad \text{and} \quad \sum y_i = 1 \tag{8.2-4}$$

In this calculation, T and P are known, but the liquid-phase mole fractions (x_i's), the vapor-phase mole fractions (y_i's) and the liquid-to-vapor split (L/V) are unknowns.

An algorithm for solving the flash problem is given in Fig. 8.2-5. It is based on making initial guesses for the equilibrium ratios $K_i = y_i/x_i$ as discussed earlier (Eq. 8.2-3), and for the fraction of liquid, L, and using the equations below obtained by simple rearrangement

$$x_i = \frac{x_i^\circ}{L + K_i(1 - L)} \tag{8.2-5}$$

and

$$y_i = K_i x_i$$

Furthermore, from Eqs. 8.2-4, we have

$$\sum_{i=1} x_i = \sum \frac{x_i^\circ}{L + K_i(1 - L)} = 1 \tag{8.2-6a}$$

and

$$\sum_{i=1} y_i = \sum_{i=1} \frac{K_i x_i^\circ}{L + K_i(1 - L)} = 1 \tag{8.2-6b}$$

or, equivalently,

$$\sum x_i - \sum y_i = \sum_{i=1} \frac{(1 - K_i)x_i^\circ}{L + K_i(1 - L)} = 0 \tag{8.2-6c}$$

It is Eqs. 8.2-5 and 6 that are used in the algorithm of Fig. 8.2-5. A BASIC language computer program, VLMU.BAS, which does bubble-point temperature and pressure, dew-point temperature and pressure and isothermal flash

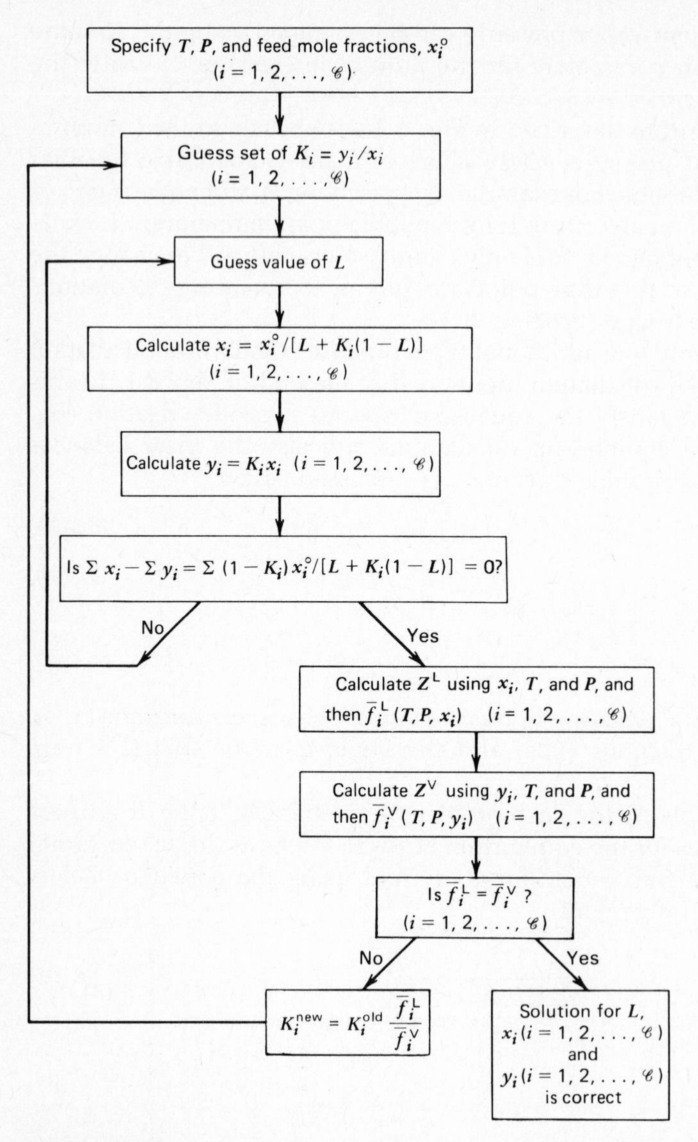

Figure 8.2-5
Flow diagram of an algorithm for the isothermal flash calculation using an equation of state.

calculations using the Peng–Robinson equation of state is discussed in Appendix A7.2 and appears on the disk which accompanies this book.

Now that the manner in which phase equilibrium calculations can be performed using equations of state has been discussed, we are in a position to consider the accuracy of such calculations. To begin, we again consider the ethane–propylene data in Fig. 8.2-1 using the isothermal flash algorithm, the Peng–Robinson equation of state, and setting the single adjustable parameter in the calculation, the binary interaction parameter k_{ij} in

$$a_{ij} = \sqrt{a_{ii}a_{jj}}\,(1 - k_{ij}) \tag{8.2-7}$$

equal to zero (so, in fact, there are no adjustable parameters). Using the BASIC language program of Appendix A7.2, the results indicated by the dashed lines in

Fig. 8.2-1 are obtained. As you can see, the predicted results are in excellent agreement with experiment. Furthermore, from the equation of state description, we get not only the compositions of the coexisting phases but also good estimates of the compressibilities or densities. Also, starting from Eqs. 4.4-27 and 28 and using the Peng–Robinson equation of state, one can show (Problem 8.2-4) that

$$\underline{H}(T, P, x_i) - \underline{H}^{IGM}(T, P, x_i) = RT(Z_m - 1) + \frac{T\left(\dfrac{da_m}{dT}\right) - a_m}{2\sqrt{2}\, b_m} \ln\left[\frac{Z_m + (\sqrt{2}+1)B_m}{Z_m - (\sqrt{2}-1)B_m}\right] \tag{8.2-8a}$$

and

$$\underline{S}(T, P, x_i) - \underline{S}^{IGM}(T, P, x_i) = R\ln(Z_m - B_m) + \frac{\dfrac{da_m}{dT}}{2\sqrt{2}\, b_m} \ln\left[\frac{Z_m + (\sqrt{2}+1)B_m}{Z_m - (\sqrt{2}-1)B_m}\right] \tag{8.2-8b}$$

where the subscript m denotes a mixture property and $\underline{H}^{IGM}(T, P, x_i)$ and $S^{IGM}(T, P, x_i)$ are the ideal gas mixture enthalpy and entropy, respectively, at the conditions of interest. Therefore, we can also compute the enthalpy and entropy of the vapor and liquid phases.

This example shows the real power of the equation of state description in that starting with relatively little information (T_C, P_C, and ω of the pure components), the phase equilibrium, phase densities, and other thermodynamic properties can be obtained. Of course, the weakness of the method is that, at present, we have equations of state that provide an accurate description only for hydrocarbons and other relatively simple substances.

For the ethane–propylene system calculation just described, we have used no adjustable parameters in the calculation. A careful examination of the results shows that the predicted compositions of ethane are systematically approximately 0.01 mole fraction too high. These predictions can be improved by using a regression procedure with the experimental data and the Peng–Robinson equation of state. Such detailed calculations have shown that setting $k_{ij} = 0.011$ in Eq. 8.2-7 improves the accuracy of the predictions, though this improvement would be barely visible on the scale of Fig. 8.2-1, and therefore is not shown. For other systems, containing species much more dissimilar in size and types of molecular interactions, the importance of the binary interaction parameter k_{ij} is more apparent. This is evident in Fig. 8.2-6, which contains experimental phase equilibrium data for the carbon dioxide–isopentane system at two temperatures, together with predictions (setting $k_{ij} = 0$) and correlations (regression of data to obtain $k_{ij} = 0.121$) using the Peng–Robinson equation of state. There we see that the predictions with the binary interaction parameter equal to zero are not nearly as good as those with $k_{ij} = 0.121$. Therefore, in engineering applications, at least one experimental data point is needed, so that a value of k_{ij} can be obtained; better still is to have several data points and choose the binary interaction parameter to give an optimum fit of all the data points. The binary interaction parameters given in Table 7.4-1 were obtained by such a regression procedure.

Although the discussion so far has been concerned with binary mixtures, the calculational procedures described here are applicable to multicomponent mixtures as well. For accurate predictions, however, one needs a value of the interaction parameter for each binary pair in the mixture.

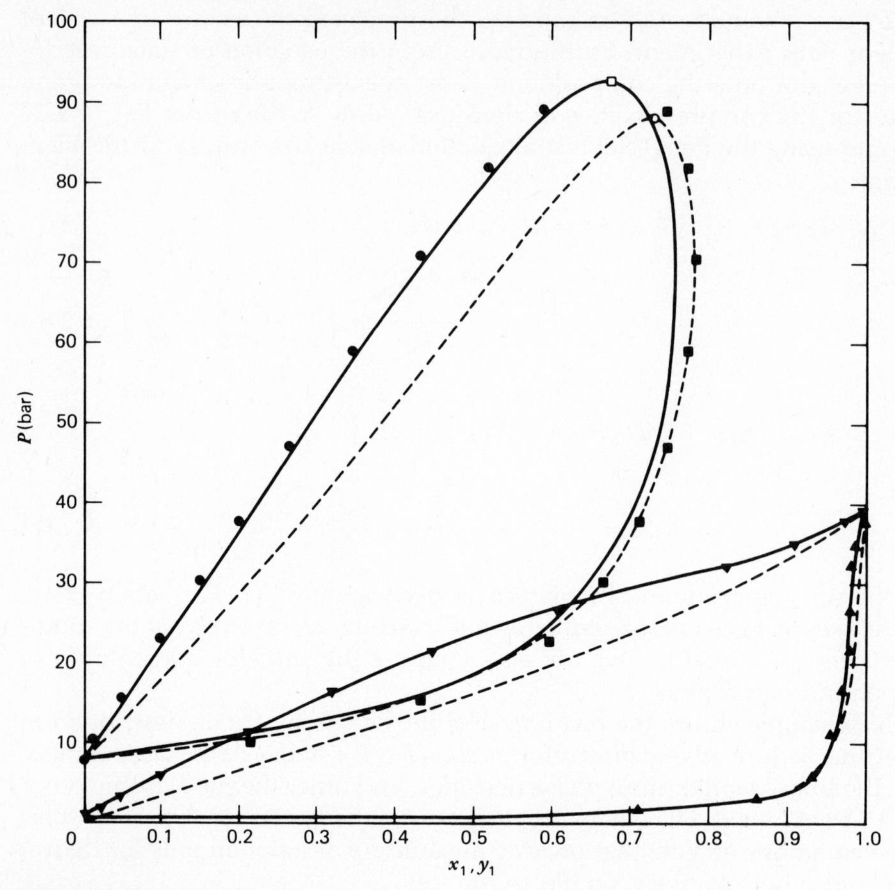

Figure 8.2-6
Vapor–liquid equilibrium of the carbon dioxide (1)–isopentane system. The experimental data of G. J. Besserer and D. B. Robinson [*J. Chem. Eng. Data* **20,** 93 (1976)] are shown at 277.59 K (▼ = liquid and ▲ = vapor) and 377.65 K (● = liquid and ■ = vapor). The dashed curves are the predictions using the Peng–Robinson equation of state with $k_{12} = 0$, and the solid lines are the correlation using the same equation of state with $k_{12} = 0.121$. The points ○ and □ are the estimated mixture critical points at 377.65 K using the same equation of state with $k_{12} = 0$ and 0.121, respectively.

As a final example of the complicated and unusual behavior that is possible in binary and multicomponent mixtures at high pressures, consider the vapor–liquid equilibrium for the ethane–*n*-heptane system shown in Fig. 8.2-7. The leftmost curve in this figure is the vapor pressure line as a function of temperature for pure ethane, and the rightmost curve is the vapor pressure curve for *n*-heptane. Between the two are the bubble-point and dew-point curves for a mixture at the constant composition of 58.71 mole percent ethane. The bubble-point curve is the locus of pairs of temperatures and pressures at which the first bubble of vapor will appear in a liquid of $x_{C_2H_6} = 0.5871$. (Note that the composition of this coexisting vapor cannot be found from the information in the figure. Why?) Similarly, the dew-point curve is the locus of temperature–pressure pairs at which the first drop of liquid appears in a vapor of $y_{C_2H_6} = 0.5871$. (Again, the composition of the coexisting liquid cannot be found from this figure.)

A number of interesting features are apparent from Fig. 8.2-7. First, the critical point of the mixture, which is the intersection of the bubble-point and

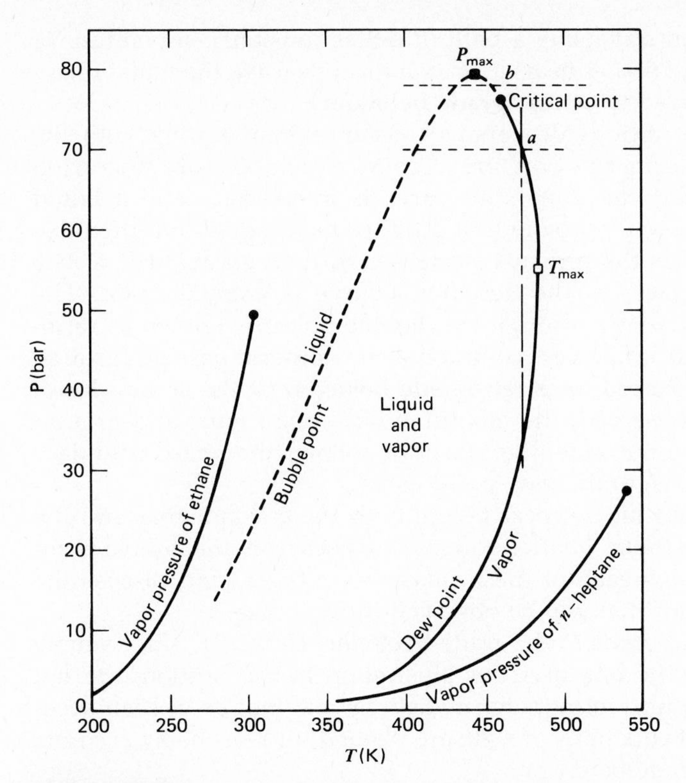

Figure 8.2-7
Vapor–liquid equilibrium for ethane, *n*-heptane, and their mixture at 58.71 mole percent ethane. [Data of W. B. Kay, *Ind. Eng. Chem.* **30,** 459 (1938).] The symbols ● denote critical points, ■ is the cricondenbar, and □ is the cricondentherm.

dew-point curves (indicated by ●), is intermediate in temperature, but at a much higher pressure, than the critical points of the pure components (also denoted by filled circles). Second, the critical point for the $x_{C_2H_6} = 0.5871$ mixture is neither the highest temperature nor the highest pressure along the phase boundary where vapor and liquid coexist. The point of maximum pressure along the phase boundary (indicated by ■) is referred to as the **cricondenbar,** and the point of maximum temperature (denoted by □) is called the **cricondentherm.**

Next note that if one has a vapor at $y_{C_2H_6} = 0.5871$, $P = 70$ bar and $T = 475$ K (denoted by point *a*) and cools it at constant pressure (following the line ----), first the dew-point curve is intersected and a drop of liquid forms. Further cooling produces additional liquid, so that both the vapor and liquid will differ from the starting composition. Finally, at about $T = 408$ K, the last bit of vapor condenses, and a liquid with $x_{C_2H_6} = 0.5871$ is obtained. This is usual phase equilibrium behavior.

However, starting with a fluid of ethane mole fraction 0.5871, $P = 78$ bar, and temperature about 455 K (point *b*), and cooling at constant pressure, the bubble-point curve is intersected. Consequently, the initial mixture was a liquid, and, at the bubble-point curve, the first bubble of vapor is produced. Reducing the temperature still further produces more vapor, until a temperature of approximately 445 K is reached. On further cooling, some of this vapor condenses until, at approximately $T = 433$ K, all the vapor is condensed and the bubble-point curve is crossed again into the region of all liquid. Vapor–liquid behavior such as this, in which, on traversing a path of either constant temperature or constant pressure, a phase first appears and then disappears so the initial phase is again obtained, is referred to as **retrograde behavior.**

Another example of retrograde behavior occurs when starting with the vapor at point *a* and reducing the pressure at constant temperature (following the line —-—). At 70 bar, the dew-point curve is intersected, and a liquid appears. Further reductions in pressure first produce more liquid, but then this liquid begins to vaporize as the pressure decreases further. At about $P = 34.5$ bar, all the liquid vaporizes, as the dew-point curve is crossed again. The behavior of an isotherm passing through two bubble points is known as **retrograde behavior of the first kind.** Less common is an isotherm passing through two dew points; this is referred to as **retrograde behavior of the second kind.** For such behavior to be observed, the mixture critical point must appear after both the cricondenbar and cricondentherm as one follows the phase boundary from the bubble-point curve to the dew-point curve.
[*Question:* The critical point may appear before both the cricondenbar and the cricondentherm, between them, or after both as one goes from the bubble-point to dew-point curves. Sketch each of these phase boundaries, and discuss the type of retrograde behavior that will be observed in each case.]

Equations of state do predict retrograde behavior. However, since simple cubic equations, such as the one used for illustration in this section, are not accurate in the critical region, retrograde predictions may not be of high accuracy. Accurate, multiterm equations of state are needed for reasonably accurate critical region and retrograde behavior.

Finally, we note from Figures 8.2-2 and 3 that the shapes of the critical loci of mixtures are complicated and that, in general, the critical temperature and/or pressure of a binary mixture is not intermediate to those of the pure fluids. This is to be contrasted with the pseudocritical mixing rules of Kay (Eqs. 7.7-1 and 2) and Prausnitz and Gunn (Eqs. 7.7-1 and 3) used in corresponding states theory that predict pseudocritical properties that are between the critical properties of the pure components. This difference arises because the pseudocritical conditions are estimates of the point at which, at fixed composition, the equation of state predicts

$$\left(\frac{\partial P}{\partial \underline{V}}\right)_{T,x} = \left(\frac{\partial^2 P}{\partial \underline{V}^2}\right)_{T,x} = 0$$

Unlike the situation for the pure fluid, this inflection point is not the mixture critical point. The mixture critical point is the point of intersection of the dew-point and bubble-point curves, and this must be determined from phase equilibrium calculations, more complicated mixture stability conditions, or experiment.

PROBLEMS FOR SECTION 8.2

8.2-1 The following mixture of hydrocarbons occurs in petroleum processing.

Component	Mole Percent
Ethane	5
Propane	57
n-Butane	38

Estimate the bubble-point temperature and the composition of the coexisting vapor for this mixture at all pressures above 1 bar.

8.2-2 Estimate the dew-point temperature and the composition of the coexisting liquid for the mixture in the previous problem at all pressures above 1 bar.

8.2-3 A liquid mixture of the composition given in Problem 8.2-1 is to be flashed at $P = 20$ bar and a collection of temperatures between the bubble-point temperature and the dew-point temperature. Determine the compositions of the coexisting vapor and liquid, and the vapor–liquid equilibrium split for several temperatures in this range.

8.2-4 Derive Eqs. 8.2-8.

8.2-5 **a** Develop an algorithm for the equation of state prediction of the dew-point pressure.

b Develop an algorithm for the equation of state prediction of the dew-point temperature.

8.2-6 **a** For the adiabatic, steady flow flash process from specified initial conditions of T_1 and P_1 to a specified final pressure P_2 shown here develop the equilibrium and balance equations to compute the final temperature, the vapor–liquid split, and the compositions of the coexisting phases.

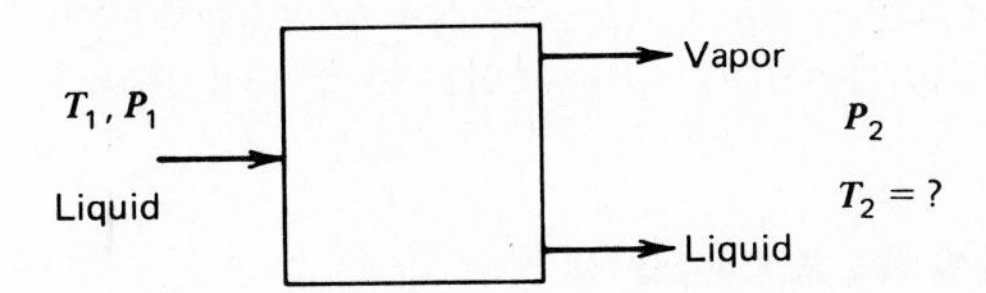

b Develop an algorithm for the calculation of part a using an equation of state.

c Incorporate this calculation into the program VLMU.BAS of Appendix A7.2.

8.2-7 **a** Make the best estimate you can of the composition of the vapor in equilibrium with a liquid containing 30.3 mole percent ethane and 69.7 mole percent ethylene at −0.01°C. Compare your results with the experimental data in the table.

b Repeat the calculation in part a at other compositions for which experimental data below are available.

Mole Percent of Ethane		
Liquid	Vapor	Pressure bar
7.8	6.2	39.73
22.8	19.7	37.07
30.3	25.5	35.60
59.0	53.1	32.13
89.0	85.4	25.45

8.2-8 Vapor–liquid equilibria in petroleum technology are usually expressed in terms of K factors $K_i = y_i/x_i$ where y_i and x_i are the mole fractions of species i in the vapor and liquid phases, respectively. Estimate the K values for methane and benzene in the benzene–methane system at 300 K and a total pressure of 30 bar.

8.2-9 In a petroleum refinery an equimolar stream containing propane and n-butane is fed to a flash separator operating at 40°C. Determine the pressure that this separator should be operated at so that an equal number of moles of liquid and vapor are produced.

8.2-10 A storage tank is known to contain the following mixture at 45°C and 15 bar:

Species	Overall Mole Fraction
Ethane	0.31
Propane	0.34
n-Butane	0.21
i-Butane	0.14

What is the composition of the coexisting vapor and liquid phases, and what fraction (by moles) of the contents of the tank are liquid?

8.3 THE SOLUBILITY OF A GAS IN A LIQUID

In the study of the solubility of a gas in a liquid one is interested in the situation in which the mixture temperature T is greater than the critical temperature of at least one of the components in the mixture. If the mixture can be described by an equation of state, no special difficulties are involved, and the calculations proceed as described in Sec. 8.2. Indeed, a number of cases encountered in Sec. 8.2 were of this type (e.g., ethane in the ethane–propylene mixture at 344.3 K); consequently, it is not necessary to consider the equation of state description of gas solubility separately from vapor-liquid equilibrium calculations.

The description of gas solubility using activity coefficient models requires some additional attention, however. The activity coefficient description is of interest because it is applicable to mixtures that are not describable by an equation of state, and also because it may be possible to make simple gas solubility estimates using an activity coefficient model, whereas a computer program is required for equation of state calculations.

To study the solubility of a gas in a liquid using an activity coefficient model we start with the equilibrium relation

$$\bar{f}_i^{\mathrm{L}}(T, P, x_i) = \bar{f}_i^{\mathrm{V}}(T, P, y_i) \qquad i = 1, 2, \ldots \tag{8.3-1}$$

which, after using the Lewis–Randall rule for the gas phase and the definition of the activity coefficient, reduces to

$$x_i\gamma_i(T, P, x_i)f_i^{\mathrm{L}}(T, P) = y_iP\left(\frac{f}{P}\right)_i \tag{8.3-2}$$

The situation of interest here is when the mixture temperature T is greater than the critical temperature of one of the components, say component 1 (i.e., $T >$

$T_{C,1}$), so that this species exists only as a gas in the pure component state. In this case the evaluation of the liquid-phase properties for this species, such as $f_1^L(T, P)$ and $\gamma_1(T, P, x_1)$, is not straightforward. (It is this complication that distinguishes gas solubility problems from those of vapor–liquid equilibrium, which were considered in Sec. 8.1.) We will refer to liquid-phase species above their critical temperatures as the solutes. For those species below their critical temperature, which we designate as the solvents, Eq. 8.3-2 is used just as in Sec. 8.1.

If the temperature of the mixture is only slightly greater than the critical temperature of the gaseous (solute) species, the (hypothetical) pure component liquid-phase fugacity $f_i^L(T, P)$ can be computed using the fugacity extrapolation scheme for nonsimple mixtures discussed in Sec. 7.8. In this case, the gas solubility problem is just like the vapor–liquid equilibrium problem of Sec. 8.1 and is treated accordingly.

However, if the temperature of the mixture is well above $T_{C,1}$, the evaluation of $f_1^L(T, P)$ is more troublesome. A number of different procedures for estimating this hypothetical liquid fugacity have been proposed. Prausnitz and Shair[7] have suggested the simple, approximate corresponding states correlation of Fig. 8.3-1 be used to evaluate the liquid fugacity at 1.013 bar total pressure,

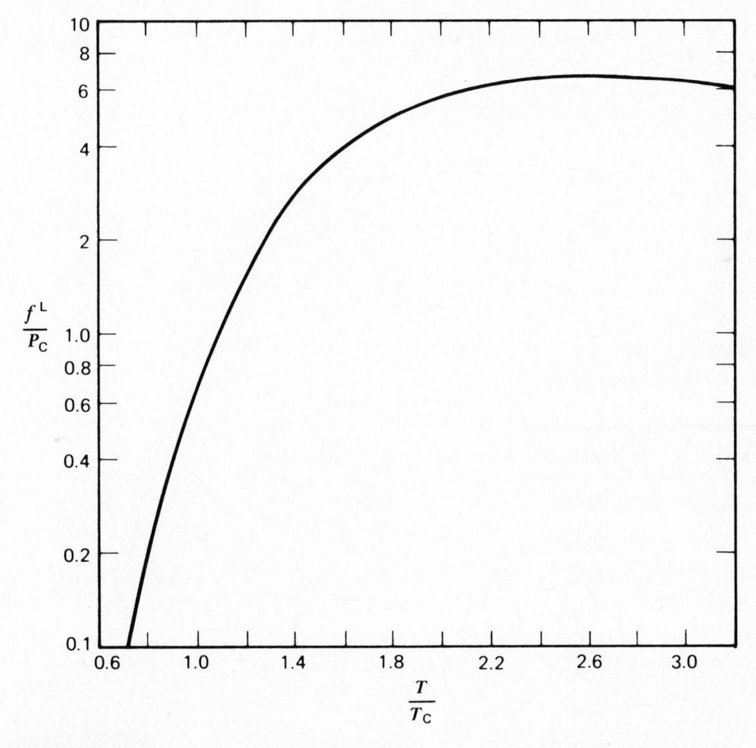

Figure 8.3-1
Extrapolated liquid-phase fugacity coefficients at 1.013 bar as a function of reduced temperature. [This figure originally appeared in J. M. Prausnitz and F. H. Shair, *A. I. Ch. E. J.* **7,** 682 (1961). It appears here courtesy of the copyright owners, the American Institute of Chemical Engineers.]

[7] J. M. Prausnitz and F. H. Shair, *A.I.Ch.E. J.* **7**, 682 (1961); also, J. M. Prausnitz, R. N. Lichtenthaler, and E. G. Azevedo, *Molecular Thermodynamics of Fluid Phase Equilibrium*, 2nd ed. Prentice–Hall, Englewood Cliffs, N.J., 1986, p. 392ff.

that is, $f_1^L(T, P = 1.013 \text{ bar})$. To compute the hypothetical liquid phase fugacity at any other pressure, a Poynting pressure correction is made to the value at 1.013 bar as follows:

$$f_1^L(T, P) = f_1^L(T, P = 1.013 \text{ bar}) \exp\left[\int_{1.013 \text{ bar}}^{P} \frac{\underline{V}_1^L}{RT}\, dP\right]$$

$$\cong f_1^L(T, P = 1.013 \text{ bar}) \exp\left[\frac{\underline{V}_1^L(P - 1.013 \text{ bar})}{RT}\right] \tag{8.3-3}$$

Unfortunately, Eq. 8.3-3 introduces another unknown quantity, the molar volume of the hypothetical liquid. These have been tabulated by Prausnitz and Shair, and are given in Table 8.3-1.

The question of evaluating the liquid-phase activity coefficient of the solute species still remains. Although experimental data for γ_1 would be preferable, such data are rarely available. Consequently, various liquid solution theories and correlations are used. If the regular solution model is used, we have

$$\ln \gamma_1(x_1) = \frac{\underline{V}_1^L \Phi_2^2(\delta_1 - \delta_2)^2}{RT} \tag{8.3-4a}$$

for pure solvents, and

$$\ln \gamma_1(x_1) = \frac{\underline{V}_1^L(\delta_1 - \bar{\delta})^2}{RT} \tag{8.3-4b}$$

with

$$\bar{\delta} = \sum_j \Phi_j \delta_j \qquad \text{and} \qquad \Phi_j = \frac{x_j \underline{V}_j^L}{\sum_l x_l \underline{V}_l^L}$$

Table 8.3-1
"LIQUID" VOLUMES AND SOLUBILITY PARAMETERS FOR GASEOUS SOLUTES AT 25°C

Gas	$\underline{V}^L$(cc/mol)	δ (cal/cc)$^{1/2}$
N_2	32.4	2.58
CO	32.1	3.13
O_2	33.0	4.0
Ar	57.1	5.33
CH_4	52	5.68
CO_2	55	6.0
Kr	65	6.4
C_2H_4	65	6.6
C_2H_6	70	6.6
Cl_2	74	8.7

Source: This table originally appeared in J. M. Prausnitz and F. H. Shair, *A.I.Ch.E. J.* **7**, 682 (1961). It appears here courtesy of the copyright owners, the American Institute of Chemical Engineers.

for mixed solvents. The Prausnitz and Shair estimates for the solubility parameters of the hypothetical liquids of several common gases at 25°C have also been tabulated in Table 8.3-1. (It is interesting to note that the values of this parameter for the hypothetical liquids at 25°C are quite different from that for the real liquids at 90 K given in Table 7.6-1.) Of course, any other solution model, for which the necessary parameters are available, can be used to evaluate γ_1. (However, as the UNIFAC model can only be used with substances that are liquid at 25°C and 1.013 bar, it cannot be used.)

Using these estimates for the liquid-phase fugacity and activity coefficient of the solute species, Eqs. 8.3-2 and 3 can be combined to give

$$x_1 = \frac{y_1 P (f/P)_1}{\gamma_1(T, P, x_i) f_1^{\mathrm{L}}(T, P = 1.013 \text{ bar}) \exp[\underline{V}_1^{\mathrm{L}}(P - 1.013 \text{ bar})/RT]} \tag{8.3-5}$$

This equation is solved together with the equilibrium relations for the solvent species

$$x_i = \frac{y_i P (f/P)_i}{\gamma_i(T, P, x_i) P_i^{\mathrm{vap}} (f/P)_{\mathrm{sat},i} \exp[\underline{V}_i^{\mathrm{L}}(P - P_i^{\mathrm{vap}})/RT]} \tag{8.3-6}$$

to compute the solute solubility in the liquid solvent, and the solvent solubility in the gas.

For ideal solutions [i.e., solutions for which $\gamma_i(x_i) = 1$], the solubility of the gas depends on its partial pressure (or gas-phase fugacity), and not on the liquid or liquid mixture into which it dissolves. This solubility is termed the ideal solubility of the gas.

Since the fugacity of the solute species, obtained either from the extrapolation of the vapor pressure or from Fig. 8.3-1, will be very large, the mole fraction x_1 of the gaseous species in the liquid is likely to be quite small. This observation may provide a useful simplification in the solution of Eqs. 8.3-5 and 6. Also, if one is merely interested in the solubility of the gas in the liquid for a given gas phase partial pressure, only Eq. 8.3-5 need be solved.

When a gas is only sparingly soluble in a liquid or liquid mixture (i.e., as $x_1 \to 0$), it is observed that the liquid-phase mole fraction of the solute species is, at fixed temperature, linearly proportional to its gas-phase fugacity, that is,

$$x_1 H_1(T, P) = \bar{f}_1^{\mathrm{V}}(T, P) = y_1 P \left(\frac{f}{P}\right)_1 \qquad \text{as } x_1 \to 0 \tag{8.3-7}$$

where H_1 is the Henry's law constant[8] (see Sec. 7.8). Gas solubility measurements are frequently reported in terms of the Henry's law constant, so that values of H for many gas–liquid pairs appear in the chemical and chemical engineering literature.

To relate the Henry's law constant to other thermodynamic quantities, we recognize that since, at equilibrium

$$\bar{f}_1^{\mathrm{L}}(T, P, x_1) = \bar{f}_1^{\mathrm{V}}(T, P, y_1) = x_1 H_1(T, P) \qquad \text{as } x_1 \to 0$$

we can take the following to be the formal definition of the Henry's law constant

$$\lim_{x_1 \to 0} \frac{\bar{f}_1^{\mathrm{L}}(T, P, x_1)}{x_1} = H_1(T, P) \tag{8.3-8}$$

[8]The Henry's law "constant," as indicated in Sec. 7.8, is independent of concentration, but a function of temperature, pressure and solvent.

Comparing Eqs. 8.3-2 and 8 yields

$$H_1(T, P) = \gamma_1(x_1 = 0)f_1^L(T, P) = \gamma_\alpha f_1^L(T, P) \tag{8.3-9}$$

where $\gamma_1(x_1 = 0)$ is the limiting value of the activity coefficient of the gas in the liquid at infinite dilution. Thus, Eqs. 8.3-3 and 4 and the correlation of Fig. 8.3-1 can be used to predict values of the Henry's law constant.

As the pressure, and hence the solute mole fraction x_1, increases, deviations from this simple limiting law are observed (see Fig. 7.8-3*a*). For appreciable concentrations of the gaseous species in the liquid phase we write instead

$$y_1 P\left(\frac{f}{P}\right)_1 = x_1\gamma_1^*(x_1)H_1(T, P) \tag{8.3-10}$$

where

$$\gamma_1^*(x_1) = \frac{\gamma_1(x_1)}{\gamma_1(x_1 = 0)}$$

is the renormalized activity coefficient defined by Eq. 7.8-5. Clearly $\gamma_1^* \to 1$ as $x_1 \to 0$, and γ_1^* departs from unity as the mole fraction of the solute increases. The regular solution theory prediction for γ_1^* (see Problem 7.9) is

$$\ln \gamma_1^*(x_1) = \ln \gamma_1(x_1) - \ln \gamma_1(x_1 = 0) = \frac{\underline{V}_1^L(\gamma_1 - \delta_2)^2(\Phi_2^2 - 1)}{RT} \tag{8.3-11}$$

ILLUSTRATION 8.3-1

Estimate the solubility and Henry's law constant for carbon dioxide in a liquid mixture of toluene and carbon disulfide as a function of CS_2 mole fraction at 25°C and a partial pressure of CO_2 of 1.013 bar (1 atm).

Data

See Tables 4.6-1, 7.5-1, and 8.3-1.

Solution

Equation 8.3-5 provides the starting point for the solution of this problem. Since the partial pressure of carbon dioxide and the vapor pressures of toluene and carbon disulfide are so low, the total pressure must be low, so

$$\left(\frac{f}{P}\right) = 1 \quad \text{and} \quad \exp\left[\frac{\underline{V}_{CO_2}^L(P - 1.013 \text{ bar})}{RT}\right] = 1$$

Next, using the regular solution model for γ, we obtain

$$x_{CO_2} = \frac{y_{CO_2}P}{f_{CO_2}^L(T, P = 1.013 \text{ bar}) \exp\left[\frac{\underline{V}_1^L(\delta_{CO_2} - \bar{\delta})^2}{RT}\right]}$$

with

$$\bar{\delta} = \sum_j \Phi_j \delta_j$$

The reduced temperature of CO_2 is $T_r = 298.2 \text{ K}/304.3 \text{ K} = 0.98$, so from the Shair–Prausnitz correlation $f^L/P_C \approx 0.82$ and $f^L \approx 0.82 \times 73.76 \text{ bar} = 60.48$ bar. To calculate the activity coefficients we will assume that CO_2 is only slightly soluble in the solvents, so that its volume fraction is small; we will then verify this assumption. Thus, as a first guess, the contribution of CO_2 to $\bar{\delta}$ will be neglected.

To compute the solubility of CO_2 in pure carbon disulfide, we note that

$$\bar{\delta} \approx \delta_{CS_2} = 10\ (\text{cal/cc})^{1/2} \quad \text{and} \quad (\delta_{CO_2} - \bar{\delta})^2 = 16\ \text{cal/cc} = 66.94\ \text{J/cc}$$

so that

$$x_{CO_2} = \frac{1.013\ \text{bar}}{60.48\ \text{bar} \times \exp\left\{\dfrac{55\ \text{cc/mol} \times 66.94\ \text{J/cc}}{8.314\ \text{J/mol K} \times 298.2\ \text{K}}\right\}} = 3.79 \times 10^{-3}$$

(The experimental value is $x_1 = 3.28 \times 10^{-3}$.)
Also

$$H = P/x_{CO_2} = 1.013\ \text{bar}/3.79 \times 10^{-3} = 267\ \text{bar}$$

The solubility of CO_2 in pure toluene is computed as follows

$$\bar{\delta} \approx \delta_T = 8.9\ (\text{cal/cc})^{1/2} \quad \text{and} \quad (\delta_{CO_2} - \bar{\delta})^2 = 8.4\ \text{cal/cc} = 35.15\ \text{J/cc}$$

so that

$$x_{CO_2} = \frac{1.013\ \text{bar}}{60.48\ \text{bar} \times \exp\left\{\dfrac{55 \times 35.15}{8.314 \times 298.2}\right\}} = 7.68 \times 10^{-3}$$

and

$$H = 131.9\ \text{bar}$$

Finally, the solubility of CO_2 in a 50 mole percent toluene, 50 mole percent CS_2 mixture is gotten from

$$\underline{V}^L_{mix} = x_{CS_2}\underline{V}^L_{CS_2} + x_T\underline{V}^L_T = 0.5 \times 61\ \frac{\text{cc}}{\text{mol}} + 0.5 \times 107\ \frac{\text{cc}}{\text{mol}} = 84\ \frac{\text{cc}}{\text{mol}}$$

$$\Phi_{CS_2} = \frac{0.5 \times 61}{84} = 0.363, \qquad \Phi_T = \frac{0.5 \times 107}{84} = 0.637$$

$$\bar{\delta} = 0.363 \times 10\left(\frac{\text{cal}}{\text{cc}}\right)^{1/2} + 0.637 \times 8.9\left(\frac{\text{cal}}{\text{cc}}\right)^{1/2}$$

$$= 9.30\ (\text{cal/cc})^{1/2}$$

and

$$(\delta - \bar{\delta})^2 = 10.88\ \text{cal/cc} = 45.52\ \text{J/cc}$$

Thus

$$x_{CO_2} = \frac{1.013}{60.48 \times \exp\left\{\dfrac{55 \times 45.52}{8.314 \times 298.2}\right\}} = 6.10 \times 10^{-3}$$

and

$$H = 166.0\ \text{bar}$$

These results are plotted in Fig. 8.3-2. In all cases x_{CO_2} is small, as had initially been assumed, so that an iterative calculation is not necessary.

Comment
The UNIFAC group contribution model is only meant to be used with substances that are liquids at 25°C and 1.013 bar. Consequently, UNIFAC cannot be used with carbon dioxide, and therefore cannot be used to estimate its solubility in either toluene or carbon disulfide. ■

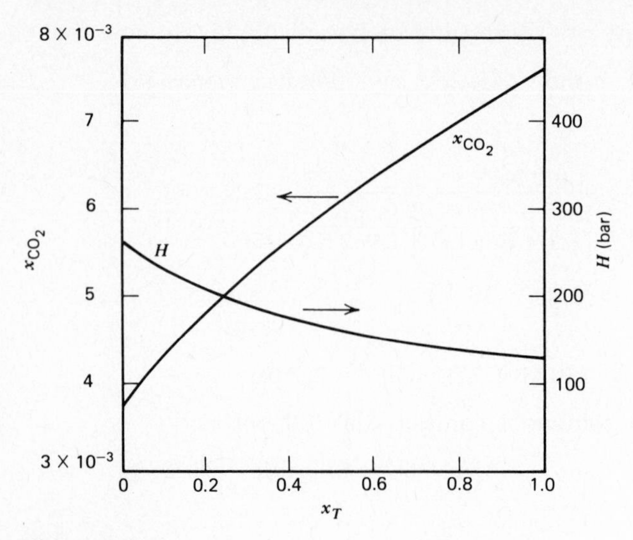

Figure 8.3-2
The Henry's law constant for carbon dioxide in carbon disulfide–toluene mixtures.

ILLUSTRATION 8.3-2

Predict the solubility of carbon dioxide in toluene at 25°C and a 1.013 bar carbon dioxide partial pressure using the Peng–Robinson equation of state.

Solution

The critical properties for both carbon dioxide and toluene are given in Table 4.6-1. The binary interaction parameter for the CO_2–toluene mixture is not given in Table 7.4-1. However, the value for CO_2–benzene is 0.077 and that for CO_2–n-heptane is 0.10. Therefore, we estimate that the CO_2–toluene interaction parameter will be 0.09. Using this value and the bubble-point pressure program of VLMU.BAS, the following values were obtained

x_{CO_2}	P_{tot}(bar)	y_{CO_2}	P_{CO_2}(bar) $= y_{CO_2}P_{tot}$
0.001	0.11	0.6579	0.072
0.002	0.20	0.7915	0.158
0.004	0.36	0.8834	0.318
0.006	0.51	0.9189	0.469
0.008	0.67	0.9378	0.628
0.010	0.83	0.9495	0.788
0.0125	1.03	0.9590	0.988
0.013	1.07	0.9605	1.028
0.015	1.23	0.9655	1.188

Therefore, using the Peng–Robinson equation state we estimate that at a partial pressure of 1.013 bar, carbon dioxide will be soluble in liquid toluene to the extent of 0.0128 mole fraction. This values differs from the value of 0.0077 computed in the last illustration using the Prausnitz–Shair correlation and regular solution theory. However, given the inaccuracies of both methods, this difference is not unreasonable.

Comment

Had we assumed that the CO_2–toluene binary interaction parameter was zero, the predicted CO_2 solubility in toluene at 1.013 bar CO_2 partial pressure would be 0.0221 mole fraction (Problem 8.3-6).

There is no binary interaction parameter reported for the CO_2–CS_2 mixture, nor for any similar mixtures. If we assume the binary interaction parameter is zero, we find that the CO_2 solubility in CS_2 at 1.013 bar CO_2 partial pressure is 0.0159 mole fraction, which is a factor of five greater than the measured value (Problem 8.3-7). However, if we set $k_{CO_2-CS_2} = 0.2$, we obtain a CO_2 solubility of 3.4×10^{-3}, which is in excellent agreement with experiment. ■

It is clear from Eq. 8.3-9 that the Henry's law constant will vary with pressure, since f_1^L and γ_1 are functions of pressure. The common method of accounting for this pressure variation is to define the Henry's law constant to be specific to a fixed pressure P_0 (frequently taken to be atmospheric pressure) and then include a Poynting pressure correction for other pressures. Independent of whether we apply the correction to the fugacity of the solute species in solution $\bar{f}_1^L(T, P, x_1 \to 0)$ or separately to the pure component fugacity and the infinite dilution activity coefficient (see Eq. 7.3-20), we obtain

$$\bar{f}_1^L(T, P, x_1 \to 0) = x_1\gamma_1(T, P_0, x_1 \to 0) f_1^L(T, P_0) \exp\left[\int_{P_0}^{P} \frac{\bar{V}_1^L(x_1 = 0)}{RT} dP\right] \quad (8.3\text{-}12)$$

where $\bar{V}_1^L(x_1 = 0)$ is the partial molar volume of the gaseous species in the liquid at infinite dilution. Using this expression in Eq. 8.3-10 yields

$$y_1 P \left(\frac{f}{P}\right)_1 = x_1\gamma_1^*(T, P_0) H_1(T, P_0) \exp\left[\int_{P_0}^{P} \frac{\bar{V}_1^L(x_1 = 0)}{RT} dP\right] \quad (8.3\text{-}13)$$

Finally, we note from Fig. 8.3-3 that the solubility in a liquid of some gases increases as the temperature increases, whereas for other gases it decreases. To

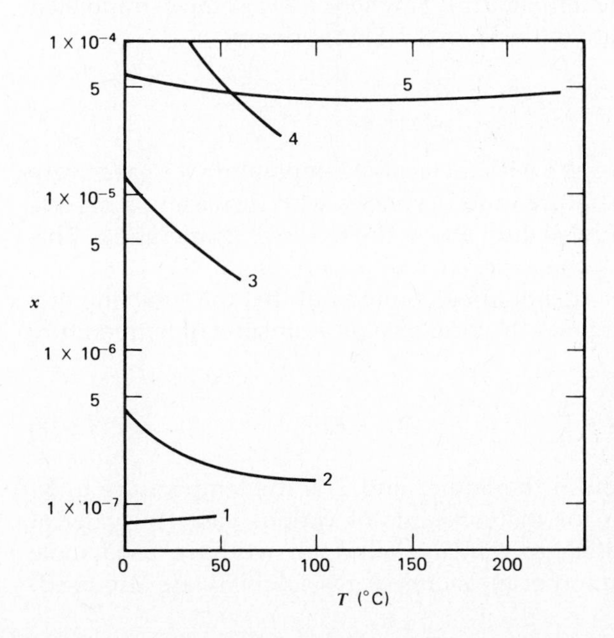

Figure 8.3-3
The solubility of several gases in liquids as a function of temperature. The solubility is expressed as mole fraction of the gas in the liquid at a gas partial pressure of 0.01 bar. Curve 1: Helium in water. Curve 2: Oxygen in water. Curve 3: Carbon dioxide in water. Curve 4: Bromine in water. Curve 5: Methane in *n*-heptane.

explain this observation we take the derivative of Eq. 8.3-2 with respect to temperature (at constant pressure and gas-phase composition) to get

$$0 = \left(\frac{\partial \ln x_1}{\partial T}\right)_P + \left(\frac{\partial \ln \gamma_1}{\partial T}\right)_P + \left(\frac{\partial \ln f_1^{\mathrm{L}}}{\partial T}\right)_P \tag{8.3-14}$$

where we have neglected the slight temperature dependence of the gas-phase fugacity coefficient. Now if we assume that the fugacity of the pure hypothetical liquid is obtained by extrapolating the vapor pressure of the real liquid, we have

$$\left(\frac{\partial \ln f_1^{\mathrm{L}}}{\partial T}\right)_P = \left(\frac{\partial \ln P_1^{\mathrm{vap}}}{\partial T}\right)_P = \frac{\Delta \underline{H}_1^{\mathrm{vap}}}{RT^2} \tag{8.3-15}$$

by the Clausius–Clapeyron equation, Eq. 5.7-5. Here, again, we have neglected the temperature dependence of the fugacity coefficient. Next, from Eq. 7.3-21 we have

$$\left(\frac{\partial \ln \gamma_1}{\partial T}\right)_P = -\frac{\overline{H}_1^{ex}(T, P, x_i)}{RT^2} \tag{8.3-16}$$

Combining Eqs. 8.3-14, 15, and 16 we get

$$\begin{aligned}\left(\frac{\partial \ln x_1}{\partial T}\right)_P &= \frac{-\Delta \underline{H}_1^{\mathrm{vap}} + \overline{H}_1^{ex}}{RT^2} = \frac{-(\underline{H}_1^{\mathrm{V}} - \underline{H}_1^{\mathrm{L}}) + (\overline{H}_1^{\mathrm{L}} - \underline{H}_1^{\mathrm{L}})}{RT^2} \\ &= \frac{-(\underline{H}_1^{\mathrm{V}} - \overline{H}_1^{\mathrm{L}})}{RT^2} \approx \frac{-\Delta \underline{H}_1^{\mathrm{vap}}}{RT^2}\end{aligned} \tag{8.3-17}$$

since $\overline{H}^{ex}$ is usually much smaller than $\Delta \underline{H}_1^{\mathrm{vap}}$, the heat of vaporization of the pure solute. [Note that $(\underline{H}_1^{\mathrm{V}} - \overline{H}_1^{\mathrm{L}})$ may be interpreted as the heat of vaporization of species 1 from the liquid *mixture*.]

For all fluids below their critical temperature, $\Delta \underline{H}^{\mathrm{vap}}$ is positive; that is, energy is absorbed in going from the liquid to the gas. For $T > T_C$, but $P < P_C$, $\Delta \underline{H}^{\mathrm{vap}}$ must be evaluated by extrapolation of the liquid-phase enthalpy into the vapor region. Here one finds that $\Delta \underline{H}^{\mathrm{vap}}$ is positive in the vicinity of the critical temperature (but below the critical pressure), though its magnitude decreases as T increases. Finally, above some temperature T, where $T \gg T_C$, the extrapolated enthalpy change becomes negative (Problem 8.3-1). Therefore,

$$\left(\frac{\partial \ln x_1}{\partial T}\right)_P \begin{cases} >0 & T \gg T_C \\ <0 & \text{otherwise} \end{cases}$$

Thus the solubility of a gas increases with increasing temperature for gases very much above their critical temperature, and decreases with increasing temperature at temperatures near or only slightly above their critical temperature. This conclusion is in agreement with the experimental data of Fig. 8.3-3.

Before leaving this section, it should be pointed out that the solubility of a gas at fixed partial pressure is frequently correlated as a function of temperature in the form

$$\ln x = A + \frac{B}{T} + C \ln T + DT + ET^2 \tag{8.3-18}$$

where x is the gas mole fraction in the liquid, and T is the temperature in K. Table 8.3-2 contains the values for the constants of various gases in water at 1.013 bar partial pressure of the gas. Finally, although we have used mole fraction throughout this section, other measures of gas solubility are also used. Some are listed in Table 8.3-3.

Table 8.3-2

MOLE FRACTION SOLUBILITY OF GASES IN WATER AT 1.013 BAR PARTIAL PRESSURE AS A FUNCTION OF TEMPERATURE

$\ln x = A + \frac{B}{T} + C \ln T + DT + ET^2$; ($T$ = K)

Gas	T Range	A	B	C	D	E
Helium	273–348	− 105.977	4 259.62	14.0094	—	—
Neon	273–348	− 139.967	6 104.94	18.9157	—	—
Argon	273–348	− 150.413	7 476.27	20.1398	—	—
Krypton	273–353	− 178.533	9 101.66	24.2207	—	—
Xenon	273–348	− 201.227	10 521.0	27.4664	—	—
Radon	273–373	− 251.751	13 002.6	35.0047	—	—
Hydrogen	274–339	− 180.054	6 993.54	26.3121	−0.0150432	—
Deuterium	278–303	− 181.251	7 309.62	26.1780	−0.0118151	—
Nitrogen	273–348	− 181.5870	8 632.129	24.79808	—	—
Oxygen	273–333	− 1 072.48902	27 609.2617	191.886028	−0.483090199	2.24445261E-4
Ozone	277–293	− 14.9645	1 965.31	—	—	—
Carbon monoxide	278–323	− 427.656023	15 259.9953	67.8429542	−0.0704595356	—
Carbon dioxide	273–373	− 4 957.824	105 288.4	933.1700	−2.854886	1.480857E-3
Methane	275–328	− 416.159289	15 557.5631	65.2552591	−0.0616975729	—
Ethane	275–323	−11 268.4007	221 617.099	2 158.42179	−7.18779402	4.05011924E-3
Ethylene	287–346	− 176.910	9 110.81	24.0436	—	—
Acetylene	274–343	− 156.509	8 160.17	21.4023	—	—
Propane	273–347	− 316.460	15 921.2	44.3243	—	—
Cyclopropane	298–361	326.902	− 13 526.8	− 50.9010	—	—
n-Butane	276–349	− 290.2380	15 055.5	40.1949	—	—
Isobutane	278–343	96.1066	− 2 472.33	− 17.3663	—	—
Neopentane	288–353	− 437.182	21 801.4	61.8894	—	—
CH_3F	273–353	− 135.910	7 600.23	18.1780	—	—
CH_3Cl	277–353	− 172.503	9 768.67	23.4241	—	—
CH_3Br	278–353	− 163.745	9 641.71	22.0397	—	—
CF_4	275–323	− 342.437	16 250.6	48.3441	—	—

Table 8.3-2 *(Continued)*

Gas	*T* Range	*A*	*B*	*C*	*D*	*E*
CH_2FCl	283–352	−138.912	8 141.19	18.5510	—	—
CHF_2Cl	297–352	−190.691	13 083.6	22.7782	0.032357	—
CHF_3	298–348	−19.1037	3 214.07	—	—	—
C_2F_6	278–303	−644.222	29 933.6	93.0269	—	—
C_2F_5Cl	298–348	−21.3965	2 802.35	—	—	—
Vinyl chloride	273–358	−2 399.543	72 028.1	408.576	−0.594578	—
C_2F_4	273–343	−184.958	10 843.3	23.1458	0.0209611	—
C_3F_8	278–288	−10 180.5	438 919	1 526.05	—	—
C_3F_6	278–343	−66.5820	4 491.54	6.90981	—	—
c-C_4F_8	278–303	−759.717	35 705.7	110.033	—	—
COS	273–303	−221.211	12 025.1	30.3659	—	—
CH_3NH_2	298–333	−9.1917	2 607.10	—	—	—
$(CH_3)_2NH$	298–333	−14.0980	4 026.14	—	—	—
$C_2H_5NH_2$	298–333	−12.6231	3 628.65	—	—	—
NH_3	273–373	−81.7466	1 096.82	16.5603	0.0602469	—
N_2O	273–313	−158.6208	8 882.80	21.2531	—	—
NO	273–353	−328.097	12 541.9	50.7616	−0.0451331	—
H_2S	273–333	−149.537	8 226.54	20.2308	0.00129405	—
SO_2	283–306	−13.0502	2 792.62	—	—	—
SF_6	275–323	−435.519	20 901.8	61.9692	—	—
Cl_2	283–313	108.389	−2 428.63	−19.1855	0.00892064	—
Cl_2O	273–293	−7.2207	1 798.85	—	—	—
ClO_2	283–333	56.7389	143.179	−10.7454	—	—
Air	273–373	−388.760	14 097.6	61.2018	−0.0617537	—

Experimental data were not corrected for nonideality of the gas phase and chemical reactions with the solvent. The quoted coefficients are valid in the temperature range given in the second column.

Source: S. Cabani and P. Gianni in H.-J. Hinz, editor, "Thermodynamic Data for Biochemistry and Biotechnology," Springer-Verlag, Berlin, 1986, p. 261–262. Used with permission.

Table 8.3-3
CONVERSION FORMULAS OF VARIOUS EXPRESSIONS OF GAS SOLUBILITY IN WATER TO MOLE FRACTION (x) OF THE DISSOLVED GAS, UNDER A GAS PARTIAL PRESSURE OF 1.013 BAR

Quantity	Symbol	Definition	Conversion Formula
Bunsen coefficient	α	Volume of gas, reduced to 273.15 K and 1.013 bar, absorbed by a unit volume of the absorbing solvent at the temperature of measurement under a gas partial pressure of 1.013 bar	$x = \left[1 + \frac{1.244 \times 10^{-3} \cdot \rho_W}{\alpha}\right]^{-1}$
Ostwald coefficient	L	Ratio of the volume of gas absorbed to the volume of absorbing liquid, both measured at the same temperature	$x = \left[1 + \frac{4.555 \times 10^{-6}\, T \cdot \rho_W}{L}\right]^{-1}$
Kuenen coefficient	S	Volume of gas (cm^3) at a partial pressure of 1.013 bar reduced to 273.15 K and 1.013 bar, dissolved by the quantity of solution containing 1 g of solvent	$x = \frac{S}{S + 1244.1}$
Henry's law constant	H	Limiting value of the ratio of the gas partial pressure to its mole fraction in solution as the latter tends to zero	$x = \frac{1.013}{H}$
Weight solubility	S_0	Grams of gas dissolved by 100 g of solvent, under a gas partial pressure of 1.013 bar	$x = \frac{S_0 \cdot 18.015}{100\, m_s + S_0 \cdot 18.105}$

The ideal behavior of the solute in the gas phase is assumed. T = temperature (K); ρ_W = water density (g m^{-3}); m_s = solute molecular weight.
The Henry's constant is expressed in bar.
Based on table in S. Cabani and P. Gianni, in H.-J. Hinz, editor, "Thermodynamic Data for Biochemistry and Biotechnology," Springer-Verlag, Berlin, 1986, p. 260. Used with permission.

PROBLEMS FOR SECTION 8.3

8.3-1 In this exercise we want to establish that the extrapolated value of $\Delta\hat{H}^{vap}$ is positive for a liquid at its critical temperature but below its critical pressure, and becomes negative as the temperature is increased much above the critical temperature. Use the data in the steam tables in Appendix III to prepare a plot of $\hat{H}^L$ and $\hat{H}^V$ for liquid water and steam at 0.1 MPa for all temperatures above 0°C, and extrapolate $\hat{H}^L$ to high temperatures from the low-pressure liquid-phase enthalpy data.

a Estimate the temperature at which $\Delta\hat{H}^{vap}$ will equal zero.

b Find the hypothetical $\Delta\hat{H}^{vap}$ of water at 1100°C and 1 bar.

8.3-2 Here we want to estimate the solubility of gaseous nitrogen in liquid carbon tetrachloride at 25°C and a partial pressure of nitrogen of 1 bar. The (hypothetical) liquid nitrogen fugacity at 25°C is 1000 bar.

a Calculate the mole fraction of nitrogen present in the liquid CCl_4 at equilibrium if the two species form an ideal solution.

b From regular solution theory it is estimated that

$$\ln \gamma_{N_2} = 0.526(1 - x_{N_2})^2$$

What is the equilibrium mole fraction of nitrogen in CCl_4 under these circumstances? What is the Henry's law constant for this system, if the excess volume for this system is zero?

8.3-3 The following data have been reported for the solubility of methane in various solvents at 25°C and 1.013 bar partial pressure of methane.[9]

Solvent	Mole Fraction of CH_4 in Saturated Liquid	
Benzene	2.07×10^{-3}	
Carbon tetrachloride	2.86×10^{-3}	
Cyclohexane	2.83×10^{-3}	
n-Hexane	3.15×10^{-3}	[A. S. McDaniel, *J. Phys. Chem.* **15**, 587 (1911)]
	4.24×10^{-3}	(D. Guerry, Jr., Thesis, Vanderbilt Univ., 1944)

a Estimate the ideal solubility of methane at 25°C and a partial pressure of 1.013 bar.

b Calculate the activity coefficient of methane in each of the solvents in the table.

c Is the datum of either McDaniel or Guerry consistent with the regular solution model?

8.3-4 **a** Estimate the vapor and liquid compositions of a nitrogen and benzene mixture in equilibrium at 75 bar and 100°C.

b Estimate the vapor and liquid compositions at 100 bar and 100°C.

Data

	Temperature (°C)	
Vapor Pressure, bar	**Benzene**	**Nitrogen**
1.013	80.1	−195.8
2.027	103.8	−189.2
5.067	142.5	−179.1
10.133	178.8	−169.8
30.399	249.5	−148.3
50.665	290.3	

[9] The data for this problem were abstracted from J. H. Hildebrand and R. L. Scott, *The Solubility of Nonelectrolytes*, 3rd ed., Table 4, p. 243, Reinhold, New York, 1950.

8.3-5 One way to remove a dissolved gas from a liquid is by vaporizing a small amount of the liquid in such a way that the vapor formed is continually withdrawn from the system. This process is called differential distillation.

a If N is the total number of moles of liquid, x the instantaneous mole fraction of dissolved gas, and y the gas mole fraction of the vapor in equilibrium with that liquid, show that

$$\frac{dx}{dN} = \frac{y - x}{N}$$

b If the mole fraction of the dissolved gas is low, then

$$y = Hx/P$$

where H is the Henry's law constant for the gas-solvent combination, and P is the total pressure above the liquid. Under this circumstance show that

$$\frac{x}{x_0} = \left(\frac{N}{N_0}\right)^{(H-P)/P} \quad \text{or, equivalently} \quad \frac{N}{N_0} = \left(\frac{x}{x_0}\right)^{P/(H-P)}$$

where N_0 and x_0 are the initial moles of solution and dissolved gas mole fraction, respectively.

c The Henry's law constant for carbon dioxide in carbon disulfide was computed (in Illustration 8.3-1) to be 267 bar at 25°C. At a total pressure of 1 bar, how much liquid should be vaporized from a solution saturated in carbon dioxide to decrease the CO_2 concentration to 1% of its equilibrium value? To 0.01% of its equilibrium value?

8.3-6 Estimate the solubility of carbon dioxide in toluene at 25°C and 1.013 bar CO_2 partial pressure using the Peng–Robinson equation of state assuming $k_{CO_2-T} = 0.0$.

8.3-7 Estimate the solubility of carbon dioxide in carbon disulfide at 25°C and 1.013 bar CO_2 partial pressure using the Peng–Robinson equation of state assuming

a $k_{CO_2-CS_2} = 0.0$

b $k_{CO_2-CS_2} = 0.2$

8.3-8 Derive the conversion factors in Table 8.3-3.

8.3-9 From Eq. 8.3-18 derive equations for the Gibbs free energy, enthalpy, entropy, and heat capacity for the transfer of the gas from the ideal gas state at 1.013 bar pressure to the (hypothetical) ideal solution at unit mole fraction. Comment on the expected accuracy of the thermodynamic properties determined in this way.

8.4 THE SOLUBILITY OF A LIQUID IN A LIQUID AND LIQUID–LIQUID–VAPOR EQUILIBRIUM

At low pressures all gases are mutually soluble in all proportions. The same is not generally true with liquids in that the equilibrium state of many binary liquid mixtures is two stable liquid phases of differing composition rather than as a single liquid phase, at least over certain ranges of temperature and composition. Our aim in this section is to obtain an understanding of why liquid–liquid phase separation occurs, and to develop thermodynamic equations that relate the properties of the two equilibrium phases. In this way we will be able to estimate the compositions of the coexisting phases when data are not available or use available liquid–liquid phase equilibrium data to get information on the activity coefficients of each species in a mixture.

Examples of liquid–liquid phase equilibrium behavior are given in Figs. 8.4-1 through 3.[10] These phase diagrams are to be interpreted as follows: If the temperature and overall composition of the mixture lie in the two-phase region, two liquid phases will form; the compositions of these phases are given by the intersection of the constant temperature (horizontal) line with the boundaries of the two-phase region. If, however, the temperature and composition are in the single-phase region, the equilibrium state is a single liquid phase. For example, if one attempts to prepare a mixture of 40 mole percent β-picoline in water at 125.7°C, he will obtain, instead, two liquid phases, one containing 12.7 mole percent β-picoline, and the other containing 61.7 mole percent β-picoline (see Fig. 8.4-1). The relative amounts of the two phases in such cases can be computed from the species mass balance

$$N_i = N_i^{\text{I}} + N_i^{\text{II}} \qquad i = 1, 2, \ldots \tag{8.4-1a}$$

Here N_i^J is the number of moles of species i in phase J, and N_i is the total number of moles of this species. An alternative form of this equation is

$$N_i = x_i^{\text{I}} N^{\text{I}} + x_i^{\text{II}} N^{\text{II}} \qquad i = 1, 2, \ldots \tag{8.4-1b}$$

where x_i^J is the mole fraction of species i in phase J, and N^J is the total number of moles of that phase.

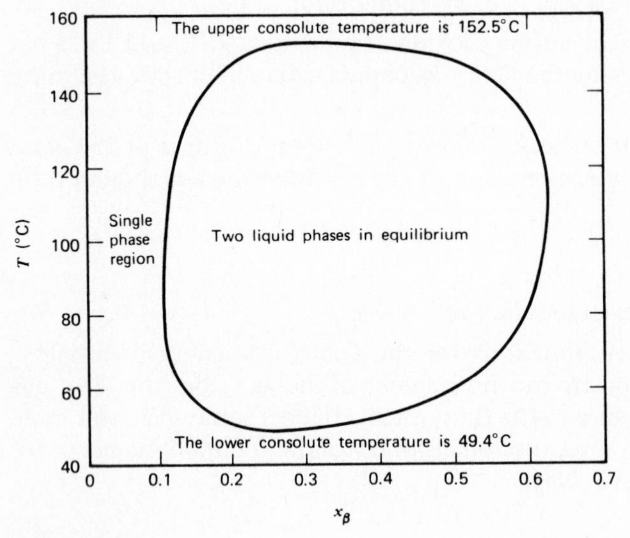

Figure 8.4-1
The liquid–liquid phase diagram of β-picoline and water.

[10]The experimental data for the partial solubility of perfluoro-n-heptane in various solvents has been plotted as a function of both mole fraction and volume fraction in Fig. 8.4-3. It is of interest to notice that these solubility data are almost symmetric functions of the volume fraction and nonsymmetric functions of the mole fraction. Such behavior has also been found with other thermodynamic mixture properties; these observations suggest the use of volume fractions, rather than mole fractions, as the appropriate concentration variables for describing nonideal mixture behavior. Indeed, this is the reason that volume fractions have been used in both the regular solution model and the Wohl expansion for liquid mixtures.

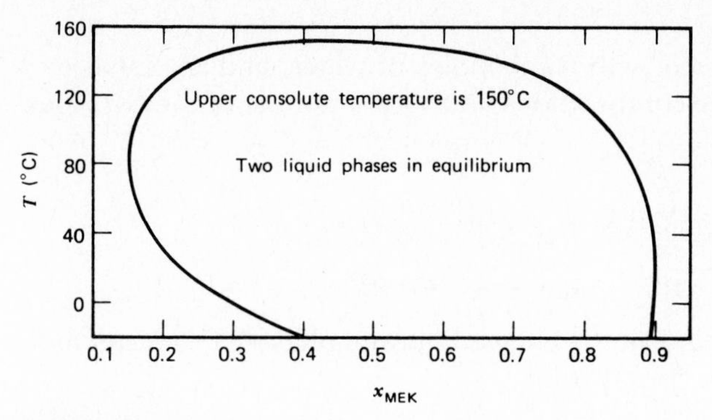

Figure 8.4-2
The liquid–liquid phase diagram for methyl ketone and water.

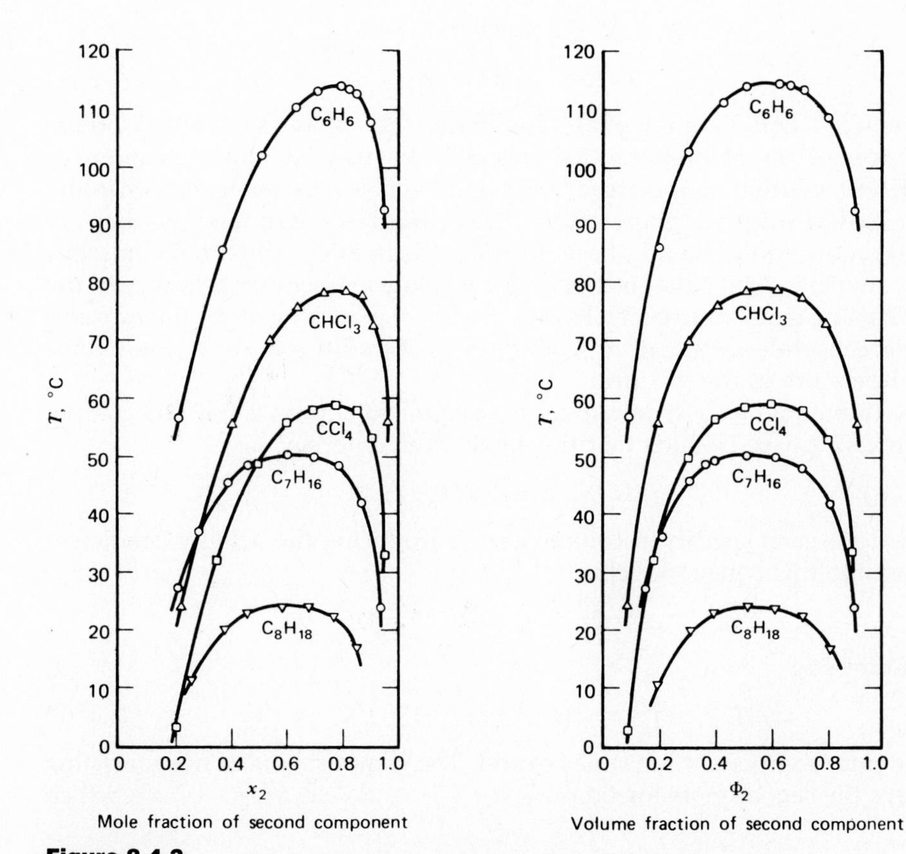

Figure 8.4-3
The solubility of perfluoro-*n*-heptane in various solvents; species 1 is perfluoro-*n*-heptane, species 2 is the solvent. [Reprinted with permission from J. H. Hildebrand, B. B. Fisher, and H. A. Benesi, *J. Am. Chem. Soc.* **72,** 4348 (1950). Copyright by the American Chemical Society.]

ILLUSTRATION 8.4-1

One mole of β-picoline is mixed with three moles of water, and the mixture is heated to 80°C. Determine the compositions and total amounts of the two coexisting liquid phases.

Solution

From the equilibrium phase diagram, Fig. 8.4-1, we have

$$x_\beta^{\mathrm{I}} = 0.107 \qquad \text{and} \qquad x_\beta^{\mathrm{II}} = 0.593$$

as the mole fractions of β-picoline in the two equilibrium phases. The water mole fractions in the phases can be found from

$$x_{\mathrm{H_2O}}^{J} = 1 - x_\beta^{J}$$

To compute the amounts of each of the phases we use the mass balances of Eq. 8.4-1b, which yield

$$1 \text{ mol } \beta\text{-picoline} = x_\beta^{\mathrm{I}} N^{\mathrm{I}} + x_\beta^{\mathrm{II}} N^{\mathrm{II}} = 0.107 N^{\mathrm{I}} + 0.593 N^{\mathrm{II}}$$

and

$$3 \text{ mol water} = (1 - x_\beta^{\mathrm{I}}) N^{\mathrm{I}} + (1 - x_\beta^{\mathrm{II}}) N^{\mathrm{II}} = 0.893 N^{\mathrm{I}} + 0.407 N^{\mathrm{II}}$$

These equations have the solution

$$N^{\mathrm{I}} = 2.82 \text{ mol}$$

$$N^{\mathrm{II}} = 1.18 \text{ mol} \quad \blacksquare$$

Generally, liquid–liquid phase equilibrium (or phase separation) occurs only over certain temperature ranges, bounded above by the **upper consolute** or **upper critical solution temperature**, and bounded below by the **lower consolute** or **lower critical solution temperature**. These critical solution temperatures are indicated on the liquid–liquid phase diagrams given here. All partially miscible mixtures should exhibit either one or both consolute temperatures; however, the lower consolute temperature may be obscured by the freezing of the mixture and the upper consolute temperature will not be observed if it is above the bubble point temperature of the mixture.

The thermodynamic requirement for phase equilibrium is that the compositions in each phase be such that the equilibrium criterion

$$\bar{f}_i^{\mathrm{I}}(T, P, \underline{x}^{\mathrm{I}}) = \bar{f}_i^{\mathrm{II}}(T, P, \underline{x}^{\mathrm{II}})$$

is satisfied for each species in the mixture. Introducing the activity coefficient definition into this equation yields

$$x_i^{\mathrm{I}} \gamma_i^{\mathrm{I}}(T, P, \underline{x}^{\mathrm{I}}) f_i(T, P) = x_i^{\mathrm{II}} \gamma_i^{\mathrm{II}}(T, P, \underline{x}^{\mathrm{II}}) f_i(T, P)$$

which reduces to

$$x_i^{\mathrm{I}} \gamma_i^{\mathrm{I}}(T, P, \underline{x}^{\mathrm{I}}) = x_i^{\mathrm{II}} \gamma_i^{\mathrm{II}}(T, P, \underline{x}^{\mathrm{II}}) \qquad i = 1, 2, \ldots \mathscr{C} \tag{8.4-2}$$[11]

since the pure component fugacities cancel. The compositions of the coexisting phases are the sets of mole fractions $x_1^{\mathrm{I}}, x_2^{\mathrm{I}}, \ldots, x_{\mathscr{C}}^{\mathrm{I}}, x_1^{\mathrm{II}}, x_2^{\mathrm{II}}, \ldots, x_{\mathscr{C}}^{\mathrm{II}}$, which

[11]From this equation it is clear that liquid–liquid phase separation is a result of solution nonideality. If the solutions were ideal, $\gamma_i^{\mathrm{I}} = \gamma_i^{\mathrm{II}} = 1$, so that $x_i^{\mathrm{I}} = x_i^{\mathrm{II}}$ for all species i, and there would be only a single phase, and not two distinguishable liquid phases, present at equilibrium.

simultaneously satisfy Eqs. 8.4-2 and

$$\sum_{i=1}^{\mathscr{C}} x_i^{\mathrm{I}} = 1 \quad \text{and} \quad \sum_{i=1}^{\mathscr{C}} x_i^{\mathrm{II}} = 1 \tag{8.4-3}$$

Equations 8.4-2 can be used with experimental phase equilibrium data to calculate the activity coefficient of a species in one phase from its known value in the second phase, or, with Eqs. 8.4-3 and experimental activity coefficient data or appropriate solution models, to compute the compositions of both coexisting liquid phases (see Illustration 8.4-2). Using the one-constant Margules equation to represent the activity coefficients, we obtain, from Eq. 8.4-2, the following relationship between the phase compositions

$$x_i^{\mathrm{I}} \exp\left[\frac{A(1-x_i^{\mathrm{I}})^2}{RT}\right] = x_i^{\mathrm{II}} \exp\left[\frac{A(1-x_i^{\mathrm{II}})^2}{RT}\right] \tag{8.4-4a}$$

whereas using the regular solution model leads to

$$x_i^{\mathrm{I}} \exp\left[\frac{\underline{V}_i^{\mathrm{L}}(\Phi_j^{\mathrm{I}})^2(\delta_1-\delta_2)^2}{RT}\right] = x_i^{\mathrm{II}} \exp\left[\frac{\underline{V}_i^{\mathrm{L}}(\Phi_j^{\mathrm{II}})^2(\delta_1-\delta_2)^2}{RT}\right] \tag{8.4-4b}$$

Alternatively, the van Laar, NRTL, and UNIQUAC activity coefficient models could be used. (The UNIFAC method can also be used to predict liquid–liquid equilibrium, but with different main group interaction parameters than are used to predict vapor–liquid equilibrium.)

ILLUSTRATION 8.4-2

Use the van Laar equations to estimate the compositions of the coexisting liquid phases in an isobutane–furfural mixture at 37.8°C and a pressure of 5 bar. (You may assume that the van Laar constants for this system given in Table 7.5-1 are applicable at this pressure.)

Solution

The compositions of the coexisting phases are the solution of the following set of algebraic equations:

$$x_1^{\mathrm{I}}\gamma_1^{\mathrm{I}} = x_1^{\mathrm{I}} \exp\left\{\frac{\alpha}{\left[1+\dfrac{\alpha x_1^{\mathrm{I}}}{\beta(1-x_1^{\mathrm{I}})}\right]^2}\right\} = x_1^{\mathrm{II}} \exp\left\{\frac{\alpha}{\left[1+\dfrac{\alpha x_1^{\mathrm{II}}}{\beta(1-x_1^{\mathrm{II}})}\right]^2}\right\} = x_1^{\mathrm{II}}\gamma_1^{\mathrm{II}} \tag{a}$$

$$x_2^{\mathrm{I}}\gamma_2^{\mathrm{I}} = x_2^{\mathrm{I}} \exp\left\{\frac{\beta}{\left[1+\dfrac{\beta x_2^{\mathrm{I}}}{\alpha(1-x_2^{\mathrm{I}})}\right]^2}\right\} = x_2^{\mathrm{II}} \exp\left\{\frac{\beta}{\left[1+\dfrac{\beta x_2^{\mathrm{II}}}{\alpha(1-x_2^{\mathrm{II}})}\right]^2}\right\} = x_2^{\mathrm{II}}\gamma_2^{\mathrm{II}} \tag{b}$$

$$x_1^{\mathrm{I}} + x_2^{\mathrm{I}} = 1 \tag{c}$$

$$x_1^{\mathrm{II}} + x_2^{\mathrm{II}} = 1 \tag{d}$$

Here isobutane will be taken as species 1, and furfural species 2; thus, from Table 7.5-1, $\alpha = 2.62$ and $\beta = 3.02$. The following procedure will be used in solving the equations:

i. A guess will be made for the value of x_1^{I}.
ii. Eq. a will then be used to compute x_1^{II}, and Eq. c to compute x_2^{I}.
iii. x_2^{II} will then be computed using Eq. b.

iv. Finally, we will check whether Eq. d is satisfied with the computed values of x_1^{II} and x_2^{II}. If it is, we have found the solution to the equations. If Eq. d is not satisfied, a new value of x_1^{I} is chosen, and the calculation is repeated.

The difficult part of this computation is the solution of Eqs. a and b for x_1^{II} and x_2^{II}. One way to solve these equations is as follows. First, the values of the product $x_1\gamma_1$ are computed for various values of x_1 (see the table that follows). From a graph of $x_1\gamma_1$ versus x_1 (see Fig. 8.4-4), we can easily obtain the value of $x_1\gamma_1$ for any choice of x_1 and then locate other values of x_1 that have the same value of $x_1\gamma_1$. Thus, given the trial value of x_1^{I}, x_1^{II} is quickly found. In a similar fashion, computing and plotting $x_2\gamma_2$ as a function of x_2 (or $x_1 = 1 - x_2$) provides an easy method of finding x_2^{II} given $x_2^{I} = 1 - x_1^{I}$. We then need to check whether Eq. d is satisfied. If not, the whole calculation is repeated with another guess for x_1^{I}.

Using this procedure, we obtain the following compositions for the coexisting liquid phases

$$x_1^{I} = 0.1128 \qquad x_1^{II} = 0.9284$$

and

$$x_2^{I} = 0.8872 \qquad x_2^{II} = 0.0716$$

At these compositions the activity coefficients are

$$\gamma_1^{I} = 8.375, \qquad \gamma_1^{II} = 1.018$$

$$\gamma_2^{II} = 1.030, \qquad \gamma_2^{II} = 12.77$$

and

$$x_1^{I}\gamma_1^{I} = 0.1128 \times 8.375 = 0.945 = x_1^{II}\gamma_1^{II} = 0.9284 \times 1.018$$

$$x_2^{I}\gamma_2^{I} = 0.8872 \times 1.030 = 0.914 = x_2^{II}\gamma_2^{II} = 0.0716 \times 12.77$$

There are no experimental liquid–liquid phase equilibrium data for the isobutane–furfural system with which we can compare our predictions. However, binary mixtures of furfural and, separately, 2,2-dimethyl pentane, 2-methyl pentane, and hexane, which are similar to the binary mixture here, all

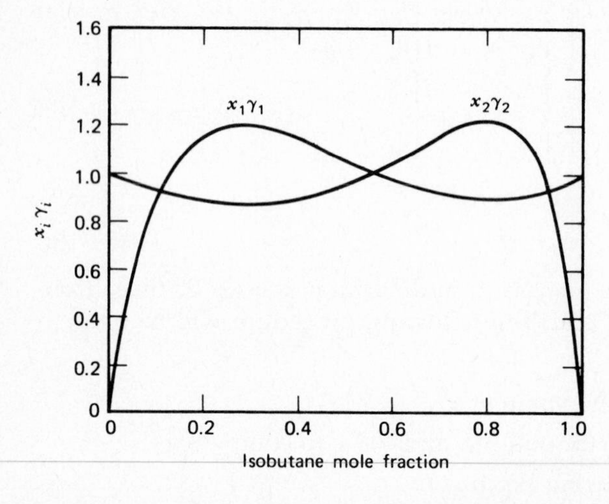

Figure 8.4-4
The product of species mole fraction and activity coefficients for the isobutane–furfural system at 37.8°C and 5 bar.

exhibit liquid–liquid phase separation,[12] with one phase containing between 89 and 90 mole percent furfural (x_2^{I}) and the other between 6 and 7 mole percent furfural (x_2^{II}).

TABLE OF $x_i\gamma_i$ VERSUS x_1

x_1	$x_1\gamma_1$	x_2	$x_2\gamma_2$
0	0	1.0	1.0
0.05	0.5491	0.95	0.9555
0.10	0.8843	0.90	0.9231
0.15	1.0761	0.85	0.8965
0.20	1.1734	0.80	0.8805
0.30	1.2072	0.70	0.8739
0.40	1.1405	0.60	0.9000
0.50	1.0598	0.50	0.9594
0.60	0.9840	0.40	1.0506
0.70	0.9322	0.30	1.1607
0.80	0.9121	0.20	1.2343
0.85	0.9161	0.15	1.2071
0.90	0.9309	0.10	1.0731
0.95	0.9582	0.05	0.7325
1.00	1.0	0.0	0.0

■

Although by starting with Eq. 8.4-2 one can proceed directly to the calculation of the liquid–liquid phase equilibrium state, this equation provides no real insight into the reason that phase separation and critical solution temperature behavior occur. To obtain this insight it is necessary to study the Gibbs free energy versus composition diagram for various mixtures. For an ideal binary mixture we have (Table 7.3-1)

$$\underline{G}^{IM} = x_1\underline{G}_1 + x_2\underline{G}_2 + RT(x_1 \ln x_1 + x_2 \ln x_2) \tag{8.4-5}$$

Now since x_1 and x_2 are always less than unity, $\ln x_1, \ln x_2 \leq 0$, and the last term in Eq. 8.4-5 is negative. Therefore, the Gibbs free energy of an ideal mixture is always less than the mole fraction weighted sum of the pure component Gibbs free energies, as illustrated by curve a of Fig. 8.4-5. For a real mixture we have

$$\underline{G} = \underline{G}^{IM} + \underline{G}^{ex} \tag{8.4-6}$$

where the excess Gibbs free energy would be determined by experiment, or approximated using a liquid solution model. For the purposes of illustration assume that the one-constant Margules equation is satisfactory, so that

$$\underline{G}^{ex} = Ax_1x_2 \tag{8.4-7}$$

[12]Reference for these data is H. Stephen and T. Stephen, eds., *Solubilities of Inorganic and Organic Compounds,* Vol. 1, *Binary Systems,* Macmillan, New York, 1963.

with $A > 0$. The total Gibbs free energy,

$$\underline{G} = x_1\underline{G}_1 + x_2\underline{G}_2 + RT(x_1 \ln x_1 + x_2 \ln x_2) + Ax_1x_2 \tag{8.4-8}$$

for two values of A is plotted as curves b and c of Fig. 8.4-5.

The equilibrium criterion for a closed system at constant temperature and pressure is that the Gibbs free energy of the system be a minimum.[13] For the mixture of curve c with an overall composition between x_α and x_β, the lowest value of $\underline{G}$ is obtained when the mixture separates into two phases, one of composition x_α and the other of composition x_β. In this case the Gibbs free energy of the mixture is a linear combination of the Gibbs free energies of two coexisting equilibrium liquid phases (the lever rule, Eqs. 5.3-1), and is represented by the dashed line (representing different amounts of the two phases) rather than the solid line in Fig. 8.4-5. If, however, the total mixture mole fraction of species 1 is less than x_α or greater than x_β, only a single phase will exist. Of course, the phase equilibrium compositions x_α and x_β can also be found directly from Eq. 8.4-2 in general, and from Eq. 8.4-4a for the case here, without reference to the figure.

The temperature range over which liquid–liquid phase separation occurs (i.e., the range of critical solution temperatures) can be found by using the

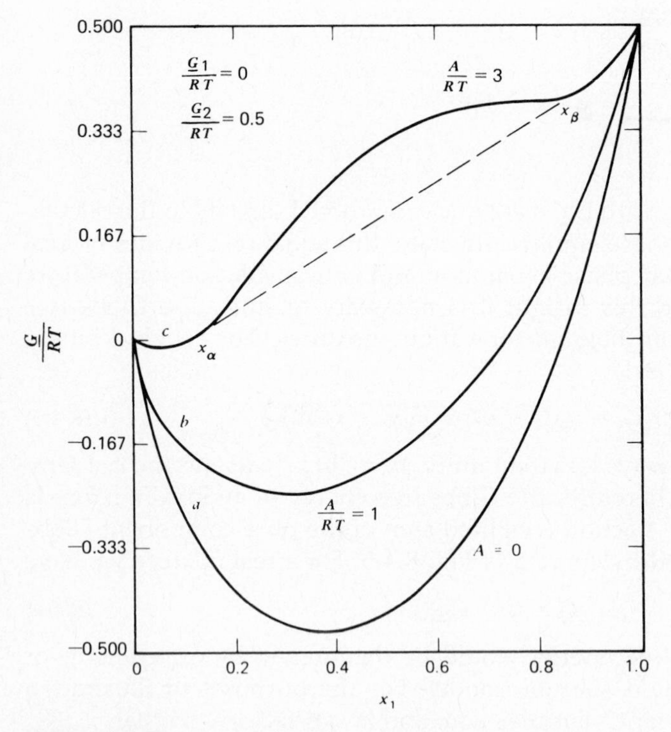

Figure 8.4-5
The molar Gibbs free energy of ideal ($A = 0$) and nonideal ($A \neq 0$) binary mixtures if phase separation does not occur (solid line) and when phase separation occurs (dashed line).

[13]Since the system is closed and nonreacting, the number of moles of each species is fixed. Therefore, in the analysis here we will consider the molar Gibbs free energy of the mixture $\underline{G}$, rather than G.

requirement for intrinsic fluid stability of Chapter 5 (see also Problem 6.22), that is,

$$d^2\underline{G} > 0 \qquad \text{at constant } N,\ T, \text{ and } P \tag{8.4-9}$$

What, in fact, we will do is to look at the second composition derivative of the Gibbs free energy. If $(\partial^2\underline{G}/\partial x_1^2)_{T,P} > 0$ (which follows from Eq. 8.4-9), for a given temperature and composition, the single phase is stable; if, however, $(\partial^2\underline{G}/\partial x_1^2)_{T,P} < 0$ at the given values of T and x_1, the single phase is unstable and phase separation occurs. The compositions at which $(\partial^2\underline{G}/\partial x_1^2)_{T,P} = 0$, which are inflection points on the $\underline{G}$ versus x_1 curve, at the limits of stability of the single phase at the given temperature. If a temperature T_{uc} exists for which

$$\left(\frac{\partial^2\underline{G}}{\partial x_1^2}\right)_{T,P} \begin{cases} =0 & \text{for some value of } x_1 \text{ at } T = T_{uc} \\ >0 & \text{for all values of } x_1 \text{ at } T > T_{uc} \end{cases} \tag{8.4-10a}$$

it is the upper consolute temperature of the mixture. Similarly, if there is a temperature T_{lc} for which

$$\left(\frac{\partial^2\underline{G}}{\partial x_1^2}\right)_{T,P} \begin{cases} =0 & \text{for some value of } x_1 \text{ at } T = T_{lc} \\ >0 & \text{for all values of } x_1 \text{ at } T < T_{lc} \end{cases} \tag{8.4-10b}$$

it is the lower consolute temperature.

To find the consolute temperatures of a mixture obeying the one-constant Margules model we start from Eqs. 8.4-8 and obtain[14]

$$\left(\frac{\partial^2\underline{G}}{\partial x_1^2}\right)_{T,P} = \frac{RT}{x_1x_2} - 2A \tag{8.4-11}$$

Consequently, $(\partial^2\underline{G}/\partial x_1^2)_{T,P} > 0$ and the single liquid phase is the equilibrium state if

$$T > \frac{2Ax_1x_2}{R} \tag{8.4-12a}$$

whereas $(\partial^2\underline{G}/\partial x_1^2)_{T,P} < 0$ and phase separation occurs if

$$T < \frac{2Ax_1x_2}{R} \tag{8.4-12b}$$

The limit of stability occurs at

$$T = \frac{2Ax_1(1 - x_1)}{R} \tag{8.4-13}$$

and the highest temperature at which phase separation is possible regardless of composition for a Margules mixture occurs at $x_1 = x_2 = 0.5$ which gives the maximum value of the product $x_1(1 - x_1)$, so that

$$T_{uc} = \frac{A}{2R} \tag{8.4-14}$$

This is the upper consolute temperature for a Margules mixture. Note that the Margules equation does not have a lower critical solution temperature (i.e., there is no solution of Eq. 8.4-11 for which Eq. 8.4-10b is satisfied). Thus the two partially miscible liquid phases of a Margules mixture cannot be made to combine by lowering the temperature.

[14]In taking this derivative remember that $(\partial x_2/\partial x_1)_{T,P} = -1$.

Equations 8.4-11 through 14 are specific to the choice of the one-constant Margules equation to represent the excess Gibbs free energy of the mixture. The use of more realistic models for $\underline{G}^{ex}$ will lead to other predictions for phase separation, such as the limit of stability at the upper consolute temperature occurring away from $x_1 = 0.5$ (Problem 8.4-1). For very nonideal mixtures, for example, aqueous solutions in which hydrogen-bonding and other associative phenomena occur, the species activity coefficients may be very different in value from unity, asymmetric in composition, and temperature dependent. A detailed analysis of such systems is beyond the scope of this book, but it is useful to note here that such mixtures may have a lower consolute temperature. Also, in very nonideal mixtures the species activity coefficients can be large, so that the two species will be relatively insoluble, as indicated in Illustration 8.4-3. (Since species activity coefficients are never infinite, it is evident from Eq. 8.4-2 that any liquid must be at least sparingly soluble in any other, so that no two liquids are truly immiscible. However, for some species the equilibrium solubilities are so small that it is convenient to refer to them as being immiscible.)

ILLUSTRATION 8.4-3

From the data in Volume III of the *International Critical Tables* (McGraw–Hill, New York, 1929) we know the equilibrium state in the carbon tetrachloride–water system at 25°C is two phases: one, an aqueous phase containing 0.083 wt% carbon tetrachloride, and the other an organic phase containing 0.011 wt% water. Estimate the activity coefficient of CCl_4 in the aqueous phase and H_2O in the organic phase.

Solution

Since the aqueous phase is 99.917 wt% water, it seems reasonable to assume that the activity coefficient of water in the aqueous phase is unity. Similarly, the activity coefficient of carbon tetrachloride in the organic phase will be taken to be unity. Therefore, from Eq. 8.4-2, we have

$$\gamma_{CCl_4}^{I} = \frac{x_{CCl_4}^{II}}{x_{CCl_4}^{I}}$$

and

$$\gamma_{H_2O}^{II} = \frac{x_{H_2O}^{I}}{x_{H_2O}^{II}}$$

where we have used the superscripts I and II to denote the aqueous and organic phases, respectively.

Clearly the first calculation to be done is the conversion of the weight fraction data to mole fractions. The activity coefficients are then easily computed. The results are given in the table:

	x_{CCl_4}	x_{H_2O}	γ_{CCl_4}	γ_{H_2O}
Aqueous phase	0.9708×10^{-4}	0.9999	1.029×10^{4}	1 (assumed)
Organic phase	0.9991	0.9403×10^{-3}	1 (assumed)	1.063×10^{3}

Comment

The values of the activity coefficients here are very large especially when compared to those we have found previously. Such activity coefficient behavior for the dilute species in a mixture of such different chemical species as carbon tetrachloride and water is, however, plausible. ■

The addition of a solute affects the activity coefficients of all species in the mixture. Thus, the addition of one solute may increase or decrease the equilibrium solubility of another solute (be it gas, liquid, or solid) in a given solvent. An increase in solubility in this way is termed **salting-in**, a decrease **salting-out** (see Fig. 8.6-1).[15] In some cases the addition of a solute (usually an electrolyte) can so increase the mutual solubility of two partially miscible liquids that a completely miscible mixture is formed.

Liquid–liquid equilibrium may occur when the species in a mixture are dissimilar. The most common situation is the one in which the species are of different chemical nature. Such mixtures are, at present, only describable by activity coefficient models, and that is the case that was considered in Illustration 8.4-3. However, liquid–liquid equilibrium may also occur when the two species are of similar chemical nature but differ greatly in size as in the methane–*n*-heptane system, or when the species differ in both size and chemical nature as in the carbon dioxide–*n*-octane system shown in Fig. 8.2-3*b*. Since both carbon dioxide and *n*-octane can be described by simple equations of state, liquid-liquid equilibrium in this system can be predicted or correlated using equations of state.

The calculation of liquid–liquid equilibrium using equations of state proceeds as in the equilibrium flash calculation described in Sec. 8.2 and illustrated in Fig. 8.2-5, with two changes. First, since both phases are liquids, the liquid root for the compressibility (which will differ in the two liquid phases) must be found in each case and used in the fugacity calculation. Second, the initial guesses for $K_i = x_i^{\mathrm{II}}/x_i^{\mathrm{I}}$ are not made using the pure component vapor pressures as in vapor–liquid equilibrium, but are chosen arbitrarily. (For example, $K_1 = 10$ and $K_2 = 0.1$ in a prediction calculation, or using the experimental data as the initial guess in a correlation to obtain the binary interaction parameter.)

ILLUSTRATION 8.4-4

The experimental data for liquid–liquid equilibrium in the CO_2–*n*-decane system appear in the following table.

T(K)	P(bar)	$x_{CO_2}^{\mathrm{I}}$	$x_{CO_2}^{\mathrm{II}}$
235.65	10.58	0.577	0.974
236.15	10.75	0.582	0.973
238.15	11.52	0.602	0.970
240.15	12.38	0.627	0.965
242.15	13.19	0.659	0.960
244.15	14.14	0.695	0.954
246.15	15.10	0.734	0.942
248.15	16.11	0.783	0.916
248.74	16.38	0.850	0.850

Source: A. A. Kulkarni, B. Y. Zarah, K. D. Luks, and J. P. Kohn, *J. Chem. Eng. Data* **19,** 92 (1974).

[15]Usually the terms salting-in and salting-out are used to describe the increase or decrease in solubility that results from the addition of a salt or electrolyte to a solute–solvent system. Their use here to describe the effects of the addition of a nonelectrolyte is slight generalization of the definition of these terms.

Make predictions for the liquid–liquid equilibrium in this system using the Peng–Robinson equation of state with the binary interaction parameter equal to 0.114 as given in Table 7.4-1, as well as several other values of this parameter.

Solution

Using the Peng–Robinson equation of state flash program modified as just described for the liquid–liquid equilibrium calculation gives the results shown in Fig. 8.4-6. Clearly, no choice for the binary interaction parameter k_{12} will result in predictions that are in complete agreement with the experimental data. In particular, the value of the binary interaction parameter determined from higher temperature vapor–liquid equilibrium data ($k_{12} = 0.114$) results in a much higher liquid–liquid critical solution temperature than is observed in the laboratory. Clearly, the Peng–Robinson equation of state prediction for liquid–liquid coexistence curve is not of the correct shape for this system.

What should be stressed, however, is not the poor accuracy of the equation of state predictions for the CO_2–n-decane system, but rather the fact that the same, simple equation of state can lead to good vapor–liquid equilibrium predictions over a wide range of temperature and pressures, as well as a qualitative description of liquid–liquid equilibrium at lower temperatures. ■

Although the discussion so far has concerned liquid–liquid equilibrium, the extension to liquid–liquid–vapor equilibrium is straightforward. As discussed in Sec. 6.7 the condition for liquid–liquid–vapor equilibrium is that

$$\bar{G}_i^{\mathrm{I}} = \bar{G}_i^{\mathrm{II}} = \bar{G}_i^{\mathrm{III}} \tag{8.4-15}$$

or equivalently

$$\bar{f}_i^{\mathrm{I}} = \bar{f}_i^{\mathrm{II}} = \bar{f}_i^{\mathrm{III}} \tag{8.4-16}$$

for each species distributed among the three phases. Consequently, one method of computing the three phases that are in equilibrium is to solve two two-phase

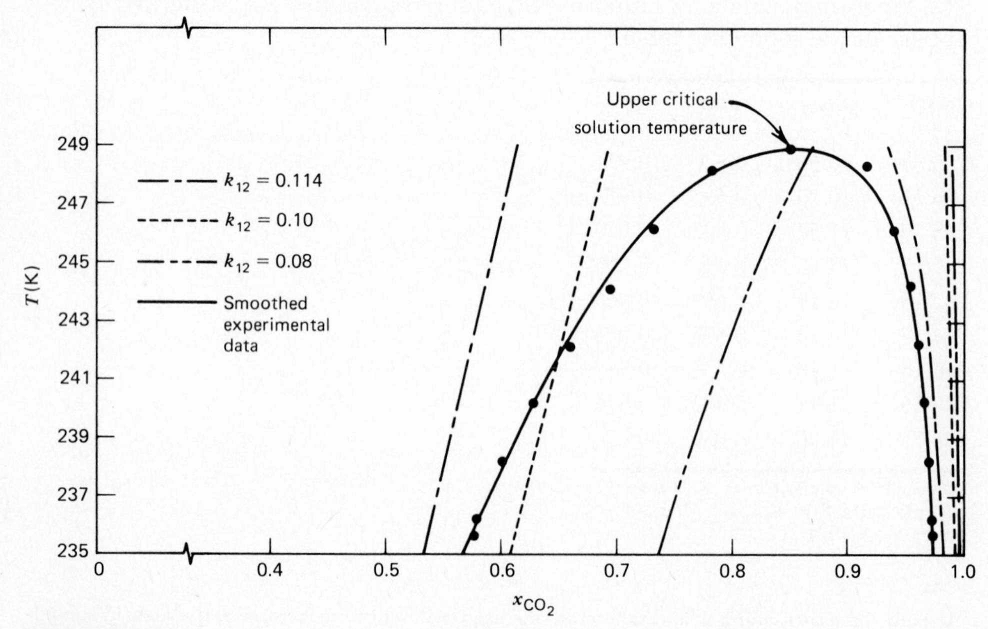

Figure 8.4-6
Liquid–liquid equilibrium for the system CO_2–n-decane. Experimental data and predictions using the Peng–Robinson equation of state.

problems. For example, one could first determine the compositions of the two liquids that are in equilibrium, and then use the methods of Secs. 8.1 or 2, as appropriate, to find the vapor that would be in equilibrium with either one of the liquids (since, by Eqs. 8.4-16, a vapor in equilibrium with one of the coexisting equilibrium liquid phases will also be in equilibrium with the other). This is illustrated in the two examples that follow.

ILLUSTRATION 8.4-5

Since liquids are not very compressible, at low and moderate pressures liquid–liquid equilibrium compositions are almost independent of pressure. Therefore, assuming that the liquid–liquid equilibrium of the isobutane(1)–furfural(2) mixture at 37.8°C calculated in Illustration 8.4-2 is unaffected by pressure, compute the pressure at which the first bubble of vapor will form (i.e., compute the bubble-point pressure of this system) and the composition of the vapor that forms.

Data

$$P_1^{\text{vap}}\ (T = 37.8°\text{C}) = 4.956\ \text{bar}$$

$$P_2^{\text{vap}}\ (T = 37.8°\text{C}) = 0.005\ \text{bar}$$

Solution

Using the van Laar activity coefficient model as in Illustration 8.4-2, and assuming the vapor phase is ideal, we have that the bubble-point pressure of liquid phase I is

$$P = x_1^{\text{I}}\gamma_1^{\text{I}}P_1^{\text{vap}} + x_2^{\text{I}}\gamma_2^{\text{I}}P_2^{\text{vap}}$$

$$= 0.1128 \times 8.375 \times 4.956 + 0.8872 \times 1.030 \times 0.005 = 4.69\ \text{bar}$$

The bubble-point pressure of liquid phase II is

$$P = x_1^{\text{II}}\gamma_1^{\text{II}}P_1^{\text{vap}} + x_2^{\text{II}}\gamma_2^{\text{II}}P_2^{\text{vap}}$$

$$= 0.9284 \times 1.018 \times 4.956 + 0.0716 \times 12.77 \times 0.005 = 4.69\ \text{bar}$$

which is the same as liquid phase I (as it must be since $x_1^{\text{I}}\gamma_1^{\text{I}} = x_1^{\text{II}}\gamma_1^{\text{II}}$ and $x_2^{\text{I}}\gamma_2^{\text{I}} = x_2^{\text{II}}\gamma_2^{\text{II}}$). The composition of the vapor (computed using either liquid phase I or II) is gotten from $x_i\gamma_i P_i^{\text{vap}} = y_i P$, so that

$$y_1 = \frac{x_1\gamma_1 P_1^{\text{vap}}}{P} = \frac{0.1128 \times 8.375 \times 4.956}{4.69} = 0.999$$

and

$$y_2 = \frac{x_2\gamma_2 P_2^{\text{vap}}}{P} = \frac{0.8872 \times 1.030 \times 0.005}{4.69} = 0.001$$

Therefore, from the van Laar model, at $T = 37.8°\text{C}$ and $P = 4.69$ bar, the isobutane–furfural mixture can have two liquid phases and a vapor coexisting at equilibrium, with the following compositions

Substance	Liquid I	Liquid II	Vapor
Isobutane	0.1128	0.9284	0.999
Furfural	0.8872	0.0716	0.001

Comment

The Gibbs phase rule for this nonreacting ($\mathcal{M} = 0$) system of two components ($\mathcal{C} = 2$) and three phases ($\mathcal{P} = 3$) establishes that the number of degrees of freedom are

$$\begin{aligned}\mathcal{F} &= \mathcal{C} + 2 - \mathcal{P} - \mathcal{M} \\ &= 2 + 2 - 3 - 0 = 1\end{aligned}$$

Therefore, at each temperature, there is only a single pressure at which three phases will coexist at equilibrium in this two-component system, and the compositions of these three phases are fixed. That is, as the feed changes within the boundaries of the three-phase region, the distribution of mass between the three phases will change, but not the composition of those phases. ■

ILLUSTRATION 8.4-6

Use the Peng–Robinson equation of state, with $k_{12} = 0.114$, to compute the bubble-point pressure and vapor composition in equilibrium with the two coexisting liquid phases in the CO_2–n-decane system of Illustration 8.4-4.

Solution

Using the bubble-point pressure program in VLMU.BAS, and the computed composition for the liquid richer in n-decane, we obtain the following results for the three-phase coexistence region

T(K)	P(bar)	$x^{I}_{CO_2}$	$x^{II}_{CO_2}$	$y_{C_{10}}$
235.65	10.84	0.539	0.997	1.2×10^{-6}
236.15	11.01	0.540	0.996	1.3×10^{-6}
238.15	11.82	0.552	0.996	1.6×10^{-6}
240.15	12.66	0.563	0.995	2.0×10^{-6}
242.15	13.56	0.575	0.995	2.4×10^{-6}
244.15	14.49	0.586	0.994	3.0×10^{-6}
246.15	15.46	0.597	0.993	3.6×10^{-6}
248.74	16.76	0.609	0.992	4.7×10^{-6}

As we can see from these results, there is very little n-decane in the vapor. This is because of the large volatility difference between carbon dioxide and n-decane. The three-phase experimental data for this system confirms this behavior. ■

The prediction of three-phase equilibria considered so far were done as two separate two-phase calculations. Although applicable to the examples here, such a procedure cannot easily be followed in a three-phase flash calculation in which the temperature or pressure of a mixture of two or more components is changed so that three phases are formed. In this case the equilibrium relations and mass balance equations for all three phases must be solved simultaneously to find the compositions of the three coexisting phases. It is left to you (Problem 8.4-5) to develop the algorithm for such a calculation.

PROBLEMS FOR SECTION 8.4

8.4-1 **a** Show that if two liquids form a regular solution the critical temperature for phase separation is

$$RT_C = \frac{2x_1x_2\underline{V}_1^2\underline{V}_2^2}{(x_1\underline{V}_1 + x_2\underline{V}_2)^3}(\delta_1 - \delta_2)^2 = \frac{2\Phi_1\Phi_2\underline{V}_1\underline{V}_2}{(x_1\underline{V}_1 + x_2\underline{V}_2)}(\delta_1 - \delta_2)^2$$

b Show that the composition at the upper consolute temperature is

$$x_1 = 1 - x_2 = \frac{(\underline{V}_1^2 + \underline{V}_2^2 - \underline{V}_1\underline{V}_2)^{1/2} - \underline{V}_1}{\underline{V}_2 - \underline{V}_1}$$

and develop an expression for the upper consolute temperature for the regular solution model.

8.4-2 Following is a portion of a phase diagram for two liquids that are only partially miscible. Note that the phase diagram contains both the coexistence curve (solid line) and a curve indicating the stability limit for each phase (dashed line). (That is, the dashed line represents the greatest concentration of the dilute species, or extent of supersaturation, that may occur in a metastable phase.) This line is called the spinodal curve. For a binary mixture for which $\underline{G}^{ex} = Ax_1x_2$, develop the equations to be used to

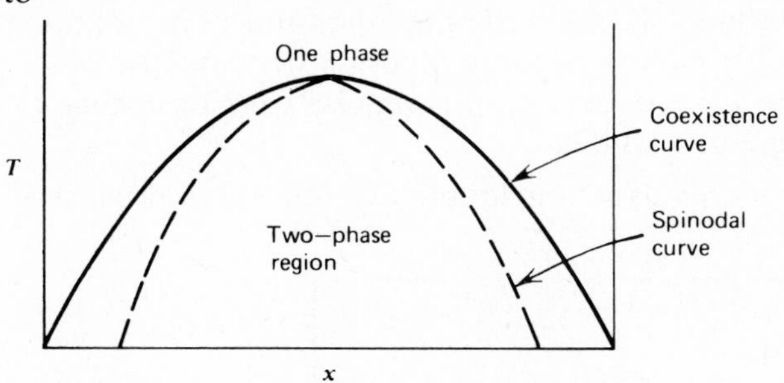

a Compute the liquid–liquid coexistence line.
b Compute the spinodal curve.

8.4-3 The two figures here have been obtained from experimental measurements of the excess Gibbs free energy and excess enthalpy for the benzene–CCl_4 and benzene–CS_2 systems, respectively, at 25°C.

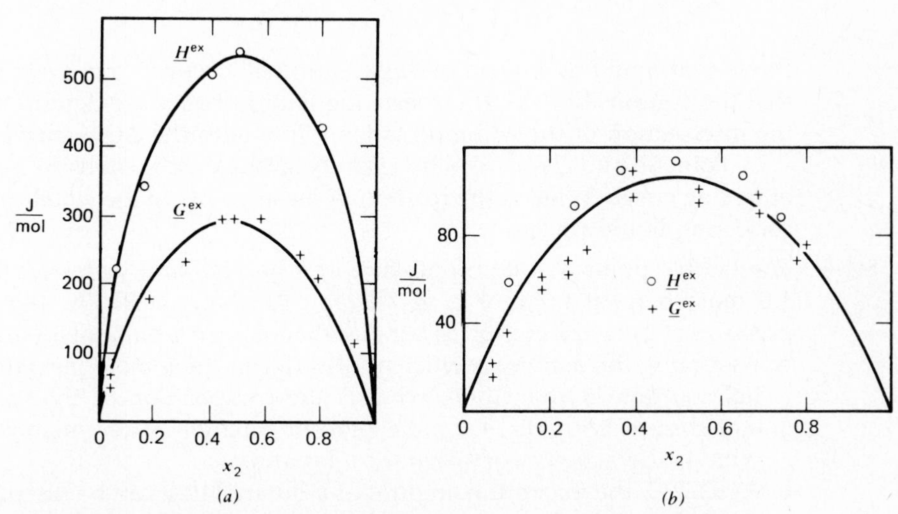

(*a*) Thermodynamic excess functions for the system carbon disulfide(1) + benzene(2). (*b*) Thermodynamic excess functions for the system benzene(1) + carbon tetrachloride(2). [From I. Prigogine and R. Defay (D. H. Everett, trans.), *Chemical Thermodynamics*. Copyright 1954. Used with permission of Longman Group Limited and Halsted Press (John Wiley & Sons, Inc.).]

a Comment on the applicability of the regular solution theory model to these two systems.

b Assuming that the excess Gibbs free energy for the C_6H_6–CS_2 system is temperature independent, estimate the upper consolute temperature of the system. Since the melting point of benzene is 5.5°C and that of CS_2 is −108.6°C, will a liquid–liquid phase separation be observed?

c At 46.5°C, the vapor pressure of CS_2 is 1.013 bar and that of C_6H_6 is 0.320 bar. Will an azeotrope occur in this system at this temperature?

8.4-4 The liquids perfluoro-*n*-heptane and benzene are only partially miscible at temperatures below their upper consolute temperature of 113.4°C. At 100°C one liquid phase is approximately 0.48 mole fraction benzene and the other 0.94 mole fraction benzene (see Fig. 8.4-3). The liquid molar volume of perfluoro-*n*-heptane at 25°C is 0.226 m^3/kmol.

a Use these data to compute the interaction parameter A in a one-constant Margules equation for the excess Gibbs free energy. Is the one-constant Margules equation consistent with the experimental data?

b Use the experimental data to compute the value of the regular solution theory solubility parameter for perfluoro-*n*-heptane [the value given in Table 7.6-1 is $\delta = 6.0\ (\text{cal/cc})^{1/2}$]. Is regular solution theory consistent with the experimental data?

8.4-5 The Gibbs free energy of mixing for highly nonideal solutions can behave as shown here:

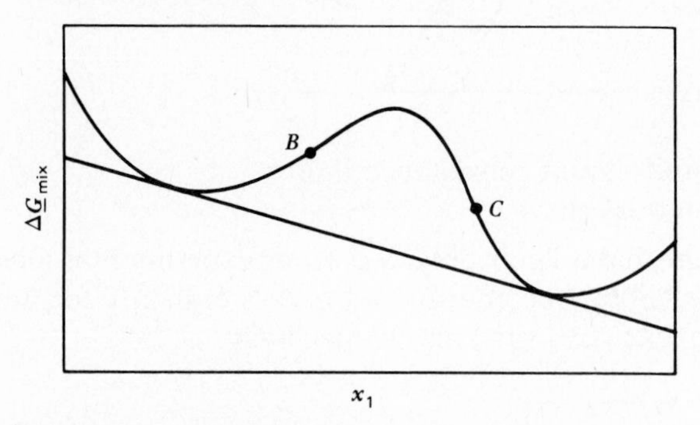

Prove that liquid–liquid phase separation will occur in this system, and that the compositions of the coexisting liquid phases are determined by the intersection of the common tangent line with the $\Delta \underline{G}_{\text{mix}}$ curve.

Points B and C are inflection points on the excess Gibbs free energy of mixing curve. What is the relation of these points to the stability of the coexisting liquid phases?

8.4-6 The bubble point of a liquid mixture of *n*-butanol and water containing 4.0 mole percent butanol is 92.7°C at 1.013 bar. At 92.7°C the vapor pressure of pure water is 0.784 bar and that of pure *n*-butanol is 0.427 bar.

a Assuming the activity coefficient of water in the 4 mole percent butanol solution is about one, what is the composition of the vapor in equilibrium with the 4.0 mole percent butanol–water mixture, and what is the activity coefficient of *n*-butanol.

b At 92.7°C, the maximum amount of *n*-butanol that can be dissolved in water is 4.0 mole percent. When larger amounts of *n*-butanol are present, a second liquid phase, containing 40.0 mole percent *n*-butanol, appears. What are the activity coefficients for *n*-butanol and water in this second liquid phase?

c If the two coexisting liquids in part b are kept at 92.7°C, what will be the pressure when the first bubble of vapor is formed?

8.4-7 Estimate the pressure and vapor-phase composition in equilibrium with a liquid whose overall composition is 50 mole percent furfural and 50 mole percent isobutane at $T = 37.8°C$.

Data

$$P_{\text{iso}}^{\text{vap}}(T = 37.8°\text{C}) = 4.909 \text{ bar}$$

$$P_{\text{furf}}^{\text{vap}}(T = 37.8°\text{C}) = 4.93 \times 10^{-3} \text{ bar}$$

Isobutane and furfural are only partially miscible in the liquid phase. At 37.8°C the two coexisting liquid phases contain 11.8 and 92.5 mole percent isobutane, and $\gamma_{\text{iso}}\ (x_{\text{iso}} = 0.925) = 1.019$, and $\gamma_{\text{furf}}\ (x_{\text{furf}} = 0.882) = 1.003$.

8.4-8 Ethyl alcohol and n-hexane are put into an evacuated, isothermal container. After equilibrium is established at 75°C, it is observed that two liquid phases and a vapor phase are in equilibrium. One of the liquid phases contains 9.02 mole percent n-hexane. Activity coefficients for n-hexane–ethyl alcohol liquid mixtures can be represented by the following equation

$$RT \ln \gamma_i = 8.163\, x_j^2 \frac{\text{kJ}}{\text{mol}}$$

a Compute the equilibrium composition of the coexisting liquid phase.
b Compute the equilibrium pressure and vapor-phase mole fractions.

Data
Vapor pressure equations (T in K, P in bar)

$$\ln P_H^{\text{vap}} = \frac{-3570.58}{T} + 10.4575$$

$$\ln P_{\text{EOH}}^{\text{vap}} = \frac{-4728.98}{T} + 13.4643$$

8.4-9 Glycerol and acetophenone are partially miscible at 140°C, one liquid phase containing 9.7 mole percent acetophenone and the other containing 90.3 mole percent acetophenone. The vapor pressure of pure glycerol at 140°C is 3.13×10^{-3} bar and that of acetophenone is 0.167 bar. Assuming that the activity coefficients for the glycerol-acetophenone system obey the 1-constant Margules equation, compute the pressure versus liquid-phase mole fraction (P-x) diagram for this system.

8.4-10 For a separation process, it is necessary to ascertain whether water and methyl ethyl ketone form an azeotrope at 73.4°C. The engineer needing this information could not find any vapor–liquid equilibrium data for this system, but the following liquid–liquid equilibrium data are available;

$T = 20°C$

	x_{MEK}	x_{H_2O}
Liquid phase 1	0.0850	0.9150
Liquid phase 2	0.6363	0.3637

In addition, at 73.4°C

$$P^{\text{vap}}_{\text{H}_2\text{O}} = 0.3603 \text{ bar}$$

$$P^{\text{vap}}_{\text{MEK}} = 0.8337 \text{ bar}$$

Calculate, as best you can, the *P-x-y* diagram for MEK and water at 73.4°C and determine whether this system has an azeotrope, and if so, at what composition the azeotrope occurs.

8.4-11 Show that the Wilson activity coefficient model of Eqs. 7.5-11 and 12 cannot predict the existence of two liquid phases for any values of its parameters.

8.5 THE SOLUBILITY OF A SOLID IN A LIQUID, GAS, OR SUPERCRITICAL FLUID

We now want to consider the extent to which a solid is soluble in a liquid, a gas, or a supercritical fluid. (This last case is of interest for supercritical extraction, a new separations method.) To analyze these phenomena we again start with the equality of the species fugacities in each phase. However, since the fluid (either liquid, gas, or supercritical fluid) is not present in the solid, two simplifications arise. First, the equilibrium criterion applies only to the solute, which we denote by the subscript 1, and second, the solid phase fugacity of the solute is that of the pure solid. Thus we have the single equilibrium relation

$$f_1^{\text{S}}(T, P) = \bar{f}_1^{\text{F}}(T, P, x_1) \tag{8.5-1}$$

where the superscripts S and F refer to the solid and fluid (liquid, gas, or supercritical fluid) phases, respectively.

We consider first the solubility of a solid in a liquid. Using Eq. 7.3-11 for the fugacity of the solute in a liquid we obtain

$$f_1^{\text{S}}(T, P) = x_1\gamma_1(T, P, x_1)f_1^{\text{L}}(T, P) \tag{8.5-2}$$

where $f_1^{\text{S}}(T, P)$ and $f_1^{\text{L}}(T, P)$ refer to the fugacity of the pure species as a solid and as a liquid, respectively, at the temperature and pressure of the mixture, and x_1 is the saturation mole fraction of the solid solute in the solvent.

If the temperature of the mixture is equal to the normal melting temperature of the solid T_m, then

$$f_1^{\text{S}}(T_m) = f_1^{\text{L}}(T_m) \tag{8.5-3}$$

by the pure component phase equilibrium condition, so that at the melting point we have

$$x_1 = 1/\gamma_1(T_m, P, x_1) \tag{8.5-4}$$

Thus, the solubility in a liquid of a solid at its melting point is just equal to the reciprocal of its activity coefficient in the solute–solvent mixture.

If, as is usually the case, the solid is below its melting point, $f_1^{\text{L}} > f_1^{\text{S}}$, so that Eq. 8.5-4 is not valid. To predict the solubility in this case Eq. 8.5-2 must be used with some estimate for the ratio $f_1^{\text{S}}/f_1^{\text{L}}$. One way to make this estimate is (1) to use the sublimation pressure for f_1^{S} (see Eq. 5.4-24), and (2) to compute the fugacity for the "subcooled" liquid f_1^{L} by extrapolation of the liquid thermodynamic properties into the solid region. This can be done graphically as indicated in Fig. 7.8-2*b* if the temperature is not too far below the melting point. Alternatively, if heat capacity data for both the solid and liquid, and heat (enthalpy) of fusion data are available, we can directly compute the Gibbs free energy of fusion $\Delta \underline{G}^{\text{fus}}(T)$, which is related to the fugacity ratio as follows

$$\frac{\Delta \underline{G}^{\text{fus}}(T, P)}{RT} = \frac{\underline{G}_1^{\text{L}}(T, P) - \underline{G}_1^{\text{S}}(T, P)}{RT} = \ln \frac{f_1^{\text{L}}(T, P)}{f_1^{\text{S}}(T, P)}$$

Combining this result with Eq. 8.5-2 gives

$$\ln [x_1 \gamma_1(T, P, x_1)] = -\frac{\Delta \underline{G}^{\text{fus}}(T, P)}{RT} \tag{8.5-5}$$

(Clearly, if $T = T_m$, $\Delta \underline{G}^{fus} = 0$. Why?)

$\Delta \underline{G}^{\text{fus}}(T)$ is computed by separately calculating $\Delta \underline{H}^{\text{fus}}(T)$ and $\Delta \underline{S}^{\text{fus}}(T)$, and then using the relation $\Delta \underline{G}^{\text{fus}}(T) = \Delta \underline{H}^{\text{fus}}(T) - T\Delta \underline{S}^{\text{fus}}(T)$. To compute the enthalpy and entropy changes of fusion, we suppose that the melting of a solid (below its normal melting point) to form a liquid is carried out in the following three-step constant pressure process:

1. The solid is heated at fixed pressure from the temperature T to its normal melting temperature T_m.
2. The solid is then melted to form a liquid, and
3. The liquid is cooled *without* solidification from T_m back to the temperature of the mixture.

The enthalpy and entropy changes for this process are

$$\begin{aligned} \Delta \underline{H}^{\text{fus}}(T) &= \int_T^{T_m} C_{\text{P}}^{\text{S}}\, dT + \Delta \underline{H}^{\text{fus}}(T_m) + \int_{T_m}^{T} C_{\text{P}}^{\text{L}}\, dT \\ &= \Delta \underline{H}^{\text{fus}}(T_m) + \int_{T_m}^{T} \Delta C_{\text{P}}\, dT \end{aligned} \tag{8.5-6}$$

and

$$\begin{aligned} \Delta \underline{S}^{\text{fus}}(T) &= \int_T^{T_m} \frac{C_{\text{P}}^{\text{S}}}{T}\, dT + \Delta \underline{S}^{\text{fus}}(T_m) + \int_{T_m}^{T} \frac{C_{\text{P}}^{\text{L}}}{T}\, dT \\ &= \Delta \underline{S}^{\text{fus}}(T_m) + \int_{T_m}^{T} \frac{\Delta C_{\text{P}}}{T}\, dT \end{aligned} \tag{8.5-7}$$

where $\Delta C_{\text{P}} = C_{\text{P}}^{\text{L}} - C_{\text{P}}^{\text{S}}$.

Note that Eqs. 8.5-6 and 7 relate the enthalpy and entropy changes of fusion at any temperature T to those changes at the melting point at the same pressure. Now since $\underline{G} = \underline{H} - T\underline{S}$, and $\Delta \underline{G}^{\text{fus}}(T = T_m) = 0$, Eq. 8.5-7 can be rewritten as

$$\Delta \underline{S}^{\text{fus}}(T) = \frac{\Delta \underline{H}^{\text{fus}}(T_m)}{T_m} + \int_{T_m}^{T} \frac{\Delta C_{\text{P}}}{T}\, dT \tag{8.5-8}$$

and, therefore,

$$\begin{aligned} \Delta \underline{G}^{\text{fus}}(T) &= \Delta \underline{H}^{\text{fus}}(T) - T\Delta \underline{S}^{\text{fus}}(T) \\ &= \Delta \underline{H}^{\text{fus}}(T_m)\left[1 - \frac{T}{T_m}\right] + \int_{T_m}^{T} \Delta C_{\text{P}}\, dT - T\int_{T_m}^{T} \frac{\Delta C_{\text{P}}}{T}\, dT \\ &\equiv RT \ln \left[\frac{f_1^{\text{L}}(T, P)}{f_1^{\text{S}}(T, P)}\right] \end{aligned} \tag{8.5-9}$$

Using this result in Eq. 8.5-5 gives

$$\ln \gamma_1 x_1 = -\frac{\Delta \underline{H}^{\text{fus}}(T_m)}{RT}\left[1 - \frac{T}{T_m}\right] - \frac{1}{RT}\int_{T_m}^{T} \Delta C_{\text{P}}\, dT + \frac{1}{R}\int_{T_m}^{T} \frac{\Delta C_{\text{P}}}{T}\, dT \qquad (8.5\text{-}10)$$

Equation 8.5-10 is the basic equation for predicting the saturation mole fraction of a solid in a liquid. As this equation now stands, it is almost exact.[16]

Two approximations can be made in Eq. 8.5-10 without introducing appreciable error. First, we assume that ΔC_{P} is independent of temperature, so that Eq. 8.5-10 becomes

$$\ln x_1 \gamma_1 = -\left\{\frac{\Delta \underline{H}^{\text{fus}}(T_m)}{RT}\left[1 - \frac{T}{T_m}\right] + \frac{\Delta C_{\text{P}}}{R}\left[1 - \frac{T_m}{T} + \ln\left(\frac{T_m}{T}\right)\right]\right\} \qquad (8.5\text{-}11)$$

Next, since the melting point temperature at any pressure and the triple-point temperature (T_{T}) are only slightly different for most solids (see Fig. 7.8-2), we can rewrite Eq. 8.4-11, without much error, as

$$\ln x_1 = -\ln \gamma_1 - \left\{\frac{\Delta \underline{H}^{\text{fus}}(T_{\text{T}})}{RT}\left[1 - \frac{T}{T_{\text{T}}}\right] + \frac{\Delta C_{\text{P}}}{R}\left[1 - \frac{T_{\text{T}}}{T} + \ln\left(\frac{T_{\text{T}}}{T}\right)\right]\right\} \qquad (8.5\text{-}12)$$

If the liquid mixture is ideal, so that $\gamma_1 = 1$, we have the case of ideal solubility of a solid in a liquid, and the solubility can be computed from only thermodynamic data ($\Delta \underline{H}^{\text{fus}}$ and ΔC_{P}) for the solid species near the melting point. For nonideal solutions γ_1 must be estimated from either experimental data or a liquid solution theory. For example, the regular solution theory estimate for this activity coefficient is

$$RT \ln \gamma_1 = \underline{V}_1^{\text{L}}(\delta_1 - \delta_2)^2 \Phi_2^2 \qquad (8.5\text{-}13)$$

where δ_1 and δ_2 are the solubility parameters of the solute as a liquid and the liquid solvent, respectively. The difficulty in using this relation is that we must be able to estimate both δ_1 and the liquid molar volume for a species whose pure component state is a solid at the mixture temperature.

Usually one can neglect the thermal expansibility of both solids and liquids, so that $\underline{V}^{\text{L}}$ can be taken to be molar volume of the liquid at the normal melting point, or it can be computed from

$$\underline{V}_1^{\text{L}} = \underline{V}_1^{\text{S}} + \Delta \underline{V}^{\text{fus}} \qquad (8.5\text{-}14)$$

where $\Delta \underline{V}^{\text{fus}}$ is the molar volume change on fusion at the triple point of the solid. To compute the solubility parameter δ_1, where

$$\delta = \left(\frac{\Delta \underline{U}^{\text{vap}}}{\underline{V}^{\text{L}}}\right)^{1/2} \qquad (8.5\text{-}15)$$

it is necessary to estimate $\Delta \underline{U}^{\text{vap}}$, the molar internal energy change on vaporization of the subcooled liquid. The enthalpy change on vaporization of the subcooled liquid is

$$\Delta \underline{H}^{\text{vap}}(T) = \Delta \underline{H}^{\text{sub}}(T) - \Delta \underline{H}^{\text{fus}}(T)$$

where $\Delta \underline{H}^{\text{sub}}(T)$ is the heat of sublimation of the solid at the temperature T; this quantity can either be found in tables of thermodynamic properties or estimated

[16]The only approximation that has been made is to neglect the Poynting pressure correction terms to the solid and liquid free energies (or fugacities).

from sublimation pressure data using the Clausius–Clapeyron equation (Eq. 5.7-5a). $\Delta \underline{H}^{\text{fus}}(T)$ is computed from Eq. 8.5-6. Thus

$$\begin{aligned}\Delta \underline{U}^{\text{vap}} &= \Delta \underline{H}^{\text{vap}} - P\Delta \underline{V}^{\text{vap}} \\ &= \Delta \underline{H}^{\text{sub}}(T) - \Delta \underline{H}^{\text{fus}}(T_m) - \Delta C_P(T - T_m) - RT \end{aligned} \tag{8.5-16}$$

where we have assumed that $P\Delta \underline{V}^{\text{vap}} = P(\underline{V}^{\text{V}} - \underline{V}^{\text{L}}) \approx P\underline{V}^{\text{V}} = RT$, and that ΔC_P is independent of temperature.

Equations 8.5-13 through 16 can be used in Eq. 8.5-12 to compute the saturation solubility for any regular solution solute–solvent pair.

ILLUSTRATION 8.5-1

Estimate the solubility of solid naphthalene in liquid n-hexane at 20°C.

Data[17]

Naphthalene ($C_{10}H_8$, molecular weight = 128.19)
Melting point: 80.2°C
Heat of fusion: 18.804 kJ/mol
Density of the solid: 1.0253 g/cc at 20°C
Density of the liquid: 0.9625 g/cc at 100°C
Vapor pressure of the solid:

$$\log_{10} P(\text{bar}) = 8.722 - \frac{3783}{T} \qquad (T = \text{K})$$

The heat capacities of liquid and solid naphthalene may be assumed to be equal.

Solution

The solubility parameter and liquid molar volume for n-hexane are given in Table 7.6-1 as $\delta_2 = 7.3$ and $\underline{V}_2^{\text{L}} = 132$ cc/mol, respectively. Since the liquid molar volume of naphthalene given in the data is for a temperature 80°C higher than the temperature of interest, and the volume change on melting of naphthalene is small, the molar volume of liquid naphthalene below its melting temperature will be taken to be that of the solid; that is,

$$\underline{V}_1^{\text{L}} = \frac{128.19 \text{ g/mol}}{1.0253 \text{ g/cc}} = 125 \text{ cc/mol}$$

The heat of sublimation of naphthalene is not given. However, we can compute this quantity from the vapor pressure curve of the solid and the Clausius–Clapeyron equation (Eq. 5.7-5a) by taking P equal to the sublimation pressure, $\Delta \underline{H}$ equal to the heat of sublimation, and setting $\Delta \underline{V} = \underline{V}^V - \underline{V}^S \cong RT/P$. Thus

$$\frac{\Delta \underline{H}^{\text{sub}}}{RT^2} = \frac{d \ln P}{dT} = 2.303 \frac{d \log_{10} P}{dT} = +2.303 \frac{(3783)}{T^2}$$

and

$$\Delta \underline{H}^{\text{sub}} = 2.303(3783)(8.314 \text{ J/mol}) = 72{,}434 \text{ J/mol}$$

[17]Reference: R. C. Weast, ed., *Handbook of Chemistry and Physics*, 68th ed., Chemical Rubber Publishing Co., Cleveland, 1987, pp. C-357, D-214.

Next

$$\Delta \underline{U}^{\text{vap}} = \Delta \underline{H}^{\text{sub}} - \Delta \underline{H}^{\text{fus}} - RT$$
$$= 72{,}434 - 18{,}804 - 8.314 \times 293.15$$
$$= 51{,}193 \text{ J/mol}$$

so that

$$\delta_1 = \left(\frac{51{,}193 \text{ J/mol}}{125 \text{ cc/mol} \times 4.184 \text{ J/cal}} \right)^{1/2} = 9.9 \text{ (cal/cc)}^{1/2}$$

Now using Eq. 8.5-12, with $\Delta C_P = 0$, and the regular solution expression for the activity coefficient we obtain

$$\ln x_1 = -\frac{\underline{V}_1^{\text{L}}(\delta_1 - \delta_2)^2 \Phi_2^2}{RT} - \frac{\Delta \underline{H}^{\text{fus}}(T_m)}{RT}\left(1 - \frac{T}{T_m}\right)$$

As a first guess assume that x_1 will be small so that

$$\Phi_2 = \frac{x_2 \underline{V}_2^{\text{L}}}{x_1 \underline{V}_1^{\text{L}} + x_2 \underline{V}_2^{\text{L}}} \approx 1$$

In this case

$$\ln x_1 = \frac{-125 \dfrac{\text{cc}}{\text{mol}} \times (9.9 - 7.3)^2 \dfrac{\text{cal}}{\text{cc}} \times 4.184 \dfrac{\text{J}}{\text{cal}}}{8.314 \dfrac{\text{J}}{\text{mol K}} \times 293.15 \text{ K}} - \frac{18804 \dfrac{\text{J}}{\text{mol}}}{8.314 \dfrac{\text{J}}{\text{mol K}} \times 293.15} \left(1 - \frac{293.15}{353.35}\right)$$
$$= -1.451 - 1.314 = -2.765$$

$$x_1 = 0.063$$

With such a large value for x_1 we must go back and correct the value of Φ_2 for the presence of the solute and repeat the computation. Thus

$$\Phi_2 = \frac{0.937 \times 132}{0.937 \times 132 + 0.063 \times 125} = 0.94$$

and

$$\ln x_1 = 1.282 - 1.314 = 2.596$$
$$x_1 = 0.0746$$

The results of the next two iterations are $x_1 = 0.0768$ and $x_1 = 0.0772$, respectively. This last prediction is in reasonable agreement with the experimental result of $x_1 = 0.09$.[18]

Comment

Note that had we assumed ideal solution behavior, $\gamma_1 = 1$ and $\ln \gamma_1 = 0$, so that

$$\ln x_1 = -1.314$$
$$x_1 = +0.269$$

which is a factor of three too large. ■

[18]G. Scatchard, *Chem. Revs.* **8**, 329 (1931).

ILLUSTRATION 8.5-2

Repeat Illustration 8.5-1 using the UNIFAC group contribution model to estimate the naphthalene activity coefficient.

Solution

Naphthalene has 8 aromatic CH (subgroup 10) and 2 aromatic C (subgroup 11) groups, and *n*-hexane has 2 CH_3 (subgroup 1) and 4 CH_2 (subgroup 2) groups. Since the output of the program UNIFAC.BAS is activity coefficients, Eq. 8.5-12 is rewritten as

$$x_1 = \frac{\exp\left(-\left\{\frac{\Delta \underline{H}^{\text{fus}}(T_{\text{T}})}{RT}\left[1 - \frac{T}{T_{\text{T}}}\right] + \frac{\Delta C_{\text{P}}}{R}\left[1 - \frac{T_{\text{T}}}{T} + \ln\left(\frac{T_{\text{T}}}{T}\right)\right]\right\}\right)}{\gamma_1} \tag{a}$$

or, using the same simplification as in the previous illustration

$$x_1 = \frac{\exp\left(-\left\{\frac{\Delta \underline{H}^{\text{fus}}(T_{\text{T}})}{RT}\left[1 - \frac{T_{\text{T}}}{T}\right]\right\}\right)}{\gamma_1} = \frac{\exp(-1.314)}{\gamma_1} \tag{b}$$

Using $x_1 = 0.07$ as the first guess, we find $\gamma_1 = 2.3888$ and, from Eq. b, $x_1 = 0.114$. Iterating several additional times yields $x_1 = 0.124$, which is 38% larger than the experimental value. ■

Although the discussion of this section has centered on the solubility of a solid in a pure liquid, the methods used can be easily extended to mixed solvents. In fact, to apply the equations developed in this section to mixed solvents we need only recognize that the measured or computed value of γ_1, the activity coefficient for the dissolved solid, used in the calculations must be appropriate to the solid and mixed solvent combination being considered. In the regular solution model, for example, this means replacing Eq. 8.5-13 with

$$RT \ln \gamma_1 = \underline{V}_1^{\text{L}}(\delta_1 - \bar{\delta})^2 \tag{8.5-17}$$

Similarly, the UNIFAC model and program are applicable to both binary and multicomponent mixtures.

To estimate the solubility of a solid in a gas we again start from Eq. 8.5-1, which, using Eq. 5.4-23 and 7.2-13, can be written as

$$P_i^{\text{sat}}(T)\left(\frac{f}{P}\right)_{\text{sat},T} \exp\left[\frac{\underline{V}_i^{\text{S}}(P - P_i^{\text{sat}})}{RT}\right] = y_i P\left(\frac{\bar{f}_i}{y_i P}\right) = y_i P \bar{\phi}_i \tag{8.5-18}$$

where (in the Poynting factor) we have assumed the solid is incompressible. At low pressures, the Poynting factor and the fugacity coefficients in the solid and fluid phases can all be taken equal to unity. In this case we obtain the following expression for the ideal solubility of a solid in a gas,

$$y_i^{\text{ID}} = \frac{P_i^{\text{sat}}(T)}{P} \tag{8.5-19}$$

ILLUSTRATION 8.5-3

Estimate the solubility of naphthalene in carbon dioxide at 1 bar and temperatures of 35.0 and 60.4°C.

Solution
Using the solid vapor pressure in Illustration 8.5-1 we have

T(°C)	P_N^{sat}(bar)	y_N (at 1 bar pressure)
35.0	2.789×10^{-4}	0.00028
60.4	2.401×10^{-3}	0.00240

■

At moderate and high pressures the solubility of the solid in the gas is computed from

$$y_i = \frac{P_i^{sat}(T)\left(\frac{f}{P}\right)_{sat,T} \exp\left[\frac{\underline{V}_i^S(P - P_i^{sat})}{RT}\right]}{P\phi_i(T, P, y_i)} \tag{8.5-20}$$

This equation usually must be solved by iteration since the solute mole fraction appears in the fugacity coefficient $\phi_i(T, P, y_i)$. However, since the vapor pressure of the solid is generally small, the term $(f/P)_{sat,T}$ is usually unity. Finally, we define an **enhancement factor E** as follows

$$E(T, P, y_i) = \frac{\left(\frac{f}{P}\right)_{sat,T} \exp\left[\frac{\underline{V}_i^S(P - P_i^{sat})}{RT}\right]}{\phi_i(T, P, y_i)} \tag{8.5-21a}$$

so that

$$y_i = \frac{P_i^{sat}(T)}{P} E(T, P, y_i) \tag{8.5-21b}$$

Note that E has contributions from both the Poynting factor and the vapor phase fugacity coefficient, both of which are important at high pressure, and that $E \rightarrow 1$ as $P \rightarrow P_i^{sat}$.

ILLUSTRATION 8.5-4

McHugh and Paulaitis [*J. Chem. Eng. Data* **25**, 326, (1980)] report the following data for the solubility of naphthalene in carbon dioxide at temperatures slightly above the CO_2 critical temperature and pressures considerably higher than its critical pressure.

T = 35.0°C		T = 60.4°C	
P(bar)	y_N	P(bar)	y_N
86.8	0.00750	108.4	0.00524
98.2	0.00975	133.8	0.01516
133.0	0.01410	152.5	0.02589
199.5	0.01709	164.2	0.04296
255.3	0.01922	192.6	0.05386
		206.0	0.06259

Assuming that the CO_2–naphthalene mixture obeys the Peng–Robinson equation of state with $k_{CO_2\text{-}N} = 0.103$, estimate the solubility of naphthalene in the CO_2 supercritical fluid. Also compute the predicted enhancement factors and the contribution of the Poynting factor to the enhancement factor.

Solution

Using the data in Illustration 8.5-1 and Table 4.6-1, and the program VLMU.BAS to calculate the vapor phase fugacities, the following results are obtained.

$T = 35.0°C$	P(bar)	y_N	y_N^{ID} (Eq. 8.5-19)	E	Poynting Factor
	86.8	1.99×10^{-3}	3.21×10^{-6}	619	1.527
	98.2	2.87×10^{-3}	2.84×10^{-6}	1010	1.615
	199.5	1.654×10^{-2}	1.40×10^{-6}	11,830	2.647
	255.3	1.940×10^{-2}	1.09×10^{-6}	17,759	3.475

$T = 60.4°C$	P(bar)	y_N	y_N^{ID}	E	Poynting Factor
	108.4	2.80×10^{-3}	2.21×10^{-5}	126	1.630
	133.8	1.05×10^{-2}	1.79×10^{-5}	584	1.828
	152.5	1.96×10^{-2}	1.57×10^{-5}	1246	1.989
	164.0	2.64×10^{-2}	1.46×10^{-5}	1806	2.096
	192.6	5.25×10^{-2}	1.25×10^{-5}	4211	2.383
	206.0	6.61×10^{-2}	1.17×10^{-5}	5671	2.531

The predicted and measured naphthalene mole fractions in supercritical carbon dioxide are plotted in Fig. 8.5-1.

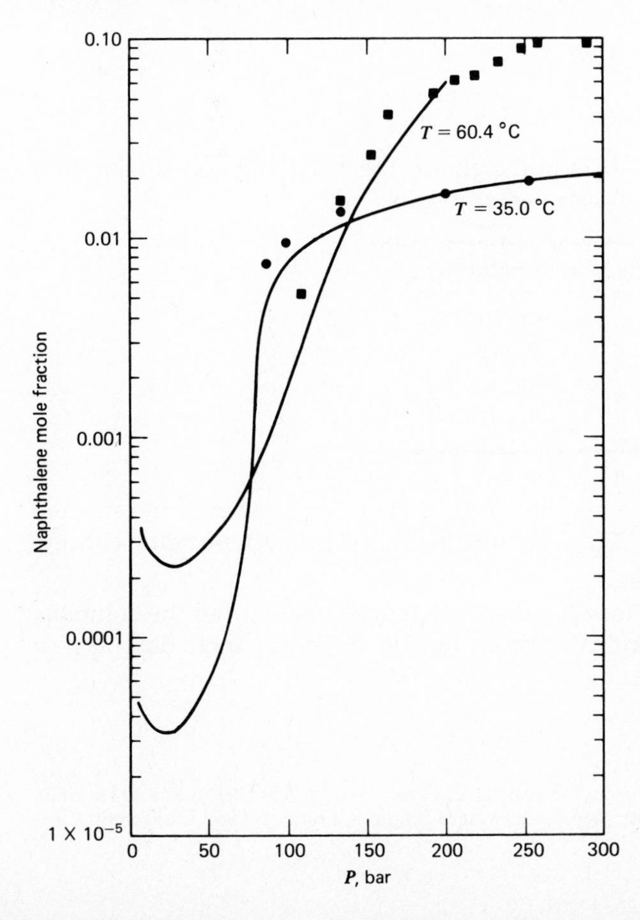

Figure 8.5-1
The solubility of napthalene in supercritical carbon dioxide as a function of pressure. The points ● and ■ are the experimental data of McHugh and Paulaitis [*J. Chem. Engr. Data* **25,** 326 (1980)] at T = 35.0 and 60.4°C, respectively. The lines are the correlations of the data using the Peng–Robinson equation of state with $k_{CO_2\text{-}N} = 0.103$. Note the sharp increase in naphthalene solubility with pressure near the CO_2 critical pressure of 73.76 bar.

Comments

1. Notice that the predictions for these extreme conditions are good. Indeed, with only one adjustable parameter ($k_{ij} = 0.103$) reasonable predictions are obtained for the solubility of naphthalene in supercritical carbon dioxide for a range of temperatures in the near critical region, and to moderately high pressures.
2. The enhancement factors here are huge, in fact, among the larger nonideal corrections encountered in chemical engineering thermodynamics. (Enhancement factors for other mixtures at cryogenic conditions of 10^9 and larger have been reported). Note that at $T = 35°C$ and $P = 255.3$ bar, the solubility of naphthalene is enhanced by a factor of more than 17,700 above its ideal value; however, its total solubility is still small at less than 2 mole percent.
3. The solubility of naphthalene in supercritical carbon dioxide increases from a mole fraction of 0.00240 at 1 bar (Illustration 8.5-3) to 0.098 at 291.3 bar. This illustrates the large increase that may occur in the solubility of a solute with increasing pressure, which is the basis of supercritical extraction to, for example, remove caffeine from coffee beans or fragrances and oils from plant material.
4. The Poynting corrections in this illustration are large, reaching values greater than 3. Consequently, the Poynting factor could not be ignored in this example. However, the main contribution to the enhancement factor arises from gas-phase nonidealities (the fugacity coefficient, ϕ_i). ■

PROBLEMS FOR SECTION 8.5

8.5-1 Estimate the solubility of naphthalene in the following solvents at 20°C and compare with the experimental results.[19]

Solvent	Measured Solubility, x_N
Chlorbenzene	0.256
Benzene	0.241
Toluene	0.224
Carbon tetrachloride	0.205

8.5-2 Estimate the solubility of naphthalene in the mixed solvent *n*-hexane and carbon tetrachloride at 20°C as a function of the (initial) *n*-hexane concentration.

8.5-3 McHugh and Paulaitis (see Illustration 8.5-4) also measured the solubility of biphenyl in supercritical carbon dioxide. Some of their data appear here.

[19]Data for this problem were taken from G. Scatchard, *Chem. Revs.* **8**, 329 (1931), and J. H. Hildebrand, J. M. Prausnitz and R. L. Scott, *Regular and Related Solutions*, Prentice–Hall, Englewood Cliffs, N.J., 1970. p. 152.

T(°C)	P(bar)	y_B
49.5	155.6	0.01782
49.5	204.5	0.02689
49.5	296.5	0.03605
49.5	379.4	0.03795
55.2	110.6	0.00447
55.2	132.6	0.01031
55.2	167.2	0.01829
55.2	252.5	0.03516
55.2	334.6	0.05615
55.2	412.8	0.07918
55.2	469.9	0.11054
55.2	482.7	0.12669
57.5	361.4	0.06365
57.5	430.3	0.09208

Choose a binary interaction parameter $k_{CO_2\text{-}B}$ that gives a reasonable correlation of these data.

8.5-4 At moderate pressures, the virial equation of state can be used to estimate the solubility of a solid in a supercritical fluid.

a Compute the solubility of naphthalene in carbon dioxide at 50°C and 60 bar using the data here:

Vapor pressure of naphthalene at 50°C = 1.11×10^{-3} bar

Virial coefficients at 50°C

$$B(CO_2\text{-}CO_2) = -0.103 \text{ m}^3/\text{kmol}$$

$$B(C_8H_{10}\text{-}CO_2) = -0.405 \text{ m}^3/\text{kmol}$$

Molar volume of naphthalene at 50°C = 0.112 m^3/kmol

b Repeat the computation in part a using the Peng–Robinson equation of state and compare the results.

8.6 THE PARTITIONING OF A SOLUTE AMONG TWO COEXISTING LIQUID PHASES; THE DISTRIBUTION COEFFICIENT

When a gas, liquid, or solid is added to two partially miscible or completely immiscible solvents, it will, depending on the amount of solute present, either partially or completely dissolve and be distributed unequally between the two liquid phases. Most chemists and chemical engineers first encounter this phenomenon in the organic chemistry laboratory where diethyl ether, which is virtually immiscible with water, is used to extract reaction products from aqueous solutions. The distribution of a solute among coexisting liquid phases is of industrial importance in purification procedures such as liquid extraction and partition chromatography, and of pharmacological interest in the distribution of drugs between lipids and body fluids.

Experimental data on the partitioning of a solute between liquid phases are usually reported in terms of a **distribution coefficient** K, defined to be the ratio of the solute concentration in the two phases

$$K = \frac{\text{concentration of solute in phase I}}{\text{concentration of solute in phase II}} \tag{8.6-1}$$

The purpose of this section is to study some aspects of the partitioning phenomenon, and to relate the distribution coefficient to more fundamental thermodynamic quantities, so that we can (1) predict the distribution coefficient for a solute among two given solvents if experimental data are not available, or (2) use experimental distribution coefficient data to obtain information on liquid-phase activity coefficients.

The simplest type of solute distribution problem occurs when N_1 moles of a solute are completely dissolved and distributed between two immiscible solvents. The equilibrium distribution of the solute is determined from the single equilibrium relation

$$\bar{f}_1^{\text{I}}(T, P, x_1^{\text{I}}) = \bar{f}_1^{\text{II}}(T, P, x_1^{\text{II}}) \tag{8.6-2}$$

and the constraint that the number of moles of solute be conserved

$$N_1 = N_1^{\text{I}} + N_1^{\text{II}} \tag{8.6-3}$$

where N_1^i is the number of moles of solute in phase i. Now, using the definition of the activity coefficient

$$\bar{f}_1(T, P, x_1) = x_1\gamma_1(T, P, x_1)f_1^{\text{L}}(T, P)$$

Eq. 8.6-2 can be rewritten as

$$x_1^{\text{I}}\gamma_1^{\text{I}}(T, P, x_1^{\text{I}}) = x_1^{\text{II}}\gamma_1^{\text{II}}(T, P, x_1^{\text{II}}) \tag{8.6-4}$$

since the pure component solute fugacity cancels. This last equation can be rearranged to

$$\frac{x_1^{\text{I}}}{x_1^{\text{II}}} = K_x = \frac{\gamma_1^{\text{II}}(T, P, x_1^{\text{II}})}{\gamma_1^{\text{I}}(T, P, x_1^{\text{I}})} \tag{8.6-5}$$

which establishes that the distribution coefficient for solute mole fractions is equal to the reciprocal of the ratio of the solute activity coefficients in the two phases. Thus, given activity coefficient information for the solute in the two phases, one can compute the distribution of the solvent among the phases, or, given information about distribution of the solute, one can compute the ratio of the solute activity coefficients.

ILLUSTRATION 8.6-1

The following data are available for the concentration distribution coefficient K_C of bromine between carbon tetrachloride and water at 25°C:[20]

[20]Reference for distribution coefficient data is *International Critical Tables,* Vol. 3, McGraw-Hill, New York, 1929.

Concentration of Bromine in CCl_4 (kmol/m³)	$K_C = \left\{ \frac{\text{kmol } Br_2}{\text{m}^3\ CCl_4 \text{ Solution}} \Big/ \frac{\text{kmol } Br_2}{\text{m}^3\ H_2O \text{ Solution}} \right\}$
0.04	26.8
0.1	27.2
0.5	29.0
1.0	30.4
1.5	31.4
2.0	33.4
2.5	35.0

Compute the ratio of the activity coefficient of bromine in water to that in carbon tetrachloride.

Data for the pure species are:

Species	Molecular Weight	Density ρ, kg m^{-3}
CCl_4	153.84	$1.595 \times 10^{+3}$
H_2O	18.0	$1.0 \times 10^{+3}$
Br_2	159.83	$3.119 \times 10^{+3}$

Solution

Since carbon tetrachloride and water are virtually immiscible (see Illustration 8.4-3), we will assume that the liquid bromine is distributed between water-free carbon tetrachloride and water free of carbon tetrachloride. Also, since solution volumetric data are not available, we will assume that there is no volume change on mixing bromine with either carbon tetrachloride or water.

In order to use Eq. 8.6-5 to evaluate the ratio of the activity coefficients, it is first necessary to convert all the concentration distribution data to mole fractions. To do this let C_B represent the concentration of bromine in kmol/m³ of solution, which is known from the data, and C_S represent the concentration of solvent, again in kmol/m³ of solution, which is also known. By the $\Delta \underline{V}_{mix} = 0$ assumption, we have that

$$\underline{V}_{mix} = x_B \underline{V}_B + x_S \underline{V}_S = x_B \frac{m_B}{\rho_B} + (1 - x_B) \frac{m_S}{\rho_S}$$

where $\underline{V}_i$ is the molar volume of species i, and m and ρ are its molecular weight and density, respectively. The moles of bromine in one mole of solution is then computed from

$$\begin{aligned}\text{moles of bromine in one mole of solution} &= C_B \underline{V}_{mix} = C_B \left[x_B \frac{m_B}{\rho_B} + (1 - x_B) \frac{m_S}{\rho_S} \right] \\ &= \text{mole fraction of bromine} = x_B\end{aligned}$$

or

$$x_B = \frac{C_B\, m_S/\rho_S}{1 + C_B \left(\dfrac{m_S}{\rho_S} - \dfrac{m_B}{\rho_B} \right)} \tag{1}$$

Therefore, to compute the mole fraction of bromine in the CCl_4 phase we use

$$x_{Br_2}^{CCl_4} = \frac{C_B\left(\dfrac{153.84\ \text{kg/mol}}{1595\ \text{kg/m}^3}\right)}{1 + C_B\left(\dfrac{153.84}{1595} - \dfrac{159.83}{3119}\right)\ \text{m}^3/\text{kmol}} = \frac{0.09645\ C_B}{1 + 0.04521\ C_B} \tag{2}$$

where C_B is the concentration entries in the distribution coefficient table. Similarly, to compute the mole fraction of bromine in the aqueous phase we again use Eq. 1, but now recognizing that if C_B is the concentration of bromine in the organic phase (i.e., the entries in the table), then C_B/K_C is the concentration of bromine in the aqueous phase. Therefore

$$x_{Br_2}^{H_2O} = \frac{\dfrac{C_B}{K_C}\left(\dfrac{18\ \text{kg/kmol}}{1000\ \text{kg/m}^3}\right)}{1 + \dfrac{C_B}{K_C}\left(\dfrac{18}{100} - \dfrac{159.83}{3119}\right)\ \text{m}^3/\text{kmol}} = \frac{0.018\ C_B/K_C}{1 - 0.03324\ C_B/K_C} \tag{3}$$

Using Eqs. 2 and 3 we obtain the following results

	Mole Fraction of Br_2 in		
Concentration Bromine in CCl_4	CCl_4 $x_{Br_2}^{CCl_4}$	H_2O $x_{Br_2}^{H_2O}$	$\dfrac{x_{Br_2}^{CCl_4}}{x_{Br_2}^{H_2O}} = \dfrac{\gamma_{Br_2}^{H_2O}}{\gamma_{Br_2}^{CCl_4}}$
0.04	3.85×10^{-3}	2.7×10^{-5}	142.6
0.10	9.60×10^{-3}	6.6×10^{-5}	145.4
0.5	4.70×10^{-2}	3.1×10^{-4}	151.6
1.0	9.23×10^{-2}	5.93×10^{-4}	155.6
1.5	0.135	8.62×10^{-4}	156.6
2.0	0.177	1.08×10^{-3}	163.9
2.5	0.217	1.29×10^{-3}	168.2

Comment

If, separately, we knew the activity coefficient behavior of bromine in carbon tetrachloride, we could use the data in the table to evaluate the activity coefficient of bromine in water (Problem 8.6-3). Since the regular solution model can be used to represent the Br_2–CCl_4 mixture, we can surmise that $\gamma_{Br_2}^{CCl_4}$ will be of the order of unity. This suggests that $\gamma_{Br_2}^{H_2O}$ will be a very large number, of the order of 100 or more. Such behavior for the activity coefficient of the minor component is not unusual in mixtures of species with such different molecular characteristics as strongly quadrupolar liquid bromine and strongly polar and hydrogen-bonded water. ■

A slightly more complicated situation than the one just considered arises when there is some undissolved solute in equilibrium with two immiscible solvents. Here we will suppose that the undissolved solute is either a solid or a gas, and that neither solvent is present in the undissolved solute.[21] In this case the equilibrium conditions are

$$f_1(T, P) = \bar{f}_1^{\mathrm{I}}(T, P, x_1^{\mathrm{I}}) \quad \text{and} \quad f_1(T, P) = \bar{f}_1^{\mathrm{II}}(T, P, x_1^{\mathrm{II}})$$

or

$$f_1(T, P) = \bar{f}_1^{\mathrm{I}}(T, P, x_1^{\mathrm{I}}) = \bar{f}_1^{\mathrm{II}}(T, P, x_1^{\mathrm{II}}) \tag{8.6-6}$$

since the undissolved solute must be in equilibrium with both liquid phases (which implies that both liquid phases are saturated with the solute). In this equation $f_1(T, P)$ is to be interpreted as the fugacity of the undissolved solute, which may be either a solid or a pure or mixed gas. Using the definition of the activity coefficient, Eq. 7.3-11, we obtain the following equations

$$\frac{f_1(T, P)}{f_1^{\mathrm{L}}(T, P)} = x_1^{\mathrm{I}}\gamma_1(T, P, x_1^{\mathrm{I}}) = x_1^{\mathrm{II}}\gamma_1(T, P, x_1^{\mathrm{II}}) \tag{8.6-7}$$

for the saturation mole fractions of the solute in each phase. If the undissolved solute is a solid, the fugacity ratio $f_1(T, P)/f_1^{\mathrm{L}}(T, P)$ is equal to the ratio $f_1^{\mathrm{S}}(T, P)/f_1^{\mathrm{L}}(T, P)$, which can be evaluated using the prescription of Sec. 8.5 (eq. 8.5-9). If, on the other hand, the solute is a gas, $f_1(T, P)$ can be computed using an equation of state or the Lewis-Randall rule, and $f_1^{\mathrm{L}}(T, P)$ is evaluated using the procedure discussed in Sec. 8.3. Note that Eqs. 8.6-5 and 7 require that the mole fraction distribution coefficient K_x be equal to the reciprocal of solute activity coefficients in the two liquid phases; this is, in fact, a general result.

Further complications arise when the two solvents are partially miscible (instead of immiscible), since in this case we have, in addition to the previous equations, the equilibrium conditions

$$\bar{f}_j^{\mathrm{I}}(T, P, x_j^{\mathrm{I}}) = \bar{f}_j^{\mathrm{II}}(T, P, x_j^{\mathrm{II}}) \tag{8.6-8}$$

or

$$x_j^{\mathrm{I}}\gamma_j(T, P, x_j^{\mathrm{I}}) = x_j^{\mathrm{II}}\gamma_j(T, P, x_j^{\mathrm{II}})$$

and mass balance constraints

$$N_j = N_j^{\mathrm{I}} + N_j^{\mathrm{II}} \tag{8.6-9}$$

for each of the solvent species, which we number 2, 3, and so forth. Thus to compute the equilibrium state when N_2 moles of solvent 2, N_3 moles of solvent 3, etc., are mixed with N_1 moles of a solute, Eqs. 8.6-8 and 9 are to be solved together with Eqs. 8.6-3 and 4 if all the solute dissolves, or with Eqs. 8.6-3 and 7 if some undissolved solute remains. These equations can be difficult to solve, first because of the complicated dependence of the activity coefficients on the mole fractions (these equations are nonlinear) and second, because even in the simplest case of a single solute in a mixture of two partially miscible solvents, there are six coupled equations to be solved.

Since the addition of any solute affects the activity coefficients of all species in a mixture, some interesting effects may accompany solute addition. For example, the addition of a solute may increase (salting in) or decrease (salting out) the mutual solubility of the two solvents; sometimes to the extent that they become

[21] When the solute is a liquid, the problem is really one of multiple liquid phase equilibrium. This is considered in Problem 8.6-1 and later in this section.

completely miscible. Also, the addition of a second solute may change the distribution coefficient for the first solute, as in Problem 8.6-4. The occurrence of these phenomena are allowed by the equations developed in this section.

If a liquid solute (rather than a gas or a solid) is added to two coexisting liquids, there is the possibility that three partially miscible phases will be present at equilibrium. Although three-phase equilibrium phenomena could be studied by a generalization of the two-phase equilibrium analysis used here (see Problem 8.6-1), this will not be done. Instead, we will merely discuss the triangular diagram method of presentation of liquid-phase equilibrium data for ternary mixtures indicated in Fig. 8.6-1. These two-dimensional representations of the three composition variables are to be interpreted as follows. The three apexes of the triangle each represent a pure species; the composition (which, depending on the figure may be either mass fraction or mole fraction) of each species decreases linearly with distance along the perpendicular bisector from the apex of that species to the opposite side of the triangle. Thus each side of the triangle denotes the absence of the species at the opposite apex. For convenience, these fractional concentrations are frequently indicated along one side of the triangle for each species. The fractional concentration of each species at a specific point on these diagrams is found by drawing a line through that point parallel to the side opposite the apex for that substance and noting the intersection of this line with the appropriate side of the triangle.

The dome-shaped regions in the diagrams in Fig. 8.6-1 denote phase separation (in some systems these regions merge, as in Fig. 8.6-1*c*), and tie lines are

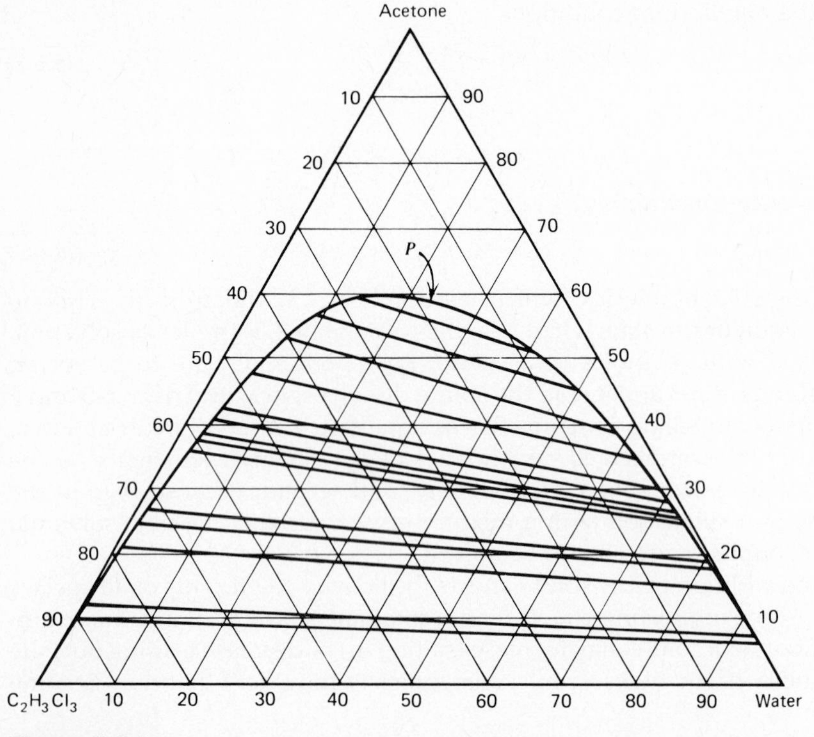

Figure 8.6-1a
Equilibrium diagram for the acetone–water–1,1,2-trichloroethane system. [Reprinted with permission from R. E. Treybal, L. D. Weber, and J. F. Daley, *Ind. Eng. Chem.* **38,** 817 (1946). Copyright by American Chemical Society.]

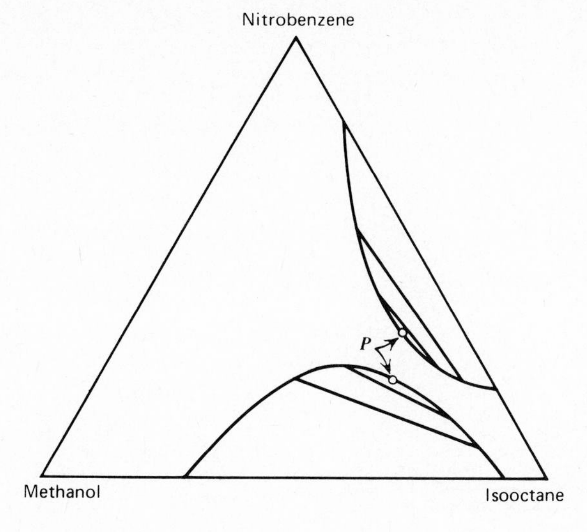

Figure 8.6-1*b*
Equilibrium diagram for the nitrobenzene–methanol–isooctane system at 15°C. (From A. W. Francis, *Liquid–Liquid Equilibriums*. Copyright 1963 by John Wiley & Sons, Inc. Used with permission.)

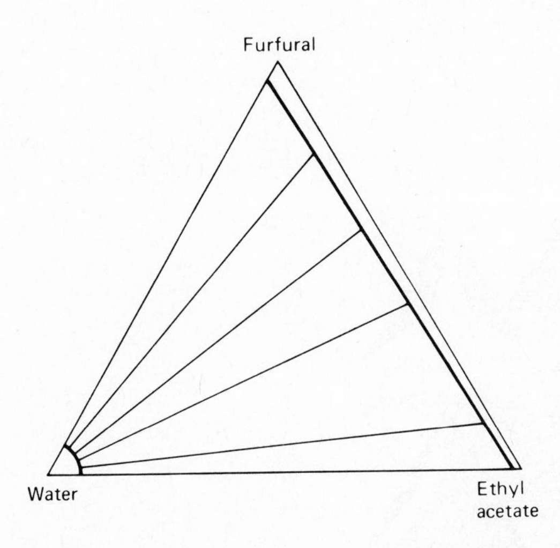

Figure 8.6-1*c*
Equilibrium diagram for the furfural–water–ethyl acetate system. (From A. W. Francis, *Liquid–Liquid Equilibriums*. Copyright 1963 by John Wiley & Sons, Inc. Used with permission.)

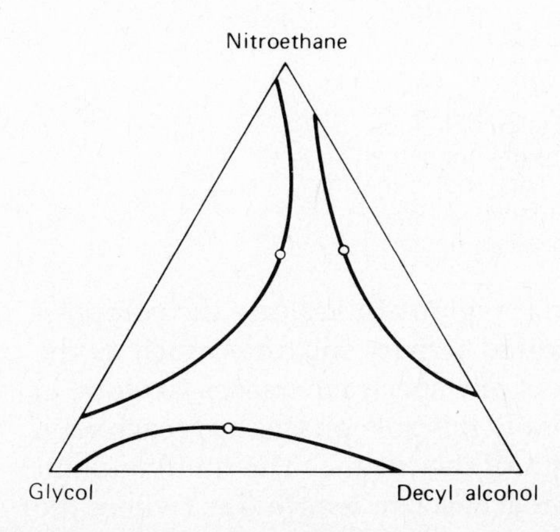

Figure 8.6-1*d*
Equilibrium diagram for the nitroethane–glycol–decyl alcohol system at 10°C. [Reprinted with permission from A. W. Francis, *J. Phys. Chem.* **60,** 20 (1956). Copyright by the American Chemical Society.]

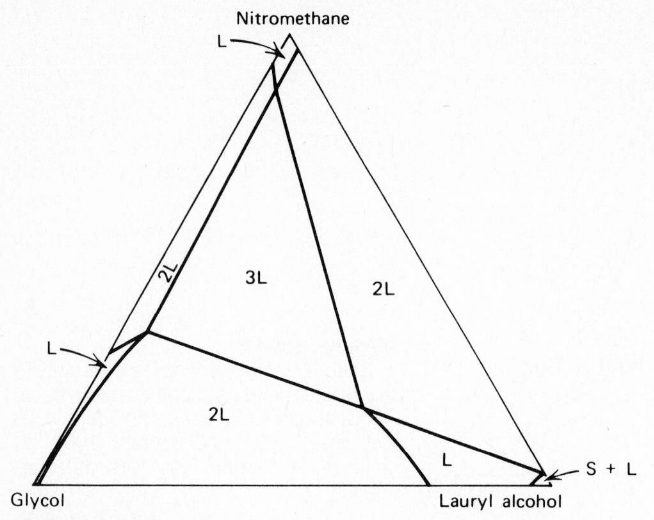

Figure 8.6-1e
Equilibrium diagram for the nitromethane–glycol–lauryl alcohol system at 20°C, showing the presence of two and three coexisting solid and liquid phases. [Reprinted with permission from A. W. Francis, *J. Phys. Chem.* **60,** 20 (1956). Copyright by the American Chemical Society.]

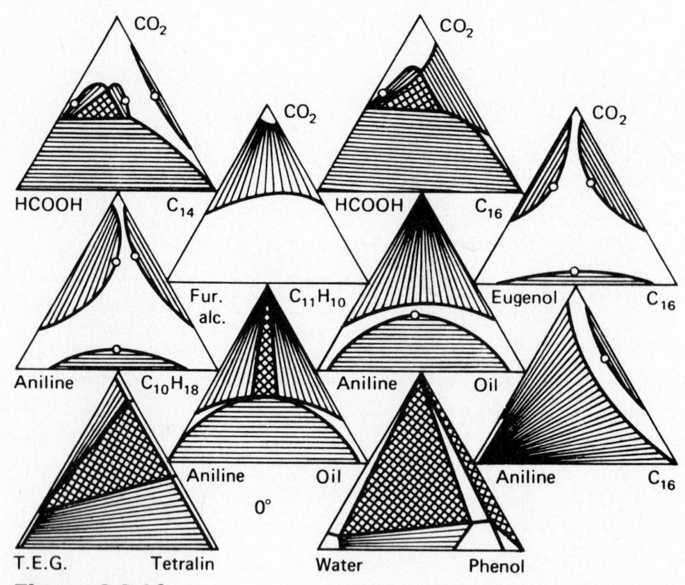

Figure 8.6-1f
Equilibrium diagrams for carbon dioxide with 10 pairs of other liquids, demonstrating the wide variety of liquid–liquid phase equilibrium that occur. Note that tie lines and plait points have been included in the diagrams. [Reprinted with permission from A. W. Francis, *J. Phys. Chem.* **58,** 1099 (1954). Copyright by the American Chemical Society.]

sometimes drawn within the phase separation regions to indicate the compositions of the coexisting phases. In uncomplicated ternary mixtures, such as the one shown in Fig. 8.6-1*a*, as the composition of one species increases (acetone in Fig. 8.6-1*a*), the compositions of the two partially miscible phases approach each other, as indicated by a decrease in the length of the tie line. The limiting point on the phase diagram where the tie line approaches zero length (i.e., where the two phases merge) is termed a critical point or **plait point** of the mixture (denoted by the point P in the figures).

In more complicated systems, such as the nitrobenzene–methanol–isooctane system of Fig. 8.6-1*b*, there may be two two-phase regions or, as in Fig. 8.6-1*c* for the furfural–water–ethyl acetate system, the two two-phase regions merge into a band. Similarly, the nitroethane–glycol–decyl alcohol system of Fig. 8.6-1*d* has three distinct two-phase regions, whereas Fig. 8.6-1*e*, for the nitromethane–glycol–lauryl alcohol system, shows the merging of these two-phase regions (denoted by 2L) into regions where three liquid phases coexist (denoted by 3L).

Finally, Fig. 8.6-1*f*, for the liquid–liquid phase equilibrium behavior of liquid carbon dioxide with pairs of other liquids, has been included to illustrate the variety of types of ternary system phase diagrams the chemist and engineer may encounter. A complete discussion of these different types of phase diagrams are given in numerous places, including A. W. Francis, *Liquid–Liquid Equilibriums*, Interscience, New York, 1963.

PROBLEMS FOR SECTION 8.6

8.6-1 Write the equations that are to be used to compute the equilibrium compositions when three liquids are mixed and form

a Two liquid phases.

b Two liquid phases in which species 1 is distributed among completely immiscible species 2 and 3.

c Three partially miscible liquid phases.

8.6-2 An important application of liquid-phase partitioning is in the process of partition chromatography. There a solvent containing several solutes is brought into contact with another solvent, usually one that is mutually immiscible with the first. When equilibrium is established, some solutes are concentrated in the first solvent and others in the second solvent. By repeating this process a number of times, it is possible to partially separate the solutes.

The concentration distribution coefficient K_C for gallic acid ($C_7H_6O_5$) between diethyl ether (phase I) and water (phase II) is 0.25, and that for *p*-hydroxybenzoic acid is 8.0. If 2 g of each of gallic acid and *p*-hydroxybenzoic acid are present in 1 liter of water, how much of each component would be left in the aqueous phase if the aqueous phase were

a Allowed to achieve equilibrium with 5 liters of diethyl ether.

b Allowed to achieve equilibrium first with one batch of 2 1/2 liters of diethyl ether, and then with a second batch of the same amount.

c Allowed to achieve equilibrium with five successive 1-liter batches of pure diethyl ether.

8.6-3 Use regular solution theory to calculate the activity coefficients of bromine in carbon tetrachloride at the mole fractions found in Illustration 8.6-1, and use this information and that in the illustration to compute the activity coefficients of bromine in water.

8.6-4 Volume 3 of the *International Critical Tables* (McGraw-Hill, New York, 1928) gives the following data for the distribution coefficient of bromine between carbon disulfide and various aqueous solutions of potassium bromide at 25°C:

Bromine Concentration in Carbon Disulfide (mol Br_2/l of solution)	Distribution Coefficient $\left(\frac{\text{mol } Br_2\text{/l of aqueous phase}}{\text{mol } Br_2\text{/l of organic phase}}\right)$		
	$\frac{1}{16}$ Molar KBr	$\frac{1}{4}$ Molar KBr	$\frac{1}{2}$ Molar KBr
0.005	0.0696	0.1735	
0.02	0.0651	0.167	0.303
0.03	0.0621	0.163	0.298
0.04		0.159	0.293
0.05		0.155	0.288
0.06		0.151	0.283
0.095			0.271

Using regular solution theory to estimate the activity coefficient of bromine in carbon disulfide, compute the activity coefficient of bromine in

a The 1/16 molar KBr solution.
b The 1/4 molar KBr solution.
c The 1/2 molar KBr solution.
d Try to infer a relationship between the salt concentration and the activity coefficient of bromine in aqueous solution.

8.7 FREEZING POINT DEPRESSION OF A SOLVENT DUE TO THE PRESENCE OF A SOLUTE; THE FREEZING POINT OF LIQUID MIXTURES

If a small amount of solute (gas, liquid, or solid) is added to a liquid solvent and the temperature of the mixture lowered, a temperature T_f is reached at which the pure solvent begins to separate out as a solid. This temperature is lower than the freezing (or melting) point of the pure solvent, T_m. Here we are interested in estimating the depression of the solvent freezing point, $\Delta T = T_m - T_f$, due to the addition of a solid solute. This equilibrium state is, in a sense, similar to the one encountered in the solubility of a solid in a liquid in that there we were interested in the equilibrium between a solid and a dilute liquid mixture of that component; here we are interested in the equilibrium between a solid and a liquid mixture that is concentrated in that component.

The starting point for the analysis of the freezing point depression phenomenon is the observation that when the first drop of pure solvent freezes, the equilibrium condition is

$$f_1^S(T_f, P) = \bar{f}_1^L(T_f, P, x_1) = x_1\gamma_1(T_f, P, x_1)f_1^L(T_f, P) \tag{8.7-1}$$

where x_1 is the original liquid solution mole fraction of the component that precipitates out as a solid. In writing this equation we have assumed that the solid phase is pure solvent and used the fact that the composition of the liquid phase is not significantly changed by the appearance of a very small amount of solid phase (in this regard the analysis here resembles that for dew-point and bubble-point phenomena considered in Secs. 8.1 and 2).

Combining Eq. 8.7-1 with Eq. 8.5-9 for $\ln[f_1^S(T_f, P)/f_1^L(T_f, P)]$ yields

$$\begin{aligned}\ln \gamma_1 x_1 &= \ln \frac{f_1^S(T_f, P)}{f_1^L(T_f, P)} \\ &= -\frac{\Delta \underline{H}^{\text{fus}}(T_m)}{R}\left[\frac{T_m - T_f}{T_m T_f}\right] - \frac{1}{RT_f}\int_{T_m}^{T_f} \Delta C_P \, dT + \frac{1}{R}\int_{T_m}^{T_f} \frac{\Delta C_P}{T} dT \\ &= -\frac{\Delta \underline{H}^{\text{fus}}(T_m)}{R}\left[\frac{T_m - T_f}{T_m T_f}\right] - \frac{\Delta C_P}{R}\left[1 - \frac{T_m}{T_f} + \ln\left(\frac{T_m}{T_f}\right)\right]\end{aligned} \quad (8.7\text{-}2)$$

where $\Delta \underline{H}^{\text{fus}}(T_m)$ is the heat of fusion of the solid solvent at its normal melting point temperature T_m. To obtain the last form of Eq. 8.7-2, we have assumed that ΔC_P is independent of temperature.

In many instances, T_m and T_f are sufficiently close that Eq. 8.7-2 can be simplified to

$$\ln(\gamma_1 x_1) \cong -\frac{\Delta \underline{H}^{\text{fus}}(T_m)}{R}\left(\frac{T_m - T_f}{T_m T_f}\right) \cong -\frac{\Delta \underline{H}^{\text{fus}}(T_m)}{RT_m^2}(T_m - T_f) \quad (8.7\text{-}3)$$

so that

$$\Delta T = T_m - T_f = -\frac{RT_m^2}{\Delta \underline{H}^{\text{fus}}(T_m)} \ln(\gamma_1 x_1) \quad (8.7\text{-}4)$$

For the case of very dilute solutes we expect that $\gamma_1 \sim 1$, and $\ln x_1 = \ln(1 - x_2) \sim -x_2$, so that the freezing point depression equation may be further simplified to

$$\Delta T = T_m - T_f \cong +\frac{RT_m^2}{\Delta \underline{H}^{\text{fus}}(T_m)} x_2 \quad (8.7\text{-}5)$$

In the case of dilute mixed solutes the equation becomes

$$\Delta T = T_m - T_f = \frac{RT_m^2}{\Delta \underline{H}^{\text{fus}}(T_m)} \sum_{i=2}^{\mathscr{C}} x_i \quad (8.7\text{-}6)$$

where the sum, which extends from 2 to $\mathscr{C}$, is over all solute species.

ILLUSTRATION 8.7-1

Determine the freezing point depression of water as a result of the addition of 0.01 g/cm³ of (a) methanol, and (b) a protein whose molecular weight is 60,000.

Solution

Since such a small amount of solute is involved, we will assume the density of the solution is the same as that of pure water, 1 g/cm³. Thus the solute mole fraction is

$$x_{\text{solute}} = \frac{\text{moles of solute in 1 cm}^3 \text{ of solution}}{(\text{moles of solute + moles of water) in 1 cm}^2 \text{ of solution}} = \frac{\dfrac{0.01}{m}}{\dfrac{0.99}{18} + \dfrac{0.01}{m}}$$

where m is the molecular weight of the solute. $\Delta \underline{H}^{\text{fus}}(T_m)$ for water is 6025 J/mol.

a. The molecular weight of methanol is 32. Thus, $x_{\text{solute}} = 0.00565$, and

$$\Delta T = \frac{8.314 \dfrac{\text{J}}{\text{mol K}} \times (273.15\ \text{K})^2 \times 5.65 \times 10^{-3}}{6025\ \text{J/mol}} = 0.58\ \text{K}$$

b. The molecular weight of the protein is 60,000, so that $x_{\text{solute}} \sim 3 \times 10^{-6}$ and

$$\Delta T = \frac{8.314 \dfrac{\text{J}}{\text{mol K}} \times (273.15\ \text{K})^2 \times 3 \times 10^{-6}}{6025\ \text{J/mol}} = 3.1 \times 10^{-4}\ \text{K}$$

This small depression of the freezing point is barely measurable. ■

One interesting application of Eq. 8.7-2 is the computation of the freezing point of a general binary liquid mixture; that is, a liquid mixture in which neither component is easily recognized to be either the solvent or the solute. This is considered in Illustration 8.7-2.

ILLUSTRATION 8.7-2

Compute the freezing point temperature versus composition curve for an ethyl benzene–toluene mixture. The physical properties for this system are given in the table.

	Toluene	Ethyl Benzene
δ (cal/cc)$^{1/2}$ at 25°C	8.9	8.8
$\underline{V}^{\text{L}}$ (cc/mol) at 25°C	107	123
Normal melting point (K)	178.16	178.2
Heat of fusion (J/mol)	6610.7	9070.9
C_{P}^{S} J/mol K	87.0	105.9
C_{P}^{L} J/mol K	135.6	157.4

Solution

The solubility parameters for toluene and ethyl benzene are nearly equal, so that the liquid mixture will be considered to be ideal ($\gamma_1, \gamma_2 = 1$). Since we do not know a priori whether pure toluene or pure ethyl benzene will freeze out as the solid component from a given mixture, the calculational procedure we will follow is to first assume that toluene appears as the solid component and compute the freezing point temperature T_f of all toluene–ethyl benzene mixtures. The calculations will then be repeated assuming ethyl benzene appears as the solid phase. The freezing point of a given mixture, and the solid that appears, will then be determined by noting which of the two calculations leads to the higher freezing point.

If we assume that toluene freezes out, T_f is found from the solution of

$$\ln x_1 = -\frac{6610.7}{8.314}\left(\frac{178.16 - T_f}{178.16\,T_f}\right) - \frac{48.6}{8.314}\left(1 - \frac{178.16}{T_f} + \ln \frac{178.16}{T_f}\right)$$

where x is the mole fraction of toluene. Since ΔC_{P} is so large, none of the terms in this equation can be neglected, and the solution is found by trial and error. The results are given in the following table. Similarly, if we assume the solid precipitate is ethyl benzene, the mixture freezing point is computed from

$$\ln x_1 = -\frac{9070.9}{8.314}\left(\frac{178.2 - T_f}{178.2\,T_f}\right) - \frac{51.5}{8.314}\left(1 - \frac{178.2}{T_f} + \ln\frac{178.2}{T_f}\right)$$

where x is now the mole fraction of ethyl benzene. The results of this calculation are also given in the following table. Both sets of freezing points are plotted in Fig. 8.7-1.

	T_f(K)	
x_T	**Toluene as the Solid**	**Ethyl Benzene as the Solid**
1.0	178.16	
0.9	174.0	122.2
0.8	169.4	137.0
0.7	164.3	146.4
0.6	158.6	153.4
0.5	151.9	159.1
0.4	144.0	163.9
0.3	134.2	168.1
0.2	121.0	171.8
0.1	99.8	175.1
0		178.2

■

From the figure it is evident that below 47.5 mole percent toluene, ethyl benzene precipitates as the solid phase, whereas at higher concentrations toluene is the solid precipitate. Furthermore, the minimum freezing point of the mixture is 155.9 K, which occurs at $x_T = 0.558$. This point is called the **eutectic point** of the mixture. At the eutectic temperature the remaining liquid mixture solidifies into a mixed solid of the same composition (55.8 mole percent toluene) as the liquid from which it was formed.

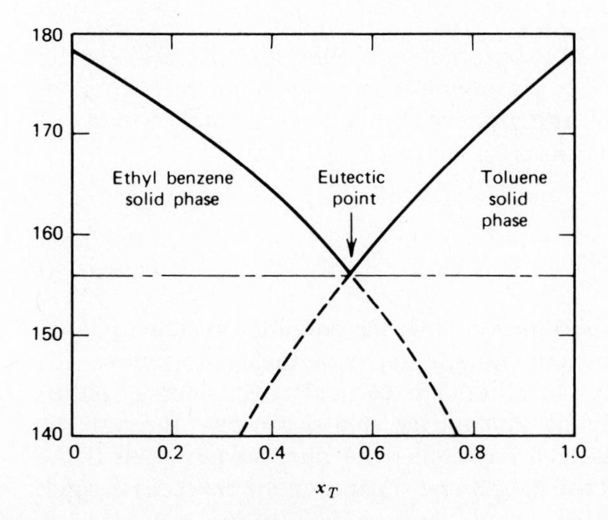

Figure 8.7-1
The liquid–solid phase diagram for ethyl benzene–toluene mixtures.

Although the phase diagram for the toluene–ethyl benzene mixture was constructed by determining the temperatures at which the first minute amount of solid appears in liquid mixtures of various compositions, it may now be used to analyze the complete solidification behavior of a liquid mixture at any composition. For example, suppose 1 mole of a 20 mole percent toluene mixture is cooled. From the phase diagram we see that at 171.8 K a minute amount of pure ethyl benzene appears as the solid. As the temperature is lowered, more pure ethyl benzene precipitates out, and the remaining liquid becomes increasingly richer in toluene. The temperature and composition of this liquid, which is in equilibrium with the solid, follows the freezing point curve in the figure; the relative amounts of the liquid and solid phases can be found by a species mass balance equation. Thus, at $T = 163.9$ K, the liquid contains 40 mole percent toluene. Letting L represent the number of moles of liquid in equilibrium with the solid, and writing a mass (mole) balance on toluene using, as a basis, 1 mole of the initial mixture, we find

$$(0.2)(1) = 0.4L$$

or

$$L = 0.5 \text{ moles}$$

So that when the temperature of the mixture is 163.9 K, half the original mixture is present as pure, solid ethyl benzene, and the remainder appears as a liquid enriched in toluene.

As the temperature is lowered further, additional pure ethyl benzene precipitates out and the liquid becomes richer in toluene, until the eutectic composition and temperature is reached. At this point all the remaining liquid solidifies as two mixed solid phases.

PROBLEMS FOR SECTION 8.7

8.7-1 The addition of a nonvolatile solute, such as a protein or a salt, to a pure solvent will raise its normal boiling point as well as lower its freezing point. Develop an expression relating the boiling-point elevation of a solvent to the solute concentration, heat of vaporization, and normal boiling temperature.

8.7-2 Chemical additives are added to automotive cooling systems to lower the freezing point of water used as the engine coolant. Estimate the number of grams of methanol, ethanol, and glycerol that, when separately added to 1 kg of pure water, will lower its freezing point to −12°C.

8.8 OSMOTIC EQUILIBRIUM AND OSMOTIC PRESSURE

As a final example of phase equilibrium, consider the equilibrium state of liquid mixtures in two cells separated by a membrane that is permeable to some of the species present and impermeable to others. In particular, consider a solute–solvent system in which the solvent, but not the solute, can pass through the membrane. We will assume that cell I contains the pure solvent, cell II the solvent–solute mixture, and that the membrane separating the two cells is rigid, so that the two cells need not be at the same pressure. The equilibrium criterion for this system is

$$\bar{f}^{\mathrm{I}}_{\text{solvent}} = \bar{f}^{\mathrm{II}}_{\text{solvent}} \tag{8.8-1}$$

or

$$f_{\text{solvent}}(T, P^{\mathrm{I}}) = x^{\mathrm{II}}_{\text{solvent}} \gamma^{\mathrm{II}}_{\text{solvent}} f_{\text{solvent}}(T, P^{\mathrm{II}}) \tag{8.8-2}$$

A similar equation is not written for the solute since it is mechanically constrained from passing through the membrane and thus need not be in thermodynamic equilibrium.

Actually, the use of Eq. 8.8-1 as the equilibrium requirement for this case deserves some discussion. We have shown that this equation is the equilibrium criterion for a system at constant temperature and pressure. From the discussion of Sec. 6.7, it is clear that it is also valid for systems subject to certain other constraints. Here we are interested in the equilibrium criterion for a system divided into two parts, each of which is maintained at constant temperature and pressure, but not necessarily the same pressure (or, for that matter, temperature). Though it is not obvious that Eq. 8.8-1 applies to this case, it can be shown to be valid (Problem 8.8-1).

Since $P^{\mathrm{I}} \neq P^{\mathrm{II}}$, the fugacities of the pure solvent at the conditions in the two cells are related by the Poynting pressure correction (see Eq. 5.4-21)

$$f_{\text{solvent}}(T, P^{\mathrm{II}}) = f_{\text{solvent}}(T, P^{\mathrm{I}}) \exp\left[\frac{\underline{V}^{\mathrm{L}}_{\text{solvent}}(P^{\mathrm{II}} - P^{\mathrm{I}})}{RT}\right] \tag{8.8-3}$$

where we have assumed that the liquid is incompressible, so that $\int_{P^{\mathrm{I}}}^{P^{\mathrm{II}}} \underline{V}^{\mathrm{L}}_{\text{solvent}}\, dP \simeq \underline{V}^{\mathrm{L}}_{\text{solvent}}(P^{\mathrm{II}} - P^{\mathrm{I}})$. Using Eq. 8.8-3 in Eq. 8.8-2 yields

$$1 = x^{\mathrm{II}}_{\text{solvent}} \gamma^{\mathrm{II}}_{\text{solvent}} \exp\left[\frac{\underline{V}^{\mathrm{L}}_{\text{solvent}}(P^{\mathrm{II}} - P^{\mathrm{I}})}{RT}\right]$$

or

$$P^{\mathrm{II}} - P^{\mathrm{I}} = -\frac{RT}{\underline{V}^{\mathrm{L}}_{\text{solvent}}} \ln\left(x^{\mathrm{II}}_{\text{solvent}} \gamma^{\mathrm{II}}_{\text{solvent}}\right) \tag{8.7-4}$$

where $P^{\mathrm{II}} - P^{\mathrm{I}}$, called the **osmotic pressure,** is the pressure difference between the two cells needed to maintain thermodynamic equilibrium.

A very large pressure difference can be required to maintain osmotic equilibrium in a system with only a small concentration difference. Consider, for example, the very simple case of osmotic equilibrium at room temperature between an ideal aqueous solution containing 98 mole percent water and pure water. Here

$$P^{\mathrm{II}} - P^{\mathrm{I}} = -\frac{RT}{\underline{V}^{\mathrm{L}}_{\text{solvent}}} \ln\left(x^{\mathrm{II}}_{\text{solvent}}\right) = -\frac{8.314 \times 10^{-5}\,\dfrac{\text{bar m}^3}{\text{mol K}} \times 298.15\text{ K}}{18 \times 10^{-6}\,\dfrac{\text{m}^3}{\text{mol}}} \ln 0.98 = 27.8 \text{ bar}$$

Thus if cell I is at atmospheric pressure, cell II would have to be maintained at 28.8 bar to prevent the migration of water from cell I to cell II. (A more accurate estimate of the osmotic pressure could be obtained using experimental data or appropriate liquid solution theories to evaluate the activity coefficient of water in aqueous solutions. See Problem 8.8-3.)

Since pressure measurements are relatively simple, Eq. 8.8-4 can be the basis for determining solvent activity coefficients in a solvent–solute system, provided a suitable leakproof membrane can be found. Osmotic pressure measurements are, however, more commonly used to determine the molecular weight of proteins and other macromolecules (for which impermeable membranes are easily found). In such cases an osmometer, such as the one shown in schematic form in Fig. 8.8-1, is used to measure the equilibrium pressure difference between the pure solvent and the solvent containing the macromolecules (which are too large to pass through the membrane); the pressure difference ΔP is equal to ρgh, where ρ is the solution density and h is the difference in liquid heights. If the solute concentration is small, we have

$$\Delta P = -\frac{RT}{\underline{V}_{\text{solvent}}} \ln (x_{\text{solvent}}) \approx \frac{RT}{\underline{V}_{\text{solvent}}} (1 - x_{\text{solvent}}) \tag{8.8-5}$$

furthermore, since $x_{\text{solvent}} + x_{\text{solute}} = 1$, this reduces to

$$\begin{aligned} \Delta P &= \frac{RT}{\underline{V}_{\text{solvent}}} x_{\text{solute}} \cong \frac{RTC_{\text{solute}}/m_{\text{solute}}}{\underline{V}_{\text{solvent}} C_{\text{solvent}}/m_{\text{solvent}}} \\ &= RTC_{\text{solute}}/m_{\text{solute}} \end{aligned} \tag{8.8-6}$$

Here the concentration C is in mass per volume, m is the molecular weight, and we have used the fact that since the solute mole fraction is low

$$x_{\text{solute}} = \frac{\text{moles solute}}{\text{moles solute} + \text{moles solvent}} \simeq \frac{\text{moles solute}}{\text{moles solvent}} = \frac{C_{\text{solute}}/m_{\text{solute}}}{C_{\text{solvent}}/m_{\text{solvent}}}$$

and

$$\frac{C_{\text{solvent}}}{m_{\text{solvent}}} = \frac{1}{\underline{V}_{\text{solvent}}}$$

Relatively small concentrations of macromolecules produce easily measurable osmotic pressure differences. For example, suppose 1 g of a protein or polymer of molecular weight 60,000 dissolved in 100 cc of water in the os-

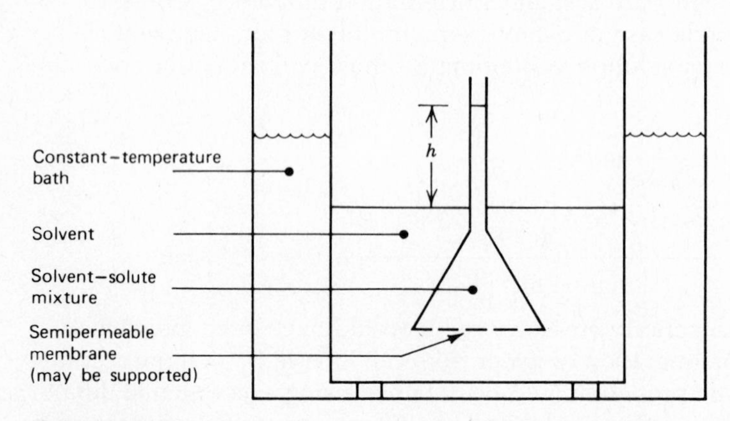

Figure 8.8-1
A schematic drawing of a simple osmometer.

mometer of Fig. 8.8-1. At a temperature of 25°C, the osmotic pressure difference would be

$$\Delta P = \frac{0.01 \frac{\text{g}}{\text{cc}} \times 298.15 \text{ K} \times 8.314 \times 10^{-5} \frac{\text{bar m}^3}{\text{mol K}} \times 10^6 \frac{\text{cc}}{\text{m}^3}}{60{,}000 \text{ g/mol}}$$

$$= 4.13 \times 10^{-3} \text{ bar} = 4.22 \text{ cm H}_2\text{O}$$

which is quite easily measurable.

To determine the molecular weight of a macromolecule, a known weight of the substance is added to the osmometer and ΔP measured. The molecular weight is then found from

$$m_{\text{solute}} = \frac{RTC_{\text{solute}}}{\Delta P} \tag{8.8-7}$$

For high accuracy, this measurement is repeated several times at varying solute concentrations, and the limiting value of $C_{\text{solute}}/\Delta P$ as C_{solute} approaches zero is used in Eq. 8.8-7. This procedure allows for the fact that C_{solute} is imperfectly known (since additional solvent passes through the membrane until equilibrium is established) and that the simplifications in Eq. 8.8-6 become exact and solvent nonidealities vanish as $C_{\text{solute}} \rightarrow 0$.

The advantage of the osmotic pressure difference method over the freezing point depression method of determining the molecular weights of macromolecules is evident from a comparison of the magnitudes of the effects to be measured. In Illustration 8.7-1 it was shown that the addition of 0.01 g/cc of a 60,000 molecular weight protein results in a 0.00031 K freezing point depression of water, whereas here we find it results in an osmotic pressure of 4.22 cm H_2O.

Finally, it should be pointed out that the Gibbs phase rule, as developed in Sec. 6.9, does not apply to osmotic equilibrium. This is because (1) the total pressure need not be the same in each phase (cell), so that Eq. 6.9-2 is not satisfied, and (2) the equality of partial molar Gibbs free energies in each phase (cell) does not apply to all species, but only to those that can pass through the membrane. A form of the Gibbs phase rule applicable to osmotic equilibrium can be easily developed (Problem 8.8-2).

PROBLEMS FOR SECTION 8.8

8.8-1 Show that the condition for equilibrium in a closed isothermal system, one part of which is maintained at P^{I} and the remainder at P^{II}, is that the function $G' = U - TS + P^{\text{I}}V^{\text{I}} + P^{\text{II}}V^{\text{II}}$ be a minimum (here V^{I} and V^{II} are the volumes of the portions of the system maintained at P^{I} and P^{II}, respectively). Then show that Eq. 8.8-1 is the condition for osmotic equilibrium.

8.8-2 Derive a form of the Gibbs phase rule that applies to osmotic equilibrium.

8.8-3 **a** The osmotic coefficient of a solvent ϕ_S is defined by the relation

$$\bar{G}_S = \underline{G}_S + \phi_S RT \ln x_S$$

where $\phi_S \rightarrow 1$ as $x_S \rightarrow 1$. Develop an expression relating the activity coefficient γ_S to the osmotic coefficient ϕ_S.

b Robinson and Wood[22] report the following interpolated values for the osmotic coefficient of seawater as a function of concentration at 25°C.

[22]R. A. Robinson and R. H. Wood, *J. Sol. Chem.* **1,** 481 (1972).

μ	ϕ_S
0.5	0.9018
1.0	0.9155
2.0	0.9619
4.0	1.0938
6.0	1.2536

Here μ is the ionic strength, defined as

$$\mu = \frac{1}{2}\sum_i z_i^2 M_i$$

where z_i is the valence of the ith ionic species and M_i is its molality. For simplicity, seawater may be considered to be a mixture of only sodium chloride and water.

Compute the water activity coefficient and the equilibrium osmotic pressure for each of the solutions in the table.

8.8-4 Bruno and coworkers [*Int. J. Thermop.*, **7**, 1033 (1986)] describe an apparatus that contains a palladium/silver membrane that is permeable to hydrogen but not other gases. In their experiments they measure the temperature T_1 in the apparatus, the pressure P_1 of pure hydrogen on one side of the semipermeable membrane, and the equilibrium composition of hydrogen x_{H_2} and other gases and pressure P_2 on the other side of the membrane. Next by equating the hydrogen fugacities on both sides of the membrane.

$$f_{H_2}(T_1, P_1) = \bar{f}_{H_2}(T_1, P_2, x_{H_2})$$

and calculating the fugacity of the pure hydrogen using known virial coefficients from

$$\ln \frac{f_{H_2}}{P_1} = \frac{BP_1}{RT} + \frac{C - B^2}{2}\left(\frac{P_1}{RT}\right)^2$$

they have a direct measure of the fugacity of hydrogen in the mixture. Some of their results are given here:

Hydrogen + Propane System at 3.45 MPa						
At 80°C	x_{H_2} =	0.2801	0.4452	0.5935	0.7298	0.8215
	ϕ_{H_2} =	1.283	1.106	1.058	1.028	1.033
At 130°C	x_{H_2} =	0.2649	0.4715	0.5449	0.7827	0.8354
	ϕ_{H_2} =	1.22	1.116	1.096	1.047	1.038
Hydrogen + Methane System at 3.45 MPa						
At 80°C	x_{H_2} =	0.2155	0.4594	0.5355	0.7901	0.8494
	ϕ_{H_2} =	1.223	1.134	1.125	1.115	1.121

$\phi_{H_2} = \bar{f}_{H_2}/x_{H_2}P$

Compare these experimental results with the predicted fugacity coefficients for hydrogen in these mixtures calculated using the Peng–Robinson equation of state. Determine the sensitivity of the predictions to the value of the binary interaction parameter.

8.8-5 It is necessary to determine the molecular weight of a soluble, but essentially involatile component. The methods that can be used include dissolution in a solvent and then measurement of (i) the freezing-point depression of a solvent, (ii) the boiling-point elevation of a solvent, or (iii) the osmotic pressure of the solvent. Comment on these alternatives for

a A solute of molecular weight of about 100.
b A solute of molecular weight of about 1000.
c A solute of molecular weight of about 1,000,000.

8.9 CONCLUDING REMARKS

In this chapter many different types of phase equilibrium were considered. It is my hope that by first presenting the thermodynamic basis of phase equilibrium (in Chapters 6 and 7) and then grouping together many types of phase equilibria here, you will appreciate the essential unity of this subject and its wide applications in chemical engineering.

It should be evident from the examples in this chapter that the evaluation of species fugacities or partial molar Gibbs free energies (or chemical potentials) are central to any phase equilibrium calculation. Two different fugacity descriptions are used. For mixtures containing only hydrocarbons and perhaps some inorganic gases, the equation of state description is preferred. For all other mixtures, activity coefficient models are used. Both had adjustable parameters; the binary interaction parameter k_{ij} for equations of state, and the parameters in the activity coefficient models. If the values of these adjustable parameters are known or estimated, the phase equilibrium state may be predicted. Equally important, however, is the observation that experimental phase equilibrium can be used to obtain these parameters. For example, in Sec. 8.1 we demonstrated how activity coefficients could be computed directly from P-T-x-y data and how activity coefficient models could be fit to such data. Similarly, in Sec. 8.2, we pointed out how fitting equation of state predictions to experimental high pressure phase equilibrium data could be used to obtain a best-fit value of the binary interaction parameter.

In a similar fashion solubility measurements (of a gas in a liquid, a liquid in a liquid, or a solid in a liquid) can be used to determine the activity coefficient of a solute in a solvent at saturation. Also, measurements of the solubility of a solid in two liquid phases can be used to relate the activity coefficient of a solid in one liquid to a known activity coefficient in another liquid, and freezing-point depression or boiling-point elevation measurements are frequently used to determine the activity of the solvent in a solute–solvent mixture. Finally, as was mentioned in the previous section, osmotic pressure measurements can be used to determine solvent activity coefficients, or to determine the molecular weight of a large polymer or protein.

The body of thermodynamic information determined in the ways just described provide a base of knowledge for making the estimates and predictions needed for engineering design, for testing equations of state, their mixing rules and liquid solution theories, and for extending our knowledge of the way molecules in mixtures interact.

ADDITIONAL PHASE EQUILIBRIUM PROBLEMS

8.9-1 Relative humidity is the ratio of the partial pressure of water in air to the partial pressure of water in air saturated with water at the same temperature, expressed as a percentage,

$$\text{Relative humidity} = \frac{\text{partial pressure of water in air}}{\text{partial pressure of water in air saturated with water at same temperature}} \times 100$$

Among aviators it is more common to express the moisture content of the air by giving the air temperature and its dew point, that is, the temperature to which the air must be cooled for the first drop of water to condense.

The following atmospheric conditions have been reported:

Atmospheric pressure = 1.011 bar

Air temperature = 25.6°C

Dew point of air = 20.6°C

What is the relative humidity of air? (See Problem 5.12 for the necessary data.)

8.9-2 One intriguing problem in atmospheric physics is the relatively long persistence of contrails emitted from high-altitude jet aircraft when the sky is clear. An explanation of this phenomenon is that as the water vapor emitted from the jet engines cools, some of it will condense to form water droplets, and, on further cooling, will form ice crystals. The claim is that the ice crystals, although they are formed when the air is saturated with respect to water, will persist as long as the air is saturated with respect to ice. Therefore, the ice crystal contrail can persist in equilibrium even though all the water droplets have either crystallized or evaporated, and the partial pressure of water in the air is less than the liquid water vapor pressure (i.e., the relative humidity of the air is less than 100%).

a Establish, by the principles of thermodynamics, the validity or fallacy of the explanation given here.

b Estimate the relative humidity above which ice crystals will be stable at −25°C and 0.5 bar pressure.

The vapor pressure of liquid water and ice are given in Problem 5.12.

8.9-3 The partial pressure of water above aqueous hydrochloric acid solutions can be represented by

$$\log_{10} P = A - \frac{B}{T}$$

where P is the pressure in bar and T is the temperature in K. The values of A and B are given in the table:

wt% HCl	A	B
10	6.12357	2295
20	6.10370	2334
30	6.12610	2422
40	6.46416	2647

Source: R. H. Perry, D. W. Green, and J. O. Maloney, eds., *The Chemical Engineers' Handbook,* 6th ed., McGraw–Hill, New York, 1984, p. 3–64.

The vapor pressure of pure water is given in Problem 5.12. Compute the activity coefficients of water in each of the hydrochloric acid solutions in the table at 25°C. (*Hint*: Hydrogen chloride can be assumed to completely ionize in aqueous solution.)

8.9-4 The following data for the partial pressure of water vapor over aqueous solutions of sodium carbonate at 30°C are given in *The Chemical Engineers' Handbook,* 5th ed. (R. H. Perry and C. H. Chilton, eds., McGraw-Hill, New York, 1973), p. 3–68.

wt% Na_2CO_3	0	5	10	15	20	25	30
P_{H_2O}, Pa	4.24	4.16	4.05	3.95	3.84	3.71	3.52

Compute the activity coefficient of water in each of these solutions. (*Hint*: Does sodium carbonate ionize in aqueous solution?)

8.9-5 The triple-point properties of a substance may be experimentally determined by measuring its temperature and pressure when the vapor, liquid, and solid phases all coexist at equilibrium. There will be an error in these measurements if air is trapped in the system; the entrapped air will be present in the vapor phase and dissolved in the liquid. Determine the error that would occur in the measurement of the triple-point temperature and pressure of water if, at equilibrium, the vapor contained 0.1333 Pa partial pressure of air.

Data

See Problem 5.12.

The Henry's law expression for air in water at $T \sim 0°C$ is

$$P_{air} = 4.3 \times 10^4 x_{air}$$

where P_{air} is the partial pressure of air in the vapor phase in bars, and x_{air} is the equilibrium mole fraction of air dissolved in water.

8.9-6 In Sec. 8.1 it was shown that the excess entropy and excess enthalpy can be gotten from various temperature derivatives of the excess Gibbs free energy. These and other excess thermodynamic functions can also be computed directly from derivatives of the activity coefficients. Show that in a binary mixture the following equations may be used for such calculations:

$$\underline{S}^{ex} = -\left(\frac{\partial \underline{G}^{ex}}{\partial T}\right)_{P,x_i} = -RT\left(x_1 \frac{\partial \ln \gamma_1}{\partial T} + x_2 \frac{\partial \ln \gamma_2}{\partial T}\right) - R(x_1 \ln \gamma_1 + x_2 \ln \gamma_2)$$

$$\underline{H}^{ex} = -T^2 \frac{\partial(\underline{G}^{ex}/T)_{P,x_i}}{\partial T} = -RT^2\left(x_1 \frac{\partial \ln \gamma_1}{\partial T} + x_2 \frac{\partial \ln \gamma_2}{\partial T}\right)$$

$$\underline{V}^{ex} = \left(\frac{\partial \underline{G}^{ex}}{\partial P}\right)_{T,x_i} = RT\left(x_1 \frac{\partial \ln \gamma_1}{\partial P} + x_2 \frac{\partial \ln \gamma_2}{\partial P}\right)$$

$$\underline{U}^{ex} = -RT\left\{T\left(x_1 \frac{\partial \ln \gamma_1}{\partial T} + x_2 \frac{\partial \ln \gamma_2}{\partial T}\right) + P\left(x_1 \frac{\partial \ln \gamma_1}{\partial P} + x_2 \frac{\partial \ln \gamma_2}{\partial P}\right)\right\}$$

$$C_P^{ex} = -2RT\left(x_1 \frac{\partial \ln \gamma_1}{\partial T} + x_2 \frac{\partial \ln \gamma_2}{\partial T}\right) - RT^2\left(x_1 \frac{\partial^2 \ln \gamma_1}{\partial T^2} + x_2 \frac{\partial^2 \ln \gamma_2}{\partial T^2}\right)$$

Note that in the activity coefficient derivatives, all variables from the set T, P, x_i, other than the one being varied, are fixed.

8.9-7 The following table gives the boiling point of liquid mixtures containing x grams of tartaric acid ($C_4H_6O_6$; molecular weight = 150) per 100 g of water at 1.013 bar, as well as the vapor pressure of pure water at the boiling point temperatures.

Boiling point (°C)	105	110	115
x (g/100 g H_2O)	87.0	177.0	272.0
Vapor pressure of pure water (bar)	1.1848	1.4050	1.6580

a What is the activity coefficient of water in these solutions?
b Assuming that the activity coefficients (be they unity or not) are independent of temperature, calculate the freezing point for the mixtures in the table.

The vapor pressures of liquid water and ice are given in Problem 5.12.

8.9-8 **a** Estimate the heat and work flows needed to reversibly and isothermally separate an equimolar mixture of two species into its pure components if the excess Gibbs free energy for the mixture is given by

$$\underline{G}^{ex} = Ax_1x_2$$

where A is independent of temperature.
b How does the temperature at which $W = 0$ compare with the upper consolute temperature of the mixture?
c How would the answers to parts a and b change if A were a function of temperature?

8.9-9 A binary liquid solution, having mole fraction x of component 1, is in equilibrium with a vapor that has a mole fraction y of that component. Show that for this mixture the effect of a change in temperature on the equilibrium pressure at fixed liquid composition is approximately

$$\left(\frac{\partial \ln P}{\partial T}\right)_x = \frac{\Delta \underline{H}}{RT^2}$$

where $\Delta \underline{H}$ is the heat of vaporization of y moles of component 1 and $(1 - y)$ moles of component 2 from a large volume of solution.

8.9-10 The change in vapor pressure of a pure liquid with temperature can be computed from the Clapyron equation:

$$\frac{dP}{dT} = \frac{\Delta \underline{H}}{T\Delta \underline{V}}$$

where $\Delta \underline{H}$ and $\Delta \underline{V}$ are the enthalpy and volume changes on going from the liquid to the vapor. The vapor pressure and temperature must, of course, be related, since the Gibbs phase rule indicates that the system is univariant.

Now consider two partially miscible liquids and the vapor in equilibrium with these two liquid phases. This system is also univariant (two components and three phases).

a Derive a "generalized" Clapyron equation relating the equilibrium pressure and temperature for this system.

b Qualitatively discuss the variation of the equilibrium pressure with the overall two-phase liquid composition for this system at fixed temperature. How does the pressure variation with composition differ in the one- and two-liquid phase regions? Also, discuss the variation of the equilibrium temperature with overall liquid phase composition at fixed pressure in the single-liquid phase and two-liquid phase regions.

8.9-11 Assuming oxygen and nitrogen form ideal liquid mixtures, and that their solid phases are immiscible, compute the complete solid–liquid–vapor phase diagram for this mixture for the temperature range of 46 to 100 K.

Data

Solid oxygen[23]

Molar volume = 0.02462 m^3/kmol

Melting point = 54.35 K

T (K)	46	48	50	52	54	54.35
Sublimation pressure (Pa)	5.252	13.02	29.86	64.25	130.1	146.4
$\Delta \underline{H}^{sub}$(kJ/mol)	8.350	8.316	8.281	8.248	8.213	8.207

Liquid oxygen

Molar volume = 0.02735 m^3/kmol

T(°C)[24]	−219.1	−213.4	−210.6	−204.1	−198.8	−188.8	−183.1
Vapor pressure (kPa)	0.1333	0.6665	1.333	5.332	13.33	53.32	101.3

[23] J. C. Mullins, W. T. Ziegler, and B. S. Kirk, *Adv. Cryog. Eng.* **8**, 126 (1963).

[24] R. H. Perry, D. W. Green, and J. O. Maloney, eds., *Chemical Engineers' Handbook*, 6th ed., McGraw-Hill, New York, 1984, p. 3–48.

Solid nitrogen[25]

Molar volume = 0.03186 m^3/kmol

Melting point = 63.2 K

Sublimation pressure: $\log_{10} P = -\frac{381.6}{T} - 0.0062372T + 7.53588$

$(P = \text{kPa}, T = \text{K})$

$\Delta \underline{H}^{\text{fus}}$ (melting point) = 720.9 J/mol

Liquid nitrogen

Molar volume = $0.02414/(1 - 0.0039T)$ m^3/kmol

Vapor pressure: $\log_{10} P = -\frac{339.8}{T} - 0.0056286T + 6.83540$

$(P = \text{kPa}, T = \text{K})$

8.9-12 Estimate the equilibrium pressure and isobutane vapor-phase mole fraction as a function of liquid composition for the isobutane–furfural system at 37.8°C. At this temperature the vapor pressure of furfural is 0.493 kPa and that of isobutane is 4.909 kPa.
(*Hint*: See Illustration 8.4-2 and Problem 8.9-10.)

8.9-13 A mixture of 80 mole percent acetone and 20 mole percent water is to be transported in a pipeline within a chemical plant. This mixture can be transported as either a liquid or a gas.

a A reciprocating pump would be used with a liquid mixture. This pump will vapor-lock and cease functioning if any vapor is present. Compute the minimum pressure that must be maintained at the pump inlet so that no vapor is formed when the liquid temperature is 100°C.

b A centrifugal pump would be used with a gas mixture. To prevent erosion of the pump blades, only vapor should be present. If the pump effluent is 100°C, compute the maximum pump effluent pressure so that no liquid is formed.

Data

Temperature (°C)	56.6	78.6	113.0
Vapor pressure of acetone (bar)	1.013	2.026	5.065

8.9-14 To evaluate the potential use of carbon dioxide in tertiary oil recovery, it is necessary to estimate the vapor–liquid equilibrium between carbon dioxide and reservoir petroleum, which we will take to be *n*-hexane, at oil-well conditions, typically 140 bar and 75°C. Make this estimate as best you can.

8.9-15 An equimolar mixture of ethanol and ethyl acetate is maintained at 75°C and 10 bar. The pressure on the system is isothermally reduced to 1.8

[25] *Kirk-Othmer Encyclopedia of Chemical Technology*, 3rd ed., John Wiley & Sons, New York, 1981, vol. 15, p. 933.

bar. How many phases are present when equilibrium is achieved under these new conditions, and what are their compositions?

Data for ethanol-ethyl acetate system:

$$\ln \gamma_i = 0.896 x_j^2$$

$$\ln P_{\mathrm{EAC}}^{\mathrm{vap}} = \frac{-3861.37}{T} + 11.0372 \qquad T\,[=]\,\mathrm{K};\ P\,[=]\,\mathrm{bar}$$

$$\ln P_{\mathrm{EOH}}^{\mathrm{vap}} = \frac{-4728.98}{T} + 13.4643$$

9

Chemical Equilibrium and the Balance Equations for Chemically Reacting Systems

Our interest in this chapter is with changes of state involving chemical reaction. Of particular concern are the prediction of the equilibrium state in both simple and complicated chemical reaction systems and the mass and energy balance equations for such systems. We consider first chemical equilibrium in a single-phase, single-reaction system to indicate the underlying chemistry and physics of reaction equilibrium. This simple case is then generalized, in steps, to the study of equilibrium in multiphase, multireaction systems. The mass and energy balances for tank and tubular reactors, and for a general "black-box" chemical reactor, are considered next, since these balance equations are an interesting application of the thermodynamic equations for reacting mixtures and the starting point for practical reactor design and analysis. For simplicity the discussion of this chapter is largely of reaction equilibrium in ideal mixtures. Chemical equilibrium computations for nonideal systems are tedious, so only a few examples are given in this chapter.

9.1 CHEMICAL EQUILIBRIUM IN A SINGLE-PHASE SYSTEM

In Chapter 5 the criteria for equilibrium were found to be

$$S = \text{maximum for a closed system at constant } U \text{ and } V$$

$$A = \text{minimum for a closed system at constant } T \text{ and } V$$

$$G = \text{minimum for a closed system at constant } T \text{ and } P \qquad (9.1\text{-}1)$$

For chemical reaction equilibrium in a single-phase, single-reaction system, these criteria led to (See Sec. 6.8)

$$\sum \nu_i \overline{G}_i = 0 \qquad (9.1\text{-}2)$$

which, *together with the set of mass balance and state variable constraints on the system*, can be used to identify the equilibrium state in a chemically reacting system. The constraints on the system are important since, as will be seen shortly, a system in a given initial state, but subject to different constraints (e.g., constant T and P, or constant T and V), may have different equilibrium states (see Illustration 9.1-4). In each case the equilibrium state will satisfy both Eq. 9.1-2 and the constraints, and it will correspond to an extreme value of the thermodynamic function appropriate to the imposed constraints.

As an introduction to the application of Eqs. 9.1-1 and 2 to chemical equilibrium problems, consider the prediction of the equilibrium state for the low pressure, gas-phase reaction

$$CO_2 + H_2 = CO + H_2O$$

occurring in a closed system at constant pressure and a temperature of 1000°C. The total Gibbs free energy for this gaseous system is

$$\begin{aligned} G &= \sum N_i \bar{G}_i(T, P, y_i) \\ &= \sum N_i \underline{G}_i(T, P) + \sum N_i \{\bar{G}_i(T, P, y_i) - \underline{G}_i(T, P)\} \\ &= \sum N_i \underline{G}_i(T, P) + RT \sum N_i \ln \{\bar{f}_i(T, P, y_i)/f_i(T, P)\} \qquad (9.1\text{-}3a) \\ &= \sum N_i \underline{G}_i(T, P) + RT \sum_i N_i \ln \left\{ \frac{y_i P \left(\dfrac{\bar{f}_i(T, P, y_i)}{y_i P} \right)}{P \left(\dfrac{f_i(T, P)}{P} \right)} \right\} \qquad (9.1\text{-}3b) \end{aligned}$$

where $\bar{G}_i(T, P, y_i)$ and $\underline{G}_i(T, P)$ are the partial molar and pure component Gibbs free energies for gaseous species i evaluated at the reaction conditions, and $(\bar{f}_i/y_iP)$ and (f_i/P) are the fugacity coefficients for species i in the mixture (Eq. 7.2-13) and as a pure component (Eq. 5.4-8), respectively. Since the pressure is low, we will assume that the gas phase is ideal, so that all fugacity coefficients will be set equal to unity and Eq. 9.1-3 reduces to

$$G = \sum N_i \underline{G}_i(T, P) + RT \sum_i N_i \ln y_i \qquad (9.1\text{-}4)$$

As the reaction proceeds, the mole numbers and mole fraction of each species, and the total Gibbs free energy of the reacting mixture, change. The number of moles of each reacting species in a closed system is not an independent variable (i.e., it cannot take on any value), but is related to the mole numbers of the other species and the initial mole numbers through the reaction stoichiometry. This is most easily taken into account using the molar extent of reaction variable of Chapter 6

$$N_i = N_{i,0} + \nu_i X$$

or

$$X = \frac{N_i - N_{i,0}}{\nu_i}$$

In this case, the initial and final mole numbers of the species are related as follows:

$$\begin{aligned} X &= (N_{CO} - N_{CO,0}) = (N_{H_2O} - N_{H_2O,0}) \\ &= -(N_{CO_2} - N_{CO_2,0}) = -(N_{H_2} - N_{H_2,0}) \qquad (9.1\text{-}5) \end{aligned}$$

where $N_{i,0}$ is the initial number of moles of species i. The number of moles of each species and the gas-phase mole fractions at any extent of reaction X are given in Table 9.1-1 for the case in which $N_{CO_2,0} = N_{H_2,0} = 1$ mol and $N_{CO,0} = N_{H_2O,0} = 0$. Balance tables such as this one will be used throughout this chapter in the solution of chemical equilibrium problems.

Table 9.1-1
MASS BALANCE FOR THE REACTION $CO_2 + H_2 = CO + H_2O$

Species	Initial Number of Moles	Final Number of Moles	Mole Fraction
CO_2	1	$1 - X$	$(1 - X)/2$
H_2	1	$1 - X$	$(1 - X)/2$
CO	0	X	$X/2$
H_2O	0	X	$X/2$
Total	2	2	

The first term on the right side of Eq. 9.1-4 is the sum of the Gibbs free energies of the pure components at the temperature and pressure of the mixture. The second term is the Gibbs free energy change on forming a mixture from the pure components; it arises here because the reaction takes place not between the pure components, but between components in a mixture. Figure 9.1-1 is a plot of the Gibbs free energy of this reacting mixture as a function of extent of reaction for the initial mole numbers given in the table. That is, the solid line is a plot of

$$\begin{aligned} G &= \sum_i N_i \underline{G}_i + RT \sum_i N_i \ln y_i \\ &= (1 - X)(\underline{G}_{CO_2} + \underline{G}_{H_2}) + X(\underline{G}_{CO} + \underline{G}_{H_2O}) \\ &\quad + 2RT[(1 - X) \ln \{(1 - X)/2\} + X \ln \{X/2\}] \end{aligned} \tag{9.1-6}$$

The dashed line in the figure is the Gibbs free energy change as a function of the extent of reaction *neglecting* the Gibbs free energy of mixing (the logarithmic terms in X).

The important feature of Fig. 9.1-1 is that, because of the Gibbs free energy of mixing term, there is an intermediate value of the reaction variable X for which the total Gibbs free energy of the mixture is a minimum; this value of X has been denoted by X^*. Since the condition for equilibrium at constant temperature and pressure is that G be a minimum, X^* is the equilibrium extent of reaction for this system.

This equilibrium state can be mathematically (rather than graphically) identified using the criterion that at equilibrium in a closed system at constant T and

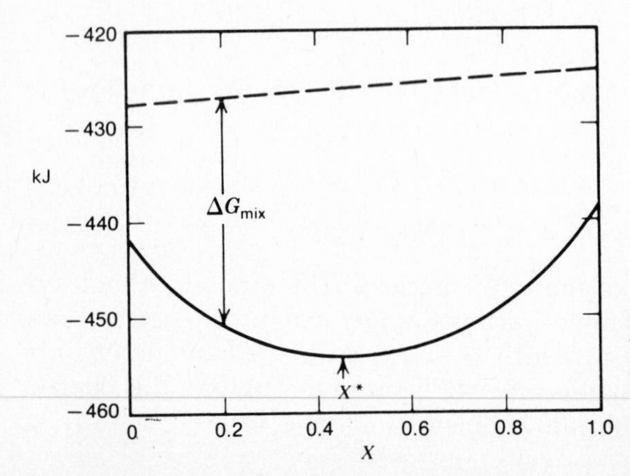

Figure 9.1-1
The Gibbs free energy for the system $CO_2 + H_2 = CO + H_2O$ at 1000 K and 1.013 bar ~0.1 MPa relative to each atomic species in its standard state at 298.15 K (25°C).

P, the Gibbs free energy G is a minimum, or $dG = 0$ for all possible mole number variations *consistent* with the reaction stoichiometry. This implies that

$$\left(\frac{\partial G}{\partial X}\right)_{T,P} = 0 \tag{9.1-7}$$

since X is the single variable describing the mole number variations consistent with the stoichiometry and the initial amounts of each species. Using this result in Eq. 9.1-6 yields

$$(\underline{G}_{CO} + \underline{G}_{H_2O} - \underline{G}_{CO_2} - \underline{G}_{H_2}) + 2RT[-\ln\{(1 - X^*)/2\} + \ln(X^*/2)] = 0$$

or

$$\frac{-(\underline{G}_{CO} + \underline{G}_{H_2O} - \underline{G}_{CO_2} - \underline{G}_{H_2})}{RT} = \ln\left[\frac{X^{*2}}{(1 - X^*)^2}\right] \tag{9.1-8}$$

To find the equilibrium mole fractions, Eq. 9.1-8 is first solved for the equilibrium extent of reaction X^*, and this is used with the initial mole numbers and stoichiometric information (i.e., Table 9.1-1) to find the mole fractions.

It is also of interest to notice that using the stoichiometric relations between X and the species mole fractions we obtain

$$\frac{-(\underline{G}_{CO} + \underline{G}_{H_2O} - \underline{G}_{CO_2} - \underline{G}_{H_2})}{RT} = \ln\left[\frac{y_{CO}y_{H_2O}}{y_{CO_2}y_{H_2}}\right] \tag{9.1-9a}$$

or, equivalently (and more generally)

$$\frac{-\sum_i \nu_i \underline{G}_i}{RT} = \sum_i \ln y_i^{\nu_i} = \ln\left[\prod_i y_i^{\nu_i}\right] \tag{9.1-9b}$$

(The symbol Π_i used here denotes a product of numbers, that is, $\Pi_i\, y_i^{\nu_i} = y_A^{\nu_A}\, y_B^{\nu_B}\, y_C^{\nu_C} \ldots$.) Thus, the equilibrium product of species mole fractions, each taken to the power of its stoichiometric coefficient, is related to the sum over the species of the stoichiometric coefficients times the *pure component* Gibbs free energies at the temperature and pressure of the reacting mixture.

Whereas Eq. 9.1-8 is specific to the reaction and initial mole numbers of the example being considered, Eq. 9.1-9b (of which Eq. 9.1-9a is a special case) is generally valid for single-phase reactions in ideal mixtures; furthermore, Eq. 9.1-9b can be obtained directly from the general equilibrium relation, Eq. 9.1-2:

$$\sum_i \nu_i \bar{G}_i(T, P, x_i) = 0$$

We show this here by first considering the more general case of chemical reactions in nonideal vapor or liquid systems for which

$$\bar{G}_i(T, P, x_i) = \underline{G}_i(T, P) + RT \ln \frac{\bar{f}_i(T, P, x_i)}{f_i(T, P)} \tag{9.1-10}$$

Thus

$$\sum_i \nu_i \bar{G}_i(T, P, x_i) = \sum_i \nu_i \underline{G}_i(T, P) + RT \sum_i \nu_i \ln\left(\frac{\bar{f}_i(T, P, x_i)}{f_i(T, P)}\right) = 0$$

or

$$\frac{-\sum_i \nu_i \underline{G}_i(T, P, x_i)}{RT} = \sum_i \nu_i \ln \frac{\bar{f}_i(T, P, x_i)}{f_i(T, P)} = \ln\left[\Pi\left(\frac{\bar{f}_i(T, P, x_i)}{f_i(T, P)}\right)^{\nu_i}\right] \tag{9.1-11}$$

For a gas-phase mixture

$$\bar{f}_i(T, P, x_i) = x_i P \left(\frac{\bar{f}_i}{x_i P}\right) \qquad \text{and} \qquad f_i(T, P) = P\left(\frac{f}{P}\right)_i$$

where the fugacity coefficients can be evaluated from an equation of state (or the principle of corresponding states). Thus Eq. 9.1-11 reduces to

$$\frac{-\sum \nu_i \underline{G}_i}{RT} = \ln\left[\Pi\left(\frac{x_i(\bar{f}_i/x_i P)}{(f/P)_i}\right)^{\nu_i}\right] \qquad \text{(reactions in gas phase)} \qquad (9.1\text{-}12a)$$

which, for an ideal gas (all fugacity coefficients equal to unity) becomes

$$\frac{-\sum \nu_i \underline{G}_i}{RT} = [\Pi\, x_i^{\nu_i}] \qquad \text{(reactions in ideal gas phase)} \qquad (9.1\text{-}12b)$$

This last equation is identical to Eq. 9.1-9b. (Note that Eq. 9.1-12a also applies to liquid mixtures describable by an equation of state.)

For a liquid mixture described by an activity coefficient model

$$\bar{f}_i(T, P, x_i) = x_i \gamma_i f_i(T, P)$$

so that

$$\frac{-\sum \nu_i \underline{G}_i}{RT} = \ln\,[\Pi(x_i\gamma_i)^{\nu_i}] \qquad \text{(reactions in liquid phase)} \qquad (9.1\text{-}13a)$$

which, for an ideal liquid mixture (all activity coefficients equal to unity) reduces to

$$\frac{-\sum \nu_i \underline{G}_i}{RT} = \ln\,(\Pi\, x_i^{\nu_i}) \qquad \text{(reactions in ideal liquid phase)} \qquad (9.1\text{-}13b)$$

For all future chemical equilibrium calculations in single-phase, single-reaction systems, we will start from Eqs. 9.1-12 or 13 as appropriate, rather than starting at Eq. 9.1-2 and repeating the derivation each time.

Figures like Fig. 9.1-1 provide some insight into the direction of progress of a chemical reaction. To be specific, the equilibrium and stability analysis of Chapter 5 establishes that a system at constant temperature and pressure evolves toward a state of minimum Gibbs free energy. Here this is the state for which Eqs. 9.1-2, 12, or 13 are satisfied. Therefore, if, at any instant, the mole fractions of the reacting species in an ideal mixture are such that

$$\prod_i x_i^{\nu_i} < \exp\left[-\frac{\sum_i \nu_i \underline{G}_i(T, P)}{RT}\right]$$

(which, for the example in Fig. 9.1-1, occurs when the molar extent of reaction is to the left of X^*), the reaction proceeds as written. That is, reactants (species with negative stoichiometric coefficients) are consumed to form the reaction products (species with positive stoichiometric coefficients), until the equilibrium composition is reached. Conversely, if the species mole fractions are such that

$$\prod_i x_i^{\nu_i} > \exp\left[-\frac{\sum_i \nu_i \underline{G}_i(T, P)}{RT}\right]$$

the reaction proceeds in a manner such that what heretofore had been designated as the reaction products would be consumed to form reactants, until equilibrium was achieved; that is, until

$$\prod_i x_i^{\nu_i} = \exp\left[-\frac{\sum_i \nu_i \underline{G}_i(T, P)}{RT}\right]$$

Thus, our choice of reactants and products in a chemical reaction is arbitrary, in that the reaction can proceed in either direction. Indeed, replacing each stoichiometric coefficient (ν_i) by its negative ($-\nu_i$) in the foregoing equations, which is equivalent to interchanging the choice of reactants and products, leads to the same equilibrium state; again a state of minimum Gibbs free energy. This freedom to arbitrarily choose which species are the reactants and which are the products is also evident from the equilibrium relation

$$\sum \nu_i \overline{G}_i(T, P, x_i) = 0$$

since multiplying this equation by the constant -1 leaves the equation unchanged. In fact, multiplying this equation by any constant, either positive or negative, affects neither the equation nor the equilibrium state computed from it. Thus, whether we choose to write a reaction as

$$\alpha A + \beta B + \cdots \rightarrow \rho R + \cdots$$

$$2\alpha A + 2\beta B + \cdots \rightarrow 2\rho R + \cdots$$

or

$$\rho R + \cdots \rightarrow \alpha A + \beta B + \cdots$$

has no effect on what is ultimately predicted to be the equilibrium state of the system. Since chemical reactions can proceed either in the direction written or in the opposite direction, they are said to be **reversible**. (Note that here the word reversible is being used in a different sense than in Chapter 3.)

A chemical reaction goes to **completion** if it proceeds until one of the reactants is completely consumed. In principle, no homogeneous (i.e., single-phase) reaction goes to completion because the balance between the Gibbs free energy of mixing and the $\Delta G_{rxn} = \Sigma_i \nu_i \underline{G}_i$ terms (see Fig. 9.1-1) forces the equilibrium value of X (that is, X^*) to lie between 0 and complete reaction. However, there are many instances when ΔG_{rxn} is so large in magnitude compared to the free energy of mixing term, which is of the order of RT (see Eq. 9.1-4), that the reaction either goes essentially to completion ($-\Delta G_{rxn}/RT \gg 1$) or does not measurably occur ($\Delta G_{rxn}/RT \gg 1$). The room temperature oxidation of hydrogen to form water

$$H_2 + \tfrac{1}{2}O_2 = H_2O$$

for which

$$\frac{\Delta G_{rxn}}{RT} = \frac{-228.59 \text{ kJ/mol}}{8.314 \text{ kJ/mol K} \times 298.15 \text{ K}} = -92.22$$

is an example of a reaction that goes essentially to completion. To see this we note that starting with stoichiometric amounts of hydrogen and oxygen we have

Species	Initial Number of Moles	Final Number of Moles	Mole Fraction
H_2	1	$1 - X$	$(1 - X)/(1.5 - 0.5X)$
O_2	0.5	$0.5 - 0.5X$	$0.5(1 - X)/(1.5 - 0.5X)$
H_2O	0	X	$X/(1.5 - 0.5X)$
Total	1.5	$1.5 - 0.5X$	

so that

$$\exp\left(-\frac{\sum \nu_i \underline{G}_i}{RT}\right) = \Pi\, x_i^{\nu_i}$$

$$\exp(+92.22) = 1.12 \times 10^{40} = \frac{x_{H_2O}}{x_{H_2} x_{O_2}^{0.5}} = \frac{X(1.5 - 0.5X)^{0.5}}{0.5(1.0 - X)^{1.5}}$$

The solution to this equation is $X \sim 1$ [actually $1.0 - (3 \times 10^{-27})$] so that, for all practical purposes, the reaction goes to completion. It is left to you to show that the room temperature oxidation of nitrogen

$$\tfrac{1}{2}N_2 + \tfrac{1}{2}O_2 = NO$$

for which

$$\frac{\Delta G_{rxn}}{RT} = \frac{86{,}688 \text{ J/mol}}{8.314 \text{ J/mol K} \times 298.15 \text{ K}} = 34.97$$

does not, for practical purposes, occur at all.

The identification of the equilibrium extent of reaction in the carbon dioxide–hydrogen reaction discussed earlier was straightforward for two reasons. First, the mixture was ideal so that there were no fugacity corrections or activity coefficients to consider. Second, it was presumed that the pure component Gibbs free energies were available for the reacting species at the same temperature, pressure, and state of aggregation as the reacting mixture. Most chemical equilibrium calculations are, unfortunately, more complicated because few mixtures (other than low pressure gas mixtures) are ideal, and pure component thermodynamic properties are usually tabulated only at a single temperature and pressure (25°C and 1 atm as in Appendix IV.)[1] Thus, for most chemical equilibrium computations one needs fugacity or activity coefficient data (or adequate mixture models) and must be able to estimate Gibbs free energies for any pure component state.

One possible starting point for the analysis of general single-phase chemical equilibrium problems is the observation that the partial molar Gibbs free energy of a molecular species can be written as

$$\begin{aligned} \overline{G}_i(T, P, x_i) &= \overline{G}_i^\circ(T = 25°\text{C}, P = 1 \text{ atm}, x_i^\circ) \\ &\quad + [\overline{G}_i(T, P, x_i) - \overline{G}_i^\circ(T = 25°\text{C}, P = 1 \text{ atm}, x_i^\circ)] \end{aligned} \tag{9.1-14}$$

where $\overline{G}_i^\circ$ denotes the Gibbs free energy of species i at 25°C, 1 atm and the composition x_i° (usually taken to be the pure component state $x_i^\circ = 1$, the infinite

[1] Note that since the enthalpy and Gibbs free energy of formation tables in Appendix IV are for 25°C and 1 atm, we will use the same conditions throughout this chapter although the unit atmosphere is not an S.I. unit. 1 atm = 1.013 bar = 0.1013 MPa ~ 0.1 MPa.

dilution state, $x_i^\circ = 0$, or the ideal 1 molal solution). However, starting from Eq. 9.1-14 requires estimates of the changes in species partial molar Gibbs free energy with temperature, pressure, and composition; a difficult task. (The formulas of Chapter 7 account mainly for the composition and pressure variations of the species fugacity and partial molar Gibbs free energy.) A more practical procedure is to choose the **standard state** of each species to be the species at composition x_i°, the *temperature of interest* T, and a pressure of 1 atm. In this case we have

$$
\begin{aligned}
\overline{G}_i(T, P, x_i) &= \overline{G}_i^\circ(T, P = 1 \text{ atm}, x_i^\circ) \\
&\quad + [\overline{G}_i(T, P, x_i) - \overline{G}_i^\circ(T, P = 1 \text{ atm}, x_i^\circ)] \\
&= \overline{G}_i^\circ(T, P = 1 \text{ atm}, x_i^\circ) + RT \ln \left\{ \frac{\bar{f}_i(T, P, x_i)}{\bar{f}_i^\circ(T, P = 1 \text{ atm}, x_i^\circ)} \right\} \\
&= \overline{G}_i^\circ(T, P = 1 \text{ atm}, x_i^\circ) + RT \ln a_i
\end{aligned} \tag{9.1-15}
$$

where we have introduced the **activity** of species i, denoted by a_i, defined to be the ratio of the species fugacity in the mixture to the fugacity in its standard state,

$$
a_i = \frac{\bar{f}_i(T, P, x_i)}{\bar{f}_i^\circ(T, P = 1 \text{ atm}, x_i^\circ)} = \exp\left(\frac{\overline{G}_i(T, P, x_i) - \overline{G}_i^\circ(T, P = 1 \text{ atm}, x_i^\circ)}{RT} \right) \tag{9.1-16}
$$

With this formulation of the Gibbs free energy function, the activity or fugacity for each species is a function of pressure and composition only, and its value can be computed using the generalized f/P charts, equations of state, or liquid solution data or models, all of which are discussed in Chapter 7. Note that once the standard state is chosen, the standard state Gibbs free energy $\overline{G}_i^\circ$ is a function of temperature only. The calculation of its value for any temperature will be considered shortly.

For convenience we have listed in Table 9.1-2 species activities for several common choices of the standard state.

With the notation introduced here, the equilibrium relation, Eq. 9.1-2, can now be written as

$$
\begin{aligned}
0 = \sum_{i=1}^{\mathscr{C}} \nu_i \overline{G}_i &= \sum_{i=1}^{\mathscr{C}} \nu_i \overline{G}_i^\circ(T, P = 1 \text{ atm}, x_i^\circ) + RT \sum_{i=1}^{\mathscr{C}} \nu_i \ln a_i \\
&= \Delta G_{rxn}^\circ + RT \sum_{i=1}^{\mathscr{C}} \nu_i \ln a_i
\end{aligned}
$$

or

$$
-\frac{\Delta G_{rxn}^\circ}{RT} = \ln \prod_{i=1}^{\mathscr{C}} a_i^{\nu_i} \tag{9.1-17}
$$

where we have used ΔG_{rxn}° to denote the quantity $\Sigma\, \nu_i \overline{G}_i^\circ(T, P = 1 \text{ atm}, x_i^\circ)$, the Gibbs free energy change on reaction with each species in its standard state or state of unit activity. Finally, the **equilibrium constant** K_a is defined by the relation

$$
K_a(T) = \exp\left(-\frac{\Delta G_{rxn}^\circ}{RT} \right) \tag{9.1-18}
$$

Table 9.1-2
SPECIES ACTIVITY BASED ON DIFFERENT CHOICES OF STANDARD STATE ACTIVITY*

State (All at Temperature T and Pressure P)	Standard State (All at Temperature T and 1 atm Pressure)	General	Simplification for Low and Moderate Pressures
Pure gas	Pure gas $\overline{G}_i^\circ = \underline{G}_i^V(T, P = 1\text{ atm})$	$a_i = \dfrac{f_i^V(T, P)}{f_i^V(T, P = 1\text{ atm})} = \dfrac{P(f/P)}{1\text{ atm}}$	$a_i = \dfrac{P}{1\text{ atm}}$
Species in a gaseous mixture	Pure gas $\overline{G}_i^\circ = \underline{G}_i^V(T, P = 1\text{ atm})$	$a_i = \dfrac{\bar{f}_i^V(T, P, y_i)}{f_i^V(T, P = 1\text{ atm})} = \dfrac{y_iP(\bar{f}_i/y_iP)}{1\text{ atm}} = \dfrac{P_i(\bar{f}_i/y_iP)}{1\text{ atm}};$	$a_i = \dfrac{y_iP}{1\text{ atm}} = \dfrac{P_i}{1\text{ atm}}$
Pure liquid†	Pure liquid $\overline{G}_i^\circ = \underline{G}_i^L(T, P = 1\text{ atm})$	$a_i = \dfrac{f_i^L(T, P)}{f_i^L(T, P = 1\text{ atm})} = \dfrac{P^{vap}(T)(f/P)_{sat}\exp\{(\underline{V}^L/RT)(P - P^{vap})\}}{P^{vap}(T)(f/P)_{sat}} = \exp\left\{\dfrac{\underline{V}^L}{RT}(P - P^{vap})\right\}$	$a_i = 1$
Species in a liquid mixture†	Pure liquid $\overline{G}_i^\circ = \underline{G}_i^L(T, P = 1\text{ atm})$	$a_i = \dfrac{x_i\gamma_iP^{vap}(f/P)_{sat}\exp\{(\overline{V}_i/RT)(P - P^{vap})\}}{P^{vap}(f/P)_{sat}} = x_i\gamma_i\exp\left\{\dfrac{\overline{V}_i}{RT}(P - P^{vap})\right\}$	$a_i = x_i\gamma_i$ ($\gamma_i \to 1$ as $x_i \to 1$)
	Species in a 1 molal ideal solution $\overline{G}_i^\circ = \overline{G}_i(T, P = 1\text{ atm}, M_i = 1)$; (see Eq. 7.8-15)	$a_i = \dfrac{M_i\mathcal{H}_i(T, P = 1\text{ atm})\gamma_i^\square}{(M_i = 1)\mathcal{H}_i(T, P = 1\text{ atm})}\exp\left\{\dfrac{\overline{V}_i^\infty}{RT}(P - 1\text{ atm})\right\} = \dfrac{M_i\gamma_i^\square}{(M_i = 1)}\exp\left\{\dfrac{\overline{V}_i^\infty}{RT}(P - 1\text{ atm})\right\}$	$a_i = M_i\gamma_i^\square/(M_i = 1)$ ($\gamma_i^\square \to 1$ as $M_i \to 0$)
	Species as a pure liquid with infinite dilution properties $\overline{G}_i^\circ = \overline{G}_i^*(T, P = 1\text{ atm}, x_i = 1)$	$a_i = \dfrac{x_iH_i(T, P = 1\text{ atm})\gamma_i^*}{H_i(T, P = 1\text{ atm})}\exp\left\{\dfrac{\overline{V}_i^\infty}{RT}(P - 1\text{ atm})\right\} = x_i\gamma_i^*\exp\left\{\dfrac{\overline{V}_i^\infty}{RT}(P - 1\text{ atm})\right\}$	$a_i = x_i\gamma_i^*$ ($\gamma_i^* \to 1$ as $x_i \to 0$)
Pure solid	Pure solid $\overline{G}_i^\circ = \underline{G}_i^S(T, P = 1\text{ atm})$	$a_i = \dfrac{f_i^S(T, P)}{f_i^S(T, P = 1\text{ atm})} = \dfrac{P^{sat}(T)(f/P)_{sat}\exp\{(\underline{V}^S/RT)(P - P^{sat})\}}{P^{sat}(T)(f/P)_{sat}} = \exp\left\{\dfrac{\underline{V}^S}{RT}(P - P^{sat})\right\}$	$a_i = 1$
Species in a solid mixture	Pure solid $\overline{G}_i^\circ = \underline{G}_i^S(T, P = 1\text{ atm})$	$a_i = \exp\left\{\dfrac{\underline{V}^S}{RT}(P - 1\text{ atm})\right\}$ as above	$a_i = 1$
Dissolved electrolyte in solution	Dissolved electrolyte, each ion at unit molality in an ideal solution $\overline{G}_i^\circ = \nu_+\overline{G}_A(T, P = 1\text{ atm}, M_A = 1) + \nu_-\overline{G}_B(T, P = 1\text{ atm}, M_B = 1) = \overline{G}_{AB,D}^\square(T, P = 1\text{ atm})$		$a_i = \dfrac{M_A^{\nu_+}M_B^{\nu_-}(\gamma_\pm)^\nu}{(M_\pm = 1)^\nu} = \dfrac{(M_\pm\gamma_\pm)^\nu}{(M_\pm = 1)^\nu}$

*In this table we have neglected the Poynting correction to the pure component standard states; also 1 atm = 1.013 bar.

†If a liquid or liquid mixture is describable by equation of state, the same equations as for gases or gaseous mixtures are used.

Thus, Eq. 9.1-17 can be rewritten as

$$K_a(T) = \prod_{i=1}^{\mathscr{C}} a_i^{\nu_i} = \frac{a_R^{\rho} \cdots}{a_A^{\alpha} a_B^{\beta} \cdots} \tag{9.1-19}$$

This equation is equivalent to Eq. 9.1-2 and can be used in the prediction of the chemical equilibrium state, provided that we can calculate a value for the equilibrium constant K_a and the species activities a_i.

If the standard state of each component is chosen to be $T = 25°C$, $P = 1$ atm, and the state of aggregation listed in Appendix IV, then, following Eq. 6.5-2,

$$\Delta G^{\circ}_{rxn}(T = 25°\text{C}) = \sum_i \nu_i \Delta \underline{G}^{\circ}_{f,i}(T = 25°\text{C})$$

where $\Delta \underline{G}^{\circ}_f$, the Gibbs free energy of formation, is given in Appendix IV.

ILLUSTRATION 9.1-1

Calculate the equilibrium extent of decomposition of nitrogen tetroxide due to the chemical reaction $N_2O_4(g) = 2NO_2(g)$ at 25°C and 1 atm (1.013 bar).

Solution

The equilibrium relation is

$$K_a = \exp\left\{-\frac{\Delta G^{\circ}_{rxn}(T = 25°\text{C}, P = 1 \text{ atm})}{RT}\right\} = \frac{a^2_{NO_2}}{a_{N_2O_4}} = \frac{\left(\dfrac{y_{NO_2}P}{1 \text{ atm}}\right)^2}{\left(\dfrac{y_{N_2O_4}P}{1 \text{ atm}}\right)} = \frac{y^2_{NO_2}}{y_{N_2O_4}}$$

Here we have assumed the gas phase is ideal. Furthermore, since the reaction and standard state pressures are both one atmosphere, the pressure cancels out of the equation. Using the entries in Appendix IV we have

$$\Delta G^{\circ}_{rxn}(T = 25°\text{C}, P = 1 \text{ atm}) = 2\Delta \underline{G}^{\circ}_{f,NO_2} - \Delta \underline{G}^{\circ}_{f,N_2O_4} = 2 \times 12.26 - 23.41 \text{ kcal/mol}$$

$$= 1.11 \text{ kcal/mol} = 4644 \text{ J/mol}$$

so that

$$K_a = \exp\left\{\frac{-4644 \text{ J/mol}}{8.314 \text{ J/mol K} \times 298.15 \text{ K}}\right\} = 0.154$$

Next, we write the mole fractions of both NO_2 and N_2O_4 in terms of a single extent of reaction variable. This is most easily done using the mass balance table:

	Initial Number of Moles of Each Species	Final Number of Moles of Each Species	Equilibrium Mole Fraction
N_2O_4	1	$1 - X$	$y_{N_2O_4} = (1 - X)/(1 + X)$
NO_2	0	$2X$	$y_{NO_2} = (2X)/(1 + X)$
Total		$1 + X$	

Therefore,

$$K_a = 0.154 = \frac{(2X/(1+X))^2}{(1-X)/(1+X)} = \frac{4X^2}{(1+X)(1-X)} = \frac{4X^2}{(1-X^2)}$$

or

$$X = \sqrt{\frac{K_a}{4+K_a}} = 0.192$$

so that the nitrogen tetroxide is 19.2% decomposed at the conditions given, and

$$y_{N_2O_4} = 0.677$$

$$y_{NO_2} = 0.323 \quad ■$$

Equilibrium compositions in a chemically reacting system are affected by changes in the state variables (i.e., temperature and pressure), the presence of diluents, or variations in the initial state of the system. These effects are considered in this discussion and illustrations in the remainder of this section.

If an inert diluent is added to a reacting mixture, it may change the equilibrium state of the system, not as a result of a change in the value of the equilibrium constant (which depends only on temperature), but rather as a result of the change in the concentration, and hence, activity of each reacting species. This effect is illustrated in the following example.

ILLUSTRATION 9.1-2

Pure nitrogen tetroxide at a low temperature is diluted with nitrogen and heated to 25°C and 1.013 bar. If the initial mole fraction of N_2O_4 in the N_2O_4–nitrogen mixture before dissociation begins is 0.20, what is the extent of the decomposition and the mole fractions of NO_2 and N_2O_4 present at equilibrium?

Solution

As in the preceding illustration,

$$K_a = 0.154 = \frac{y_{NO_2}^2}{y_{N_2O_4}}$$

Here, however, we have

	Initial Number of Moles	Final Number of Moles	Equilibrium Mole Fraction
N_2O_4	0.2	$0.2 - X$	$(0.2 - X)/(1 + X)$
NO_2	0	$2X$	$2X/(1 + X)$
N_2	0.8	0.8	$0.8/(1 + X)$
Total		$1 + X$	

Therefore,

$$0.154 = \frac{4X^2}{(0.2 - X)(1 + X)}$$

so that

$$X = 0.0725 \qquad y_{N_2O_4} = 0.119 \qquad y_{NO_2} = 0.135 \qquad \text{and} \qquad y_{N_2} = 0.746$$

Comment

The fractional decomposition of N_2O_4, which is equal to

$$\frac{\text{number of moles of } N_2O_4 \text{ reacted}}{\text{initial number of moles of } N_2O_4} = \frac{X}{0.2} = 0.363$$

is higher here than in the case of undiluted nitrogen tetroxide (preceding illustration). At a fixed extent of reaction, the presence of the inert diluent nitrogen decreases the mole fractions of NO_2 and N_2O_4 equally. However, since the mole fraction of NO_2 appears in the equilibrium relation to the second power, the equilibrium must shift to the right (more dissociation of nitrogen tetroxide). (How would the presence of an inert diluent effect an association reaction, i.e., $2A \rightarrow B$? How would the presence of an inert diluent effect a gas phase reaction in which $\Sigma\, \nu_i = 0$?) ■

To compute the equilibrium constant K_a at any temperature T, given the Gibbs free energies of formation at 25°C, we start with the observation that

$$\frac{\partial}{\partial T}\left(\frac{\overline{G}_i}{T}\right)_P = \frac{1}{T}\left(\frac{\partial \overline{G}_i}{\partial T}\right)_P - \frac{\overline{G}_i}{T^2} = -\frac{\overline{S}_i}{T} - \frac{\overline{H}_i}{T^2} + \frac{\overline{S}_i}{T} = -\frac{\overline{H}_i}{T^2} \tag{9.1-20a}$$

and use the fact that $\ln K_a = -\Sigma\, \nu_i \Delta \underline{G}^\circ_{f,i}/RT$ to obtain

$$\left(\frac{\partial \ln K_a}{\partial T}\right)_P = -\frac{1}{R}\frac{\partial}{\partial T}\left[\frac{\sum_i \nu_i \Delta \underline{G}^\circ_{f,i}}{T}\right] = \frac{1}{RT^2}\sum_i \nu_i \Delta \underline{H}^\circ_{f,i} = \frac{\Delta H^\circ_{rxn}(T)}{RT^2} \tag{9.1-20b}$$

Here $\Delta H^\circ_{rxn} = \Sigma\, \nu_i \Delta \underline{H}^\circ_{f,i}$ is the heat of reaction in the standard state, that is, the heat of reaction if the reaction took place with each species in its standard state at the reaction temperature. Equation 9.1-20b is known as the **van't Hoff equation.** If a reaction is **exothermic,** that is, if energy is released on reaction so that ΔH°_{rxn} is negative, the equilibrium constant and the equilibrium conversion from reactants to products decrease with increasing temperature. Conversely, if energy is absorbed as the reaction proceeds, so that ΔH°_{rxn} is positive, the reaction is said to be **endothermic,** and both the equilibrium constant and the equilibrium extent of reaction increase with increasing temperature. These facts are easily remembered by noting that reactions that release energy are favored at lower temperatures, and reactions that absorb energy are favored at higher temperatures.

The standard state heat of reaction ΔH°_{rxn} at 25°C and 1 atm can be computed, as was pointed out in Chapter 6, from

$$\Delta H^\circ_{rxn}(T = 25°\text{C}) = \sum_i \nu_i \Delta \underline{H}^\circ_{f,i}(T = 25°\text{C})$$

and the heats of formation data, which appear in Appendix IV. At temperatures other than 25°C we start from

$$\underline{H}_i(T) = \underline{H}_i(T = 25°\text{C}) + \int_{T=25°\text{C}}^{T} C_{P,i}(T')\, dT'$$

(where T' is a dummy variable in integration), and obtain

$$\Delta H^\circ_{rxn}(T) = \sum \nu_i \underline{H}^\circ_i(T) = \Delta H^\circ_{rxn}(T = 25°\text{C}) + \int_{\substack{T=25°\text{C}\\(298.15\text{ K})}}^{T} \Delta C^\circ_P(T')\, dT' \tag{9.1-21}$$

with $\Delta C^\circ_P = \Sigma\, \nu_i C^\circ_{P,i}$ where $C^\circ_{P,i}$ is the heat capacity of species i in its standard state. Note that in this integration ΔC°_P may be a function of temperature.

Equation 9.1-20b can be integrated between any two temperatures T_1 and T_2 to give

$$\ln \frac{K_a(T_2)}{K_a(T_1)} = \int_{T_1}^{T_2} \frac{\Delta H^\circ_{rxn}(T)}{RT^2}\, dT \tag{9.1-22a}$$

so that if K_a is known at one temperature, usually 25°C, its value at any other temperature can be computed if the standard state heat of reaction is known as a function of temperature. If ΔH°_{rxn} is temperature independent, or if T_1 and T_2 are so close that ΔH°_{rxn} may be assumed to be constant over the temperature range, we obtain

$$\ln \frac{K_a(T_2)}{K_a(T_1)} = -\frac{\Delta H^\circ_{rxn}}{R}\left(\frac{1}{T_2} - \frac{1}{T_1}\right) \tag{9.1-22b}$$

Equation 9.1-22b suggests that the logarithm of the equilibrium constant should be a linear function of the reciprocal of the absolute temperature if the heat of reaction is independent of temperature and, presumably, an almost linear function of $1/T$ even if ΔH°_{rxn} is a function of temperature. (Compare this behavior with that of the vapor pressure of a pure substance in Sec. 5.7.) Consequently, it is common practice to plot the logarithm of the equilibrium constant versus the reciprocal of temperature. Figure 9.1-2 gives the equilibrium constants for a number of reactions as a function of temperature plotted in this way. (Can you identify those reactions that are endothermic and those that are exothermic?)

For the general case in which ΔH°_{rxn} is a function of temperature, we start from the observation that the constant-pressure heat capacity is usually given in the form[2]

$$C_{P,i} = a_i + b_i T + c_i T^2 + d_i T^3 + e_i T^{-2}$$

(see Appendix II), and obtain, from Eq. 9.1-21 with $T_1 = 298.15$ K, that

$$\Delta H^\circ_{rxn}(T) = \Delta H^\circ_{rxn}(T_1) + \Delta a(T - T_1) + \frac{\Delta b}{2}(T^2 - T_1^2) + \frac{\Delta c}{3}(T^3 - T_1^3) + \frac{\Delta d}{4}(T^4 - T_1^4) - \Delta e(T^{-1} - T_1^{-1}) \tag{9.1-23a}$$

Further, from Eq. 9.1-22a, we obtain

$$\begin{aligned}\ln \frac{K_a(T_2)}{K_a(T_1)} &= \frac{\Delta a}{R}\ln\frac{T_2}{T_1} + \frac{\Delta b}{2R}(T_2 - T_1) + \frac{\Delta c}{6R}(T_2^2 - T_1^2) \\ &+ \frac{\Delta d}{12R}(T_2^3 - T_1^3) + \frac{\Delta e}{2R}(T_2^{-2} - T_1^{-2}) \\ &+ \frac{1}{R}\left[-\Delta H^\circ_{rxn}(T_1) + \Delta a T_1 + \frac{\Delta b}{2}T_1^2 + \frac{\Delta c}{3}T_1^3 + \frac{\Delta d}{4}T_1^4 - \frac{\Delta e}{T_1}\right] \times \left[\frac{1}{T_2} - \frac{1}{T_1}\right]\end{aligned} \tag{9.1-23b}$$

where $\Delta a = \Sigma\, \nu_i a_i$, $\Delta b = \Sigma\, \nu_i b_i$, and so on. The BASIC language program CHEMEQ.BAS of Appendix A9.1 can be used for calculations using Eqs. 9.1-22 and 23 for reactions involving compounds in its data base.

[2]The last term, $e_i T^{-2}$, is usually present in the heat capacity of solids and is included here for generality.

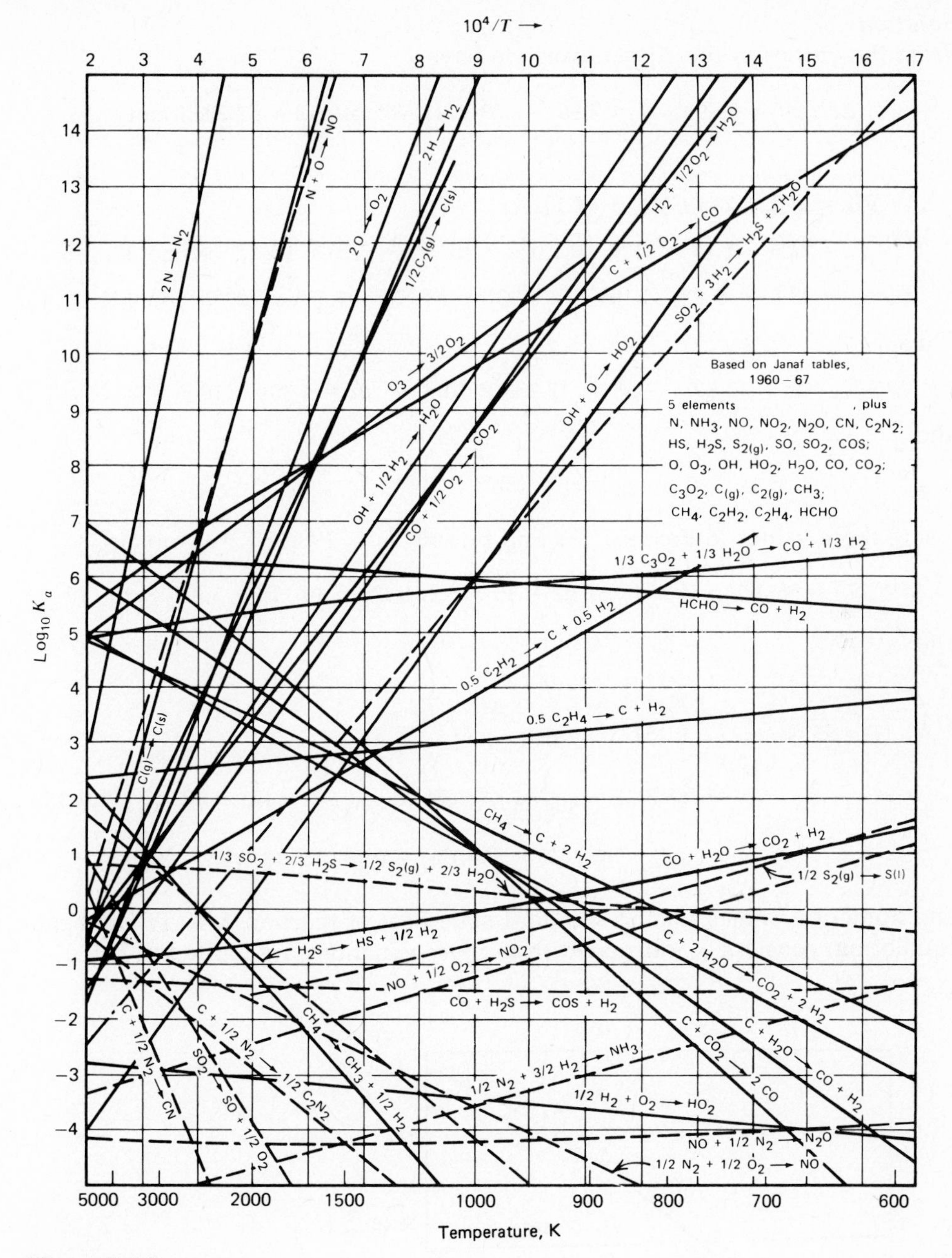

Figure 9.1-2
Chemical equilibrium constants as a function of temperature. (From M. Modell and R. C. Reid, *Thermodynamics and Its Applications* © 1974, p. 396. Reprinted by permission of Prentice–Hall, Inc., Englewood Cliffs, N.J.)

ILLUSTRATION 9.1-3

Compute the equilibrium extent of decomposition of pure nitrogen tetroxide due to the chemical reaction $N_2O_4(g) = 2NO_2(g)$ over the temperature range of 200 to 400 K, at pressures of 0.1, 1, and 10 atmospheres.

Data

See Appendixes II and IV.

Solution

From the entries in the Appendixes we have

$$\Delta H^\circ_{rxn}(T = 25°\text{C}) = 2 \times 7.96 - 2.23 = 13.69 \text{ kcal/mol} = 57{,}279 \text{ J/mol}$$

and

$$\begin{aligned}\Delta C_P = \Delta C_P^* &= 2 \times C_P^*|_{NO_2} - C_P^*|_{N_2O_4} \\ &= 3.06 - 1.73 \times 10^{-2}T + 1.028 \times 10^{-5}T^2 + 3.76 \times 10^{-9}T^3 \text{ cal/mol K} \\ &= 12.80 - 7.238 \times 10^{-2}T + 4.301 \times 10^{-5}T^2 + 1.573 \times 10^{-8}T^3 \text{ J/mol K}\end{aligned}$$

Thus

$$\Delta a = 12.80, \qquad \Delta b = -7.238 \times 10^{-2}, \qquad \Delta c = 4.301 \times 10^{-5}$$

and

$$\Delta d = 1.573 \times 10^{-8}$$

Using these values in Eqs. 9.1-23a and b (with $T_1 = 298.15$ K) we obtain

$$\Delta H^\circ_{rxn}(T) = 56{,}271 + 12.80T - 3.62 \times 10^{-2}T^2 + 1.435 \times 10^{-5}T^3 + 3.933 \times 10^{-9}T^4$$

and

$$\begin{aligned}\ln\left(\frac{K_a(T)}{K_a(T = 25°\text{C})}\right) &= \ln\left(\frac{K_a(T)}{0.154}\right) = \int_{298.15\text{ K}}^{T} \frac{\Delta H^\circ_{rxn}(T)}{RT^2}\, dT \\ &= -6768.5\left(\frac{1}{T} - \frac{1}{298.15}\right) + 1.54 \ln\frac{T}{298.15} - 0.435 \times 10^{-2}(T - 298.15) \\ &\quad + 0.862 \times 10^{-6}(T^2 - 298.15^2) + 0.1577 \times 10^{-9}(T^3 - 298.15^3)\end{aligned}$$

The numerical values for the standard state heat of reaction $\Delta H^\circ_{rxn}(T)$ and the equilibrium constant K_a calculated from these equations are plotted in Fig. 1.

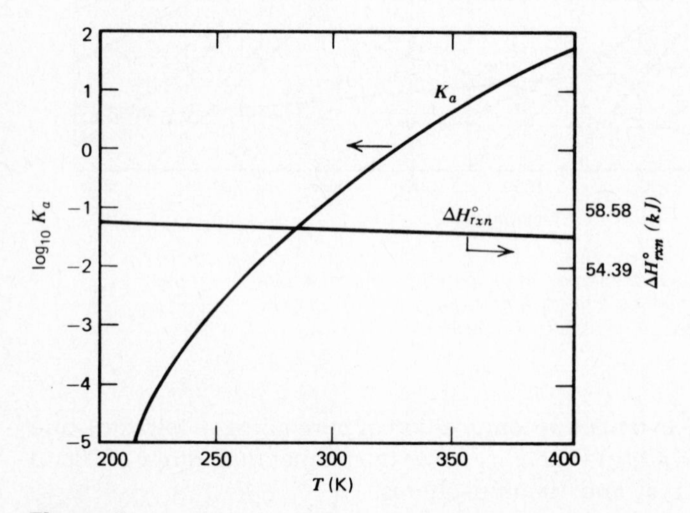

Figure 1
The standard heat of reaction and equilibrium constant for the reaction $N_2O_4(g) = 2NO_2(g)$.

Now assuming that the gas phase is ideal, we have that

$$K_a = \frac{a_{NO_2}^2}{a_{N_2O_4}} = \frac{(y_{NO_2}P/P = 1 \text{ atm})^2}{(y_{N_2O_4}P/P = 1 \text{ atm})} = \frac{y_{NO_2}^2}{y_{N_2O_4}}\left(\frac{P}{1 \text{ atm}}\right)$$

where, as in Illustration 9.1-1, $y_{NO_2} = 2X/(1 + X)$, and $y_{N_2O_4} = (1 - X)/(1 + X)$, so that

$$K_a = \frac{4X^2}{(1 - X^2)}\left(\frac{P}{1 \text{ atm}}\right) \qquad K_a\left(\frac{1 \text{ atm}}{P}\right) = \frac{4X^2}{1 - X^2}$$

and

$$X = \sqrt{\frac{K_a/P}{1 + K_a/P}} \qquad \text{for } P \text{ in atm}$$

The extent of reaction X and the mole fraction of nitrogen dioxide as a function of temperature and pressure are plotted in Fig. 2. (Note that the equilibrium constant is independent of pressure, but the equilibrium activity ratio increases linearly with pressure. Therefore, the extent of reaction for this reaction decreases as the pressure increases.) ■

In the next illustration, the effects on the equilibrium composition of both feed composition and maintaining reactor volume (rather than reactor pressure) constant are considered.

ILLUSTRATION 9.1-4

Nitrogen and hydrogen react to form ammonia in the presence of a catalyst,

$$\tfrac{1}{2}N_2 + \tfrac{3}{2}H_2 \rightarrow NH_3$$

The reactor in which this reaction is to be run is maintained at 450 K and has a sufficiently long residence time so that equilibrium is achieved at the reactor exit.

a. What will be the mole fractions of nitrogen, hydrogen, and ammonia exiting the reactor if stoichiometric amounts of nitrogen and hydrogen enter the reactor which is kept at 4 atm pressure?

b. What will be the exit mole fractions if the reactor operates at 4 atm pressure and the feed consists of equal amounts of nitrogen, hydrogen, and an inert diluent?

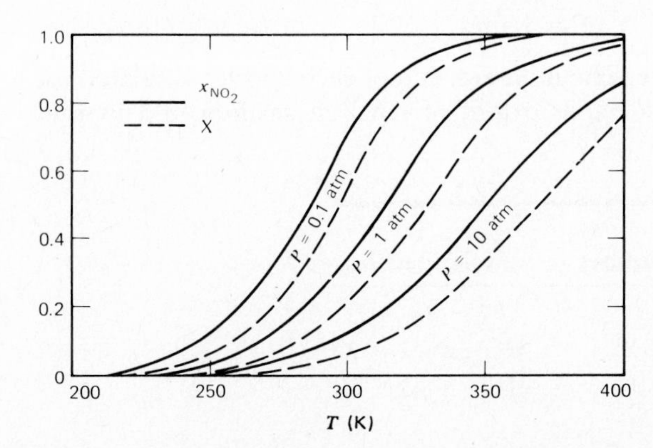

Figure 2 Equilibrium mole fraction x_{NO_2} and molar extent of reaction X for the reaction $N_2O_4 = 2NO_2$ as a function of temperature and pressure.

c. The reaction is to be run in an isothermal, constant-volume reaction vessel with a feed consisting of stoichiometric amounts of nitrogen and hydrogen. The initial pressure of the reactant mixture (before any reaction has occurred) is 4 atm. What is the pressure in the reactor and the species mole fractions when equilibrium is achieved?

Solution

a. The starting point for this problem is the evaluation of the equilibrium constant for the ammonia production reaction at 450 K. From Appendixes II and IV we have

$$\Delta \underline{H}^\circ_{f,NH_3}(T = 25°C, P = 1 \text{ atm}) = -10{,}960 \text{ cal/mol } NH_3 = -45{,}857 \text{ J/mol } NH_3 = \Delta H^\circ_{rxn}$$

$$\Delta \underline{G}^\circ_{f,NH_3}(T = 25°C, P = 1 \text{ atm}) = -3{,}903 \text{ cal/mol } NH_3 = -16{,}330 \text{ J/mol } NH_3 = \Delta G^\circ_{rxn}$$

and

$$\Delta C_P^* = -30.522 + 2.928 \times 10^{-2}T - 1.387 \times 10^{-7}T^2 - 3.9463 \times 10^{-9}T^3 \text{ J/mol K}$$

Thus,

$$\Delta H^\circ_{rxn}(T) = -45{,}857 + \int_{298.15\text{ K}}^{T} \Delta C_P^*\, dT$$

$$= -38{,}049 - 30.522T + 1.464 \times 10^{-2}T^2 - 4.623 \times 10^{-8}T^3 - 9.866 \times 10^{-10}T^4 \text{ J/mol}$$

and

$$\ln \frac{K_a(T = 450\text{ K})}{K_a(T = 298.15\text{ K})} = \int_{298.15\text{ K}}^{450\text{ K}} \frac{\Delta H^\circ_{rxn}(T)}{RT^2}\, dT = -6.426$$

Also,

$$\ln K_a(T = 298.15\text{ K}) = -\frac{\Delta G^\circ_{rxn}}{RT} = \frac{16{,}330}{8.314 \times 298.15} = 6.588$$

so that

$$\ln K_a(T = 450\text{ K}) = 6.588 - 6.426 = 0.162$$

and

$$K_a(T = 450\text{ K}) = 1.175 = \frac{a_{NH_3}}{a_{N_2}^{1/2} a_{H_2}^{3/2}}$$

At the low pressure here we will assume the gas phase is ideal so that

$$a_i = \frac{y_i P_{rxn}}{P = 1 \text{ atm}}$$

where P_{rxn} is the reaction pressure. The mole fraction of each species is related to the inlet mole numbers and the molar extent of reaction as indicated in the following table.

	Initial Mole Number	Final Mole Number	Mole Fraction
NH_3	0	X	$X/(2 - X)$
N_2	$\frac{1}{2}$	$\frac{1}{2}(1 - X)$	$\frac{1}{2}(1 - X)/(2 - X)$
H_2	$\frac{3}{2}$	$\frac{3}{2}(1 - X)$	$\frac{3}{2}(1 - X)/(2 - X)$
Total		$2 - X$	

Therefore

$$K_a = 1.175 = \frac{X(2 - X)}{\left\{\frac{1}{2}(1 - X)\right\}^{1/2} \left\{\frac{3}{2}(1 - X)\right\}^{3/2} \left(\frac{P_{rxn}}{1 \text{ atm}}\right)} = \frac{X(2 - X)}{\left(\frac{1}{2}\right)^{1/2} \left(\frac{3}{2}\right)^{3/2} (1 - X)^2 \left(\frac{P_{rxn}}{1 \text{ atm}}\right)}$$

and at $P_{rxn} = 4$ atm

$$K' \equiv \left(\frac{1}{2}\right)^{1/2} \left(\frac{3}{2}\right)^{3/2} \left(\frac{P_{rxn}}{1 \text{ atm}}\right) K_a = 6.105 = \frac{X(2 - X)}{(1 - X)^2}$$

This equation has the solution

$$X = 1 - \sqrt{\frac{1}{1 + K'}} = 1 - \sqrt{\frac{1}{7.105}} = 0.6248$$

so that

$$y_{NH_3} = 0.455, \qquad y_{N_2} = 0.136, \qquad \text{and} \qquad y_{H_2} = 0.409$$

b. Here we have

	Initial Mole Numbers	Final Mole Numbers	Mole Fractions
NH_3	0	X	$X/(3 - X)$
N_2	1	$1 - \frac{1}{2}X$	$(1 - \frac{1}{2}X)/(3 - X)$
H_2	1	$1 - \frac{3}{2}X$	$(1 - \frac{3}{2}X)/(3 - X)$
Diluent	1	1	$1/(3 - X)$
Total		$3 - X$	

and

$$K_a = \frac{X(3 - X)}{\left(1 - \frac{1}{2}X\right)^{1/2} \left(1 - \frac{3}{2}X\right)^{3/2} \left(\frac{P_{rxn}}{1 \text{ atm}}\right)}$$

with $K_a = 1.175$ and $P_{rxn} = 4$ atm. By trial and error we find $X = 0.4028$, and

$$y_{NH_3} = 0.155 \qquad y_{H_2} = 0.152$$

$$y_{N_2} = 0.307 \qquad y_{Dil} = 0.385$$

c. As in part a we have

$$\left(\frac{1}{2}\right)^{1/2} \left(\frac{3}{2}\right)^{1/2} K_a \left(\frac{P_{rxn}}{1 \text{ atm}}\right) = \frac{X(2 - X)}{(1 - X)^2}$$

since the equilibrium criterion $\sum \nu_i \overline{G}_i = 0$ holds for reactions at constant T and V, just as it does for reactions at constant T and P (see Problem 6.4). Here, however, P_{rxn} is not fixed at 4 atm, but depends on the number of moles of gas through the ideal gas law. Since the volume and temperature are fixed,

$$P_{rxn} = \frac{N_{eq}}{N_0} P_0 = \frac{2 - X}{2} P_0 = 2(2 - X) \text{ atm}$$

where P_0 is the initial state pressure, 4 atm. Thus,

$$2\left(\frac{1}{2}\right)^{1/2}\left(\frac{3}{2}\right)^{3/2} K_a = 3.0527 = \frac{X}{(1 - X)^2}$$

since $K_a = 1.175$. This equation has the solution $X = 0.5685$, so that

$$y_{NH_3} = 0.397$$

$$P = 2.863 \text{ atm} \quad \text{and} \quad y_{N_2} = 0.151$$

$$y_{H_2} = 0.452$$

This solution should be compared with that obtained in part a for a reactor maintained at constant pressure. Can you explain why the answers differ? ■

The algebra involved in solving for the molar extent of reaction in general chemical equilibrium calculations can be tedious, especially if several reactions occur simultaneously, because of the nonlinear equations that arise. It is frequently possible, however, to make judicious simplifications based on the magnitude of the equilibrium constant. This is demonstrated in the next illustration.

ILLUSTRATION 9.1-5

At high temperatures, hydrogen sulfide dissociates into molecular hydrogen and sulfur:

$$2H_2S = 2H_2 + S_2$$

At 700°C all species are gases and the equilibrium constant for this reaction, K_a, is equal to 2.17×10^{-5}, based on standard states of the pure gases at the reaction temperature and a pressure of 1 atm.

a. Estimate the extent of dissociation of pure hydrogen sulfide at 700°C and $P = 1$ atm.

b. Show that the extent of dissociation is proportional to $P^{-1/3}$, and that if N moles of an inert diluent are added to 1 mole of hydrogen sulfide

$$\frac{X(\text{with diluent})}{X(\text{without diluent})} = (1 + N)^{1/3}$$

Solution

The starting point for the solution of this problem is the construction of a species balance table relating the mole fractions of each species to the molar extent of reaction. For brevity, this table is presented in a form that is applicable to parts a and b of this illustration.

Species	Initial Number of Moles	Final Number of Moles	Mole Fraction $N = 0$	Mole Fraction $N \neq 0$
H_2S	1	$1 - 2X$	$(1 - 2X)/(1 + X)$	$(1 - 2X)/(1 + X + N)$
H_2	0	$2X$	$2X/(1 + X)$	$2X/(1 + X + N)$
S_2	0	X	$X/(1 + X)$	$X/(1 + X + N)$
Diluent	N	N		
Total		$1 + X + N$		

The chemical equilibrium relation is

$$K_a = 2.17 \times 10^{-5} = \frac{a_{S_2} a_{H_2}^2}{a_{H_2S}^2} = \frac{y_{S_2} y_{H_2}^2}{y_{H_2S}^2} (P/1 \text{ atm}) = \frac{4X^3(P/1 \text{ atm})}{(1 - 2X)^2(1 + X + N)} \tag{1}$$

since $a_i = y_i P/(1 \text{ atm})$ for low pressure gas mixtures. As K_a is so small, we expect that $X \ll 1$. Therefore, it is reasonable to assume that

$$1 - 2X \approx 1$$

and

$$1 + X + N \approx 1 + N$$

so that

$$\frac{4X^3 P}{(1 - 2X)^2(1 + X + N)} \approx \frac{4X^3 P}{(1 - N)} = 2.17 \times 10^{-5}$$

or

$$X = \left(\frac{2.17 \times 10^{-5}(1 + N)}{4P}\right)^{1/3}$$

Thus, it is clear that the equilibrium extent of reaction X is proportional to $P^{-1/3}$, and that

$$\frac{X(N \neq 0)}{X(N = 0)} = (1 + N)^{1/3}$$

At $P = 1$ atm and $N = 0$ we have

$$X = \left(\frac{2.17 \times 10^{-5}}{4}\right)^{1/3} = 0.0176$$

so that the fraction of H_2S dissociated, $2X/1$, is equal to 0.035. (Had we not made the simplification, the solution, by trial and error, would be $X = 0.0173$.)

Comment

At high temperatures K_a can become large and we can expect X to be close to 0.5. In this case we may assume that

$$1 + X + N \approx 1.5 + N$$

$$X^3 \approx (0.5)^3 = 0.125$$

However, $(1 - 2X)$, which appears in the denominator of Eq. 1, cannot be set equal to zero since the equilibrium relation is not satisfied in this case. Instead, at high temperatures (large values of K_a) we obtain an estimate of X by solving the equation

$$(1 - 2X)^2 \approx \frac{4(0.125)P}{(1.5 + N)K_a}$$

or

$$X \approx 0.5 - \sqrt{\frac{0.125P}{(1.5 + N)K_a}} \quad \blacksquare$$

A number of other equilibrium ratios that are more easily measured than K_a frequently appear in the scientific literature. For example, the concentration equilibrium ratio K_c defined to be

$$K_c = \prod_i C_i^{\nu_i} \tag{9.1-24a}$$

where C_i is the concentration of species i (in kmol/m^3 or similar units), the mole fraction equilibrium ratios

$$K_x = \prod_i x_i^{\nu_i} \quad \text{and} \quad K_y = \prod_i y_i^{\nu_i} \tag{9.1-24b}$$

and the partial pressure equilibrium ratio

$$K_p = \prod_i P_i^{\nu_i} \tag{9.1-24c}$$

are frequently used. In Table 9.1-3 each of these quantities is related to the equilibrium constant K_a, and the quantities

$$K_\nu = \prod_i \left(\frac{\bar{f}_i}{y_i P}\right)^{\nu_i} \quad \text{and} \quad K_\gamma = \prod_i \gamma_i^{\nu_i} \tag{9.1-24d}$$

which account for gas-phase and solution nonidealities, respectively. [Note that if the Lewis–Randall rule, $\bar{f}_i(T, P, y_i) = y_i f_i(T, P)$, is used, then

$$K_\nu = \prod_i \left(\frac{f_i}{P}\right)^{\nu_i} \tag{9.1-24e}$$

This approximate expression, together with Fig. 5.4-1 for f/P is useful for making rapid estimates of the importance of gas-phase nonidealities.]

There are important differences between the equilibrium constant K_a and the equilibrium ratios defined here. First, K_a depends only on temperature and the standard states of the reactants and products; the equilibrium ratios, however, depend on mixture nonidealities (through K_ν and K_γ) and on the total pressure or the total molar concentration. Consequently, while the thermodynamic equilibrium constant K_a can be used to study the same reaction with different diluents or solvents, the ratios K_c or K_y only have meaning in the situation in which they were obtained. Finally, K_a is nondimensional, whereas K_p has units of (pressure)$^{\Sigma \nu_i}$ and K_c has units of (concentration)$^{\Sigma \nu_i}$.

ILLUSTRATION 9.1-6

Compare the numerical values of K_a, K_p, and K_y for the ammonia production reaction of Illustration 9.1-4 at $P = 4$ atm and $T = 450$ K.

Solution

From Illustration 9.1-4, $K_a = 1.175$ for the pure gas, $T = 450$ K, $P = 1$ atm standard state. From Table 9.1-3

$$K_a = 1.175 = (P = 1 \text{ atm})^{-\Sigma \nu_i} K_p = \left(\frac{P}{P = 1 \text{ atm}}\right)^{\Sigma \nu_i} K_y$$

where we have set $K_\nu = 1$. For the reaction being considered

$$\sum \nu_i = 1 - \frac{1}{2} - \frac{3}{2} = -1$$

so that

$$K_p = \frac{P_{NH_3}}{P_{N_2}^{1/2} P_{H_2}^{3/2}} = K_a\,(1 \text{ atm})^{-1} = 1.175 \text{ atm}^{-1}$$

and

$$K_y = K_a \left(\frac{P}{1 \text{ atm}}\right)^{-\Sigma \nu_i} = 1.175 \left(\frac{4}{1}\right)^1 = 4.700 \quad \text{(unitless)} \quad \blacksquare$$

Table 9.1-3
CHEMICAL EQUILIBRIUM RATIOS

Gaseous Mixture at Moderate or High Density
Standard State: Pure gases at $P = 1$ atm

$$K_a = \prod_i a_i^{\nu_i} = \prod_i \left[\frac{y_i P \left(\frac{\bar{f}_i}{y_i P} \right)}{1 \text{ atm}} \right]^{\nu_i} = \prod_i \left[\frac{P_i \left(\frac{\bar{f}_i}{y_i P} \right)}{1 \text{ atm}} \right]^{\nu_i}$$

$$= (1 \text{ atm})^{-\Sigma \nu_i} K_p K_\nu = \left(\frac{P}{1 \text{ atm}} \right)^{+\Sigma \nu_i} K_y K_\nu$$

Gaseous Mixture at Low Density
Standard State: State of unit activity

$$\left(\frac{f}{P} \right) \cong 1 \qquad K_\nu \cong 1$$

and

$$K_a = (1 \text{ atm})^{-\Sigma \nu_i} K_p = \left(\frac{P}{1 \text{ atm}} \right)^{\Sigma \nu_i} K_y$$

Liquid mixture
Standard State: State of unit activity*

$$K_a = \prod_i a_i^{\nu_i} = \prod_i (x_i \gamma_i)^{\nu_i} = K_x K_\gamma$$

Using

$$x_i = C_i / C$$

where C_i is the molar concentration of species i, and C is the total molar concentration of the mixture, we have

$$K_a = \prod_i (x_i \gamma_i)^{\nu_i} = \prod_i \left(\frac{C_i}{C} \gamma_i \right)^{\nu_i} = C^{-\Sigma \nu_i} K_c K_\gamma$$

For an ideal mixture $\gamma_i = 1$, and

$$K_a = C^{-\Sigma \nu_i} K_c$$

*The expressions here have been written assuming that the standard state of each component is the pure component state. Analogous expressions can be written using either of the Henry's law standard states for each component or, more generally, for the case in which the standard state of some species in the reaction is the pure component state and for others it is the infinite dilution or ideal 1 molal states.

The prediction of the reaction equilibrium state for nonideal gas-phase reactions is somewhat more complicated than for ideal gas-phase reactions. This is demonstrated in the following illustration, which also shows that there is a small effect of pressure on reaction equilibrium due to gas-phase nonideality (as a result of the fugacity coefficient ratio K_ν), in addition to the primary pressure effect found in Illustration 9.1-3.

ILLUSTRATION 9.1-7

Compute the equilibrium mole fraction of each of the species in the gas-phase reaction

$$CO_2 + H_2 = CO + H_2O$$

at 1000 K and (a) 1 atm total pressure, (b) 500 atm total pressure. The equilibrium constant for this reaction K_a experimentally has been found to be equal to 0.693 at 1000 K (standard state is the pure gases at $T = 1000$ K, $P = 1$ atm), and initially there are equal amounts of carbon dioxide and hydrogen present.

Solution

For this reaction, we have

$$K_a = 0.693 = \frac{a_{CO}a_{H_2O}}{a_{CO_2}a_{H_2}} = \frac{y_{CO}y_{H_2O}}{y_{CO_2}y_{H_2}} \times \frac{\left(\frac{\bar{f}_{CO}}{y_{CO}P}\right)\left(\frac{\bar{f}_{H_2O}}{y_{H_2O}P}\right)}{\left(\frac{\bar{f}_{CO_2}}{y_{CO_2}P}\right)\left(\frac{\bar{f}_{H_2}}{y_{H_2}P}\right)} \times \left(\frac{P}{1 \text{ atm}}\right)^{\Sigma \nu_i}$$

$$= K_y K_\nu \left(\frac{P}{1 \text{ atm}}\right)^{\Sigma \nu_i}$$

Since $\Sigma \nu_i = 0$, there is no primary effect of pressure on the extent of reaction; there is, however, a secondary effect through the pressure dependence of K_ν. Eliminating the species mole fractions in terms of the extent of reaction, we obtain

	Initial Mole Numbers	Final Mole Numbers	Mole Fractions
CO_2	1	$1 - X$	$(1 - X)/2$
H_2	1	$1 - X$	$(1 - X)/2$
CO	0	X	$X/2$
H_2O	0	X	$X/2$
Total		2	

and

$$K_a = 0.693 = \frac{X^2}{(1 - X)^2} K_\nu$$

a. At 1 atm $K_\nu = 1$, so that to predict the equilibrium state we need solve

$$0.693 = X^2/(1 - X)^2$$

The solution to this equation is $X = 0.4543$, so that

$$y_{CO} = y_{H_2O} = 0.227 \quad \text{and} \quad y_{CO_2} = y_{H_2} = 0.273.$$

b. At 500 atm, K_ν cannot be assumed to be equal to unity since each of the fugacity coefficients will have values different from unity; instead, the value of K_ν will be computed by using the Peng–Robinson equation of state to calculate each of the fugacity coefficients. However, since, in the fugacity coefficient calculation, the mole fractions are needed (first to compute the a and b parameters in the mixture, and then for the evaluation of the fugacity coefficients after the compressibility factor has been found), the computation of the equilibrium state will be iterative. That is, a value of K_ν will be assumed (unity for the first iteration), and the equilibrium compositions will be computed. The mole fractions that result will then be used in the Peng–Robinson equation to compute the species fugacity coefficients and a new value of K_ν, which is then used to compute new equilibrium mole fractions. This procedure is repeated until the calculation has converged.

The final result is

$$\left(\frac{\bar{f}}{yP}\right)_{H_2O} = 0.9609; \qquad \left(\frac{\bar{f}}{yP}\right)_{CO} = 1.1804; \qquad \left(\frac{\bar{f}}{yP}\right)_{CO_2} = 1.1282$$

$$\left(\frac{\bar{f}}{yP}\right)_{H_2} = 1.0991$$

and

$$K_\nu = 0.9147$$

Thus $X = 0.4654$, so that

$$y_{CO} = y_{H_2} = 0.2327 \qquad \text{and} \qquad y_{CO_2} = y_{H_2O} = 0.2673,$$

which differs slightly from the low pressure result.

Note that $K_p = \Pi_i P^{\nu_i} = P^{\Sigma \nu_i}$ is equal to unity for the reaction being considered since $\Sigma \nu_i = 0$. Therefore, the only effect of pressure on the reaction equilibrium is as a result of gas-phase nonidealities. Such an effect is much smaller than the direct effect of pressure when $\Sigma \nu_i \neq 0$ (and $K_p \neq 1$) as found in Illustration 9.1-4. ■

The discussion so far has been restricted to gas-phase reactions. One reason for this is that a large number of reactions, including many high temperature reactions (except metallurgical reactions), occur in the gas phase. Also, the identification of the equilibrium state is easiest for gases as nonidealities play only a minor role. However, many reactions of interest to engineers occur in the liquid phase. The prediction of the equilibrium state in such cases can be complicated if the only information available is K_a or ΔG°_{rxn}, since liquid solutions are rarely ideal, so that K_γ will not be unity. Furthermore, some liquid-phase reactions, especially those in aqueous solution, involve dissociation of the reactants into ions (which form highly nonideal solutions), and then reaction of the ions (see Illustration 9.1-9).

Consequently, for many liquid-phase reactions one is more likely to use experimental measurements of the equilibrium concentration ratio

$$K_c = \prod_i C_i^{\nu_i} = C^{\Sigma \nu_i} K_a / K_\gamma \tag{9.1-25}$$

if such data are available, then to try to calculate the equilibrium state from K_a or Gibbs free energy of formation data. However, the equilibrium ratio K_c is a function of the solvent and reactant concentrations, both through the $C^{\Sigma \nu_i}$ term and the activity coefficient ratio K_γ. Therefore, at a given temperature and pressure, several values of K_c may be given, each corresponding to different solvents or molar concentrations of reactants and diluents. (Frequently, as in Illustration 9.1-9, we will be satisfied with relating the value of K_c in one solution to its value in another, rather than trying to predict its value a priori.)

Given a value of K_c for the reaction conditions of interest, equilibrium calculations for liquid-phase reactions become straightforward, as next indicated.

ILLUSTRATION 9.1-8

The ester ethyl acetate is produced by the reaction

$$CH_3COOH + C_2H_5OH = CH_3COOC_2H_5 + H_2O$$

In aqueous solution at 100°C, the equilibrium ratio K_c for this reaction is 2.92 (which is unitless since $\Sigma\, \nu_i = 0$). Compute the equilibrium concentrations of each species in an aqueous solution that initially contains 250 kg of acetic acid and 500 kg of ethyl alcohol for each 1 m³ of solution. The density of the solution may be assumed to be constant and equal to 1040 kg/m³.

Solution

The initial concentration of each species is

$$C_A = \frac{250 \text{ kg/m}^3}{60 \text{ g/mol}} = 4.17 \text{ kmol/m}^3$$

$$C_E = \frac{500 \text{ kg/m}^3}{46 \text{ g/mol}} = 10.9 \text{ kmol/m}^3$$

$$C_W = \frac{(1040 - 250 - 500) \text{ kg/m}^3}{18 \text{ g/mol}} = 16.1 \text{ kmol/m}^3$$

and the concentration of each species at an extent of reaction $\hat{X}$, in units of kmol/m³, is

$$C_A = 4.17 - \hat{X}$$

$$C_E = 10.9 - \hat{X}$$

$$C_W = 16.1 + \hat{X}$$

$$C_{EA} = \hat{X}$$

Therefore, at equilibrium, we have

$$K_c = 2.92 = \frac{(16.1 + \hat{X})\hat{X}}{(10.9 - \hat{X})(4.17 - \hat{X})}$$

This equation has the solution $\hat{X} = 2.39$ kmol/m³, so that

$$C_A = 1.78 \text{ kmol/m}^3$$

$$C_E = 8.51 \text{ kmol/m}^3$$

$$C_W = 18.49 \text{ kmol/m}^3$$

$$C_{EA} = 2.39 \text{ kmol/m}^3$$

Consequently, 57.3% of the acid has reacted at equilibrium. ■

The partial ionization of weak acids, bases, and salts in solution, most commonly in aqueous solution, can be considered to be a chemical equilibrium process. Examples include the ionization of acetic acid

$$CH_3COOH = CH_3COO^- + H^+$$

the ionization of water

$$H_2O = H^+ + OH^-$$

and the dissociation of organic acids and alcohols. In the discussion here the ionization reaction will be designated as

$$A_{\nu_+}B_{\nu_-} = \nu_+ A^{z+} + \nu_- B^{z-} \tag{9.1-26}$$

where ν_+ and ν_- are the stoichiometric coefficients of the cation and anion, respectively, and z_+ and z_- are their valences.

The chemical equilibrium relation for this reaction is

$$\begin{aligned} K_a &= \exp\left[-\frac{1}{RT}\{\nu_+ \Delta \underline{G}^\circ_{f,A^{z+}} \text{ (ideal 1 molal solution)} \right. \\ &\quad + \nu_- \Delta \underline{G}^\circ_{f,B^{z-}} \text{ (ideal 1 molal solution)} \\ &\quad \left. - \Delta \underline{G}^\circ_{f,A_{\nu_+}B_{\nu_-}} \text{ (ideal 1 molal solution)}\}\right] \\ &= \frac{a_{A^{z+}}^{\nu_+} a_{B^{z-}}^{\nu_-}}{a_{A_{\nu_+}B_{\nu_-}}} = \frac{\left(\dfrac{M_{A^{z+}}\gamma^{\square}_{A^{z+}}}{1 \text{ molal}}\right)^{\nu_+}\left(\dfrac{M_{B^{z-}}\gamma^{\square}_{B^{z-}}}{1 \text{ molal}}\right)^{\nu_-}}{\left(\dfrac{M_{A_{\nu_+}B_{\nu_-}}\gamma^{\square}_{A_{\nu_+}B_{\nu_-}}}{1 \text{ molal}}\right)} \\ &= \frac{\left(\dfrac{M_{A^{z+}}}{1 \text{ molal}}\right)^{\nu_+}\left(\dfrac{M_{B^{z-}}}{1 \text{ molal}}\right)^{\nu_-}(\gamma_\pm)^{\nu_+ + \nu_-}}{\left(\dfrac{M_{A_{\nu_+}B_{\nu_-}}\gamma^{\square}_{A_{\nu_+}B_{\nu_-}}}{1 \text{ molal}}\right)} \end{aligned} \tag{9.1-27}$$

Here the standard state for the ionic species is a 1 molal ideal solution; the enthalpies and Gibbs free energies of formation for some ions in this standard state at 25°C are given in Table 9.1-4. In Eq. 9.1-27 the standard state for the undissociated molecule has also been chosen to be the ideal 1 molal solution (see Eq. 7.8-15), although the pure component state could have been used as well (with appropriate changes in $\Delta \underline{G}^\circ_{f,A_{\nu_+}B_{\nu_-}}$ and $a_{A_{\nu_+}B_{\nu_-}}$). Finally, we have used the mean molal activity coefficient, $\gamma_\pm$ of Eq. 7.9-11. Also remember that for the 1 molal standard state, $\gamma^{\square} \to 1$ as the solution becomes very dilute.

From electrical conductance measurements it is possible to measure the total ionic concentration in solution of weak electrolytes and thus determine their degrees of ionization. This information is usually summarized in terms of the equilibrium concentration ratio

$$K_c = \frac{(C_{A^{z+}})^{\nu_+}(C_{B^{z-}})^{\nu_-}}{C_{A_{\nu_+}B_{\nu_-}}} \tag{9.1-28}$$

Historically, the equilibrium ratio K_c for ionization reactions has been called the **ionization constant** or the **dissociation constant**; clearly, its value will depend on both the total electrolyte concentration and the solvent.

Comparing Eqs. 9.1-27 and 28, and neglecting the difference between concentration and molality, yields

$$K_c = K_a \left(\frac{\gamma^{\square}_{A_{\nu_+}B_{\nu_-}}}{(\gamma_\pm)^{\nu_+ + \nu_-}}\right)(1 \text{ molal})^{\nu_+ + \nu_- - 1} \tag{9.1-29}$$

Table 1
THE HEATS AND GIBBS FREE ENERGIES OF FORMATION FOR IONS IN AN IDEAL ONE MOLAL SOLUTION AT 25°C

Ion	$\Delta \underline{H}_f^\circ$(kcal/mol)	$\Delta \underline{G}_f^\circ$(kcal/mol)
Ag^+	25.31	18.43
Al^{+++}	1307.44	
Ba^{++}	−128.67	−134.0
Ca^{++}	−129.77	−132.18
Cl^-	−40.023	−31.350
Cs^+	−59.2	−67.41
Cu^+	12.4	12.0
Cu^{++}	15.39	15.53
F^-	−78.66	−66.08
Fe^{++}	−21.0	−20.30
Fe^{+++}	−11.4	−2.52
Hg^{++}		39.38
HSO_3^-	−150.09	−126.03
HSO_4^-	−211.70	−179.94
I^-	−13.37	−12.35
K^+	−60.04	−67.466
Li^+	−66.54	−70.22
Mg^{++}	−110.41	−108.99
Mn^{++}	−52.3	−53.4
Na^+	−57.28	−62.59
NH_4^+	−31.74	−19.00
Ni^{++}	−15.3	−11.1
OH^-	−54.957	−37.595
Pb^{++}	0.39	−5.81
SO_3^{--}	−149.2	−118.8
SO_4^{--}	−216.9	−177.34
Sr^{++}	−130.38	−133.2
Tl^+	1.38	−7.755
Tl^{+++}	27.7	50.0
Zn^{++}	−36.43	−35.184

Source: Adapted from G. N. Lewis, M. Randall, K. S. Pitzer, and L. Brewer, *Thermodynamics*, 2nd ed. Copyright 1961 by McGraw-Hill, Inc. Used with permission of McGraw-Hill Book Company.
Note: kcal/mol = 4.184 kJ/mol

Generally we will be interested in very dilute solutions, so that $\gamma^{\square}_{A_{\nu+}B_{\nu-}}$ can be taken to be unity. Also, we define an equilibrium constant K_c° by

$$K_c^\circ = K_a(1 \text{ molal})^{\nu_+ + \nu_- - 1} \tag{9.1-30}$$

and obtain

$$\ln K_c = \ln K_c^\circ - (\nu_+ + \nu_-) \ln \gamma_\pm \tag{9.1-31}$$

(Note that K_c° would be the ionization constant if the ions formed an ideal solution.) Depending on the total ionic strength of the solution, one of Eqs. 7.9-15, 17, 18, or 19 will be used to compute $\ln \gamma_\pm$. At very low ionic strengths, the Debye–Hückel relation applies, so that

$$\ln K_c = \ln K_c^\circ + (\nu_+ + \nu_-)|z_+z_-|\alpha\sqrt{\mu} \tag{9.1-32}$$

where the ionic strength μ is given by

$$\mu = \frac{1}{2}\sum_{\text{ions}} z_i^2 M_i$$

The important feature of Eq. 9.1-31 is that it can be used to predict the ionization constant of a molecule when data on the free energy of formation of its ion fragments are available (so that K_a and K_c° can be computed), or, when such data are not available, it can at least be used to interrelate the ionization constants for the same molecule at different ionic strengths, as in Illustration 9.1-9.

ILLUSTRATION 9.1-9

MacInnes and Shedlovsky [*J. Am. Chem. Soc.* **54,** 1429 (1932)] report the following data for ionization of acetic acid in water at 25°C:

Total Amount of Acetic Acid Added, C_T, kmol/m^3	CH_3COO^- Concentration kmol/m^3
0.028×10^{-3}	0.1511×10^{-4}
0.1532×10^{-3}	0.4405×10^{-4}
1.0283×10^{-3}	1.273×10^{-4}
2.4140×10^{-3}	2.001×10^{-4}
5.9115×10^{-3}	3.193×10^{-4}
20.000×10^{-3}	5.975×10^{-4}

Establish that, at low acetic acid concentrations, these data satisfy Eq. 9.1-32, and compute K_c° and the standard state Gibbs free energy change for this reaction.

Solution

The ionization reaction is

$$CH_3COOH = CH_3COO^- + H^+$$

so that $\nu_+ = \nu_- = 1$, $z_+ = 1$, $z_- = -1$ and

$$\ln K_c = \ln K_c^\circ + 2.356\sqrt{\mu}$$

Also, $C_{H^+} = C_{CH_3COO^-}$, $C_{CH_3COOH} = C_T - C_{CH_3COO^-}$, and

$$\mu = \frac{1}{2}\sum z_i^2 C_i = \frac{1}{2}\{C_{H^+} + C_{CH_3COO^-}\} = C_{CH_3COO^-}$$

The dissociation constant K_c is related to the ionic strength as follows

$$K_c = \frac{C_{CH_3COO^-}C_{H^+}}{C_{CH_3COOH}} = \frac{(C_{CH_3COO^-})^2}{C_T - C_{CH_3COO^-}} = \frac{\mu^2}{C_T - \mu}$$

The dissociation constant and ionic strength are tabulated and plotted here:

Total Amount Acetic Acid Added C_T, kmol/m³	$\mu = C_{CH_3COO^-}$ kmol/m³	K_c	ln K_c
0.028×10^{-3}	0.1511×10^{-4}	1.768×10^{-5}	−10.943
0.1532×10^{-3}	0.4405×10^{-4}	1.778×10^{-5}	−10.938
1.0283×10^{-3}	1.273×10^{-4}	1.799×10^{-5}	−10.926
2.4140×10^{-3}	2.001×10^{-4}	1.809×10^{-5}	−10.920
5.9115×10^{-3}	3.193×10^{-4}	1.823×10^{-5}	−10.912
20.000×10^{-3}	5.975×10^{-4}	1.840×10^{-5}	−10.903

From Fig. 9.1-3 we see that Eq. 9.1-32 does fit the low ionic strength data well, and that $\ln K_c^\circ = -10.9515$. Thus, $K_c^\circ = 1.753 \times 10^{-5}$ kmol/m³ and $K_a = 1.753 \times 10^{-5}$. Finally, since $\Delta G_{rxn}^\circ = -RT \ln K_a$, we have that at 25°C

$$\Delta G_{rxn}^\circ = \overline{G}_{CH_3COO^-}(\text{ideal 1 molal solution}) + \overline{G}_{H^+}(\text{ideal 1 molal solution}) - \overline{G}_{CH_3COOH}(\text{ideal 1 molal solution}) = 27.15 \text{ kJ/mol} \quad \blacksquare$$

A complete chemical equilibrium stability analysis is beyond the scope of this book.[3] It is useful to note, however, that generally states of chemical equilibrium are thermodynamically stable in that if the value of a state variable or constraint on the system is changed, the equilibrium will shift, but return to its previous state, after sufficient time, if the altered state variable or constraint is restored to its initial value. By direct calculation involving either the equilibrium constant (for changes in temperature), activities (for changes in pressure and species concentration), or the system constraints, we can determine the shift in the equilibrium state in response to any external change. For example, Illustra-

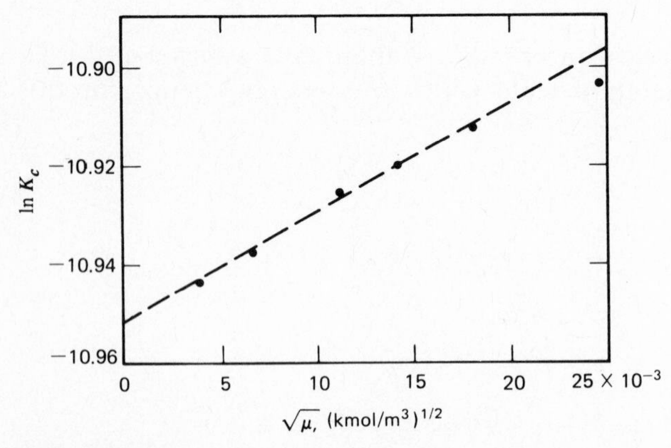

Figure 9.1-3
The ln K_c versus $\sqrt{\mu}$ for the dissociation of acetic acid.

[3]See Chapter 9 of *Thermodynamics and Its Applications*, 2nd ed. by M. Modell and R. C. Reid, Prentice–Hall, Englewood Cliffs, N.J., 1983, and *Chemical Thermodynamics* by I. Prigogine and R. Defay (Transl. by D. H. Everett), Longmans Green, New York, 1954.

tion 9.1-3 shows that both the equilibrium constant K_a and the molar extent of reaction X increase with increasing temperature for an endothermic ($\Delta H_{rxn} > 0$) reaction, whereas Illustration 9.1-4 shows that K_a decreases with increasing temperature for an exothermic ($\Delta H_{rxn} < 0$) reaction. Illustration 9.1-3 can also be interpreted as demonstrating that an increase in pressure decreases the extent of reaction for a reaction in which ΔV_{rxn} is positive and would increase X if ΔV_{rxn} were negative (note that in an ideal gas-phase reaction ΔV_{rxn} is proportional to $\Sigma \nu_i$). Adding an inert diluent to a reaction at constant temperature and pressure reduces the concentration of the reactant species, and its effect can be found accordingly. (In a gas-phase reaction the partial pressure of each species is reduced, so that the direction of the shift in equilibrium is the same as that which accompanies a reduction in total pressure.) Adding a diluent at constant temperature and volume, however, has no effect on the equilibrium, except through mixture nonidealities.

These observations and others on the direction of the shift in equilibrium in response to a given change are usually summarized by a statement referred to as the Principle of LeChatelier[4] and Braun[5] (but also known as the Principle of Moderation or the Principle of Spite[6]):

> A system in chemical equilibrium responds to an imposed change in any of the factors governing the equilibrium (for example, temperature, pressure, or the concentration of one of the species) in such a way that, had this same response occurred without the imposed variation, the factor would have changed in the opposite direction.

Although this simple statement is a good rule-of-thumb for inferring the effects of changes in an equilibrium system, it is not universally valid, and exceptions to it do occur (Problem 9.17). A more general, universally valid statement of this principle is best given within the context of a complete thermodynamic stability analysis for a multicomponent reacting system. For many situations, however, it may be more useful, and even more expeditious, to ascertain the equilibrium shift by direct computation of the new equilibrium state, rather than merely surmising the direction of that shift by a detailed stability analysis.

9.2 HETEROGENEOUS CHEMICAL REACTIONS

The discussion of the previous section was concerned with chemical reactions that occur in a single phase. Here our interest is with reactions that occur among species in different phases but that do not involve combined chemical and phase equilibrium. Several examples of such reactions are

$$CaF_2(\text{s, fluorspar}) + H_2SO_4(\text{l, pure}) \rightarrow CaSO_4(\text{s}) + 2HF(\text{g}) \tag{9.2-1}$$

$$CH_4 \rightarrow C(\text{s}) + 2H_2 \tag{9.2-2}$$

and

$$CaCO_3(\text{s}) \rightarrow CaO(\text{s}) + CO_2 \tag{9.2-3}$$

[4] H. LeChatelier, "Recherches sur les equilibres chimique" (Paris, 1888); *Annales des mines* **13,** 200 (1888).
[5] F. Braun, *Zeit. Physik. Chem.* **1,** 259 (1887).
[6] This term was used by J. Kestin in *A Course of Thermodynamics,* Vol. 2 Blaisdell, (John Wiley & Sons, Inc.), 1968, p. 304.

In the first of these reactions, the solubility of hydrogen fluoride in sulfuric acid is negligible, and in all these reactions the gaseous and solid species can be considered to be mutually insoluble. Consequently, the determination of the equilibrium state for each of these systems involves considerations of chemical, but not phase, equilibrium.

The most important characteristic that differentiates the foregoing heterogeneous reactions from the homogeneous reactions considered in the previous section is that in the calculation of the equilibrium state for heterogeneous reactions, the activity of a species is affected only by other components that appear in the same phase. Thus, to analyze the production of hydrogen fluoride gas by the reaction of Eq. 9.2-1, we have

$$K_a = \frac{a_{CaSO_4} a_{HF}^2}{a_{CaF_2} a_{H_2SO_4}}$$

which, at low pressures and using the 1 atm pure component gaseous standard state for HF, the pure component liquid standard state for H_2SO_4, and pure component solid standard states for CaF_2 and $CaSO_4$, reduces to

$$K_a = \left(\frac{y_{HF} P}{1 \text{ atm}}\right)^2 = \left(\frac{P_{HF}}{1 \text{ atm}}\right)^2$$

since the activities of the pure liquid (H_2SO_4) and the unmixed solids ($CaSO_4$ and CaF_2) are unity (see Table 9.1-2), and

$$a_{HF} = \frac{y_{HF} P}{(P = 1 \text{ atm})}$$

Thus the equilibrium pressure of hydrogen fluoride is easily calculated; clearly much more easily than if all reaction species were present in the gas or liquid phase.

ILLUSTRATION 9.2-1

Carbon black is to be produced from methane in a reactor maintained at a pressure of 1 atm and a temperature of 700°C (the reaction of Eq. 9.2-2). Compute the equilibrium gas-phase conversion and the fraction of pure methane charged that is reacted. The equilibrium constant K_a for this reaction at $T = 700°C$ is 7.403 based on the pure component standard states (gaseous for CH_4 and H_2, and solid for C) at the temperature of the reaction and 1 atm.

Solution

Basis of the calculation: 1 mole of methane

Species	Initial Amount	Final Amount	Present in Gas Phase	Gas Phase Mole Fraction
CH_4	1	$1 - X$	$1 - X$	$(1 - X)/(1 + X)$
C	0	X		
H_2	0	$2X$	$2X$	$(2X)/(1 + X)$
Total			$1 + X$	

Thus,

$$K_a = 7.403 = \frac{a_C a_{H_2}^2}{a_{CH_4}} = \frac{y_{H_2}^2}{y_{CH_4}} = \frac{(2X)^2}{(1+X)(1-X)}$$

since the activity of solid carbon is unity, and the reaction and standard state pressures are both 1 atm. Therefore

$$X = \sqrt{\frac{K_a}{4+K_a}} = 0.806 \quad \text{and} \quad \begin{matrix} y_{CH_4} = 0.108 \\ y_{H_2} = 0.892 \end{matrix} \quad \blacksquare$$

An important result of the equilibrium calculation just performed, or, for that matter, of any chemical equilibrium calculation in which reactants and products both appear in the same gaseous or liquid phase, is that the reaction will not go to completion (i.e., completely consume one of the reacting species) unless K_a is infinite. This behavior results from the contribution to the total Gibbs free energy from the free energy of mixing reactant and product species, as was the case in homogeneous liquid or gas phase chemical reactions (see Sec. 9.1 and Fig. 9.1-1)

The decomposition of calcium carbonate (Eq. 9.2-3), or any other reaction in which the reaction products and reactants do not mix in the gas or liquid phase represents a fundamentally different situation from that just considered, and the reaction *may* go to completion. To see why this occurs, consider the reaction of Eq. 9.2-3 in a constant temperature and pressure reaction vessel, and let $N_{CaCO_3,0}$ and $N_{CO_2,0}$ represent the number of moles of calcium carbonate and carbon dioxide, respectively, before the decomposition has started. Then,

$$\begin{pmatrix} \text{Initial Gibbs} \\ \text{free energy of} \\ \text{the system} \end{pmatrix} = G_0 = N_{CaCO_3,0}\underline{G}_{CaCO_3}(T, P) + N_{CO_2,0}\underline{G}_{CO_2}(T, P)$$

and

$$\begin{aligned} \begin{pmatrix} \text{Gibbs free energy of} \\ \text{the system at molar} \\ \text{extent of reaction } X \end{pmatrix} &= (N_{CaCO_3,0} - X)\underline{G}_{CaCO_3}(T, P) + X\underline{G}_{CaO}(T, P) \\ &\quad + (N_{CO_2,0} + X)\underline{G}_{CO_2}(T, P) \\ &= G_0 + X \sum \nu_i \underline{G}_i \\ &= G_0 + X\left\{\Delta G^\circ_{rxn} + RT \ln\left(\frac{P_{CO_2}}{P = 1 \text{ atm}}\right)\right\} \end{aligned} \tag{9.2-4}$$

where ΔG°_{rxn} is the standard state Gibbs free energy change of reaction, that is, the Gibbs free energy change if the reaction took place between the pure components at the temperature T and a pressure of 1 atmosphere. Notice that the Gibbs free energy is a function of the molar extent of reaction both explicitly and, since the system is closed and the CO_2 partial pressure depends on X, implicitly.

If the initial partial pressure of carbon dioxide is so low that

$$\Delta G^\circ_{rxn} + RT \ln\left(\frac{P_{CO_2}}{P = 1 \text{ atm}}\right) < 0$$

the minimum value of the total Gibbs free energy of the system occurs when enough calcium carbonate has decomposed to raise the partial pressure of carbon dioxide so that

$$\Delta G^\circ_{rxn} + RT \ln\left(\frac{P_{CO_2}}{P = 1 \text{ atm}}\right) = 0$$

or, if there is insufficient calcium carbonate present to do this, when all the calcium carbonate has been consumed and the reaction has gone to completion (i.e., $X = N_{CaCO_3}$). That is, if

$$-\frac{\Delta G^\circ_{rxn}}{RT} > \ln\left(\frac{P_{CO_2}}{P = 1 \text{ atm}}\right)$$

or, equivalently, if

$$K_a = \exp\left(-\frac{\Delta G^\circ_{rxn}}{RT}\right) > a_{CO_2}$$

where $a_{CO_2} = P_{CO_2}/1$ atm, the reaction will proceed as written until either the equilibrium partial pressure of CO_2 is achieved or all the calcium carbonate is consumed.

On the other hand, if initially P_{CO_2} is so large that

$$\Delta G^\circ_{rxn} + RT \ln\left(\frac{P_{CO_2}}{P = 1 \text{ atm}}\right) > 0$$

or

$$K_a = \exp\left\{-\frac{\Delta G^\circ_{rxn}}{RT}\right\} < a_{CO_2}$$

the minimum Gibbs free energy occurs when $X = 0$ (i.e., when there is no decomposition of calcium carbonate). In systems containing both $CaCO_3$ and CaO, when the partial pressure of carbon dioxide is such that

$$K_a = a_{CO_2} = \left\{\frac{P_{CO_2}}{P = 1 \text{ atm}}\right\}$$

equilibrium is obtained; that is, calcium carbonate will neither decompose to, nor be formed from, calcium oxide.

These various possibilities are shown in Fig. 9.2-1 where we plot, as a function of the imposed partial pressure of carbon dioxide, the quantity

$$\Delta G = \begin{pmatrix}\text{Gibbs free energy of} \\ \text{system at molar extent} \\ \text{of reaction } X\end{pmatrix} - \begin{pmatrix}\text{Initial Gibbs free} \\ \text{energy of system}\end{pmatrix}$$

$$= \left[G_0 + X\left\{\Delta G^\circ_{rxn} + RT \ln\left(\frac{P_{CO_2}}{P = 1 \text{ atm}}\right)\right\}\right] - G_0$$

$$= X\left\{\Delta G^\circ_{rxn} + RT \ln\left(\frac{P_{CO_2}}{P = 1 \text{ atm}}\right)\right\}$$

at $T = 1200$ K. At this temperature $\Delta G^\circ_{rxn} = -4908$ J/mol (see Illustration 9.2-2). In Fig. 9.2-1 we see that for carbon dioxide partial pressures of less than 1.635 atm, the Gibbs free energy change on reaction, and thus the system Gibbs free energy, decrease as the molar extent of reaction increases. Therefore, the reaction will proceed until all the calcium carbonate decomposes to calcium oxide. Conversely, for carbon dioxide partial pressures above 1.635 atm, the Gibbs free energy change on reaction, and the system Gibbs free energy, increase as the reaction proceeds. Therefore, the state of minimum Gibbs free energy of the system, which is the equilibrium state, is when $X = 0$ and no dissociation of the calcium carbonate occurs. However, if the partial pressure of carbon dioxide is maintained exactly at 1.635 atm and the system temperature at 1200 K, there is no

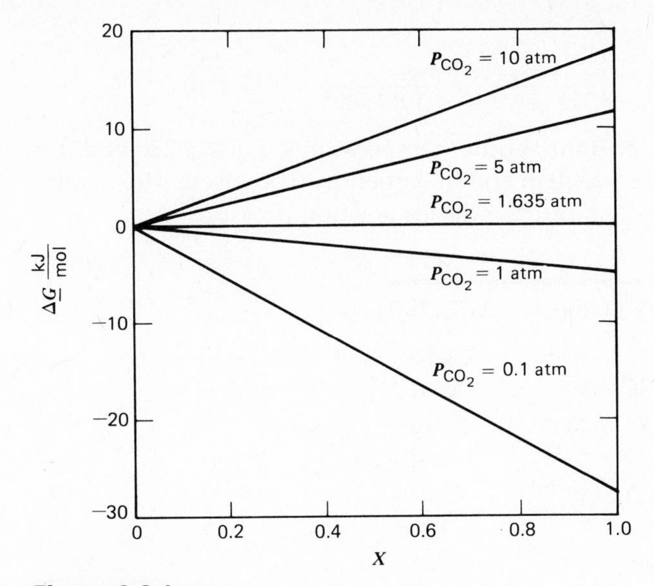

Figure 9.2-1
Gibbs free energy change for the heterogeneous reaction $CaCO_3 \rightarrow CaO + CO_2$ as a function of extent of reaction.

Gibbs free energy change on reaction, and any extent of reaction is allowed. Thus, independent of whether the reaction cell, because of previous history, contained calcium carbonate in an undecomposed or partially or fully decomposed state, that state would remain as there is no driving force for change. (You should compare Figs. 9.1-1 and 9.2-1 and understand the differences between the two.)

The equilibrium partial pressure of a gaseous species that results from the dissociation of a solid is called the **decomposition pressure** of the solid. In the foregoing discussion, this is 1.635 atm for calcium carbonate at 1200 K. As is evident from Illustration 9.2-2, the decomposition pressure of a solid is a strong function of temperature.

ILLUSTRATION 9.2-2

Compute the decomposition pressure of calcium carbonate over the temperature range of 298.15 K to 1400 K.

Data

$$\begin{aligned} C_{P,CaCO_3} &= 82.34 + 0.04975T - 1{,}287{,}000/T^2 \\ C_{P,CaO} &= 41.84 + 0.02025T - 451{,}870/T^2 \end{aligned} \quad \frac{J}{mol\ K}$$

Solution

From the preceding discussion we have

$$K_a = a_{CO_2} = P_{CO_2}/(P = 1 \text{ atm})$$

so that here the calculation of the decomposition pressure, P_{CO_2}, is equivalent to the calculation of the equilibrium constant as a function of temperature. From Appendix IV we have

$$\Delta G^\circ_{rxn}(T = 25°C) = \sum \nu_i \Delta \underline{G}^\circ_{f,i} = 32.24 \text{ kcal/mol} = 134.89 \text{ kJ/mol}$$

and

$$K_a(T = 25°\text{C}) = \exp\left\{-\frac{134{,}890\ \text{J/mol}}{8.314\ \text{J/mol K} \times 298.15\ \text{K}}\right\} = 2.33 \times 10^{-24}$$

To calculate the equilibrium constant at other temperatures, Eq. 9.1-23b and the heat capacity data given in this problem and in Appendix II are used. The results for both the equilibrium constant and the decomposition pressure are:

$T(K)$	K_a	P_{CO_2}(atm)	ΔG°_{rxn} (kJ)
298.15	2.33×10^{-24}	2.33×10^{-24}	134.89
400	3.34×10^{-16}	3.34×10^{-16}	118.51
600	2.78×10^{-8}	2.78×10^{-8}	86.78
800	2.33×10^{-4}	2.33×10^{-4}	55.64
1000	4.90×10^{-2}	4.90×10^{-2}	25.07
1200	1.64	1.64	−4.908
1400	18.98	18.98	−34.26

Note that the decomposition pressure changes by 25 orders of magnitude over the temperature range. ■

Another example of heterogeneous chemical equilibrium is the dissolution and dissociation of a weak salt in solution, most commonly aqueous solution. This process can be represented as

$$A_{\nu_+}B_{\nu_-}(\text{s}) = \nu_+ A^{z+}(\text{aq}) + \nu_- B^{z-}(\text{aq})$$

where the notation (s) and (aq) indicate solid and aqueous solution, respectively, and z_+ and z_- are the valences of the cation and anion. The equilibrium relation for this ionization process is

$$\begin{aligned} K_a &= \exp\left[-\frac{1}{RT}\{\nu_+\Delta\underline{G}^\circ_{f,A^{z+}}(\text{ideal, 1 molal}) + \nu_-\Delta\underline{G}^\circ_{f,B^{z-}}(\text{ideal, 1 molal}) \right. \\ &\qquad \left. - \Delta\underline{G}^\circ_{f,A_{\nu_+}B_{\nu_-}}(\text{solid})\}\right] \\ &= \frac{(a_{A^{z+}})^{\nu_+}(a_{B^{z-}})^{\nu_-}}{(a_{A_{\nu_+}B_{\nu_-}})} = (a_{A^{z+}})^{\nu_+}(a_{B^{z-}})^{\nu_-} \\ &= \frac{(M_{A^{z+}}\gamma^{\square}_{A^{z+}})^{\nu_+}(M_{B^{z-}}\gamma^{\square}_{B^{z-}})^{\nu_-}}{(M = 1\ \text{molal})^{\nu_+ + \nu_-}} = \frac{(M_{A^{z+}})^{\nu_+}(M_{B^{z-}})^{\nu_-}\gamma_\pm^{(\nu_+ + \nu_-)}}{(M = 1\ \text{molal})^{\nu_+ + \nu_-}} \end{aligned} \tag{9.2-5}$$

since $a_{A_{\nu_+}B_{\nu_-}}$, the activity of the pure, undissociated solid is unity, and we have taken the standard state of the ions to be an ideal one molal solution. Also, the mean activity coefficient, discussed in Sec. 7.9, has been used.

Common notation is to define the **solubility product** K_s by

$$K_s = (C_{A^{z+}})^{\nu_+}(C_{B^{z-}})^{\nu_-}$$

so that K_s and K_a are related as follows:[7]

$$K_a = K_s\gamma_\pm^{(\nu_+ + \nu_-)}/(M = 1\ \text{molal})^{(\nu_+ + \nu_-)}$$

[7]In writing this expression we have neglected the difference between concentration in kmol/m³ = moles/liter and molality.

or

$$K_s = K_a(M = 1 \text{ molal})^{(\nu_+ + \nu_-)} / \gamma_\pm^{(\nu_+ + \nu_-)}$$

If we now define K_s° to be the solubility in an ideal solution, we have

$$K_s^\circ = K_a(M = 1 \text{ molal})^{(\nu_+ + \nu_-)}$$

and

$$K_s = K_s^\circ / \gamma_\pm^{(\nu_+ + \nu_-)}$$

or

$$\ln K_s = \ln K_s^\circ - (\nu_+ + \nu_-) \ln \gamma_\pm \tag{9.2-6}$$

where one of Eqs. 7.9-15, 17, 18, or 19 is used for $\ln \gamma_\pm$, depending on the ionic strength. (You should compare the relation in Eq. 9.2-6 with Eq. 9.1-31.) At low ionic strengths, the Debye–Hückel limiting law applies, so that

$$\ln K_s = \ln K_s^\circ + (\nu_+ + \nu_-)|z_+ z_-|\alpha\sqrt{\mu} \tag{9.2-7}$$

where the ionic strength μ is given by

$$\mu = \frac{1}{2} \sum_{\text{ions}} z_i^2 M_i$$

The important feature of Eq. 9.2-7 is that it can be used to predict the solubility product of molecules when data on the free energy of formation of the ions are available (so that K_a and K_s° can be computed) or, when such data are not available, it can be used to interrelate the solubility products for the same molecule at different ionic strengths. These two types of calculations are demonstrated in Illustration 9.2-3.

ILLUSTRATION 9.2-3

The following data give the solubility of silver chloride in aqueous solutions of potassium nitrate at 25°C:

Concentration of KNO_3 (kmol/m^3)	**Concentration of AgCl at Saturation (kmol/m^3)**
0.0	1.273×10^{-5}
0.000509	1.311
0.009931	1.427
0.016431	1.469
0.040144	1.552

Source: S. Popoff and E. W. Neumann, *J. Phys. Chem.* **34,** 1853 (1930); E. W. Neumann, *J. Am. Chem. Soc.* **54,** 2195 (1932).

a. Compute the solubility product for silver chloride in each of these solutions.

b. Make an independent prediction of the solubility product of silver chloride in the absence of any potassium nitrate.

c. Show that the silver chloride solubility data satisfy Eq. 9.2-7, at least at low potassium nitrate concentrations, and find the numerical value of K_s°.

Solution

a. The solubility product for silver chloride is

$$K_s = C_{Ag^+}C_{Cl^-} = (C_{AgCl})^2$$

since here the molar concentration of silver ions and chloride ions are each equal to the molar concentration of dissolved silver chloride; that is, $C_{Ag^+} = C_{Cl^-} = C_{AgCl}$. The values of K_s and $\ln K_s$ are given in the table that follows.

b. To independently predict the solubility product for silver chloride, we first compute the thermodynamic equilibrium constant K_a from Eq. 9.2-5, Appendix IV, and Table 9.1-4. Thus

$$K_a = \exp\left[-\frac{4.184 \times \{18{,}430 + (-31{,}350) - (-25{,}980)\}}{8.314 \times 298.15}\right] = 2.670 \times 10^{-10}$$

and $K_s^\circ = 2.670 \times 10^{-10}$ (kmol/m^3)^2. Now using Eq. 9.2-7 we have

$$\ln K_s = 2 \ln C_{Ag^+} = \ln(2.670 \times 10^{-10}) + 2 \times 1.178\sqrt{C_{Ag^+}}$$

since, for this case $\mu = \frac{1}{2}(C_{Ag^+} + C_{Cl^-}) = C_{Ag^+}$, and $K_s = C_{Ag^+}C_{Cl^-} = (C_{Ag^+})^2$. This equation has the solution that, at saturation, $C_{Ag^+} = C_{Cl^-} = 1.634 \times 10^{-5}$ kmol/m^3, and $K_s = 2.67 \times 10^{-10}$ (kmol/m^3)^2. Thus our prediction leads to a silver-ion concentration and a chloride-ion concentration, at saturation, that are each about 30% too large, and a solubility product that is about 65% too large.

c. For the situation here

$$\mu = \frac{1}{2}\sum z_i^2 C_i = \frac{1}{2}\{C_{Ag^+} + C_{Cl^-} + C_{K^+} + C_{NO_3^-}\}$$

$$= \frac{1}{2}\{2C_{AgCl} + 2C_{KNO_3}\} = C_{AgCl} + C_{KNO_3}$$

and

$$\sqrt{\mu} = \sqrt{C_{AgCl} + C_{KNO_3}}$$

C_{KNO_3}, kmol/m^3	C_{AgCl}, kmol/m^3	K_s	$\ln K_s$	$\sqrt{\mu}$, (kmol/m^3)$^{1/2}$
0	1.273×10^{-5}	1.621×10^{-10}	−22.543	3.568×10^{-3}
0.000509	1.311	1.719	−22.484	2.285×10^{-2}
0.009931	1.427	2.036	−22.315	9.973×10^{-2}
0.016431	1.469	2.158	−22.257	1.282×10^{-1}
0.040144	1.552	2.409	−22.147	2.004×10^{-1}

The $\ln K_s$ is plotted versus $\sqrt{\mu}$ in Fig. 9.2-2, together with the dashed line, which has a slope of 2α, that was drawn to pass through the $C_{KNO_3} = 0$ datum point. Clearly, Eq. 9.2-7 is valid up to an ionic strength of about 0.0225 kmol/m^3; at higher ionic strengths there are deviations from the Debye–Hückel limiting law (Eq. 7.9-15) and, therefore, from Eq. 9.2-7.

Using Eq. 2 and the $C_{KNO_3} = 0.0$ datum point, we find that $\ln K_s^\circ = -22.5514$, so that the ideal solution solubility product is 1.607×10^{-10} (kmol/m^3)^2. ■

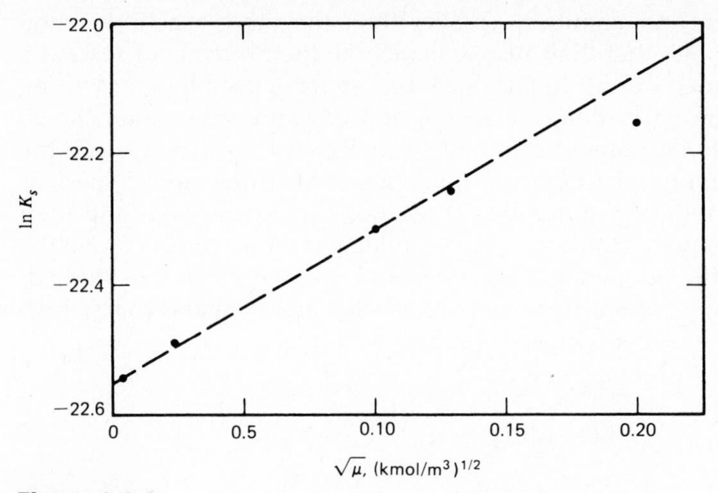

Figure 9.2-2
The solubility product of silver chloride as a function of the square root of the ionic strength in aqueous potassium nitrate solutions.

9.3 CHEMICAL EQUILIBRIUM WHEN SEVERAL REACTIONS OCCUR IN A SINGLE PHASE

The state of chemical equilibrium for $\mathcal{M}$ independent reactions occurring in a single phase is that state which satisfies the constraints on the system, the set of stoichiometric relations

$$N_i = N_{i,0} + \sum_{j=1}^{\mathcal{M}} \nu_{ij} X_j \qquad i = 1, 2, \ldots, \mathscr{C}$$

and the $\mathcal{M}$ equilibrium relations (cf. Sec. 6.8)

$$\sum_{i=1}^{\mathscr{C}} \nu_{ij} \overline{G}_i = 0 \qquad j = 1, 2, \ldots, \mathcal{M} \tag{9.3-1}$$

Using the notation introduced in this chapter, this last equation can be rewritten as

$$K_{a,j} = \prod_{i=1}^{\mathscr{C}} a_i^{\nu_{ij}} \qquad j = 1, 2, \ldots, \mathcal{M} \tag{9.3-2}$$

where $K_{a,j}$ is the equilibrium constant for the jth reaction. Since a collection of equations (which are nonlinear, and usually coupled) are to be solved here, rather than merely one equation as when only a single reaction occurs, equilibrium calculations for multiple reaction systems can be complicated.

Before proceeding to a sample calculation of a multireaction equilibrium state, it is useful to consider the simplifications that can be made in the analysis. First, the number of reactions that must be considered (and hence the number of simultaneous equations that must be solved) can frequently be reduced by taking into account the accuracy desired in the calculations and eliminating from consideration reactions that occur to such a small extent as to produce products with concentrations that are below our level of interest. Such reactions are usually recognized by very small equilibrium constants, equilibrium constants that are several orders of magnitude smaller than the other reactions being considered, or by the fact that their reactants include the products of reactions that occur only to a small extent. In fact, such reasoning is used here in writing only five reactions among carbon, hydrogen, and oxygen, rather than the 20 reactions among these substances that appear in Fig. 9.1-2 or, worse, all the reactions that occur in organic chemistry involving these three atomic species.

We can further reduce the number of reactions by identifying, and then only including in the equilibrium analysis, the independent reactions among the species. To see that the independent reactions and no others need be studied, consider the following set of reactions that occur between steam and coal (which we take to be pure carbon) at temperatures below 2000 K:

$$C + 2H_2O \rightarrow CO_2 + 2H_2 \qquad (rxn\ 1)$$

$$C + H_2O \rightarrow CO + H_2 \qquad (rxn\ 2)$$

$$C + CO_2 \rightarrow 2CO \qquad (rxn\ 3)$$

$$C + 2H_2 \rightarrow CH_4 \qquad (rxn\ 4)$$

$$CO + H_2O \rightarrow CO_2 + H_2 \qquad (rxn\ 5)$$

There are only three independent chemical reactions among these five reactions (see Problem 6.9c). This is easily demonstrated using Denbigh's method (cf. Sec. 6.3); we start by writing the reactions

$$2H + O \rightarrow H_2O \qquad (a)$$

$$C + 2O \rightarrow CO_2 \qquad (b)$$

$$C + O \rightarrow CO \qquad (c)$$

$$C + 4H \rightarrow CH_4 \qquad (d)$$

$$2H \rightarrow H_2 \qquad (e)$$

Since neither atomic hydrogen nor atomic oxygen are present below several thousand degrees, these two species are to be eliminated from the equations. Using first reaction (e) to eliminate atomic hydrogen (i.e., $H = \frac{1}{2}H_2$), and then reactions (a) and (e) to eliminate atomic oxygen (i.e., $O = H_2O - H_2$) yields

$$\begin{aligned} C + 2H_2O &= CO_2 + 2H_2 \\ C + H_2O &= CO + H_2 \\ C + 2H_2 &= CH_4 \end{aligned} \qquad (9.3\text{-}3)$$

These three reactions form a set of independent reactions among the six chemical species being considered.

Now the claim is that only these three reactions (or, equivalently, any other set of three independent reactions) need be considered in computing the equilibrium state for this system. This is easily verified by an examination of the five equilibrium relations

$$K_{a,1} = \exp\left[-\frac{(\Delta \underline{G}^\circ_{f,CO_2} + 2\Delta \underline{G}^\circ_{f,H_2} - \Delta \underline{G}^\circ_{f,C} - 2\Delta \underline{G}^\circ_{f,H_2O})}{RT}\right] = \frac{a_{CO_2} a_{H_2}^2}{a_C a_{H_2O}^2} \qquad (9.3\text{-}4a)$$

$$K_{a,2} = \exp\left[-\frac{(\Delta \underline{G}^\circ_{f,CO} + \Delta \underline{G}^\circ_{f,H_2} - \Delta \underline{G}^\circ_{f,C} - \Delta \underline{G}^\circ_{f,H_2O})}{RT}\right] = \frac{a_{CO} a_{H_2}}{a_C a_{H_2O}} \qquad (9.3\text{-}4b)$$

$$K_{a,3} = \exp\left[-\frac{(2\Delta \underline{G}^\circ_{f,CO} - \Delta \underline{G}^\circ_{f,C} - \Delta \underline{G}^\circ_{f,CO_2})}{RT}\right] = \frac{a^2_{CO}}{a_C a_{CO_2}} \tag{9.3-4c}$$

$$K_{a,4} = \exp\left[-\frac{(\Delta \underline{G}^\circ_{f,CH_4} - \Delta \underline{G}^\circ_{f,C} - 2\Delta \underline{G}^\circ_{f,H_2})}{RT}\right] = \frac{a_{CH_4}}{a_C a^2_{H_2}} \tag{9.3-4d}$$

and

$$K_{a,5} = \exp\left[-\frac{(\Delta \underline{G}^\circ_{f,CO_2} + \Delta \underline{G}^\circ_{f,H_2} - \Delta \underline{G}^\circ_{f,CO} - \Delta \underline{G}^\circ_{f,H_2O})}{RT}\right] = \frac{a_{CO_2} a_{H_2}}{a_{CO} a_{H_2O}} \tag{9.3-4e}$$

It is clear that if Eqs. 9.3-4a, b, and d are satisfied, Eqs. 9.3-4c and e will also be satisfied because these two equations are merely ratios of the other three; that is,

$$K_{a,3} = \frac{(K_{a,2})^2}{(K_{a,1})} = \frac{\left(\dfrac{a_{CO} a_{H_2}}{a_C a_{H_2O}}\right)^2}{\left(\dfrac{a_{CO_2} a^2_{H_2}}{a_C a^2_{H_2O}}\right)} = \frac{a^2_{CO}}{a_C a_{CO_2}} \tag{9.3-5a}$$

and

$$K_{a,5} = \frac{K_{a,1}}{K_{a,2}} = \frac{\left(\dfrac{a_{CO_2} a^2_{H_2}}{a_C a^2_{H_2O}}\right)}{\left(\dfrac{a_{CO} a_{H_2}}{a_C a_{H_2O}}\right)} = \frac{a_{CO_2} a_{H_2}}{a_{CO} a_{H_2O}} \tag{9.3-5b}$$

Thus we are led to the conclusion that it is not necessary to consider all five reactions in computing the equilibrium state of the reaction system, but merely the three independent reactions of Eqs. 9.3-3. In fact, this result could have been anticipated by observing that reactions 3 and 5 are linear combinations of the other reactions, that is

$$rxn\ 3 = 2\ (rxn\ 2) - rxn\ 1$$

and

$$rxn\ 5 = rxn\ 1 - rxn\ 2$$

In all the chemical equilibrium calculations that follow, we limit our attention to the independent chemical reactions among the reacting species.

ILLUSTRATION 9.3-1

Compute the equilibrium mole fractions of H_2O, CO, CO_2, H_2, and CH_4 in the steam–carbon system at a total pressure of 1 atm and over the temperature range of 600 to 1600 K.

Data

The equilibrium constants (based on the pure component standard states at $P =$ 1 atm and the reaction temperature) for the set of independent reactions

$$C + 2H_2O = CO_2 + 2H_2 \qquad (rxn\ 1)$$

$$C + H_2O = CO + H_2 \qquad (rxn\ 2)$$

$$C + 2H_2 = CH_4 \qquad (rxn\ 3)$$

are given in Fig. 9.3-1 or can be calculated using the BASIC program CHEMEQ.BAS of Appendix A9.1.

Solution

X_1, X_2, and X_3 will be used as the molar extents of reaction for the three reactions being considered. Basing all calculations on 1 mole of steam, we can construct the following mole balance table for the *gaseous* species (as long as there is sufficient solid carbon present to ensure equilibrium, it need not be considered, since it does not appear in the gas phase and, therefore, does not affect the activities of the other species).

	Number of Moles in the Gas Phase		Equilibrium Mole Fraction
	Initial	Final	
H_2O	1	$1 - 2X_1 - X_2$	$(1 - 2X_1 - X_2)/\Sigma$
CO_2	0	X_1	X_1/Σ
CO	0	X_2	X_2/Σ
H_2	0	$2X_1 + X_2 - 2X_3$	$(2X_1 + X_2 - 2X_3)/\Sigma$
CH_4	0	X_3	X_3/Σ
Total	1	$\Sigma = 1 + X_1 + X_2 - X_3$	

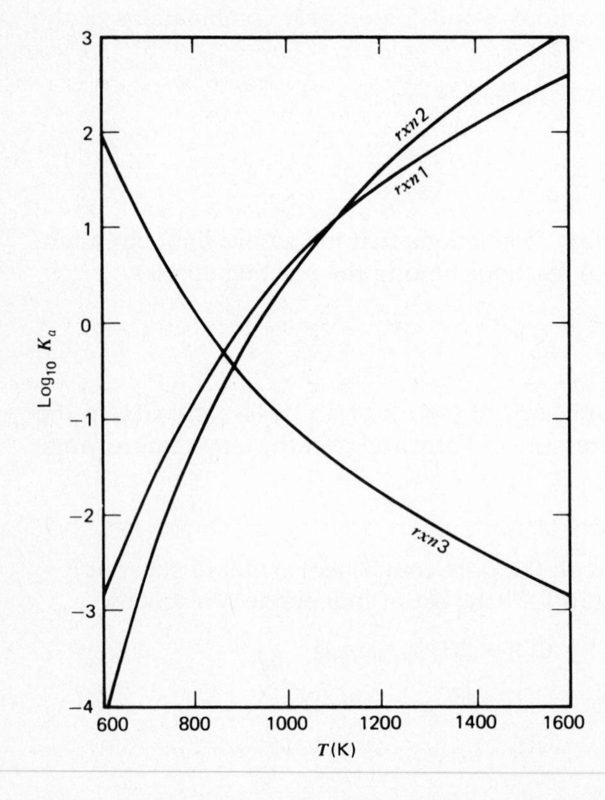

Figure 9.3-1
The equilibrium constants for the three independent reactions between coal and steam.

To solve for the three molar extents of reaction at 1 atm pressure we use the three equilibrium equations

$$K_{a,1} = \frac{a_{CO_2}a_{H_2}^2}{a_C a_{H_2O}^2} = \frac{(y_{CO_2})(y_{H_2})^2}{(y_{H_2O})^2} = \frac{X_1(2X_1 + X_2 - 2X_3)^2}{(1 - 2X_1 - X_2)^2(1 + X_1 + X_2 - X_3)} \quad \text{(a)}$$

$$K_{a,2} = \frac{a_{CO}a_{H_2}}{a_C a_{H_2O}} = \frac{y_{CO}y_{H_2}}{y_{H_2O}} = \frac{X_2(2X_1 + X_2 - 2X_3)}{(1 - 2X_1 - X_2)(1 + X_1 + X_2 - X_3)} \quad \text{(b)}$$

and

$$K_{a,3} = \frac{a_{CH_4}}{a_C a_{H_2}^2} = \frac{y_{CH_4}}{y_{H_2}^2} = \frac{X_3(1 + X_1 + X_2 - X_3)}{(2X_1 + X_2 - 2X_3)^2} \quad \text{(c)}$$

Since these equations are nonlinear, there will be more than one set of solutions for the molar extents of reaction. The only acceptable solution to this problem is one for which

$$2X_1 + X_2 \leq 1$$

and

$$0 \leq 2X_3 \leq 2X_1 + X_2$$

The first of these restrictions ensures that we do not use more steam than was supplied, and the second that no more hydrogen is used than has been produced.

Clearly, finding the solution to this problem is a nontrivial computational task, involving sophisticated numerical analytic techniques and a digital computer. The results of the calculation are given in Fig. 9.3-2. ■

The dissolution and ionization of a mixture of electrolytes provides another example of equilibrium in a multireaction system. To be specific, suppose two electrolytes $A_{\nu_A}B_{\nu_B}$ and $G_{\nu_G}H_{\nu_H}$ ionize in solution as follows:

$$A_{\nu_A}B_{\nu_B} = \nu_A A^{z_A} + \nu_B B^{z_B}$$

$$G_{\nu_G}H_{\nu_H} = \nu_G G^{z_G} + \nu_H H^{z_H}$$

The equilibrium relations for these ionization processes (see Eqs. 9.1-31 and 9.2-6) are

$$\ln K_{AB} = \ln (C_A^{\nu_A}C_B^{\nu_B}) = \ln K_{AB}^{\circ} - (\nu_A + \nu_B) \ln \gamma_{\pm} \quad (9.3\text{-}6)$$

and

$$\ln K_{GH} = \ln (C_G^{\nu_G}C_H^{\nu_H}) = \ln K_{GH}^{\circ} - (\nu_G + \nu_H) \ln \gamma_{\pm} \quad (9.3\text{-}7)$$

where K_{AB} and K_{GH} are the equilibrium ratios of Eqs. 9.1-31 and 9.2-6. The difficulty in solving these equations to find the equilibrium state is that even if there is no common ion among the electrolytes, the equations are coupled by the fact that the activity coefficient $\gamma_{\pm}$ is a function of the total ionic strength μ,

$$\mu = \frac{1}{2} \sum_{\text{ions}} z_i^2 C_i = \frac{1}{2} [z_A^2 C_A + z_B^2 C_B + z_G^2 C_G + z_H^2 C_H] \quad (9.3\text{-}8)$$

and thus depends on the concentration of all the ions present.

If there is an ion that is common to both electrolytes, the solubility and extent of ionization of each may be much more affected by the presence of the

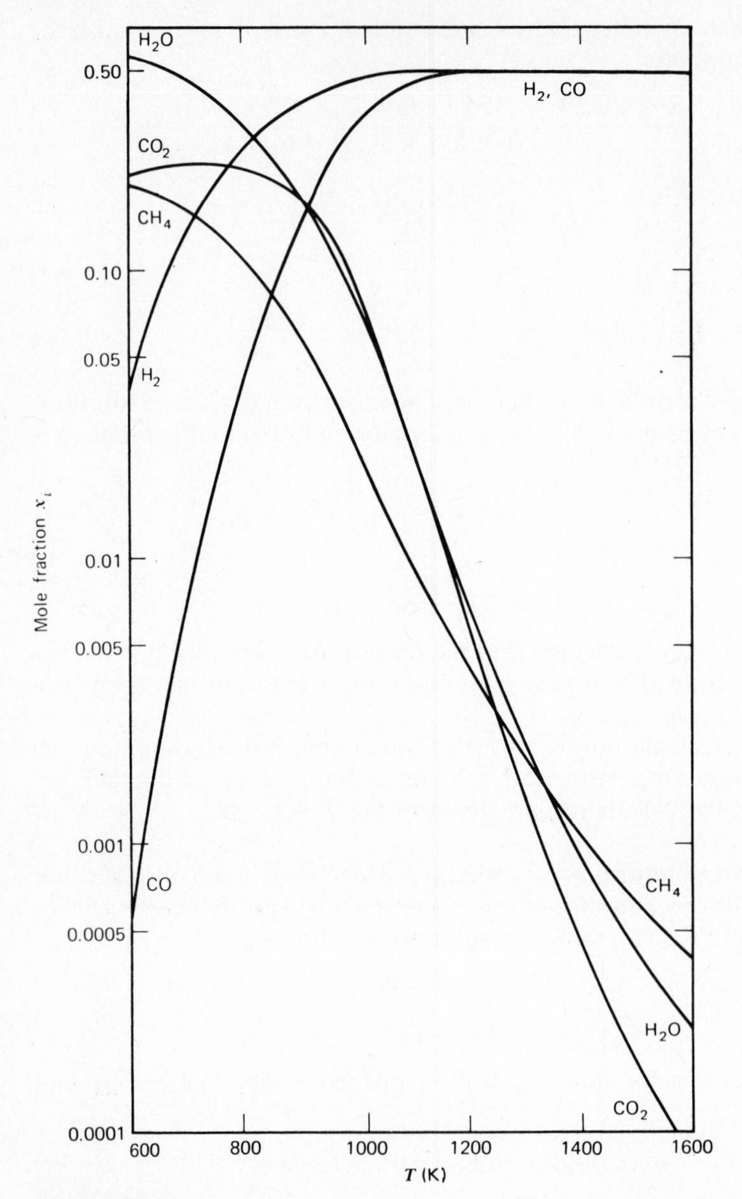

Figure 9.3-2
The equilibrium mole fractions for the coal–steam reactions considered in Illustration 9.3-1.

other than would be the case with only an ionic-strength coupling. This phenomenon, known as the **common ion effect**, is demonstrated in Illustration 9.3-2.

ILLUSTRATION 9.3-2

The solubility of silver chloride in water at 25°C is 1.273×10^{-5} kmol/m^3, and that of thallium chloride is 0.144 kmol/m^3. Estimate the simultaneous solubility of AgCl and TlCl in water.

Solution

The first step in the calculation is to compute the constants K°_{AgCl} and K°_{TlCl} using the solubility data for each of the pure salts. Recognizing that $\nu_{Ag} = \nu_{Tl} = \nu_{Cl} = 1$ and that $z_{Ag} = z_{Tl} = +1$ and $z_{Cl} = -1$, we write

$$\ln K^\circ_{AgCl} = \ln (C_{Ag}C_{Cl}) + 2 \ln \gamma_\pm = 2 \ln C_{AgCl} + 2 \ln \gamma_\pm \tag{a}$$

and

$$\ln K^\circ_{TlCl} = \ln (C_{Tl}C_{Cl}) + 2 \ln \gamma_\pm = 2 \ln C_{TlCl} + 2 \ln \gamma_\pm \tag{b}$$

Also, for the pure salts, μ(silver chloride) $= 1.273 \times 10^{-5}$ kmol/m^3 and μ(thallium chloride) $= 0.144$ kmol/m^3. Because of the high ionic strengths, Eq. 7.9-18 will be used to compute $\gamma_\pm$; that is,

$$\begin{aligned}\ln \gamma_\pm &= \frac{-1.178|z_+z_-|\sqrt{\mu}}{1 + \sqrt{\mu}} + 0.1|z_+z_-|\mu \\ &= \frac{-1.178\sqrt{\mu}}{1 + \sqrt{\mu}} + 0.1\mu \end{aligned} \tag{c}$$

Using Eq. c and the experimental solubility data in Eqs. a and b yields

$$K^\circ_{AgCl} = 1.607 \times 10^{-10}\ (\text{kmol/m}^3)^2$$

and

$$K^\circ_{TlCl} = 1.116 \times 10^{-2}\ (\text{kmol/m}^3)^2$$

To find the simultaneous solubility of silver chloride and thallium chloride, we must solve the equations

$$\ln (C_{Ag}C_{Cl}) = \ln K^\circ_{AgCl} - 2 \ln \gamma_\pm$$

and

$$\ln (C_{Tl}C_{Cl}) = \ln K^\circ_{TlCl} - 2 \ln \gamma_\pm$$

or, using Eq. c, the equations

$$1.607 \times 10^{-10} \exp \left[\frac{2.356\sqrt{\mu}}{1 + \sqrt{\mu}} - 0.2\mu\right] = C_{Ag}C_{Cl}$$

$$1.116 \times 10^{-2} \exp \left[\frac{2.356\sqrt{\mu}}{1 + \sqrt{\mu}} - 0.2\mu\right] = C_{Tl}C_{Cl}$$

where

$$\mu = \tfrac{1}{2}(C_{Ag} + C_{Tl} + C_{Cl}) = C_{AgCl} + C_{TlCl}$$

since $C_{Ag} + C_{Tl} = C_{Cl}$. These nonlinear equations can be solved by trial and error.

Because of the low solubility of silver chloride, as a first guess, we will assume that the solubility of thallium chloride is unaffected by the presence of silver chloride, and that the ionization of silver chloride has little effect on either μ or the total chloride ion concentration. In this case we have

$$C_{Tl} = C_{Cl} = 0.144\ \text{kmol/m}^3$$

$$\mu = 0.144\ \text{kmol/m}^3 \quad \text{and} \quad \sqrt{\mu} = 0.3975\ (\text{kmol/m}^3)^{1/2}$$

The silver ion concentration (and the solubility of silver chloride) can then be computed from

$$1.607 \times 10^{-10} \exp\left[\frac{2.356 \times 0.3975}{1.3975} - 0.2 \times 0.144\right] = C_{Ag}(0.144)$$

so that

$$C_{Ag} = C_{AgCl} = 2.119 \times 10^{-9} \text{ kmol/m}^3$$

Therefore, as we assumed, AgCl is so slightly soluble in the aqueous TlCl solution that neither the ionic strength nor the solubility of thallium chloride is greatly affected by its presence. On the other hand, owing to the common ion effect, the solubility of silver chloride is reduced by almost four orders of magnitude from the value when only AgCl is present.

Comment

If, instead of thallium chloride, some other salt that does not have a common ion with AgCl, for example, $NaNO_3$, had been introduced into the aqueous solution, the solubility of silver chloride would have increased because of the ionic strength dependence of the ionic activity coefficient. Thus, if $NaNO_3$ were present in the same concentration as, but instead of, thallium chloride, we would have

$$C_{Ag}C_{Cl} = 1.607 \times 10^{-10} \exp\left\{\frac{2.356 \times 0.3975}{1.3975} - 0.2 \times 0.144\right\} = C_{Ag}^2 = C_{Cl}^2$$

or

$$C_{Ag} = C_{Cl} = C_{AgCl} = 1.75 \times 10^{-5} \text{ kmol/m}^3$$

which is an increase of 37% in the solubility of silver chloride over the value when only AgCl is present. ■

9.4 COMBINED CHEMICAL AND PHASE EQUILIBRIUM

In Sec. 6.8 we established that the conditions for equilibrium when chemical reaction and phase equilibrium occur simultaneously are that (1) each species must be in phase equilibrium among all the phases; that is,

$$\bar{G}_i^{\text{I}} = \bar{G}_i^{\text{II}} = \cdots = \bar{G}_i^{\mathcal{P}} = \bar{G}_i \qquad i = 1, 2, \ldots, \mathcal{C} \tag{9.4-1}$$

(2) each chemical reaction must be in chemical equilibrium in each phase

$$\sum_{i=1}^{\mathcal{C}} \nu_{ij}\bar{G}_i = 0 \qquad \text{for all independent reactions } j = 1, 2, \ldots, \mathcal{M} \tag{9.4-2}$$

and (3) the stoichiometric and state variable constraints on the system must be satisfied. In fact, the equality of the partial molar Gibbs free energy of each species in all phases at equilibrium (Eq. 9.4-1) assures that if the chemical equilibrium criterion is satisfied in any one phase, it will be satisfied in all phases. Thus, in computations it is only necessary to seek a solution for which

$$\bar{f}_i^{\text{I}} = \bar{f}_i^{\text{II}} = \cdots = \bar{f}_i^{\mathcal{P}} = \bar{f}_i \qquad i = 1, 2, \ldots, \mathcal{C} \tag{9.4-3}$$

in all phases (which follows from Eq. 9.4-1), and for which

$$K_{a,j} = \prod_{i=1}^{\mathscr{C}} a_i^{\nu_{ij}} \qquad j = 1, 2, \ldots, \mathscr{M} \tag{9.4-4}$$

is satisfied in any one phase[8] (which follows from Eq. 9.4-2).

The prediction of a state of combined chemical and phase equilibrium using these equations can be complicated because of the large number of nonlinear equations that must be solved simultaneously. Frequently, the equations involved can be simplified or reduced in number by recognizing that some species are only slightly soluble in certain phases, and that some reactions may go virtually to completion or do not measurably proceed at all in some phases. However, even with such simplifications, the calculations are likely to be tedious, as indicated in Illustration 9.4-1.

ILLUSTRATION 9.4-1

One mole of nitrogen, 3 moles of hydrogen, and 5 moles of water are placed in a closed container maintained at 25°C and 13.33 kPa and, using the appropriate catalyst and stirring, allowed to attain phase and chemical equilibrium. Assuming that the liquid and vapor phases are ideal, compute the amount and composition of each phase at equilibrium (neglecting the aqueous phase reaction of ammonia to form NH_4OH, and its subsequent ionization). The following data are available:

Vapor pressure of water at 25°C = 3.167 kPa

Henry's law constants

$$\begin{aligned} N_2 \text{ in } H_2O: &\quad H_{N_2} = 9.224 \times 10^7 \text{ kPa} \\ H_2 \text{ in } H_2O: &\quad H_{H_2} = 7.158 \times 10^7 \text{ kPa} \\ NH_3 \text{ in } H_2O: &\quad H_{NH_3} = 97.58 \text{ kPa} \end{aligned}$$

(The Henry's law constant for ammonia was computed from experimental data using the equation $P_{NH_3} = H_{NH_3} x_{NH_3}$ and assuming all the ammonia asorbed to be present as NH_3.)

Solution

The only reaction that can occur between the species at 25°C is the formation of ammonia

$$\tfrac{1}{2}N_2 + \tfrac{3}{2}H_2 = NH_3$$

for which

$$\begin{aligned} \Delta G_{rxn} &= \sum \nu_i \Delta \underline{G}^\circ_{f,i} \\ &= \Delta \underline{G}^\circ_{f,NH_3} - \tfrac{1}{2}\Delta \underline{G}^\circ_{f,N_2} - \tfrac{3}{2}\Delta \underline{G}^\circ_{f,H_2} = -3903 \text{ cal/mol} = -16330 \text{ J/mol} \end{aligned}$$

and

$$K_a = \exp\left[-\frac{\Delta G_{rxn}}{RT}\right] = 726.2$$

[8]In certain instances it may be convenient to choose different states of aggregation as the standard states for the various species in a reaction, usually because of the availability of ΔG_f° data. In such cases the activity of each species is evaluated in that phase which corresponds most closely to the state of aggregation of the standard state. (See the discussion of Eqs. 9.4-5 and 9.)

Consequently, at chemical equilibrium the equation[9]

$$K_a = \frac{a_{NH_3}}{a_{N_2}^{1/2} a_{H_2}^{3/2}} = \frac{y_{NH_3}}{y_{N_2}^{1/2} y_{H_2}^{3/2}(P/1 \text{ atm})} = 726.2 \tag{1}$$

must be satisfied. In writing the equation in this way we have assumed, since the pressure is so low, that the gas phase is ideal.

Also, the phase equilibrium relation

$$\bar{f}_i^V = \bar{f}_i^L$$

must be satisfied for each species. For N_2, H_2, and NH_3, this leads to the following equations

$$P_i = y_i P = H_i x_i$$

or

$$y_i = H_i x_i / P$$

Thus

$$y_{N_2} = 6.92 \times 10^5 x_{N_2} \tag{2}$$

$$y_{H_2} = 5.37 \times 10^5 x_{H_2} \tag{3}$$

and

$$y_{NH_3} = 7.32 x_{NH_3} \tag{4}$$

For water we have

$$y_{H_2O} P = x_{H_2O} P_{H_2O}^{vap}$$

or

$$y_{H_2O} = \frac{P_{H_2O}^{vap} x_{H_2O}}{P} = 0.2376 x_{H_2O} \tag{5}$$

The mass balance constraints are most easily taken into account using the mole numbers of each species in each phase as the independent variables rather than the mole fractions. Thus we have

$$N_{H_2O} = 5 \text{ mol} = N_{H_2O}^L + N_{H_2O}^V \tag{6}$$

$$N_{NH_3} = X = N_{NH_3}^L + N_{NH_3}^V \tag{7}$$

$$N_{N_2} = (1 - X/2) = N_{N_2}^L + N_{N_2}^V \tag{8}$$

and

$$N_{H_2} = 3(1 - X/2) = N_{H_2}^L + N_{H_2}^V \tag{9}$$

where N_i^J is the number of moles of species i in phase J. Also letting L equal the total number of moles present in the liquid phase at equilibrium, and V equal the total number of moles in the vapor phase, we have

$$L = N_{H_2O}^L + N_{NH_3}^L + N_{N_2}^L + N_{H_2}^L \tag{10}$$

[9] 1 atm = 1.013 bar = 101.3 kPa.

and

$$V = N_{H_2O}^V + N_{NH_3}^V + N_{N_2}^V + N_{H_2}^V \tag{11}$$

Using Eqs. 6 through 11 in Eqs. 2 through 5 yields

$$N_{N_2}^V = 6.92 \times 10^5\, N_{N_2}^L\, V/L \tag{12}$$

$$N_{H_2}^V = 5.37 \times 10^5\, N_{H_2}^L\, V/L \tag{13}$$

$$N_{NH_3}^V = 7.32 N_{NH_3}^L\, V/L \tag{14}$$

and

$$N_{H_2O}^V = 0.2376 N_{H_2O}^L\, V/L \tag{15}$$

From these equations it is evident that the number of moles of both nitrogen and hydrogen in the liquid phase must be small. For what follows we will assume that $N_{H_2}^L = N_{N_2}^L \approx 0$, and later verify that this is, in fact, the case.

Recognizing that nitrogen and hydrogen appear only in the vapor phase, we then have, by stoichiometry, that $N_{H_2}^V = 3N_{N_2}^V$, so that Eq. 1 becomes

$$\frac{N_{NH_3}^V}{3^{3/2}(N_{N_2}^V)^2(P/1\text{ atm})} = 726.2$$

or since $P = 13.33$ kPa

$$N_{N_2}^V = 0.0449\sqrt{N_{NH_3}^V\, V} \tag{16}$$

Equations 6 through 11 and 14 through 16 are to be solved for the nine unknowns $N_{H_2O}^V$, $N_{N_2}^V$, $N_{H_2}^V$, $N_{NH_3}^V$, $N_{NH_3}^L$, $N_{H_2O}^L$, L, V, and X. We will solve these equations by trial and error using the procedure indicated in Fig. 9.4-1.

Since K_a is so large in numerical value, our first guess is that the reaction goes to completion so that $X = 2$. Thus,

$$L + V = 5 + X + 4(1 - X/2) = 9 - X = 7\text{ mol} \tag{17}$$

Also

$$N_{NH_3} = 2\text{ mol} \qquad N_{H_2} = 0\text{ mol}$$

$$N_{H_2O} = 2\text{ mol} \qquad N_{N_2} = 0\text{ mol}$$

and

$$N_{NH_3} = 2\text{ mol} = N_{NH_3}^V + N_{NH_3}^L = (1 + 7.32V/L)N_{NH_3}^L \tag{18}$$

$$N_{H_2O} = 5\text{ mol} = N_{H_2O}^V + N_{H_2O}^L = (1 + 0.2376V/L)N_{H_2O}^L \tag{19}$$

Our first guess for V is 2 moles, thus $L = 5$ moles. Using this guess in Eqs. 18 and 19 yields

$$N_{NH_3}^L = 0.5092\text{ mol} \qquad N_{H_2O}^L = 4.5662\text{ mol}$$

$$N_{NH_3}^V = 1.4908\text{ mol} \qquad N_{H_2O}^V = 0.4338\text{ mol}$$

and

$$L = 0.5092 + 4.5662 = 5.0754\text{ mol}$$

$$V = 1.4908 + 0.4338 = 1.9246\text{ mol}$$

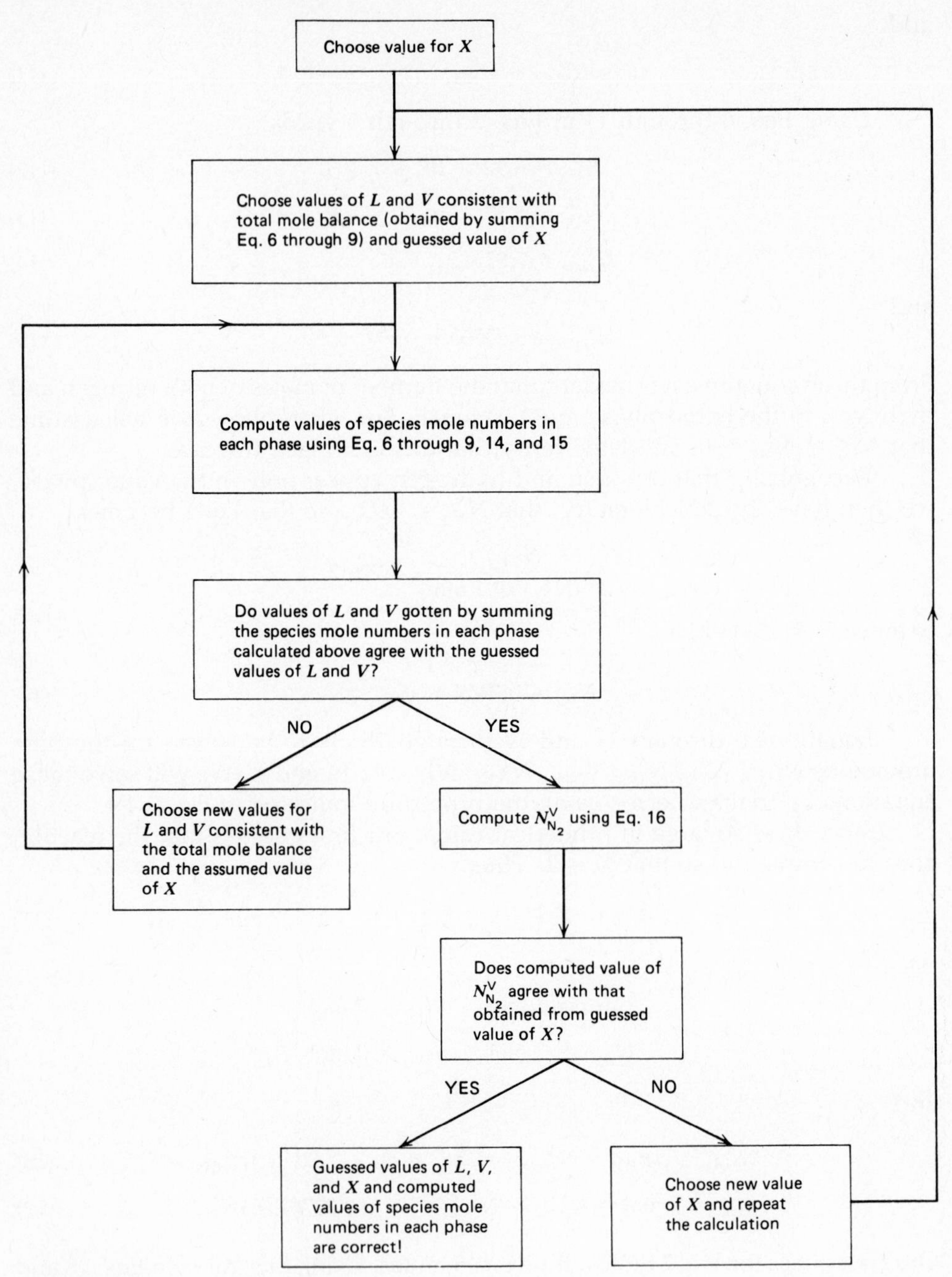

Figure 9.4-1
Logic diagram for solution of Eqs. 6 through 11 and 14 through 16.

and

Our next guess is $V = 1.85$ mol. In this case $L = 5.15$ mol, and

$$N_{NH_3}^L = 0.5510 \text{ mol} \qquad N_{H_2O}^L = 4.6068 \text{ mol}$$

$$N_{NH_3}^V = 1.4490 \text{ mol} \qquad N_{H_2O}^V = 0.3932 \text{ mol}$$

and

$$L = 5.1578 \text{ mol}$$

$$V = 1.8422 \text{ mol}$$

These values of L and V are sufficiently close to the assumed values that they will be taken as being correct for the guessed value of X. It is necessary to check the guess for X. Using Eq. 16 yields

$$N_{N_2}^V = (1 - \tfrac{1}{2}X) = 0.0449\sqrt{1.4490 \times 1.8422} \text{ mol} = 0.0734 \text{ mol}$$

so that $X = 1.8532$ mol. We now have to correct the guess for X and repeat the calculation.

After a number of iterations we find that for assumed values of

$$X = 1.84 \text{ mol}$$

$$V = 2.20 \text{ mol}$$

$$L = 4.96 \text{ mol}$$

we obtain

$$N_{NH_3}^L = 0.433 \text{ mol} \qquad N_{NH_3}^V = 1.407 \text{ mol}$$

$$N_{H_2O}^L = 4.524 \text{ mol} \qquad N_{H_2O}^V = 0.477 \text{ mol}$$

$$N_{N_2}^V = 0.080 \text{ mol}$$

$$N_{H_2}^V = 0.240 \text{ mol}$$

so that $L = 4.956$ mol, $V = 2.204$ mol, and, from Eq. 16, $X = 1.842$ mol. Since these are close to the assumed values, another iteration is not needed.

It is also necessary to check the assumption that $N_{N_2}^L$ and $N_{H_2}^L$ are negligible. Using Eqs. 12 and 13 and the foregoing results, this is clearly the case, so that the solution is complete. The results are summarized in the following table.

	Liquid Phase		Vapor Phase	
Species	**Moles**	**Mole Fraction**	**Moles**	**Mole Fraction**
H_2O	4.523	0.913	0.477	0.216
NH_3	0.433	0.087	1.407	0.638
N_2	0.26×10^{-6}	0	0.080	0.036
H_2	0.10×10^{-5}	0	0.240	0.109
Total	4.956		2.204	

■

For relatively simple combined chemical and phase equilibrium problems, such as the one just given, hand calculations are possible. More complicated problems, especially those involving multiple reactions (even in a single phase) are best solved by digital computation. In fact, several prepared computer programs are available[10] to both generate all the thermodynamic data for the reactant and product species and identify the state of thermodynamic equilibrium. The computation algorithm used in these programs is different than the procedures discussed here in that the chemical equilibrium constant concept is not used. Instead, a general expression is obtained for the Gibbs free energy in terms of all the species and phases that may be present, and this function is minimized, by direct search, subject to the state variable and mass balance constraints on the system. Computationally, this method is more efficient than setting up and solving a large collection of nonlinear equilibrium equations. The two procedures are theoretically equivalent and must lead to the same result. (For the reaction considered in Fig. 9.1-1, the direct search would involve an iterative calculation of X^* by searching for the value of X for which G was a minimum, whereas in the equilibrium constant approach X^* is found as the solution to a nonlinear algebraic equation that results from analytically identifying the state of minimum Gibbs free energy.)

We conclude this section by noting that, in some cases, states involving both chemical and phase equilibrium can be considered to be chemical equilibrium problems only, but with different states of aggregation for the standard states of the various species present. Consider, for example, the dissolution of gaseous ammonia in water, and its subsequent reaction to form ammonium hydroxide. This process can be considered either to occur in two steps, the first involving phase equilibrium

$$\mathrm{NH_3(g) = NH_3(aqueous\ solution)}$$

and the second single-phase chemical equilibrium

$$\mathrm{NH_3(aqueous\ solution) + H_2O(l) = NH_4OH(aqueous\ solution)}$$

or to occur as a single-step multiphase chemical equilibrium process

$$\mathrm{NH_3(g) + H_2O(l) = NH_4OH(aqueous\ solution)}$$

We will now establish that the same equilibrium state will be found independent of which of the two descriptions is used.

In the second case we take the standard state of ammonia to be the pure gas, of water to be the pure liquid, and of ammonium hydroxide to be the ideal 1 molal solution, and obtain:

$$K_{A,\mathrm{II}} = \exp\left\{-\frac{1}{RT}\left[\Delta\underline{G}^\circ_{f,\mathrm{NH_4OH}}(\text{ideal, 1 molal}) - \Delta\underline{G}^\circ_{f,\mathrm{H_2O}}(\text{liquid}) - \Delta\underline{G}^\circ_{f,\mathrm{NH_3}}(\text{gas, 1 atm})\right]\right\}$$

$$= \frac{a_{\mathrm{NH_4OH}}}{a_{\mathrm{NH_3}}a_{\mathrm{H_2O}}} = \frac{\left(\dfrac{M_{\mathrm{NH_4OH}}\gamma^{\Box}_{\mathrm{NH_4OH}}}{M = 1\ \text{molal}}\right)}{\left(\dfrac{P_{\mathrm{NH_3}}}{1\ \text{atm}}\right)(x_{\mathrm{H_2O}}\gamma_{\mathrm{H_2O}})} \tag{9.4-5}$$

[10]For a general discussion see *Chemical Reaction Equilibrium Analysis: Theory and Algorithms* by W. R. Smith and R. W. Missen, J. Wiley, New York, 1982. Also *Fortran IV Computer Program for Calculation of Thermodynamic and Transport Properties of Complex Chemical Systems* by R. A. Svehla and B. J. McBride, National Aeronautics and Space Administration Technical Note D-7056, January 1973; *Rand's Chemical Composition Program*, The Rand Corporation, Santa Monica, California; and others.

In the two-step description, the standard state of ammonia would be taken to be the ideal 1 molal solution so that

$$K_{A,\mathrm{I}} = \exp\left\{-\frac{1}{RT}\left[\Delta \underline{G}^{\circ}_{f,\mathrm{NH_4OH}}(\text{ideal, 1 molal}) - \Delta \underline{G}^{\circ}_{f,\mathrm{H_2O}}(\text{liquid}) - \Delta \underline{G}^{\circ}_{f,\mathrm{NH_3}}(\text{ideal, 1 molal})\right]\right\}$$

$$= \frac{a_{\mathrm{NH_4OH}}}{a_{\mathrm{NH_3}}a_{\mathrm{H_2O}}} = \frac{\left(\dfrac{M_{\mathrm{NH_4OH}}\gamma^{\square}_{\mathrm{NH_4OH}}}{M = 1\text{ molal}}\right)}{\left(\dfrac{M_{\mathrm{NH_3}}\gamma^{\square}_{\mathrm{NH_3}}}{M = 1\text{ molal}}\right)(x_{\mathrm{H_2O}}\gamma_{\mathrm{H_2O}})} \tag{9.4-6}$$

Clearly, $K_{A,\mathrm{I}} \neq K_{A,\mathrm{II}}$. However, Eq. 9.4-6 must be solved together with the phase equilibrium requirement

$$\bar{f}^{\mathrm{V}}_{\mathrm{NH_3}} = \bar{f}^{\mathrm{L}}_{\mathrm{NH_3}}$$

or equivalently,

$$\bar{G}^{\mathrm{V}}_{\mathrm{NH_3}} = \bar{G}^{\mathrm{L}}_{\mathrm{NH_3}}$$

we will use the latter of these two relations. Next, we note that

$$\bar{G}^{\mathrm{V}}_{\mathrm{NH_3}}(T, P, y_{\mathrm{NH_3}}) = \bar{G}^{\mathrm{V}}_{\mathrm{NH_3}}(T, P_{\mathrm{NH_3}}) = \underline{G}^{\mathrm{V}}_{\mathrm{NH_3}}(T, P = 1\text{ atm}) + RT\ln\left[\frac{P_{\mathrm{NH_3}}}{1\text{ atm}}\right] \tag{9.4-7}$$

$$\bar{G}^{\mathrm{L}}_{\mathrm{NH_3}}(T, M_{\mathrm{NH_3}}) = \bar{G}^{\mathrm{L}}_{\mathrm{NH_3}}(T, \text{ideal 1 molal}) + RT\ln\left[\frac{M_{\mathrm{NH_3}}\gamma^{\square}_{\mathrm{NH_3}}}{M_{\mathrm{NH_3}} = 1}\right] \tag{9.4-8}$$

and that

$$\begin{aligned}
&\Delta \underline{G}^{\circ}_{f,\mathrm{NH_3}}(T, \text{ideal 1 molal}) + \underline{G}^{\mathrm{V}}_{\mathrm{NH_3}}(T, P = 1\text{ atm}) - \bar{G}^{\mathrm{L}}_{\mathrm{NH_3}}(T, \text{ideal 1 molal})\\
&= \bar{G}^{\mathrm{L}}_{\mathrm{NH_3}}(T, \text{ideal 1 molal}) - \tfrac{1}{2}\underline{G}^{\mathrm{V}}_{\mathrm{N_2}}(T, P = 1\text{ atm}) - \tfrac{3}{2}\underline{G}^{\mathrm{V}}_{\mathrm{H_2}}(T, P = 1\text{ atm})\\
&\quad + \underline{G}^{\mathrm{V}}_{\mathrm{NH_3}}(T, P = 1\text{ atm}) - \bar{G}^{\mathrm{L}}_{\mathrm{NH_3}}(T, \text{ideal 1 molal})\\
&= \Delta \underline{G}^{\circ}_{f,\mathrm{NH_3}}(\text{gas}, T, P = 1\text{ atm})
\end{aligned} \tag{9.4-9}$$

Using Eqs. 9.4-7, 8, and 9 in the equilibrium relation of Eq. 9.4-6 yields Eq. 9.4-5. Thus, precisely the same equilibrium relation is found between the ammonia partial pressure in the gas phase and the ammonium hydroxide concentration in the liquid phase independent of the manner in which we presume the absorption-reaction process to take place. Consequently, it is a matter of convenience whether we consider multiphase reactions such as the one here to be chemical equilibrium problems or problems of combined chemical and phase equilibrium.[11]

9.5 THE BALANCE EQUATIONS FOR A TANK-TYPE CHEMICAL REACTOR

The design of a chemical reactor, that is, choosing its type, size, shape, and conditions of operation, is largely determined by the kinetics of the chemical

[11] In general, considering the process to be a combined chemical and phase equilibrium problem will result in more complicated calculations, but will yield somewhat more information; here the concentration of ammonia in the liquid phase.

reactions[12] (i.e., the rates at which reactions occur and whether a catalyst is needed) and the rate at which heat is produced or absorbed during the reaction. Thermodynamics can be useful in reactor design. In particular, the multicomponent mass and energy balances of thermodynamics can provide useful information on the energy requirements and temperature programming in reactor operations. Also, through the use of computational techniques discussed earlier in this chapter, thermodynamics can provide information on the maximum (equilibrium) conversions that can be obtained with any reactor at the given operating conditions. Although it is not our intention to study reactor design in great detail here, we will try, in this section and the next, to establish the relationship between the thermodynamic balance equations and those of reaction engineering.

In this section we are concerned with analysis of tank-type reactors used for liquid-phase reactions. A schematic diagram of a tank reactor is given in Fig. 9.5-1. To develop a quantitative description of such reactors, we will assume that its contents are well mixed, as is the case with many industrial reactors, so that the species concentrations and the temperature are uniform throughout the reactor. *We will not assume that the reactor exit stream is in chemical equilibrium, since industrial reactors generally do not operate in such a manner.* Using the entries of Table 6.4-1, the species mass and total energy balances for a reactor with one inlet

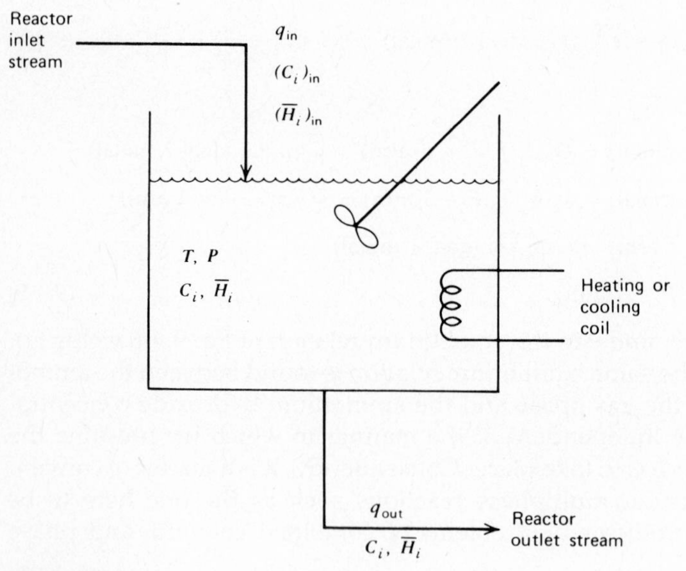

Figure 9.5-1
A schematic diagram of a simple stirred tank reactor.

[12]The relationship between the rate of a chemical reaction and the species concentrations cannot be predicted and must be determined from experiment. Detailed discussions of the analysis of experimental reaction rate data to get the constitutive equation relating the reaction rate to species concentrations are given in *Kinetics and Mechanism*, 3rd ed., by J. W. Moore and R. G. Pearson, J. Wiley & Sons, New York, 1981, Chaps. 2 and 3; *Chemical Reaction Engineering*, 2nd ed., by O. Levenspiel, J. Wiley & Sons, New York, 1972, Chap. 3; and *Introduction to Chemical Engineering Analysis*, by T. W. F. Russell and M. M. Denn, J. Wiley & Sons, New York, 1972, Chap. 5.

and one outlet stream are, respectively

$$\frac{dN_i}{dt} = (\dot{N}_i)_{\text{in}} - (\dot{N}_i)_{\text{out}} + \sum_{j=1}^{\mathcal{M}} \nu_{ij} \frac{dX_j}{dt} \tag{9.5-1}$$

and

$$\frac{dU}{dt} = \sum_{i}^{\mathscr{C}} (\dot{N}_i\overline{H}_i)_{\text{in}} - \sum_{i}^{\mathscr{C}} (\dot{N}_i\overline{H}_i)_{out} + \dot{Q} \tag{9.5-2}$$

(The generalization of these equations to multiple feed streams is simple, and is left to you.) In writing the energy balance equation, the kinetic and potential energy, shaft work, and $P(dV/dt)$ terms have been neglected, because these terms are usually of little importance compared to the thermal energy and chemical reaction terms. Also, since the contents of the reactor are of uniform temperature and composition (by the "well mixed" assumption), the species concentrations and temperature of the exit stream are the same as those of the reactor contents.

With several small changes in notation the mass and energy balances of Eqs. 9.5-1 and 2 can be made to look more like those commonly used in reactor analysis. By letting q_{in} and q_{out} represent the volumetric flow rates into and out of the reactor, C_i be the molar concentration of species i, V be the fluid volume in the reactor, and $r_j = (1/V)(dX_j/dt)$ be the specific reaction rate for jth reaction, Eqs. 9.5-1 and 2 can be rewritten as

$$\frac{d}{dt}(C_iV) = (C_i)_{\text{in}}q_{\text{in}} - C_iq_{\text{out}} + V\sum_{j=1}^{\mathcal{M}} \nu_{ij}r_j \tag{9.5-3}$$

and

$$\frac{d}{dt}\left(V\sum^{\mathscr{C}} C_i\overline{U}_i\right) = \left(\sum_{i}^{\mathscr{C}} C_i\overline{H}_i\right)_{\text{in}} q_{\text{in}} - \left(\sum_{i}^{\mathscr{C}} C_i\overline{H}_i\right) q_{\text{out}} + \dot{Q} \tag{9.5-4}$$

Here $(C_i)_{\text{in}}$ and $(\underline{H}_i)_{\text{in}}$ are the concentration and partial molar enthalpy, respectively, of species i in the inlet stream, and C_i is the concentration of species i and $\overline{H}_i$ its partial molar enthalpy both in the reactor and in the outlet stream.

It is possible that the volume of fluid in a reactor is a function of time, in which case a total mass balance for the reactor would have to be written to obtain this time dependence. However, few reactors are operated in this manner, so we will neglect this complication, as well as volume changes that may occur on mixing and reaction, and assume that $q_{\text{in}} = q_{\text{out}} = q$. Also, since the tank-type reactor is used almost exclusively for liquid-phase reactions, we may safely assume that $\overline{U}_i = \overline{H}_i$ (since $P\overline{V}_i \ll RT$ for liquids). Thus Eqs. 9.5-3 and 4 can be rewritten as

$$V\frac{dC_i}{dt} = q\{(C_i)_{\text{in}} - C_i\} + V\sum_{j=1}^{\mathcal{M}} \nu_{ij}r_j \tag{9.5-5}$$

and

$$V\frac{d}{dt}\left(\sum C_i\overline{H}_i\right) = q\left(\sum_{i} (C_i\overline{H}_i)_{\text{in}} - \sum_{i} C_i\overline{H}_i\right) + \dot{Q} \tag{9.5-6}$$

which are the equations usually used in the design of tank reactors.

An important special case of these equations is their application to the steady-state operation of a continuous flow reactor. At steady-state the contents of the reactor do not change with time so that $dC_i/dt = 0$ and $dU/dt = dH/dt = 0$, and the design equations reduce to

$$C_i = (C_i)_{\text{in}} + \frac{V}{q}\sum_{j=1}^{\mathcal{M}} \nu_{ij} r_j \tag{9.5-7}$$

and

$$\dot{Q} = q\left(\sum_i C_i \overline{H}_i - \sum_i (C_i \overline{H}_i)_{\text{in}}\right) \tag{9.5-8}$$

The first of these equations relates the exit composition of the reactor to its volume, the inlet composition and flow rate, and the reaction kinetics (i.e., constitutive relations for the reaction rates), whereas the second equation is used to determine the heat load for steady isothermal operation of the reactor.

Actually, the form of Eq. 9.5-8 is a little deceptive because one expects the heat of reaction to play an important role in determining the reactor heat load, yet it does not appear explicitly in the equation. As was indicated in Chapter 6, the heat of reaction is implicit in this equation. To see this, we use Eq. 9.5-7 to eliminate either C_i or $(C_i)_{\text{in}}$ from Eq. 9.5-8 and obtain

$$\dot{Q} = q\sum_i (C_i)_{\text{in}}\{\overline{H}_i - (\overline{H}_i)_{\text{in}}\} + V\sum_j r_j\, \Delta H_{rxn,j} \tag{9.5-9a}$$

if C_i is eliminated, or, if $(C_i)_{\text{in}}$ is eliminated

$$\dot{Q} = q\sum_i C_i\{\overline{H}_i - (\overline{H}_i)_{\text{in}}\} + V\sum_j r_j(\Delta H_{rxn,j})_{\text{in}} \tag{9.5-9b}$$

Here $\Delta H_{rxn,j} = \Sigma_i\, \nu_{ij}\overline{H}_i$, and $(\Delta H_{rxn,j})_{\text{in}} = \Sigma_i\, \nu_{ij}(\overline{H}_i)_{\text{in}}$ are both heats of reaction, the first being the heat reaction at the reactor outlet conditions (temperature and composition), and the second at the reactor inlet conditions.

One simplification that is usually made in writing the energy balance for a reacting system is to neglect solution nonidealities with respect to the energy changes that accompany the chemical reaction and temperature changes in the system [i.e., to neglect the difference between $\overline{H}_i(T, P, x_i)$ and $\underline{H}_i(T, P)$]. With this assumption we have

$$\Delta H_{rxn,j}(T, P, x_1, \ldots, x_{\mathscr{C}-1}) = \sum_i \nu_{ij}\overline{H}_i(T, P, x_1, \ldots, x_{\mathscr{C}-1}) = \sum_i \nu_{ij}\underline{H}_i(T, P)$$
$$= \Delta H_{rxn,j}(T, P)$$

so that Eq. 9.5-9a becomes

$$\dot{Q} = q\sum_{i=1}^{\mathscr{C}} (C_i)_{\text{in}}\{\underline{H}_i(T) - \underline{H}_i(T_{\text{in}})\} + V\sum_{j=1}^{\mathcal{M}} \Delta H_{rxn,j}(T, P) r_j \tag{9.5-10a}$$

The two terms in this equation have simple physical interpretations. The first term represents the rate of flow of energy into the reactor required to heat the inlet fluid from T_{in} to T without reaction, and the second term is the energy requirement for all chemical reactions to occur isothermally at the reactor exit temperature. Similarly, neglecting solution nonidealities in Eq. 9.5-9b yields

$$\dot{Q} = V\sum_{j=1}^{\mathcal{M}} \Delta H_{rxn,j}(T_{\text{in}}) r_j + q\sum_i C_i\{\underline{H}_i(T) - H_i(T_{\text{in}})\} \tag{9.5-10b}$$

Here the first term is the energy required for isothermal chemical reaction at the reactor feed temperature, and the second term is the energy required to heat the reaction mixture, without further reaction, from the inlet temperature to the reactor operating temperature.

The form of Eq. 9.5-10a suggests that the first process in the reactor is the heating of the feed to the reactor operating temperature, and then isothermal chemical reactions occur, whereas Eq. 9.5-10b indicates the chemical reactions occur, followed by fluid heating. Of course, chemical reaction and fluid heating occur simultaneously. These two alternative, and seemingly contradictory, descriptions of the process illustrate again that the enthalpy change between two given states is independent of path, and, therefore, any convenient path may be used for its calculation. In this regard the equations here are similar to those used in Sec. 4.4, where the change in thermodynamic properties accompanying a change in state of a real fluid were computed along any convenient path.

ILLUSTRATION 9.5-1

The ester ethyl acetate is produced by the reversible reaction

$$CH_3COOH + C_2H_5OH \underset{k'}{\overset{k}{\rightleftharpoons}} CH_3COOC_2H_5 + H_2O$$

in the presence of a catalyst such as sulfuric or hydrochloric acid. The rate of ethyl acetate production has been found, from the analysis of chemical kinetic data, to be given by the following equation

$$\frac{dC_{EA}}{dt} = kC_AC_E - k'C_{EA}C_W$$

where the subscripts EA, A, E, and W denote ethyl acetate, acetic acid, ethanol, and water, respectively. The values of the reaction rate constants at 100°C and the catalyst concentration of interest are

$$k = 4.76 \times 10^{-4} \text{ m}^3/\text{kmol min}$$

and

$$k' = 1.63 \times 10^{-4} \text{ m}^3/\text{kmol min}$$

The feed stream is an aqueous solution containing 250 kg of acetic acid and 500 kg of ethyl alcohol per m^3 of solution (including catalyst). The density of the solution is 1040 kg/m^3 and constant. The reaction will be carried out at 100°C, the feed is at 100°C, and the reactor will be operated at a sufficiently high pressure that a negligible amount of reactants or products vaporize.

If a continuous flow stirred tank reactor is used for the reaction, determine

a. The size of the reactor needed to produce 1250 kg/h of the ester, if 37.2% of the acid reacts.

b. The heat load on the reactor for this extent of reaction.

Solution

a. Using the equations in Illustration 9.1-8, and the fact that 37.2% of the acid (or $0.372 \times 4.17 = 1.55$ kmol/m^3) reacts, we have

$$C_A = 4.17 - 1.55 = 2.62 \text{ kmol/m}^3$$

$$C_E = 10.9 - 1.55 = 9.35 \text{ kmol/m}^3$$

$$C_{EA} = 0 + 1.55 = 1.55 \text{ kmol/m}^3$$

$$C_W = 16.1 + 1.55 = 17.65 \text{ kmol/m}^3$$

Next, using Eq. 9.5-7 yields

$$C_{EA,in} = \frac{V}{q} r = \frac{V}{q} \{kC_AC_E - k'C_{EA}C_W\}$$

$$r = \{4.76 \times 10^{-4} \times 2.62 \times 9.35 - 1.63 \times 10^{-4} \times 1.55 \times 17.65\}$$

$$= 7.20 \times 10^{-3} \frac{\text{kmol}}{\text{m}^3 \text{ min}} = 7.20 \frac{\text{mol}}{\text{m}^3 \text{ min}}$$

and

$$1.55 \frac{\text{kmol}}{\text{m}^3} = \frac{V}{q} 7.20 \times 10^{-3} \frac{\text{kmol}}{\text{m}^3 \text{ min}}$$

so that

$$\tau = \frac{V}{q} = 215.3 \text{ min} = 3.588 \text{ h}$$

is required to obtain the desired extent of reaction. (Note that τ has the units of time and can be interpreted to be an average residence time for fluid in the reactor.) The volumetric flow rate into and out of the reactor must be such that

$$q \times C_{EA} \times \text{mol wt of ester} = 1250 \text{ kg/h}$$

or

$$q = \frac{1250 \text{ kg/h}}{1.55 \frac{\text{kmol}}{\text{m}^3} \times 88 \frac{\text{g}}{\text{mol}}} = 9.164 \text{ m}^3\text{/h}$$

Therefore, the reactor volume should be

$$V = \tau q = 3.588 \text{ h} \times 9.164 \text{ m}^3\text{/h} = 32.88 \text{ m}^3$$

to obtain the desired production rate.

b. To determine the heat load on the reactor, we can use either Eqs. 9.5-9 or 10. Assuming that liquid-phase nonidealities are unimportant, and that the reactor effluent is at the reactor temperature (100°C), these equations reduce to

$$Q = V\Delta H_{rxn} r$$

since the inlet stream and reactor temperatures are equal. From part a, $V = 32.88$ m^3 and $r = 7.20$ mol/m^3 min, and from Appendix IV we have that

$$\Delta \underline{H}^\circ_{f,A}(T = 25°\text{C}) = -116.2 \text{ kcal/mol} = -486.18 \text{ kJ/mol}$$

$$\Delta \underline{H}^\circ_{f,E}(T = 25°\text{C}) = -66.35 \text{ kcal/mol} = -277.61 \text{ kJ/mol}$$

$$\Delta \underline{H}^\circ_{f,EA}(T = 25°\text{C}) = -110.72 \text{ kcal/mol} = -463.25 \text{ kJ/mol}$$

$$\Delta \underline{H}^\circ_{f,W}(T = 25°\text{C}) = -68.3 \text{ kcal/mol} = -285.77 \text{ kJ/mol}$$

Therefore

$$\Delta H^\circ_{rxn}(T = 25°\text{C}) = \sum \nu_i \Delta \underline{H}^\circ_{f,i} = (-463.25) + (-285.77) - (-486.18) - (-277.61)$$

$$= 14.75 \text{ kJ/mol ester produced}$$

and

$$\Delta H^{\circ}_{rxn}(T = 100°\text{C}) = \Delta H^{\circ}_{rxn}(T = 25°\text{C}) + \int_{T=25°\text{C}}^{T=100°\text{C}} \Delta C_{\text{P}}\, dT$$

where $\Delta C_{\text{P}} = -13.39$ J/mol K. Therefore

$$\begin{aligned}\Delta H^{\circ}_{rxn}(T = 100°\text{C}) &= 14{,}750 \text{ J/mol} - 13.39 \text{ (J/mol K)} \times 75 \text{ K}\\ &= 14{,}750 - 1004 = 13{,}746 \text{ J/mol}\end{aligned}$$

Neglecting solution nonidealities, we have $\Delta H_{rxn}(T = 100°\text{C}) = \Delta H^{\circ}_{rxn}(T = 100°\text{C})$, so that

$$\dot{Q} = 32.88 \text{ m}^3 \times 7.20 \text{ mol/m}^3 \text{ min} \times 13.746 \text{ kJ/mol} = 3254 \text{ kJ/min}.$$

Note: The fact that $\dot{Q}$ is positive indicates that the reaction is endothermic, so that heat must be supplied to the reactor to maintain isothermal operation. ■

Another tank-type reactor is the batch reactor. In the batch reactor the total charge is introduced initially, the reaction then proceeds without mass flows into or out of the reactor until some later time, when the contents of the reactor are discharged. The fluid leaving the reactor may or may not be in chemical equilibrium, and the reactor may or may not be operated in either an isothermal and/or an isobaric manner.

The mass (mole) and energy balances for the batch reactor are (see Table 6.4-1):

$$\frac{dN_i}{dt} = \sum_{j=1}^{\mathcal{M}} \nu_{ij} \frac{dX_j}{dt} \tag{9.5-11}$$

and

$$\frac{d\underline{U}}{dt} = \frac{d}{dt}\left(\sum_{i=1}^{\mathscr{C}} N_i\bar{U}_i\right) = \dot{Q} \tag{9.5-12}$$

where V is the fluid volume in the reactor. Note that here, as before, the kinetic energy and shaft work terms have been neglected. The term $P\,(dV/dt)$ has also been neglected since it is small for liquid-phase reactions and identically zero for gaseous reactions where the reactant mixture fills the total volume of the reactor. Making the same substitutions as were used in going from Eqs. 9.5-5 and 6 to Eqs. 9.5-7 and 8 here yields

$$\frac{dC_i}{dt} = \sum_{j=1}^{\mathcal{M}} \nu_{ij} r_j \tag{9.5-13}$$

and

$$V \frac{d}{dt} \sum_{i=1}^{\mathscr{C}} (C_i\bar{U}_i) = \dot{Q} \tag{9.5-14}$$

The term on the left side of Eq. 9.5-14 can be written as follows

$$V \frac{d(\Sigma\, C_i \overline{U}_i)}{dt} = V \sum_{i=1}^{\mathscr{C}} \overline{U}_i \frac{dC_i}{dt} + V \sum_{i=1}^{\mathscr{C}} C_i \frac{d\overline{U}_i}{dt}$$

$$= V \sum_{i=1}^{\mathscr{C}} \overline{U}_i \left(\sum_{j=1}^{\mathscr{M}} \nu_{ij} r_j \right) + V \sum_{i=1}^{\mathscr{C}} C_i \overline{C}_{V,i} \frac{dT}{dt}$$

$$= V \sum_{j=1}^{\mathscr{M}} \Delta U_{rxn,j}(T, P, x_i) r_j + \mathbb{C}_V \frac{dT}{dt}$$

where $\Delta U_{rxn,j}(T, P, x_i) = \Sigma_i \nu_{ij} \underline{U}_i(T, P, x_i)$ is the internal energy change for the jth reaction at the reaction conditions (i.e., the temperature, pressure, and composition of reactant mixture), and $\mathbb{C}_V = V\, \Sigma\, C_i \overline{C}_{V,i}$ is the total heat capacity of the fluid in the reactor. Thus, Eq. 9.5-14 becomes

$$\mathbb{C}_V \frac{dT}{dt} = -V \sum_{j=1}^{\mathscr{M}} \Delta U_{rxn,j}(T) r_j + \dot{Q} \tag{9.5-15}$$

For a liquid mixture, $\underline{U} \approx \underline{H}$, and $\mathbb{C}_V \approx \mathbb{C}_P$, so that this equation can be rewritten as

$$\mathbb{C}_P \frac{dT}{dt} = -V \sum_{j=1}^{\mathscr{M}} \Delta H_{rxn,j}(T) r_j + \dot{Q} \tag{9.5-16}$$

Equations 9.5-13 and 16 are used by chemical engineers for the design of batch reactors.

ILLUSTRATION 9.5-2

Ethyl acetate is to be produced in a batch reactor operating at 100°C and at 37.2% conversion of acetic acid. If the reactor charge of the preceding example is used and 20 minutes are needed to discharge, clean, and charge the reactor, what size reactor is required to produce, on the average, 1250 kg of the ester per hour? What heat program should be followed to ensure the reactor remains at 100°C?

Data

See preceding example.

Solution

The starting point here is Eq. 9.5-13, which, for the present case, can be written

$$\frac{dC_{EA}}{dt} = kC_A C_E - k' C_{EA} C_W \tag{a}$$

where, at any extent of reaction $\hat{X} = X/V$ we have

	Initial Concentration	Concentration* at the Extent of Reaction $\hat{X}$
Acetic acid	4.17 kmol/m^3	$4.17 - \hat{X}$ kmol/m^3
Ethanol	10.9	$10.9 - \hat{X}$
Ethyl acetate	0	$\hat{X}$
Water	16.1	$16.1 + \hat{X}$

*Note that $\hat{X} = X/V$ is the molar extent of reaction per unit volume and has units of concentration.

Thus, Eq. a becomes

$$\frac{d\hat{X}}{dt} = k(4.17 - \hat{X})(10.9 - \hat{X}) - k'(16.1 + \hat{X})\hat{X}$$

$$= 4.76 \times 10^{-4}(4.17 - \hat{X})(10.9 - \hat{X}) - 1.63 \times 10^{-4}(16.1 + \hat{X})\hat{X}$$

$$= 2.163 \times 10^{-2}(1 - 0.4528\hat{X} + 0.01447\hat{X}^2) \text{ kmol/m}^3 \text{ min} \tag{b}$$

which can be rearranged to

$$2.163 \times 10^{-2}\, dt = \frac{d\hat{X}}{1 - 0.4528\hat{X} + 0.01447\hat{X}^2}$$

Integrating this equation between $t = 0$ and the time t yields[13]

$$2.163 \times 10^{-2} \int_0^t dt = 2.163 \times 10^{-2} \frac{\text{kmol}}{\text{m}^3 \text{ min}} \times t = \int_0^{\hat{X}} \frac{d\hat{X}}{1 - 0.4528\hat{X} + 0.01447\hat{X}^2}$$

or

$$\ln\left\{\frac{0.02894\hat{X} - 0.8364}{0.02894\hat{X} - 0.0692}\right\} - \ln\left(\frac{0.8364}{0.0692}\right) = 0.8297 \times 10^{-2} t \tag{c}$$

and for $\hat{X} = 1.55$ (37.2% conversion of the acid) we have $0.8297 \times 10^{-2} t = 0.9896$ min or $t = 119.3$ min.

Thus, the total cycle time for the batch reactor is $119.3 + 20 = 139.3$ min = 2.322 hr. Therefore, 2.322 h × 1250 kg/h = 2902.5 kg of ester must be produced in each reactor cycle. Since, in this example, the conversion and feed in the batch reactor are the same as those in the previous illustration, the effluent concentration of the ester is again

$$C_{EA} = 1.55 \text{ kmol m}^3 = 136.4 \text{ kg/m}^3$$

Therefore, the reactor volume is

$$V = \frac{2902.5 \text{ kg}}{136.4 \text{ kg/m}^3} = 21.28 \text{ m}^3$$

To compute the reactor heat program, Eq. 9.5-16 is used, noting that for the isothermal case $dT/dt = 0$, so that

$$\dot{Q} = V(\Delta H_{rxn})r = V(\Delta H_{rxn})\frac{d\hat{X}}{dt} \tag{d}$$

where $\frac{d\hat{X}}{dt}$ is given by Eq. b and $\hat{X}$ is given as a function of time by Eq. c.

The instantaneous values of $\hat{X}$ and $\dot{Q}$ as a function of time are plotted in Fig. 9.5-2. ■

[13]Note that

$$\int \frac{dX}{a + bX + cX^2} = \frac{1}{\sqrt{2}} \ln\left[\frac{2cX + b - \sqrt{-q}}{2cX + b + \sqrt{-q}}\right]$$

where $q = 4ac - b^2$.

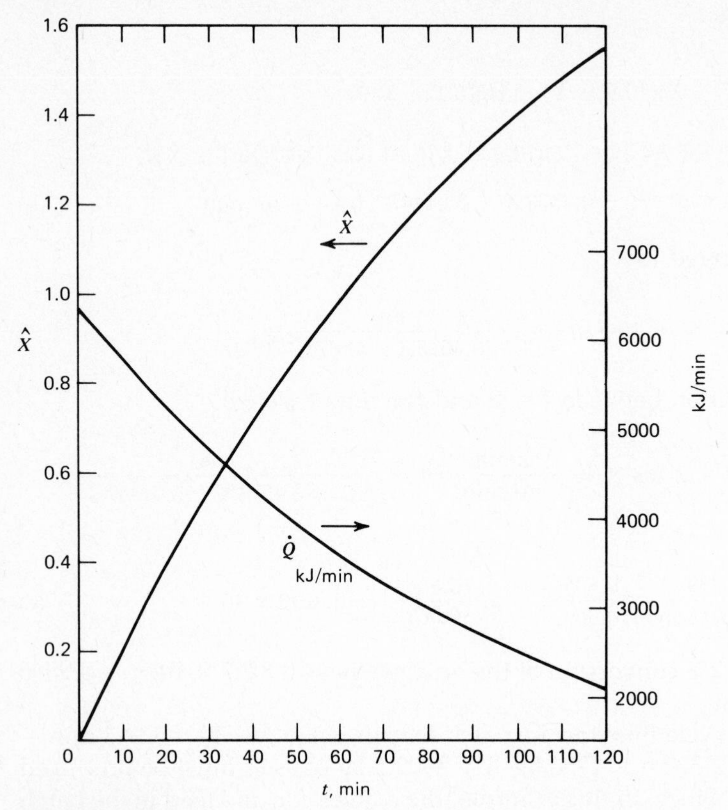

Figure 9.5-2
The heat flow rate $\dot{Q}$ and extent of reaction $\hat{X}$ for ethyl acetate production in a batch reactor.

9.6 THE BALANCE EQUATIONS FOR A TUBULAR REACTOR

Tubular reactors are commonly used in the chemical process industry. This reactor, as the name implies, is a tube that may be packed with catalytic or other material and through which the reactant mixture flows (see Fig. 9.6-1*a*). Such reactors are frequently used for gas-phase reactions and for reactions in which good heat transfer or contact with a heterogeneous catalyst is needed. Our interest here will be in the "plug-flow" tubular reactor; that is, a reactor in which the fluid flow is sufficiently turbulent that there are no radial gradients of concentration, temperature, or fluid velocity. Also, we will assume that the fluid velocity is much greater than the species diffusion velocity, so that diffusion along the tube axis can be neglected.

Since the temperature and concentration vary along the tube axis, we cannot analyze the performance of a tubular reactor using balance equations for the whole reactor. Instead, we write balance equations for the differential reactor element Δz shown in Fig. 9.6-1*b*, and then integrate the equations so obtained over the reactor length. In writing these balance equations Δz will be taken to be so small (in fact, we will be interested in the limit of Δz going to zero) that we can assume the species concentrations and the temperature to be uniform within the

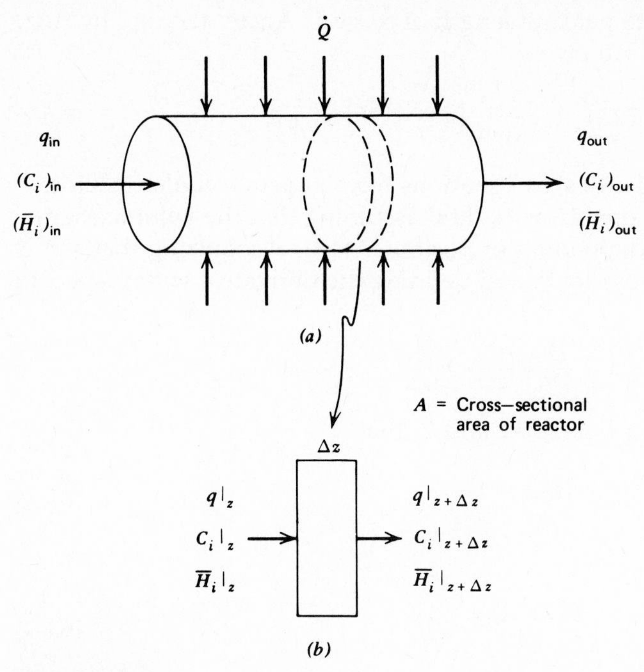

Figure 9.6-1
The tubular reactor.

element. The balance equation for the amount of species i contained within the reactor element shown in Fig. 9.6-1b is

$$\begin{pmatrix}\text{rate of}\\ \text{accumulation}\\ \text{of moles of}\\ \text{species } i \text{ in}\\ \text{reactor element}\\ \text{of length } \Delta z\end{pmatrix} = \begin{pmatrix}\text{rate at}\\ \text{which moles}\\ \text{of species } i\\ \text{enter the}\\ \text{reactor}\\ \text{element}\end{pmatrix} - \begin{pmatrix}\text{rate at}\\ \text{which moles}\\ \text{of species } i\\ \text{leave the}\\ \text{reactor}\\ \text{element}\end{pmatrix} + \begin{pmatrix}\text{rate at}\\ \text{which species } i \text{ is}\\ \text{produced in}\\ \text{the reactor}\\ \text{element by chemical}\\ \text{reaction}\end{pmatrix}$$

$$A\Delta z \frac{\partial C_i}{\partial t} = (qC_i)_z - (qC_i)_{z+\Delta z} + A(\Delta z) \sum_{j=1}^{\mathcal{M}} \nu_{ij} r_j$$

where A is the cross-sectional area of the tubular reactor and q is the volumetric flow rate in the reactor. Now dividing by $A\Delta z$ and taking the limit as $\Delta z \to 0$ gives

$$\frac{\partial C_i}{\partial t} = -\frac{1}{A}\frac{\partial (qC_i)}{\partial z} + \sum_{j=1}^{\mathcal{M}} \nu_{ij} r_j \tag{9.6-1}$$

Similarly, the energy balance for the differential element of reactor volume is

$$\begin{pmatrix}\text{rate of accumulation}\\ \text{of energy in reactor}\\ \text{element of length } \Delta z\end{pmatrix} = \begin{pmatrix}\text{rate at which}\\ \text{energy enters}\\ \text{reactor element}\\ \text{due to flow}\end{pmatrix} - \begin{pmatrix}\text{rate at which}\\ \text{energy leaves}\\ \text{reactor element}\\ \text{due to flow}\end{pmatrix} + \begin{pmatrix}\text{rate of}\\ \text{energy input}\\ \text{by heat flow}\end{pmatrix}$$

$$A\Delta z \frac{\partial}{\partial t}\left(\sum C_i \overline{U}_i\right) = q\left(\sum_i C_i \overline{H}_i\right)\Big|_z - q\left(\sum_i C_i \overline{H}_i\right)\Big|_{z+\Delta z} + \mathcal{Q}\Delta z$$

where $\dot{\mathcal{Q}}$ is the heat flow rate per unit length of reactor. Again dividing by $A\Delta z$, and taking the limit as $\Delta z \to 0$ gives

$$\frac{\partial}{\partial t}\left(\sum_i C_i\overline{U}_i\right) = -\frac{1}{A}\frac{\partial}{\partial z}\left(q\sum_i C_i\overline{H}_i\right) + \frac{1}{A}\dot{\mathcal{Q}} \tag{9.6-2}$$

Equations 9.6-1 and 2 are the design equations for a general tubular reactor.

For liquid-phase reactions, $\overline{U}_i = \overline{H}_i$, and assuming that the volume change on reaction is small, q is independent of position. Also, recognizing that q/A is equal to the mass average velocity v, and defining the derivative with respect to the fluid motion D/Dt by

$$\frac{D}{Dt} = \frac{\partial}{\partial t} + v\frac{\partial}{\partial z}$$

(see Sec. 2.7) we have, from Eqs. 9.6-1 and 2, that

$$\frac{DC_i}{Dt} = \sum_{j=1}^{\mathcal{M}} \nu_{ij} r_j \tag{9.6-3}$$

and

$$\frac{D(\Sigma\, C_i\overline{U}_i)}{Dt} = \frac{1}{A}\dot{\mathcal{Q}} \tag{9.6-4}$$

Equations 9.6-3 and 4 bear a striking resemblance to the mass and energy balances for a batch reactor, Eqs. 9.5-13 and 14. There is, in fact, good reason why these equations should look very much alike. Our model of a plug flow reactor, which neglects diffusion and does not allow for velocity gradients, in essence requires that each element of fluid travel through the reactor with no interaction with the fluid elements before or after it. Therefore, if we could follow a small fluid element in a tubular reactor, we would find that it had precisely the same time behavior as is found in a batch reactor. This similarity of physical situation is mirrored in the similarity of the descriptive equations.

For gas-phase reactions this tubular reactor–batch reactor analogy may not be valid since the volumetric flow rate of the gas can vary along the reactor length as a result mole number changes accompanying chemical reaction, the thermal expansibility of the gas if the reactor temperature varies, and the expansion of the gas accompanying the hydrodynamic pressure drop in the reactor. Consequently, for gas-phase reactions there may not be a simple relationship between distance traversed down the reactor and residence time. Thus, it is not surprising that in this case Eqs. 9.6-1 and 2 cannot be rewritten in the form of Eqs. 9.6-3 and 4

ILLUSTRATION 9.6-1

Ethyl acetate is to be produced at a rate of 1250 kg/h in a tubular reactor operating at 100°C. If the feed is the same as that used in the previous examples, how large should the reactor be to produce the desired amount of ester by achieving the same conversion as the batch reactor? What are the heat transfer specifications for this reactor?

Solution

The steady-state mass and energy balances for the tubular reactor are

$$\frac{q}{A}\frac{dC_i}{dz} = r_i \tag{a}$$

and

$$\frac{q}{A}\frac{d}{dz}\sum_i C_i\overline{H}_i = \frac{q}{A}\left[\sum_i \frac{dC_i}{dz}\overline{H}_i + \sum_i C_i\frac{d\overline{H}_i}{dz}\right] = \frac{\dot{\mathcal{Q}}}{A} \tag{b}$$

Since we are neglecting solution nonidealities, and the temperature is constant, $d\overline{H}_i/dz = 0$, so that Eq. b can be rewritten as

$$\frac{q}{A}\sum_i \overline{H}_i\frac{dC_i}{dz} = \frac{\dot{\mathcal{Q}}}{A} \tag{c}$$

Setting $\tau = Az/q$, we obtain

$$\frac{dC_{EA}}{d\tau} = r_{EA} = kC_A C_E - k'C_{EA}C_W \tag{d}$$

and using Eq. a in Eq. c yields

$$\sum \frac{dC_i}{d\tau}\overline{H}_i = \sum r_i\overline{H}_i = \frac{d\hat{X}}{d\tau}\sum \nu_i\overline{H}_i = \Delta H_{rxn}\frac{d\hat{X}}{d\tau} = \frac{\dot{\mathcal{Q}}}{A} \tag{e}$$

If we equate the time variable t with the distance variable τ, Eqs. d and e become identical to Eqs. a and c of the preceding illustration. That is, the composition in a batch reactor t minutes after start-up will be the same as that at a distance z feet down a tubular reactor, where z is the distance traversed by the fluid in the time t. That is, $z = qt/A$ or $t = Az/q$.

Thus, to convert 37.2% of the acid to ester, the reaction time τ in the tubular reactor must be such that $\tau = AL/q = V/q = 119.3$ min, where L is the total reactor length, and $V = AL$ its volume. Therefore, $V = (119.3/139.3)\, V_{\substack{\text{batch}\\\text{reactor}}} = 18.22\ \text{m}^3$, since we do not have to allow for dumping, cleaning, and charging of the tubular reactor, as we did with the batch reactor. Also

$$\frac{\dot{\mathcal{Q}}}{A} = \Delta H_{rxn}\frac{d\hat{X}}{d\tau} \qquad \text{or} \qquad \dot{\mathcal{Q}} = A\Delta H_{rxn}\frac{d\hat{X}}{d\tau}$$

Since $\hat{X}$ is known as a function of t from the previous illustration, we can use the same figure to obtain $\hat{X}$, and hence $d\hat{X}/d\tau$, as a function of τ (or distance down the reactor). Thus, the heat flux at each point along the tubular reactor needed to maintain isothermal conditions can be computed from the heat program as a function of time in a batch reactor. ■

9.7 OVERALL REACTOR BALANCE EQUATIONS AND THE ADIABATIC REACTION TEMPERATURE

The mass and energy balance equations developed in Secs. 9.5 and 6 are the basic equations used in reactor design and analysis. In many cases, however, our needs are much more modest than in engineering design. In particular, we may not be interested in such details as the type of reactor used and the concentration and temperature profiles or time history in the reactor, but merely in the species mass and total energy balances for the reactor. In such situations one can use the general "black-box" equations of Table 6.4-1:

$$\frac{dN_i}{dt} = (\dot{N}_i)_{\text{in}} - (\dot{N}_i)_{\text{out}} + \sum_{j=1}^{\mathcal{M}} \nu_{ij} \frac{dX_j}{dt}$$

$$= (\dot{N}_i)_{\text{in}} - (\dot{N}_i)_{\text{out}} + V \sum_{j=1}^{\mathcal{M}} \nu_{ij} r_j \qquad (9.7\text{-}1)$$

and

$$\frac{d(\Sigma_i N_i \overline{U}_i)}{dt} = \sum_{i=1}^{\mathscr{C}} (\dot{N}_i \overline{H}_i)_{\text{in}} - \sum_{i=1}^{\mathscr{C}} (\dot{N}_i \overline{H}_i)_{\text{out}} + \dot{Q} + \dot{W} \qquad (9.7\text{-}2)$$

The difficulty that arises is that to evaluate each of the terms in these equations we need chemical reaction rate data, reactor heat programs, and so forth; information we may not have. We can avoid this difficulty by instead of using these differential equations directly, using the balance equations obtained by integrating these equations over the time interval t_1 to t_2, chosen so that the reactor is in the same state at t_2 as it was at t_1. If the reactor is a flow reactor this corresponds to any period of steady-state operation, whereas if it is a batch reactor the time interval is such that the reactor is charged, the reaction run, and the reactor contents dumped over the time interval. In either of these cases the results of the integrations are

$$(N_i)_{\text{out}} = (N_i)_{\text{in}} + \sum \nu_{ij} X_j \qquad (9.7\text{-}3)$$

and

$$Q + W = \sum (N_i \overline{H}_i)_{\text{out}} - \sum (N_i \overline{H}_i)_{\text{in}} \qquad (9.7\text{-}4)$$

where Q and W are the total heat and work flows into the reactor over the time interval t_1 to t_2, and X is the total molar extent of reaction in this time period. Comparing Eqs. 9.7-1 and 2 with 9.7-3 and 4 we find

$$Q = \int_{t_1}^{t_2} \dot{Q}\, dt \qquad W = \int_{t_1}^{t_2} \dot{W}\, dt$$

$$X_j = \int_{t_1}^{t_2} \frac{dX_j}{dt}\, dt = \int_{t_1}^{t_2} \int_V \frac{d\hat{X}_j}{dt}\, dV\, dt$$

$$(N_i)_{\text{in}} = \int_{t_1}^{t_2} (\dot{N}_i)_{\text{in}}\, dt \qquad \text{and} \qquad (N_i)_{\text{out}} = \int_{t_1}^{t_2} (\dot{N}_i)_{\text{out}}\, dt$$

The important feature of Eqs. 9.7-3 and 4 is that they contain only the total mass, heat, and work flows into the system, and the total molar extents of reaction, rather than the flow rates and rates of change of these quantities. Therefore, although Q, W, and X_j can be evaluated from the integrals here, Eqs. 9.7-3 and 4 can also be used to interrelate Q, W, and X_j when the detailed information needed to do these integrations are not available. This is demonstrated in Illustration 9.7-1.

ILLUSTRATION 9.7-1

Ethyl acetate is to be produced from the acetic acid and ethyl alcohol feed used in the illustrations of Secs. 9.5 and 9.6. If the feed and the product streams are both at 100°C, and 37.2% of the acid is converted to the ester, determine the total heat input into any reactor to produce 2500 kg of the ester. Show that this heat input is equivalent to that obtained in the illustrations in the previous sections.

Solution
From Eqs. 9.7-3 and 4 we can immediately write

$$(N_i)_{\text{out}} = (N_i)_{\text{in}} + \nu_i X$$

and

$$\begin{aligned} Q &= \sum (N_i \underline{H}_i)_{\text{out}} - \sum (N_i \underline{H}_i)_{\text{in}} \\ &= \sum_i (N_i)_{\text{in}} [\underline{H}_i(T_{\text{out}}) - \underline{H}_i(T_{\text{in}})] + X \sum \nu_i \underline{H}_i(T_{\text{out}}) \\ &= X \Delta H_{rxn}(T_{\text{out}}) = 13.746X \text{ kJ} \end{aligned}$$

In writing this last equation we have neglected solution nonidealities; that is, we have set $\overline{H}_i = \underline{H}_i$, and noted from Illustration 9.5-1, that $\Delta H_{rxn}(T = 100°\text{C}) =$ 13.746 kJ/mol.

The total molar extent of reaction can be computed by writing the species mass balance, Eq. 9.7-3, for ethyl acetate to obtain

$$\begin{aligned} X &= \frac{(N_i)_{\text{out}} - (N_i)_{\text{in}}}{\nu_i} = \frac{2500 \text{ kg} \times 1000 \text{ g/kg} \times 1 \text{ mol/88 g} - 0}{+1} \\ &= 2.841 \times 10^4 \text{ mol} \end{aligned}$$

Thus $Q = 2.841 \times 10^4 \text{ mol} \times 13.746 \text{ kJ/mol} = 3.905 \times 10^5 \text{ kJ}$.

From Illustration 9.5-1 we have that for the continuous flow stirred tank reactor $\dot{Q} = 3254$ kJ/min and that 120.0 min are required to produce 2500 kg of ethyl acetate. Therefore, the total heat load is

$$Q = 3.254 \times 10^3 \text{ kJ/min} \times 1.20 \times 10^2 \text{ min} = 3.905 \times 10^5 \text{ kJ}$$

To obtain the total input to either the batch or tubular reactor, we must integrate the area under the $\dot{Q}$ versus t curve given in Illustration 9.5-2. The result is

$$Q = 3.905 \times 10^5 \text{ kJ}$$

Thus, the energy required to produce a given amount of product from specified amounts of reactants all at the same temperature is independent of the type of reactor used to accomplish the transformation. This is still another demonstration that the change in thermodynamic state properties of a system, here ΔH, between any two states is independent of the path between the states.

Comment
This example illustrates both the advantages and disadvantages of the black-box style of thermodynamic analysis. If we are interested in merely computing the total heat requirement for the reaction, the black-box analysis is clearly the most expeditious and does not require any kinetic data. On the other hand, black-box thermodynamics also gives us no information about the reactor size or heat program. The decision whether to use the black-box or more detailed thermodynamic analysis will largely depend on the amount of kinetic information available and the degree of detail desired in the final solution. ■

To put Eq. 9.7-4 into a form analogous to Eq. 9.5-9b, we first multiply Eq. 9.7-3 by $(\overline{H}_i)_{\text{in}}$, sum over all species i, and then use the result in Eq. 9.7-4 to obtain

$$Q + W = \sum_i (N_i)_{\text{out}}[(\overline{H}_i)_{\text{out}} - (\overline{H}_i)_{\text{in}}] + \sum_i \sum_j (\overline{H}_i)_{\text{in}} \nu_{ij} X_j$$

$$= \sum_i (N_i)_{\text{out}}[(\overline{H}_i)_{\text{out}} - (\overline{H}_i)_{\text{in}}] + \sum_j (\Delta\overline{H}_{rxn,j})_{\text{in}} X_j \tag{9.7-5}$$

Similarly, multiplying Eq. 9.7-3 by $(\overline{H}_i)_{\text{out}}$, and following the same procedure as the one that led to Eq. 9.7-15, yields the analog of Eq. 9.5-9a

$$Q + W = \sum_i (N_i)_{\text{in}}[(\overline{H}_i)_{\text{out}} - (\overline{H}_i)_{\text{in}}] + \sum_j (\Delta\overline{H}_{rxn,j})_{\text{out}} X_j \tag{9.7-6}$$

For simplicity we will again neglect the effects of solution nonidealities with respect to the heats of reaction. With this assumption we have

$$\overline{H}_i(T, P, x_i) = \underline{H}_i(T, P)$$

and

$$(\overline{H}_i)_{\text{out}} - (\overline{H}_i)_{\text{in}} = (\underline{H}_i)_{\text{out}} - (\underline{H}_i)_{\text{in}} = \int_{T_{\text{in}}}^{T_{\text{out}}} C_{\text{P},i}\, dT$$

so that

$$(\Delta\overline{H}_{rxn,j})_{\text{out}} = \Delta H_{rxn,j}(T_{\text{out}})$$

and

$$(\Delta\overline{H}_{rxn,j})_{\text{in}} = \Delta H_{rxn,j}(T_{\text{in}})$$

Using these results in Eqs. 9.7-5 and 6 yields

$$Q + W = \sum_{j=1}^{\mathcal{M}} X_j \Delta H_{rxn,j}(T_{\text{in}}) + \sum_{i=1}^{\mathscr{C}} (N_i)_{\text{out}} \int_{T_{\text{in}}}^{T_{\text{out}}} C_{\text{P},i}\, dT \tag{9.7-7}$$

and

$$Q + W = \sum_{i=1}^{\mathscr{C}} (N_i)_{\text{in}} \int_{T_{\text{in}}}^{T_{\text{out}}} C_{\text{P},i}\, dT + \sum_{j=1}^{\mathcal{M}} X_j \Delta H_{rxn,j}(T_{\text{out}}) \tag{9.7-8}$$

Equations 9.7-7 and 8 have the same dual interpretation concerning the occurrence of fluid heating and chemical reaction as Eqs. 9.5-10, and the comments made previously about those equations are equally valid here.

Equations 9.7-7 and 8 can be used to compute the steady-state outlet temperature of an *adiabatic* reactor whose effluent is in chemical equilibrium; this temperature is called the **adiabatic reaction temperature**[14] and will be designated here by T_{ad}. By definition, the adiabatic reaction temperature must satisfy (1) the equilibrium relations

$$K_{a,j}(T_{\text{ad}}) = \prod_{i=1}^{\mathscr{C}} a_i^{\nu_{ij}} \qquad j = 1, 2, \ldots, \mathcal{M} \tag{9.7-9}$$

(2) one of the energy balances

$$0 = \sum_{i=1}^{\mathscr{C}} (N_i)_{\text{in}} \int_{T_{\text{in}}}^{T_{\text{ad}}} C_{\text{P},i}\, dT + \sum_{j=1}^{\mathcal{M}} [\Delta H_{rxn,j}(T_{\text{ad}})] X_j \tag{9.7-10a}$$

[14]The computation of the temperature of an adiabatic combustion flame is the most common adiabatic reaction temperature calculation (see Problems 9.19 and 23).

or

$$0 = \sum_{j=1}^{\mathcal{M}} [\Delta H_{rxn,j}(T_{in})] X_j + \sum_{i=1}^{\mathscr{C}} (N_i)_{out} \int_{T_{in}}^{T_{ad}} C_{P,i}\, dT \tag{9.7-10b}$$

and (3) the mass balance and state variable constraints on the system. Since the equilibrium constants $K_{a,j}$ are nonlinear functions of temperature through the van't Hoff relation

$$\ln \frac{K_{a,j}(T_{ad})}{K_{a,j}(T_0)} = \int_{T_0}^{T_{ad}} \frac{\Delta H^{\circ}_{rxn,j}(T)}{RT^2}\, dT \tag{9.7-11}$$

the computation of T_{ad} can be tedious. The graphical procedure illustrated next can be used in solving Eqs. 9.7-9 through 11 for the adiabatic reaction temperature when only a single chemical reaction is involved. For multiple reactions T_{ad} is usually found by computer calculation.

ILLUSTRATION 9.7-2

An equimolar gaseous mixture of benzene and ethylene at 300 K is fed into a reactor where ethyl benzene is formed. If the reactor is operated adiabatically at a pressure of 1 atm, and the reactor effluent is in chemical equilibrium, find the reactor effluent temperature and species concentrations.

Data

See Appendixes II and IV.

Solution

Since the reactor effluent is in equilibrium, we can write

$$K_a(T_{ad}) = \frac{a_{EB}}{a_E a_B} = \frac{y_{EB}\left(\frac{P}{1\text{ atm}}\right)}{y_E\left(\frac{P}{1\text{ atm}}\right) y_B \left(\frac{P}{1\text{ atm}}\right)} = \frac{y_{EB}}{y_E y_B}$$

where we have assumed ideal gas mixture behavior and used the fact that the pressure in the reactor is 1 atm. Using the following table, the three mole fractions in this equation can be replaced by the single molar extent of reaction variable X:

	Inlet	Outlet	Outlet Mole Fraction
C_6H_6	1	$1 - X$	$(1 - X)/(2 - X)$
C_2H_4	1	$1 - X$	$(1 - X)/(2 - X)$
$C_6H_5C_2H_5$	0	X	$X/(2 - X)$
		$2 - X$	

Thus, we have

$$K_a(T_{ad}) = \frac{X(2 - X)}{(1 - X)^2}$$

so that

$$X = 1 - \sqrt{\frac{1}{1 + K(T_{ad})}}$$

With the data given in the appendixes, and Eq. 9.1-19b, it is possible to compute the value of K_a at any temperature. Once K_a is known, X can be computed. The equilibrium values of X calculated in this manner for various values of T_{ad} are plotted in Fig. 9.7-1.

The adiabatic reaction temperature also must satisfy the energy balance equation; using Eq. 9.7-10a we have

$$0 = \int_{T_{in}}^{T_{ad}} (C_{P,B} + C_{P,E})\, dT + \Delta H_{rxn}(T_{ad}) X$$

or

$$X = \frac{-\int_{T_{in}}^{T_{ad}} (C_{P,B} + C_{P,E})\, dT}{\Delta H_{rxn}(T_{ad})}$$

where the heat capacity data for benzene and ethylene are given in Appendix II. Using this equation, the molar extent of reaction needed to satisfy the energy balance for fixed values of T_{in} and T_{ad} can be computed. Since the inlet temperature is fixed at 300 K, we will use this equation to compute values of the molar extent of reaction X for a number of different choices of T_{ad}; the results of this calculation are plotted in Fig. 9.7-1. For illustration, energy balance X-T_{ad} curves for several other values of the reactor inlet temperature are given as dashed lines.

The adiabatic reaction temperature and the equilibrium extent of reaction X that satisfy both the equilibrium and the energy balance relations for a given value of T_{in} are found at the intersection of the equilibrium and energy balance curves in the figure. For the reactor feed at 300 K this occurs at

$$T_{ad} \simeq 588 \text{ K}$$

$$X \simeq 0.955$$

thus

$$y_B = y_E = 0.045/1.045 = 0.043$$

$$y_{EB} = 0.955/1.045 = 0.914$$

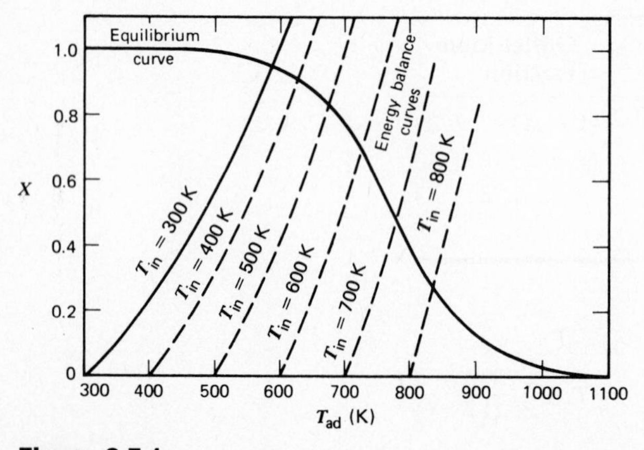

Figure 9.7-1
Graph for computing the adiabatic reaction temperature for the formation of ethyl benzene from ethylene and benzene.

The results for other reactor feed temperatures T_{in} can be computed in a similar fashion from the data in the figure. ■

In Illustration 9.7-2 the reaction was exothermic ($\Delta H_{rxn} < 0$). This implies that (1) energy is released on reaction, which must appear as an increase in temperature of the reactor effluent since the reactor is adiabatic, and (2) the equilibrium conversion of reactants to products decreases with increasing temperature. Consequently, the equilibrium extent of reaction achieved in an adiabatic reactor is less than that which would be obtained had the reactor been operating isothermally at the reactor inlet conditions. For an endothermic reaction ($\Delta H_{rxn} > 0$), the equilibrium conversion decreases with decreasing temperature and, since energy is absorbed on reaction, the reactor effluent is at a lower temperature than the feed stream. Thus, the equilibrium and energy balance curves are mirror images of those in Illustration 9.7-2 (see Problem 9.16), and again the equilibrium extent of reaction is less than would be obtained with an isothermal reactor operating at the feed temperature.

Another interesting application of the multicomponent balance equations for reacting systems is in the estimation of the maximum work that can be obtained from an isothermal chemical reaction that occurs in a steady-state fuel cell or other work-producing device. The starting point here is the steady-state isothermal energy and entropy balances of Table 6.4-1

$$0 = \sum_{i=1}^{\mathscr{C}} (\dot{N}_i \overline{H}_i)_{in} - \sum_{i=1}^{\mathscr{C}} (\dot{N}_i \overline{H}_i)_{out} + \dot{Q} + \dot{W}_s \tag{9.7-12}$$

and

$$0 = \sum_{i=1}^{\mathscr{C}} (\dot{N}_i \overline{S}_i)_{in} - \sum_{i=1}^{\mathscr{C}} (\dot{N}_i \overline{S}_i)_{out} + \frac{\dot{Q}}{T} + \dot{S}_{gen} \tag{9.7-13}$$

Solving the second of these equations for the heat flow $\dot{Q}$

$$\dot{Q} = T \sum_{i=1}^{\mathscr{C}} (\dot{N}_i \overline{S}_i)_{out} - T \sum_{i=1}^{\mathscr{C}} (\dot{N}_i \overline{S}_i)_{in} - T\dot{S}_{gen} \tag{9.7-14}$$

and then eliminating the heat flow from Eq. 9.7-12 yields

$$\begin{aligned} \dot{W}_s &= \sum_{i=1}^{\mathscr{C}} [\dot{N}_i(\overline{H}_i - T\overline{S}_i)]_{out} - \sum_{i=1}^{\mathscr{C}} [\dot{N}_i(\overline{H}_i - T\overline{S}_i)]_{in} + T\dot{S}_{gen} \\ &= \sum_{i=1}^{\mathscr{C}} [\dot{N}_i \overline{G}_i]_{out} - \sum_{i=1}^{\mathscr{C}} [\dot{N}_i \overline{G}_i]_{in} + T\dot{S}_{gen} \end{aligned} \tag{9.7-15}$$

Since a fuel cell is a work-producing device, $\dot{W}_s$ will be negative. The maximum work (i.e., the largest negative value of $\dot{W}_s$ for given inlet conditions) is obtained when the terms $\Sigma_{i=1}^{\mathscr{C}} [\dot{N}_i \overline{G}_i]_{out}$ and $T\dot{S}_{gen}$ are as small as possible. Since $\dot{S}_{gen} \geqslant 0$, with the equality holding only for reversible processes, one condition for obtaining the maximum work is that the chemical reaction and mechanical energy production process be carried out reversibly, so that $\dot{S}_{gen}$ is equal to zero. The first term on the right side of Eq. 9.7-15, $\Sigma_{i=1}^{\mathscr{C}} [\dot{N}_i \overline{G}_i]_{out}$, is the flow of Gibbs free energy accompanying the mass flow out of the system. The minimum value of this term for a given mass flow at fixed temperature occurs when the exit stream is in chemical equilibrium (so that the Gibbs free energy per unit mass is a minimum). Thus, another condition for obtaining maximum work from an isothermal flow reactor is that the exit stream be in chemical equilibrium. When both these conditions are met

$$\dot{W}_{max} = \sum_{i=1}^{\mathscr{C}} (\dot{N}_i\overline{G}_i)_{out} - \sum_{i=1}^{\mathscr{C}} (\dot{N}_i\overline{G}_i)_{in} \tag{9.7-16a}$$

and

$$W_{max} = \sum_{i=1}^{\mathscr{C}} (N_i\overline{G}_i)_{out} - \sum_{i=1}^{\mathscr{C}} (N_i\overline{G}_i)_{in} \tag{9.7-16b}$$

It is also possible to compute the maximum work attainable from an isothermal, constant-volume batch reactor. It is left to you to show that

$$W_{max} = \sum_{i=1}^{\mathscr{C}} (N_i\overline{A}_i)_{final} - \sum_{i=1}^{\mathscr{C}} (N_i\overline{A}_i)_{initial} \tag{9.7-17}$$

ILLUSTRATION 9.7-3

The November 1972 issue of *Fortune* magazine discusses the possibility of the future energy economy of the United States being based on hydrogen. In particular, the use of hydrogen fuel in automobiles is considered. Assuming that pure hydrogen gas and twice the stoichiometric amount of dry air, each at 400 K and 1 atm, are fed into a catalytic fuel cell, and that the reaction products leave at the same temperature and pressure, compute the maximum amount of work that can be obtained from each mole of hydrogen.

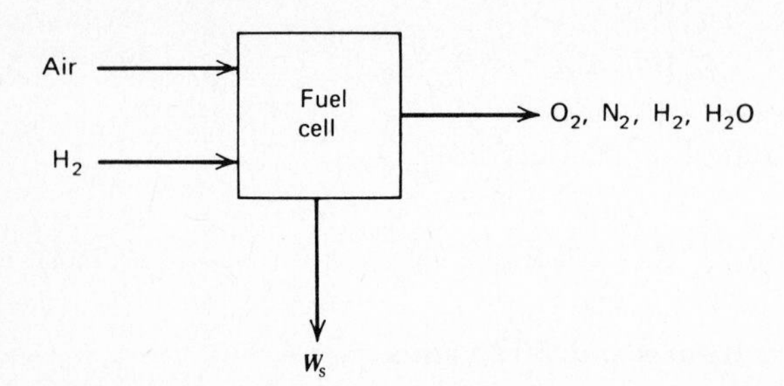

Data

Air may be considered to be 21 mole percent oxygen and 79 mole percent nitrogen; the heats and free energies of formation are given in Appendix IV, and the heat capacities in Appendix II.

Solution

From Eq. 9.7-16 we have

$$W_{max} = \sum_{i=1}^{\mathscr{C}} \{(N_i\overline{G}_i)_{out} - (N_i\overline{G}_i)_{in}\}$$

where the outlet compositions are related by the fact that chemical equilibrium exists. Here, since both the reactants and products are gases (presumably ideal gases at the reaction conditions), we have

$$\overline{G}_i(T, P, y_i) = \underline{G}_i(T, P) + RT \ln y_i$$

Since hydrogen enters the fuel cell as a separate, pure stream, we have

Species	$(y_i)_{in}$	$(N_i)_{in}$ mol	$(N_i)_{out}$ mol	$(y_i)_{out}$
H_2	1	1	$1 - X$	$(1 - X)/\Sigma$
O_2	0.21	1	$1 - \frac{1}{2}X$	$(1 - \frac{1}{2}X)/\Sigma$
N_2	0.79	$\frac{0.79}{0.21} \times 1 = 3.762$	3.762	$3.762/\Sigma$
H_2O	0		X	X/Σ
Total			$\Sigma = 5.762 - \frac{1}{2}X$	

Now, from the equilibrium relation, we have

$$K_a = \frac{a_{H_2O}}{a_{H_2}(a_{O_2})^{1/2}} = \frac{y_{H_2O}}{y_{H_2}(y_{O_2})^{1/2}(P/1\text{ atm})^{1/2}} = \frac{X(5.762 - \frac{1}{2}X)^{1/2}}{(1 - X)(1 - \frac{1}{2}X)^{1/2}}$$

Using the data in the problem statement and the van't Hoff equation (Eqs. 9.1-16 and 18), we find that $K_a = 1.745 \times 10^{+29}$; a huge number. Therefore, $X \sim 1$. Also, using Eqs. 9.1-16a and the heat capacity data, the values of the pure component Gibbs free energies at 400 K and 1 atm, relative to the reference states in Appendix IV, can be computed; the results are:

Species	$\underline{G}_i(T = 400\text{ K}, P = 1\text{ atm})$	$(N_i)_{out}$	$(y_i)_{out}$
H_2	−453.1 J/mol	0	0
O_2	−466.9	0.5	0.095
N_2	−457.5	3.762	0.715
H_2O	−224,600	1	0.190

Therefore,

$$\begin{aligned}
W_{max}(T = 400\text{ K}) &= \sum_i (N_i\overline{G}_i)_{out} - \sum_i (N_i\overline{G}_i)_{in} \\
&= \sum_i [N_i(\underline{G}_i + RT \ln y_i)]_{out} - \sum_i [N_i(\underline{G}_i + RT \ln y_i)]_{in} \\
&= \{0 + 0.5[-466.9 + 8.314 \times 400 \ln (0.095)] \\
&\quad + 3.762[-457.5 + 8.314 \times 400 \ln (0.715)] \\
&\quad + 1(-224{,}600 + 8.314 \times 400 \ln (0.190)]\} \\
&\quad - \{1[-453.1 + 8.314 \times 400 \ln (1.)] \\
&\quad + 1[-466.9 + 8.314 \times 400 \ln (0.21)] \\
&\quad + 3.762[-457.5 + 8.314 \times 400 \ln (0.79)] + 0\} \\
&= (0 - 4147.5 - 5918.2 - 230122.9) - (-453.1 - 5657.0 - 4670.1)\text{ J/mol} \\
&= -229408.4\text{ J/mol} = -229.4\text{ kJ/mol}
\end{aligned}$$

Comments

1. In this illustration the Gibbs free energy of mixing terms make a relatively small contribution to $W_{\max}$. This is only because for the $H_2 + \frac{1}{2}O_2 = H_2O$ reaction the dominant term in the calculation is ΔG_{rxn}. For many reactions ΔG_{rxn} is not so large, and the Gibbs free energy of mixing terms can be important.
2. Although the simple black-box thermodynamic analysis here permits us to calculate the maximum work obtainable from the fuel cell, it does not provide any indication as to how to design the fuel cell (i.e., as an internal combustion engine, electrolytic cell, etc.) to obtain this work. The design of efficient fuel cells is currently an important unsolved engineering problem, much like the design of heat engines at the time of Sadi Carnot. ■

A BASIC Language Program for the Calculation of Chemical Equilibrium Constants as a Function of Temperature, CHEMEQ.BAS

This appendix describes a BASIC language program for the calculation of chemical equilibrium constants, and standard state heats and free energies as a function of temperature. The program uses a datafile, REACT.DTA, which contains species names, $\Delta \underline{G}_f^\circ$, $\Delta \underline{H}_f^\circ$, and heat capacity constants for 99 compounds. The user need only supply the stoichiometric coefficients for the species in the reaction and the temperature range of interest.

As written, the program runs using BASICA (advanced BASIC) on an IBM PC or compatible computers. For faster execution, the program may be compiled using a BASIC compiler. To obtain a listing of the program, insert the disk accompanying this book into drive A of an IBM-PC compatible computer with a printer and type

```
PRINT A:CHEMEQ.BAS
```

To obtain a listing of the data file, type

```
PRINT A:REACT.DTA
```

The structure of the program is indicated below.

Statement Range	Function
100–650	Program information, loading of data file REACT.DTA.
670–1150	Prints component names and stoichiometric coefficients, computes heat capacity constants, $\Delta G_{rxn}^\circ(T = 25°C)$, $\Delta H_{rxn}^\circ(T = 25°C)$, and $K_a(T = 25°C)$. Since data in REACT1.DAT is in units of calories, units conversion occurs.
1160–1240	Input for temperature range over which equilibrium constant calculations are to be done.
1250–1430	Output of equilibrium constant results over temperature range.
1440–1530	Calculation of K_a, ΔG_{rxn}° and ΔH_{rxn}° as a function of temperature.
1540–1580	Program exit.
1590–2340	Routine to display names of components in data bank and allow user to enter stoichiometric coefficients for species in reaction using cursor keys.

PROBLEMS

9.1 Isopropyl alcohol is to be dehydrogenated in the gas phase to form propionaldehyde according to the reaction:

$$(CH_3)_2CHOH(g) = CH_3CH_2CHO(g) + H_2(g)$$

For this reaction

$$\Delta G^{\circ}_{rxn}(T = 298.15 \text{ K}) = 17.74 \text{ kJ/mol } i\text{-propanol reacted}$$

$$\Delta H^{\circ}_{rxn}(T = 298.15 \text{ K}) = 55.48 \text{ kJ/mol } i\text{-propanol reacted}$$

and

$$\Delta C_{P} = \sum_{i} \nu_i C_{P,i} = 16.736 \text{ J/kmol of } i\text{-propanol reacted}$$

Compute the equilibrium fraction of isopropyl alcohol that would be dehydrogenated at 500 K and 1.013 bar.

9.2 The dissociation pressure of calcium oxalate in the reaction

$$CaC_2O_4(s) = CaCO_3(s) + CO$$

at various temperatures is:

T(°C)	375	388	403	410	416	418
Dissociation pressure (kPa)	1.09	4.00	17.86	33.33	78.25	91.18

Source: J. H. Perry, ed., *Chemical Engineers' Handbook*, 4th ed., McGraw-Hill, New York, 1963, p. 3–69.

Compute the standard state Gibbs free energy change, enthalpy change, and entropy change for this reaction for the temperature range listed in the table.

9.3 Carbon dioxide can react with graphite to form carbon monoxide

$$C(\text{graphite}) + CO_2(g) = 2CO(g)$$

and the carbon monoxide formed can further react to form oxygen

$$2CO(g) = 2C(s) + O_2(g)$$

Determine the equilibrium composition when pure carbon dioxide is passed over a hot carbon bed maintained at 1 atm and (a) 2000 K or (b) 1000 K.

9.4 The extent of reaction generally depends on pressure as well as temperature. For the reaction (or phase transition)

$$C(\text{graphite}) = C(\text{diamond})$$

the standard state free energy change at 25°C is 2866 J/mol. The density of graphite is 2.25 g/cc and that of diamond is approximately 3.51 g/cc; both solids may be considered to be incompressible. To help chemical engineering students pay their tuition, we are thinking of setting up equipment to run this reaction in the senior laboratory. Estimate what pressure, at room temperature ($T \sim 25$°C), must be used to convert old pencil "leads" (graphite) into diamonds.

9.5 The production of NO by the direct oxidation of nitrogen

$$\tfrac{1}{2}N_2 + \tfrac{1}{2}O_2 = NO$$

occurs naturally in internal combustion engines. This reaction is also used to commercially produce nitric oxide in electric arcs in the Berkeland–Eyde process. If air is used as the feed, compute the equilibrium conversion of oxygen at 1 atm (1.013 bar) total pressure over the temperature range of 1500 to 3000°C. Air contains 21 mole percent oxygen and 79 mole percent nitrogen.

9.6 Crystalline sodium sulfate, in the presence of water vapor, may form a decahydrate

$$Na_2SO_4(s) + 10H_2O(g) = Na_2SO_4 \cdot 10H_2O(s)$$

a Estimate the minimum partial pressure of water at which the decahydrate will form at 25°C.

b Make a rough estimate of the minimum water partial pressure for decahydrate formation at 15°C.

9.7 Carbon disulfide is produced from the high temperature reaction of carbon and sulfur

$$C(s) + S_2(g) = CS_2(g)$$

This reaction is carried out in a retort at low pressure, and in the absence of oxygen and other species that may react with either the carbon or the sulfur. Compute the equilibrium percentage conversion of sulfur at 750°C and 1000°C.

9.8 The data in the following table give the solubility of silver chloride in various aqueous solutions at 25°C. Show that these data can be plotted on the same ln K_s versus $\sqrt{\mu}$ curve as used in Illustration 9.2-3.

Electrolyte Added	Concentration of Added Salt (mol/m^3)	Concentration of Silver Chloride (mol/m^3)
$Ba(NO_3)_2$	0.2111	1.309×10^{-2}
	0.7064	1.339
	4.402	1.450
	5.600	1.467
$La(NO_3)_3$	0.1438	1.317×10^{-2}
	0.5780	1.367
	1.660	1.432
	2.807	1.477

Source: E. W. Neuman, *J. Am. Chem. Soc.* **54,** 2195 (1932).

9.9 The following data are available for the solubility of barium sulfate in water:

Temperature (°C)	5	10	15	20	25
Solubility ($mol\ m^{-3}$)	0.0156	0.0167	0.0183	0.0198	0.0216

The mean activity coefficient $\gamma_\pm$ for ions of a salt at low ionic strength is given by

$$\ln \gamma_\pm = -\alpha|z_+z_-|\sqrt{\mu}$$

where values for the parameter α for water are given in Table 7.9-1

a Compute K_s°, the ideal solution solubility product, for barium sulfate at each of the temperatures in the table.

b At each of the temperatures in the table calculate the Gibbs free energy change for the reaction

$$BaSO_4(s) = Ba^{++}(aq, M = 1, \text{ideal}) + SO_4^{=}(aq, M = 1, \text{ideal})$$

Here (s) denotes the pure solid state, and (aq, M = 1, ideal) indicates the ion in an ideal aqueous solution at 1 molal concentration.

c Compute the entropy and enthalpy changes for the reaction in part b at 5°C, 15°C, and 25°C.

9.10 Hydrogen gas can be produced by the following reactions between propane and steam in the presence of a nickel catalyst

$$C_3H_8 + 3H_2O = 3CO + 7H_2$$

$$C_3H_8 + 6H_2O = 3CO_2 + 10H_2$$

a Compute the standard heat of reaction and Gibbs free energy change on reaction for each of the reactions at 1000 and 1100 K.

b What is the equilibrium composition of the product gas at 1000 K and 1 atm if the inlet to the catalytic reactor is pure propane and steam in a 1 to 10 ratio?

c Repeat calculation b at 1100 K.

9.11 An important step in the manufacture of sulfuric acid is the gas-phase oxidation reaction

$$SO_2 + \tfrac{1}{2}O_2 = SO_3$$

Compute the equilibrium conversion of sulfur dioxide to sulfur trioxide over the temperature range of 0 to 1400°C for a reactant mixture consisting of initially pure sulfur dioxide and a stoichiometric amount of air at a total pressure of 1.013 bar (Air contains 21 mole percent oxygen and 79 mole percent nitrogen.)

9.12 Ethylene dichloride is produced by the direct chlorination of ethylene using small amounts of ethylene dibromide as a catalyst:

$$C_2H_4 + Cl_2 = C_2H_4Cl_2$$

If stoichiometric amounts of ethylene and chlorine are used, and the reaction is carried out at 50°C and 1 atm, what is the equilibrium conversion of ethylene? (*Note:* The normal boiling point of ethylene dichloride is 83.47°C).

9.13 Polar molecules interact more strongly at large distances than do nonpolar molecules and generally form nonideal solutions. One model for solution nonidealities in a binary mixture consisting of a nonpolar species, which we denote by A, and a polar substance, designated by the symbol B, is based on the supposition that the polar substance partially dimerizes

$$2B \rightleftharpoons B_2$$

Thus, although the mixture is considered to be a binary mixture with mole fractions

$$x_A = \frac{N_A^\circ}{N_A^\circ + N_B^\circ} \qquad \text{and} \qquad x_B = \frac{N_B^\circ}{N_A^\circ + N_B^\circ}$$

where N_A° and N_B° are the initial number of moles of A and B, respectively, it is, according to the supposition, really a ternary mixture with mole fractions

$$x_A^\ddagger = \frac{N_A^\circ}{N_A^\circ + N_B + N_{B_2}} \qquad x_B^\ddagger = \frac{N_B}{N_A^\circ + N_B + N_{B_2}} \qquad \text{and} \qquad x_{B_2}^\ddagger = \frac{N_{B_2}}{N_A^\circ + N_B + N_{B_2}}$$

where, by conservation of B molecules

$$N_B^\circ = N_B + 2N_{B_2}$$

It is further assumed that the ternary mixture is ideal, and that the apparent nonidealities in mixture properties result from considering the A-B solution to be a binary mixture with mole fractions x_i, rather than a ternary mixture with true fractions $x_i^\ddagger$.

Show that the apparent activity coefficients for the binary mixture that result from this model are

$$\gamma_A = 2k/\delta \qquad \gamma_B = \left(\frac{2}{x_B}\right) \frac{(-x_A + (x_A^2 + 2kx_Ax_B + kx_B^2)^{1/2})}{\delta}$$

and

$$\delta = (2k - 1)x_A + kx_B + (x_A^2 + 2kx_Ax_B + kx_B^2)^{1/2}$$

where $k = 4K_a + 1$, and K_a is the equilibrium constant for the dimerization reaction.

9.14 Acetaldehyde is produced from ethanol by the following gas-phase reactions

$$C_2H_5OH + \tfrac{1}{2}O_2 = CH_3CHO + H_2O \qquad \text{(a)}$$

$$C_2H_5OH = CH_3CHO + H_2 \qquad \text{(b)}$$

The reactions are carried out at 540°C and 1 atm pressure using a silver gauze catalyst and air as an oxidant. If 50% excess air (sufficient air so that 50% more oxygen is present than is needed for all the ethanol to react by reaction a) is used, calculate the equilibrium composition of the reactor effluent.

9.15 When propane is heated to high temperatures it pyrolyzes or decomposes. Assume that the only reactions that occur are

$$C_3H_8 = C_3H_6 + H_2$$

$$C_3H_8 = C_2H_4 + CH_4$$

and that these reactions take place in the gas phase.

a Calculate the composition of the equilibrium mixture of propane and its pyrolysis products at a pressure of 1 atm and over a temperature range of 1000 to 2000 K.

b Calculate the composition of the equilibrium mixture of propane and its pyrolysis products over a temperature range of 1000 to 2000 K if pure propane at 25°C and 1 atm is loaded into a constant-volume (bomb) reactor and heated.

9.16 Equal amounts of pure nitrogen and pure oxygen, each at 5000 K and 1 atm, are continuously fed into a chemical reactor, and the reactor effluent, consisting of the reaction product nitric oxide and unreacted nitrogen and oxygen, is continually withdrawn. Assuming that the reactor is adiabatic and that the reactor effluent is in chemical equilibrium, determine the temperature and composition of the effluent.

9.17 The simple statement of the LeChatelier–Braun principle given in Sec. 9.1 leads one to expect that if the concentration of a reactant were increased, the reaction would proceed so as to consume the added reactant. This, however, is not always true. Consider the gas-phase reaction

$$N_2 + 3H_2 = 2NH_3$$

Show that if the mole fraction of nitrogen is less than 0.5, the addition of a small amount of nitrogen to the system at constant temperature and pressure results in the reaction of nitrogen and hydrogen to form ammonia, *whereas* if the mole fraction of nitrogen is greater than 0.5 the addition of a small amount of nitrogen leads to the dissociation of some ammonia to form more nitrogen and hydrogen. Why does this occur?

9.18 By catalytic dehydrogenation, 1-butene can be produced from *n*-butane,

$$C_4H_{10} = C_4H_8 + H_2$$

However, butene may also be dehydrogenated to form 1,3-butadiene,

$$C_4H_8 = C_4H_6 + H_2$$

Compute the equilibrium conversion of *n*-butane to 1-butene and 1,3-butadiene at 1 atm and

a 900 K.
b 1000 K.

Data
For 1,3-butadiene at 25°C and 1 atm

$$\Delta \underline{H}_f^\circ = 111.92 \text{ kJ/mol}$$

$$\Delta \underline{G}_f^\circ = 152.42 \text{ kJ/mol}$$

9.19 The flame temperature attained in a torch or a burner can be computed using the adiabatic reaction temperature analysis of Sec. 9.7 if it is assumed that the radiant heat loss from the flame is negligible. Compute the flame temperature and exit composition in a hydrogen torch if:

a A stoichiometric amount of pure oxygen is used as the oxidant.
b Oxygen is the oxidant, but a 100% excess is used.
c Twice the stoichiometric amount of air is used as the oxidant.

In each case the hydrogen and oxidant entering the torch are at room temperature (298.15 K), and the torch pressure is 1.013 bar.

9.20 The Soviet Venera VII probe, which reached Venus on December 15, 1970, found that the conditions on the planet surface are:

$$T \approx 747 \pm 20 \text{ K}$$

$$P \approx 90 \pm 15 \text{ (earth) atmospheres}$$

Interferometric measurements indicate that there is only a 10 to 20 K temperature variation across the planet, and that the coldest region lies at the equator. Radar astronomy measurements suggest that the dielectric constant of the Venus surface is typical of mineral silicates. Furthermore, from spectroscopic observations it has been concluded that the approximate atmospheric composition of Venus is:

	Species	Mole Percent
	CO_2	99.9−
	H_2O	0.01
	CO	0.01
	HCl	1×10^{-1}
	HF	1×10^{-6}
	O_2	trace
Sulfur-bearing constituents	H_2S, COS, SO_2, SO_3	0

Since the temperature of Venus is so high, and the planet is quite old, it may be assumed that all chemical reactions occurring between the atmosphere and the surface minerals are in chemical equilibrium. The following reactions are thought to occur:

$$CaCO_3 + SiO_2 = CaO{\cdot}SiO_2 + CO_2$$

$$3FeO{\cdot}SiO_2 + CO_2 = 3SiO_2 + Fe_3O_4 + CO$$

$$3FeO{\cdot}SiO_2 + \tfrac{1}{2}O_2 = Fe_3O_4 + 3SiO_2$$

Determine whether the chemical equilibrium assumption is consistent with the reported data.

Data

For $FeO{\cdot}SiO_2$, $\Delta\underline{G}_f^\circ = -1.060$ MJ/mol and $\Delta\underline{H}_f^\circ = -1.144$ MJ/mol; other heat and Gibbs free energy of formation data are given in Appendix IV. The heat capacity of the solid species are given here:[15]

$$C_P = a + bT + e/T^2 \text{ (J/mol K); T[=]K}$$

Species	a	$b \times 10^2$	$e \times 10^{-4}$
$CaO{\cdot}SiO_2$	116.94	0.8602	−3.120
$CaCO_3$	82.34	4.975	−1.287
Fe_3O_4	172.26	7.874	−4.098
$FeO{\cdot}SiO_2$	98.28	4.269	−1.215
SiO_2	45.48	3.645	−1.009

Since the temperature is so high, you may assume that the gas phase is ideal.

[15]Reference: R. H. Perry, D. W. Green, and J. O. Maloney, eds., *Chemical Engineers' Handbook*, 6th ed., McGraw-Hill, New York, 1984, pp. 3-129 to 135.

9.21 Styrene can be hydrogenated to ethyl benzene at moderate conditions in both the liquid and the gas phase. Calculate the equilibrium compositions in the vapor and liquid phases of hydrogen, styrene, and ethyl benzene at each of the following conditions:

a 3 atm pressure and 25°C, with a starting mole ratio of hydrogen to styrene of 2 to 1.

b 3 atm pressure and 150°C, with a starting mole ratio of hydrogen to styrene of 2 to 1.

Data

Reaction stoichiometry:

$$C_6H_5\text{—}HC{=}CH_2 + H_2 \longrightarrow C_6H_5\text{—}CH_2\text{—}CH_3$$

Physical properties:

Species	δ (cal/cc)$^{1/2}$	$\underline{V}^L$ cc/mol	$\Delta \underline{G}_f^\circ$ (gas, 1 atm and 25°C)
Styrene C_8H_8	9.3	116	213.8 kJ/mol
Ethyl benzene C_8H_{10}	10.1	123	130.57 kJ/mol
Hydrogen	3.25	3.1	0.0

Heat capacity: See Appendix II.

Vapor pressure:

	1.333 kPa	13.33	53.32	101.3	202.6	506.5
C_8H_8	30.8°C	82.0	122.5	145.2		
C_8H_{10}	25.9	74.1	113.8	136.2	163.5	207.5

9.22 If 1 mole of a gas in a constant volume system is heated, and both the heat flow and the gas temperature are measured as a function of time, the constant-volume heat capacity can be computed from

$$C_V = \left(\frac{\partial U}{\partial T}\right)_V = \frac{\dot{Q}}{\left(\frac{\partial T}{\partial t}\right)_V}$$

This equation can also be used to calculate the effective heat capacity $C_{V,\text{eff}}$ of a gas that is undergoing a chemical reaction, such as nitrogen tetroxide dissociating to form nitrogen dioxide

$$N_2O_4 \rightleftharpoons 2NO_2$$

In such cases $C_{V,\text{eff}}$ can be much larger than the heat capacity of the non-reacting gas.

a Develop an expression for $C_{v,eff}$ for dissociating nitrogen tetroxide, and comment on the dependence of $C_{v,eff}$ on the internal energy change on chemical reaction.

b Compute the molar effective heat capacity $C_{v,eff}$ as a function of temperature for nitrogen tetroxide over the temperature range of 300 to 600 K, if pure N_2O_4 is loaded into a constant-volume reactor at 300 K at a pressure of 1 atm (1.013 bar).

9.23 Compute the flame temperature of an oxyacetylene torch using pure acetylene and 50% more pure oxygen than is needed to convert all the acetylene to carbon dioxide and water. Both the oxygen and acetylene are initially at room temperature and atmospheric pressure. The following reactions may occur:

$$HCCH + \tfrac{3}{2}O_2 = 2CO + H_2O$$

$$HCCH + \tfrac{5}{2}O_2 = 2CO_2 + H_2O$$

$$CO + \tfrac{1}{2}O_2 = CO_2$$

$$CO + H_2O = CO_2 + H_2$$

$$HCCH = 2C(s) + H_2$$

9.24 One mole of ethylene and one mole of benzene are fed to a constant-volume batch reactor and heated to 600 K. On addition of a catalyst, an equilibrium mixture of ethylbenzene, benzene, and ethylene is formed:

$$C_6H_6(g) + C_2H_4(g) \rightleftarrows C_6H_5C_2H_5(g)$$

The pressure in the reactor, before addition of the catalyst (i.e., before any reaction has occurred), is 1.013 bar. Calculate the equilibrium conversion and the heat that must be removed to maintain the reactor temperature constant at 600 K.

9.25 When pure hydrogen iodide gas enters an evacuated cylinder, the following reactions may occur

$$HI(g) = \tfrac{1}{2}H_2(g) + \tfrac{1}{2}I_2(g)$$

$$I_2(g) = I_2(s)$$

[Note that since Gibbs free energies of formation data are available for iodine in both the gaseous and solid phases, it is more convenient to think of the solid–vapor iodine phase equilibrium as a chemical equilibrium.] If the reaction mixture is gradually compressed at 25°C, a pressure is reached at which the first bit of solid iodine appears. What is the pressure at which this occurs, and what is the vapor composition at this pressure?

9.26 The calcination of sodium bicarbonate takes place according to the reaction

$$2NaHCO_3(s) = Na_2CO_3(s) + CO_2(g) + H_2O(g)$$

When this reaction was run in the laboratory by placing sodium bicarbonate in an initially evacuated cylinder, it was observed that the equilibrium total pressure was 0.826 kPa at 30°C and 166.97 kPa at 110°C. The heat of reaction for the calcination can be assumed to be independent of temperature.

a What is the heat of reaction for this reaction?

b Develop an equation for the equilibrium constant at any temperature.

c At what temperature will the partial pressure of carbon dioxide in the reaction vessel be exactly 1 bar.

9.27 Carbon is deposited on a catalytic reactor bed as a result of the cracking of hydrocarbons. Periodically, hydrogen gas is passed through the reactor in an effort to remove the carbon and to preserve the reduced state of the catalyst. It has been found, by experiment, that the effluent gas contains about 10 mole percent methane and 90 mole percent hydrogen when the temperature is 1000 K and the pressure is 1 bar. Is this conversion thermodynamically limited? Can a higher concentration of methane be produced by reducing the rate of hydrogen flow through the reactor, thereby removing more carbon for a given amount of hydrogen?

9.28 A gas mixture containing equimolar quantities of carbon dioxide and hydrogen is to be "reformed" by passing it over a catalyst. The pressure in the reformer will be determined by the possibility of solid carbon deposition. Although a large number of reactions are possible, only the following are believed to occur:

$$C + H_2O = CO + H_2$$

$$C + 2H_2O = CO_2 + 2H_2$$

$$CO_2 + C = 2CO$$

$$CO + H_2O = CO_2 + H_2$$

a At temperatures between 600 and 1000 K over what range of pressure will carbon deposit if each of the reactions is assumed to achieve equilibrium?

b For this feed, what pressure should be maintained for exactly 30% of the carbon present in the feed to precipitate as solid carbon at each temperature between 600 and 1000 K?

9.29 Assume two species can associate in the vapor phase according to the reactions

$$iA_1 = A_i \qquad i = 1, 2, \text{etc.}$$

$$jB_1 = B_j \qquad j = 1, 2, \text{etc.} \qquad \begin{pmatrix}\text{but not all}\\ \text{integers need}\\ \text{be included}\end{pmatrix}$$

$$iA_1 + jB_1 = A_iB_j$$

Using the notation that $\overline{G}_A$ and $\overline{G}_B$ are the partial molar Gibbs free energies of total species A and B (in all its forms, unassociated and associated), prove that

$$\overline{G}_A = \overline{G}_{A_1} \qquad \text{and} \qquad \overline{G}_B = \overline{G}_{B_1}$$

where $\overline{G}_{A_1}$ and $\overline{G}_{B_1}$ are the partial molar Gibbs free energies of the monomeric species, respectively. Also prove that

$$\bar{f}_A = \bar{f}_{A_1} \qquad \text{and} \qquad \bar{f}_B = \bar{f}_{B_1}$$

9.30 The description of components that associate or hydrogen bond is difficult. An alternative model to the one considered in the previous problem is the continuous association model, in which

$$A_1 + A_1 = A_2$$

$$A_1 + A_2 = A_3 \qquad \begin{pmatrix}\text{here all integers to } \infty\\ \text{are included}\end{pmatrix}$$

$$A_1 + A_n = A_{n+1}$$

etc.

We will assume that this associating fluid is described by the van der Waals equation, and in the equation of state representation the parameters of the j-mer are gotten from

$$a_j = j^2 a_1$$

and

$$b_j = j b_1$$

Further, the value of the equilibrium constant for the association

$$K_{j+1} = \frac{a_{j+1}}{a_j a_1}$$

will be assumed independent of the degree of association, that is,

$$K_2 = K_3 = \cdots = K_n = K$$

Using the following notation

N_0 = number of moles if no association occurs

N_T = number of moles when association occurs

and assuming the binary interaction parameters are all zero since the species are so similar, do the following

a Obtain expressions for a and b for the mixture in terms of a_1, b_1, N_0, and N_T only.

b Show that the ratio

$$\frac{P\phi_j\phi_1}{\phi_{j+1}}$$

is independent of index j and obtain an explicit expression for this ratio. (Note that ϕ_j is the fugacity coefficient of species j.)

c Obtain an expression for the equation of state of this associating fluid that contains only P, V, T, N_0, a_1, b_1, and the equilibrium constant K. (Does your equation reduce to the van der Waals equation for a nonassociating one-component system in the limit of $K \rightarrow 0$?)

9.31 The behavior of hydrogen fluoride is unusual! For example, here are the critical properties of various hydrogen halides:

	MW	T_C**(K)**	P_C**(bar)**	Z_C	ω
HF	20	461.0	64.88	0.12	0.372
HCl	36.46	324.6	83.07	0.249	0.12
HBr	80.91	363.2	85.50	0.283	0.063
HI	127.9	424.0	83.07	0.309	0.05

We see that HF has a very high critical temperature and acentric factor for its molecular weight; it also has the lowest reported critical compressibility of any species.

Experimental data for the vapor pressure and the apparent molecular weight of saturated HF vapor are given here.

T(K)	P^{vap}(bar)	$(MW)_{apparent}$
227.3	0.0519	92.8
243.9	0.1265	85.0
256.4	0.2328	79.4
277.8	0.5780	69.8
303.0	1.4353	58.4
322.6	2.6178	50.3

Apparent molecular weight at 0.993 bar

T(K)	$(MW)_{apparent}$	
227.3	117.6	Calculated
250.0	110.7	Calculated
270.3	95.7	Calculated
285.7	74.6	Calculated
294.1	59.8	Measured
303.0	43.0	Measured
312.5	28.8	Measured
322.6	21.8	Measured

In each case the apparent molecular weight has been found by measuring the mass density of the vapor and comparing that with an ideal gas of molecular weight 20.

One possible explanation for this behavior is that hydrogen fluoride associates according to the following set of reactions

$$2HF = (HF)_2$$

$$6HF = (HF)_6$$

$$8HF = (HF)_8$$

Describe how you would use the foregoing data to develop a model for HF so that you could determine the vapor–liquid equilibrium of HF and a component that did not associate.

Conversion Factors to SI Units

To Convert from	To	Multiply by
Atmosphere (standard)	Pa	1.03×10^5
bar	Pa	1×10^5
British thermal unit*	J	1.054×10^3
BTU/lb	J kg^{-1}	2.324×10^3
BTU/lb °F	J $kg^{-}K^{-1}$	4.184×10^3
calorie*	J	4.184
cal/g	J kg^{-1}	4.184×10^3
cal/g °C	J $kg^{-1}K^{-1}$	4.184×10^3
cm of mercury (0°C)	Pa	1.33×10^3
cm^3	m^3	1×10^{-6}
dyne	N	1×10^{-5}
dyne cm	N m	1×10^{-7}
dyne/cm^2	Pa	1×10^{-1}
erg	J	1×10^{-7}
foot	m	3.048×10^{-1}
gallon (U.S., liquid)	m^3	3.785×10^{-3}
g/cm^3	kg m^{-3}	1×10^3
g/liter	kg m^{-3}	1
horsepower	W	7.457×10^2
inch	m	2.540×10^{-2}
inch water (60°F)	Pa	2.488×10^2
kilogram-force	N	9.807
kilowatt hour	J	3.600×10^6
liter	m^3	1×10^{-3}
millibar	Pa	1×10^2
millimeter of mercury (0°C)	Pa	1.333×10^2
ounce (avoirdupois)	kg	2.835×10^{-2}
pounds (avoirdupois)	kg	4.536×10^{-1}
lb/ft^3	kg m^{-3}	1.602×10^1
lb/gal (U.S.)	kg m^{-3}	1.198×10^2
lb-force/ft^2	Pa	4.788×10^1
lb-force/in^2 (psi)	Pa	6.895×10^3
ton (refrigeration)	W	3.517×10^3
ton (2000 lbs)	kg	9.072×10^2
torr (mm Hg, 0°C)	Pa	1.333×10^2
yard	m	9.144×10^{-1}
degree Celsius	K	$T(K) = T(°C) + 273.15$
degree Fahrenheit	K	$T(K) = [T(°F) + 459.67]/1.8$
degree Rankine	K	$T(K) = T(°R)/1.8$

*Thermochemical unit that is used, for example, in the tables in the *Chemical Engineers Handbook*. For International Table units use the constant 4.1868 instead of 4.184, and for mean calorie use 4.19002.

APPENDIX II

The Molar Heat Capacities of Gases in the Ideal Gas (Zero Pressure) State*

		a	$b \times 10^2$	$c \times 10^5$	$d \times 10^9$	Temperature Range, K
Paraffinic Hydrocarbons						
Methane	CH_4	4.750	1.200	0.3030	−2.630	273–1500
Ethane	C_2H_6	1.648	4.124	−1.530	1.740	273–1500
Propane	C_3H_8	−0.966	7.279	−3.755	7.580	273–1500
n-Butane	C_4H_{10}	0.945	8.873	−4.380	8.360	273–1500
i-Butane	C_4H_{10}	−1.890	9.936	−5.495	11.92	273–1500
n-Pentane	C_5H_{12}	1.618	10.85	−5.365	10.10	273–1500
n-Hexane	C_6H_{14}	1.657	13.19	−6.844	13.78	273–1500
Monoolefinic Hydrocarbons						
Ethylene	C_2H_4	0.944	3.735	−1.993	4.220	273–1500
Propylene	C_3H_6	0.753	5.691	−2.910	5.880	273–1500
1-Butene	C_4H_8	−0.240	8.650	−5.110	12.07	273–1500
i-Butene	C_4H_8	1.650	7.702	−3.981	8.020	273–1500
cis-2-Butene	C_4H_8	−1.778	8.078	−4.074	7.890	273–1500
trans-2-Butene	C_4H_8	2.340	7.220	−3.403	6.070	273–1500
Cycloparaffinic Hydrocarbons						
Cyclopentane	C_5H_{10}	−12.957	13.087	−7.447	16.41	273–1500
Methylcyclopentane	C_6H_{12}	−12.114	15.380	−8.915	20.03	273–1500
Cyclohexane	C_6H_{12}	−15.935	16.454	−9.203	19.27	273–1500
Methylcyclohexane	C_7H_{14}	−15.070	18.972	−10.989	24.09	273–1500
Aromatic Hydrocarbons						
Benzene	C_6H_6	−8.650	11.578	−7.540	18.54	273–1500
Toluene	C_7H_8	−8.213	13.357	−8.230	19.20	273–1500
Ethylbenzene	C_8H_{10}	−8.398	15.935	−10.003	23.95	273–1500
Styrene	C_8H_8	−5.968	14.354	−9.150	22.03	273–1500
Cumene	C_9H_{12}	−9.452	18.686	−11.869	28.80	273–1500

		a	$b \times 10^2$	$c \times 10^5$	$d \times 10^9$	Temperature Range, K
Oxygenated Hydrocarbons						
Formaldehyde	CH_2O	5.447	0.9739	0.1703	−2.078	273–1500
Acetaldehyde	C_2H_4O	4.19	3.164	−0.515	−3.800	273–1000
Methanol	CH_4O	4.55	2.186	−0.291	−1.92	273–1000
Ethanol	C_2H_6O	4.75	5.006	−2.479	4.790	273–1500
Ethylene oxide	C_2H_4O	−1.12	4.925	−2.389	3.149	273–1000
Ketene	C_2H_2O	4.11	2.966	−1.793	4.22	273–1500
Miscellaneous Hydrocarbons						
Cyclopropane	C_3H_6	−6.481	8.206	−5.577	15.61	273–1000
Isopentane	C_5H_{12}	−2.273	12.434	−7.097	15.86	273–1500
Neopentane	C_5H_{12}	−3.865	13.305	−8.018	18.83	273–1500
o-Xylene	C_8H_{10}	−3.789	14.291	−8.354	18.80	273–1500
m-Xylene	C_8H_{10}	−6.533	14.905	−8.831	20.05	273–1500
p-Xylene	C_8H_{10}	−5.334	14.220	−7.984	17.03	273–1500
C_3 Oxygenated Hydrocarbons						
Carbon suboxide	C_3O_2	8.203	3.073	−2.081	5.182	273–1500
Acetone	C_3H_6O	1.625	6.661	−3.737	8.307	273–1500
i-Propyl alcohol	C_3H_8O	0.7936	8.502	−5.016	11.56	273–1500
n-Propyl alcohol	C_3H_8O	−1.307	9.235	−5.800	14.14	273–1500
Allyl alcohol	C_3H_6O	0.5203	7.122	−4.259	9.948	273–1500
Chloroethenes						
Chloroethene	C_2H_3Cl	2.401	4.270	−2.751	6.797	273–1500
1,1-Dichloroethene	$C_2H_2Cl_2$	5.899	4.383	−3.182	8.516	273–1500
cis-1,2-Dichloroethene	$C_2H_2Cl_2$	4.336	4.691	−3.397	9.010	273–1500
trans-1,2-Dichloroethene	$C_2H_2Cl_2$	5.661	4.295	−3.022	7.891	273–1500
Trichloroethene	C_2HCl_3	9.200	4.517	−3.600	10.10	273–1500
Tetrachloroethene	C_2Cl_4	15.11	3.799	−3.179	9.089	273–1500
Nitrogen Compounds						
Ammonia	NH_3	6.5846	0.61251	0.23663	−1.5981	273–1500
Hydrazine	N_2H_4	3.890	3.554	−2.304	5.990	273–1500
Methylamine	CH_5N	2.9956	3.6101	−1.6446	2.9505	273–1500
Dimethylamine	C_2H_7N	−0.275	6.6152	−3.4826	7.1510	273–1500
Trimethylamine	C_3H_9N	−2.098	9.6187	−5.5488	12.432	273–1500

APPENDIX II THE MOLAR HEAT CAPACITIES OF GASES IN THE IDEAL GAS (ZERO PRESSURE) STATE *(Continued)*

		a	$b \times 10^2$	$c \times 10^5$	$d \times 10^9$	Temperature Range, K
Halogens and Halogen Acids						
Fluorine	F_2	6.115	0.5864	−0.4186	0.9797	273–2000
Chlorine	Cl_2	6.8214	0.57095	−0.5107	1.547	273–1500
Bromine	Br_2	8.051	0.2462	−0.2128	0.6406	273–1500
Iodine	I_2	8.504	0.13135	−0.10684	0.3125	273–1800
Hydrogen fluoride	HF	7.201	−0.1178	0.1576	−0.3760	273–2000
Hydrogen chloride	HCl	7.244	−0.1820	0.3170	−1.036	273–1500
Hydrogen bromide	HBr	7.169	−0.1604	0.3314	−1.161	273–1500
Hydrogen iodide	HI	6.702	0.04546	0.1216	−0.4813	273–1900
Chloromethanes						
Methyl chloride	CH_3Cl	3.05	2.596	−1.244	2.300	273–1500
Methylene chloride	CH_2Cl_2	4.20	3.419	−2.3500	6.068	273–1500
Chloroform	$CHCl_3$	7.61	3.461	−2.668	7.344	273–1500
Carbon tetrachloride	CCl_4	12.24	3.400	−2.995	8.828	273–1500
Phosgene	$COCl_2$	10.35	1.653	−0.8408	——	273–1000
Thiophosgene	$CSCl_2$	10.80	1.859	−1.045	——	273–1000
Cyanogens						
Cyanogen	$(CN)_2$	9.82	1.4858	−0.6571	——	273–1000
Hydrogen cyanide	HCN	6.34	0.8375	−0.2611	——	273–1500
Cyanogen chloride	CNCl	7.97	1.0745	−0.5265	——	273–1000
Cyanogen bromide	CNBr	8.82	0.9084	−0.4367	——	273–1000
Cyanogen iodide	CNI	9.69	0.7213	−0.3265	——	273–1000
Acetonitrile	CH_3CN	5.09	2.7634	−0.9111	——	273–1200
Acrylic nitrile	CH_2CHCN	4.55	4.1039	−1.6939	——	273–1000
Oxides of Nitrogen						
Nitric oxide	NO	6.461	0.2358	−0.07705	0.08729	273–3800
Nitric oxide	NO	7.008	−0.02244	0.2328	−1.000	273–1500
Nitrous oxide	N_2O	5.758	1.4004	−0.8508	2.526	273–1500
Nitrogen dioxide	NO_2	5.48	1.365	−0.841	1.88	273–1500
Nitrogen tetroxide	N_2O_4	7.9	4.46	−2.71	——	273–600

		a	$b \times 10^2$	$c \times 10^5$	$d \times 10^9$	Temperature Range, K
Acetylenes and Diolefins						
Acetylene	C_2H_2	5.21	2.2008	−1.559	4.349	273–1500
Methylacetylene	C_3H_4	4.21	4.073	−2.192	4.713	273–1500
Dimethylacetylene	C_4H_6	3.54	5.838	−2.760	4.974	273–1500
Propadiene	C_3H_4	2.43	4.693	−2.781	6.484	273–1500
1,3-Butadiene	C_4H_6	−1.29	8.350	−5.582	14.24	273–1500
Isoprene	C_5H_8	−0.44	10.418	−6.762	16.93	273–1500
Combustion Gases (Low Range)						
Nitrogen	N_2	6.903	−0.03753	0.1930	−0.6861	273–1800
Oxygen	O_2	6.085	0.3631	−0.1709	0.3133	273–1800
Air		6.713	0.04697	0.1147	−0.4696	273–1800
Hydrogen	H_2	6.952	−0.04576	0.09563	−0.2079	273–1800
Carbon monoxide	CO	6.726	0.04001	0.1283	−0.5307	273–1800
Carbon dioxide	CO_2	5.316	1.4285	−0.8362	1.784	273–1800
Water vapor	H_2O	7.700	0.04594	0.2521	−0.8587	273–1800
Combustion Gases (High Range)†						
Nitrogen	N_2	6.529	0.1488	−0.02271	——	273–3800
Oxygen	O_2	6.732	0.1505	−0.01791	——	273–3800
Air		6.557	0.1477	−0.02148	——	273–3800
Hydrogen	H_2	6.424	0.1039	−0.007804	——	273–3800
Carbon monoxide	CO	6.480	0.1566	−0.02387	——	273–3800
Water vapor	H_2O	6.970	0.3464	−0.04833	——	273–3800
Sulfur Compounds						
Sulfur	S_2	6.499	0.5298	−0.3888	0.9520	273–1800
Sulfur dioxide	SO_2	6.157	1.384	−0.9103	2.057	273–1800
Sulfur trioxide	SO_3	3.918	3.483	−2.675	7.744	273–1300
Hydrogen sulfide	H_2S	7.070	0.3128	0.1364	−0.7867	273–1800
Carbon disulfide	CS_2	7.390	1.489	−1.096	2.760	273–1800
Carbonyl sulfide	COS	6.222	1.536	−1.058	2.560	273–1800

Source: O. Hougen, K. Watson, and R. A. Ragatz, *Chemical Process Principles,* Part 1. John Wiley & Sons, New York, 1954. Used with permission.

*Constants are for the equation $C_p = a + bT + cT^2 + dT^3$, where T is in degrees Kelvin and C_p in cal (mol K)$^{-1}$. Note that 1 cal (mol K)$^{-1}$ = 4.184 J (mol K)$^{-1}$.

†The equation for CO_2 in the temperature range of 273–3800 K is $C_p = 18.036 - 4.474 \times 10^{-5}T - 158.08/\sqrt{T}$.

APPENDIX III

The Thermodynamic Properties of Water and Steam[1]

THERMODYNAMIC PROPERTIES OF STEAM

SATURATED STEAM: TEMPERATURE TABLE

		Specific Volume		Internal Energy			Enthalpy			Entropy		
Temp. °C T	Press. kPa P	Sat. Liquid $\hat{V}^L$	Sat. Vapor $\hat{V}^V$	Sat. Liquid $\hat{U}^L$	Evap. $\Delta\hat{U}$	Sat. Vapor $\hat{U}^V$	Sat. Liquid $\hat{H}^L$	Evap. $\Delta\hat{H}$	Sat. Vapor $\hat{H}^V$	Sat. Liquid $\hat{S}^L$	Evap. $\Delta\hat{S}$	Sat. Vapor $\hat{S}^V$
0.01	0.6113	0.001 000	206.14	.00	2375.3	2375.3	.01	2501.3	2501.4	.0000	9.1562	9.1562
5	0.8721	0.001 000	147.12	20.97	2361.3	2382.3	20.98	2489.6	2510.6	.0761	8.9496	9.0257
10	1.2276	0.001 000	106.38	42.00	2347.2	2389.2	42.01	2477.7	2519.8	.1510	8.7498	8.9008
15	1.7051	0.001 001	77.93	62.99	2333.1	2396.1	62.99	2465.9	2528.9	.2245	8.5569	8.7814
20	2.339	0.001 002	57.79	83.95	2319.0	2402.9	83.96	2454.1	2538.1	.2966	8.3706	8.6672
25	3.169	0.001 003	43.36	104.88	2304.9	2409.8	104.89	2442.3	2547.2	.3674	8.1905	8.5580
30	4.246	0.001 004	32.89	125.78	2290.8	2416.6	125.79	2430.5	2556.3	.4369	8.0164	8.4533
35	5.628	0.001 006	25.22	146.67	2276.7	2423.4	146.68	2418.6	2565.3	.5053	7.8478	8.3531
40	7.384	0.001 008	19.52	167.56	2262.6	2430.1	167.57	2406.7	2574.3	.5725	7.6845	8.2570
45	9.593	0.001 010	15.26	188.44	2248.4	2436.8	188.45	2394.8	2583.2	.6387	7.5261	8.1648
50	12.349	0.001 012	12.03	209.32	2234.2	2443.5	209.33	2382.7	2592.1	.7038	7.3725	8.0763
55	15.758	0.001 015	9.568	230.21	2219.9	2450.1	230.23	2370.7	2600.9	.7679	7.2234	7.9913
60	19.940	0.001 017	7.671	251.11	2205.5	2456.6	251.13	2358.5	2609.6	.8312	7.0784	7.9096
65	25.03	0.001 020	6.197	272.02	2191.1	2463.1	272.06	2346.2	2618.3	.8935	6.9375	7.8310
70	31.19	0.001 023	5.042	292.95	2176.6	2469.6	292.98	2333.8	2626.8	.9549	6.8004	7.7553
75	38.58	0.001 026	4.131	313.90	2162.0	2475.9	313.93	2321.4	2635.3	1.0155	6.6669	7.6824
80	47.39	0.001 029	3.407	334.86	2147.4	2482.2	334.91	2308.8	2643.7	1.0753	6.5369	7.6122
85	57.83	0.001 033	2.828	355.84	2132.6	2488.4	355.90	2296.0	2651.9	1.1343	6.4102	7.5445
90	70.14	0.001 036	2.361	376.85	2117.7	2494.5	376.92	2283.2	2660.1	1.1925	6.2866	7.4791
95	84.55	0.001 040	1.982	397.88	2102.7	2500.6	397.96	2270.2	2668.1	1.2500	6.1659	7.4159
100	0.101 35	0.001 044	1.6729	418.94	2087.6	2506.5	419.04	2257.0	2676.1	1.3069	6.0480	7.3549
105	0.120 82	0.001 048	1.4194	440.02	2072.3	2512.4	440.15	2243.7	2683.8	1.3630	5.9328	7.2958
110	0.143 27	0.001 052	1.2102	461.14	2057.0	2518.1	461.30	2230.2	2691.5	1.4185	5.8202	7.2387
115	0.169 06	0.001 056	1.0366	482.30	2041.4	2523.7	482.48	2216.5	2699.0	1.4734	5.7100	7.1833
120	0.198 53	0.001 060	0.8919	503.50	2025.8	2529.3	503.71	2202.6	2706.3	1.5276	5.6020	7.1296
125	0.2321	0.001 065	0.7706	524.74	2009.9	2534.6	524.99	2188.5	2713.5	1.5813	5.4962	7.0775
130	0.2701	0.001 070	0.6685	546.02	1993.9	2539.9	546.31	2174.2	2720.5	1.6344	5.3925	7.0269
135	0.3130	0.001 075	0.5822	567.35	1977.7	2545.0	567.69	2159.6	2727.3	1.6870	5.2907	6.9777
140	0.3613	0.001 080	0.5089	588.74	1961.3	2550.0	589.13	2144.7	2733.9	1.7391	5.1908	6.9299
145	0.4154	0.001 085	0.4463	610.18	1944.7	2554.9	610.63	2129.6	2740.3	1.7907	5.0926	6.8833
150	0.4758	0.001 091	0.3928	631.68	1927.9	2559.5	632.20	2114.3	2746.5	1.8418	4.9960	6.8379
155	0.5431	0.001 096	0.3468	653.24	1910.8	2564.1	653.84	2098.6	2752.4	1.8925	4.9010	6.7935
160	0.6178	0.001 102	0.3071	674.87	1893.5	2568.4	675.55	2082.6	2758.1	1.9427	4.8075	6.7502
165	0.7005	0.001 108	0.2727	696.56	1876.0	2572.5	697.34	2066.2	2763.5	1.9925	4.7153	6.7078
170	0.7917	0.001 114	0.2428	718.33	1858.1	2576.5	719.21	2049.5	2768.7	2.0419	4.6244	6.6663
175	0.8920	0.001 121	0.2168	740.17	1840.0	2580.2	741.17	2032.4	2773.6	2.0909	4.5347	6.6256

[1]From G. J. Van Wylen and R. E. Sontagg, *Fundamentals of Classical Thermodynamics, S. I. Version*, 2nd ed., John Wiley & Sons, New York, 1978. Used with permission.

Temp. °C T	Press. ~~kPa~~ MPa P	*Specific Volume* Sat. Liquid $\hat{V}^L$	Sat. Vapor $\hat{V}^V$	*Internal Energy* Sat. Liquid $\hat{U}^L$	Evap. $\Delta\hat{U}$	Sat. Vapor $\hat{U}^V$	*Enthalpy* Sat. Liquid $\hat{H}^L$	Evap. $\Delta\hat{H}$	Sat. Vapor $\hat{H}^V$	*Entropy* Sat. Liquid $\hat{S}^L$	Evap. $\Delta\hat{S}$	Sat. Vapor $\hat{S}^V$
180	1.0021	0.001 127	0.194 05	762.09	1821.6	2583.7	763.22	2015.0	2778.2	2.1396	4.4461	6.5857
185	1.1227	0.001 134	0.174 09	784.10	1802.9	2587.0	785.37	1997.1	2782.4	2.1879	4.3586	6.5465
190	1.2544	0.001 141	0.156 54	806.19	1783.8	2590.0	807.62	1978.8	2786.4	2.2359	4.2720	6.5079
195	1.3978	0.001 149	0.141 05	828.37	1764.4	2592.8	829.98	1960.0	2790.0	2.2835	4.1863	6.4698
200	1.5538	0.001 157	0.127 36	850.65	1744.7	2595.3	852.45	1940.7	2793.2	2.3309	4.1014	6.4323
205	1.7230	0.001 164	0.115 21	873.04	1724.5	2597.5	875.04	1921.0	2796.0	2.3780	4.0172	6.3952
210	1.9062	0.001 173	0.104 41	895.53	1703.9	2599.5	897.76	1900.7	2798.5	2.4248	3.9337	6.3585
215	2.104	0.001 181	0.094 79	918.14	1682.9	2601.1	920.62	1879.9	2800.5	2.4714	3.8507	6.3221
220	2.318	0.001 190	0.086 19	940.87	1661.5	2602.4	943.62	1858.5	2802.1	2.5178	3.7683	6.2861
225	2.548	0.001 199	0.078 49	963.73	1639.6	2603.3	966.78	1836.5	2803.3	2.5639	3.6863	6.2503
230	2.795	0.001 209	0.071 58	986.74	1617.2	2603.9	990.12	1813.8	2804.0	2.6099	3.6047	6.2146
235	3.060	0.001 219	0.065 37	1009.89	1594.2	2604.1	1013.62	1790.5	2804.2	2.6558	3.5233	6.1791
240	3.344	0.001 229	0.059 76	1033.21	1570.8	2604.0	1037.32	1766.5	2803.8	2.7015	3.4422	6.1437
245	3.648	0.001 240	0.054 71	1056.71	1546.7	2603.4	1061.23	1741.7	2803.0	2.7472	3.3612	6.1083
250	3.973	0.001 251	0.050 13	1080.39	1522.0	2602.4	1085.36	1716.2	2801.5	2.7927	3.2802	6.0730
255	4.319	0.001 263	0.045 98	1104.28	1496.7	2600.9	1109.73	1689.8	2799.5	2.8383	3.1992	6.0375
260	4.688	0.001 276	0.042 21	1128.39	1470.6	2599.0	1134.37	1662.5	2796.9	2.8838	3.1181	6.0019
265	5.081	0.001 289	0.038 77	1152.74	1443.9	2596.6	1159.28	1634.4	2793.6	2.9294	3.0368	5.9662
270	5.499	0.001 302	0.035 64	1177.36	1416.3	2593.7	1184.51	1605.2	2789.7	2.9751	2.9551	5.9301
275	5.942	0.001 317	0.032 79	1202.25	1387.9	2590.2	1210.07	1574.9	2785.0	3.0208	2.8730	5.8938
280	6.412	0.001 332	0.030 17	1227.46	1358.7	2586.1	1235.99	1543.6	2779.6	3.0668	2.7903	5.8571
285	6.909	0.001 348	0.027 77	1253.00	1328.4	2581.4	1262.31	1511.0	2773.3	3.1130	2.7070	5.8199
290	7.436	0.001 366	0.025 57	1278.92	1297.1	2576.0	1289.07	1477.1	2766.2	3.1594	2.6227	5.7821
295	7.993	0.001 384	0.023 54	1305.2	1264.7	2569.9	1316.3	1441.8	2758.1	3.2062	2.5375	5.7437
300	8.581	0.001 404	0.021 67	1332.0	1231.0	2563.0	1344.0	1404.9	2749.0	3.2534	2.4511	5.7045
305	9.202	0.001 425	0.019 948	1359.3	1195.9	2555.2	1372.4	1366.4	2738.7	3.3010	2.3633	5.6643
310	9.856	0.001 447	0.018 350	1387.1	1159.4	2546.4	1401.3	1326.0	2727.3	3.3493	2.2737	5.6230
315	10.547	0.001 472	0.016 867	1415.5	1121.1	2536.6	1431.0	1283.5	2714.5	3.3982	2.1821	5.5804
320	11.274	0.001 499	0.015 488	1444.6	1080.9	2525.5	1461.5	1238.6	2700.1	3.4480	2.0882	5.5362
330	12.845	0.001 561	0.012 996	1505.3	993.7	2498.9	1525.3	1140.6	2665.9	3.5507	1.8909	5.4417
340	14.586	0.001 638	0.010 797	1570.3	894.3	2464.6	1594.2	1027.9	2622.0	3.6594	1.6763	5.3357
350	16.513	0.001 740	0.008 813	1641.9	776.6	2418.4	1670.6	893.4	2563.9	3.7777	1.4335	5.2112
360	18.651	0.001 893	0.006 945	1725.2	626.3	2351.5	1760.5	720.5	2481.0	3.9147	1.1379	5.0526
370	21.03	0.002 213	0.004 925	1844.0	384.5	2228.5	1890.5	441.6	2332.1	4.1106	.6865	4.7971
374.14	22.09	0.003 155	0.003 155	2029.6	0	2029.6	2099.3	0	2099.3	4.4298	0	4.4298

$\hat{V}$ [=] m³/kg; $\hat{U}$, $\hat{H}$ [=] J/g = kJ/kg; $\hat{S}$ [=] kJ/kg K

SATURATED STEAM: PRESSURE TABLE

		Specific Volume		Internal Energy			Enthalpy			Entropy		
Press. kPa P	Temp. °C T	Sat. Liquid $\hat{V}^L$	Sat. Vapor $\hat{V}^V$	Sat. Liquid $\hat{U}^L$	Evap. $\Delta\hat{U}$	Sat. Vapor $\hat{U}^V$	Sat. Liquid $\hat{H}^L$	Evap. $\Delta\hat{H}$	Sat. Vapor $\hat{H}^V$	Sat. Liquid $\hat{S}^L$	Evap. $\Delta\hat{S}$	Sat. Vapor $\hat{S}^V$
0.6113	0.01	0.001 000	206.14	0.00	2375.3	2375.3	0.01	2501.3	2501.4	0.0000	9.1562	9.1562
1.0	6.98	0.001 000	129.21	29.30	2355.7	2385.0	29.30	2484.9	2514.2	0.1059	8.8697	8.9756
1.5	13.03	0.001 001	87.98	54.71	2338.6	2393.3	54.71	2470.6	2525.3	0.1957	8.6322	8.8279
2.0	17.50	0.001 001	67.00	73.48	2326.0	2399.5	73.48	2460.0	2533.5	0.2607	8.4629	8.7237
2.5	21.08	0.001 002	54.25	88.48	2315.9	2404.4	88.49	2451.6	2540.0	0.3120	8.3311	8.6432
3.0	24.08	0.001 003	45.67	101.04	2307.5	2408.5	101.05	2444.5	2545.5	0.3545	8.2231	8.5776
4.0	28.96	0.001 004	34.80	121.45	2293.7	2415.2	121.46	2432.9	2554.4	0.4226	8.0520	8.4746
5.0	32.88	0.001 005	28.19	137.81	2282.7	2420.5	137.82	2423.7	2561.5	0.4764	7.9187	8.3951
7.5	40.29	0.001 008	19.24	168.78	2261.7	2430.5	168.79	2406.0	2574.8	0.5764	7.6750	8.2515
10	45.81	0.001 010	14.67	191.82	2246.1	2437.9	191.83	2392.8	2584.7	0.6493	7.5009	8.1502
15	53.97	0.001 014	10.02	225.92	2222.8	2448.7	225.94	2373.1	2599.1	0.7549	7.2536	8.0085
20	60.06	0.001 017	7.649	251.38	2205.4	2456.7	251.40	2358.3	2609.7	0.8320	7.0766	7.9085
25	64.97	0.001 020	6.204	271.90	2191.2	2463.1	271.93	2346.3	2618.2	0.8931	6.9383	7.8314
30	69.10	0.001 022	5.229	289.20	2179.2	2468.4	289.23	2336.1	2625.3	0.9439	6.8247	7.7686
40	75.87	0.001 027	3.993	317.53	2159.5	2477.0	317.58	2319.2	2636.8	1.0259	6.6441	7.6700
50	81.33	0.001 030	3.240	340.44	2143.4	2483.9	340.49	2305.4	2645.9	1.0910	6.5029	7.5939
75	91.78	0.001 037	2.217	384.31	2112.4	2496.7	384.39	2278.6	2663.0	1.2130	6.2434	7.4564
MPa												
0.100	99.63	0.001 043	1.6940	417.36	2088.7	2506.1	417.46	2258.0	2675.5	1.3026	6.0568	7.3594
0.125	105.99	0.001 048	1.3749	444.19	2069.3	2513.5	444.32	2241.0	2685.4	1.3740	5.9104	7.2844
0.150	111.37	0.001 053	1.1593	466.94	2052.7	2519.7	467.11	2226.5	2693.6	1.4336	5.7897	7.2233
0.175	116.06	0.001 057	1.0036	486.80	2038.1	2524.9	486.99	2213.6	2700.6	1.4849	5.6868	7.1717
0.200	120.23	0.001 061	0.8857	504.49	2025.0	2529.5	504.70	2201.9	2706.7	1.5301	5.5970	7.1271
0.225	124.00	0.001 064	0.7933	520.47	2013.1	2533.6	520.72	2191.3	2712.1	1.5706	5.5173	7.0878
0.250	127.44	0.001 067	0.7187	535.10	2002.1	2537.2	535.37	2181.5	2716.9	1.6072	5.4455	7.0527
0.275	130.60	0.001 070	0.6573	548.59	1991.9	2540.5	548.89	2172.4	2721.3	1.6408	5.3801	7.0209
0.300	133.55	0.001 073	0.6058	561.15	1982.4	2543.6	561.47	2163.8	2725.3	1.6718	5.3201	6.9919

SATURATED STEAM: PRESSURE TABLE *(Continued)*

		Specific Volume		*Internal Energy*			*Enthalpy*			*Entropy*		
Press. MPa P	Temp. °C T	Sat. Liquid $\hat{V}^L$	Sat. Vapor $\hat{V}^V$	Sat. Liquid $\hat{U}^L$	Evap. $\Delta\hat{U}$	Sat. Vapor $\hat{U}^V$	Sat. Liquid $\hat{H}^L$	Evap. $\Delta\hat{H}$	Sat. Vapor $\hat{H}^V$	Sat. Liquid $\hat{S}^L$	Evap. $\Delta\hat{S}$	Sat. Vapor $\hat{S}^V$
0.325	136.30	0.001 076	0.5620	572.90	1973.5	2546.4	573.25	2155.8	2729.0	1.7006	5.2646	6.9652
0.350	138.88	0.001 079	0.5243	583.95	1965.0	2548.9	584.33	2148.1	2732.4	1.7275	5.2130	6.9405
0.375	141.32	0.001 081	0.4914	594.40	1956.9	2551.3	594.81	2140.8	2735.6	1.7528	5.1647	6.9175
0.40	143.63	0.001 084	0.4625	604.31	1949.3	2553.6	604.74	2133.8	2738.6	1.7766	5.1193	6.8959
0.45	147.93	0.001 088	0.4140	622.77	1934.9	2557.6	623.25	2120.7	2743.9	1.8207	5.0359	6.8565
0.50	151.86	0.001 093	0.3749	639.68	1921.6	2561.2	640.23	2108.5	2748.7	1.8607	4.9606	6.8213
0.55	155.48	0.001 097	0.3427	655.32	1909.2	2564.5	655.93	2097.0	2753.0	1.8973	4.8920	6.7893
0.60	158.85	0.001 101	0.3157	669.90	1897.5	2567.4	670.56	2086.3	2756.8	1.9312	4.8288	6.7600
0.65	162.01	0.001 104	0.2927	683.56	1886.5	2570.1	684.28	2076.0	2760.3	1.9627	4.7703	6.7331
0.70	164.97	0.001 108	0.2729	696.44	1876.1	2572.5	697.22	2066.3	2763.5	1.9922	4.7158	6.7080
0.75	167.78	0.001 112	0.2556	708.64	1866.1	2574.7	709.47	2057.0	2766.4	2.0200	4.6647	6.6847
0.80	170.43	0.001 115	0.2404	720.22	1856.6	2576.8	721.11	2048.0	2769.1	2.0462	4.6166	6.6628
0.85	172.96	0.001 118	0.2270	731.27	1847.4	2578.7	732.22	2039.4	2771.6	2.0710	4.5711	6.6421
0.90	175.38	0.001 121	0.2150	741.83	1838.6	2580.5	742.83	2031.1	2773.9	2.0946	4.5280	6.6226
0.95	177.69	0.001 124	0.2042	751.95	1830.2	2582.1	753.02	2023.1	2776.1	2.1172	4.4869	6.6041
1.00	179.91	0.001 127	0.194 44	761.68	1822.0	2583.6	762.81	2015.3	2778.1	2.1387	4.4478	6.5865
1.10	184.09	0.001 133	0.177 53	780.09	1806.3	2586.4	781.34	2000.4	2781.7	2.1792	4.3744	6.5536
1.20	187.99	0.001 139	0.163 33	797.29	1791.5	2588.8	798.65	1986.2	2784.8	2.2166	4.3067	6.5233
1.30	191.64	0.001 144	0.151 25	813.44	1777.5	2591.0	814.93	1972.7	2787.6	2.2515	4.2438	6.4953
1.40	195.07	0.001 149	0.140 84	828.70	1764.1	2592.8	830.30	1959.7	2790.0	2.2842	4.1850	6.4693
1.50	198.32	0.001 154	0.131 77	843.16	1751.3	2594.5	844.89	1947.3	2792.2	2.3150	4.1298	6.4448
1.75	205.76	0.001 166	0.113 49	876.46	1721.4	2597.8	878.50	1917.9	2796.4	2.3851	4.0044	6.3896
2.00	212.42	0.001 177	0.099 63	906.44	1693.8	2600.3	908.79	1890.7	2799.5	2.4474	3.8935	6.3409
2.25	218.45	0.001 187	0.088 75	933.83	1668.2	2602.0	936.49	1865.2	2801.7	2.5035	3.7937	6.2972
2.5	223.99	0.001 197	0.079 98	959.11	1644.0	2603.1	962.11	1841.0	2803.1	2.5547	3.7028	6.2575
3.0	233.90	0.001 217	0.066 68	1004.78	1599.3	2604.1	1008.42	1795.7	2804.2	2.6457	3.5412	6.1869

$\hat{V}$ [=] m^3/kg; $\hat{U}$, $\hat{H}$ [=] J/g = kJ/kg; $\hat{S}$ [=] kJ/kg K

SATURATED STEAM: PRESSURE TABLE *(Continued)*

		Specific Volume		*Internal Energy*			*Enthalpy*			*Entropy*		
Press. MPa P	Temp. °C T	Sat. Liquid $\hat{V}^L$	Sat. Vapor $\hat{V}^V$	Sat. Liquid $\hat{U}^L$	Evap. $\Delta\hat{U}$	Sat. Vapor $\hat{U}^V$	Sat. Liquid $\hat{H}^L$	Evap. $\Delta\hat{H}$	Sat. Vapor $\hat{H}^V$	Sat. Liquid $\hat{S}^L$	Evap. $\Delta\hat{S}$	Sat. Vapor $\hat{S}^V$
3.5	242.60	0.001 235	0.057 07	1045.43	1558.3	2603.7	1049.75	1753.7	2803.4	2.7253	3.4000	6.1253
4	250.40	0.001 252	0.049 78	1082.31	1520.0	2602.3	1087.31	1714.1	2801.4	2.7964	3.2737	6.0701
5	263.99	0.001 286	0.039 44	1147.81	1449.3	2597.1	1154.23	1640.1	2794.3	2.9202	3.0532	5.9734
6	275.64	0.001 319	0.032 44	1205.44	1384.3	2589.7	1213.35	1571.0	2784.3	3.0267	2.8625	5.8892
7	285.88	0.001 351	0.027 37	1257.55	1323.0	2580.5	1267.00	1505.1	2772.1	3.1211	2.6922	5.8133
8	295.06	0.001 384	0.023 52	1305.57	1264.2	2569.8	1316.64	1441.3	2758.0	3.2068	2.5364	5.7432
9	303.40	0.001 418	0.020 48	1350.51	1207.3	2557.8	1363.26	1378.9	2742.1	3.2858	2.3915	5.6772
10	311.06	0.001 452	0.018 026	1393.04	1151.4	2544.4	1407.56	1317.1	2724.7	3.3596	2.2544	5.6141
11	318.15	0.001 489	0.015 987	1433.7	1096.0	2529.8	1450.1	1255.5	2705.6	3.4295	2.1233	5.5527
12	324.75	0.001 527	0.014 263	1473.0	1040.7	2513.7	1491.3	1193.6	2684.9	3.4962	1.9962	5.4924
13	330.93	0.001 567	0.012 780	1511.1	985.0	2496.1	1531.5	1130.7	2662.2	3.5606	1.8718	5.4323
14	336.75	0.001 611	0.011 485	1548.6	928.2	2476.8	1571.1	1066.5	2637.6	3.6232	1.7485	5.3717
15	342.24	0.001 658	0.010 337	1585.6	869.8	2455.5	1610.5	1000.0	2610.5	3.6848	1.6249	5.3098
16	347.44	0.001 711	0.009 306	1622.7	809.0	2431.7	1650.1	930.6	2580.6	3.7461	1.4994	5.2455
17	352.37	0.001 770	0.008 364	1660.2	744.8	2405.0	1690.3	856.9	2547.2	3.8079	1.3698	5.1777
18	357.06	0.001 840	0.007 489	1698.9	675.4	2374.3	1732.0	777.1	2509.1	3.8715	1.2329	5.1044
19	361.54	0.001 924	0.006 657	1739.9	598.1	2338.1	1776.5	688.0	2464.5	3.9388	1.0839	5.0228
20	365.81	0.002 036	0.005 834	1785.6	507.5	2293.0	1826.3	583.4	2409.7	4.0139	0.9130	4.9269
21	369.89	0.002 207	0.004 952	1842.1	388.5	2230.6	1888.4	446.2	2334.6	4.1075	0.6938	4.8013
22	373.80	0.002 742	0.003 568	1961.9	125.2	2087.1	2022.2	143.4	2165.6	4.3110	0.2216	4.5327
22.09	374.14	0.003 155	0.003 155	2029.6	0.0	2029.6	2099.3	0.0	2099.3	4.4298	0.0	4.4298

$\hat{V}$ [=] m^3/kg; $\hat{U}$, $\hat{H}$ [=] J/g = kJ/kg; $\hat{S}$ [=] kJ/kg K

T°C	P = 0.010 MPa (45.81)				P = 0.050 MPa (81.33)				P = 0.10 MPa (99.63)			
	$\hat{V}$	$\hat{U}$	$\hat{H}$	$\hat{S}$	$\hat{V}$	$\hat{U}$	$\hat{H}$	$\hat{S}$	$\hat{V}$	$\hat{U}$	$\hat{H}$	$\hat{S}$
Sat.	14.674	2437.9	2584.7	8.1502	3.240	2483.9	2645.9	7.5939	1.6940	2506.1	2675.5	7.3594
50	14.869	2443.9	2592.6	8.1749	—	—	—	—	—	—	—	—
100	17.196	2515.5	2687.5	8.4479	3.418	2511.6	2682.5	7.6947	1.6958	2506.7	2676.2	7.3614
150	19.512	2587.9	2783.0	8.6882	3.889	2585.6	2780.1	7.9401	1.9364	2582.8	2776.4	7.6134
200	21.825	2661.3	2879.5	8.9038	4.356	2659.9	2877.7	8.1580	2.172	2658.1	2875.3	7.8343
250	24.136	2736.0	2977.3	9.1002	4.820	2735.0	2976.0	8.3556	2.406	2733.7	2974.3	8.0333
300	26.445	2812.1	3076.5	9.2813	5.284	2811.3	3075.5	8.5373	2.639	2810.4	3074.3	8.2158
400	31.063	2968.9	3279.6	9.6077	6.209	2968.5	3278.9	8.8642	3.103	2967.9	3278.2	8.5435
500	35.679	3132.3	3489.1	9.8978	7.134	3132.0	3488.7	9.1546	3.565	3131.6	3488.1	8.8342
600	40.295	3302.5	3705.4	10.1608	8.057	3302.2	3705.1	9.4178	4.028	3301.9	3704.7	9.0976
700	44.911	3479.6	3928.7	10.4028	8.981	3479.4	3928.5	9.6599	4.490	3479.2	3928.2	9.3398
800	49.526	3663.8	4159.0	10.6281	9.904	3663.6	4158.9	9.8852	4.952	3663.5	4158.6	9.5652
900	54.141	3855.0	4396.4	10.8396	10.828	3854.9	4396.3	10.0967	5.414	3854.8	4396.1	9.7767
1000	58.757	4053.0	4640.6	11.0393	11.751	4052.9	4640.5	10.2964	5.875	4052.8	4640.3	9.9764
1100	63.372	4257.5	4891.2	11.2287	12.674	4257.4	4891.1	10.4859	6.337	4257.3	4891.0	10.1659
1200	67.987	4467.9	5147.8	11.4091	13.597	4467.8	5147.7	10.6662	6.799	4467.7	5147.6	10.3463
1300	72.602	4683.7	5409.7	11.5811	14.521	4683.6	5409.6	10.8382	7.260	4683.5	5409.5	10.5183

T°C	P = 0.20 MPa (120.23)				P = 0.30 MPa (133.55)				P = 0.40 MPa (143.63)			
	$\hat{V}$	$\hat{U}$	$\hat{H}$	$\hat{S}$	$\hat{V}$	$\hat{U}$	$\hat{H}$	$\hat{S}$	$\hat{V}$	$\hat{U}$	$\hat{H}$	$\hat{S}$
Sat.	0.8857	2529.5	2706.7	7.1272	0.6058	2543.6	2725.3	6.9919	0.4625	2553.6	2738.6	6.8959
150	0.9596	2576.9	2768.8	7.2795	0.6339	2570.8	2761.0	7.0778	0.4708	2564.5	2752.8	6.9299
200	1.0803	2654.4	2870.5	7.5066	0.7163	2650.7	2865.6	7.3115	0.5342	2646.8	2860.5	7.1706
250	1.1988	2731.2	2971.0	7.7086	0.7964	2728.7	2967.6	7.5166	0.5951	2726.1	2964.2	7.3789
300	1.3162	2808.6	3071.8	7.8926	0.8753	2806.7	3069.3	7.7022	0.6548	2804.8	3066.8	7.5662
400	1.5493	2966.7	3276.6	8.2218	1.0315	2965.6	3275.0	8.0330	0.7726	2964.4	3273.4	7.8985
500	1.7814	3130.8	3487.1	8.5133	1.1867	3130.0	3486.0	8.3251	0.8893	3129.2	3484.9	8.1913
600	2.013	3301.4	3704.0	8.7770	1.3414	3300.8	3703.2	8.5892	1.0055	3300.2	3702.4	8.4558
700	2.244	3478.8	3927.6	9.0194	1.4957	3478.1	3927.1	8.8319	1.1215	3477.9	3926.5	8.6987
800	2.475	3663.1	4158.2	9.2449	1.6499	3662.9	4157.8	9.0576	1.2372	3662.4	4157.3	8.9244
900	2.706	3854.5	4395.8	9.4566	1.8041	3854.2	4395.4	9.2692	1.3529	3853.9	4395.1	9.1362
1000	2.937	4052.5	4640.0	9.6563	1.9581	4052.3	4639.7	9.4690	1.4685	4052.0	4639.4	9.3360
1100	3.168	4257.0	4890.7	9.8458	2.1121	4256.8	4890.4	9.6585	1.5840	4256.5	4890.2	9.5256
1200	3.399	4467.5	5147.3	10.0262	2.2661	4467.2	5147.1	9.8389	1.6996	4467.0	5146.8	9.7060
1300	3.630	4683.2	5409.3	10.1982	2.4201	4683.0	5409.0	10.0110	1.8151	4682.8	5408.8	9.8780

‡Note: number in parenthesis is temperature of saturated steam at the specified pressure.
$\hat{V}$ [=] m^3/kg; $\hat{U}$, $\hat{H}$ [=] J/g = kJ/kg; $\hat{S}$ [=] kJ/kg K

T°C	P = 0.50 MPa (151.86)				P = 0.60 MPa (158.85)				P = 0.80 MPa (170.43)			
	$\hat{V}$	$\hat{U}$	$\hat{H}$	$\hat{S}$	$\hat{V}$	$\hat{U}$	$\hat{H}$	$\hat{S}$	$\hat{V}$	$\hat{U}$	$\hat{H}$	$\hat{S}$
Sat.	0.3749	2561.2	2748.7	6.8213	0.3157	2567.4	2756.8	6.7600	0.2404	2576.8	2769.1	6.6628
200	0.4249	2642.9	2855.4	7.0592	0.3520	2638.9	2850.1	6.9665	0.2608	2630.6	2839.3	6.8158
250	0.4744	2723.5	2960.7	7.2709	0.3938	2720.9	2957.2	7.1816	0.2931	2715.5	2950.0	7.0384
300	0.5226	2802.9	3064.2	7.4599	0.4344	2801.0	3061.6	7.3724	0.3241	2797.2	3056.5	7.2328
350	0.5701	2882.6	3167.7	7.6329	0.4742	2881.2	3165.7	7.5464	0.3544	2878.2	3161.7	7.4089
400	0.6173	2963.2	3271.9	7.7938	0.5137	2962.1	3270.3	7.7079	0.3843	2959.7	3267.1	7.5716
500	0.7109	3128.4	3483.9	8.0873	0.5920	3127.6	3482.8	8.0021	0.4433	3126.0	3480.6	7.8673
600	0.8041	3299.6	3701.7	7.3522	0.6697	3299.1	3700.9	8.2674	0.5018	3297.9	3699.4	8.1333
700	0.8969	3477.5	3925.9	8.5952	0.7472	3477.0	3925.3	8.5107	0.5601	3476.2	3924.2	8.3770
800	0.9896	3662.1	4156.9	8.8211	0.8245	3661.8	4156.5	8.7367	0.6181	3661.1	4155.6	8.6033
900	1.0822	3853.6	4394.7	9.0329	0.9017	3853.4	4394.4	8.9486	0.6761	3852.8	4393.7	8.8153
1000	1.1747	4051.8	4639.1	9.2328	0.9788	4051.5	4638.8	9.1485	0.7340	4051.0	4638.2	9.0153
1100	1.2672	4256.3	4889.9	9.4224	1.0559	4256.1	4889.6	9.3381	0.7919	4255.6	4889.1	9.2050
1200	1.3596	4466.8	5146.6	9.6029	1.1330	4466.5	5146.3	9.5185	0.8497	4466.1	5145.9	9.3855
1300	1.4521	4682.5	5408.6	9.7749	1.2101	4682.3	5408.3	9.6906	0.9076	4681.8	5407.9	9.5575

T°C	P = 1.00 MPa (179.91)				P = 1.20 MPa (187.99)				P = 1.40 MPa (195.07)			
	$\hat{V}$	$\hat{U}$	$\hat{H}$	$\hat{S}$	$\hat{V}$	$\hat{U}$	$\hat{H}$	$\hat{S}$	$\hat{V}$	$\hat{U}$	$\hat{H}$	$\hat{S}$
Sat.	0.194 44	2583.6	2778.1	6.5865	0.163 33	2588.8	2784.8	6.5233	0.140 84	2592.8	2790.0	6.4693
200	0.2060	2621.9	2827.9	6.6940	0.169 30	2612.8	2815.9	6.5898	0.143 02	2603.1	2803.3	6.4975
250	0.2327	2709.9	2942.6	6.9247	0.192 34	2704.2	2935.0	6.8294	0.163 50	2698.3	2927.2	6.7467
300	0.2579	2793.2	3051.2	7.1229	0.2138	2789.2	3045.8	7.0317	0.182 28	2785.2	3040.4	6.9534
350	0.2825	2875.2	3157.7	7.3011	0.2345	2872.2	3153.6	7.2121	0.2003	2869.2	3149.5	7.1360
400	0.3066	2957.3	3263.9	7.4651	0.2548	2954.9	3260.7	7.3774	0.2178	2952.5	3257.5	7.3026
500	0.3541	3124.4	3478.5	7.7622	0.2946	3122.8	3476.3	7.6759	0.2521	3121.1	3474.1	7.6027
600	0.4011	3296.8	3697.9	8.0290	0.3339	3295.6	3696.3	7.9435	0.2860	3294.4	3694.8	7.8710
700	0.4478	3475.3	3923.1	8.2731	0.3729	3474.4	3922.0	8.1881	0.3195	3473.6	3920.8	8.1160
800	0.4943	3660.4	4154.7	8.4996	0.4118	3659.7	4153.8	8.4148	0.3528	3659.0	4153.0	8.3431
900	0.5407	3852.2	4392.9	8.7118	0.4505	3851.6	4392.2	8.6272	0.3861	3851.1	4391.5	8.5556
1000	0.5871	4050.5	4637.6	8.9119	0.4892	4050.0	4637.0	8.8274	0.4192	4049.5	4636.4	8.7559
1100	0.6335	4255.1	4888.6	9.1017	0.5278	4254.6	4888.0	9.0172	0.4524	4254.1	4887.5	8.9457
1200	0.6798	4465.6	5145.4	9.2822	0.5665	4465.1	5144.9	9.1977	0.4855	4464.7	5144.4	9.1262
1300	0.7261	4681.3	5407.4	9.4543	0.6051	4680.9	5407.0	9.3698	0.5186	4680.4	5406.5	9.2984

	P = 1.60 MPa (201.41)				P = 1.80 MPa (207.15)				P = 2.00 MPa (212.42)			
T°C	$\hat{V}$	$\hat{U}$	$\hat{H}$	$\hat{S}$	$\hat{V}$	$\hat{U}$	$\hat{H}$	$\hat{S}$	$\hat{V}$	$\hat{U}$	$\hat{H}$	$\hat{S}$
Sat.	0.123 80	2596.0	2794.0	6.4218	0.110 42	2598.4	2797.1	6.3794	0.099 63	2600.3	2799.5	6.3409
225	0.132 87	2644.7	2857.3	6.5518	0.116 73	2636.6	2846.7	6.4808	0.103 77	2628.3	2835.8	6.4147
250	0.141 84	2692.3	2919.2	6.6732	0.124 97	2686.0	2911.0	6.6066	0.111 44	2679.6	2902.5	6.5453
300	0.158 62	2781.1	3034.8	6.8844	0.140 21	2776.9	3029.2	6.8226	0.125 47	2772.6	3023.5	6.7664
350	0.174 56	2866.1	3145.4	7.0694	0.154 57	2863.0	3141.2	7.0100	0.138 57	2859.8	3137.0	6.9563
400	0.190 05	2950.1	3254.2	7.2374	0.168 47	2947.7	3250.9	7.1794	0.151 20	2945.2	3247.6	7.1271
500	0.2203	3119.5	3472.0	7.5390	0.195 50	3117.9	3469.8	7.4825	0.175 68	3116.2	3467.6	7.4317
600	0.2500	3293.3	3693.2	7.8080	0.2220	3292.1	3691.7	7.7523	0.199 60	3290.9	3690.1	7.7024
700	0.2794	3472.7	3919.7	8.0535	0.2482	3471.8	3918.5	7.9983	0.2232	3470.9	3917.4	7.9487
800	0.3086	3658.3	4152.1	8.2808	0.2742	3657.6	4151.2	8.2258	0.2467	3657.0	4150.3	8.1765
900	0.3377	3850.5	4390.8	8.4935	0.3001	3849.9	4390.1	8.4386	0.2700	3849.3	4389.4	8.3895
1000	0.3668	4049.0	4635.8	8.6938	0.3260	4048.5	4635.2	8.6391	0.2933	4048.0	4634.6	8.5901
1100	0.3958	4253.7	4887.0	8.8837	0.3518	4253.2	4886.4	8.8290	0.3166	4252.7	4885.9	8.7800
1200	0.4248	4464.2	5143.9	9.0643	0.3776	4463.7	5143.4	9.0096	0.3398	4463.3	5142.9	8.9607
1300	0.4538	4679.9	5406.0	9.2364	0.4034	4679.5	5405.6	9.1818	0.3631	4679.0	5405.1	9.1329
	P = 2.50 MPa (223.99)				P = 3.00 MPa (233.90)				P = 3.50 MPa (242.60)			
Sat.	0.079 98	2603.1	2803.1	6.2575	0.066 68	2604.1	2804.2	6.1869	0.057 07	2603.7	2803.4	6.1253
225	0.080 27	2605.6	2806.3	6.2639	—	—	—	—	—	—	—	—
250	0.087 00	2662.6	2880.1	6.4085	0.070 58	2644.0	2855.8	6.2872	0.058 72	2623.7	2829.2	6.1749
300	0.098 90	2761.6	3008.8	6.6438	0.081 14	2750.1	2993.5	6.5390	0.068 42	2738.0	2977.5	6.4461
350	0.109 76	2851.9	3126.3	6.8403	0.090 53	2843.7	3115.3	6.7428	0.076 78	2835.3	3104.0	6.6579
400	0.120 10	2939.1	3239.3	7.0148	0.099 36	2932.8	3230.9	6.9212	0.084 53	2926.4	3222.3	6.8405
450	0.130 14	3025.5	3350.8	7.1746	0.107 87	3020.4	3344.0	7.0834	0.091 96	3015.3	3337.2	7.0052
500	0.139 98	3112.1	3462.1	7.3234	0.116 19	3108.0	3456.5	7.2338	0.099 18	3103.0	3450.9	7.1572
600	0.159 30	3288.0	3686.3	7.5960	0.132 43	3285.0	3682.3	7.5085	0.113 24	3282.1	3678.4	7.4339
700	0.178 32	3468.7	3914.5	7.8435	0.148 38	3466.5	3911.7	7.7571	0.126 99	3464.3	3908.8	7.6837
800	0.197 16	3655.3	4148.2	8.0720	0.164 14	3653.5	4145.9	7.9862	0.140 56	3651.8	4143.7	7.9134
900	0.215 90	3847.9	4387.6	8.2853	0.179 80	3846.5	4385.9	8.1999	0.154 02	3845.0	4384.1	8.1276
1000	0.2346	4046.7	4633.1	8.4861	0.195 41	4045.4	4631.6	8.4009	0.167 43	4044.1	4630.1	8.3288
1100	0.2532	4251.5	4884.6	8.6762	0.210 98	4250.3	4883.3	8.5912	0.180 80	4249.2	4881.9	8.5192
1200	0.2718	4462.1	5141.7	8.8569	0.226 52	4460.9	5140.5	8.7720	0.194 15	4459.8	5139.3	8.7000
1300	0.2905	4677.8	5404.0	9.0291	0.242 06	4676.6	5402.8	8.9442	0.207 49	4675.5	5401.7	8.8723

$\hat{V}$ [=] m^3/kg; $\hat{U}$, $\hat{H}$ [=] J/g = kJ/kg; $\hat{S}$ [=] kJ/kg K

SUPERHEATED VAPOR *(Continued)*

	P = 4.0 MPa (250.40)				P = 4.5 MPa (257.49)				P = 5.0 MPa (263.99)			
T°C	$\hat{V}$	$\hat{U}$	$\hat{H}$	$\hat{S}$	$\hat{V}$	$\hat{U}$	$\hat{H}$	$\hat{S}$	$\hat{V}$	$\hat{U}$	$\hat{H}$	$\hat{S}$
Sat.	0.049 78	2602.3	2801.4	6.0701	0.044 06	2600.1	2798.3	6.0198	0.039 44	2597.1	2794.3	5.9734
275	0.054 57	2667.9	2886.2	6.2285	0.047 30	2650.3	2863.2	6.1401	0.041 41	2631.3	2838.3	6.0544
300	0.058 84	2725.3	2960.7	6.3615	0.051 35	2712.0	2943.1	6.2828	0.045 32	2698.0	2924.5	6.2084
350	0.066 45	2826.7	3092.5	6.5821	0.058 40	2817.8	3080.6	6.5131	0.051 94	2808.7	3068.4	6.4493
400	0.073 41	2919.9	3213.6	6.7690	0.064 75	2913.3	3204.7	6.7047	0.057 81	2906.6	3195.7	6.6459
450	0.080 02	3010.2	3330.3	6.9363	0.070 74	3005.0	3323.3	6.8746	0.063 30	2999.7	3316.2	6.8186
500	0.086 43	3099.5	3445.3	7.0901	0.076 51	3095.3	3439.6	7.0301	0.068 57	3091.0	3433.8	6.9759
600	0.098 85	3279.1	3674.4	7.3688	0.087 65	3276.0	3670.5	7.3110	0.078 69	3273.0	3666.5	7.2589
700	0.110 95	3462.1	3905.9	7.6198	0.098 47	3459.9	3903.0	7.5631	0.088 49	3457.6	3900.1	7.5122
800	0.122 87	3650.0	4141.5	7.8502	0.109 11	3648.3	4139.3	7.7942	0.098 11	3646.6	4137.1	7.7440
900	0.134 69	3843.6	4382.3	8.0647	0.119 65	3842.2	4380.6	8.0091	0.107 62	3840.7	4378.8	7.9593
1000	0.146 45	4042.9	4628.7	8.2662	0.130 13	4041.6	4627.2	8.2108	0.117 07	4040.4	4625.7	8.1612
1100	0.158 17	4248.0	4880.6	8.4567	0.140 56	4246.8	4879.3	8.4015	0.126 48	4245.6	4878.0	8.3520
1200	0.169 87	4458.6	5138.1	8.6376	0.150 98	4457.5	5136.9	8.5825	0.135 87	4456.3	5135.7	8.5331
1300	0.181 56	4674.3	5400.5	8.8100	0.161 39	4673.1	5399.4	8.7549	0.145 26	4672.0	5398.2	8.7055

T°C	P = 6.0 MPa (275.64)				P = 7.0 MPa (285.88)				P = 8.0 MPa (295.06)			
	$\hat{V}$	$\hat{U}$	$\hat{H}$	$\hat{S}$	$\hat{V}$	$\hat{U}$	$\hat{H}$	$\hat{S}$	$\hat{V}$	$\hat{U}$	$\hat{H}$	$\hat{S}$
Sat.	0.032 44	2589.7	2784.3	5.8892	0.027 37	2580.5	2772.1	5.8133	0.023 52	2569.8	2758.0	5.7432
300	0.036 16	2667.2	2884.2	6.0674	0.029 47	2632.2	2838.4	5.9305	0.024 26	2590.9	2785.0	5.7906
350	0.042 23	2789.6	3043.0	6.3335	0.035 24	2769.4	3016.0	6.2283	0.029 95	2747.7	2987.3	6.1301
400	0.047 39	2892.9	3177.2	6.5408	0.039 93	2878.6	3158.1	6.4478	0.034 32	2863.8	3138.3	6.3634
450	0.052 14	2988.9	3301.8	6.7193	0.044 16	2978.0	3287.1	6.6327	0.038 17	2966.7	3272.0	6.5551
500	0.056 65	3082.2	3422.2	6.8803	0.048 14	3073.4	3410.3	6.7975	0.041 75	3064.3	3398.3	6.7240
550	0.061 01	3174.6	3540.6	7.0288	0.051 95	3167.2	3530.9	6.9486	0.045 16	3159.8	3521.0	6.8778
600	0.065 25	3266.9	3658.4	7.1677	0.055 65	3260.7	3650.3	7.0894	0.048 45	3254.4	3642.0	7.0206
700	0.073 52	3453.1	3894.2	7.4234	0.062 83	3448.5	3888.3	7.3476	0.054 81	3443.9	3882.4	7.2812
800	0.081 60	3643.1	4132.7	7.6566	0.069 81	3639.5	4128.2	7.5822	0.060 97	3636.0	4123.8	7.5173
900	0.089 58	3837.8	4375.3	7.8727	0.076 69	3835.0	4371.8	7.7991	0.067 02	3832.1	4368.3	7.7351
1000	0.097 49	4037.8	4622.7	8.0751	0.083 50	4035.3	4619.8	8.0020	0.073 01	4032.8	4616.9	7.9384
1100	0.105 36	4243.3	4875.4	8.2661	0.090 27	4240.9	4872.8	8.1933	0.078 96	4238.6	4870.3	8.1300
1200	0.113 21	4454.0	5133.3	8.4474	0.097 03	4451.7	5130.9	8.3747	0.084 89	4449.5	5128.5	8.3115
1300	0.121 06	4669.6	5396.0	8.6199	0.103 77	4667.3	5393.7	8.5473	0.090 80	4665.0	5391.5	8.4842

T°C	P = 9.0 MPa (303.40)				P = 10.0 MPa (311.06)				P = 12.5 MPa (327.89)			
	$\hat{V}$	$\hat{U}$	$\hat{H}$	$\hat{S}$	$\hat{V}$	$\hat{U}$	$\hat{H}$	$\hat{S}$	$\hat{V}$	$\hat{U}$	$\hat{H}$	$\hat{S}$
Sat.	0.020 48	2557.8	2742.1	5.6772	0.018 026	2544.4	2724.7	5.6141	0.013 495	2505.1	2673.8	5.4624
325	0.023 27	2646.6	2856.0	5.8712	0.019 861	2610.4	2809.1	5.7568	—	—	—	—
350	0.025 80	2724.4	2956.6	6.0361	0.022 42	2699.2	2923.4	5.9443	0.016 126	2624.6	2826.2	5.7118
400	0.029 93	2848.4	3117.8	6.2854	0.026 41	2832.4	3096.5	6.2120	0.020 00	2789.3	3039.3	6.0417
450	0.033 50	2955.2	3256.6	6.4844	0.029 75	2943.4	3240.9	6.4190	0.022 99	2912.5	3199.8	6.2719
500	0.036 77	3055.2	3386.1	6.6576	0.032 79	3045.8	3373.7	6.5966	0.025 60	3021.7	3341.8	6.4618
550	0.039 87	3152.2	3511.0	6.8142	0.035 64	3144.6	3500.9	6.7561	0.028 01	3125.0	3475.2	6.6290
600	0.042 85	3248.1	3633.7	6.9589	0.038 37	3241.7	3625.3	6.9029	0.030 29	3225.4	3604.0	6.7810
650	0.045 74	3343.6	3755.3	7.0943	0.041 01	3338.2	3748.2	7.0398	0.032 48	3324.4	3730.4	6.9218
700	0.048 57	3439.3	3876.5	7.2221	0.043 58	3434.7	3870.5	7.1687	0.034 60	3422.9	3855.3	7.0536
800	0.054 09	3632.5	4119.3	7.4596	0.048 59	3628.9	4114.8	7.4077	0.038 69	3620.0	4103.6	7.2965
900	0.059 50	3829.2	4364.8	7.6783	0.053 49	3826.3	4361.2	7.6272	0.042 67	3819.1	4352.5	7.5182
1000	0.064 85	4030.3	4614.0	7.8821	0.058 32	4027.8	4611.0	7.8315	0.046 58	4021.6	4603.8	7.7237
1100	0.070 16	4236.3	4867.7	8.0740	0.063 12	4234.0	4865.1	8.0237	0.050 45	4228.2	4858.8	7.9165
1200	0.075 44	4447.2	5126.2	8.2556	0.067 89	4444.9	5123.8	8.2055	0.054 30	4439.3	5118.0	8.0987
1300	0.080 72	4662.7	5389.2	8.4284	0.072 65	4460.5	5387.0	8.3783	0.058 13	4654.8	5381.4	8.2717

$\hat{V}$ [=] m^3/kg; $\hat{U}$, $\hat{H}$ [=] J/g = kJ/kg; $\hat{S}$ [=] kJ/kg K

SUPERHEATED VAPOR *(Continued)*

	P = 15.0 MPa (342.24)				P = 17.5 MPa (354.75)				P = 20.0 MPa (365.81)			
T°C	$\hat{V}$	$\hat{U}$	$\hat{H}$	$\hat{S}$	$\hat{V}$	$\hat{U}$	$\hat{H}$	$\hat{S}$	$\hat{V}$	$\hat{U}$	$\hat{H}$	$\hat{S}$
Sat.	0.010 337	2455.5	2610.5	5.3098	0.007 920	2390.2	2528.8	5.1419	0.005 834	2293.0	2409.7	4.9269
350	0.011 470	2520.4	2692.4	5.4421	—	—	—	—	—	—	—	—
400	0.015 649	2740.7	2975.5	5.8811	0.012 447	2685.0	2902.9	5.7213	0.009 942	2619.3	2818.1	5.5540
450	0.018 445	2879.5	3156.2	6.1404	0.015 174	2844.2	3109.7	6.0184	0.012 695	2806.2	3060.1	5.9017
500	0.020 80	2996.6	3308.6	6.3443	0.017 358	2970.3	3274.1	6.2383	0.014 768	2942.9	3238.2	6.1401
550	0.022 93	3104.7	3448.6	6.5199	0.019 288	3083.9	3421.4	6.4230	0.016 555	3062.4	3393.5	6.3348
600	0.024 91	3208.6	3582.3	6.6776	0.021 06	3191.5	3560.1	6.5866	0.018 178	3174.0	3537.6	6.5048
650	0.026 80	3310.3	3712.3	6.8224	0.022 74	3296.0	3693.9	6.7357	0.019 693	3281.4	3675.3	6.6582
700	0.028 61	3410.9	3840.1	6.9572	0.024 34	3398.7	3824.6	6.8736	0.021 13	3386.4	3809.0	6.7993
800	0.032 10	3610.9	4092.4	7.2040	0.027 38	3601.8	4081.1	7.1244	0.023 85	3592.7	4069.7	7.0544
900	0.035 46	3811.9	4343.8	7.4279	0.030 31	3804.7	4335.1	7.3507	0.026 45	3797.5	4326.4	7.2830
1000	0.038 75	4015.4	4596.6	7.6348	0.033 16	4009.3	4589.5	7.5589	0.028 97	4003.1	4582.5	7.4925
1100	0.042 00	4222.6	4852.6	7.8283	0.035 97	4216.9	4846.4	7.7531	0.031 45	4211.3	4840.2	7.6874
1200	0.045 23	4433.8	5112.3	8.0108	0.038 76	4428.3	5106.6	7.9360	0.033 91	4422.8	5101.0	7.8707
1300	0.048 45	4649.1	5376.0	8.1840	0.041 54	4643.5	5370.5	8.1093	0.036 36	4638.0	5365.1	8.0442

T°C	P = 25.0 MPa $\hat{V}$	$\hat{U}$	$\hat{H}$	$\hat{S}$	P = 30.0 MPa $\hat{V}$	$\hat{U}$	$\hat{H}$	$\hat{S}$	P = 35.0 MPa $\hat{V}$	$\hat{U}$	$\hat{H}$	$\hat{S}$
375	0.001 973	1798.7	1848.0	4.0320	0.001 789	1737.8	1791.5	3.9305	0.001 700	1702.9	1762.4	3.8722
400	0.006 004	2430.1	2580.2	5.1418	0.002 790	2067.4	2151.1	4.4728	0.002 100	1914.1	1987.6	4.2126
425	0.007 881	2609.2	2806.3	5.4723	0.005 303	2455.1	2614 2	5.1504	0.003 428	2253.4	2373.4	4.7747
450	0.009 162	2720.7	2949.7	5.6744	0.006 735	2619.3	2821.4	5.4424	0.004 961	2498.7	2672.4	5.1962
500	0.011 123	2884.3	3162.4	5.9592	0.008 678	2820.7	3081.1	5.7905	0.006 927	2751.9	2994.4	5.6282
550	0.012 724	3017.5	3335.6	6.1765	0.010 168	2970.3	3275.4	6.0342	0.008 345	2921.0	3213.0	5.9026
600	0.014 137	3137.9	3491.4	6.3602	0.011 446	3100.5	3443.9	6.2331	0.009 527	3062.0	3395.5	6.1179
650	0.015 433	3251.6	3637.4	6.5229	0.012 596	3221.0	3598.9	6.4058	0.010 575	3189.8	3559.9	6.3010
700	0.016 646	3361.3	3777.5	6.6707	0.013 661	3335.8	3745.6	6.5606	0.011 533	3309.8	3713.5	6.4631
800	0.018 912	3574.3	4047.1	6.9345	0.015 623	3555.5	4024.2	6.8332	0.013 278	3536.7	4001.5	6.7450
900	0.021 045	3783.0	4309.1	7.1680	0.017 448	3768.5	4291.9	7.0718	0.014 883	3754.0	4274.9	6.9886
1000	0.023 10	3990.9	4568.5	7.3802	0.019 196	3978.8	4554.7	7.2867	0.016 410	3966.7	4541.1	7.2064
1100	0.025 12	4200.2	4828.2	7.5765	0.020 903	4189.2	4816.3	7.4845	0.017 895	4178.3	4804.6	7.4057
1200	0.027 11	4412.0	5089.9	7.7605	0.022 589	4401.3	5079.0	7.6692	0.019 360	4390.7	5068.3	7.5910
1300	0.029 10	4626.9	5354.4	7.9342	0.024 266	4616.0	5344.0	7.8432	0.020 815	4605.1	5333.6	7.7653
	P = 40.0 MPa				**P = 50.0 MPa**				**P = 60.0 MPa**			
375	0.001 641	1677.1	1742.8	3.8290	0.001 559	1638.6	1716.6	3.7639	0.001 503	1609.4	1699.5	3.7141
400	0.001 908	1854.6	1930.9	4.1135	0.001 731	1788.1	1874.6	4.0031	0.001 634	1745.4	1843.4	3.9318
425	0.002 532	2096.9	2198.1	4.5029	0.002 007	1959.7	2060.0	4.2734	0.001 817	1892.7	2001.7	4.1626
450	0.003 693	2365.1	2512.8	4.9459	0.002 486	2159.6	2284.0	4.5884	0.002 085	2053.9	2179.0	4.4121
500	0.005 622	2678.4	2903.3	5.4700	0.003 892	2525.5	2720.1	5.1726	0.002 956	2390.6	2567.9	4.9321
550	0.006 984	2869.7	3149.1	5.7785	0.005 118	2763.6	3019.5	5.5485	0.003 956	2658.8	2896.2	5.3441
600	0.008 094	3022.6	3346.4	6.0114	0.006 112	2942.0	3247.6	5.8178	0.004 834	2861.1	3151.2	5.6452
650	0.009 063	3158.0	3520.6	6.2054	0.006 966	3093.5	3441.8	6.0342	0.005 595	3028.8	3364.5	5.8829
700	0.009 941	3283.6	3681.2	6.3750	0.007 727	3230.5	3616.8	6.2189	0.006 272	3177.2	3553.5	6.0824
800	0.011 523	3517.8	3978.7	6.6662	0.009 076	3479.8	3933.6	6.5290	0.007 459	3441.5	3889.1	6.4109
900	0.012 962	3739.4	4257.9	6.9150	0.010 283	3710.3	4224.4	6.7882	0.008 508	3681.0	4191.5	6.6805
1000	0.014 324	3954.6	4527.6	7.1356	0.011 411	3930.5	4501.1	7.0146	0.009 480	3906.4	4475.2	6.9127
1100	0.015 642	4167.4	4793.1	7.3364	0.012 496	4145.7	4770.5	7.2184	0.010 409	4124.1	4748.6	7.1195
1200	0.016 940	4380.1	5057.7	7.5224	0.013 561	4359.1	5037.2	7.4058	0.011 317	4338.2	5017.2	7.3083
1300	0.018 229	4594.3	5323.5	7.6969	0.014 616	4572.8	5303.6	7.5808	0.012 215	4551.4	5284.3	7.4837

$\hat{V}$ [=] m³/kg; $\hat{U}$, $\hat{H}$ [=] J/g = kJ/kg; $\hat{S}$ [=] kJ/kg K

COMPRESSED LIQUID

T°C	P = 5 MPa (263.99)				P = 10 MPa (311.06)				P = 15 MPa (342.24)			
	$\hat{V}$	$\hat{U}$	$\hat{H}$	$\hat{S}$	$\hat{V}$	$\hat{U}$	$\hat{H}$	$\hat{S}$	$\hat{V}$	$\hat{U}$	$\hat{H}$	$\hat{S}$
Sat.	0.001 285 9	1147.8	1154.2	2.9202	0.001 452 4	1393.0	1407.6	3.3596	0.001 658 1	1585.6	1610.5	3.6848
0	0.000 997 7	0.04	5.04	0.0001	0.000 995 2	0.09	10.04	0.0002	0.000 992 8	0.15	15.05	0.0004
20	0.000 999 5	83.65	88.65	0.2956	0.000 997 2	83.36	93.33	0.2945	0.000 995 0	83.06	97.99	0.2934
40	0.001 005 6	166.95	171.97	0.5705	0.001 003 4	166.35	176.38	0.5686	0.001 001 3	165.76	180.78	0.5666
60	0.001 014 9	250.23	255.30	0.8285	0.001 012 7	249.36	259.49	0.8258	0.001 010 5	248.51	263.67	0.8232
80	0.001 026 8	333.72	338.85	1.0720	0.001 024 5	332.59	342.83	1.0688	0.001 022 2	331.48	346.81	1.0656
100	0.001 041 0	417.52	422.72	1.3030	0.001 038 5	416.12	426.50	1.2992	0.001 036 1	414.74	430.28	1.2955
120	0.001 057 6	501.80	507.09	1.5233	0.001 054 9	500.08	510.64	1.5189	0.001 052 2	498.40	514.19	1.5145
140	0.001 076 8	586.76	592.15	1.7343	0.001 073 7	584.68	595.42	1.7292	0.001 070 7	582.66	598.72	1.7242
160	0.001 098 8	672.62	678.12	1.9375	0.001 095 3	670.13	681.08	1.9317	0.001 091 8	667.71	684.09	1.9260
180	0.001 124 0	759.63	765.25	2.1341	0.001 119 9	756.65	767.84	2.1275	0.001 115 9	753.76	770.50	2.1210
200	0.001 153 0	848.1	853.9	2.3255	0.001 148 0	844.5	856.0	2.3178	0.001 143 3	841.0	858.2	2.3104
220	0.001 186 6	938.4	944.4	2.5128	0.001 180 5	934.1	945.9	2.5039	0.001 174 8	929.9	947.5	2.4953
240	0.001 226 4	1031.4	1037.5	2.6979	0.001 218 7	1026.0	1038.1	2.6872	0.001 211 4	1020.8	1039.0	2.6771
260	0.001 274 9	1127.9	1134.3	2.8830	0.001 264 5	1121.1	1133.7	2.8699	0.001 255 0	1114.6	1133.4	2.8576
280					0.001 321 6	1220.9	1234.1	3.0548	0.001 308 4	1212.5	1232.1	3.0393
300					0.001 397 2	1328.4	1342.3	3.2469	0.001 377 0	1316.6	1337.3	3.2260
320									0.001 472 4	1431.1	1453.2	3.4247
340									0.001 631 1	1567.5	1591.9	3.6546

	P = 20 MPa (365.81)				P = 30 MPa				P = 50 MPa			
T°C	$\hat{V}$	$\hat{U}$	$\hat{H}$	$\hat{S}$	$\hat{V}$	$\hat{U}$	$\hat{H}$	$\hat{S}$	$\hat{V}$	$\hat{U}$	$\hat{H}$	$\hat{S}$
Sat.	0.002 036	1785.6	1826.3	4.0139								
0	0.000 990 4	0.19	20.01	0.0004	0.000 985 6	0.25	29.82	0.0001	0.000 976 6	0.20	49.03	−0.0014
20	0.000 992 8	82.77	102.62	0.2923	0.000 988 6	82.17	111.84	0.2899	0.000 980 4	81.00	130.02	0.2848
40	0.000 999 2	165.17	185.16	0.5646	0.000 995 1	164.04	193.89	0.5607	0.000 987 2	161.86	211.21	0.5527
60	0.001 008 4	247.68	267.85	0.8206	0.001 004 2	246.06	276.19	0.8154	0.000 996 2	242.98	292.79	0.8052
80	0.001 019 9	330.40	350.80	1.0624	0.001 015 6	328.30	358.77	1.0561	0.001 007 3	324.34	374.70	1.0440
100	0.001 033 7	413.39	434.06	1.2917	0.001 029 0	410.78	441.66	1.2844	0.001 020 1	405.88	456.89	1.2703
120	0.001 049 6	496.76	517.76	1.5102	0.001 044 5	493.59	524.93	1.5018	0.001 034 8	487.65	539.39	1.4857
140	0.001 067 8	580.69	602.04	1.7193	0.001 062 1	576.88	608.75	1.7098	0.001 051 5	569.77	622.35	1.6915
160	0.001 088 5	665.35	687.12	1.9204	0.001 082 1	660.82	693.28	1.9096	0.001 070 3	652.41	705.92	1.8891
180	0.001 112 0	750.95	773.20	2.1147	0.001 104 7	745.59	778.73	2.1024	0.001 091 2	735.69	790.25	2.0794
200	0.001 138 8	837.7	860.5	2.3031	0.001 130 2	831.4	865.3	2.2893	0.001 114 6	819.7	875.5	2.2634
220	0.001 169 3	925.9	949.3	2.4870	0.001 159 0	918.3	953.1	2.4711	0.001 140 8	904.7	961.7	2.4419
240	0.001 204 6	1016.0	1040.0	2.6674	0.001 192 0	1006.9	1042.6	2.6490	0.001 170 2	990.7	1049.2	2.6158
260	0.001 246 2	1108.6	1133.5	2.8459	0.001 230 3	1097.4	1134.3	2.8243	0.001 203 4	1078.1	1138.2	2.7860
280	0.001 296 5	1204.7	1230.6	3.0248	0.001 275 5	1190.7	1229.0	2.9986	0.001 241 5	1167.2	1229.3	2.9537
300	0.001 359 6	1306.1	1333.3	3.2071	0.001 330 4	1287.9	1327.8	3.1741	0.001 286 0	1258.7	1323.0	3.1200
320	0.001 443 7	1415.7	1444.6	3.3979	0.001 399 7	1390.7	1432.7	3.3539	0.001 338 8	1353.3	1420.2	3.2868
340	0.001 568 4	1539.7	1571.0	3.6075	0.001 492 0	1501.7	1546.5	3.5426	0.001 403 2	1452.0	1522.1	3.4557
360	0.001 822 6	1702.8	1739.3	3.8772	0.001 626 5	1626.6	1675.4	3.7494	0.001 483 8	1556.0	1630.2	3.6291
380					0.001 869 1	1781.4	1837.5	4.0012	0.001 588 4	1667.2	1746.6	3.8101

$\hat{V}$ [=] m^3/kg; $\hat{U}$, $\hat{H}$ [=] J/g = kJ/kg; $\hat{S}$ [=] kJ/kg K

SATURATED SOLID-VAPOR

	Specific Volume			Internal Energy			Enthalpy			Entropy		
Temp. °C T	Press. kPa P	Sat. Solid $\hat{V}^S \times 10^3$	Sat. Vapor $\hat{V}^V$	Sat. Solid $\hat{U}^S$	Subl. $\Delta\hat{U}$	Sat. Vapor $\hat{U}^V$	Sat. Solid $\hat{H}^S$	Subl. $\Delta\hat{H}$	Sat. Vapor $\hat{H}^V$	Sat. Solid $\hat{S}^S$	Subl. $\Delta\hat{S}$	Sat. Vapor $\hat{S}^V$
0.01	0.6113	1.0908	206.1	−333.40	2708.7	2375.3	−333.40	2834.8	2501.4	−1.221	10.378	9.156
0	0.6108	1.0908	206.3	−333.43	2708.8	2375.3	−333.43	2834.8	2501.3	−1.221	10.378	9.157
−2	0.5176	1.0904	241.7	−337.62	2710.2	2372.6	−337.62	2835.3	2497.7	−1.237	10.456	9.219
−4	0.4375	1.0901	283.8	−341.78	2711.6	2369.8	−341.78	2835.7	2494.0	−1.253	10.536	9.283
−6	0.3689	1.0898	334.2	−345.91	2712.9	2367.0	−345.91	2836.2	2490.3	−1.268	10.616	9.348
−8	0.3102	1.0894	394.4	−350.02	2714.2	2364.2	−350.02	2836.6	2486.6	−1.284	10.698	9.414
−10	0.2602	1.0891	466.7	−354.09	2715.5	2361.4	−354.09	2837.0	2482.9	−1.299	10.781	9.481
−12	0.2176	1.0888	553.7	−358.14	2716.8	2358.7	−358.14	2837.3	2479.2	−1.315	10.865	9.550
−14	0.1815	1.0884	658.8	−362.15	2718.0	2355.9	−362.15	2837.6	2475.5	−1.331	10.950	9.619
−16	0.1510	1.0881	786.0	−366.14	2719.2	2353.1	−366.14	2837.9	2471.8	−1.346	11.036	9.690
−18	0.1252	1.0878	940.5	−370.10	2720.4	2350.3	−370.10	2838.2	2468.1	−1.362	11.123	9.762
−20	0.1035	1.0874	1128.6	−374.03	2721.6	2347.5	−374.03	2838.4	2464.3	−1.377	11.212	9.835
−22	0.0853	1.0871	1358.4	−377.93	2722.7	2344.7	−377.93	2838.6	2460.6	−1.393	11.302	9.909
−24	0.0701	1.0868	1640.1	−381.80	2723.7	2342.0	−381.80	2838.7	2456.9	−1.408	11.394	9.985
−26	0.0574	1.0864	1986.4	−385.64	2724.8	2339.2	−385.64	2838.9	2453.2	−1.424	11.486	10.062
−28	0.0469	1.0861	2413.7	−389.45	2725.8	2336.4	−389.45	2839.0	2449.5	−1.439	11.580	10.141
−30	0.0381	1.0858	2943	−393.23	2726.8	2333.6	−393.23	2839.0	2445.8	−1.455	11.676	10.221
−32	0.0309	1.0854	3600	−396.98	2727.8	2330.8	−396.98	2839.1	2442.1	−1.471	11.773	10.303
−34	0.0250	1.0851	4419	−400.71	2728.7	2328.0	−400.71	2839.1	2438.4	−1.486	11.872	10.386
−36	0.0201	1.0848	5444	−404.40	2729.6	2325.2	−404.40	2839.1	2434.7	−1.501	11.972	10.470
−38	0.0161	1.0844	6731	−408.06	2730.5	2322.4	−408.06	2839.0	2430.9	−1.517	12.073	10.556
−40	0.0129	1.0841	8354	−411.70	2731.3	2319.6	−411.70	2838.9	2427.2	−1.532	12.176	10.644

$\hat{V}$ [=] m^3/kg; $\hat{U}$, $\hat{H}$ [=] J/g = kJ/kg; $\hat{S}$ [=] kJ/kg K

Heats and Free Energies of Formation

Compound	State†	Heat of formation ‡ § at 25°C., kcal./mole	Free energy of formation ‖ ¶ at 25°C., kcal./mole
Aluminum:			
Al	c	0.00	0.00
$AlBr_3$	c	−123.4	
	aq	−209.5	−189.2
Al_4C_3	c	−30.8	−29.0
$AlCl_3$	c	−163.8	
	aq, 600	−243.9	−209.5
AlF_3	c	−329	
	aq	−360.8	−312.6
AlI_3	c	−72.8	
	aq	−163.4	−152.5
AlN	c	−57.7	−50.4
$Al(NH_4)(SO_4)_2$	c	−561.19	−486.17
$Al(NH_4)(SO_4)_2.12H_2O$	c	−1419.36	−1179.26
$Al(NO_3)_3.6H_2O$	c	−680.89	−526.32
$Al(NO_3)_3.9H_2O$	c	−897.59	
Al_2O_3	c, corundum	−399.09	−376.87
$Al(OH)_3$	c	−304.8	−272.9
$Al_2O_3.SiO_2$	c, sillimanite	−648.7	
$Al_2O_3.SiO_2$	c, disthene	−642.4	
$Al_2O_3.SiO_2$	c, andalusite	−642.0	
$3Al_2O_3.2SiO_2$	c, mullite	−1874	
Al_2S_3	c	−121.6	
$Al_2(SO_4)_3$	c	−820.99	−739.53
	aq	−893.9	−759.3
$Al_2(SO_4)_3.6H_2O$	c	−1268.15	−1103.39
$Al_2(SO_4)_3.18H_2O$	c	−2120	
Antimony:			
Sb	c	0.00	0.00
$SbBr_3$	c	−59.9	
$SbCl_3$	c	−91.3	−77.8
$SbCl_5$	l	−104.8	
SbF_3	c	−216.6	
SbI_3	c	−22.8	
Sb_2O_3	c, I, orthorhombic	−165.4	−146.0
	c, II, octahedral	−166.6	
Sb_2O_4	c	−213.0	−186.6
Sb_2O_5	c	−230.0	−196.1
Sb_2S_3	c, black	−38.2	−36.9
Arsenic:			
As	c	0.00	0.00
$AsBr_3$	c	−45.9	
$AsCl_3$	l	−80.2	−70.5
AsF_3	l	−223.76	−212.27
AsH_3	g	43.6	37.7
AsI_3	c	−13.6	
As_2O_3	c	−154.1	−134.8
As_2O_5	c	−217.9	−183.9
As_2S_3	c	−20	−20
	amorphous	−34.76	
Barium:			
Ba	c	0.00	0.00
$BaBr_2$	c	−180.38	
	aq, 400	−185.67	−183.0
$BaCl_2$	c	−205.25	
	aq, 300	−207.92	−196.5
$Ba(ClO_3)_2$	c	−176.6	
	aq, 1600	−170.0	−134.4
$Ba(ClO_4)_2$	c	−210.2	
	aq, 800	...	−155.3
$Ba(CN)_2$	c	−48	
$Ba(CNO)_2$	c	−212.1	
	aq	...	−180.7
$BaCN_2$	c	−63.6	
$BaCO_3$	c, witherite	−284.2	−271.4
$BaCrO_4$	c	−342.2	
BaF_2	c	−287.9	
	aq, 1600	−284.6	−265.3
BaH_2	c	−40.8	−31.5
$Ba(HCO_2)_2$	aq	−459	−414.4
BaI_2	c	−144.6	
	aq, 400	−155.17	−158.52
$Ba(IO_3)_2$	c	−264.5	
	aq	−237.50	−198.35
$BaMoO_4$	c	−370	
Ba_3N_2	c	−90.7	
$Ba(NO_2)_2$	c	−184.5	
	aq	−179.05	−150.75
Barium (*Cont.*):			
$Ba(NO_3)_2$	c	−236.99	−189.94
	aq, 600	−227.74	
BaO	c	−133.0	
$Ba(OH)_2$	c	−225.9	
	aq, 400	−237.76	−209.02
$BaO.SiO_2$	c	−363	
$Ba_3(PO_4)_2$	c	−992	
$BaPtCl_6$	c	−284.9	
BaS	c	−111.2	
$BaSO_3$	c	−282.5	
$BaSO_4$	c	−340.2	−313.4
$BaWO_4$	c	−402	
Beryllium:			
Be	c	0.00	0.00
$BeBr_2$	c	−79.4	
	aq	−142	−127.9
$BeCl_2$	c	−112.6	
	aq	−163.9	−141.4
BeI_2	c	−39.4	
	aq	−112	−103.4
Be_3N_2	c	−134.5	−122.4
BeO	c	−145.3	−138.3
$Be(OH)_2$	c	−215.6	
BeS	c	−56.1	
$BeSO_4$	c	−284	
	aq	...	−254.8
Bismuth:			
Bi	c	0.00	0.00
$BiCl_3$	c	−90.5	−76.4
	aq	−101.6	
BiI_3	c	−24	
	aq	−27	
BiO	c	−49.5	−43.2
Bi_2O_3	c	−137.1	−117.9
$Bi(OH)_3$	c	−171.1	
Bi_2S_3	c	−43.9	−39.1
$Bi_2(SO_4)_3$	c	−607.1	
Boron:			
B	c	0.00	0.00
BBr_3	l	−52.7	
	g	−44.6	−50.9
BCl_3	g	−94.5	−90.8
BF_3	g	−265.2	−261.0
B_2H_6	g	7.5	19.9
BN	c	−32.1	−27.2
B_2O_3	c	−302.0	−282.9
	gls	−297.6	−280.3
$B(OH)_3$	c	−260.0	−229.4
B_2S_3	c	−56.6	
Bromine:			
Br_2	l	0.00	0.00
	g	7.47	0.931
BrCl	g	3.06	−0.63
Cadmium:			
Cd	c	0.00	0.00
$CdBr_2$	c	−75.8	−70.7
	aq, 400	−76.6	−67.6
$CdCl_2$	c	−92.149	−81.889
	aq, 400	−96.44	−81.2
$Cd(CN)_2$	c	36.2	
$CdCO_3$	c	−178.2	−163.2
CdI_2	c	−48.40	
	aq, 400	−47.46	−43.22
Cd_3N_2	c	39.8	
$Cd(NO_3)_2$	aq, 400	−115.67	−71.05
CdO	c	−62.35	−55.28
$Cd(OH)_2$	c	−135.0	−113.7
CdS	c	−34.5	−33.6
$CdSO_4$	c	−222.23	
	aq, 400	−232.635	−194.65
Calcium:			
Ca	c	0.00	0.00
$CaBr_2$	c	−162.20	
	aq, 400	−187.19	−181.86
CaC_2	c	−14.8	−16.0
$CaCl_2$	c	−190.6	−179.8
	aq	−209.15	−195.36

*For footnotes see end of table.

Source: Adapted from the R. H. Perry and C. H. Chilton, eds., *Chemical Engineers' Handbook*, Copyright © 1973 McGraw–Hill Inc. Used with permission of McGraw–Hill Book Company.

1 kcal = 4.184 kJ

Compound	State†	Heat of formation ‡ § at 25°C., kcal./mole	Free energy of formation ‖ ¶ at 25°C., kcal./mole
Calcium (*Cont.*):			
$CaCN_2$	c	−85	
$Ca(CN)_2$	c	−43.3	
	aq	....	−54.0
$CaCO_3$	c, calcite	−289.5	−270.8
	c, aragonite	−289.54	−270.57
$CaCO_3.MgCO_3$	c	−558.8	
CaC_2O_4	c	−332.2	
$Ca(C_2H_3O_2)_2$	c	−356.3	
	aq	−364.1	−311.3
CaF_2	c	−290.2	
	aq	−286.5	−264.1
CaH_2	c	−46	−35.7
CaI_2	c	−128.49	
	aq, 400	−156.63	−157.37
Ca_3N_2	c	−103.2	−88.2
$Ca(NO_3)_2$	c	−224.05	−177.38
	aq, 400	−228.29	
$Ca(NO_3)_2.2H_2O$	c	−367.95	−293.57
$Ca(NO_3)_2.3H_2O$	c	−439.05	−351.58
$Ca(NO_3)_2.4H_2O$	c	−509.43	−409.32
CaO	c	−151.7	−144.3
$Ca(OH)_2$	c	−235.58	−213.9
	aq, 800	−239.2	−207.9
$CaO.SiO_2$	c, II, wollastonite	−377.9	−357.5
	c, I, pseudowollastonite	−376.6	−356.6
CaS	c	−114.3	−113.1
$CaSO_4$	c, insoluble form	−338.73	−311.9
	c, soluble form α	−336.58	−309.8
	c, soluble form β	−335.52	−308.8
$CaSO_4.\frac{1}{2}H_2O$	c	−376.13	
$CaSO_4.2H_2O$	c	−479.33	−425.47
$CaWO_4$	c	−387	
Carbon:			
C	c, graphite	0.00	0.00
	c, diamond	0.453	0.685
CO	g	−26.416	−32.808
CO_2	g	−94.052	−94.260
CH_4 methane	g	−17.889	−12.140
C_2H_6 ethane	g	−20.236	−7.860
C_3H_8 propane	g	−24.820	−5.614
C_4H_{10} n-butane	g	−29.812	−3.754
C_4H_{10} isobutane	g	−31.452	−4.296
C_5H_{12} n-pentane	g	−35.00	−1.96
	l	−41.36	−2.21
C_5H_{12} 2-methylbutane	g	−36.92	−3.50
	l	−42.85	−3.59
C_5H_{12} 2,2-dimethylpropane	g	−39.67	−3.64
C_6H_{14} n-hexane	g	−39.96	0.05
	l	−47.52	−0.91
C_6H_{14} 2-methylpentane	g	−41.66	−0.96
	l	−48.82	−1.73
C_6H_{14} 3-methylpentane	g	−41.02	−0.29
	l	−48.28	−1.12
C_6H_{14} 2,2-dimethylbutane	g	−44.35	−2.35
	l	−51.00	−2.88
C_6H_{14} 2,3-dimethylbutane	g	−42.49	−0.73
	l	−49.48	−1.44
C_7H_{16} n-heptane	g	−44.89	2.09
	l	−53.63	0.42
C_7H_{16} 2-methylhexane	g	−46.60	0.98
	l	−54.93	−0.47
C_7H_{16} 3-methylhexane	g	−45.96	1.10
	l	−54.35	−0.39
C_7H_{16} 3-ethylpentane	g	−45.34	2.59
	l	−53.77	1.06
C_7H_{16} 2,2-dimethylpentane	g	−49.29	0.09
	l	−57.05	−1.08
C_7H_{16} 2,3-dimethylpentane	g	−47.62	0.16
	l	−55.81	−1.27
C_7H_{16} 2,4-dimethylpentane	g	−48.30	0.72
	l	−56.17	−0.49
C_7H_{16} 3,3-dimethylpentane	g	−48.17	0.63
	l	−56.07	−0.69
C_7H_{16} 2,2,3-trimethylbutane	g	−48.96	0.76
	l	−56.63	−0.43
C_8H_{18} n-octane	g	−49.82	4.14
	l	−59.74	1.77
C_8H_{18} 2-methylheptane	g	−51.50	3.06
	l	−60.98	0.92
C_8H_{18} 3-methylheptane	g	−50.82	3.29
	l	−60.34	1.12
C_8H_{18} 4-methylheptane	g	−50.69	4.00
	l	−60.17	1.86
C_8H_{18} 3-ethylhexane	g	−50.40	3.95
	l	−59.88	1.80
Carbon (*Cont.*):			
C_8H_{18} 2,2-dimethylhexane	g	−53.71	2.56
	l	−62.63	−0.72
C_8H_{18} 2,3-dimethylhexane	g	−51.13	4.23
	l	−60.40	2.17
C_8H_{18} 2,4-dimethylhexane	g	−52.44	2.80
	l	−61.47	0.89
C_8H_{18} 2,5-dimethylhexane	g	−53.21	2.50
	l	−62.26	0.59
C_8H_{18} 3,3-dimethylhexane	g	−52.61	3.17
	l	−61.58	1.23
C_8H_{18} 3,4-dimethylhexane	g	−50.91	4.97
	l	−60.23	2.86
C_8H_{18} 2-methyl-3-ethylpentane	g	−50.48	5.08
	l	−59.69	3.03
C_8H_{18} 3-methyl-3-ethylpentane	g	−51.38	4.76
	l	−60.46	2.69
C_8H_{18} 2,2,3-trimethylpentane	g	−52.61	4.09
	l	−61.44	2.22
C_8H_{18} 2,2,4-trimethylpentane	g	−53.57	3.13
	l	−61.97	1.51
C_8H_{18} 2,3,3,-trimethylpentane	g	−51.73	4.52
	l	−60.63	2.54
C_8H_{18} 2,3,4-trimethylpentane	g	−51.97	4.32
	l	−60.98	2.34
C_8H_{18} 2,2,3,3,-tetramethylbutane	g	−53.99	4.88
	c	−64.23	2.74
C_2H_4 ethylene	g	12.496	16.282
C_3H_6 propylene	g	4.879	14.964
C_4H_8 1-butene	g	0.280	17.217
C_4H_8 cis-2-butene	g	−1.362	16.007
C_4H_8 trans-2-butene	g	−2.405	15.323
C_4H_8 2-methyl-2-propene	g	−3.343	14.574
C_5H_{10} 1-pentene	g	−5.000	18.787
C_5H_{10} cis-2-pentene	g	−6.710	17.173
C_5H_{10} trans-2-pentene	g	−7.590	16.575
C_5H_{10} 2-methyl-1-butene	g	−8.680	15.509
C_5H_{10} 3-methyl-1-butene	g	−6.920	17.874
C_5H_{10} 2-methyl-2-butene	g	−10.170	14.267
C_2H_2 acetylene	g	54.194	50.000
C_3H_4 methylacetylene	g	44.319	46.313
C_4H_6 1-butyne	g	39.70	48.52
C_4H_6 2-butyne	g	35.374	44.725
C_5H_8 1-pentyne	g	34.50	50.17
C_5H_8 2-pentyne	g	30.80	46.41
C_5H_8 3-methyl-1-butyne	g	32.60	49.12
C_6H_6 benzene	g	19.820	30.989
C_7H_8 toluene	l	11.718	29.756
	g	11.950	29.228
	l	2.867	27.282
C_8H_{10} ethylbenzene	g	7.120	31.208
	l	−2.977	28.614
C_8H_{10} o-xylene	g	4.540	29.177
	l	−5.841	26.370
C_8H_{10} m-xylene	g	4.120	28.405
	l	−6.075	25.730
C_8H_{10} p-xylene	g	4.290	28.952
	l	−5.838	26.310
C_9H_{12} n-propylbenzene	g	1.870	32.810
	l	−9.178	29.600
C_9H_{12} isopropylbenzene	g	0.940	32.738
	l	−9.848	29.708
C_9H_{12} 1-methyl-2-ethylbenzene	g	0.290	31.323
	l	−11.110	27.973
C_9H_{12} 1-methyl-3-ethylbenzene	g	−0.460	30.217
	l	−11.670	26.977
C_9H_{12} 1-methyl-4-ethylbenzene	g	−0.780	30.281
	l	−11.920	27.041
C_9H_{12} 1,2,3-trimethylbenzene	g	−2.290	29.319
	l	−14.013	25.679
C_9H_{12} 1,2,4-trimethylbenzene	g	−3.330	27.912
	l	−14.785	24.462
C_9H_{12} 1,3,5-trimethylbenzene	g	−3.840	28.172
	l	−15.184	24.832
C_5H_{10} cyclopentane	g	−18.46	9.23
	l	−25.31	8.70
C_6H_{12} methylcyclopentane	g	−25.50	8.55
	l	−33.08	7.53
C_7H_{14} ethylcyclopentane	g	−30.38	10.59
	l	−39.09	8.84

Compound	State†	Heat of formation ‡ § at 25°C., kcal./mole	Free energy of formation ‖ ¶ at 25°C., kcal./mole
Carbon (*Cont.*):			
C_6H_{12} cyclohexane	g	−29.43	7.59
	l	−37.34	6.39
C_7H_{14} methylcyclohexane	g	−37.00	6.52
	l	−45.46	4.86
C_8H_{16} ethylcyclohexane	g	−41.06	9.38
	l	−50.73	6.96
CH_4O methanol	g	−48.08	−38.62
	l	−57.04	−39.80
C_2H_6O ethanol	g	−52.23	−40.23
	l	−66.35	−41.76
C_3H_8O *n*-propanol	g	−61.17	−38.83
	l	−71.87	−39.84
C_3H_8O isopropanol	g	−62.41	−38.20
	l	−74.32	−38.83
$C_4H_{10}O$ *n*-butanol	g	−67.81	−38.88
	l	−79.61	−40.37
$C_4H_{10}O$ isobutanol	g	−69.05	−38.25
	l	−81.06	−39.36
$C_2H_6O_2$ ethylene glycol	g	−92.53	−71.26
	l	−107.91	−76.44
$C_3H_8O_3$ glycerol	g		
	l	−159.16	−113.65
C_6H_6O phenol	g	−21.71	−6.26
	l	−37.80	−11.02
C_7H_8O cresol	g	....	−13.17
C_2H_4O ethylene oxide	g	−16.1	−6.94
C_2H_6O dimethyl ether	g	−43.06	−26.06
	l	−51.3	
$C_4H_{10}O$ diethyl ether	l	−65.2	−27.75
CH_2O formaldehyde	g	−28.29	−26.88
C_2H_4O acetaldehyde	g	−39.72	−31.46
C_3H_4O acrolein	g	−20.50	−15.57
	l	−27.97	−16.17
C_3H_6O propionaldehyde	g	−49.15	−33.96
C_4H_8O *n*-butyraldehyde	g	−52.40	−73.24
C_7H_6O bensaldehyde	g	−9.57	5.85
	l	−21.23	2.24
C_8H_8O *p*-toluic aldehyde	g	−17.78	4.09
	l	−29.79	0.97
C_2H_2O ketene	g	−14.78	−14.30
	l	−18.78	−13.32
C_3H_6O acetone	g	−51.79	−36.45
	l	−59.32	−37.16
$C_5H_{10}O$ diethylketone	l	−73.8	
CH_2O_2 formic acid	g	−86.67	−80.24
	l	−97.8	−82.7
½$(CH_2O_2)_2$ bimolecular formic acid	g	−93.85	−81.90
$C_2H_4O_2$ acetic acid	g	−104.72	−91.24
	l	−116.2	−93.56
$C_3H_6O_2$ propionic acid	g	−108.75	−88.27
	l	−121.7	−91.65
$C_2H_4O_3$ hydroxyacetic acid	l	−155.33	−125.57
$C_6H_{10}O_4$ adipic acid	g	−216.19	−163.96
	l	−235.51	−177.17
$C_2H_4O_2$ methyl formate	g	−84.69	−71.37
	l	−95.26	−71.53
$C_4H_6O_2$ methyl acrylate	g	−70.10	−56.78
	l	−82.76	−58.13
$C_4H_8O_2$ ethyl acetate	g	−102.02	−74.93
	l	−110.72	−76.11
$C_5H_{10}O_2$ ethyl propionate	g	−112.36	−77.37
	l	−122.16	−79.16
$C_4H_6O_3$ acetic anhydride	g	−148.82	−119.29
	l	−155.16	−121.75
$C_6H_{10}O_3$ propionic anhydride	g	−147.32	−109.78
	l	−161.53	−113.66
CS_2 carbon disulfide	g	28.11	16.13
COS carbonyl sulfide	g	−33.83	−40.85
C_2N_2 cyanogen	g	73.82	71.02
HCN hydrogen cyanide	g	31.1	27.94
	l	25.2	29.0
	aq, 100	25.2	26.8
C_2H_3N acetonitrile	g	19.81	
CH_5N methylamine	g	−6.7	6.6
C_2H_7N ethylamine	g	−12.24	10.01
C_3H_9N propylamine	g	−16.45	14.38
$C_4H_{11}N$ butylamine	g	−15.60	19.55
$C_6H_{13}N$ hexamethyleneimine	g	−14.37	31.52
	l	−24.90	28.84
CH_2N_2 cyanamide	l	11.18	24.30
	c	9.15	24.18
$C_6H_8N_2$ adiponitrile	g	33.34	61.43
	l	19.19	54.63
$C_6H_{16}N_2$ hexamethylenediamine	g	−30.57	28.91
CH_5N_3 guanidine	l	−27.48	7.34
	c	−30.68	6.33

Compound	State†	Heat of formation ‡ § at 25°C., kcal./mole	Free energy of formation ‖ ¶ at 25°C., kcal./mole
Carbon (*Cont.*):			
$C_3H_6N_6$ melamine	l	−19.33	40.80
CH_3NO formamide	g	−44.64	−36.60
C_2H_7NO ethanolamine	l	−62.52	27.50
CH_4N_2O urea	l	−77.55	−46.45
	c	−79.634	−47.118
Cerium:			
Ce	c	0.00	0.00
CeN	c	−78.2	−70.8
Cesium:			
Cs	c	0.00	0.00
CsBr	c	−97.64	
	aq, 500	−91.39	−94.86
CsCl	c	−106.31	
	aq, 400	−102.01	−101.61
Cs_2CO_3	c	−271.88	
CsF	c	−131.67	
	aq, 400	−140.48	−135.98
CsH	c	−12	−7.30
$CsHCO_3$	c	−230.6	
	aq, 2000	−226.6	−210.56
CsI	c	−83.91	
	aq, 400	−75.74	−82.61
$CsNH_2$	c	−28.2	
$CsNO_3$	c	−121.14	
	aq, 400	−111.54	−96.53
Cs_2O	c	−82.1	
CsOH	c	−100.2	
	aq, 200	−117.0	−107.87
Cs_2S	c	−87	
Cs_2SO_4	c	−344.86	
	aq	−340.12	−316.66
Chlorine:			
Cl_2	g	0.00	0.00
ClF	g	−25.7	
ClO	g	33	
ClO_2	g	24.7	29.5
ClO_3	g	37	
Cl_2O	g	18.20	22.40
Cl_2O_7	g	63	
Chromium:			
Cr	c	0.00	0.00
$CrBr_3$	aq	....	−122.7
Cr_3C_2	c	−21.008	−21.20
Cr_4C	c	−16.378	−16.74
$CrCl_3$	c	−103.1	−93.8
	aq	....	−102.1
CrF_2	c	−152	
CrF_3	c	−231	
CrI_3	c	−63.7	
	aq	....	−64.1
CrO_3	c	−139.3	
Cr_2O_3	c	−268.8	−249.3
$Cr_2(SO_4)_3$	aq	....	−626.3
Cobalt:			
Co	c	0.00	0.00
$CoBr_2$	c	−55.0	
	aq	−73.61	−61.96
Co_2C	c	9.49	7.08
$CoCl_2$	c	−76.9	−66.6
	aq, 400	−95.58	−75.46
$CoCO_3$	c	−172.39	−155.36
CoF_2	aq	−172.98	−144.2
CoI_2	c	−24.2	
	aq	−43.15	−37.4
$Co(NO_3)_2$	c	−102.8	
	aq	−114.9	−65.3
CoO	c	−57.5	
Co_3O_4	c	−196.5	
$Co(OH)_2$	c	−131.5	−108.9
$Co(OH)_3$	c	−177.0	−142.0
CoS	c	−22.3	−19.8
Co_2S_3	c	−40.0	
$CoSO_4$	c	−216.6	
	aq, 400	....	−188.9
Columbium:			
Cb	c	0.00	0.00
Cb_2O_5	c	−462.96	
Copper:			
Cu	c	0.00	0.00
CuBr	c	−26.7	−23.8
$CuBr_2$	c	−34.0	
	aq	−42.4	−33.25
CuCl	c	−31.4	−24.13
$CuCl_2$	c	−48.83	
	aq, 400	−64.7	
$CuClO_4$	aq	−28.3	1.34
$Cu(ClO_3)_2$	aq, 400	....	15.4
$Cu(ClO_4)_2$	aq	....	−5.5

Compound	State†	Heat of formation ‡ § at 25°C., kcal./mole	Free energy of formation ‖ ¶ at 25°C., kcal./mole
Copper (*Cont.*):			
CuI	c	−17.8	−16.66
CuI_2	c	−4.8	
	aq	−11.9	−8.76
Cu_3N	c	17.78	
$Cu(NO_3)_2$	c	−73.1	
	aq, 200	−83.6	−36.6
CuO	c	−38.5	−31.9
Cu_2O	c	−43.00	−38.13
$Cu(OH)_2$	c	−108.9	−85.5
CuS	c	−11.6	−11.69
Cu_2S	c	−18.97	−20.56
$CuSO_4$	c	−184.7	−158.3
	aq, 800	−200.78	−160.19
Cu_2SO_4	c	−179.6	
	aq	....	−152.0
Erbium:			
Er	c	0.00	0.00
$Er(OH)_3$	c	−326.8	
Fluorine:			
F_2	g	0.00	0.00
F_2O	g	5.5	9.7
Gallium:			
Ga	c	0.00	0.00
$GaBr_3$	c	−92.4	
$GaCl_3$	c	−125.4	
GaN	c	−26.2	
Ga_2O	c	−84.3	
Ga_2O_3	c	−259.9	
Germanium:			
Ge	c	0.00	0.00
Ge_3N_4	c	−15.7	
GeO_2	c	−128.6	
Gold:			
Au	c	0.00	0.00
$AuBr$	c	−3.4	
$AuBr_3$	c	−14.5	
	aq	−11.0	24.47
$AuCl$	c	−8.3	
$AuCl_3$	c	−28.3	
	aq	−32.96	4.21
AuI	c	0.2	−0.76
Au_2O_3	c	11.0	18.71
$Au(OH)_3$	c	−100.6	
Hafnium:			
Hf	c	0.00	0.00
HfO_2	c	−271.1	−258.2
Hydrogen:			
H_3AsO_3	aq	−175.6	−153.04
H_3AsO_4	c	−214.9	
	aq	−214.8	−183.93
HBr	g	−8.66	−12.72
	aq, 400	−28.80	−24.58
$HBrO$	aq	−25.4	−19.90
$HBrO_3$	aq	−11.51	5.00
HCl	g	−22.063	−22.778
	aq, 400	−39.85	−31.330
HCN	g	31.1	27.94
	aq, 100	24.2	26.55
$HClO$	aq, 400	−28.18	−19.11
$HClO_3$	aq	−23.4	−0.25
$HClO_4$	aq, 660	−31.4	−10.70
$HC_2H_3O_2$	l	−116.2	−93.56
	aq, 400	−116.74	−96.8
$H_2C_2O_4$	c	−196.7	
	aq, 300	−194.6	−165.64
$HCOOH$	l	−97.8	−82.7
	aq, 200	−98.0	−85.1
H_2CO_3	aq	−167.19	−149.0
HF	g	−64.2	−64.7
	aq, 200	−75.75	
HI	g	6.27	0.365
	aq, 400	−13.47	−12.35
HIO	aq	−38	−23.33
HIO_3	c	−56.77	
	aq	−54.8	−32.25
HN_3	g	70.3	78.50
HNO_3	g	−31.99	−17.57
	l	−41.35	−19.05
	aq, 400	−49.210	
$HNO_3.H_2O$	l	−112.91	−78.36
$HNO_3.3H_2O$	l	−252.15	−193.70
H_2O	g	−57.7979	−54.6351
	l	−68.3174	−56.6899
H_2O_2	l	−45.16	−28.23
	aq, 200	−45.80	−31.47
H_3PO_2	c	−145.5	
	aq	−145.6	−120.0
H_3PO_3	c	−232.2	
	aq	−232.2	−204.0
Hydrogen (*Cont.*):			
H_3PO_4	c	−306.2	
	aq, 400	−309.32	−270.0
H_2S	g	−4.77	−7.85
	aq, 2000	−9.38	
H_2S_2	l	−3.6	
H_2SO_3	aq, 200	−146.88	−128.54
H_2SO_4	l	−193.69	
	aq, 400	−212.03	
H_2Se	g	20.5	17.0
	aq	18.1	18.4
H_2SeO_3	c	−126.5	
	aq	−122.4	−101.36
H_2SeO_4	c	−130.23	
	aq, 400	−143.4	
H_2SiO_3	c	−267.8	−247.9
H_4SiO_4	c	−340.6	
H_2Te	g	36.9	33.1
H_2TeO_3	c	−145.0	−115.7
	aq	−145.0	
H_2TeO_4	aq	−165.6	
Indium:			
In	c	0.00	0.00
$InBr_3$	c	−97.2	
	aq	−112.9	−97.2
$InCl_3$	c	−128.5	
	aq	−145.6	−117.5
InI_3	c	−56.5	
	aq	−67.2	−60.5
InN	c	−4.8	
In_2O_3	c	−222.47	
Iodine:			
I_2	c	0.00	0.00
	g	14.88	4.63
IBr	g	10.05	1.24
ICl	g	4.20	−1.32
ICl_3	c	−21.8	−6.05
I_2O_5	c	−42.5	
Iridium:			
Ir	c	0.00	0.00
$IrCl$	c	−20.5	−16.9
$IrCl_2$	c	−40.6	−32.0
$IrCl_3$	c	−60.5	−46.5
IrF_6	l	−130	
IrO_2	c	−40.14	
Iron:			
Fe	c, α	0.00	0.00
$FeBr_2$	c	−57.15	
	aq, 540	−78.7	−69.47
$FeBr_3$	aq	−95.5	−76.26
Fe_3C	c	5.69	4.24
$Fe(CO)_5$	l	−187.6	
$FeCO_3$	c, siderite	−172.4	−154.8
$FeCl_2$	c	−81.9	−72.6
	aq	−100.0	−83.0
$FeCl_3$	c	−96.4	
	aq, 2000	−128.5	−96.5
FeF_2	aq, 1200	−177.2	−151.7
FeI_2	c	−24.2	
	aq	−47.7	−45
FeI_3	aq	−49.7	−39.5
Fe_4N	c	−2.55	0.862
$Fe(NO_3)_2$	aq	−118.9	−72.8
$Fe(NO_3)_3$	aq, 800	−156.5	−81.3
FeO	c	−64.62	−59.38
Fe_2O_3	c	−198.5	−179.1
Fe_3O_4	c	−266.9	−242.3
$Fe(OH)_2$	c	−135.9	−115.7
$Fe(OH)_3$	c	−197.3	−166.3
$FeO.SiO_2$	c	−273.5	
Fe_2P	c	−13	
$FeSi$	c	−19.0	
FeS	c	−22.64	−23.23
FeS_2	c, pyrites	−38.62	−35.93
	c, marcasite	−33.0	
$FeSO_4$	c	−221.3	−195.5
	aq, 400	−236.2	−196.4
$Fe_2(SO_4)_3$	aq, 400	−653.3	−533.4
$FeTiO_3$	c, ilmenite	−295.51	−277.06
Lanthanum:			
La	c	0.00	0.00
$LaCl_3$	c	−253.1	
	aq	−284.7	
La_3H_8	c	−160	
LaN	c	−72.0	−64.6
La_2O_3	c	−539	
LaS_2	c	−148.3	
La_2S_3	c	−351.4	
$La_2(SO_4)_3$	aq	−972	

Compound	State†	Heat of formation ‡ § at 25°C., kcal./mole	Free energy of formation ‖ ¶ at 25°C., kcal./mole
Lead:			
Pb	c	0.00	0.00
$PbBr_2$	c	−66.24	−62.06
	aq	−56.4	−54.97
$PbCO_3$	c, cerussite	−167.6	−150.0
$Pb(C_2H_3O_2)_2$	c	−232.6	
	aq, 400	−234.2	−184.40
PbC_2O_4	c	−205.3	
$PbCl_2$	c	−85.68	−75.04
	aq	−82.5	−68.47
PbF_2	c	−159.5	−148.1
PbI_2	c	−41.77	−41.47
$Pb(NO_3)_2$	c	−106.88	
	aq, 400	−99.46	−58.3
PbO	c, red	−51.72	−45.53
	c, yellow	−50.86	−43.88
PbO_2	c	−65.0	−52.0
Pb_3O_4	c	−172.4	−142.2
$Pb(OH)_2$	c	−123.0	−102.2
PbS	c	−22.38	−21.98
$PbSO_4$	c	−218.5	−192.9
Lithium:			
Li	c	0.00	0.00
LiBr	c	−83.75	
	aq, 400	−95.40	−95.28
$LiBrO_3$	aq	−77.9	−65.70
Li_2C_2	c	−13.0	
LiCN	aq	−31.4	−31.35
LiCNO	aq	−101.2	−94.12
$LiC_2H_3O_2$	aq	−183.9	−160.00
Li_2CO_3	c	−289.7	−269.8
	aq, 1900	−293.1	−267.58
LiCl	c	−97.63	
	aq, 278	−106.45	−102.03
$LiClO_3$	aq	−87.5	−70.95
$LiClO_4$	aq	−106.3	−81.4
LiF	c	−145.57	
	aq, 400	−144.85	−136.40
LiH	c	−22.9	
$LiHCO_3$	aq, 2000	−231.1	−210.98
LiI	c	−65.07	
	aq, 400	−80.09	−83.03
$LiIO_3$	aq	−121.3	−102.95
Li_3N	c	−47.45	−37.33
$LiNO_3$	c	−115.350	
	aq, 400	−115.88	−96.95
Li_2O	c	−142.3	
Li_2O_2	c	−151.9	−138.0
	aq	−159	
LiOH	c	−116.58	−106.44
	aq, 400	−121.47	−108.29
$LiOH.H_2O$	c	−188.92	
$Li_2O.SiO_2$	gls	−374	
Li_2Se	c	−84.9	
	aq	−95.5	−105.64
Li_2SO_4	c	−340.23	−314.66
	aq, 400	−347.02	
$Li_2SO_4.H_2O$	c	−411.57	−375.07
Magnesium:			
Mg	c	0.00	0.00
$Mg(AsO_4)_2$	c	−731.3	
	aq	−749	−630.14
$MgBr_2$	c	−123.9	
	aq, 400	−167.33	−156.94
$Mg(CN)_2$	aq	−39.7	−29.08
$MgCN_2$	c	−61	
$Mg(C_2H_3O_2)_2$	aq	−344.6	−286.38
$MgCO_3$	c	−261.7	−241.7
$MgCl_2$	c	−153.220	−143.77
	aq, 400	−189.76	
$MgCl_2.H_2O$	c	−230.970	−205.93
$MgCl_2.2H_2O$	c	−305.810	−267.20
$MgCl_2.4H_2O$	c	−453.820	−387.98
$MgCl_2.6H_2O$	c	−597.240	−505.45
MgF_2	c	−263.8	
MgI_2	c	−86.8	
	aq, 400	−136.79	−132.45
$MgMoO_4$	c	−329.9	
Mg_3N_2	c	−115.2	−100.8
$Mg(NO_3)_2$	c	−188.770	−140.66
	aq, 400	−209.927	−160.28
$Mg(NO_3)_2.2H_2C$	c	−336.625	
$Mg(NO_3)_2.6H_2O$	c	−624.48	−496.03
MgO	c	−143.84	−136.17
$MgO.SiO_2$	c	−347.5	−326.7
$Mg(OH)_2$	c, ppt.	−221.90	−200.17
	c, brucite	−223.9	−193.3
MgS	c	−84.2	
	aq	−108	
Magnesium (*Cont.*):			
$MgSO_4$	c	−304.94	−277.7
	aq, 400	−325.4	−283.88
MgTe	c	−25	
$MgWO_4$	c	−345.2	
Manganese:			
Mn	c, α	0.00	0.00
$MnBr_2$	c	−91	
	aq	−106	−97.8
Mn_3C	c	1.1	1.26
$Mn(C_2H_3O_2)_2$	c	−270.3	
	aq	−282.7	−227.2
$MnCO_3$	c	−211	−192.5
MnC_2O_4	c	−240.9	
$MnCl_2$	c	−112.0	−102.2
	aq, 400	−128.9	
MnF_2	aq, 1200	−206.1	−180.0
MnI_2	c	−49.8	
	aq	−76.2	−73.3
Mn_5N_2	c	−57.77	−46.49
$Mn(NO_3)_2$	c	−134.9	
	aq, 400	−148.0	−101.1
$Mn(NO_3)_2.6H_2O$	c	−557.07	−441.2
MnO	c	−92.04	−86.77
MnO_2	c	−124.58	−111.49
Mn_2O_3	c	−229.5	−209.9
Mn_3O_4	c	−331.65	−306.22
$MnO.SiO_2$	c	−301.3	−282.1
$Mn(OH)_2$	c	−163.4	−143.1
$Mn(OH)_3$	c	−221	−190
$Mn_3(PO_4)_2$	c	−736	
MnSe	c	−26.3	−27.5
MnS	c, green	−47.0	−48.0
$MnSO_4$	c	−254.18	−228.41
	aq, 400	−265.2	
$Mn_2(SO_4)_3$	c	−635	
	aq	−657	
Mercury:			
Hg	l	0.00	0.00
HgBr	g	23	18
$HgBr_2$	c	−40.68	−38.8
	aq	−38.4	−9.74
$Hg(C_2H_3O_2)_2$	c	−196.3	
	aq	−192.5	−139.2
$HgCl_2$	c	−53.4	−42.2
	aq	−50.3	−23.25
HgCl	g	19	14
Hg_2Cl_2	c	−63.13	
$Hg(CN)_2$	c	62.8	
	aq, 1110	66.25	
HgC_2O_4	c	−159.3	
HgH	g	57.1	52.25
HgI_2	c, red	−25.3	−24.0
HgI	g	33	23
Hg_2I_2	c	−28.88	−26.53
$Hg(NO_3)_2$	aq	−56.8	−13.09
$Hg_2(NO_3)_2$	aq	−58.5	−15.65
HgO	c, red	−21.6	−13.94
	c, yellow ppt.	−20.8	
Hg_2O	c	−21.6	−12.80
HgS	c, black	−10.7	−8.80
$HgSO_4$	c	−166.6	
Hg_2SO_4	c	−177.34	−149.12
Molybdenum:			
Mo	c	0.00	0.00
Mo_2C	c	4.36	2.91
Mo_2N	c	−8.3	
MoO_2	c	−130	−118.0
MoO_3	c	−180.39	−162.01
MoS_2	c	−56.27	−54.19
MoS_3	c	−61.48	−57.38
Nickel:			
Ni	c	0.00	0.00
$NiBr_2$	c	−53.4	
	aq	−72.6	−60.7
Ni_3C	c	9.2	8.88
$Ni(C_2H_3O_2)_2$	aq	−249.6	−190.1
$Ni(CN)_2$	aq	230.9	66.3
$NiCl_2$	c	−75.0	
	aq, 400	−94.34	−74.19
NiF_2	c	−157.5	
	aq	−171.6	−142.9
NiI_2	c	−22.4	
	aq	−42.0	−36.2
$Ni(NO_3)_2$	c	−101.5	
	aq, 200	−113.5	−64.0
NiO	c	−58.4	−51.7
$Ni(OH)_2$	c	−129.8	−105.6
$Ni(OH)_3$	c	−163.2	

Compound	State†	Heat of formation ‡ § at 25°C., kcal./mole	Free energy of formation ‖ ¶ at 25°C., kcal./mole
Nickel (*Cont.*):			
NiS	c	−20.4	
$NiSO_4$	c	−216	
	aq, 200	−231.3	−187.6
Nitrogen:			
N_2	g	0.00	0.00
NF_3	g	−27	
NH_3	g	−10.96	−3.903
	aq, 200	−19.27	
NH_4Br	c	−64.57	
	aq	−60.27	−43.54
$NH_4C_2H_3O_2$	c	−148.1	
	aq, 400	−148.58	−108.26
NH_4CN	c	−0.7	
	aq	3.6	20.4
NH_4CNS	c	−17.8	
	aq	−12.3	4.4
$(NH_4)_2CO_3$	aq	−223.4	−164.1
$(NH_4)_2C_2O_4$	c	−266.3	
	aq	−260.6	−196.2
NH_4Cl	c	−75.23	−48.59
	aq, 400	−71.20	
NH_4ClO_4	c	−69.4	
	aq	−63.2	−21.1
$(NH_4)_2CrO_4$	c	−276.9	
	aq	−271.3	−209.3
NH_4F	c	−111.6	
	aq	−110.2	−84.7
NH_4I	c	−48.43	
	aq	−44.97	−31.3
NH_4NO_3	c	−87.40	
	aq, 500	−80.89	
NH_4OH	aq	−87.59	
$(NH_4)_2S$	aq, 400	−55.21	−14.50
$(NH_4)_2SO_4$	c	−281.74	−215.06
	aq, 400	−279.33	−214.02
N_2H_4	l	12.06	
$N_2H_4.H_2O$	l	−57.96	
$N_2H_4.H_2SO_4$	c	−232.2	
N_2O	g	19.55	24.82
NO	g	21.600	20.719
NO_2	g	7.96	12.26
N_2O_4	g	2.23	23.41
N_2O_5	c	−10.0	
$NOBr$	l	11.6	19.26
$NOCl$	g	12.8	16.1
Osmium:			
Os	c	0.00	0.00
OsO_4	c	−93.6	−70.9
	g	−80.1	−68.1
Oxygen:			
O_2	g	0.00	0.00
O_3	g	33.88	38.86
Palladium:			
Pd	c	0.00	0.00
PdO	c	−20.40	
Phosphorus:			
P	c, white ("yellow")	0.00	0.00
	c, red ("violet")	−4.22	−1.80
P	g	150.35	141.88
P_2	g	33.82	24.60
P_4	g	13.2	5.89
PBr_3	l	−45	
PBr_5	c	−60.6	
PCl_3	g	−70.0	−65.2
	l	−76.8	−63.3
PCl_5	g	−91.0	−73.2
PH_3	g	2.21	−1.45
PI_3	c	−10.9	
P_2O_5	c	−360.0	
$POCl_3$	g	−138.4	−127.2
Platinum:			
Pt	c	0.00	0.00
$PtBr_4$	c	−40.6	
	aq	−50.7	
$PtCl_2$	c	−34	
$PtCl_4$	c	−62.6	
	aq	−82.3	
PtI_4	c	−18	
$Pt(OH)_2$	c	−87.5	−67.9
PtS	c	−20.18	−18.55
PtS_2	c	−26.64	−24.28
Potassium:			
K	c	0.00	0.00
K_3AsO_3	aq	−323.0	
K_3AsO_4	aq	−390.3	−355.7
KH_2AsO_4	c	−271.2	−236.7
KBr	c	−94.06	−90.8
	aq, 400	−89.19	−92.0
Potassium (*Cont.*):			
$KBrO_3$	c	−81.58	−60.30
	aq, 1667	−71.68	
$KC_2H_3O_2$	c	−173.80	
	aq, 400	−177.38	−156.73
KCl	c	−104.348	−97.76
	aq, 400	−100.164	−98.76
$KClO_3$	c	−93.5	−69.30
	aq, 400	−81.34	
$KClO_4$	c	−103.8	−72.86
	aq, 400	−101.14	
KCN	c	−28.1	
	aq, 400	−25.3	−28.08
$KCNO$	c	−99.6	
	aq	−94.5	−90.85
$KCNS$	c	−47.0	
	aq, 400	−41.07	−44.08
K_2CO_3	c	−274.01	
	aq, 400	−280.90	−264.04
$K_2C_2O_4$	c	−319.9	
	aq, 400	−315.5	−293.1
K_2CrO_4	c	−333.4	
	aq, 400	−328.2	−306.3
$K_2Cr_2O_7$	c	−488.5	
	aq, 400	−472.1	−440.9
KF	c	−134.50	
	aq, 180	−138.36	−133.13
$K_3Fe(CN)_6$	c	−48.4	
	aq	−34.5	
$K_4Fe(CN)_6$	c	−131.8	
	aq	−119.9	
KH	c	−10	−5.3
$KHCO_3$	c	−229.8	
	aq, 2000	−224.85	−207.71
KI	c	−78.88	−77.37
	aq, 500	−73.95	−79.76
KIO_3	c	−121.69	−101.87
	aq, 400	−115.18	−99.68
KIO_4	aq	−98.1	
$KMnO_4$	c	−192.9	−169.1
	aq, 400	−182.5	−168.0
K_2MoO_4	aq, 880	−364.2	−342.9
KNH_2	c	−28.25	
KNO_2	aq	−86.0	−75.9
KNO_3	c	−118.08	−94.29
	aq, 400	−109.79	−93.68
K_2O	c	−86.2	
$K_2O.Al_2O_3.S_iO_2$)	c, leucite	−1379.6	
	gls	−1368.2	
$K_2O.Al_2O_3.S_iO_2$	c, adularia	−1784.5	
	c, microcline	−1784.5	
	gls	−1747	
KOH	c	−102.02	
	aq, 400	−114.96	−105.0
K_3PO_3	aq	−397.5	
K_3PO_4	aq	−478.7	−443.3
KH_2PO_4	c	−362.7	−326.1
K_2PtCl_4	c	−254.7	
	aq	−242.6	−226.5
K_2PtCl_6	c	−299.5	−263.6
	aq, 9400	−286.1	
K_2Se	c	−74.4	
	aq	−83.4	−99.10
K_2SeO_4	aq	−267.1	−240.0
K_2S	c	−121.5	
	aq, 400	−110.75	−111.44
K_2SO_3	c	−267.7	
	aq	−269.7	−251.3
K_2SO_4	c	−342.65	−314.62
	aq, 400	−336.48	−310.96
$K_2SO_4.Al_2(SO_4)_3$	c	−1178.38	−1068.48
$K_2SO_4.Al_2(SO_4)_3.24H_2O$	c	−2895.44	−2455.68
$K_2S_2O_8$	c	−418.62	
Rhenium:			
Re	c	0.00	0.00
ReF_6	g	−274	
Rhodium:			
Rh	c	0.00	0.00
RhO	c	−21.7	
Rh_2O	c	−22.7	
Rh_2O_3	c	−68.3	
Rubidium:			
Rb	c	0.00	0.00
$RbBr$	c	−95.82	
	g	−45.0	−52.50
	aq, 500	−90.54	−93.38
$RbCN$	aq	−25.9	
Rb_2CO_3	c	−273.22	
	aq, 220	−282.61	−263.78

Compound	State†	Heat of formation ‡ § at 25°C., kcal./mole	Free energy of formation ‖ ¶ at 25°C., kcal./mole
Rubidium (*Cont.*):			
$RbCl$	c	−105.06	−98.48
	g	−53.6	−57.9
	aq, ∞	−101.06	−100.13
RbF	c	−133.23	
	aq, 400	−139.31	−134.5
$RbHCO_3$	c	−230.01	
	aq, 2000	−225.59	−209.07
RbI	c	−81.04	
	g	−31.2	−40.5
	aq, 400	−74.57	−81.13
$RbNH_2$	c	−27.74	
$RbNO_3$	c	−119.22	
	aq, 400	−110.52	−95.05
Rb_2O	c	−82.9	
Rb_2O_2	c	−107	
$RbOH$	c	−101.3	
	aq, 200	−115.8	−106.39
Ruthenium:			
Ru	c	0.00	0.00
RuS_2	c	−46.99	−44.11
Selenium:			
Se	c, I, hexagonal	0.00	0.00
	c, II, red, monoclinic	0.2	
Se_2Cl_2	l	−22.06	−13.73
SeF_6	g	−246	−222
SeO_2	c	−56.33	
Silicon:			
Si	c	0.00	0.00
$SiBr_4$	l	−93.0	
SiC	c	−28	−27.4
$SiCl_4$	l	−150.0	−133.9
	g	−142.5	−133.0
SiF_4	g	−370	−360
SiH_4	g	−14.8	−9.4
SiI_4	c	−29.8	
Si_3N_4	c	−179.25	−154.74
SiO_2	c, cristobalite, 1600° form	−202.62	
	c, cristobalite, 1100° form	−202.46	
	c, quarts	−203.35	−190.4
	c, tridymite	−203.23	
Silver:			
Ag	c	0.00	0.00
$AgBr$	c	−23.90	−23.02
Ag_2C_2	c	84.5	
$AgC_2H_3O_2$	c	−95.9	
	aq	−91.7	−70.86
$AgCN$	c	33.8	38.70
Ag_2CO_3	c	−119.5	−103.0
$Ag_2C_2O_4$	c	−158.7	
$AgCl$	c	−30.11	−25.98
AgF	c	−48.7	
	aq, 400	−53.1	−47.26
AgI	c	−15.14	−16.17
$AgIO_3$	c	−42.02	−24.08
$AgNO_2$	c	−11.6	3.76
	aq	−2.9	9.99
$AgNO_3$	c	−29.4	−7.66
	aq, 6500	−24.02	−7.81
Ag_2O	c	−6.95	−2.23
Ag_2S	c	−5.5	−7.6
Ag_2SO_4	c	−170.1	−146.8
	aq	−165.8	−139.22
Sodium:			
Na	c	0.00	0.00
Na_3AsO_3	aq, 500	−314.61	
Na_3AsO_4	c	−366	
	aq, 500	−381.97	−341.17
$NaBr$	c	−86.72	
	aq, 400	−86.33	−87.17
$NaBrO$	aq	−78.9	
$NaBrO_3$	aq, 400	−68.89	−57.59
$NaC_2H_3O_2$	c	−170.45	
	aq, 400	−175.450	−152.31
$NaCN$	c	−22.47	
	aq, 200	−22.29	−23.24
$NaCNO$	c	−96.3	
	aq	−91.7	−86.00
$NaCNS$	c	−39.94	
	aq, 400	−38.23	−39.24
Na_2CO_3	c	−269.46	−249.55
	aq, 1000	−275.13	−251.36
$NaCO_2NH_2$	c	−142.17	
$Na_2C_2O_4$	c	−313.8	
	aq, 600	−309.92	−283.42
$NaCl$	c	−98.321	−91.894
	aq, 400	−97.324	−93.92
Sodium (*Cont.*):			
$NaClO_3$	c	−83.59	
	aq, 400	−78.42	−62.84
$NaClO_4$	c	−101.12	
	aq, 476	−97.66	−73.29
Na_2CrO_4	c	−319.8	
	aq, 800	−323.0	−296.58
$Na_2Cr_2O_7$	aq, 1200	−465.9	−431.18
NaF	c	−135.94	−129.0
	aq, 400	−135.711	−128.29
NaH	c	−14	−9.30
$NaHCO_3$	c	−226.0	−202.66
	aq	−222.1	−202.87
NaI	c	−69.28	
	aq, ∞	−71.10	−74.92
$NaIO_3$	aq, 400	−112.300	−94.84
Na_2MoO_4	c	−364	
	aq	−358.7	−333.18
$NaNO_2$	c	−86.6	
	aq	−83.1	−71.04
$NaNO_3$	c	−111.71	−87.62
	aq, 400	−106.880	−88.84
Na_2O	c	−99.45	−90.06
Na_2O_2	c	−119.2	−105.0
$Na_2O.SiO_2$	c	−383.91	−361.49
$Na_2O.Al_2O_3.3SiO_2$	c, natrolite	−1180	
$Na_2O.Al_2O_3.4SiO_2$	c	−1366	
$NaOH$	c	−101.96	−90.60
	aq, 400	−112.193	−100.18
Na_3PO_3	aq, 1000	−389.1	
Na_3PO_4	c	−457	
	aq, 400	−471.9	−428.74
Na_2PtCl_4	aq	−237.2	−216.78
Na_2PtCl_6	c	−272.1	
	aq	−280.9	
Na_2Se	c	−59.1	
	aq, 440	−78.1	−89.42
Na_2SeO_4	c	−254	
	aq, 800	−261.5	−230.30
Na_2S	c	−89.8	
	aq, 400	−105.17	−101.76
Na_2SO_3	c	−261.2	−240.14
	aq, 800	−264.1	−241.58
Na_2SO_4	c	−330.50	−302.38
	aq, 1100	−330.82	−301.28
$Na_2SO_4.10H_2O$	c	−1033.85	−870.52
Na_2WO_4	c	−391	
	aq	−381.5	−345.18
Strontium:			
Sr	c	0.00	0.00
$SrBr_2$	c	−171.0	
	aq, 400	−187.24	−182.36
$Sr(C_2H_3O_2)_2$	c	−358.0	
	aq	−364.4	−311.80
$Sr(CN)_2$	aq	−59.5	−54.50
$SrCO_3$	c	−290.9	−271.9
$SrCl_2$	c	−197.84	
	aq, 400	−209.20	−195.86
SrF_2	c	−289.0	
$Sr(HCO_3)_2$	aq	−459.1	−413.76
SrI_2	c	−136.1	
	aq, 400	−156.70	−157.87
Sr_3N_2	c	−91.4	−76.5
$Sr(NO_3)_2$	c	−233.2	
	aq, 400	−228.73	−185.70
SrO	c	−140.8	−133.7
$SrO.SiO_2$	gls	−364	
SrO_2	c	−153.3	−139.0
Sr_2O	c	−153.6	
$Sr(OH)_2$	c	−228.7	
	aq, 800	−239.4	−208.27
$Sr_3(PO_4)_2$	c	−980	
	aq	−985	−881.54
SrS	c	−113.1	
	aq	−120.4	−109.78
$SrSO_4$	c	−345.3	
	aq, 400	−345.0	−309.30
$SrWO_4$	c	−393	
Sulfur:			
S	c, rhombic	0.00	0.00
	c, monoclinic	−0.071	−0.023
	l, λ	0.257	0.072
	l, λμ equilibrium	....	0.071
	g	53.25	43.57
S_2	g	31.02	19.36
S_6	g	27.78	13.97
S_8	g	27.090	12.770
S_2Br_2	l	−4	
SCl_4	l	−13.7	

Compound	State†	Heat of formation ‡ § at 25°C., kcal./mole	Free energy of formation ‖ ¶ at 25°C., kcal./mole
Sulfur (*Cont.*):			
S_2Cl_2	l	−14.2	−5.90
S_2Cl_4	l	−24.1	
SF_6	g	−262	−237
SO	g	19.02	12.75
SO_2	g	−70.94	−71.68
SO_3	g	−94.39	−88.59
	l	−103.03	−88.28
	c, α	−105.09	−88.22
	c, β	−105.92	−88.34
	c, γ	−109.34	−88.98
SO_2Cl_2	g	−82.04	−74.06
	l	−89.80	−75.06
Tantalum:			
Ta	c	0.00	0.00
TaN	c	−51.2	−45.11
Ta_2O_5	c	−486.0	−453.7
Tellurium:			
Te	c	0.00	0.00
$TeBr_4$	c	−49.3	
$TeCl_4$	c	−77.4	−57.4
TeF_6	g	−315	−292
TeO_2	c	−77.56	−64.66
Thallium:			
Tl	c	0.00	0.00
TlBr	c	−41.5	−39.43
	aq	−28.0	−32.34
TlCl	c	−49.37	−44.46
	aq	−38.4	−39.09
$TlCl_3$	c	−82.4	
	aq	−91.0	−44.25
TlF	aq	−77.6	−73.46
TlI	c	−31.1	−31.3
	aq	−12.7	−20.09
$TlNO_3$	c	−58.2	−36.32
	aq	−48.4	−34.01
Tl_2O	c	−43.18	
Tl_2O_3	c	−120	
TlOH	c	−57.44	−45.54
	aq	−53.9	−45.35
Tl_2S	c	−22	
Tl_2SO_4	c	−222.8	−197.79
	aq, 800	−214.1	−191.62
Thorium:			
Th	c	0.00	0.00
$ThBr_4$	c	−281.5	
	aq	−352.0	−295.31
ThC_2	c	−45.1	
$ThCl_4$	c	−335	
	aq	−392	−322.32
ThI_4	aq	−292.0	−246.33
Th_3N_4	c	−309.0	−282.3
ThO_2	c	−291.6	−280.1
$Th(OH)_4$	c, "soluble"	−336.1	
$Th(SO_4)_2$	c	−632	
	aq	−668.1	−549.2
Tin:			
Sn	c, II, tetragonal	0.00	0.00
	c, III, "gray," cubic	0.6	1.1
$SnBr_2$	c	−61.4	
	aq	−60.0	−55.43
$SnBr_4$	c	−94.8	
	aq	−110.6	−97.66
$SnCl_2$	c	−83.6	
	aq	−81.7	−68.94
$SnCl_4$	l	−127.3	−110.4
	aq	−157.6	−124.67
SnI_2	c	−38.9	
	aq	−33.3	−30.95
Tin (*Cont.*):			
SnO	c	−67.7	−60.75
SnO_2	c	−138.1	−123.6
$Sn(OH)_2$	c	−136.2	−115.95
$Sn(OH)_4$	c	−268.9	−226.00
SnS	c	−18.61	
Titanium:			
Ti	c	0.00	0.00
TiC	c	−110	−109.2
$TiCl_4$	l	−181.4	−165.5
TiN	c	−80.0	−73.17
TiO_2	c, III, rutil	−225.0	−211.9
	amorphous	−214.1	−201.4
Tungsten:			
W	c	0.00	0.00
WO_2	c	−130.5	−118.3
WO_3	c	−195.7	−177.3
WS_2	c	−84	
Uranium:			
U	c	0.00	0.00
UC_2	c	−29	
UCl_3	c	−213	
UCl_4	c	−251	
U_3N_4	c	−274	−249.6
UO_2	c	−256.6	−242.2
$UO_2(NO_3)_2.6H_2O$	c	−756.8	−617.8
UO_3	c	−291.6	
U_3O_8	c	−845.1	
Vanadium:			
V	c	0.00	0.00
VCl_2	c	−147	
VCl_3	l	−187	
VCl_4	l	−165	
VN	c	−41.43	−35.08
V_2O_2	c	−195	
V_2O_3	c	−296	−277
V_2O_4	c	−342	−316
V_2O_5	c	−373	−342
Zinc:			
Zn	c	0.00	0.00
ZnSb	c	−3.6	−3.88
$ZnBr_2$	c	−77.0	−72.9
	aq, 400	−93.6	
$Zn(C_2H_3O_2)_2$	c	−259.4	
	aq, 400	−269.4	−214.4
$Zn(CN)_2$	c	17.06	
$ZnCO_3$	c	−192.9	−173.5
$ZnCl_2$	c	−99.9	−88.8
	aq, 400	−115.44	
ZnF_2	aq	−192.9	−166.6
ZnI_2	c	−50.50	−49.93
	aq	−61.6	
$Zn(NO_3)_2$	aq, 400	−134.9	−87.7
ZnO	c, hexagonal	−83.36	−76.19
$ZnO.SiO_2$	c	−282.6	
$Zn(OH)_2$	c, rhombic	−153.66	
ZnS	c, wurtzite	−45.3	−44.2
$ZnSO_4$	c	−233.4	
	aq, 400	−252.12	−211.28
Zirconium:			
Zr	c	0.00	0.00
ZrC	c	−29.8	−34.6
$ZrCl_4$	c	−268.9	
ZrN	c	−82.5	−75.9
ZrO_2	c, monoclinic	−258.5	−244.6
$Zr(OH)_4$	c	−411.0	
$ZrO(OH)_2$	c	−337	−307.6

† The physical state is indicated as follows: *c*, crystal (solid); *l*, liquid; *g*, gas; *gls*, glass or solid supercooled liquid; *aq*, in aqueous solution. A number following the symbol *aq* applies only to the values of the heats of formation (not to those of free energies of formation); and indicates the number of moles of water per mole of solute; when no number is given, the solution is understood to be dilute. For the free energy of formation of a substance in aqueous solution, the concentration is always that of the hypothetical solution of unit molality.

‡ The increment in heat content, ΔH, in the reaction of forming the given substance from its elements in their standard states. When ΔH is negative, heat is evolved in the process, and, when positive, heat is absorbed.

§ The heat of solution in water of a given solid, liquid, or gaseous compound is given by the difference in the value for the heat of formation of the given compound in the solid, liquid, or gaseous state and its heat of formation in aqueous solution. The following two examples serve as an illustration of the procedure: (1) For NaCl(*c*) and NaCl(*aq*, $400H_2O$), the values of ΔH(formation) are, respectively, −98.321 and −97.324 kg.-cal. per mole. Subtraction of the first value from the second gives $\Delta H = 0.998$ kg.-cal. per mole for the reaction of dissolving crystalline sodium chloride in 400 moles of water. When this process occurs at a constant pressure of 1 atm., 0.998 kg.-cal. of energy are absorbed. (2) For HCl(*g*) and HCl(*aq*, $400H_2O$), the values for ΔH(formation) are, respectively, −22.06 and −39.85 kg.-cal. per mole. Subtraction of the first from the second gives $\Delta H = -17.79$ kg.-cal per mole for the reaction of dissolving gaseous hydrogen chloride in 400 moles of water. At a constant pressure of 1 atm. 17.79 kg.-cal. of energy are evolved in this process.

‖ The increment in the free energy, ΔF, in the reaction of forming the given substance in its standard state from its elements in their standard states. The standard states are: for a gas, fugacity (approximately equal to the pressure) of 1 atm.; for a pure liquid or solid, the substance at a pressure of 1 atm.; for a substance in aqueous solution, the hypothetical solution of unit molality, which has all the properties of the infinitely dilute solution except the property of concentration.

¶ The free energy of solution of a given substance from its normal standard state as a solid, liquid, or gas to the hypothetical one molal state in aqueous solution may be calculated in a manner similar to that described in footnote § for calculating the heat of solution.

APPENDIX V

Heats of Combustion

Compound	Formula	State	Heat of combustion, $-\Delta H_c^\circ$, at 25°C. and constant pressure, to form: H_2O (liq.) and CO_2 (gas)			H_2O (gas) and CO_2 (gas)		
			Kcal./mole	Cal./g.	B.t.u./lb.	Kcal./mole	Cal./g.	B.t.u./lb.
Hydrogen	H_2	gas	68.3174	33,887.6	60,957.7	57.7979	28,669.6	51,571.4
Carbon	C	solid, graph.	94.0518	7,831.1	14,086.8			
Carbon monoxide	CO	gas	67.6361	2,414.7	4,343.6			
Paraffins								
Methane	CH_4	gas	212.798	13,265.1	23,861	191.759	11,953.6	21,502
Ethane	C_2H_6	gas	372.820	12,399.2	22,304	341.261	11,349.6	20,416
Propane	C_3H_8	gas	530.605	12,033.5	21,646	488.527	11,079.2	19,929
Propane	C_3H_8	liq.*	526.782	11,946.8	21,490	484.704	10,992.5	19,774
n-Butane	C_4H_{10}	gas	687.982	11,837.3	21,293	635.384	10,932.3	19,665
n-Butane	C_4H_{10}	liq.*	682.844	11,748.9	21,134	630.246	10,843.9	19,506
2-Methylpropane (Isobutane)	C_4H_{10}	gas	686.342	11,809.1	21,242	633.744	10,904.1	19,614
2-Methylpropane (Isobutane)	C_4H_{10}	liq.*	681.625	11,727.9	21,096	629.027	10,822.9	19,468
n-Pentane	C_5H_{12}	gas	845.16	11,714.6	21,072	782.04	10,839.7	19,499
n-Pentane	C_5H_{12}	liq.	838.80	11,626.4	20,914	775.68	10,751.5	19,340
2-Methylbutane (Isopentane)	C_5H_{12}	gas	843.24	11,688.0	21,025	780.12	10,813.1	19,451
2-Methylbutane (Isopentane)	C_5H_{12}	liq.	837.31	11,605.8	20,877	774.19	10,730.9	19,303
2,2-Dimethylpropane (Neopentane)	C_5H_{12}	gas	840.49	11,649.8	20,956	777.37	10,775.0	19,382
2,2-Dimethylpropane (Neopentane)	C_5H_{12}	liq.	835.18	11,576.2	20,824	772.06	10,701.4	19,250
n Hexane	C_6H_{14}	gas	1,002.57	11,634.5	20,928	928.93	10,780.0	19,391
n-Hexane	C_6H_{14}	liq.	995.01	11,546.8	20,771	921.37	10,692.2	19,233
2-Methylpentane	C_6H_{14}	gas	1,000.87	11,614.8	20,893	927.23	10,760.2	19,356
2-Methylpentane	C_6H_{14}	liq.	993.71	11,531.7	20,743	920.07	10,677.1	19,206
3-Methylpentane	C_6H_{14}	gas	1,001.51	11,622.2	20,906	927.87	10,767.6	19,369
3-Methylpentane	C_6H_{14}	liq.	994.25	11,538.0	20,755	920.61	10,683.4	19,218
2,2-Dimethylbutane	C_6H_{14}	gas	998.17	11,583.5	20,837	924.53	10,728.9	19,299
2,2-Dimethylbutane	C_6H_{14}	liq.	991.52	11,506.3	20,698	917.88	10,651.7	19,161
2,3-Dimethylbutane	C_6H_{14}	gas	1,000.04	11,605.2	20,876	926.40	10,750.6	19,338
2,3-Dimethylbutane	C_6H_{14}	liq.	993.05	11,524.0	20,730	919.41	10,669.5	19,192
n-Heptane	C_7H_{16}	gas	1,160.01	11,577.2	20,825	1,075.85	10,737.2	19,314
n-Heptane	C_7H_{16}	liq.	1,151.27	11,489.9	20,668	1,067.11	10,650.0	19,157
2-Methylhexane	C_7H_{16}	gas	1,158.30	11,560.1	20,795	1,074.14	10,720.2	19,284
2-Methylhexane	C_7H_{16}	liq.	1,149.97	11,477.0	20,645	1,065.81	10,637.0	19,134
3-Methylhexane	C_7H_{16}	gas	1,158.94	11,566.5	20,806	1,074.78	10,726.6	19,295
3-Methylhexane	C_7H_{16}	liq.	1,150.55	11,482.8	20,655	1,066.39	10,642.8	19,145
3-Ethylpentane	C_7H_{16}	gas	1,159.56	11,572.7	20,817	1,075.40	10,732.7	19,306
3-Ethylpentane	C_7H_{16}	liq.	1,151.13	11,488.6	20,666	1,066.97	10,648.6	19,155
2,2-Dimethylpentane	C_7H_{16}	gas	1,155.61	11,533.3	20,746	1,071.45	10,693.3	19,235
2,2-Dimethylpentane	C_7H_{16}	liq.	1,147.85	11,455.8	20,607	1,063.69	10,615.9	19,096
2,3-Dimethylpentane	C_7H_{16}	gas	1,157.28	11,549.9	20,776	1,073.12	10,710.0	19,265
2,3-Dimethylpentane	C_7H_{16}	liq.	1,149.09	11,468.2	20,629	1,064.93	10,628.3	19,118
2,4-Dimethylpentane	C_7H_{16}	gas	1,156.60	11,543.1	20,764	1,072.44	10,703.2	19,253
2,4-Dimethylpentane	C_7H_{16}	liq.	1,148.73	11,464.6	20,623	1,064.57	10,624.7	19,112
3,3-Dimethylpentane	C_7H_{16}	gas	1,156.73	11,544.4	20,766	1,072.57	10,704.5	19,255
3,3-Dimethylpentane	C_7H_{16}	liq.	1,148.83	11,465.6	20,625	1,064.67	10,625.7	19,114
2,2,3-Trimethylbutane	C_7H_{16}	gas	1,155.94	11,536.6	20,752	1,071.78	10,696.6	19,241
2,2,3-Trimethylbutane	C_7H_{16}	liq.	1,148.27	11,460.0	20,614	1,064.11	10,620.1	19,104
n-Octane	C_8H_{18}	gas	1,317.45	11,533.9	20,747	1,222.77	10,705.0	19,256
n-Octane	C_8H_{18}	liq.	1,307.53	11,447.1	20,591	1,212.85	10,618.2	19,100
2-Methylheptane	C_8H_{18}	gas	1,315.76	11,519.1	20,721	1,221.08	10,690.2	19,230
2-Methylheptane	C_8H_{18}	liq.	1,306.28	11,436.1	20,572	1,211.60	10,607.2	19,080
3-Methylheptane	C_8H_{18}	gas	1,316.44	11,525.1	20,732	1,221.76	10,696.2	19,240
3-Methylheptane	C_8H_{18}	liq.	1,306.92	11,441.7	20,582	1,212.24	10,612.8	19,091
4-Methylheptane	C_8H_{18}	gas	1,316.57	11,526.2	20,734	1,221.89	10,697.3	19,243
4-Methylheptane	C_8H_{16}	liq.	1,307.09	11,443.2	20,584	1,212.41	10,614.3	19,093
3-Ethylhexane	C_8H_{18}	gas	1,316.87	11,528.8	20,738	1,222.19	10,699.9	19,247
3-Ethylhexane	C_8H_{18}	liq.	1,307.39	11,445.8	20,589	1,212.71	10,616.9	19,098
2,2-Dimethylhexane	C_8H_{18}	gas	1,313.56	11,499.9	20,686	1,218.88	10,671.0	19,195
2,2-Dimethylhexane	C_8H_{18}	liq.	1,304.64	11,421.8	20,546	1,209.96	10,592.9	19,055
2,3-Dimethylhexane	C_8H_{18}	gas	1,316.13	11,522.4	20,727	1,221.45	10,693.5	19,236
2,3-Dimethylhexane	C_8H_{18}	liq.	1,306.86	11,441.2	20,581	1,212.18	10,612.3	19,090
2,4-Dimethylhexane	C_8H_{18}	gas	1,314.83	11,511.0	20,706	1,220.15	10,682.1	19,215
2,4-Dimethylhexane	C_8H_{18}	liq.	1,305.80	11,431.9	20,564	1,211.12	10,603.0	19,073
2,5-Dimethylhexane	C_8H_{18}	gas	1,314.05	11,504.2	20,694	1,219.37	10,675.3	19,203
2,5-Dimethylhexane	C_8H_{18}	liq.	1,305.00	11,424.9	20,551	1,210.32	10,596.0	19,060
3,3-Dimethylhexane	C_8H_{18}	gas	1,314.65	11,509.4	20,703	1,219.97	10,680.5	19,212
3,3-Dimethylhexane	C_8H_{18}	liq.	1,305.68	11,430.9	20,562	1,211.00	10,602.0	19,071
3,4-Dimethylhexane	C_8H_{18}	gas	1,316.36	11,524.4	20,730	1,221.68	10,695.5	19,239
3,4-Dimethylhexane	C_8H_{18}	liq.	1,307.04	11,442.8	20,583	1,212.36	10,613.9	19,092
2-Methyl-3-ethylpentane	C_8H_{18}	gas	1,316.79	11,528.1	20,737	1,222.11	10,699.2	19,246
2-Methyl-3-ethylpentane	C_8H_{18}	liq.	1,307.58	11,447.5	20,592	1,212.90	10,618.6	19,101
3-Methyl-3-ethylpentane	C_8H_{18}	gas	1,315.88	11,520.2	20,723	1,221.20	10,691.3	19,232
3-Methyl-3-ethylpentane	C_8H_{18}	liq.	1,306.80	11,440.7	20,580	1,212.12	10,611.8	19,089
2,2,3-Trimethylpentane	C_8H_{18}	gas	1,314.66	11,509.5	20,703	1,219.98	10,680.6	19,212
2,2,3-Trimethylpentane	C_8H_{18}	liq.	1,305.83	11,432.2	20,564	1,211.15	10,603.3	19,073
2,2,4-Trimethylpentane	C_8H_{18}	gas	1,313.69	11,501.0	20,688	1,219.01	10,672.1	19,197
2,2,4-Trimethylpentane	C_8H_{18}	liq.	1,305.29	11,427.5	20,556	1,210.61	10,598.6	19,065
2,3,3-Trimethylpentane	C_8H_{18}	gas	1,315.54	11,517.2	20,717	1,220.86	10,688.3	19,226
2,3,3-Trimethylpentane	C_8H_{18}	liq.	1,306.64	11,439.3	20,577	1,211.96	10,610.4	19,086
2,3,4-Trimethylpentane	C_8H_{18}	gas	1,315.29	11,515.0	20,713	1,220.61	10,686.1	19,222
2,3,4-Trimethylpentane	C_8H_{18}	liq.	1,306.28	11,436.1	20,572	1,211.60	10,607.2	19,080
2,2,3,3-Tetramethylbutane	C_8H_{18}	gas	1,313.27	11,497.3	20,682	1,218.59	10,668.4	19,191

Source: R. H. Perry and C. H. Chilton, eds., *Chemical Engineers' Handbook*. Copyright © 1973 McGraw–Hill Inc. Used with permission of McGraw–Hill Book Company.

Compound	Formula	State	Heat of combustion, $-\Delta Hc°$, at 25°C. and constant pressure, to form					
			H_2O (liq.) and CO_2 (gas)			H_2O (gas) and CO_2 (gas)		
			Kcal./mole	Cal./g.	B.t.u./lb.	Kcal./mole	Cal./g.	B.t.u./lb.
2,2,3,3-Tetramethylbutane	C_8H_{18}	solid	1,303.03	11,407.7	20,520	1,208.35	10,578.8	19,029
n-Nonane	C_9H_{20}	gas	1,474.90	11,500.2	20,687	1,369.70	10,680.0	19,211
n-Nonane	C_9H_{20}	liq.	1,463.80	11,413.6	20,531	1,358.60	10,593.4	19,056
n-Decane	$C_{10}H_{22}$	gas	1,632.34	11,473.0	20,638	1,516.63	10,659.7	19,175
n-Decane	$C_{10}H_{22}$	liq.	1,620.06	11,386.7	20,483	1,504.35	10,573.4	19,020
n-Undecane	$C_{11}H_{24}$	gas	1,789.78	11,450.8	20,598	1,663.55	10,643.2	19,145
n-Undecane	$C_{11}H_{24}$	liq.	1,776.32	11,364.7	20,443	1,650.09	10,557.0	18,990
n-Dodecane	$C_{12}H_{26}$	gas	1,947.23	11,432.2	20,564	1,810.48	10,629.4	19,120
n-Dodecane	$C_{12}H_{26}$	liq.	1,932.59	11,346.3	20,410	1,795.84	10,543.4	18,966
n-Tridecane	$C_{13}H_{28}$	gas	2,104.67	11,416.5	20,536	1,957.40	10,617.6	19,099
n-Tridecane	$C_{13}H_{28}$	liq.	2,088.85	11,330.6	20,382	1,941.58	10,531.8	18,945
n-Tetradecane	$C_{14}H_{30}$	gas	2,262.11	11,402.9	20,512	2,104.32	10,607.5	19,081
n-Tetradecane	$C_{14}H_{30}$	liq.	2,245.11	11,317.2	20,358	2,087.32	10,521.8	18,927
n-Pentadecane	$C_{15}H_{32}$	gas	2,419.55	11,391.2	20,491	2,251.24	10,598.7	19,065
n-Pentadecane	$C_{15}H_{32}$	liq.	2,401.37	11,305.6	20,337	2,233.06	10,513.2	18,911
n-Hexadecane	$C_{16}H_{34}$	gas	2,577.00	11,380.9	20,472	2,398.17	10,591.1	19,052
n-Hexadecane	$C_{16}H_{34}$	liq.	2,557.64	11,295.4	20,318	2,378.81	10,505.6	18,898
n-Heptadecane	$C_{17}H_{36}$	gas	2,734.44	11,371.8	20,456	2,545.09	10,584.3	19,039
n-Heptadecane	$C_{17}H_{36}$	liq.	2,713.90	11,286.4	20,302	2,524.55	10,498.9	18,886
n-Octadecane	$C_{18}H_{38}$	gas	2,891.88	11,363.7	20,441	2,692.01	10,578.3	19,028
n-Octadecane	$C_{18}H_{38}$	liq.	2,870.16	11,278.4	20,288	2,670.29	10,493.0	18,875
n-Nonadecane	$C_{19}H_{40}$	gas	3,049.33	11,356.5	20,428	2,838.94	10,572.9	19,019
n-Nonadecane	$C_{19}H_{40}$	liq.	3,026.43	11,271.2	20,275	2,816.04	10,487.7	18,865
n-Eicosane	$C_{20}H_{42}$	gas	3,206.77	11,350.0	20,416	2,985.86	10,568.1	19,010
n-Eicosane	$C_{20}H_{42}$	liq.	3,182.69	11,264.7	20,263	2,961.78	10,482.8	18,857
Alkyl benzenes								
Benzene	C_6H_6	gas	789.08	10,102.4	18,172	757.52	9,698.4	17,446
Benzene	C_6H_6	liq.	780.98	9,998.7	17,986	749.42	9,594.7	17,259
Methylbenzene (toluene)	C_7H_8	gas	943.58	10,241.4	18,422	901.50	9,784.7	17,601
Methylbenzene (toluene)	C_7H_8	liq.	934.50	10,142.8	18,245	892.42	9,686.1	17,424
Ethylbenzene	C_8H_{10}	gas	1,101.13	10,372.4	18,658	1,048.53	9,876.9	17,767
Ethylbenzene	C_8H_{10}	liq.	1,091.03	10,277.2	18,487	1,038.43	9,781.7	17,596
1,2-Dimethylbenzene (o-xylene)	C_8H_{10}	gas	1,098.54	10,348.0	18,614	1,045.94	9,852.5	17,723
1,2-Dimethylbenzene (o-xylene)	C_8H_{10}	liq.	1,088.16	10,250.2	18,438	1,035.56	9,754.7	17,547
1,3-Dimethylbenzene (m-xylene)	C_8H_{10}	gas	1,098.12	10,344.0	18,607	1,045.52	9,848.5	17,716
1,3-Dimethylbenzene (m-xylene)	C_8H_{10}	liq.	1,087.92	10,247.9	18,434	1,035.32	9,752.4	17,543
1,4-Dimethylbenzene (p-xylene)	C_8H_{10}	gas	1,098.29	10,345.6	18,610	1,045.69	9,850.1	17,719
1,4-Dimethylbenzene (p-xylene)	C_8H_{10}	liq.	1,088.16	10,250.2	18,438	1,035.56	9,754.7	17,547
n-Propylbenzene	C_9H_{12}	gas	1,258.24	10,469.1	18,832	1,195.12	9,943.9	17,887
n-Propylbenzene	C_9H_{12}	liq.	1,247.19	10,377.2	18,667	1,184.07	9,852.0	17,722
Isopropylbenzene (cumene)	C_9H_{12}	gas	1,257.31	10,461.4	18,818	1,194.19	9,936.2	17,873
Isopropylbenzene (cumene)	C_9H_{12}	liq.	1,246.52	10,371.6	18,657	1,183.40	9,846.4	17,712
1-Methyl-2-ethylbenzene	C_9H_{12}	gas	1,256.66	10,456.0	18,808	1,193.54	9,930.8	17,864
1-Methyl-2-ethylbenzene	C_9H_{12}	liq.	1,245.26	10,361.1	18,638	1,182.14	9,835.9	17,693
1-Methyl-3-ethylbenzene	C_9H_{12}	gas	1,255.92	10,449.8	18,797	1,192.80	9,924.6	17,853
1-Methyl-3-ethylbenzene	C_9H_{12}	liq.	1,244.71	10,356.5	18,630	1,181.59	9,831.3	17,685
1-Methyl-4-ethylbenzene	C_9H_{12}	gas	1,255.59	10,447.1	18,792	1,192.47	9,921.9	17,848
1-Methyl-4-ethylbenzene	C_9H_{12}	liq.	1,244.45	10,354.4	18,626	1,181.33	9,829.2	17,681
1,2,3-Trimethylbenzene (hemimellitene)	C_9H_{12}	gas	1,254.08	10,434.5	18,770	1,190.96	9,909.3	17,825
1,2,3-Trimethylbenzene (hemimellitene)	C_9H_{12}	liq.	1,242.36	10,337.0	18,594	1,179.24	9,811.8	17,650
1,2,4-Trimethylbenzene (pseudocumene)	C_9H_{12}	gas	1,253.04	10,425.8	18,754	1,189.92	9,900.7	17,809
1,2,4-Trimethylbenzene (pseudocumene)	C_9H_{12}	liq.	1,241.58	10,330.5	18,583	1,178.46	9,805.3	17,638
1,3,5-Trimethylbenzene (mesitylene)	C_9H_{12}	gas	1,252.53	10,421.6	18,747	1,189.41	9,896.4	17,802
1,3,5-Trimethylbenzene (mesitylene)	C_9H_{12}	liq.	1,241.19	10,327.2	18,577	1,178.07	9,802.1	17,632
n-Butylbenzene	$C_{10}H_{14}$	gas	1,415.44	10,546.3	18,971	1,341.80	9,997.6	17,984
n-Butylbenzene	$C_{10}H_{14}$	liq.	1,403.46	10,457.0	18,810	1,329.82	9,908.4	17,823
Alkyl cyclopentanes								
Cyclopentane	C_5H_{10}	gas	793.39	11,313.1	20,350	740.79	10,563.1	19,001
Cyclopentane	C_5H_{10}	liq.	786.54	11,215.5	20,175	733.94	10,465.4	18,825
Methylcyclopentane	C_6H_{12}	gas	948.72	11,273.4	20,279	885.60	10,523.3	18,930
Methylcyclopentane	C_6H_{12}	liq.	941.14	11,183.3	20,117	878.02	10,433.2	18,768
Ethylcyclopentane	C_7H_{14}	gas	1,106.21	11,266.9	20,267	1,032.57	10,516.9	18,918
Ethylcyclopentane	C_7H_{14}	liq.	1,097.50	11,178.2	20,108	1,023.86	10,428.2	18,758
n-Propylcyclopentane	C_8H_{16}	gas	1,263.56	11,260.9	20,256	1,179.40	10,510.8	18,907
n-Propylcyclopentane	C_8H_{16}	liq.	1,253.74	11,173.4	20,099	1,169.58	10,423.3	18,750
n-Butylcyclopentane	C_9H_{18}	gas	1,421.10	11,257.7	20,250	1,326.42	10,507.6	18,901
n-Butylcyclopentane	C_9H_{18}	liq.	1,410.10	11,170.5	20,094	1,315.42	10,420.5	18,745
Alkyl cyclohexanes								
Cyclohexane	C_6H_{12}	gas	944.79	11,226.7	20,195	881.67	10,476.7	18,846
Cyclohexane	C_6H_{12}	liq.	936.88	11,132.7	20,026	873.76	10,382.7	18,676
Methylcyclohexane	C_7H_{14}	gas	1,099.59	11,199.5	20,146	1,025.95	10,449.5	18,797
Methylcyclohexane	C_7H_{14}	liq.	1,091.13	11,113.3	19,991	1,017.49	10,363.3	18,642
Ethylcyclohexane	C_8H_{16}	gas	1,257.90	11,210.4	20,166	1,173.74	10,460.4	18,816
Ethylcyclohexane	C_8H_{16}	liq.	1,248.23	11,124.3	20,011	1,164.07	10,374.3	18,661
n-Propylcyclohexane	C_9H_{18}	gas	1,415.12	11,210.3	20,165	1,320.44	10,460.3	18,816
n-Propylcyclohexane	C_9H_{18}	liq.	1,404.34	11,124.9	20,012	1,309.66	10,374.9	18,663
n-Butylcyclohexane	$C_{10}H_{20}$	gas	1,572.74	11,213.0	20,170	1,467.54	10,463.0	18,821
n-Butylcyclohexane	$C_{10}H_{20}$	liq.	1,560.78	11,127.8	20,017	1,455.58	10,377.8	18,668
Monoolefins								
Ethene (ethylene)	C_2H_4	gas	337.234	12,021.7	21,625	316.195	11,271.7	20,276
Propene (propylene)	C_3H_6	gas	491.987	11,692.3	21,032	460.428	10,942.3	19,683
1-Butene	C_4H_8	gas	649.757	11,581.3	20,833	607.679	10,831.3	19,484
cis-2-Butene	C_4H_8	gas	648.115	11,552.0	20,780	606.037	10,802.0	19,431
trans-2-Butene	C_4H_8	gas	647.072	11,533.4	20,747	604.994	10,783.4	19,397
2-Methylpropene (isobutene)	C_4H_8	gas	646.134	11,516.7	20,716	604.056	10,766.7	19,367
1-Pentene	C_5H_{10}	gas	806.85	11,505.1	20,696	754.25	10,755.1	19,346

Compound	Formula	State	Heat of combustion, $-\Delta Hc^\circ$, at 25°C. and constant pressure, to form					
			H_2O (liq.) and CO_2 (gas)			H_2O (gas) and CO_2 (gas)		
			Kcal./mole	Cal./g.	B.t.u./lb.	Kcal./mole	Cal./g.	B.t.u./lb.
cis-2-Pentene	C_5H_{10}	gas	805.34	11,483.5	20,657	752.74	10,733.5	19,308
trans-2-Pentene	C_5H_{10}	gas	804.26	11,468.1	20,629	751.66	10,718.1	19,280
2-Methyl-1-butene	C_5H_{10}	gas	803.17	11,452.6	20,601	750.57	10,702.6	19,252
3-Methyl-1-butene	C_5H_{10}	gas	804.93	11,477.7	20,646	752.33	10,727.7	19,297
2-Methyl-2-butene	C_5H_{10}	gas	801.68	11,431.3	20,563	749.08	10,681.3	19,214
Acetylenes								
Ethyne (acetylene)	C_2H_2	gas	310.615	11,930.2	21,460	300.096	11,526.2	20,734
Propyne (methylacetylene)	C_3H_4	gas	463.109	11,559.8	20,794	442.070	11,034.6	19,849
1-Butyne (ethylacetylene)	C_4H_6	gas	620.86	11,478.7	20,648	589.302	10,895.2	19,599
2-Butyne (dimethylacetylene)	C_4H_6	gas	616.533	11,398.7	20,504	584.974	10,815.2	19,455
1-Pentyne	C_5H_8	gas	778.03	11,422.5	20,547	735.95	10,804.7	19,436
2-Pentyne	C_5H_8	gas	774.33	11,368.2	20,449	732.25	10,750.4	19,338
3-Methyl-1-butyne	C_5H_8	gas	776.13	11,394.6	20,497	734.05	10,776.8	19,386

°Saturation pressure.

Index